THE GREAT BARRIER REEF

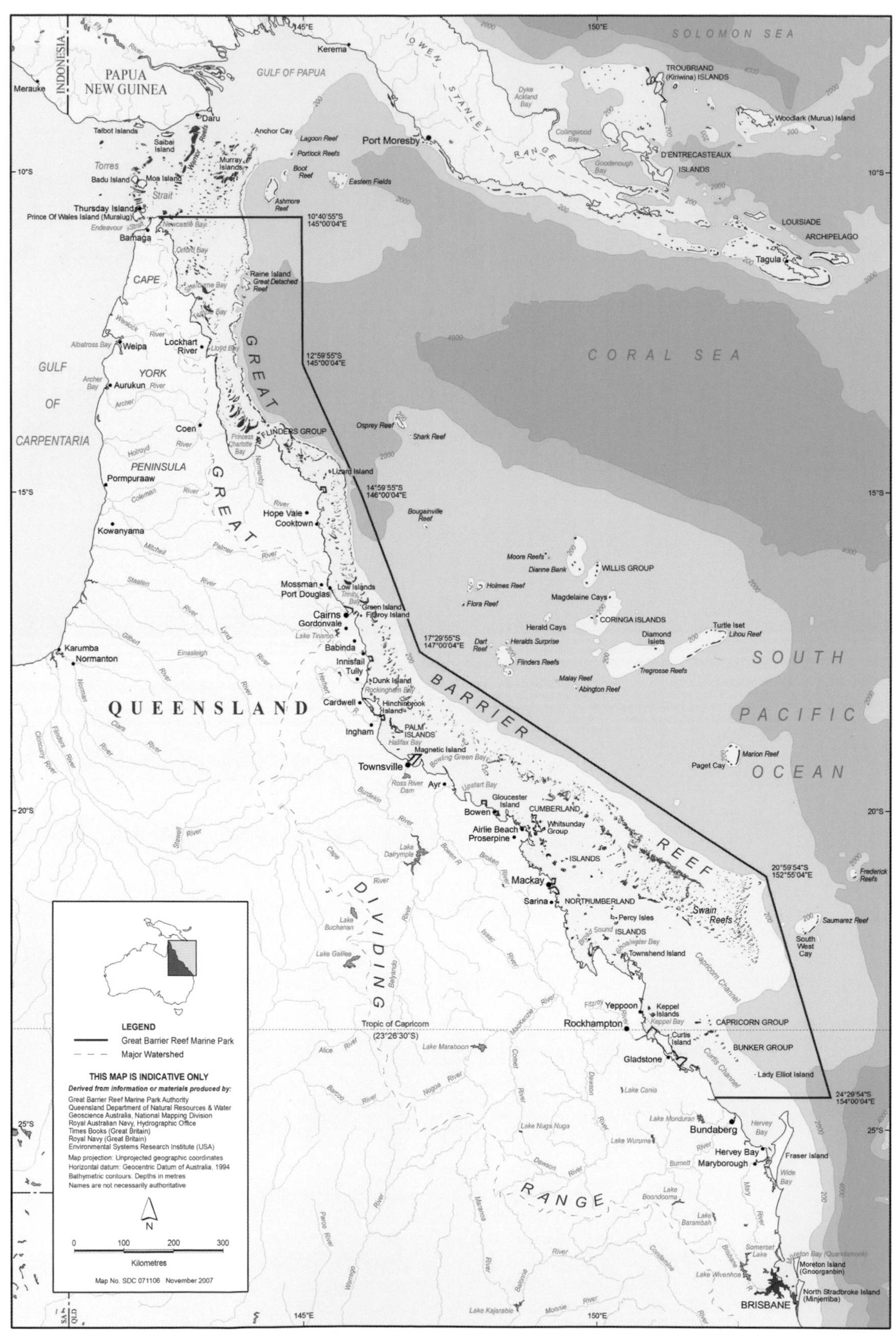

PAPUA NEW GUINEA
INDONESIA
Merauke
GULF OF PAPUA
Kerema
OWEN STANLEY RANGE
SOLOMON SEA
TROUBRIAND (Kiriwina) ISLANDS
Woodlark (Murua) Island
D'ENTRECASTEAUX ISLANDS
LOUISIADE ARCHIPELAGO
Tagula
Port Moresby
Daru
Talbot Islands
Saibai Island
Anchor Cay
Lagoon Reef
Portlock Reefs
Boot Reef
Eastern Fields
Ashmore Reef
Murray Islands
Torres Strait
Badu Island
Moa Island
Thursday Island
Prince Of Wales Island (Muralug)
Endeavour Strait
Bamaga
Newcastle Bay
10°40'55"S
145°00'04"E
Raine Island
Great Detached Reef
CAPE YORK PENINSULA
Weipa
Lockhart River
Albatross Bay
Archer Bay
Aurukun
GULF OF CARPENTARIA
12°59'55"S
145°00'04"E
CORAL SEA
GREAT BARRIER REEF
Coen
FLINDERS GROUP
Princess Charlotte Bay
Osprey Reef
Shark Reef
Lizard Island
Pormpuraaw
14°59'55"S
146°00'04"E
Hope Vale
Cooktown
Bougainville Reef
Kowanyama
GREAT DIVIDING RANGE
Moore Reefs
Dianne Bank
WILLIS GROUP
Holmes Reef
Flora Reef
Magdelaine Cays
Mossman
Low Islands
Port Douglas
Trinity Bay
Green Island
Cairns
Fitzroy Island
Gordonvale
Herald Cays
CORINGA ISLANDS
Turtle Islet
Lihou Reef
17°29'55"S
147°00'04"E
Babinda
Dart Reef
Heralds Surprise
Diamond Islets
Karumba
Normanton
Innisfail
Tully
Flinders Reefs
Tregrosse Reefs
SOUTH PACIFIC OCEAN
Dunk Island
Rockingham Bay
Malay Reef
Abington Reef
QUEENSLAND
Cardwell
Hinchinbrook Island
Ingham
PALM ISLANDS
Halifax Bay
Magnetic Island
Townsville
Bowling Green Bay
Marion Reef
Paget Cay
Ross River Dam
Ayr
Upstart Bay
Gloucester Island
Bowen
CUMBERLAND
Whitsunday Group
Airlie Beach
Proserpine
ISLANDS
Lake Dalrymple
20°59'54"S
152°55'04"E
Frederick Reefs
Mackay
Sarina
NORTHUMBERLAND
Percy Isles
ISLANDS
Swain Reefs
Saumarez Reef
South West Cay
Lake Buchanan
Lake Galilee
Broad Sound
Shoalwater Bay
Townshend Island
Capricorn Channel
Yeppoon
Keppel Islands
Keppel Bay
Rockhampton
CAPRICORN GROUP
Tropic of Capricorn
(23°26'30"S)
Curtis Island
BUNKER GROUP
Lake Maraboon
Gladstone
Curtis Channel
Lady Elliot Island
24°29'54"S
154°00'04"E
Lake Callide
Lake Monduran
Bundaberg
Hervey Bay
Lake Nuga Nuga
Lake Wuruma
Fraser Island
Hervey Bay
Maryborough
Wide Bay
Lake Boondooma
Lake Barambah
Somerset Lake
Moreton Bay (Quandamooka)
Moreton Island (Gnoorganbin)
Lake Wivenhoe
North Stradbroke Island (Minjerriba)
BRISBANE
Lake Kajarabie
145°E
150°E
10°S
15°S
20°S
25°S
SA
QLD
LEGEND
Great Barrier Reef Marine Park
Major Watershed
THIS MAP IS INDICATIVE ONLY
Derived from information or materials produced by:
Great Barrier Reef Marine Park Authority
Queensland Department of Natural Resources & Water
Geoscience Australia, National Mapping Division
Royal Australian Navy, Hydrographic Office
Times Books (Great Britain)
Royal Navy (Great Britain)
Environmental Systems Research Institute (USA)
Map projection: Unprojected geographic coordinates
Horizontal datum: Geocentric Datum of Australia, 1994
Bathymetric contours: Depths in metres
Names are not necessarily authoritative
N
0
100
200
300
Kilometres
Map No. SDC 071106 November 2007

THE GREAT BARRIER REEF

Biology, Environment and Management

Second Edition

Editors
Pat Hutchings
Michael Kingsford
and Ove Hoegh-Guldberg

PUBLISHING

CRC Press
Taylor & Francis Group
Boca Raton London New York

CRC Press is an imprint of the
Taylor & Francis Group, an **informa** business
A BALKEMA BOOK

A catalogue record for this book is available from the National Library of Australia.

Published exclusively in Australia and New Zealand by:

CSIRO Publishing
Locked Bag 10
Clayton South VIC 3169
Australia

Telephone: +61 3 9545 8400
Email: publishing.sales@csiro.au
Website: www.publish.csiro.au

Published exclusively throughout the world (excluding Australia and New Zealand) by CRC Press/Balkema, with ISBN 978-0-367-17428-6

CRC Press/Balkema
Schipholweg 107C
2316XC Leiden
The Netherlands
Tel: +31 71 524 3080
Website: www.crcpress.com

Front cover: Great Barrier Reef (Photo: Tchami/Flickr, CC BY-SA 2.0)
Back cover: (left to right) crinoids (*Cenometra bella*) attached to a gorgonian, filtering the water current as it flows past; a stingray (*Taeniurops meyeni*) gliding through the water column; banded coral shrimp (*Stenopus hispidus*) cleaning a giant moray eel (*Gymnothorax javanicus*) (Photos by David Wachenfeld)
Map (p. ii): Great Barrier Reef (Courtesy of the Spatial Data Centre, Great Barrier Reef Marine Park Authority © Commonwealth of Australia (GBRMPA))

Set in 10.5/14 Palatino and Optima
Edited by Peter Storer
Cover design by James Kelly
Typeset by Thomson Digital
Index by Max McMaster
Printed in China by Asia Pacific Offset

Royalties from the sales of this book go to the Australian Coral Reef Society.

Nov18_01

Foreword

The Great Barrier Reef is one of the most spectacular examples of natural ecosystems anywhere, which easily justified its inclusion as a UNESCO World Heritage Area in 1981. Arguably, the Reef is also the best managed marine parks to be found anywhere. And for Australians going back thousands of years, it has been a central part of our identity, community and economy.

Preserving this magnificent piece of Australia's natural wealth has grown in importance as our understanding and knowledge of the system has increased. Developing a multi-use park system in 1975, Australia pioneered the idea that areas as large as the Great Barrier Reef Marine Park should not be shut away, but rather should be part of a harmonious balance between preservation and use.

This approach, however, has not always been easy. As the 1970s taught us, communities can see things in very different lights. At one end there were people who wanted to exploit the resources of the Reef without regard; at the other, people who wanted to simply lock it up. Fortunately, however, Australia chose the middle ground which requires us to understand and regulate the use of the park, while keeping it in good condition in perpetuity.

As a test of that dialogue and transparency of process, the Park underwent a systematic review of its zoning at the beginning of the Millennium, which resulted in extensive consultation with Australians and the international community. Led by the Hon Virginia Chadwick AO, who wrote the Preface to the first edition of this book, the review led to the largest and most comprehensive review of a Park system anywhere.

The results of the rezoning of the Park were spectacular. As a result of that consultation and process, the Park went from around 5% of its zones as 'no take' to no take areas comprising more than 33% of the total Park area. A key feature of these modifications to the Great Barrier Reef was that they involved consultation with thousands, and that people discussed and owned the outcome. As a result of the changes, the *Great Barrier Reef Act* passed relatively easily through Federal Parliament in 2004.

This was a miraculous achievement in many minds, but heavily depended on having the necessary science in place to inform management decisions. And in this regard, it has been important that we have built our understanding and knowledge of the reef, so as to underpin the sorts of management decisions that were typical of re-zoning process.

The world's oceans, including the Great Barrier Reef, are being challenged at an unprecedented level. The ongoing challenges of dealing with climate change, overfishing and impacts from the land continue. Like coral reefs elsewhere, new methods of extraction are also posing additional challenges for the GBR, especially for lesser known areas such as the deep mesophotic reefs. These challenges, however, make it more, not less, important to base decisions relating to resource use and conservation on sound knowledge and understanding of the complex processes involved.

In its second edition, this book on the Reef builds on the tradition of the first edition and has expanded the already amazing array of authors and expertise to tackle a number of new and current issues to do with the Reef. Stretching from the geology, biology and oceanography of the Reef, to modern issues such as water quality, climate change and socio-economics of the Reef, this book enables the reader to get a total insight to how the world's largest coral reef functions. This is a book that will be useful for school students through to experts, and has relevance for people interested in coral reefs and related ecosystems worldwide.

For these reasons, I am extremely pleased to see the publication of the second edition of the book, the Great Barrier Reef. To put it bluntly, understanding how this vast system operates is in everyone's interest. This is why books such as *The Great Barrier Reef* are so important.

Honourable Robert Hill AC*
Former Minister for the Environment and for Defence
Ambassador for Australia to the United Nations

* The Honourable Robert Hill was responsible for establishing the Commencement of Representative Areas Program (1998) that led to the rezoning of the GBR in 2004. He also initiated what would become the world's first national Oceans Policy just before that time, although he had moved on from Environment Department before it was implemented.

Contents

Abbreviations

ABRS: Australian Biological Resources Study
ACRS: Australian Coral Reef Society
AIMS: Australian Institute of Marine Science
ARC: Australian Research Council
AUV: autonomous underwater vehicle
CFC: chlorofluorocarbon
COTS: crown-of-thorns starfish
CRC: Co-operative Research Centre
CSIRO: Commonwealth Scientific and Industrial Research Organisation
DDM: day-to-day management program
EAC: East Australian Current
ENCORE: Enrichment of Nutrients on Coral Reefs (project funded by ARC and GBRMPA)
ENSO El Niño–Southern Oscillation
EPICA: European Project for Ice Coring in Antarctica
GBR: Great Barrier Reef
GBRCA: Great Barrier Reef Catchment Area
GBRMP: Great Barrier Reef Marine Park
GBRMPA: Great Barrier Reef Marine Park Authority
GBRWHA: Great Barrier Reef World Heritage Area
IPCC: Intergovernmental Panel on Climate Change
IUCN: International Union for Conservation of Nature
JCU: James Cook University
MAGNT: Museum and Art Gallery of the Northern Territory
MODIS: moderate resolution image spectroradiometer on the Terra and Aqua satellites (see http://modis.gsfc.nasa.gov/).
MST: Marine Science and Technology (grant awarded by the federal government)
NASA National Aeronautics and Space Administration
NGOs: non-government organisations
NIWA: National Institute of Water and Atmospheric Research (New Zealand)
NOAA: National Oceanic and Atmospheric Administration (USA)
NQAIF: North Queensland Algal Identification/ Culturing Facility
QDNR&M: Queensland Department of Natural Resources and Management
QDPI: Queensland Department of Primary Industries
QPWS: Queensland Parks and Wildlife Service
OSCAR-NOAA: A NOAA project mapping, using satellite altimetry, the water surface elevation of the ocean, from which the near-surface water currents are calculated
RAP: Representative Areas Program
ROV: remote operated vehicle
RWQPP: Reef Water Quality Protection Plan
SeaWiFS: Sea-viewing Wide Field-of-view Sensor (see http://oceancolor.gsfc.nasa.gov/ SeaWiFS).
SOI: Southern Oscillation Index
UQ: University of Queensland

Author biographies

S. T. Ahyong
Shane Ahyong is a Principal Research Scientist and Manager of the Department of Marine Invertebrates at the Australian Museum, Sydney. He is an international authority on crustacean systematics, especially of decapods and stomatopods. Research interests include invertebrate taxonomy and phylogenetics, invasive species, biosecurity, and both freshwater and marine faunas. He has published more than 250 refereed articles and book chapters, and is a co-author of *The Biology of Squat Lobsters*, published by CSIRO Publishing.

P. Alderslade
Phil Alderslade has been researching the identification of octocorals since the early 1970s. Commencing in 1981 he spent 25 years as the Curator of Coelenterates at the Museum and Art Gallery of the Northern Territory in Darwin and during this time collaborated with Katharina Fabricius to co-author the book *Soft Corals and Sea Fans* – the only comprehensive field guide to the shallow water octocorals of the Red Sea, Indo-Pacific and central west Pacific regions of the world. Phil became semi-retired in 2006 and is now working as a research scientist (octocoral taxonomy) for CSIRO Oceans and Atmosphere in Hobart, Tasmania.

T. J. Anderson
Tara Anderson is a research scientist (benthic ecologist), at the National Institute of Water and Atmosphere in New Zealand. Her research over 25 years has focused primarily on the relationships between benthic habitats and biota, including bio-physical sea floor mapping and the role of biogenic habitat-formers in supporting fish assemblages, in a broad range of coastal and offshore marine ecosystems in New Zealand, Australia and along the west coast of the United States.

P. Bock
Philip Bock became fascinated by bryozoans while geological mapping in south-west Victoria, where the sediments are often packed with their skeletons. He took up study of the living bryozoans of southern Australia in order to gain an understanding of their environmental variation. After working at the Royal Melbourne Institute of Technology University for over 30 years, he retired in 1997, and maintains an active interest, including keeping the Bryozoa Home Page website available at http://www.bryozoa.net/. He also is a taxonomic editor for the World Register of Marine Species.

P. Bongaerts
Pim Bongaerts is Curator of Aquatic Biology at the California Academy of Sciences (USA). Before that, he held positions as ARC DECRA Research Fellow at the Global Change Institute (The University of Queensland) and as Lead Scientist for the deep-reef work of the 'XL Catlin Seaview Survey'. His research over the past 10 years has focused primarily on mesophotic coral ecosystems, using molecular techniques to study their biodiversity and connectivity with shallow-water reefs.

T. Bridge
Tom Bridge is Senior Curator of Corals at the Queensland Museum, based at the Museum of Tropical Queensland campus in Townsville, and at the Australian Research Council Centre of Excellence for Coral Reef Studies at James Cook University (JCU). He commenced his career with a BSc (Hons) at the University of Sydney, before moving to JCU to complete a PhD (2007–2011). From 2012 to 2016, he was a Postdoctoral Research Fellow at the ARC Centre of Excellence for Coral Reef Studies. Tom has worked on coral reefs across the Indo-Pacific. He has a particular interest in mesophotic

coral reef ecosystems and the application of remote imaging techniques such as autonomous underwater vehicles to survey habitats beyond the depths of conventional SCUBA diving.

J. Brodie
Jon Brodie is a Professorial Research Fellow with the ARC Centre of Excellence for Coral Reef Studies at James Cook University (JCU). He is also a Partner in C_2O Consulting. He has held positions as: Chief Research Scientist, Centre for Tropical Water and Aquatic Ecosystem Research, JCU; Director, Institute of Applied Science, University of the South Pacific, Fiji; and Director, Water Quality and Coastal Development Group, Great Barrier Reef Marine Park Authority (GBRMPA). Jon has collaborated with a wide range of research and environmental management colleagues from the Australian Institute of Marine Science, GBRMPA, CSIRO the Queensland Government, JCU and many Australian and international universities to establish the effects of changed terrestrial runoff and potential management solutions on Great Barrier Reef ecosystems and tropical marine environments globally.

M. Byrne
Maria Byrne is Professor of Marine and Developmental Biology at the University of Sydney. She received her BSc from the National University of Ireland Galway and PhD from University of Victoria, Canada. As a postdoctoral fellow at the Smithsonian Institution, she investigated Caribbean echinoderms and returned to Ireland working on sea urchin aquaculture and fisheries. Her research on marine invertebrate biology largely involves echinoderms and molluscs. In recent years her research has involved the quantification of the impacts of climate change stressors, ocean warming and ocean acidification on fundamental biological processes including growth, physiology, development and calcification. For 12 years Maria was the director of One Tree Island Research Station in the southern Great Barrier Reef. She has produces over 300 peer reviewed scientific articles and recently co-edited and wrote the authoritative book on Australian Echinoderms.

D. Cameron
Darren Cameron is Director – Reef Interventions with the Great Barrier Reef Marine Park Authority in Townsville. He originally worked as a fisheries biologist and manager for the Queensland Government. He was part of the team that designed and implemented the 2004 rezoning of the Great Barrier Reef Marine Park. His main interests are in coral reef ecosystem and fisheries science and management. In particular, the effective design, implementation and adaptive management of multiple-use marine protected areas, which incorporate and contribute to ecologically sustainable fisheries management. He has over 30 years professional experience, having worked in over 16 countries, mostly in the Western and Central Pacific region.

A. Chin
Andrew Chin is a research fellow with the Centre for Sustainable Tropical Fisheries and Aquaculture at James Cook University. His research focuses on fisheries science and coastal systems, and he specialises in the biology, ecology and sustainable management of sharks and rays. He has worked on blacktip reef sharks, citizen science projects, hammerhead sharks and Indigenous fisheries, and he has specific interests in coastal fisheries and the diverse values of sharks in the Indo-Pacific. Before coming to James Cook University in 2008, Andrew worked at the Great Barrier Reef Marine Park Authority connecting research to priority management needs. Now in his current role, Andrew continues to focuses on applied research that addresses end user needs.

J. H. Choat
Howard Choat is an Adjunct Professor of Marine Biology at James Cook University and has worked extensively on coral reef fishes over the Indian, Pacific and Atlantic Oceans.

B. C. Congdon
Brad Congdon is a 'Reader in Ecology' at James Cook University, Cairns. He is a field ecologist who applies ecological and evolutionary theory to the management and conservation of animal and plant species. He has a special interest in seabird conservation and has worked extensively with seabirds both in Australia and overseas. His current research is focused on understanding how changing ocean conditions impact seabird breeding success throughout the Great Barrier Reef and Coral Sea ecosystems. His research group was the first to demonstrate that seabirds are sensitive indicators of multiple climate-change impacts on top predators in these areas and have established rising sea-surface temperatures as a major conservation issue for seabirds of the Great Barrier Reef.

A. L. Crowther
Andrea Crowther is Senior Collection Manager, Marine Invertebrates at South Australian Museum. She completed her PhD at the University of Kansas, USA, where she studied shallow, tropical sea anemones. She has worked and published on sea anemones and other Cnidaria from around Australia with the Queensland Museum and she now works on sea anemones from the deep Southern Ocean.

J. C. Day
Jon Day was a terrestrial park planner and park ranger for the first 11 years of his professional career. In 1986, he joined GBRMPA, the agency responsible for the Great Barrier Reef (GBR). Over the following 28 years, Jon was involved in many aspects of planning and managing the GBR, including seven years in field management. In 1998, he was appointed as one of the Directors of GBRMPA, and for the next 16 years was variously responsible for conservation, planning, heritage (particularly World Heritage), Indigenous Partnerships, the GBR rezoning program and commencing the first 5-yearly Outlook Report. Jon retired from GBRMPA in 2014 to undertake a post-career PhD at the ARC Centre for Coral Reef Studies at JCU. He continues to publish widely on the GBR and World Heritage matters.

G. Diaz-Pulido
Guillermo Diaz-Pulido grew up in Colombia. He completed his BSc (Hons) in Marine Biology in Colombia in 1995 and his PhD in Marine Botany at James Cook University in 2002. He has done pioneering work on the ecology and diversity of reef algae from the Caribbean Sea and the Great Barrier Reef. His current research focuses on the ecology of coral–algal interactions, the impacts of ocean acidification and warming on macroalgae, and the potential of macroalgae to adapt to climate change. He is Associate Professor at the Griffith School of Environment at Griffith University.

P. J. Doherty
Peter Doherty is a post-retirement Fellow of the Australian Institute of Marine Science (AIMS), gained his PhD in 1980 by describing the population dynamics of damselfishes at One Tree Island. In 1989 he joined AIMS and led a research group in tropical fisheries ecology. In 1998, he joined the Co-operative Research Centre for the Great Barrier Reef World Heritage Area. One of his achievements was to facilitate the $9 million research collaboration known as the GBR Seabed Biodiversity Project. His fondest memories are more than 100 days spent at sea on the *RV Lady Basten* working the back deck from midnight to midday.

S. Dove
Sophie Dove obtained an undergraduate degree in Mathematics and Philosophy from the University of Edinburgh, and a PhD in Biological Sciences from the University of Sydney. She is presently an Associate Professor within the School of Biological Sciences at the University of Queensland, as well as a Chief Investigator within the Australia Research Council Centre for Excellence in Coral Reef Studies. Her research interests predominantly lie in the area of coral reef dynamics under climate change and cover a range of organisms (e.g. macro-algae, sponges and corals) that contribute to the carbonate balance of reefs. The goal of this research is to gain a broader understanding of how different combinations of elevated sea surface temperature

and acidification may influence the coastal protective properties of shallow tropical reefs.

N. C. Duke
Norm Duke is a mangrove ecologist of more than 40 years standing, specialising in mangrove floristics, biogeography, genetics, climate change ecology, vegetation mapping, plant–animal relationships, pollution and habitat restoration. As Professorial Research Fellow, he currently leads an active research and teaching group on marine tidal wetlands at James Cook University TropWATER Centre. With his detailed knowledge and understanding of tidal wetland processes, he regularly advises on effective management and mitigation of disturbed and damaged ecosystems. He has published more than 230 peer-reviewed articles and technical reports, including his authoritative popular book *Australia's Mangroves*, and innovative smart device apps on Mangrove ID for all mangrove plant taxa in Australia and the World.

M. Ekins
Merrick Ekins was previously a plant pathologist but then decided he would rather go SCUBA diving for a living than walking through fields of sunflowers. He has been the Collection Manager of Sessile Marine Invertebrates at the Queensland Museum for the last 14 years and still loves discovering new species – one of the joys of science. His main research is describing new species of sponges, octocorals, ascidians and jellyfish. As well as taxonomic identification, Merrick has also explored the population genetics and connectivity of rock sponges on ocean deep seamounts.

K. Fabricius
Katharina Fabricius is a coral reef ecologist, and holds the position of a Senior Principal Research Scientist at the Australian Institute of Marine Science. Her main research interest is to better understand the roles of disturbances, including changing water quality, ocean acidification and the cumulative effects of multiple disturbances on the biodiversity and functions of coral reefs around the world. Katharina was awarded a PhD in 1995 for her work on octocoral ecology. She has published over 140 journal articles, book chapters and a book on Indo-Pacific octocorals, jointly produced with Phil Alderslade.

L. Gershwin
Lisa-ann Gershwin is an international authority on medusae and ctenophores. Her research interests include taxonomy, systematics, biogeography, biodiversity, toxinology, bloom dynamics, marine stinger safety and prediction, and mesopelagic invertebrates. She has worked on medusae and ctenophores around the world since 1992, with particular focus on Australian species since 1998. While working in Australia, she has collected many thousands of specimens, with at least 212 species new to science, including numerous new genera, families and even a new suborder. She is the cocreator of *The Jellyfish App* and author of the best-selling books *Stung! On Jellyfish Blooms and the Future of the Ocean* and *Jellyfish: A Natural History*, both of which have been translated into multiple languages.

D. P. Gordon
Dennis Gordon FLS has been studying bryozoans for 50 years. He is a past President of the International Bryozoology Association and is currently an Emeritus Researcher at the National Institute of Water and Atmospheric Research, Wellington, New Zealand.

H. Heatwole
Harold Heatwole is an ecologist and herpetologist, who in the past 54 years successively held academic posts at the University of Puerto Rico, University of New England and North Carolina State University. His main research interests are sea snakes, island ecology, ants and tardigrades. He is the author of over 350 scientific articles and 11 books. He also produces videos for educational purposes. He holds a DSc and PhDs in Zoology and Botany, and Geography. The last two dealt respectively with vegetation dynamics on the small cays of the Great

Barrier Reef and the role of paleogeography in the evolution and dispersal of sea kraits. He is a Fellow of the Explorer's Club. Currently he is Professor Emeritus in biology at North Carolina State University, USA, and Adjunct Professor of Zoology at the University of New England, NSW, Australia.

O. Hoegh-Guldberg
Ove Hoegh-Guldberg is Professor of Marine Science and Director of the Global Change Institute (GCI) at the University of Queensland. After completing his BSc (Hons) Ove travelled to the United States to complete his PhD at the University of California at Los Angeles (UCLA). After postdoctoral work at the University of Southern California (USC) and a lectureship at UCLA, Ove, returned to University of Sydney in 1992 where he spent 8 years before taking the Chair at the University of Queensland. Here, he led the Centre for Marine Studies until 2010, when he became Director of the GCI. His research group focuses on the physiological ecology of corals and coral reefs, particularly regarding global warming and ocean acidification. In addition to his work producing published science (>300 articles), Ove was the coordinating lead author for Chapter 30, ('The Ocean') for the fifth assessment report of the Intergovernmental Panel on Climate Change (IPCC). He is currently also a Coordinating Lead Author for the Special IPCC Report on the implications of 1.5°C as a climate target. In 1999, he was awarded the Eureka Prize for his scientific research, and the Queensland Smart State Premier's Fellow (2008–2013). In 2012, he received a Thomson Reuters Citation Award in recognition of his contribution to research and an Australian Research Council Laureate Fellowship. He is member of the Australian Academy of Science, and received the Climate Change Prize from HSH Prince Albert II of Monaco in 2014 and the Banksia Foundational International Award in 2016.

M. O. Hoogenboom
Mia Hoogenboom received her PhD from James Cook University, Australia after which she conducted postdoctoral research at the Centre Scientifique de Monaco and the University of Glasgow, Scotland. She currently teaches coral reef ecology in the College of Science and Engineering at James Cook University, and is also a Program Leader in the ARC Centre of Excellence for Coral Reef Studies. She is a coral physiologist and her research focuses on chronic and acute threats to coral health and coral community structure.

J. N. A. Hooper
John Hooper, Head of the Biodiversity & Geosciences Programs, Queensland Museum, is an international authority on sponges (Phylum Porifera) with specific research interests in taxonomy, systematics, biogeography, biodiversity and conservation biology, and collaborating with 'biodiscovery' agencies over the past three decades in the search for new therapeutic pharmaceutical compounds (and discovering thousands of new species along the way).

D. Hopley
David Hopley is a coastal geomorphologist, holding the position of Professor Emeritus in the College of Science and Engineering at James Cook University. He held a personal chair in marine science and has an association with the University exceeding 50 years. He worked on coral reef evolution, and changing sea levels, especially on the Great Barrier Reef. He has published more than 150 scientific papers on this and related topics along with two major books, *The Geomorphology of the Great Barrier Reef: Quaternary Evolution of Coral Reefs* (Wiley Interscience, 1982) and, with Scott Smithers and Kevin Parnell, *The Geomorphology of the Great Barrier Reef: Development, Diversity and Change* (Cambridge University Press, 2007).

P. A. Hutchings
Pat Hutchings is a Senior Fellow, Australian Museum Research Institute, having retired in 2016 as a Senior Principal Research Scientist at the Australian Museum. She has spent her research career working on the systematics and ecology of polychaetes. In addition, she has studied the process of

bioerosion, not only on the Great Barrier Reef but also in French Polynesia. More recently she has become interested in bioerosion on fossil reefs and how this process changed with major climatic changes affecting reefs and if this can be helpful in understanding the changes currently impacting on our modern day reefs. As well as publishing extensively, she has been active in the Australian Coral Reef Society, which was recognised by her being given honorary life membership. She has also been very active in commenting on management and zoning plans for Australian coral reefs.

G. P. Jones
Geoff Jones is one of the world's most influential authors in the fields of coral and temperate reef fish ecology, and marine conservation biology, with ~270 publications in peer-reviewed journals and books. Geoff graduated with a PhD from the University of Auckland in 1981 and has held postdoctoral fellowships at the University of Melbourne and Sydney. He is currently a Distinguished Professor in Marine Biology, College of Science and Engineering at James Cook University, and a Chief Investigator in the Australian Research Council Centre of Excellence for Coral Reef Studies. His special interests are in the processes determining the structure and dynamics of reef fish populations, and human impacts on and conservation of threatened marine habitats and species. He has worked extensively on the rocky reefs of New Zealand and Australia, and on the coral reefs around the world, including the Great Barrier Reef, Papua New Guinea and the Pacific. Geoff and collaborators were the first to tag and recapture marine fish larvae. He has since become a world leader in the field of marine population connectivity and its implications for the ecology, conservation and management of reef fish populations. He teaches in marine conservation biology at James Cook University and has supervised over 160 graduate research students.

R. Kelley
Russell Kelley is a science communication consultant, author of the Indo Pacific Coral Finder and the Reef Finder, and manager of BYOGUIDES (www.byoguides.com; www.russellkelley.info). His outputs include scientific papers, book chapters, educational tools and international broadcast television productions.

M. J. Kingsford
Michael Kingsford is a Distinguished Professor in the Marine Biology and Aquaculture group of the College of Science and Engineering at James Cook University, Australia. The College is a recognised world leader in tropical marine biology and ecology. He has published extensively on the ecology of reef fishes, jellyfishes, biological oceanography and climate change. His projects have encompassed a range of latitudes and he has edited two books on tropical and temperate ecology. He is a Chief Investigator with the ARC Centre of Excellence for Innovative Coral Reef Studies. A major focus of his research has been on connectivity of reef fish populations, the utility of Marine Protected Areas, environmental records in corals and fishes and the ecology of deadly box jellyfishes. In addition to research and multiple senior leadership roles, he teaches undergraduate students and supervises many postgraduate students. He has over 30 years' research experience on the Great Barrier Reef.

A. W. D. Larkum
Anthony Larkum has worked in many fields – from molecules to ecosystems. His early interests were in the way plants absorb nutrients. However, an interest in SCUBA diving stimulated an interest in algae and in how algae are adapted to light fields underwater. This led to a lifelong interest in the physiology and ecology of algae and seagrasses. He has edited two books on the biology of seagrasses. He was instrumental in setting up the University of Sydney's One Tree Island Research Station and has been fascinated with the various roles of algae in the coral reef ecosystem. He initiated the ENCORE (Enrichment of Nutrients on Coral Reefs) Project at One Tree Island that looked at the effect of raising the local levels of nitrogen and phosphorus on coral reef organisms. He is also currently working on the potential effects of global

climate change, especially coral bleaching. His most recent interests are in the cyanobacteria of stromatolites and the role of chlorophyll-*d* and -*f* in these ecosystems.

V. Lukoschek
Vimoksalehi Lukoschek is a molecular ecologist whose overarching interest is the application of molecular genetic techniques to understand the ecology and evolution of marine vertebrates and invertebrates, and the application these findings to conservation and management. She has an ongoing research interest in sea snakes, which were the focus of her PhD at James Cook University (JCU) and a post-doctoral research fellowship at UC Irvine with Professor John Avise. She was the founding co-chair of the IUCN Sea Snake Specialist Group. Vimoksalehi has held a Research Fellowship with Professor Scott Baker at the University of Auckland working on cetacean genetics. Since 2010, Vimoksalehi has been a member of the ARC Centre of Excellence for Coral Reef Studies at JCU working on connectivity of broadcast spawning corals in collaboration with researchers at the Australian Institute of Marine Science, University of Melbourne and University of Queensland. Her research has been supported by a Queensland Government Smart Futures Fellowship, the Australian Academy of Science Endangered Species Fund, SeaWorld Research and Rescue Foundation, and Australia's Commonwealth Department of Environment. She currently holds an ARC Discovery Early Career Researcher Award.

H. Marsh
Helene Marsh is a marine conservation biologist based at James Cook University, with some 40 years' experience in research into species conservation, management and policy with particular reference to tropical coastal and riverine megafauna, especially marine mammals. She is a fellow of the Australian Academy of Science and the Australian Academy of Technological Sciences and Engineering and her research has been recognised by awards from the Pew Foundation, the Society for Conservation Biology and the American Society of Mammalogists. The policy outcomes of her research include significant contributions to the science base of the conservation of dugongs in Australia and internationally at a global scale (International Union for the Conservation of Nature, United Nations Environment Program, Convention for Migratory Species) and by providing advice to the governments of some 14 countries. Helene chairs the Australian Threatened Species Scientific Committee, a statutory committee that makes recommendations to the federal Minister for Environment and is a member of the Australian delegation to the World Heritage Committee. She is past President of the International Society of Marine Mammalogy, Co-chair of the IUCN Sirenia Specialist Group and is on the editorial boards of Conservation Biology, Endangered Species Research and Oecologia. Helene is proud of the accomplishments of the 55 PhD candidates that she has supervised to graduation, all of whom have taught her a lot. For further information see https://research.jcu.edu.au/portfolio/helene.marsh.

A.D. McKinnon
David McKinnon is an Associate Scientist at the Australian Institute of Marine Science (AIMS) and an adjunct Professor at James Cook University of North Queensland. Over a 40 year professional career, he published over 115 peer-reviewed scientific articles in several fields of marine science, including zooplankton ecology, coral reef trophodynamics and biological oceanography. David McKinnon has described over 70 copepod species, including three new genera. Since retiring from his position as a Principal Research Scientist at AIMS in 2015, he has continued to indulge his passion for marine science according to his whim.

J. M. Pandolfi
John Pandolfi is Professor in the School of Biological Sciences and the Centre for Marine Science at the University of Queensland, Australia. His research integrates paleoecological, ecological, historical and climate data to provide critical insights

into how marine communities are assembled and structured in the face of environmental variability and human impacts over extended periods of time.

C. R. Pitcher
Roland Pitcher is a Senior Principal Research Scientist at CSIRO Oceans & Atmosphere and was Principal Investigator of the Great Barrier Reef Seabed Biodiversity Project. His research over ~35 years on seabed habitats and biota, including distribution and abundance mapping, effects of prawn trawling, recovery and dynamics, and modelling provides an objective foundation to assist management in achieving sustainability of the seabed environment.

M. S. Pratchett
Morgan Pratchett is a Professorial Research fellow in the ARC Centre of Excellence for Coral Reef Studies at James Cook University (JCU). He also studied marine biology at JCU, but had several overseas postdoctoral appointments (University of Perpignan, Nova South-eastern University Oceanographic Centre and the University of Oxford) before taking up his current position. Morgan is a behavioural ecologist and conservation biologist working almost exclusively on coral reef systems and species. His current research focus is on major disturbances that impact coral reef ecosystems, including outbreaks of crown-of-thorns starfish and climate-induced coral bleaching.

B. C. Russell
Barry Russell is Curator Emeritus of Fishes at the Museum and Art Gallery of the Northern Territory in Darwin. He has over 45 years' research experience on the systematics, ecology and behaviour of tropical demersal fishes of the Indo-West Pacific. His current research interests include the taxonomy and phylogenetics of threadfin breams (Nemipteridae) and lizardfishes (Synodontidae). He is co-chair of the IUCN Species Survival Commission Snapper, Seabream and Grunt Specialist Group.

R. Saunders
Richard Saunders is a fish biologist whose current focus is on research that informs the sustainable management of Queensland's fisheries. Richard received a BSc (Hons) studying invertebrate ecology and PhD studying fish biology both from The University of Adelaide. He worked for 7 years in fisheries research at the South Australian Research and Development Institute and since 2012 has worked as a Senior Fisheries Biologist with the Queensland Department of Agriculture and Fisheries. Richard is also an adjunct Research Fellow with James Cook University in the Centre for Sustainable Tropical Fisheries and Aquaculture.

T. L. Sih
Tiffany Sih is a fish ecologist studying tropical fisheries and the deeper reefs of the Great Barrier Reef. Tiffany's research incorporates underwater video and advanced otolith chemistry to spatially refine fish populations at mesophotic depths. Tiffany received a BS in Biological Sciences from the University of Southern California, USA, and a Masters in Applied Science and PhD in Marine Biology from James Cook University, Australia, affiliated with the ARC Centre of Excellence for Coral Reef Studies and the Australian Institute of Marine Science.

S. Smithers
Scott Smithers is an Associate Professor in Tropical Geomorphology at James Cook University, Townsville. He has more than 25 years' experience of the geomorphology of coral reefs in the Pacific and Indian Oceans, as well as in the Caribbean Sea. Recent research on the Great Barrier Reef includes investigations of: Holocene sea-level change; the styles and rates of Holocene reef growth in different settings, with a particular focus on inshore turbid zone reefs; reef island evolution and morphodynamics; and reef carbonate budgets. Scott is a passionate advocate of the value of geomorphology to ecological understandings and management of the Great Barrier Reef (and other reefs). For

example, Scott's team delivered the geomorphological investigations that underpin recent sand movements on Raine Island undertaken to improve nesting and hatching outcomes for Endangered green sea turtles.

M. Srinivasan
Maya Srinivasan received her PhD from James Cook University (JCU), Australia in 2006. She is originally from Malaysia, and now lives in Australia, undertaking research and teaching at JCU. She has been conducting research on coral reef fishes for 20 years and has taught several marine biology subjects at JCU over the last 15 years. Her research focuses on population connectivity of coral reef fishes as well as conservation related questions, such as the impacts of climate change and the effectiveness of marine reserves. She has worked on coral reefs in many locations, including the Great Barrier Reef, Malaysia and Papua New Guinea. She has a voluntary role as board member and scientific advisor for Mahonia Na Dari Research and Conservation, a conservation and environmental education NGO in Papua New Guinea.

C. Syms
Craig Syms is a lecturer at the School of Marine and Tropical Biology at James Cook University. He has published a range of papers on the relationships between reef fishes and their habitats. His current research examines the role of different spatial and temporal scales of habitat variability in structuring communities. In addition to research, he has also advised extensively on marine resource management and evaluation of marine reserves in California. He teaches postgraduate sampling and experimental design and statistics, and supervises postgraduate students in a range of different marine projects.

C. C. Wallace
Carden Wallace is Principal Scientist Emeritus, Biodiversity and Geosciences Program, Queensland Museum. She has researched biodiversity and evolution of staghorn corals on reefs around the world and is author of the monograph *Staghorn Corals of the World*. She was a member of the group from James Cook University who described the coral mass spawning phenomenon and has a particular interest in the corals and sea anemones of Moreton Bay. She now works on the fossil history and evolution of family Acroporidae with colleagues from many countries.

J. M. Webster
Jody Webster is Co-Coordinator of the Geocoastal Research Group in the School of Geosciences at The University of Sydney. His research in sedimentology and stratigraphy focuses on carbonate sedimentology, climate change, and tectonics around the world (e.g. the Great Barrier Reef, Tahiti, Hawaii, Papua New Guinea, Seychelles and Brazil). Jody is particularly interested in coral reef and carbonate platform systems, both modern and ancient, and their associated sedimentary systems (i.e. slopes and canyons) as tools to address fundamental questions in paleoclimate variability and tectonics, and in turn the influence of these factors on the geometry, composition and evolution of these sedimentary systems. Jody is also heavily involved in several large international research programs including the International Ocean Discovery Program, which is focused on recovering sediment cores from the sea bed to understand past sea level and climate changes.

R. C. Willan
Richard Willan is a molluscan taxonomist, presently Senior Curator of Molluscs at the Museum and Art Gallery of the Northern Territory in Darwin. Formerly he was on the staff of the Zoology Department at the University of Queensland in Brisbane, from where he studied the molluscs of the Great Barrier Reef. During that time he visited research stations on the reef, studying sea slugs and bivalves. He is the authority on invasive marine molluscs in Australia. As the result of many visits

to the Heron Island Research Station, he collaborated with Julie Marshall to write *Nudibranchs of Heron Island, Great Barrier Reef.*

E. Wolanski
Eric Wolanski is a coastal oceanographer and holds positions at James Cook University and the Australian Institute of Marine Science. He has 402 publications. He is a fellow of the Australian Academy of Technological Sciences and Engineering, the Institution of Engineers Australia (ret.), and l'Académie Royale des Sciences d'Outre-Mer (Belgium). He was awarded an Australian Centenary medal, a Doctorate Honoris Causa from the Catholic University of Louvain and from Hull University, a Queensland Information Technology and Telecommunication award for excellence, and the 1st Lifetime Achievement Award from the Estuarine and Coastal Science Association. He is a member of the Scientific and Policy Committee of EMECS (Japan) and of the Scientific and Technical Advisory Board of Danubius-pp (Europe). He is an Erasmus Mundus scholar and is listed in Australia's *Who's Who.*

1

Introduction to the Great Barrier Reef

P. A. Hutchings, M. J. Kingsford and O. Hoegh-Guldberg

The Great Barrier Reef (GBR) is one of the world's most spectacular natural features and one of the few biological structures visible from space (Fig. 1.1). The sheer size of the GBR Marine Park (over 360 000 km^2), as well as its beauty and biodiversity, draws people from all over the world. The reef stretches over 2200 km from subtropical waters (~27°S) to the tropical waters of Torres Strait (8°S) and as far as 400 km from the coast to the outer shelf slope. For Australians, the reef is a source of much pride and enjoyment. The GBR is one of the most prominent icons of Australia, with the majority of visitors coming specifically to Australia to see it. This drawcard for visitors underpins substantial income from industries such as tourism and commercial and recreational fisheries. What is perhaps surprising about the value generated by this extraordinary ecosystem for Australia is how little we know about the reefs that make up the GBR. We are still struggling to describe the myriad of species and processes that define the GBR – all with an urgency now that is heightened by the unprecedented local and global pressures that currently face it.

The frequency, and in some cases intensity, of major events affecting the GBR have increased in recent years. The first major recorded coral bleaching incident was in the early 1980s, followed by events in 1987, 1998, 2002, 2010, 2015 and 2016. The last two bleaching incidents resulted in an 80% loss of coral in northern regions of the GBR, while southern regions were largely unaffected. Although the number of cyclones affecting the reef has if anything dropped, the intensity has increased. In the last 12 years the reef has been subjected to five category 5 cyclones, one of which (Cyclone Hamish) wreaked havoc to habitats on outer reefs of the GBR over ~1000 km. Threats from crown-of-thorns starfish, overfishing, run off from the land and a warming planet are impacts that organisms of the GBR have to deal with. Accordingly, these threats and challenges sharpen our resolve to understand how these factors can affect habitats and how a knowledge of pattern and process can provide the solutions that are so desperately needed if we are to avoid a heavily compromised GBR that is no longer a coral-dominated paradise.

Consequently, the underlying concept behind this book is to describe the patterns, processes, human interactions and organisms that underpin large reef ecosystems such as the GBR. Although much of the content of this book is focused on the GBR, we consider it highly relevant to coral reefs in other parts of Australia and the rest of the world. There has been no other comprehensive introduction to the biology, environment and management of the GBR, especially with regard to the major processes that underpin it or how issues such as deteriorating coastal water quality and climate change affect it. Extending our knowledge and understanding of these processes is vital if we are to sustainably manage the Reef, and reassure its future, especially during the coming century of rapid, anthropogenic climate change. Only by understanding and managing the GBR wisely do we have a chance of keeping the GBR 'great'.

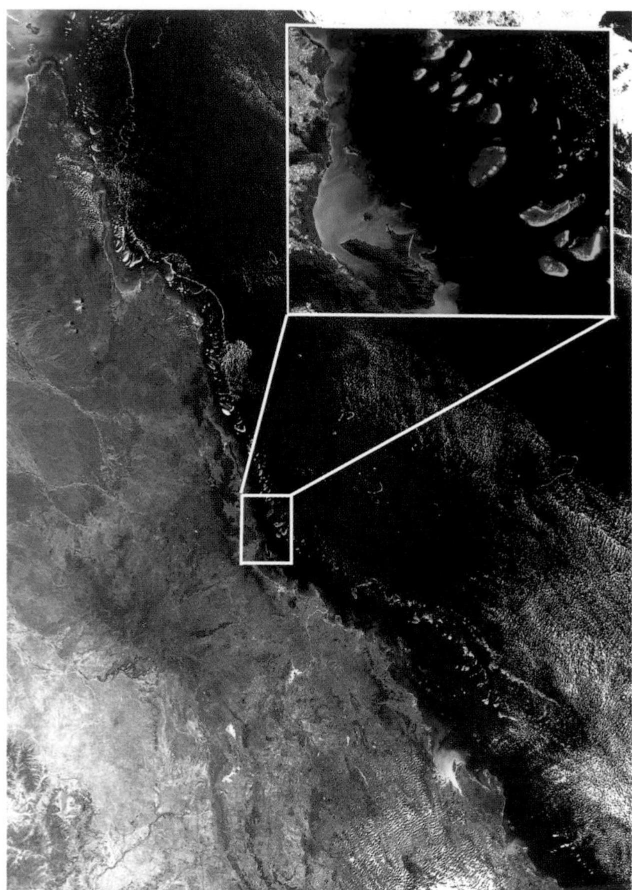

Fig. 1.1. Quasi-true colour image from space showing the Great Barrier Reef along the continental shelf of north-eastern Australia from Cape York to Gladstone (~2000 km). A mosaic of reefs can be seen parallel to the mainland and extending hundreds of kilometres across the GBR. (Source: Image generated from the Moderate Resolution Imaging Spectroradiometer (MODIS), data courtesy of NASA/GSFC, image courtesy of Dr Scarla Weeks, Centre for Marine Studies at The University of Queensland. Inset shows a closer cross-shelf view of the reefs out from Hinchinbrook, near Townsville).

This book is aimed at undergraduate and post-graduate students, the informed public, as well as researchers and managers who would like to familiarise themselves with the complexity of coral reefs such as the GBR. The project arose out of an advanced undergraduate course that has been held on the GBR for almost 20 years and which generated extensive discussions on the need for a book such as this at the Australian Coral Reef Society (ACRS). Through the Society and the course, we were able invite the appropriate international experts to contribute to this book. The ACRS (which started out as the Great Barrier Reef Committee) is the oldest coral reef society in the world and most of the authors are members and associates of this society. It was also intimately involved in the establishment of the world's largest marine park, the GBR Marine Park in 1975. This huge park, with some of the largest no-take areas in the global ocean, was enlarged and rezoned in 2004 based on our much increased scientific understanding of coral reefs and their challenges and solutions.

The book is divided into three sections. The first section focuses on the geomorphology, paleobiology and oceanography of the GBR. Here, various habitats of the GBR are discussed, not only from the point of view of coral-dominated ecosystems, but also with regard to the important associated inter-reefal areas (Chapters 2–9). These components of the GBR, along with catchments and off-shore deeper waters, are highly interconnected (Fig. 1.2). The second section of the book focuses on the major processes that are affecting the reef and includes the description of organisms and processes that contribute to photosynthetic activity and primary production, as well as the flow of energy and nutrients within coral reef ecosystems. Other chapters deal with the major forces within and around the reef, illustrating its inherent dynamic nature. This section also reviews our current understanding of how local challenges (i.e. declining water quality and over exploitation of fisheries) as well as aspects of climate change (e.g. ocean warming, acidification and intensifying storms) have changed the circumstances under which coral reefs have otherwise prospered for thousands if not tens of millions of years (Chapters 10–14).

The third and final section of this book deals with the diversity of organisms that live in and around coral reefs (Fig. 1.3). In this section, the reader is introduced to broad categories of organisms (e.g. plankton) as well as the basic taxonomy of the major groups of organisms and their biology and ecology (Chapters 15–32). We believe that this provides the reader with a fascinating journey

Fig. 1.2. Crossing the Blue Highway. Designed and written by science communicator Russell Kelley and published by the ACRS, the Blue Highway poster portrays the reefs of the Great Barrier Reef as part of a larger supporting system that includes the coastal catchments and connects numerous interacting ecosystems from ridge to reef. The poster illustrates how natural nutrient loads from runoff and ocean upwelling fuel a connected mosaic of ecosystems and the role inter-reef habitats play in supporting migrating species as they move from inshore nursery grounds to the outer reefs. The model species of fish is *Lutjanus sebae* (red emperor snapper) that spawn near the shelf edge and recruit to estuaries as larvae. Juveniles move from recruitment habitat to reefs and inter-reefal habitats, before they mature. Printed copies of this poster are available from the GBRMPA (Artwork: G. Ryan)

through the unique and spectacular biodiversity of coral reefs. By weaving the basic taxonomy of these groups together with fascinating details of their lives, it is hoped that the interest of the reader will be inspired to explore this incredible diversity.

Throughout this book we come back to the major challenges that reefs face in our changing world. For this reason, our book is unique in that it reviews the past, current and future trajectories and possible management responses of the GBR. Globally, coral reefs are at risk and knowledge-based management is absolutely critical. The GBR is at the forefront of this, with a growing risk of being severely degraded in the next few years if we don't understand the problems and apply evidence-based solutions.

It has been 11 years since the first edition of this book was published (2008). With a new decade, the pressures on the GBR have increased and trends such as increasing temperatures and ocean pH continue to warrant responses at a global and local level; you will find examples of this in all the chapters. We have added two extra chapters one on mesophotic reefs (Chapter 7) and another on fisheries of the GBR (Chapter 10). Chapter 7 reveals amazing diversity in habitats and associated fishes

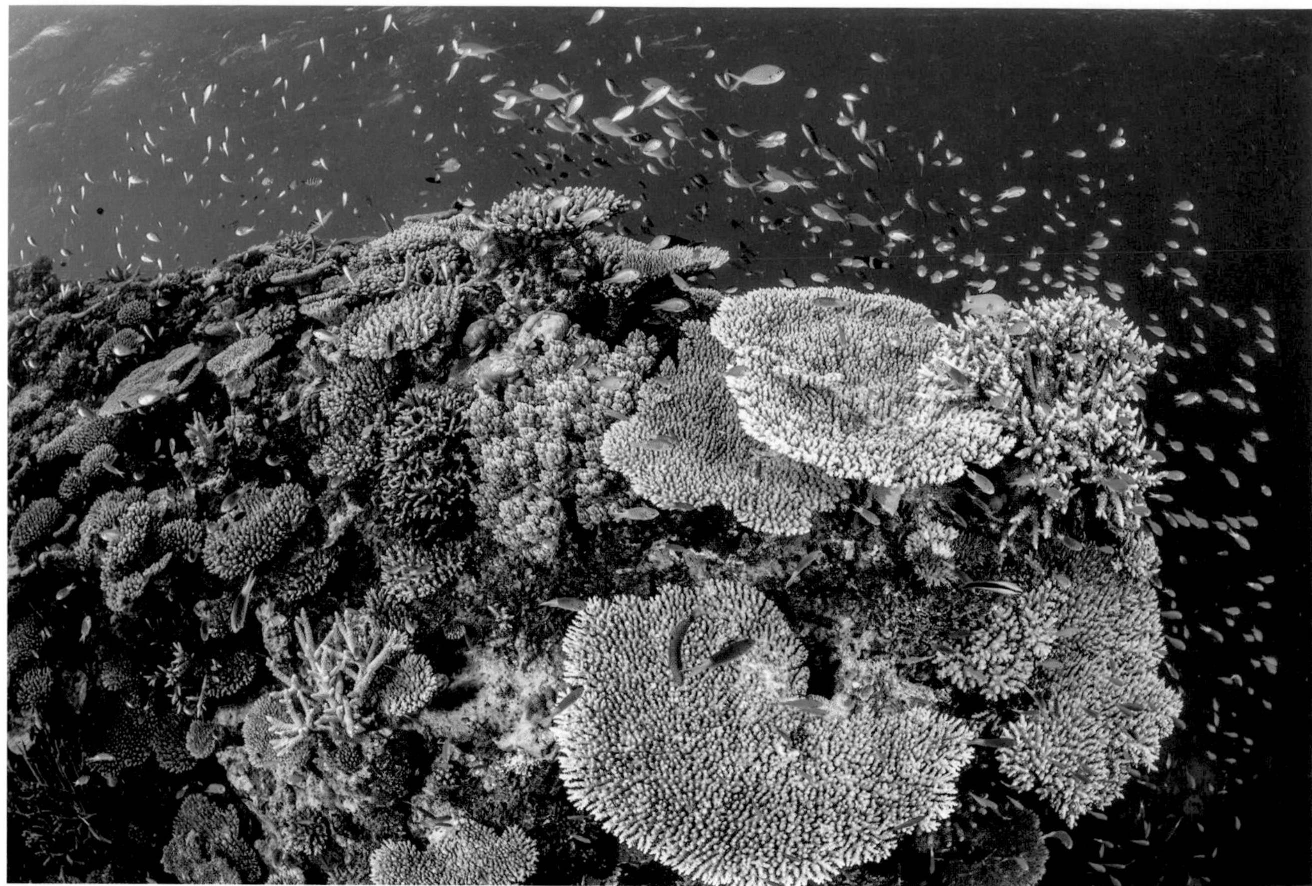

Fig. 1.3. A rich diversity of coral species in shallow water on the GBR. A large aggregation of plankton-feeding fish hover close to the complex architecture of the reef that provides them with shelter (Photo: D. Wachenfeld).

on the shelf slopes of the GBR that have been largely unknown until now (Fig. 1.4). Fishing methods and markets have changed with time and Chapter 10 describes changes in time and new challenges that relate to managing GBR fisheries with methods other than zoning plans.

Impacts of coral reefs are a global issue and most, or all, of the impacts we describe for the GBR are also true of other reef systems of the world that include the Caribbean, Indo-Pacific and Red Sea. Furthermore, many of the habitat-forming organisms (e.g. corals), those that influence them, such as herbivorous fishes, starfish and bioeroders, as well as the organisms that simply use different habitats on the GBR are similar in other coral reef systems. Detailed accounts of other coral reef ecosystems include Riegl and Dodge on reefs of the USA and that of the Japanese Coral Reef Society (MEJCRS).

For most chapters, citations are not included in the text, but a list of helpful references is provided in 'Further reading' at the end of the chapter. However, for some chapters, citations are inserted within the text because they refer to specific (sometimes controversial) points.

Finally, it is our hope that this book will help develop a better understanding of coral reefs, engage the next generation of coral reef scientists, and assist in maintaining the ecological resilience of coral reefs such as the GBR, in order to allow them to survive the challenges of the future. This said, we hope that you will enjoy using this book to discover the intricacies of the world's most diverse marine ecosystem. As editors, we would like to thank the generous contributions from the many authors that have contributed to this book, the production of which would have not been possible otherwise.

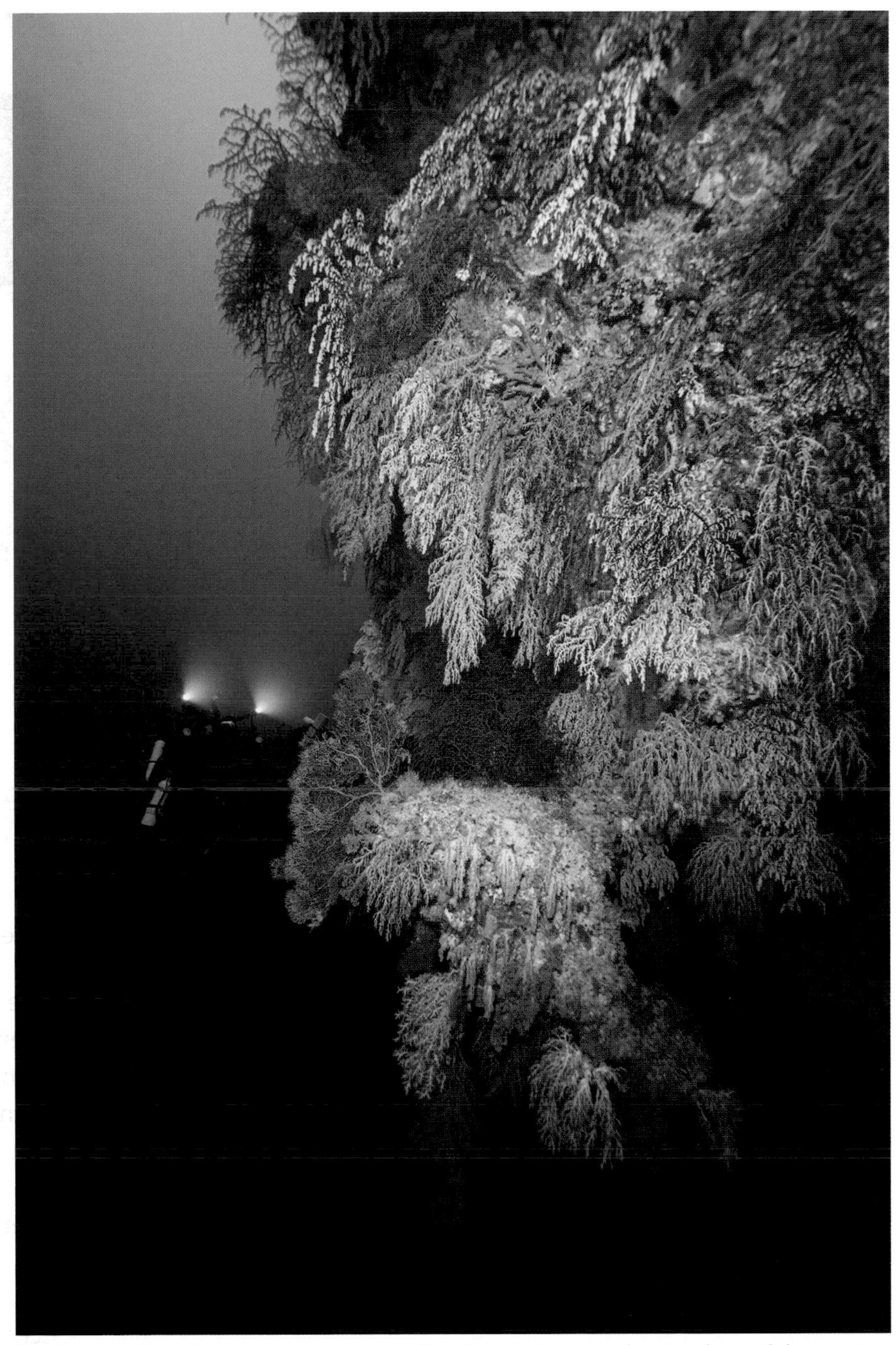

Fig. 1.4. Deep reefs of tropical environments are a new frontier and our understanding of their uniqueness and connectedness with shallower environments is in its infancy. Deep Reef ~70 m at 'Deep Arcade' on the north-western wall of Osprey Reef (Image: Simon Mitchell).

Further reading

Iguchi A, Hongo C (Eds) (2018) *Coral Reef Studies of Japan*. Coral Reefs of the World, Vol. 13. Springer, Singapore.

Omori M, Takahashi K, Moriwake N, Osada K, Kimura T, Kinoshita F, *et al.* (2004) *Coral Reefs of Japan*. Japanese Coral Reef Society, Ministry of Environment, Japan.

Riegl BM, Dodge RE (2008) *Coral Reefs of the World Vol. 1 Coral Reefs of the USA*. Springer, Netherlands.

SECTION 1

Nature of the reef

2

Geomorphology of coral reefs with special reference to the Great Barrier Reef

D. Hopley and S. Smithers

Coral reef geomorphology is the study of the morphological development of coral reef structures and associated landforms across a range of temporal and spatial scales. It integrates ecological, physical and geological information to understand controls on coral reef formation, morphological diversity and change – past, present and future. Coral reefs are biogenic limestone structures built by corals and other carbonate-producing organisms in shallow tropical and subtropical marine settings, where they grow upwards or towards sea level as landforms able to resist wave action. Most coral reefs have long and punctuated histories of development, comprising episodes of reef growth when global sea levels are high during interglacials interspersed with periods of emergence and erosion during the ice ages when global sea levels were up to 125 m lower. As such, many coral reefs consist of sequences of carbonate materials deposited by former reefs, with successive units separated by solutional unconformities developed when low sea levels exposed them to subaerial erosion. These limestone structures form the substrates on which many modern coral reefs now grow as relatively thin living veneers of corals and other reef organisms. The distribution, size and shape of these substrates significantly influences the geomorphology of many modern reefs.

Coral reef landforms include both coral reefs as described earlier and geomorphic features comprised of reefal sediments such as reef islands, sediment aprons and beaches. Unlike most other landforms, those on coral reefs are the result of interactions of ecological processes that influence the occurrence and productivity of calcium carbonate producers, and physical processes such as waves, currents and sea-level fluctuations that also influence ecological processes and can redistribute carbonate material across a reef system. Importantly, feedbacks occur between the ecological, physical and geological processes operating on coral reefs that are captured in geomorphological investigations. For example, as a coral reef grows towards the surface and into the wave base, the ecology of the reef may change to reflect increased hydrodynamic stresses. These ecological changes are recorded in a reef's geomorphology as changes in the composition of contributing organisms, reef accretion rates, and reef structure and fabric. However, as a reef grows upwards it also modifies waves and currents, influencing the distribution of carbonate-producing species and their products across it, evidence of which is also preserved geomorphologically and structurally. Coral reefs are thus 'living' and dynamic landforms that develop geomorphologies that both respond to and modify the environment in which they grow, all of which changes through time as growth proceeds.

Corals are the major reef-builders on most coral reefs (up to 10 kg $CaCO_3/m^2/year$). However, many other organisms secrete calcium carbonate skeletons and can make important contributions to reef construction, including coralline algae, molluscs, foraminifera, bryozoans and the green alga *Halimeda*. Even on healthy reefs, other plants and animals such as clionid sponges, urchins, chitons, clams and parrotfish simultaneously work to break reef framework and sediments into smaller fragments through bioerosion (see Chapter 9). Bioerosion rates can equal those of production and, together with chemical and physical erosion processes such as the mechanical destruction of reef materials by waves, constitute a suite of often synergistic destructive activities long-recognised to operate on most coral reefs. Importantly, the geomorphological development of coral reef landforms is thus the net result of constructive and destructive processes. Understanding this balance is important because significant reefal structures will not develop everywhere that corals grow. Similarly, the geomorphological development of many reefs record episodes of coral reef growth 'turn off' (sometimes for several centuries) and later 'turn on'. The nature and causes of these hiatuses, together with a reef's geomorphological response, provides important context for improved understandings of the potential outcomes and significance of challenges facing coral reefs at present and into the future (Chapter 12).

The distribution and growth of reef-building corals and thus coral reefs are constrained by strict environmental requirements. Sea-surface temperature is important; corals grow happily at various locations in sea-surface temperatures between 18° and 36°C but 26–27°C is considered optimal (Chapter 8). Extremes beyond this range may be briefly tolerated, such as at low tide in reef flat pools, but survival is limited if conditions persist. Individual corals may adapt to their ambient environmental conditions, generally also through the tolerance of their zooxanthellae. However, temperature excursions, either above or below the normal range, even if withstood by the same species elsewhere, may locally result in expulsion of the zooxanthellae, coral bleaching and possible mortality.

Salinity of ~36‰ (the open ocean level) is ideal for most reef-builders; the documented range is between 23‰ and 43‰. Low salinity produced by heavy rainfall or runoff can only be endured for short periods and is generally the reason reefs are absent close to major river mouths. Fresh water floats on salt water because it is less dense; the lowering of low salinity surface water onto a reef as the tide ebbs can be devastating, as can be the ponding of low salinity water in poorly flushed lagoons during tropical deluges.

Reef-building corals rely on photosynthesis by symbiotic zooxanthellae to rapidly produce calcium carbonate and are thus limited to the shallow photic zone where light is available (Chapter 8). The depth of the photic zone varies with latitude and with water quality, both of which affect light penetration into the water column. In highly turbid waters, coral growth may be limited to within a few metres of the surface but in clear oceanic waters, such as on the outer shelf of the Great Barrier Reef (GBR) on Myrmidon Reef off Townsville, corals and *Halimeda* continue to grow at 100 m and 125 m depth, respectively.

Sedimentation is a major control on reef growth, with poor larval settlement and coral recruitment typical where sedimentation rates and/or mobility are high. Unstable settlement substrata and burial are the major constraints. Turbidity, often confused with sedimentation, is another major control on reef growth. Turbidity is the suspension of sediments in the water column. It can be high even when sedimentation is low where sediments remain suspended and are not deposited due to the nature of the sediments, the water column, the energy regime, or all three. In some settings, turbidity is more transient and surprisingly high levels may be tolerated. The fringing reefs at Cape Tribulation and Magnetic Island experience turbidity primarily caused by resuspension of sediments during rough weather (and not flood plumes) and, as long as sediment does not settle on the reef and is kept moving by wave action, turbidity levels of more than 100 mg/L can be withstood. Furthermore, corals from inshore environments are far more tolerant of high suspended sediment levels than exactly the same species on mid and outer shelves. Strategies for success include corals switching to heterotrophy and gaining nutrition from ingestion of sediments (which can be done 24 h a day) rather than relying on autotrophy and photosynthesis only accessible during the daytime (Chapter 8). Turbid zone reefs are often surprisingly diverse. For example, 141 species from 50 genera of hard corals have been recorded from the Cape Tribulation reefs, and more than 80 species have been recorded at Middle Reef, a small nearshore reef located between Townsville and Magnetic Island. Core records from these and other inshore reefs on the GBR also indicate that these reefs have always grown in turbid water conditions and have accreted at rates equal to, and often exceeding, their clear-water counterparts, in part due to the incorporation of terrigenous sediments into their reef structures. These findings might suggest that increases in mean annual fine sediment delivery to the GBR of as much as 800% associated with European land use change are insignificant compared to the billions of tonnes delivered to the nearshore zone and resuspended by wave and current activity over the past 6000 years since sea level stabilised around its present position. However, where new sediment is reaching existing reefs it does have detrimental effects. For example, the fringing reef on High Island, south of Cairns, extends down to 20 m depth and new sediment settling below the wave base is not resuspended and is negatively impacting coral cover. Increased nutrients are in many cases also associated with increased sediment yields, and may further add to reef decline.

Geomorphological zonation

Conspicuous geomorphological and ecological zonation parallel to the reef front is a feature of most coral reefs. The zones develop in response to windward to leeward gradients in key environmental factors such as hydrodynamic energy, depth and light availability. Ecological and geomorphological zones are often clearly related, and are most strongly and distinctively developed where environmental gradients are steepest and spatial changes in conditions are most acute. Geomorphological zones commonly observed from seaward to leeward across a reef include:

- The windward coral-covered reef front slopes seawards into deeper water, and is generally steeper in higher-energy settings. A distinctive 'spur and groove' zone comprised of seaward protruding coralline algae or coral-covered buttresses (spurs) separated by channels (grooves), within which sediment may move, commonly occupies the upper reef slope in these areas, where it may merge with the reef crest at the sea surface.
- The reef crest commences with a living coral zone just below the low tide level. An intertidal algal pavement follows, the exact composition of which varies but on the GBR is rarely similar to the prominent coralline algal ridge of mid-oceanic atolls. Coralline algae are present, but this zone is usually dominated by turf algae. The pavement appearance of this zone is often mistakenly assumed to indicate low carbonate productivity, but this zone is an important habitat for large benthic foraminiferans that contribute significantly to the sand budgets of many reef islands. Large reef blocks commonly to 4 m in diameter and often larger can be scattered across this pavement; these blocks are torn from the reef front in cyclones or by tsunamis.
- The highest part of many reef flats is a rubble zone of coral shingle and larger fragments, forming a shingle rampart. It is formed of material from the reef front and deposited by waves as they lose energy passing over the reef flat. These ramparts can pond water over reef flats above the open water low tide level, allowing corals and other benthos to survive above their open water counterparts. These organisms are, however, vulnerable to water-level changes associated with subsequent breaching or attrition of the ramparts.

- The aligned coral or 'striated' zone forms leeward of the pavement zone, and is characterised by coral colonies growing in lines perpendicular to the reef front and separated by narrow sandy channels through which currents surge. Beyond this a sandy reef flat zone may occur, depending on the width, energy regime and morphology of the setting.
- The sheltered back-reef area, often in the form of a sand slope or lagoonal floor, typically contains the most fragile branching colonies and, if very deep, may be dominated by carbonate producers such as *Halimeda*. Sand deposits in this zone may be heavily bioturbated, and seagrass is also common.

Distinctive changes to coral colonial morphology also occur in response to environmental gradients expressed across many reefs (Fig. 2.1). Light, wave energy and emersion are the major controls on ecomorphological variation. At greatest depth on the reef front, light is the most limiting control and corals may adopt a globose or plate-like shape. Higher up, in the optimal photic zone, but below wave base, colonies can be intricately branching. Shallowing up into the wave zone stronger colonial structures dominate, with encrusting forms dominating reef crests in higher energy areas.

Exposure is the key limiting factor on the reef flat, where a variety of ecomorphotypes (ecomorphs) develop. The upper limit to coral growth

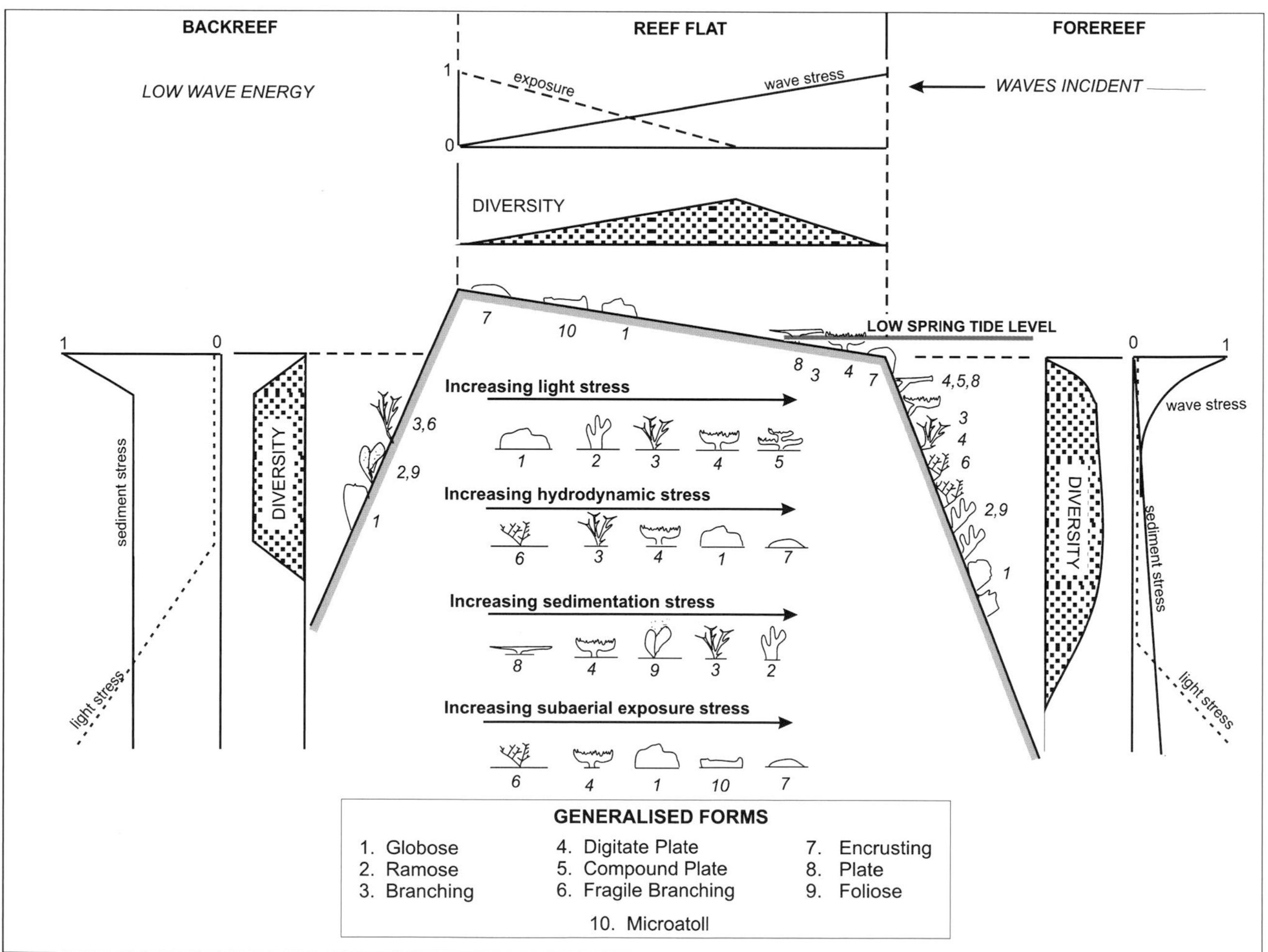

Fig. 2.1. Geomorphological and ecomorphological zonation on a reef in response to light, sediment and wave stress, and subaerial exposure (Source: reprinted by permission from Macmillan Publishers Ltd from: Chappell, J. (1980) p. 250).

is about low tide level, but varies with species and with local environmental conditions related, for example, to wave action or the time of day at which the critical low tides of the year occur. However, as indicated earlier, in some instances the low tide level on reef flats may be controlled by ponding or moating behind shingle ramparts or algal ridges. Under these conditions, the same water level is achieved on every low tide while even the rampart remains intact. Corals respond by growing to these levels and form distinctive microatolls: disc-shaped colonies with flat dead tops and living vertical sides. They form when upward coral growth is constrained by prolonged emersion at low tides, but the colony sides continue to grow laterally. Microatoll thickness varies depending on the depth of the moat, but colonies up to half a metre thick are common. Both massive and branching corals form microatolls, though they are most commonly formed by massive *Porites* colonies. Microatolls can grow to more than 10 m in diameter, and as their upper surface is determined by water level, their upper surfaces may record changes in this level. The surface morphology of ponded or moated microatolls, for example, has been used to reconstruct water-level changes associated with the cyclonic destruction of shingle ramparts, whereas the surface morphology of open water microatolls have been used to reconstruct interannual changes in sea level. Annual growth rings preserved within the skeletons of microatolls allow chronologies for surface morphology/water-level changes to be established. Because microatolls form only when corals are constrained by subaerial exposure at the sea surface, they have been widely used to determine when coral reefs have reached sea level and the age of reef flats on many reefs. When radiometrically dated, the elevations of fossil microatolls can be compared with the elevations of modern specimens growing in many cases on the same reef to establish patterns of relative sea-level change since the mid-Holocene.

In nearshore turbid settings, the whole ecomorphological zonation may be vertically compressed, and light can be a critical factor in determining the colony morphology. In clear water, colonies that are light-limited commonly adopt a plate-like form to maximise light capture, but in turbid settings plate corals are vulnerable to sedimentation, which would also impede growth. Turbid zone reefs are thus dominated by ecomorphs with forms such as whorled and foliose colonies that maximise light but also encourage sediment shedding.

The Great Barrier Reef

The GBR (Fig. 2.2) is the largest contiguous coral reef ecosystem in the world, and possibly the largest geomorphological structure ever created by living organisms. It has great geomorphological diversity, lacking only open ocean atolls. The GBR extends through almost 15 degrees of latitude between 9°15′S at the top of Torres Strait to Lady Elliott Island at 24°07′S on the southern end of the reef, a distance of ~2300 km. The outer reefs vary greatly in distance from the mainland coast, approaching within 23 km at 14°S to a maximum of 260 km at 21°S. On the wider shelf, reefs are established on the outer third, with a broad area of almost reefless sea floor at depths of between 20 m and 40 m extending landward in their lee. In the north, reefs occupy a much greater proportion of the shelf, but a narrow 'shipping channel' where reefs are largely absent can still be recognised. The outer edge of the GBR lies on the shoulder of the continental shelf in water depths of up to 100 m, beyond which the sea floor falls away almost vertically in the north, more gently in the central and southernmost GBR (Bunker-Capricorn Group), and in a step-like fashion off the southern-central GBR in the vicinity of the Pompey Reefs.

The dimensions of the GBR are often related to those of the Great Barrier Reef Marine Park (345 500 km^2) and the GBR World Heritage Area (348 000 km^2), both of which end at 10°45′S and thus do not include the reefs of Torres Strait. Within the GBRMP the area of continental shelf is ~224 000 km^2, of which reefs and shoals of near the surface cover 20 055 km^2 comprising ~9% of the total area. An additional 37 000 km^2 of shelf area

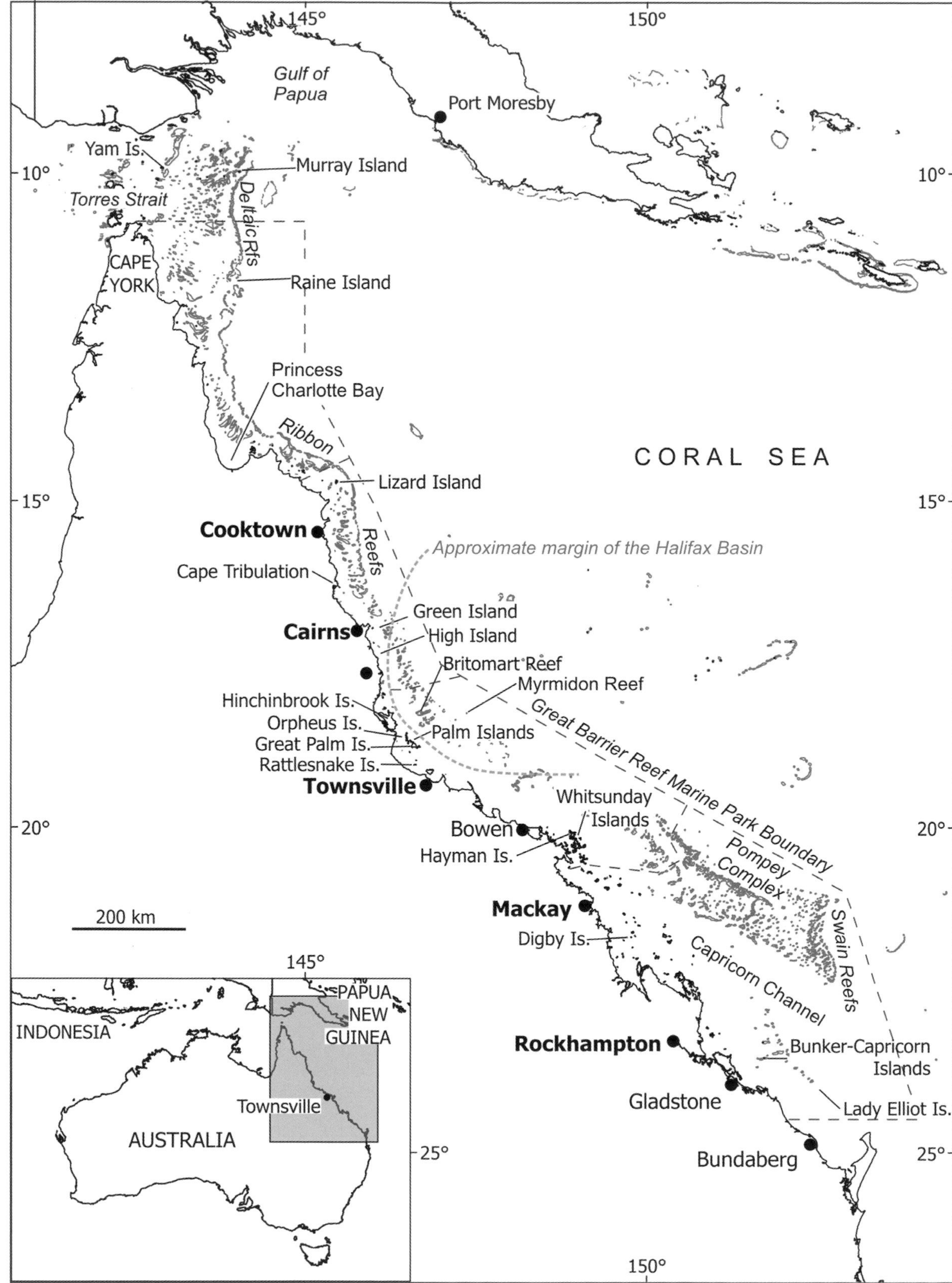

Fig. 2.2. The Great Barrier Reef – locations (Source: Scott Smithers).

and a further 750 reefs covering an area of ~6000 km^2 lie in Torres Strait. A figure of 2900 reefs, based on mapping of reefs visible at the surface on aerial photographs, has for a long time been widely quoted for the Marine Park. However, it is difficult to count separate reefs in complex parts of the GBR, and it is only recently with new technologies that the true number and distribution of more deeply submerged shoals has become apparent. Based on improved characterisation of deeper sea floor and sophisticated habitat modelling, it now seems likely that both the number reefs and area of 'coral reef' habitat on the GBR are much larger than previously recognised, with recent research suggesting that the number and area of coral reefs on the GBR may be as much as double long-believed estimates.

The most recent evolution of the GBR

The early establishment of the GBR is discussed in Chapter 3 where the importance of Pleistocene sea level changes is indicated. The last time sea level was at or above its present level was ~125 000 years ago at which time the GBR was at least as extensive as it is today. Subsequently, a series of interstadial high sea levels between 60 000 and 70 000 years ago and only 15 m to 25 m below present level were capable of adding further growth to the reefs of the outer shelf. Sea level reached its lowest point of at least –125 m during the peak of the ice age around 20 000 years ago. At this time, the whole of the GBR, including the shoulder of the continental shelf, was exposed and rivers such as the Burdekin flowed across it to the shelf edge.

The modern GBR developed as the postglacial marine transgression flooded the continental shelf. This began ~10 000 years ago although the first flooding may not have favoured coral growth as Pleistocene soils and regolith were reworked, producing turbid, eutrophic conditions. In some areas, extensive meadows of *Halimeda* were the first carbonate structures to develop. Coring shows that carbonate units rich in *Halimeda* leaflets lie beneath coral-dominated units on several reefs. *Halimeda* bioherms, or mounds, up to 18.5 m thick are found inside the ribbon reefs from Cooktown to Torres Strait. Although these structures have been known for many decades, they have recently been mapped in detail using airborne LiDAR and multibeam swathe mapping. This mapping revealed that these structures cover more than 6000 km^2 on the northern GBR, more than tripling the previously known extent. The mapping also revealed a range of intricate morphologies and pointed to the importance of such structures as habitat and for maintaining connectivity between coral reefs. Spatial patterns of development have driven speculation that *Halimeda* meadows and bioherm development is promoted where deep nutrient-rich water is jetted through the passes separating adjacent ribbon reefs.

By 9000 years ago, sea level was ~20 m below the present position and the older Pleistocene reefs, then present as limestone outcrops above a broad coastal plain, were being inundated by rapidly rising seas. The oldest radiocarbon ages from the Holocene reefs on the GBR are just over 9000 years, marking the earliest accretion of the modern GBR. Once inundated, recolonisation and vertical accretion above these Pleistocene substrates was rapid. The bulk of the modern Holocene reef was laid down between 8500 and 5500 years ago, when vertical accretion rates were up to 13 m per thousand years (mean ~6 m per thousand years). By around 7000 years, ago the transgression was effectively complete, but shelf warping as the result of hydro-isostasy (depression of the outer shelf and compensatory upwarping of the inner shelf due to the weight of the water) produced some cross-shelf contrasts. Thus, although modern sea level was reached by ~7000 years ago on the inner shelf (documented by raised reefs of up to 1.5 m on inner fringing reefs and the inner shelf low wooded island reefs of the northern GBR), the age becomes progressively younger towards the outer shelf where no emergence took place.

There is also evidence of tectonic shelf-warping in the central GBR where the Halifax Basin of the Coral Sea impinges upon the shelf. Downwarping is suggested by the extensive line of submerged reefs on the outer shelf south of Cairns. There are also regional patterns in the depths to the Pleistocene foundations of modern reefs. Although some variation may be expected due to original morphology or subsequent erosion, the pattern as shown by reef drilling also supports the downwarping on the outer half of the continental shelf. In the Torres Strait and the far northern region south to Cooktown there is a gradual increase in minimal depth from –5 m in the north to between –15 m and –17 m in the south. South of Cairns the depth to the Pleistocene foundation increases by more than –5 m and everywhere, except Britomart Reef (where it is at –8 m), it is more than 20 m below the sea surface. On a cross shelf transect just south of Mackay the variation is from 0 m at Digby Island, between –8 m and –13 m on mid-shelf reefs and –17.5 m beneath Cockatoo Reef in the Pompey Complex closer to the shelf edge.

Reef types of the GBR

Charles Darwin's reef classification that transforms fringing reefs attached to subsiding volcanic islands into barrier reefs and finally atolls elegantly explains the distribution and development of mid-ocean reefs, but contributes little to our understanding of reef development across the relatively stable, continental shelf that underlies the GBR. Nor is there any great latitudinal contrast on the GBR except for that related to shelf width. Instead, cross-shelf environmental gradients divide the GBR into three major zones: fringing and nearshore reefs; mid-shelf reefs; and outer shelf reefs.

Fringing and nearshore reefs

Some 545 fringing reefs with recognisable reef flats and 213 incipient fringing reefs (shore-attached reefs lacking reef flats) have been recognised in the GBR Marine Park. Most are small (average 1 km^2) and in total they cover only 350 km^2. In addition, there may be several hundred small inshore reefs and shoals, only a few of which have been mapped because of the turbidity of the waters they inhabit. Indeed, these poorly known reefs are often referred to as 'marginal' due to the difficult conditions in which they grow. However, recent investigations of inshore shoals in Central GBR indicate that they vary in age and continuity of growth (some are less than 1000 years old, others substantially formed 6000 years ago, and others have a punctuated growth history), can show remarkable rates of reef growth well above many clear water reefs, and that many have high diversity, coral cover and apparent resilience to bleaching.

Fringing reefs attached to the mainland occur in the Whitsunday area and north of Cairns, probably the best-known being those of Cape Tribulation. However, the majority are attached to the high continental islands of the inner shelf and their location puts them within the zero isobase of hydro-isostatic uplift. Some of the oldest reef flat dates on the GBR are found on these reefs (some >7000 years) and inner flats are frequently composed of raised microatolls indicative of a 1.5 m higher sea level. Similar to the outer reefs, the fringing reefs had commenced Holocene growth by 9000 years ago and accumulated much of their structure by 5500 years ago.

Fringing reefs have developed over a wide range of substrata (Fig. 2.3):

(i) Rocky foreshores – formed as sea level rose against the rocky slopes of offshore islands, especially on the windward side. These reefs are generally narrow and not as common as might be expected.
(ii) Pleistocene reefs – unlike mid- and outer shelf reefs, older Pleistocene fringing reef foundations are not common. The large fringing reef of Hayman Island is the best example. At Digby Island, a Pleistocene reef is exposed at the surface.
(iii) Pre-existing positive sedimentary structures such as alluvial fans (Magnetic and Great Palm Islands), transgressionary sedimentary accumulations (Pioneer Bay, Orpheus Island), leeside island sand spits (Rattlesnake Island), deltaic gravels (Myall Reef, Cape Tribulation) and boulder spits (Iris Point, Orpheus Island).

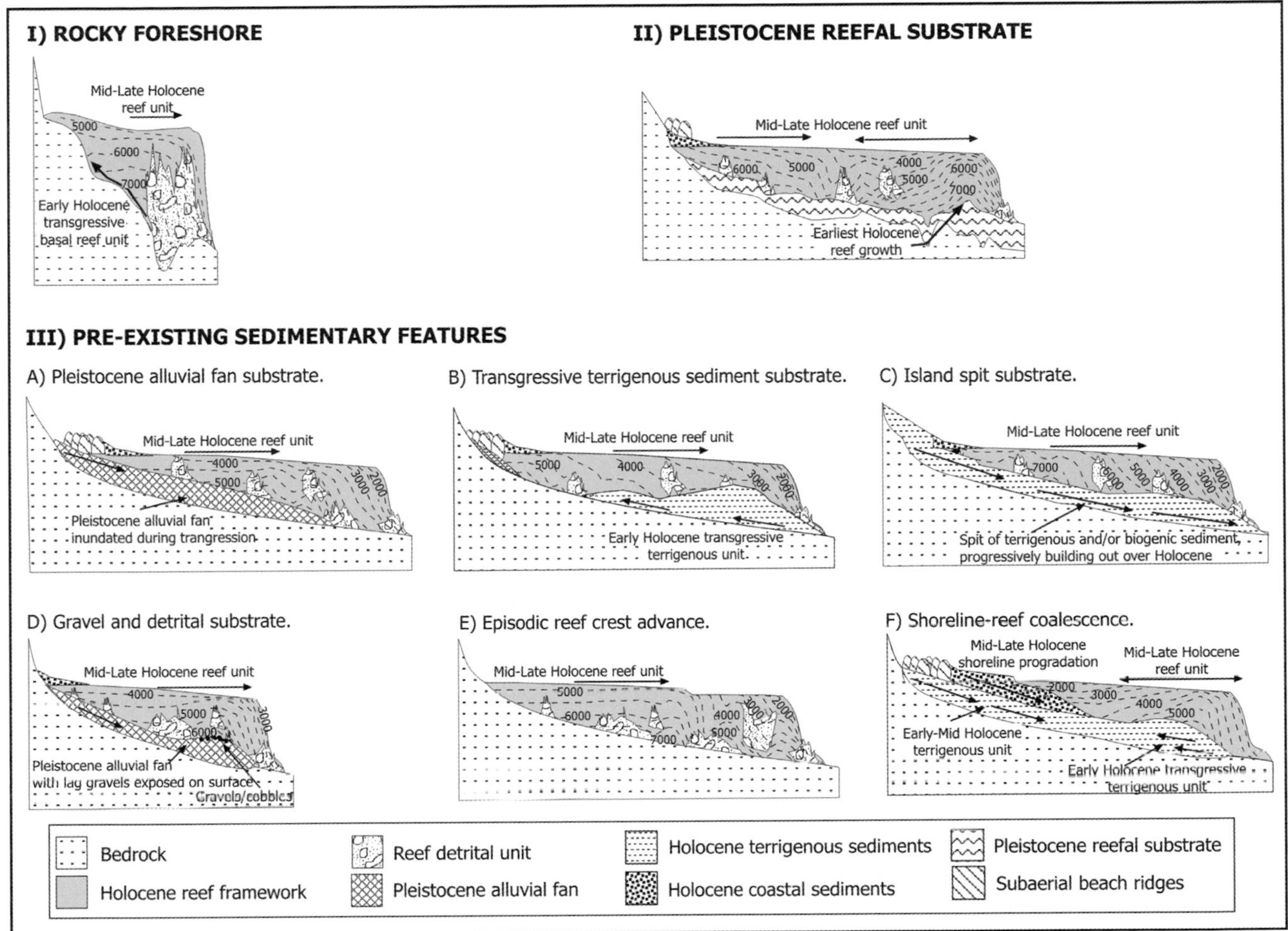

Fig. 2.3. Fringing reef classification based on foundation type and accretion patterns (Source: after Partain and Hopley 1989, p. 39; see also Hopley *et al.* 2007, Fig. 7.7).

Various modes of reef growth have been recognised, including the growth of offshore structure and subsequent backfilling to form the reef flat (e.g. Hayman Island and Yam Island in Torres Strait) or the development of a nearshore reef subsequently attached to the mainland by shoreline progradation. However, the most common mode of development is seaward progradation from the shoreline, usually episodically, due to sea level fall or other environmental factors. The landward areas of fringing reefs (that are mostly concentrated on the inner shelf) that have grown in this manner are typically emergent, with the reef flat sloping to the reef edge where live coral is constrained. This is because of the hydro-isostatic shelf flexure described earlier and its influence on mid-late Holocene relative sea-level histories across the GBR, with the effect that many fringing reefs have experienced relative sea-level fall as they prograded seaward since the mid-Holocene.

Mid-shelf reefs

Except on the narrower shelf north of Cairns, the mid-shelf reefs lie outside the hydro-isostatic zero isobase. This means that the geomorphologies of these reefs do not record the higher mid-Holocene sea level because most did not reach present sea level until after the high stand. They can be described via an evolutionary classification (Figs 2.4, 2.5A–D), commencing with reef growth initiating over older Pleistocene reefal foundations at depths between –5 to more than –25 m. The younger stages generally grow from the deepest foundations and include:

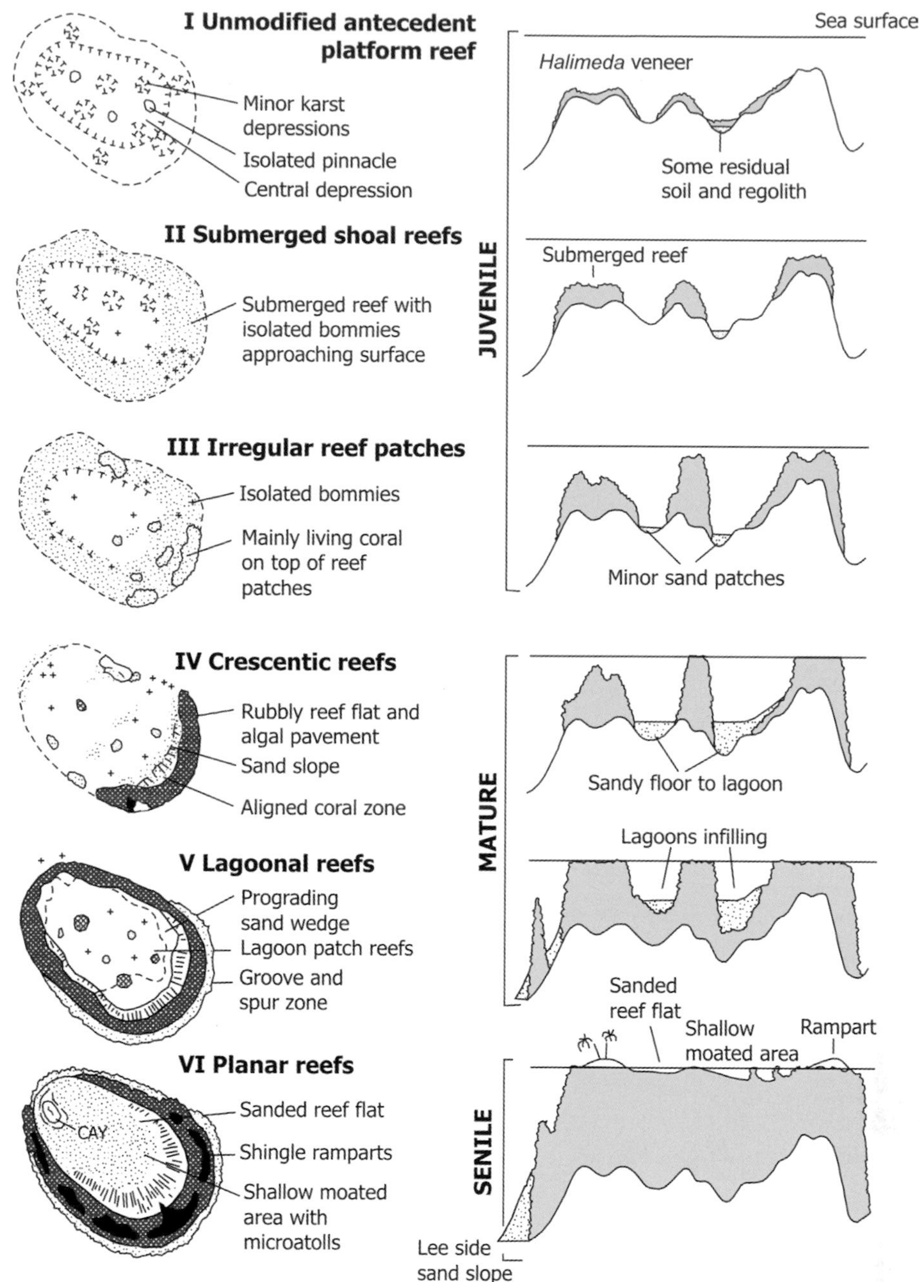

Fig. 2.4. An evolutionary classification for mid-shelf reefs growing from Pleistocene reefal foundations (Source: Hopley 1982).

A juvenile stage capturing growth between initiation and reaching sea level consisting of:

(i) the drowned Pleistocene reef with possibly only *Halimeda* growth
(ii) submerged shoal reefs, with coral growth usually restricted to the higher parts of the older reef foundation
(iii) irregular reef patches, formed when coral growth first reaches sea level.

A mature stage where reef flats are present consisting of:

(i) crescentic reefs formed when reef patches coalesce into more extensive reef flat on the windward margins, with a hard-line reef front

(ii) lagoonal reefs that form as the outer reef flat extends around the margins enclosing a lagoon that is slowly infilled by patch reef growth and by transport of sediment from the windward margins into the lagoon

(iii) a final senile stage consisting of planar reefs, with the lagoon infilled and reef flat extending across the entire reef. Sediment and seagrass beds or mangroves may dominate the reef flat and coral cays are common on such reefs, including low wooded islands.

Two dominant factors determine the stage a mid-shelf reef has reached. The first is the depth to the Pleistocene antecedent reef (that influences the depth of lagoon to be infilled), with reefs growing from shallow foundations most quickly reaching modern sea level and progressing to a mature or senile stage. The second is the size of the reef, because this determines the ratio of highly productive perimeter to the volume of lagoon to be infilled. Larger reefs, once they reach sea level, progress much more slowly. It was shown earlier that the depth of Pleistocene foundations on the GBR was not random, with shallow depths in the north and far south, a cross-shelf gradient in the southern-central GBR, and generally depths greater than 20 m on the central GBR shelf. Reef types reflect this variation with submerged reefs, patches and crescentic reefs dominating the central GBR and most planar reefs found north of Cairns or south of Mackay. Understanding this sequence, its controls, and the speed with which reefs may proceed from juvenile to senile states is critical for conservation and management of the GBR, as habitats for a vast number of organisms become available and are lost as reefs progress through this geomorphological sequence.

Outer shelf reefs

Most reefs on the shoulder of the continental shelf (with the exception of the detached reefs in the north) are linear, parallel to the shelf edge and have originated as fringing reefs at earlier low sea levels. Shelf edge morphology and tidal range are the

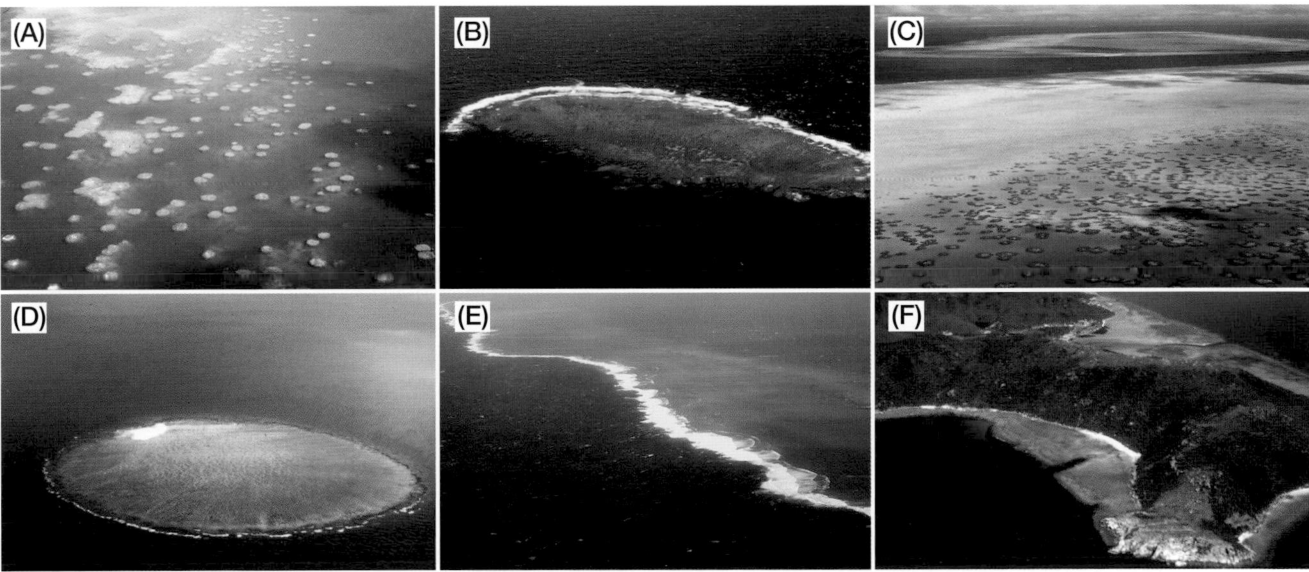

Fig. 2.5. Reef types of the Great Barrier Reef. **(A)** Reef patches, the youngest stage of reef development: Barnett Patches off Hinchinbrook Island. **(B)** Crescentic reef with hard-line growth on the windward side: Eagle Reef, off Bowen. **(C)** Enclosed lagoonal reefs, infilling from internal patch reef growth and transport of sediment from the productive margins: Heron and Wistari Reefs, Bunker Capricorn Group. **(D)** A small senile planar reef with flat top (still with a living coral veneer) and no lagoon: Wheeler Reef off Townsville (note the small unvegetated sand cay on the leeward side). **(E)** The continuous line of outer shelf ribbon reefs: Yonge Reef near Lizard Island. **(F)** Typical high island fringing reef: Great Palm Island (Photos: D. Hopley).

greatest influences on reef type, the major classes of which are described in the following sections.

Northern detached reefs. These reefs grow on isolated pinnacles of fragmented continental crust seaward of the main barrier. Ribbon-like reefs occur on the windward side of some, with large *Halimeda* meadows to the lee. On Raine Reef, the top of the Pleistocene is 11 m below reef flat level, with the reef flat developing ~6000 years ago. Raine Island, the most important green turtle rookery in the world, began to form on this reef flat by ~4500 years ago.

Northern deltaic reefs. The northernmost 96 km of shelf edge of the GBR south of the Gulf of Papua consists of short (up to 4 km) narrow reefs parallel to the shelf edge, separated by passages up to 200 m wide. Water depths at the shelf edge are only 30 m to 40 m but the seaward slope steeply drops to 700 m. Complex delta-like lobes have formed on the inner western side of each passage, deposited as sedimentary structures in response to strong flood tide currents. These structures form the foundations for deltaic reef growth.

Ribbon reefs (Fig. 2.5E). The ribbon reefs form a classic shelf edge barrier reef system between 11°S and 17°S (~700 km). Individual reefs are up to 28 km long, separated by narrow passages that are generally less than 1 km wide. Currents through the passes are weaker than opposite Torres Strait and although the ends of the reefs curve inwards there are no deltaic structures. However, tidal jets carrying nutrient-rich water upwelled from the deeper water offshore surge through the channels and maintain the extensive *Halimeda* bioherms described earlier. The outside of Ribbon 5 has been explored by submersible and shows dead reefs at –50 m and –70 m above an almost vertical limestone cliff down to 200 m. The Pleistocene is at –15 m on Ribbon 5 and –19 m on Yonge Reef. Holocene growth commenced ~8000 years ago and the ribbon reefs appear to have reached modern sea level between 6000 and 5000 years ago. Because of the high energy conditions, the ribbon reefs display a strong zonation.

The submerged reefs of the central GBR. South of Cairns, where the shelf widens, reefs are set back from the shelf edge, which is in 70 m to 80 m of water. Beyond these reefs the shelf edge slopes gently, reaching just 200 m depth over a distance of ~3 km. Reefs and shoals are scattered through this area. All those examined show the Pleistocene antecedent surface at depths below 20 m, and, although reefs quickly initiated upon these substrates once they were drowned by rising Holocene sea level (basal dates >8500 years), they did not reach modern sea level until ~4000 years ago, despite maintaining vertical accretion rates of 6 m per thousand years. The most significant feature of this 800 km section of the GBR is the presence of up to three lines of submerged reefs rising from water depths down to –100 m. In places these reefs may be as much as 70 m deep but elsewhere they approach the surface and many remain unmapped. These features are composite, formed as the sea lapped on and off this deeper shelf edge during successive transgressive and regressive events. Many appear to be dead, perhaps drowned by rapidly rising postglacial sea level, buried by sediment shed from upslope, or knocked over by poor water quality when the shelf margin was first drowned. Importantly, however, not all reefs in this area are dead; dense stands of coral have been observed on ROV dives. This area of the shelf has experienced hydro-isostatic subsidence, but the impingement of the active Halifax Basin onto the shelf may be the most important influence on the distribution of the submerged reefs on this section of the outer GBR.

The Pompey Complex. The Pompey Complex parallels the shelf edge for around 140 km and is between 10 and 15 km wide. It includes some of the largest reefs of the GBR, many between 50 km^2 and 100 km^2. The reefs themselves are not on the shelf edge, which lies a further 20 km seaward, separated from the Pompey Reefs by a series of probably fault-controlled steps. The most prominent steps occur at –70 m and –80 m, the edges of which are the location of submerged reefs some of which come to within 10 m of the surface. Narrow channels cut through the Pompey Reefs reaching depths of over 100 m, up to 40 m deeper than the shelf on either side. These reefs are on the outer edge of the widest part of the GBR shelf, in the area of highest tidal range that reaches 10 m on the mainland and exceeds 4 m within the reefs. Tidal currents greater than 4 m/s have been recorded and are responsible for both the scouring of the channels and the deltaic reefs formed at both ends of them. Unlike the northern deltaic reefs, water depths are sufficiently shallow for ebb tide deltas to develop on the seaward side of the reefs as well as the western side, producing extremely complex morphology. Intricate lagoons, with blue hole sinkholes (to 90 m deep) formed through solution processes during low sea levels also typify the Pompey Reefs. Indeed, the morphology of reefs across large parts of the Pompeys is still strongly controlled by karstic topography formed by erosion during lower sea levels.

The Swain Reefs. These reefs form a contrasting area of numerous small planar and lagoonal reefs to the south of the Pompeys. A steep drop-off occurs around the northern margin, but to the south and west adjacent to the Capricorn Channel the slope is far gentler.

The Bunker-Capricorn Group. The southernmost reefs of the GBR are set back from the shelf margin, which is poorly defined. The reefs grow over shallow Pleistocene foundations and are either planar or lagoonal. Many support cays. Corals are not found in the shelf sediments south of 24°S and it is possible that there was no reef growth at this latitude at the height of the glacial periods.

Coral reef islands

Coral reef islands (cays) are low-lying largely unconsolidated accumulations of reef-derived biogenic sediments deposited on reef platforms at or close to sea level. Cays form where waves and wave-driven currents deliver sediments produced by coral reefs to particular areas of the reef flat, where they accumulate to form cays of various types. They are generally absent on larger reefs or those with a geometry not conducive to centripetal wave refraction, or where either very low or very high hydrodynamic energy impedes accumulation. Centripetal action of refracted waves moves sediment over the reef, depositing it when they can no longer carry it further. Capacity to carry larger sediments, such as rubble and shingle, is quickly lost in normal weather conditions and shingle cays and ramparts are thus found on windward reef flats. Sand may be carried greater distances and sand cays typically form towards the leeward reef edge.

Cay shape depends on the wave refraction pattern. The depositional envelope is fairly compact on oval-shaped reefs and the cay is oval. On elongate reefs, the depositional area where wave trains from each side of the reef meet is linear, and resulting islands are longer and narrower. Linear cays, particularly their ends, can be very mobile in response to changing weather patterns. On older more stable cays (or parts of cays) mature soils, perhaps with guano deposits, may develop and support climax woodland vegetation such as *Pisonia grandis* forest. Reef islands that support *Pisonia* forest usually also support earlier stages of the vegetational succession, reflecting the active disturbance regimes and the dynamic geomorphological conditions experienced by most reef islands. Vegetated cays are typically more stable than unvegetated cays, with the relative proportion of a cay's footprint covered by vegetation a good guide to the stability of the landform, noting that vegetation itself may also improve or maintain cay stability in various ways. Stability may also be improved when unconsolidated sediments become lithified through a range of possible processes. Like vegetation, lithification both improves and indicates stability because it cannot proceed in mobile sediments. Beachrock is a common feature on many reef islands on the GBR, formed by cementation within the beach. The cementation of cay sediments by phosphate solutions derived from guano forms phosphate rock: a lithified deposit best developed on cays that support significant populations of nesting birds.

Classification and description of reef islands is based on sediment type (sand or shingle) and vegetation cover (from unvegetated to highly complex including reef flat mangroves). Island shape and size are also important because there is a threshold island width of ~120 m to support a freshwater lens on which mature vegetation depends. Some reefs have more than one island – a windward shingle cay and a leeward sand cay – but the most complex are the low wooded islands found on the GBR north of Cairns. These comprise both windward shingle and leeward say cays, but also include significant mangrove forest development over the reef top. The mangroves initially establish immediately behind the shingle rampart where they are protected from waves and currents before expanding across the reef. On some reefs, such as Bewick Island near Princess Charlotte Bay, mangroves may almost entirely cover the reef platform.

Cays on the GBR form upon reef flats; they must therefore be younger than the flats upon which they develop. Reef flats on the GBR can be up to 7000 years old, and so reef islands must all be younger than this. As discussed earlier, reefs in various parts of the GBR grow from different depths, over different sized substrates, and have different relative sea-level histories that influence when reef flats first formed. These patterns also produce a distinctive pattern in the distribution of island types. Unvegetated islands occur throughout the GBR but are least common on the central GBR where reefs are relatively immature (only recently reached sea level and limited lagoonal infill), and vegetated cays are totally absent (none between Green Island (16°45′S) and Bushy Island (25°57′S)). Vegetated cays are most numerous in the far south and far north and occur on six of the Swain Reefs. Low wooded islands are limited to the inner shelf north of Cairns. Within the GBRMP are 213 unvegetated cays, 43 vegetated cays and 44 low wooded islands (total 300), though changes are continually taking place, especially to those with ephemeral vegetation.

The growth and maintenance of cays on the GBR critically depends not only on conditions being suitable for sediment accumulation (reef flat at sea level, reef shape to focus sediment deposition, accommodation space into which the reef island can accrete, suitable hydrodynamic conditions), but also on the production of those sediments. The sand-sized sediments that dominate most reef islands are derived from the skeletons of a diverse range of organisms that live on the surrounding reef. However, the production of sediment from these skeletons can take two different pathways. The first is where framework structure or larger sediments such as coral blocks are broken down into detrital sediments by physical and biological processes. Sands produced in this way include coral sands and those derived from large mollusc shells. These sands may be produced centuries to millennia after the original secretion of the skeleton by the living organism. For example, long dead corals may produce coral sands, often at a higher rate than living colonies because bioerosion rates on dead substrates are usually higher. The second sand production pathway involves sands being initially secreted by organisms as sand-sized clasts. Sands produced this way are often available for transport to a cay immediately post mortem, with no delay associated with subsequent breakdown into smaller transportable sediments. Sands produced by a range of large benthic foraminiferans (e.g. *Baculgypsina* sp., *Amphistegina* sp.) are produced in this way and significantly contribute to many GBR reef islands (Fig. 2.6). For example, more than 50% of the sand on Raine Island is produced by the foraminiferan *Baculgyspina*, which recent investigations show are quickly transported on death from their preferred turf-covered reef pavement habitats that encircle the reef platform margin to the cay shoreline. Significantly, sediment supply on such islands is tightly coupled to the ecological condition and productivity of the contributing organisms, and disruptions or declines in productivity will quickly manifest as erosion as sediment supply is reduced. Furthermore, changes in the community composition and of surrounding reef flats can significantly influence both the composition and supply of sediments and affect the formation and stability of reef

Fig. 2.6. Sands produced by foraminiferans are important components of many reef islands. **(A)** Turf-covered algal pavement typical of the most productive habitat for many sand-producing foraminiferans. **(B)** Close-up of living foraminiferans growing on the reef flat pavement. Individual foraminiferans ~1 mm across. **(C)** Close-up of individual *Baculgypsina sphaerulata* foraminiferans harvested alive from reef flat, complete with spines. Scale bar is 1 mm. **(D)** Sands on cay supplied by foraminiferans (orange sediment grains). Note that many sediments lack spines, lost during transport to and around the island shoreline (Photos: A, B, Scott Smithers; C, D, John Dawson).

islands. Modelling studies have indicated that for many reef islands changes in sediment supply, linked to changes in reef flat community composition and productivity, sediment durability or transport may be stronger controls of reef island stability than sea level rise. Understanding the processes that influence sediment supply and reef island sediment budgets, including how they may vary in response to projected changes in reef ecology and growth, and how these changes affect the geomorphology and geomorphological processes on reef islands is an urgent research priority.

Geomorphology and the future of the Great Barrier Reef

Just 10 000 years ago the GBR as we know it did not exist. Only after the continental shelf was flooded by the postglacial transgression did it re-establish over limestone outcrops built by its ancestors during earlier phases of higher sea level. Since then it has changed remarkably as a consequence of both coral reef growth and changing environmental conditions, some driven by external factors but others the result of the growth of the reef itself. In terms of coral cover, ecological variety, and reef accretion rates, the GBR was probably at its peak around 7000 years ago. Since then reefs have progressed towards more senile stages, but new discoveries from parts of the reef that we have only recently been able to observe show more reefs and coral cover than we have hitherto known, most of which is submerged and firmly within Hopley's juvenile reef growth stage. Why these reefs remain submerged while others, often in very close proximity and presumably exposed to equivalent environmental conditions raced to the sea surface in just a few thousand years remains uncertain, and is currently the subject of investigation. Nonetheless, their existence in deeper water that may offer refuge or protection from some of the anthropogenic pressures now challenging reefs, together with depth-related accommodation space being available into which they may grow are cause for some optimism. Global climate change impacts appear to negatively affect mainly the cover and carbonate productivity of reef benthos, although it has been suggested that even a small sea level rise may renew carbonate productivity over reef flats presently constrained by sea level that have been largely senescent for 5000 years or more. Predicting the potential outcomes is, however, difficult because increases in coral mortality from bleaching, diseases, and the fragility of many surviving corals as oceans become more acidic may accelerate sediment production and lagoonal infill, particularly where transport efficiency improves as water depths over reef flats increase (Chapter 12). In the short term, these same processes may counteract the rise of sea level against reef islands dominated by sands yielded from the breakdown of existing framework. However, the impacts are likely to be negative for reef islands dependent on more fragile sands such as those produced by foraminiferans, or into the longer term future where limited growth of new corals will limit sediment supply. The details in many cases are not yet known, but it is clear that changing climate will have major effects on both ecological and geomorphological processes on the GBR.

Further reading

There is a large database on the geomorphology of the GBR that has been used in this chapter but is too large to acknowledge here. Three major references by the authors include over 1000 references on processes, morphology, and evolution and are a source for further information. They are:

Hopley D (1982) *Geomorphology of the Great Barrier Reef: Quaternary Development of Coral Reefs*. John Wiley, Interscience, New York, USA.

Hopley D, Smithers SG, Parnell KE (2007) *The Geomorphology of the Great Barrier Reef: Development, Diversity and Change*. Cambridge University Press, Cambridge, UK.

Smithers SG, Harvey N, Hopley D, Woodroffe CD (2007) Vulnerability of geomorphological features on the Great Barrier Reef to climate change. In *Climate Change and the Great Barrier Reef: A Vulnerability Assessment*. (Eds JE Johnson and PA Marshall) pp. 667–716. Great Barrier Reef Marine Park Authority and the Australian Greenhouse Office, Townsville.

In addition, excellent reviews of coral reef geomorphology more generally are provided in the following two references.

Kench PS, Perry CT, Spencer T (2009) Coral reefs. In *Geomorphology and Global Environmental Change*. (Eds O Slaymaker, T Spencer and P Embleton-Hamman) pp. 180–213. Cambridge University Press, Cambridge, UK.

Perry CT, Kench PS, Smithers SG, Riegl B, Yamano H, O'Leary MJ (2011) Implications of reef ecosystem change for the stability and maintenance of coral reef islands. *Global Change Biology* **17**, 3679–3696. doi:10.1111/j.1365-2486.2011.02523.x

Figures were sourced from:

Chappell J (1980) Coral morphology diversity and reef growth. *Nature* **286**, 249–252. doi:10.1038/286249a0

Hopley D (1982) *Geomorphology of the Great Barrier Reef: Quaternary Development of Coral Reefs*. John Wiley, Interscience, New York, USA.

Partain BR, Hopley D (1989) *Morphology and Development of the Cape Tribulation Fringing Reefs, Great Barrier Reef, Australia*. Technical Memoir No. 21, Great Barrier Reef Marine Park Authority, Townsville.

3

The Great Barrier Reef in time and space: geology and paleobiology

J. M. Pandolfi and R. Kelley

Introduction

Reefs in their many forms are found throughout the fossil record and represent some of the earliest structure-forming ecosystems on Earth. Since the explosion of metazoans in the Cambrian around 540 Mya (million years ago) many groups of organisms have formed 'reef-like' features on the sea floor, making reef communities difficult to singularly characterise. Following the greatest extinction of all time, the Permo-Triassic event (251 Mya), scleractinian corals, bivalve molluscs and crustose coralline algae have dominated the construction of wave resistant organic carbonate structures on the planet – commonly called reefs.

Although the definition of just what is a 'reef' has a long and tortuous history of debate in the scientific literature, there is no dispute about the importance of the reef ecosystems to the coastlines, continental shelves and ocean provinces of the tropical realm. In this chapter we take a broad spatial and temporal view of the largest of the world's platform reef provinces – the Great Barrier Reef (GBR). In particular we look at the boundary conditions and mechanisms that underpin the perpetuation of the GBR province in space and time and how environmental change has influenced the reef biota.

The last two decades have seen an increasing appreciation of the importance of understanding the GBR from a total system perspective. Over the same time, there has also been increasing recognition of the need for a temporal perspective in every aspect of ecology, especially where it seeks to relate to natural resource management. Understanding the historical context for the ecosystems we live in and exploit over medium-term timescales allows natural resource managers to measure the success of their operations in reducing human impacts and enhancing sustainability.

But what about the long-term view? Geological evidence accumulated over the last 40 years shows that the 'reef' part of the GBR is a relatively young feature – less than 1 million years old. Because reefs are built mainly during rising sea levels, and the highest sea levels are associated with interglacial periods, much of our attention is attracted towards GBR reef growth during the interglacial high sea level episodes of the last 500 000 years. But the GBR, like reefs elsewhere in the tropics, survived during lower sea level stands as well. In this chapter we explore the life and times of the GBR and the dynamic ecological response of coral reefs to environmental change.

Origins of the Great Barrier Reef

The history of the GBR is influenced by the post Gondwanan continental drift history of the

Australian continent and repeated episodes of global environmental change associated with the late Tertiary and Pleistocene ice ages. The 'reefal' GBR is a relatively young geological structure that was slow to respond to favourable environmental conditions early on. In fact, the central Queensland continental shelf has enjoyed warm tropical waters that could well have supported coral growth for the past 15 million years (My). However, the best evidence indicates that the initiation of the GBR did not occur until around 600 thousand years ago (kya), and the regional province of reef systems as we now know them probably did not occur until around 365–452 kya. This is coincident with Marine Isotope Stage (MIS) 11, perhaps the warmest interglacial of the past 450 000 years (ky), and one with climatic conditions most similar to those we are now experiencing. Some workers believe that the 'switching-on' of the GBR was related to the mid-Pleistocene transition from 41 ky to 100 ky-long climatic cycles, and to the development during MIS 11 of a marked high stand that enabled sustenance of both a cyclone corridor and a reef tract along a relatively wide and deeper water continental shelf (Fig. 3.1A).

Cores drilled through Ribbon Reef 5 have shown that the GBR has been able to re-establish itself

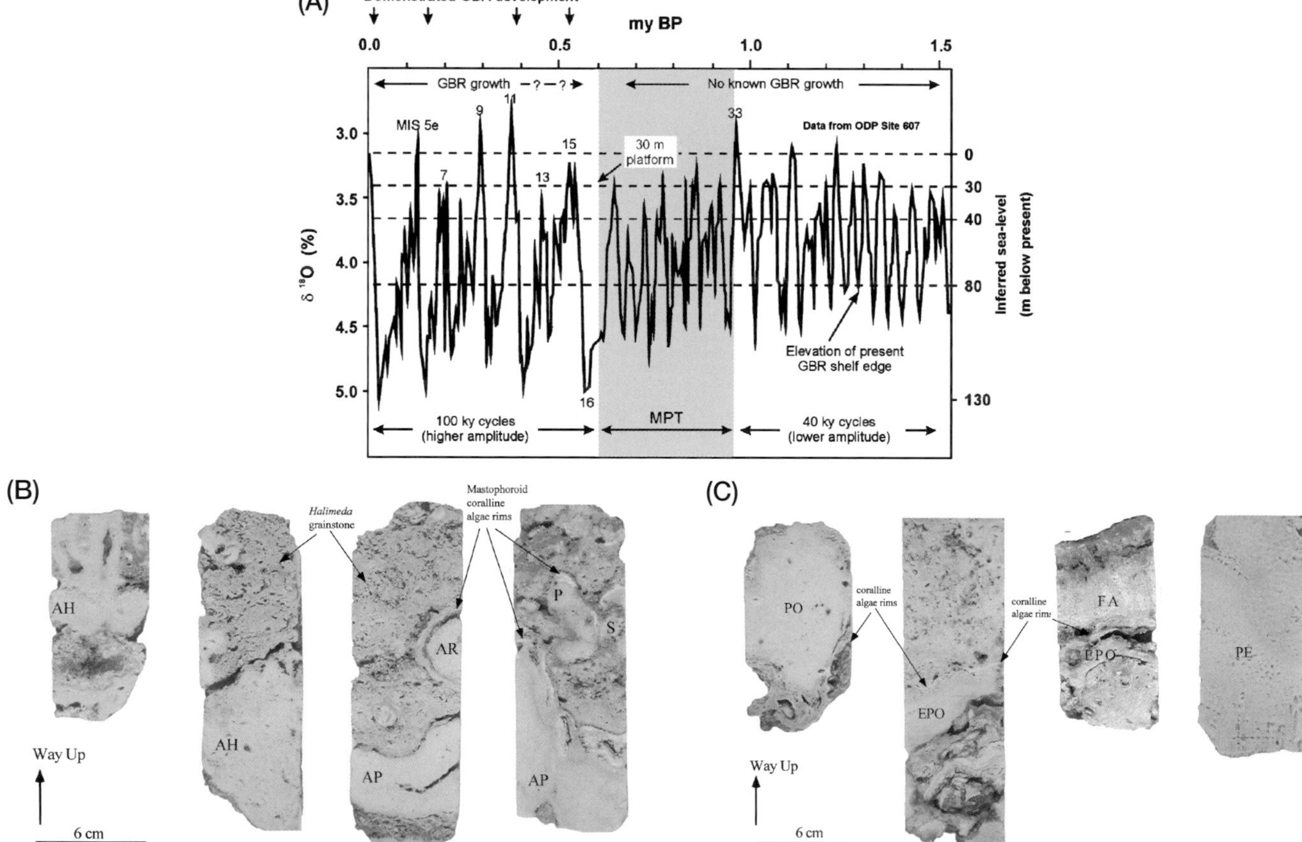

Fig. 3.1. **(A)** Plot for the past 1.5 million years showing the change in frequency and amplitude of the climatic, and by inference, sea-level fluctuations after the mid-Pleistocene transition (MPT) at 0.9–0.6 Mya (MPT, a shift in the periodicity of radiative forcing by atmospheric carbon dioxide that caused higher amplitude climate periodicities). Growth of the GBR occurred after the MPT (Source: after Larcombe and Carter 2004). **(B)** Photographs from the Ribbon Reef 5 core showing the major coral components of 'Assemblage A'. These include robust branching corals of species from the *Acropora humilis* group (AH), the *Acropora robusta* group (AR), *Acropora palifera* (AP), *Stylophora pistillata* (S) and *Pocillopora* (P). Much of the coral framework is encrusted with coralline algae. **(C)** Photographs from the Ribbon Reef 5 core showing the major coral components of 'Assemblage B1' and 'B2'. These include massive *Porites* (e.g. *Porites* cf *lutea*) (PO), encrusting *Porites* (EPO) and massive faviids such as *Favites* (FA) and *Plesiastrea versipora* (PE). Again, extensive coralline algal rims encrust much of the coral framework (Images: B and C from Webster and Davies 2003).

repeatedly during high sea level episodes associated with major environmental fluctuations in sea level, temperature and carbon dioxide (CO_2) over the past several hundred thousand years. Moreover, these reefs have maintained a similar coral and algal species composition during their repeated formation (see later section on Paleoecology) (Fig. 3.1B, C).

Reef growth and global sea level change

The growth and decay of ice sheets in the northern hemisphere were controlled by 10^4- to 10^5-year scale climate changes forced by natural cyclic changes in several parameters of Earth's orbit (so called Milankovitch cycles). These cycles influence the amount of energy from the sun received by the Earth. The cycles include obliquity (changes in the angle of Earth's axis of rotation with respect to the sun), eccentricity (changes in the circularity of Earth's orbit around the sun) and precession of the equinoxes (changes in the position of the Earth in its orbit around the sun at the time of the equinox). The cycles are 41 000, 100 000 and 23 000 years, respectively. During the last 500 000 years, global sea level underwent at least 17 such cycles of rise and fall. Average rates of sea level change

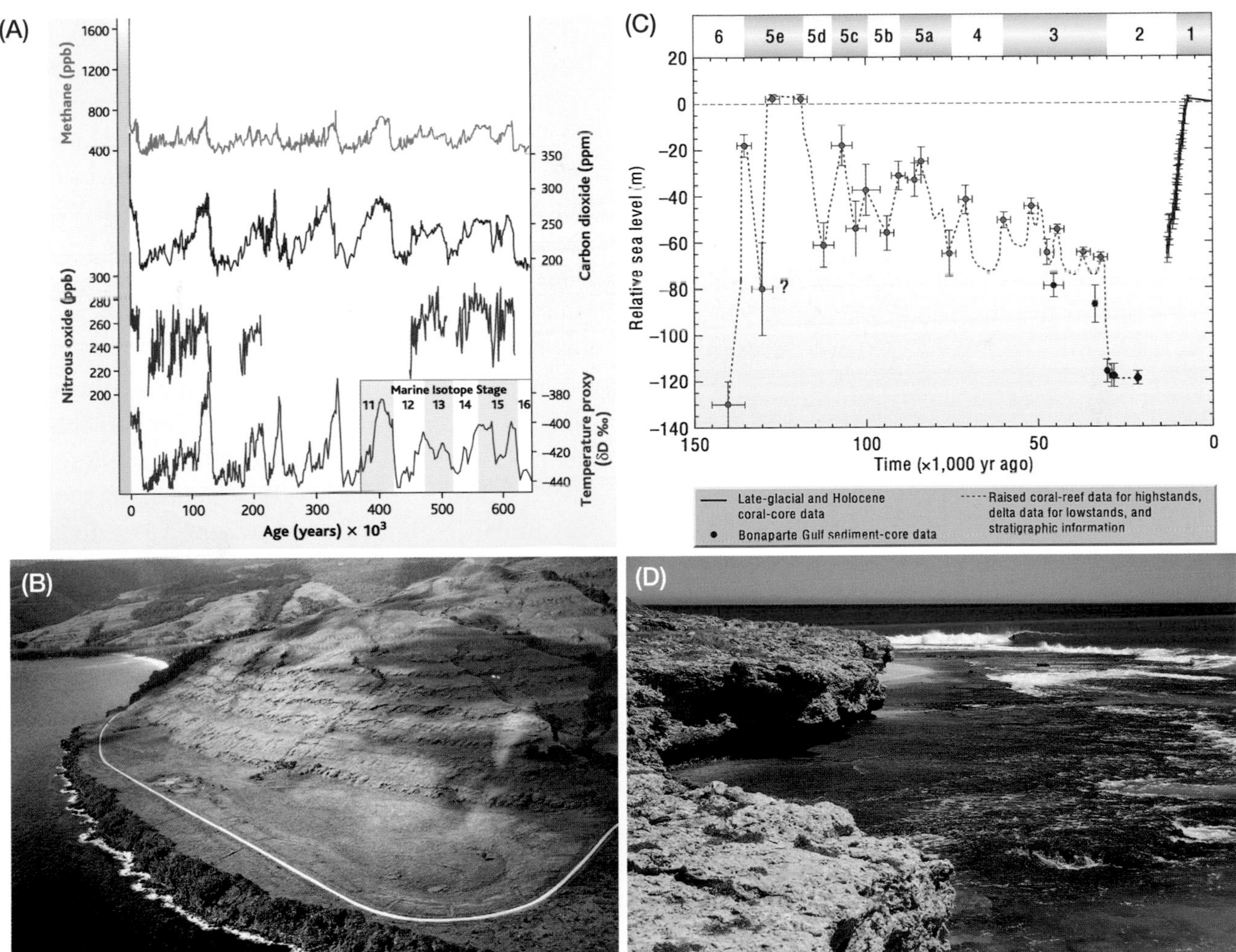

Fig. 3.2. **(A)** Sea level, temperature and greenhouse gas fluctuations over the past 650 ky from the EPICA ice core from Antarctica (Source: Brook 2005). **(B)** View of the Pleistocene and Holocene raised reef terraces at Huon Peninsula, PNG (Photo: R. Kelley). **(C)** Sea level curve for the past 150 ky derived from Huon Peninsula, supplemented with observations from Bonaparte Gulf, Australia (Source: from Lambeck *et al.* 2002). **(D)** Pleistocene reef terrace from the 125 kya reef at Exmouth, Ningaloo, Western Australia (Photo: R. Kelley).

between glacial and interglacial intervals approached 5 m per thousand years with the possibility of greater rates associated with Heinrich events (abrupt climatic episodes associated with ice rafted detritus during the last glacial). The magnitude of sea level change from one interglacial period to the next is on the order of 120 m: a major repetitive >100 m rhythm to the late Pleistocene ice ages with which all marine life contends (Fig. 3.2A).

The GBR is very similar to other reefs around the world in its growth during rises in sea level, or transgressions, associated with the deglaciation part of the cycle. One of the best examples of transgressive reef growth that has been clearly related to the oxygen isotope record for the late Pleistocene occurs at the Huon Peninsula, Papua New Guinea (PNG) (Fig. 3.2B, C). In this remarkable tectonically active locality, ongoing uplift during the last several hundred thousand years has left a record of transgressive reef terraces like 'bath rings' along over 80 km of coast. Here, nine transgressive reef growth phases are recorded between 125 kya and 30 kya. Overall the record of dated transgressive reef growth episodes extends back to at least 340 kya.

During rising seas, reefs can accumulate at rates exceeding 10 m per thousand years. This involves a huge bulk of cemented biological framework, principally coral and coralline algae, and even larger quantities of associated sediments. However, once reefs reach sea level, or sea level rise slows and stabilises, this growth also slows. From here the interplay between the growth of the bound biological framework, the production of reef associated skeletal sediment and their destruction by bioerosion and physical forces becomes of critical importance to the maintenance of reef growth (Chapter 9).

The Great Barrier Reef and global environmental change

The geology, geomorphology and age structure of the GBR is described in detail in Chapter 2. Although there is evidence of Pleistocene age reef growth older than 140 kya, we will focus on what the more recent evidence can tell us about the GBR ecosystem in time and space. Here we discuss the GBR during its most recent 'life cycle' – from the previous to the current interglacial cycle and spanning the last ice age.

The superbly exposed and documented record from the Huon Peninsula, PNG, provides a template for expected expressions of transgression within the physical GBR province. Specifically, we should find evidence of reef growth leading to a still stand (i.e. when sea level has ceased to rise or fall) in the previous (128–118 kya: –10 to +5 m a.s.l. (above sea level)) and the present (10 kya to present: –35 to +1 to 0 m a.s.l.) interglacials (Fig. 3.2C). There is extensive physical evidence from drill cores of reef growth leading into both of these interglacials. Chapter 2 discusses the dating literature associated with the Holocene transgression from the Last Glacial Maximum (18 kya) to the current high sea level stand. There is also radiometric evidence from the GBR that the Holocene reef growth was superimposed upon relic Pleistocene reef topography from the last interglacial age.

In other tectonically stable parts of the world this last interglacial reef is well documented at about +2–6 m a.s.l. For example, the last interglacial reef (125 kya, ~5 m a.s.l.) is emergent along the West Australian Ningaloo coast, where it is extensively preserved in near desert conditions (Fig. 3.2D). Whereas the north Queensland coast is well endowed with evidence of Pleistocene shorelines in the form of beach rock, dunes and beach ridges (e.g. Cowley beach near Innisfail), there are relatively few occurrences providing surficial expression of last interglacial reef framework from the GBR (only Digby Island) despite extensive evidence for it elsewhere (e.g. Stradbroke Island, Evan's Head, Lord Howe Island, Saibai Island in the Torres Strait Islands, and the Ningaloo coast of west Australia). One possible explanation for this is that the north-east coast's moist airflow from onshore trade winds has weathered and eroded the emergent 125 kya GBR reef below the current high sea level such that they now only exist as a base for Holocene reef growth. But perhaps the last interglacial reef did not grow everywhere to the high sea level. Vertical movements in the form of hydro-isostasy or tectonic lowering could also be factors.

Because of the Milankovitch cycles discussed earlier, sea level fluctuations are not confined to 'glacial' (i.e. Last Glacial Maximum (LGM) 18 kya) and 'interglacial' (i.e. Last Interglacial (LI) 125 kya) periods; smaller scale fluctuations are referred to as 'stadial' (temporary ice advance) and 'interstadial' (temporary ice retreat) times. Abundant studies carried out in tropical seas correlate the growth of 'wave-resistant organic structures' such as coral reefs with sea level transgressions. We therefore might expect to see transgressive reef deposits developed during the smaller scale sea level changes between the high stand LI and the low stand LGM (Fig. 3.2C). But is there any evidence for GBR reef-building during these lower sea levels?

Evidence for the existence of reef building during low sea level stands comes from the Huon Gulf in PNG. In contrast to their high sea level stand counterparts on the nearby raised reef terraces of the Huon Peninsula, PNG, reef coral assemblages varied through time, even while coral diversity and the taxonomic composition of benthic foraminifera and calcareous red algae assemblages remained constant over a 286 ky interval. Regardless of the patterns in community persistence, reefs must be seen as dynamic and fluid, reacting to sea level throughout the major and minor Pleistocene fluctuations in sea level (Fig. 3.2A).

On the GBR, there is a history of investigation of the terraces and positive-relief features on the continental shelf and margin for evidence of lower sea levels. 'Wave-cut' terraces have been recorded in the southern GBR at –175 m and in the central GBR at –113 m, –88 m and –75 m, where they were interpreted to correspond to postglacial shorelines. Submerged reefs, terraces and notches have been consistently recorded on single beam echo sounder transects across the southern GBR shelf edge, but so far there is insufficient evidence on the spatial distribution of these features to make accurate comparisons against sea level curves.

Recent investigations by marine geologists using multibeam echo sounders have revealed that drowned reefs extend for hundreds of kilometres along the GBR outer shelf edge in –40 m to –70 m depth. They appear to be submerged platform reefs resembling 'barrier' or 'fringing reef' structures ~200 m wide, and are comprised of two parallel ridges of eroded limestone pinnacles (Fig. 3.3A). These drowned shelf-edge reefs might be an important archive of past climate and sea level changes, and potentially provide predictive tools for GBR coral community response to future climate changes. It is now also possible to map shelf depth paleo-drainage in greater detail than ever before (Fig. 3.3B). Very recent work has extended the occurrence of these submerged shelf-edge reefs as far south as the northern end of the Swain Reefs.

In previous decades, the inherent difficulty of remote underwater exploration has restricted the usefulness of this work. A submarine terrace might represent a constructional feature – an interstadial reef – but it might also represent an erosional feature – a wave cut cliff or bench. Modern acoustic techniques involving multibeam sonar hold great promise for finally illuminating the inter-reef and shelf-edge stories by combining high resolution 3-dimensional structure with an ability to map its regional extent.

Despite the limitations of technology, four decades of exploration combined with the new hydrographic charts do tell us one thing. The interstadial GBR does not have as grand or extensive an expression as its modern interglacial sibling in terms of accumulated bound carbonate features. There are some obvious factors that may ultimately explain this. The first is the sea floor slope. Worldwide, continental shelves typically have very shallow gradients from the coast to the shelf-slope break, where the gradient markedly increases. Here, a 1 m rise in sea level can result in kilometres of shoreline displacement. During times of rapidly rising sea level, rates of reef growth from 4 m/ky to 10 m/ky are common and environmental gradients are

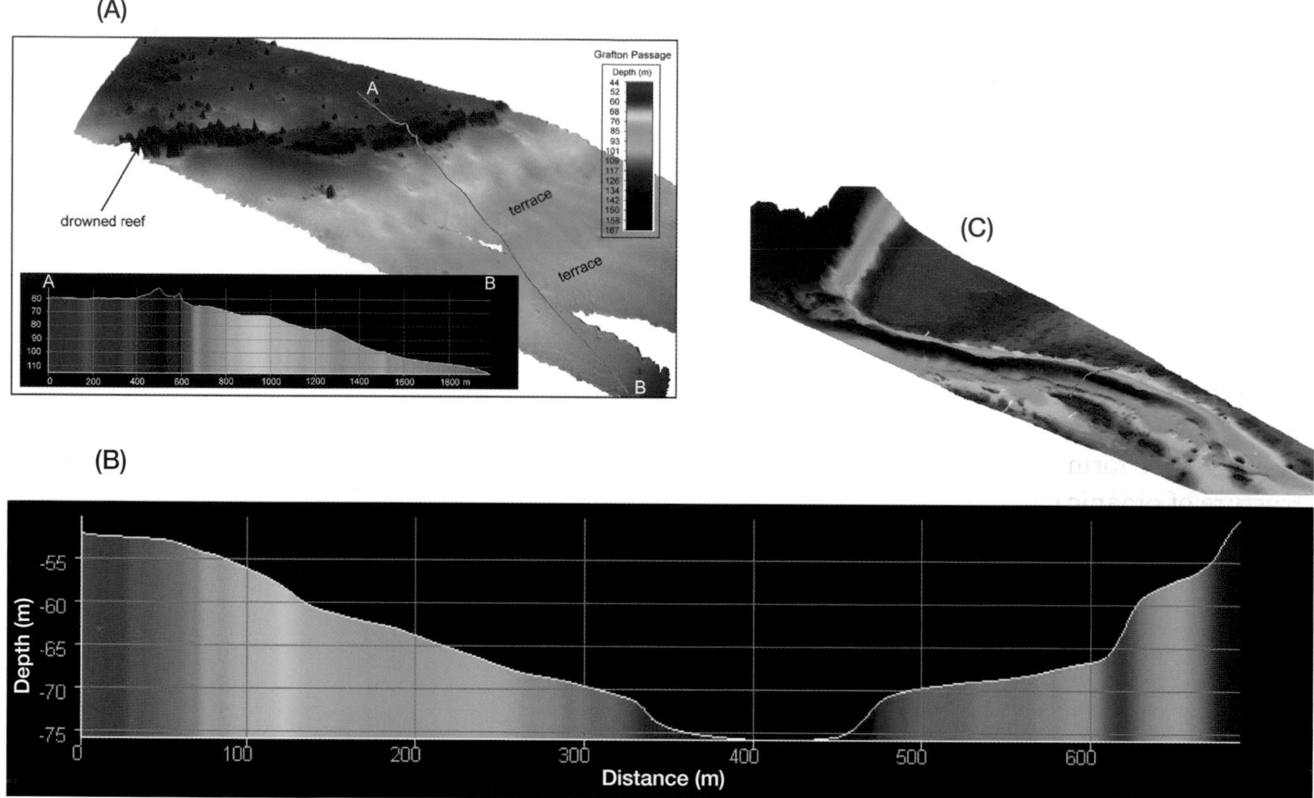

Fig. 3.3. **(A)** Drowned shelf-edge reef at Grafton Passage. Recent investigations by marine geologists at the James Cook University College of Science and Engineering using multibeam echo sounders have revealed drowned reefs that extend for hundreds of kilometres along the GBR outer shelf edge in –40 m to –70 m depth. This submerged 'barrier reef' near Grafton Passage is ~200 m wide (Source: Beaman R, Webster JM, Wust RAJ (2008) New evidence for drowned shelf edge reefs in the Great Barrier Reef, Australia. *Marine Geology* 247, 17–34). **(B, C)** Paleochannel near Cruiser Passage, North Queensland. During the last glacial maximum, sea level was over 100 m lower than today. During these times, rivers deposited floodplain and channel sediments on the continental shelf and upper slope (Images: R. Beaman).

shallow, broad and dynamic. This means that on a shallow continental shelf environmental conditions have the potential to change very rapidly, both in a 'turn on' (increased oceanic circulation/ reduced shoreline terrigenous influence) and 'turn off' (decreasing circulation, water depth, increasing sedimentation) mode. By contrast, steeper gradients, seen in the steep drop-offs on the GBR ribbon reefs and in atoll settings, are more like dipsticks, where the shoreline recedes little during sea-level rise and environmental gradients are steep, narrow and less dynamic.

A further consideration influencing reef development is the effect of sea level when still stand is achieved. Rivers that flow across the continental shelf during ice ages have their floodplain sediments remobilised during the next transgression. These materials are moved inshore by the wave climate and end up, in the eastern Australian case, coming onshore in spectacular dune fields. Geological studies of a dune island barrier system enclosing Moreton Bay, southern Queensland, showed that when sea level rise stops, the onshore movement of sediments into the nearshore sediment profile slows and coastal dune building decreases in size and extent. In the Moreton Bay example, a few thousand years of still stand also led to the development of sedimentary deposits (coastal plains and tidal deltas) and their inshore environmental correlates (mangroves, seagrasses, etc.). This restricted back-barrier circulation increased the estuarine nature of these environments with negative consequences for mid-Holocene back-barrier coral communities.

In the Moreton Bay example, many millions of tonnes of coral carbonate was deposited throughout the entire bay in sequences up to 8 m thick immediately after sea level stabilised between 6 kya and 4 kya. So extensive were these deposits they supported a cement mining operation for over six decades. Although corals are still found in Moreton Bay today, the reduced circulation and increasingly estuarine conditions experienced after 4 ky has reduced their extent, growth and diversity.

These corals are of interest to the GBR context for what they are not, as much as for what they are. They did not form 'reefs' in the 'wave-resistant structure of organic origin' sense of the definition. Rather, the corals were flourishing mounds or banks of corals in a back-barrier setting with open circulation. Unpublished radiocarbon dates from Moreton Bay show these *Acropora* dominated coral communities (~40 spp.) grew and accumulated carbonate at rates of up to 5 m/ky, similar to those known from GBR reefs. The Moreton Bay back-barrier model for coral communities is therefore a diverse, fast acting and geologically significant vehicle for corals over time. A significant difference between these communities and true reefs is that, if sea level had continued to rise, these uncemented carbonate deposits would most likely have been eroded away.

Likewise non-reef forming coral communities from turbid and inter-reefal environments are becoming increasingly well known, both as a source of recruits for areas where high mortality

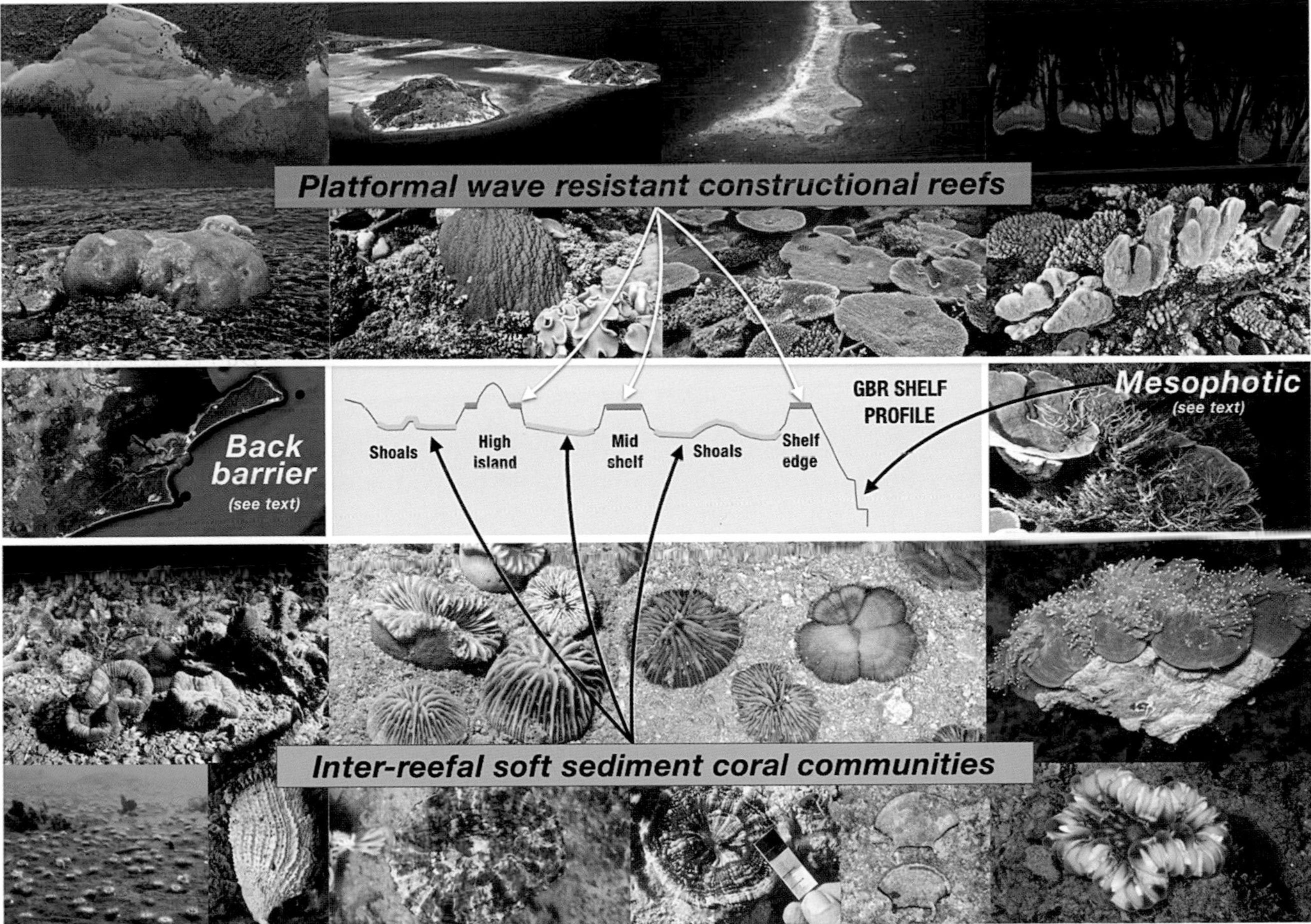

Fig. 3.4. Coral community snapshots from the Great Barrier Reef (GBR): high sea level constructional platform reefs (top two rows), inter-reefal soft sediment coral communities (bottom two rows), lightly calcified deep water mesophotic coral communities (middle row, right) and map of back-barrier coral communities from Moreton Bay (middle row, left). Also shown is a generalised GBR shelf profile where these reef communities can be found.

from bleaching has reduced coral populations and as a source of coral species of interest to the aquarium trade (Fig. 3.4). We feel it is helpful to differentiate these 'coral communities' from 'coral reefs' because they provide an alternative to high sea level platform reefs during other phases of the GBR 'life cycle'. As sea levels rise and fall, inter-reefal and Moreton Bay-style back-barrier coral communities may play an important role in the absence of a contiguous platform reef record (Fig. 3.4).

Having discussed the geological boundary conditions and some of the processes that frame the GBR in space and time, we can now better understand our notion of a single interglacial to interglacial 'life cycle' of the GBR and also better scrutinise some of our assumptions about the system. A review of the drilling data shows that the majority of framework growth associated with the current interglacial GBR grew between 9 kya and 4–5 kya window at depths shallower than 30 m. If we assume a similar window for the previous interglacial GBR, then the extensive matrix of high sea level platform reefs we know as the GBR was probably active for ~10% of the ecological time between the last two interglacials. Moreover, it may only have been actively growing for ~5% of that time.

A model of the GBR in time and space also needs to account for environmental gradients. Today, there is no single locality that supports all of the roughly 400 coral species found in the GBR region. The richest reefs are in the far northern to central region. Areas such as Princess Charlotte Bay, the Palm and Whitsunday Islands provide important ecological space for 'inshore' or turbid water coral communities. These communities collectively contain most species present in the GBR coral fauna, with only a small pool of species apparently restricted to offshore reefs. There are, nevertheless, substantial differences in species abundance in respect of the major environmental gradients, resulting in more or less characteristic community types across and along the GBR. In particular, there are major differences in species composition between the wave washed, clear water reef crest communities of the seaward slopes of outer barrier reefs and their highly sheltered, turbid water, inshore counterparts, most notably those of the deeper reef slopes of leeward sides of continental islands. These communities are at opposite ends of the physico-chemical spectrum and environmental gradients for the GBR.

So do wave resistant high-stand reefs adequately represent a model for the ecological and geological propagation of the GBR in time and space? Clearly, only partially. If sea level were to fall by 10 m, 20 m and then 30 m, would the corals of the GBR, and the thousands of coral connected species, go charging out to the Queensland Plateau to form a clear water 'reef'? Probably not. Just as understanding the workings of the modern GBR benefits from a wider whole-of-system approach, grappling with the GBR in time and space requires a broader conception of coral communities than just the clear water platform reefs where we might prefer to go diving.

What can we learn from paleoecological responses in ancient reefs?

Like tropical marine communities throughout the world, the corals, coral communities and coral reefs of the GBR are fundamentally influenced by their response to climate change and associated environmental parameters, including magnitude and rates of sea level change, CO_2, temperature and turbidity. There is a growing recognition that the integration of paleoecological and climate data on the GBR provides essential insight into how natural communities are assembled and structured in the face of environmental variability over extended periods of time. Given our ability to discriminate among the various kinds of coral and reef development on the GBR, we next consider some examples of the ecological dynamics of coral reef communities over long timeframes.

Although less true for the GBR, many reef organisms are sufficiently preserved in fossil sequences around the world as to provide generic or even species-level information on community structure, including corals, molluscs, echinoderms, coralline algae and foraminifera. We will discuss these examples to help illustrate what the long-term ecology of GBR reefs might have been.

In the Indo-Pacific, recent evolution of corals has been rather slow, with less than 20% of new taxa appearing in the past 2–3 My. In the Caribbean, only two species have gone extinct in the past 125 ky. As such, paleoecological patterns from Quaternary reefs (past 1.8 to 2.6 My) can be investigated from what are essentially modern faunas. For example, during the last interglacial (128–118 kya), sea level was 2–6 m higher than present levels. This has left a fossilised remnant reef in a large number of locations through the tropics (Fig. 3.2D), giving global insight into the ecological nature of reefs in the recent geological past.

One of the best archives for understanding the ecological effects of sea level fluctuations on coral reefs is contained in Pleistocene reef sequences from several tectonically active sites around the world, the most famous of which is the Huon Peninsula in PNG, where nine such reefs were developed between 125 kya and 30 kya (Fig. 3.2B). This series of coral reef terraces, formed by the interaction between Quaternary sea level fluctuations and local tectonic uplift, allows investigation of the assembly of coral reefs during successive sea level rises. Here, ecological trends over millennial time scales point to high levels of persistence in community structure, regardless of the magnitudes of change in environmental variables. In the Caribbean, similar coral community structure was noted among four reef-building episodes ranging in age from 104 kya to 220 kya on Barbados. Remarkably, the high similarity in community composition derived from surveys of common species was also characteristic of separate surveys targeting rare taxa. These studies point to persistence in coral community structure over successive high sea level stand reefs that grew optimally during rising sea level, and are consistent with the rare glimpses we have of the GBR that also show that recurrent associations of coral reef communities are the norm (Fig. 3.5A).

Current concern over the deteriorating condition of coral reefs worldwide has focused intense attention upon the relationship between past 'natural' levels of disturbance and community change versus modern human-induced agents of decline. To understand the impact of humans, our only recourse is to study the fossil record. The uplifted Holocene reef at the Huon Peninsula, PNG, age-equivalent to the GBR, has been studied to determine the frequency of disturbance in fossil sequences with little or no human impacts. Rates of mass coral mortality were far lower (averaging one in 500 years) than are presently being experienced in living reefs (multiple events per decade) (Fig. 3.5B). Recovery from disturbance was swift and complete, and the history of communities provides predictive power for the nature of their recovery. The stark contrast between living and fossil reefs provides novel insight to the abnormally high disturbance frequencies now occurring.

But what happens when sea level falls or stands still, and how do reefs respond to habitat reduction caused by lowered or lowering sea level? Some spectacular sequences of drowned coral reefs occur at significant water depths in Hawaii and the Gulf of Papua. Although the scale of resolution is much less for these drowned reefs, the overall picture, at least for the corals, contrasts with the community similarity observed through large intervals of geological time during successive high stand reefs. When reef growth occurred during sea level fall in the Gulf of Papua, there was a much lower degree of predictability in the coral composition of the reefs compared with high stand reefs of the Huon Peninsula, PNG.

Any discussion of the development of GBR reefs through time must mention the interruption of natural reef ecological dynamics with those imposed through human agency. Precisely dated (using U-series radiometric age dating of corals at the Radioisotope Facility at The University of Queensland) reef sediment cores derived from present-day reef slopes show dramatic changes in coral community structure associated with land-use changes brought on by the European colonisation of the Queensland coastline adjacent to the GBR. Fast-growing branching corals, such as *Acropora*

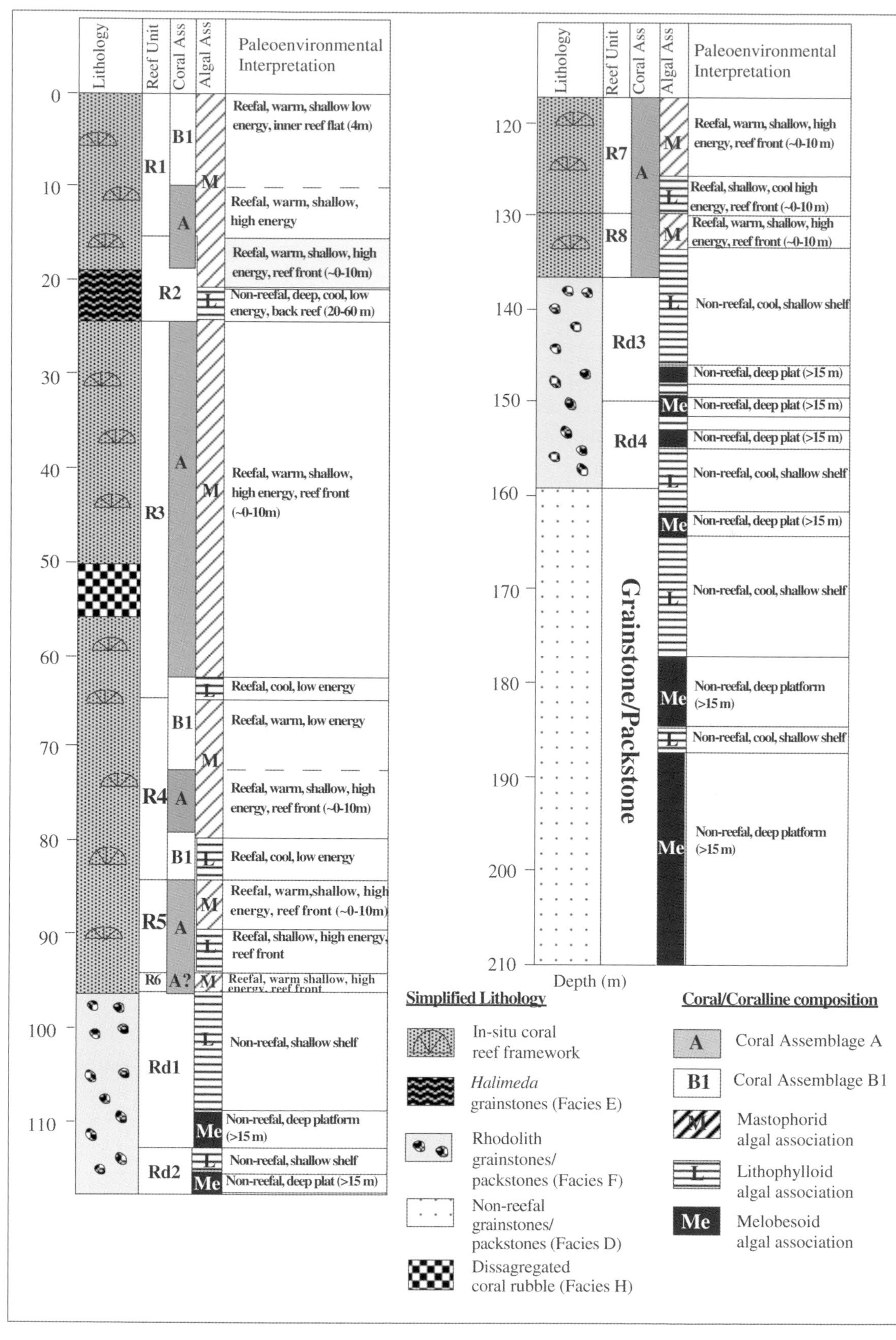
(A)
Lithology
Reef Unit
Coral Ass
Algal Ass
Paleoenvironmental Interpretation
Reefal, warm, shallow low energy, inner reef flat (4m)
Reefal, warm, shallow, high energy
Reefal, warm, shallow, high energy, reef front (~0-10m)
Non-reefal, deep, cool, low energy, back reef (20-60 m)
Reefal, warm, shallow, high energy, reef front (~0-10m)
Reefal, cool, low energy
Reefal, warm, low energy
Reefal, warm, shallow, high energy, reef front (~0-10m)
Reefal, cool, low energy
Reefal, warm,shallow, high energy, reef front (~0-10m)
Reefal, shallow, high energy, reef front
Reefal, warm shallow, high energy, reef front
Non-reefal, shallow shelf
Non-reefal, deep platform (>15 m)
Non-reefal, shallow shelf
Non-reefal, deep plat (>15 m)
R1
R2
R3
R4
R5
R6
Rd1
Rd2
B1
A
A?
M
L
Me
Reefal, warm, shallow, high energy, reef front (~0-10 m)
Reefal, shallow, cool high energy, reef front (~0-10 m)
Reefal, warm, shallow, high energy, reef front (~0-10 m)
Non-reefal, cool, shallow shelf
Non-reefal, deep plat (>15 m)
Non-reefal, deep plat (>15 m)
Non-reefal, deep plat (>15 m)
Non-reefal, cool, shallow shelf
Non-reefal, deep plat (>15 m)
Non-reefal, cool, shallow shelf
Non-reefal, deep platform (>15 m)
Non-reefal, cool, shallow shelf
Non-reefal, deep platform (>15 m)
R7
R8
Rd3
Rd4
Grainstone/Packstone
Depth (m)
Simplified Lithology
In-situ coral reef framework
Halimeda grainstones (Facies E)
Rhodolith grainstones/ packstones (Facies F)
Non-reefal grainstones/ packstones (Facies D)
Dissagregated coral rubble (Facies H)
Coral/Coralline composition
Coral Assemblage A
Coral Assemblage B1
Mastophorid algal association
Lithophylloid algal association
Melobesoid algal association

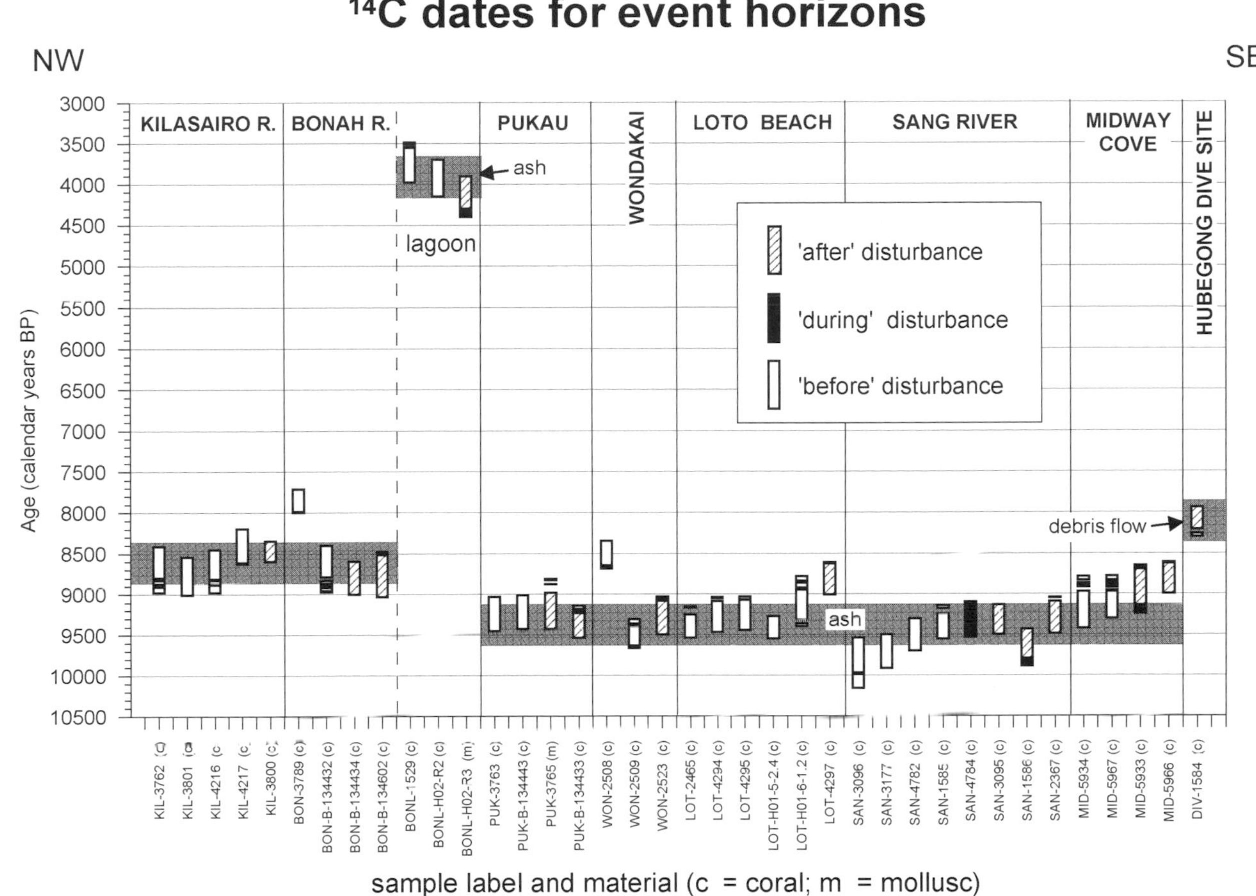

Fig. 3.5. **(A)** Lithologic and biologic variation in the Ribbon Reef 5 core through the past ~600 ky. Ancient environments and coral assemblages remained constant through several cycles during the growth and development of the GBR (Source: from Webster and Davies 2003). **(B)** [14]C dates of coral mass mortality along 27 km of the Holocene raised reef terrace from the Huon Peninsula, PNG. Two widespread disturbance events were dated at ~9100–9400 years BP (before present) and ~8500 years BP. Isolated examples of coral mortality were observed in the Bonah River lagoon and the Hubegong dive site. Mortality events are shaded, and labelled with their likely cause where this has been deduced ('ash', associated with a volcanic ash layer, or, 'debris flow', associated with a submarine debris flow). Shown here are the 2-sigma age ranges in calendar years BP (Source: from Pandolfi *et al.* 2006).

have suffered disproportionately compared to other coral taxa, resulting in coral phase shifts among the inshore reefs of the GBR.

The future of the Great Barrier Reef

Global environmental change has had a profound influence over the development of tropical coral reefs since time immemorial, and their effects are no less profound for the GBR. Coral growth culminating in today's GBR has been shaped by the natural variations in sea level, temperature, CO_2 and other climate variables that control light levels, rainfall, turbidity and ocean acidification. It is no wonder then that the ecology of coral and coral reef communities must be seen as dynamic and fluid in their response to Pleistocene sea level and climatic fluctuations.

During the last interglacial to interglacial cycle, the GBR has experienced large scale platform reef accretion during two short intervals near the peak of each transgression. However, for more extensive periods, the GBR has been doing 'something else', including intervals of restricted shelf edge fringing reef development on the eastern slopes coupled with, we speculate, other non-reefal modes of coral community exploitation of the deeper southern regions of the GBR and Queensland Plateau during the regressive and glacial intervals. Further work into the comparative response of coral communities and reef growth to these two end member phases should provide important insight into prediction of reef response to future climatic changes.

Now the GBR, like many coral reefs around the world, is changing dramatically in response to human interaction. But living ecological systems provide few clues as to the extent of their degradation. The only recourse into understanding the natural state of living reefs is the fossil record. Our knowledge of past ecosystems on the GBR contributes to formulating sound approaches to the conservation and sustainability of the GBR; specifically in ensuring that policy makers and managers use geological contexts and perspectives in setting realistic goals and measuring their success.

Further reading

Sea-level change

Beaman RJ, Webster JM, Wust RJA (2008) New evidence for drowned shelf edge reefs in the Great Barrier Reef, Australia. *Marine Geology* **247**, 17–34. doi:10.1016/j.margeo.2007.08.001

Brook EJ (2005) Tiny Bubbles Tell All. *Science*, 310(5752), 1285–1287, http://doi.org/10.1126/science.1121535

Chappell J, Omura A, Esat T, McCulloch M, Pandolfi J, Ota Y, Pillans B (1996) Reconciliation of late Quaternary sea levels derived from coral terraces at Huon Peninsula with deep sea oxygen isotope records. *Earth and Planetary Science Letters* **141**, 227–236. doi:10.1016/0012-821X(96)00062-3

Lambeck K, Esat TM, Potter E-K (2002) Links between climate and sea levels for the past three million years. *Nature* **419**, 199–206. doi:10.1038/nature01089

Reef development

Larcombe P, Carter RM (2004) Cyclone pumping, sediment partitioning and the development of the Great Barrier Reef shelf system: a review. *Quaternary Science Reviews* **23**, 107–135. doi:10.1016/j.quascirev.2003.10.003

Pickett JW, Ku TL, Thompson CH, Roman D, Kelley RA, Huang YP (1989) A review of age determinations on Pleistocene corals in eastern Australia. *Quaternary Research* **31**, 392–395. doi:10.1016/0033-5894(89)90046-X

Roff G, Clark TR, Reymond CE, Zhao J-X, Feng Y, McCook LJ, *et al.* (2013) Palaeoecological evidence of a historical collapse of corals at Pelorus Island, inshore Great Barrier Reef, following European settlement. *Proceedings of the Royal Society B: Biological Sciences* **280**(1750), 2012–2100. doi.org/10.1098/rspb.2012.2100

Tager D, Webster JM, Potts DC, Renema W, Braga JC, Pandolfi JM (2010) Community dynamics of Pleistocene coral reefs during alternative climatic regimes. *Ecology* **91**, 191–200. doi:10.1890/08-0422.1

Webster JA, Davies PJ (2003) Coral variation in two deep drill cores: significance for the Pleistocene development of the Great Barrier Reef. *Sedimentary Geology* **159**, 61–80. doi:10.1016/S0037-0738(03)00095-2

Reef systems

Bridge TCL, Done TJ, Beaman RJ, Friedman A, Williams SB, Pizarro O, *et al.* (2011) Topography, substratum and benthic macrofaunal relationships on a tropical mesophotic shelf margin, central Great Barrier Reef, Australia. *Coral Reefs* **30**, 143–153. doi:10.1007/s00338-010-0677-3.

Cappo M, Kelley R (2001) Connectivity in the Great Barrier Reef World Heritage Area – an overview of pathways and processes. In *Oceanographic Processes of Coral Reefs: Physical and Biological Links in the Great Barrier Reef.* (Ed. E Wolanski) pp. 161–187. CRC Press, New York, USA.

Hopley D, Smithers S, Parnell K (2007) *The Geomorphology of the Great Barrier Reef: Development, Diversity and Change.* Cambridge University Press, Cambridge, UK.

Pandolfi JM, Tudhope A, Burr G, Chappell J, Edinger E, Frey M, *et al.* (2006) Mass mortality following disturbance in Holocene coral reefs from Papua New Guinea. *Geology* **34**, 949–952. doi:10.1130/G22814A.1

4
Oceanography

M. J. Kingsford and E. Wolanski

Oceanography affects organisms of the GBR and the nature of contemporary geological processes. The coral reefs that form the GBR are scattered over the continental shelf, which is shallow and fringed by the deep water of the Coral Sea (Fig. 4.1). The oceanography of the GBR influences the transport of sediment and the deposition of material to the substratum, as well as contemporary geological processes by eroding and shaping reefs; it controls the dispersion of organisms of the GBR. Indeed, organisms of all sizes are affected by oceanography: waterborne plankton, eggs and larvae have limited control of their horizontal movements and, therefore, transport and dispersion will be influenced by currents. Cyclonic waves and currents destroy reef structures, kill organisms and alter the nature of habitats. Habitat-forming corals are becoming increasingly impacted by climate change and the bleaching that results when they are exposed to very warm waters. Changes in habitats can in turn affect organisms that typically 'respond' to different habitat types and the influence of oceanography on habitats can influence broad-scale patterns of biogeography. The richness of inter-reefal and reef based flora and fauna are strongly influenced by nutrient input from rivers and their dispersal by the oceanography, and upwelling over the shelf break.

Oceanography affects the GBR at all spatial scales, from large to small. At the largest spatial scales (thousands of kilometres) major oceanic currents of the Coral Sea bathe the GBR, affecting patterns of connectivity between reefs and the likelihood of coral bleaching. At moderate scales (km to hundreds of metres) the currents are strongly steered by the reefs. Finally, at very small spatial scales (centimetres to metres) small-scale turbulence affects the settlement patterns of organisms such as corals.

Currents have a great effect on highly mobile pelagic organisms. Nekton are attracted to oceanographic features such as convergences for the purposes of feeding and reproduction. For example, flying fish lay their demersal eggs on flotsam in convergences and this may provide suitable conditions for larvae. Concentrations of prey and variation in sea water temperature can also be critical for the survival of larval phases through to their settlement on reefs or other habitats. Some plankton will only spawn while aggregated. Many jellyfish and larvaceans, for example, primarily spawn when concentrated in convergence zones so that the chances of fertilisation are greatest. Currents can also move plankton to favourable or unfavourable environments and therefore influence the connectivity of populations. This is particularly important for larval forms that must settle on coral reefs; unfavourable currents may expatriate them from the GBR and favourable currents may sweep them towards reefs, either toward other coral reefs (meaning the reefs are then connected) or their natal reefs (meaning the reef is self-seeded).

Fig. 4.1. 3-D rendering of the bathymetry of the GBR and the adjoining Coral Sea. The South Equatorial Current transports waters from the oceanic surface well mixed layer that overlies waters that are 4000 m deep to the GBR on a continental shelf 30 m to 100 m deep (Source: After Wolanski 2001).

Large-scale net currents

The strong, westward flowing, South Equatorial Current (SEC), that generates the East Australian Current (EAC), prevails in the Coral Sea and is directed towards the GBR (Fig. 4.2A). The strength of this flow is modulated by the Southern Oscillation Index (SOI). The SOI is the difference in atmospheric pressure between Tahiti and Darwin. The SOI also influences the strength of the other major tropical boundary current around Australia, namely the Leeuwin Current on the west coast (Box 4.1). On meeting the GBR continental shelf, the SEC splits into two oceanic currents, namely the southward flowing East Australian Current (EAC) south of the bifurcation point and the northward flowing North Queensland Coastal Current (NQCC) north of the bifurcation point (Fig. 4.2A). The bifurcation point fluctuates between Townsville and Lizard Island. The EAC does enter the reef mosaic of the GBR and generates the Coral Sea Lagoonal Current (CSLC) from the Central GBR to the Swains (Fig. 4.2A). This flow is weak and episodic compared with the EAC flow on the outside of the reef that can reach speeds of 0.5 m/s

Box 4.1. Tropical boundary currents

A warm, southward flowing tropical boundary current is also found on the west coast of Australia. This is the Leeuwin Current (LC) that is the continuation of the 'Indonesian Throughflow' from the Indonesian seas, and flows south down the coast of Western Australia. It turns east at Cape Leeuwin, where it becomes the South Australian Current, and flows eastwards below most of South Australia to Tasmania, where it is known as the Zeehan Current. The LC creeps closest to the coast near Ningaloo, ~1200 km north of Perth, where the current may be only 10 km off the coast. The thermal signal of the LC can be readily detected at Port Lincoln South Australia.

The EAC and LC both carry warm and clear waters that are nutrient poor (oligotrophic, Chapter 8). The direction of transport and the oligotrophic waters have a great influence on the biogeography of the east and west coasts. Corals are found as far south as Rottnest Island in the west and extend to Sydney in the east. The larvae of tropical species are advected south. A host of newly settled tropical fishes and invertebrates often arrive at high latitude towards the end of the summer, but few survive the cold of winter. As Wernberg and colleagues showed in 2016, the northern extent of macroalgae (e.g. *Macrocystis* and *Ecklonia*) is also affected by warm currents, as these algae cannot persist in warm and nutrient poor waters.

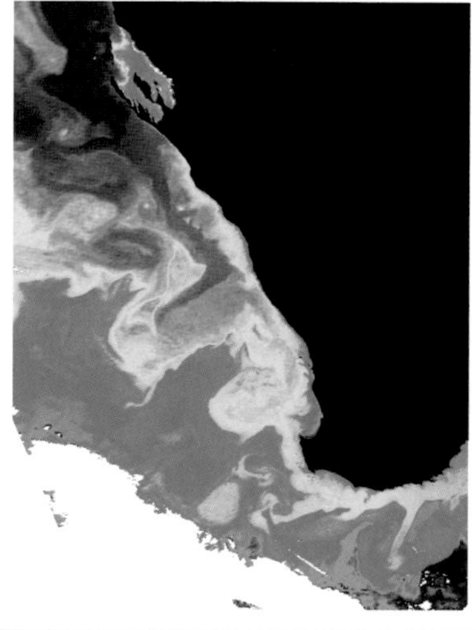

Image of sea surface temperature showing the warm Leeuwin Current flowing southwards and into the Great Australian Bight (Image: CSIRO).

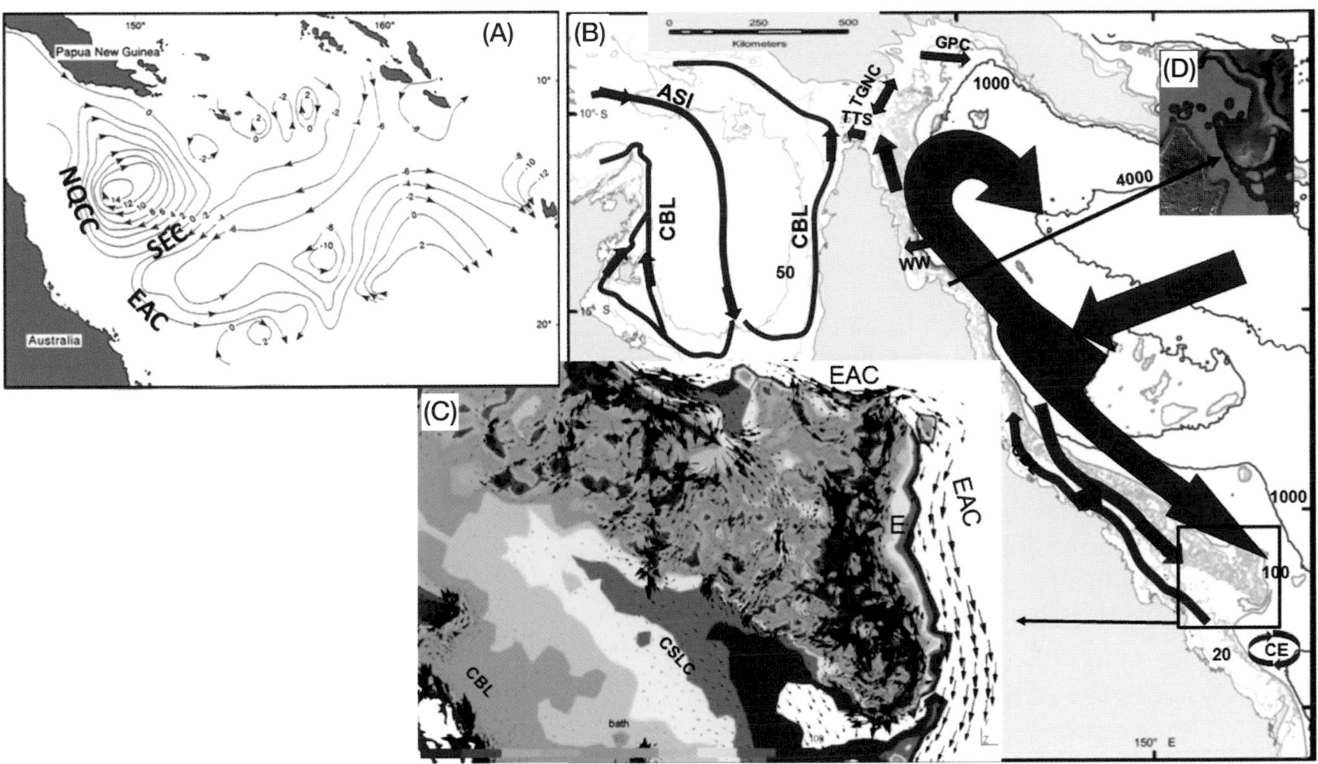

Fig. 4.2. **(A)** Large scale circulation in the Coral Sea derived from ship-borne CTD data (plotted as contours of volume transport (in Sv; 1 Sv, 1 million m³/s) for the top 1000 m (positive for clockwise flow, negative for counter-clockwise flow). (Source: Adapted from Andrews JC, Clegg S (1989) Coral Sea circulation and transport from modal information models. *Deep-Sea Research* **36**, 957–974). **(B)** The net circulation in the south-east trade wind season in the GBR and surrounding seas (Source: Adapted from Wolanski E, Lambrechts J, Thomas C, Deleersnijder E (2013) The net water circulation through Torres strait. *Continental Shelf Research* **64**, 66–74). **(C)** The SLIM model-predicted average circulation in the southern GBR. The symbols are various net currents that are explained in the text. **(D)** Satellite image of the GBR near 14°S showing the effective blockage by reefs forming a barrier that largely blocks the currents and restricts connectivity between the central and the northern regions of the GBR (Source: Map data from Google, SIO, NOAA, US Navy, NGO, GEBCO).

(Fig. 4.2C). The EAC rotates around the Swain Reefs and creates an eddy (E in Fig. 4.2C), that occasionally sheds to help create the transient Capricorn Eddy (CE in Fig. 4.2B). In turn, the Capricorn Eddy modulates the strength of the CSLC. In the southern GBR, the wind also drives an inflowing northward longshore coastal boundary layer (CBL) current (Fig. 4.2B, C) that turns into a Cross-Current (CC) that feeds into the CSLC (Fig. 4.2B). The influence of the EAC extends over a 1000 km south of the GBR and veers eastwards, forming a temperature discontinuity called the Tasman Front.

There is little oceanographic exchange of water between the northern and the central GBR, because of the blockage of the GBR shelf by reefs near 14°S (Fig. 4.2D). Thus the NQCC does not spread over the northern GBR, where the net circulation is a wind-driven current (WC) that is facilitated by the oceanic waves breaking over the outer reef crest; this generates an inflow of Coral Sea water into the northern GBR, which is the WW current shown in Fig. 4.2B.

The water circulation becomes extremely complex on reaching Torres Strait where the waters are very shallow and the net currents are diminished by friction that is enhanced by interacting with strong tidal currents. In the south-east wind season, northern GBR waters flows out to the Gulf of Carpentaria through the Through Torres Strait Current (TTS in Fig. 4.2B) and to the Gulf of Papua through the Great North-east Channel Current (TGNC in Fig. 4.2B); this water is circulated in the Gulf of Carpentaria by the Arufura Sea Inflow (AIS) and the wind-driven Coastal Boundary Current (CBL) and in the Gulf of Papua by the Gulf of Papua Current (GPC).

This wind-driven circulation in Torres Strait reverses sign during the monsoonal season when the north-westerly wind makes the mean sea level on the western side up to 0.8 m higher than on the eastern side – this was measured by satellite altimetry (Fig. 4.3A). This slope, together with the monsoonal wind, generates an inflow of Gulf of Carpentaria waters into Torres Strait (Fig. 4.3B) and this water then moves into the northern GBR.

During the 2016 ENSO event, the temperature of the intruding Gulf of Carpentaria waters was 32–33°C (Fig. 4.3C); this intrusion of hot water contributed to exceptional heat stress and subsequent mass coral bleaching and mortality of reefs in the Torres Strait and the Northern GBR. Additional causes for heating the Northern GBR waters included the ENSO shutdown of the NQCC, which would otherwise have flushed and cooled the northern GBR, and local solar heating. Historical data show that, during ENSO events, the lower cloud cover and warmer atmosphere more effectively heats up the shallow waters of the Gulf of Carpentaria (Fig. 4.3C) and these waters reside for longer than normal in Torres Strait and the NGBR due to the smaller mean sea level slope across Torres Strait.

Reef oceanography

The oceanography of the GBR is affected not just by the net oceanic currents, but also by the tides that flow on and off the shelf, and the wind that varies in direction and strength with time of year. The resulting currents interact with the complex bathymetry to generate jets, eddies, stagnation zones and convergences. Furthermore, the wind that varies in direction and strength with time of year. Variation in vertical physical structure (i.e. thermoclines and haloclines) and the input of freshwater and mud from rivers also affect water density and related currents. Across the shelf, nearshore waters are often impacted by freshwater whereas outer reefs are more affected by upwelling.

Tides

Tides have a great influence on currents within the GBR lagoon. The tides flow on and off the shelf, resulting in a considerable east–west movement of water. Maximum amplitudes of tides, by area, on the GBR range from 2.5 m to 7 m. The tides are generally semidiurnal (twice per day) and spring tides generate greatest currents (cf. neap tides). In regions that include the southern GBR and northern regions, tidal currents of 1 m/s are generated through channels between coral reefs, although in very narrow channels the peak tidal currents can be much stronger. Typically waterborne particles can travel up to 7 km westwards on incoming tides and 7 km eastwards on outgoing tides.

The interaction of tidal currents with the EAC depends on the reef region and distance across the shelf. Where the EAC enters the GBR in the central region and forms the lagoonal current (CSLC) that flows south, particles move on an east–west axis with the tide, while there is some movement to the south as a result of the EAC. The influence of the NQCC (flowing north) on inter-reefal waters of the northern GBR is smaller and the residence time of water masses tends to be greater in this area; therefore, the effect of tides generally dominates here.

Influence of the bathymetry

The bathymetry has a great influence on currents of the GBR and each one of the 2500 reefs of the GBR generate great complexities in flow. Complex bathymetry and channels generate water jets, eddies and convergence zones such as thermal fronts and internal waves. All of these features have a great influence on the transport and aggregation of plankton, connectivity of larvae among reefs and the aggregated plankton is often an attractant for nekton.

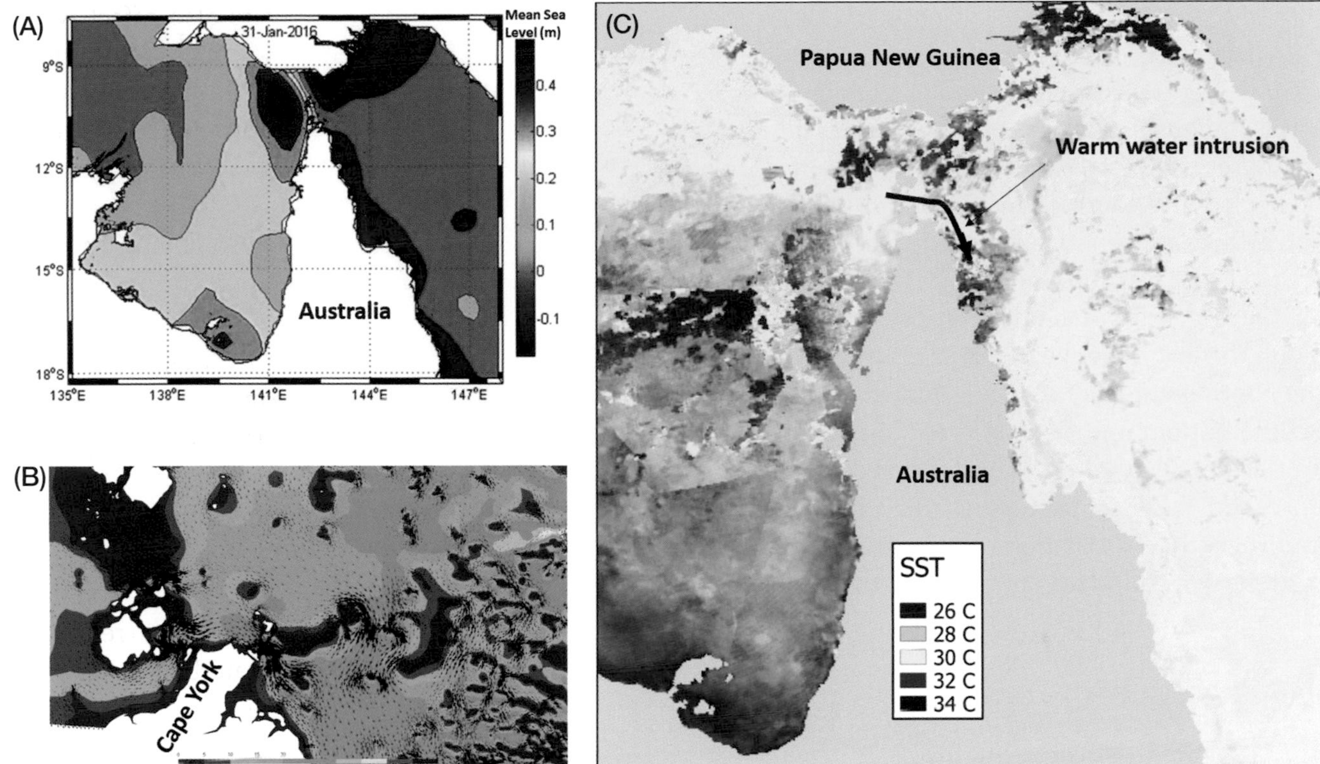

Fig. 4.3. **(A)** The observed distribution of the mean sea level in the northern GBR and surrounding seas on 31 January 2016 (Source: Reproduced from Wolanski *et al.* 2017). **(B)** The SLIM model predicted net currents in the southern Torres Strait and the northern GBR on that same date. The colour bar at the bottom indicates the depth (in metres). **(C)** The satellite observation of the sea surface temperature (SST) in the northern GBR and surrounding seas on 20 February 2016.

Eddies

Eddies vary greatly in size from small-scale features that are tens to hundreds of metres wide to those that are tens of kilometres wide. Eddies form on the down current side of reefs and will influence the trajectory of particles. These eddies are highly unsteady and recent studies have shown the complex currents that they generate at high resolutions down to 12 m (Fig. 4.4A). Detailed field studies showed that reef-generated eddies are three-dimensional structures with a shape like a doughnut (Fig. 4.4B). Particles are subducted at convergence zones around the edge of the eddy. Rotation of the eddy facilitates upwelling up through the core and sea level is lowest in the core of the eddy. These eddies are common around reefs and islands (Fig. 4.4C). It takes an hour or two for them to generate after a change of the tide and particles may only do a few orbits of the doughnut before the eddy moves off with a change of the tide and dissipates. Some reefs that are in the Coral Sea have topographically stable eddies for periods of much greater than a day because the current does not change direction. These eddies are not influenced by the Earth rotation; they are only a few km in size. They are quite different from meso-scale oceanic eddies (>100 km) that are modulated by the Earth rotation and where upwelling depends on whether the rotation is cyclonic or anti-cyclonic. Intermediate-scale eddies are also found on or near the GBR, perhaps the most conspicuous is the eddy 'E' in the lee of the Swains (Fig. 4.2C) that is driven by the EAC. The EAC forms many topographically unconstrained eddies that are transported south. Eddies of all sizes are of biological important because they can retain particles (including larvae) near individual reefs or regions through transport or through aggregation in convergence zones and they can influence local upwelling of nutrient-rich water.

Jets

At falling tide, water jets seawards through the Ribbon Reefs of the GBR and generates complex three-dimensional structures that extend hundreds of metres to over a kilometre from the edge of the reef; also the thermocline is deformed upwards and eddies are generated above (Fig. 4.5A). Floating particles are captured and aggregated along slicks by these circulations at the convergence zone or along the leading edge of the jet. The water leaving the continental shelf forms a buoyant jet that lifts off the bottom and vertically entrains deeper oceanic water (Fig. 4.5A). Nutrient-rich deep water is upwelled by jet entrainment and this upwelling, and related accumulation of prey, may explain the aggregation of black marlin in front of these passages.

On the rising tide (Fig. 4.5B), a mushroom jet is formed over the GBR shelf. On the oceanic side, deep nutrient-rich water mass is upwelled by a 'Bernoulli-effect' upwelling, and this water is entrained by the tidal jet on the GBR shelf as a bottom-tagging layer (Fig. 4.5B); the primary beneficiary may be the *Halimeda* algae that form large meadows near these passages.

The 'sticky water' effect

The tides propagating through a dense reef mosaic generate very strong tidal currents through the formation of tidal jets. In shallow waters, through non-linear hydrodynamic effects, these result in greatly increasing the bottom friction of the net currents. The prevailing net oceanic currents upstream of the reef matrix are steered sideways

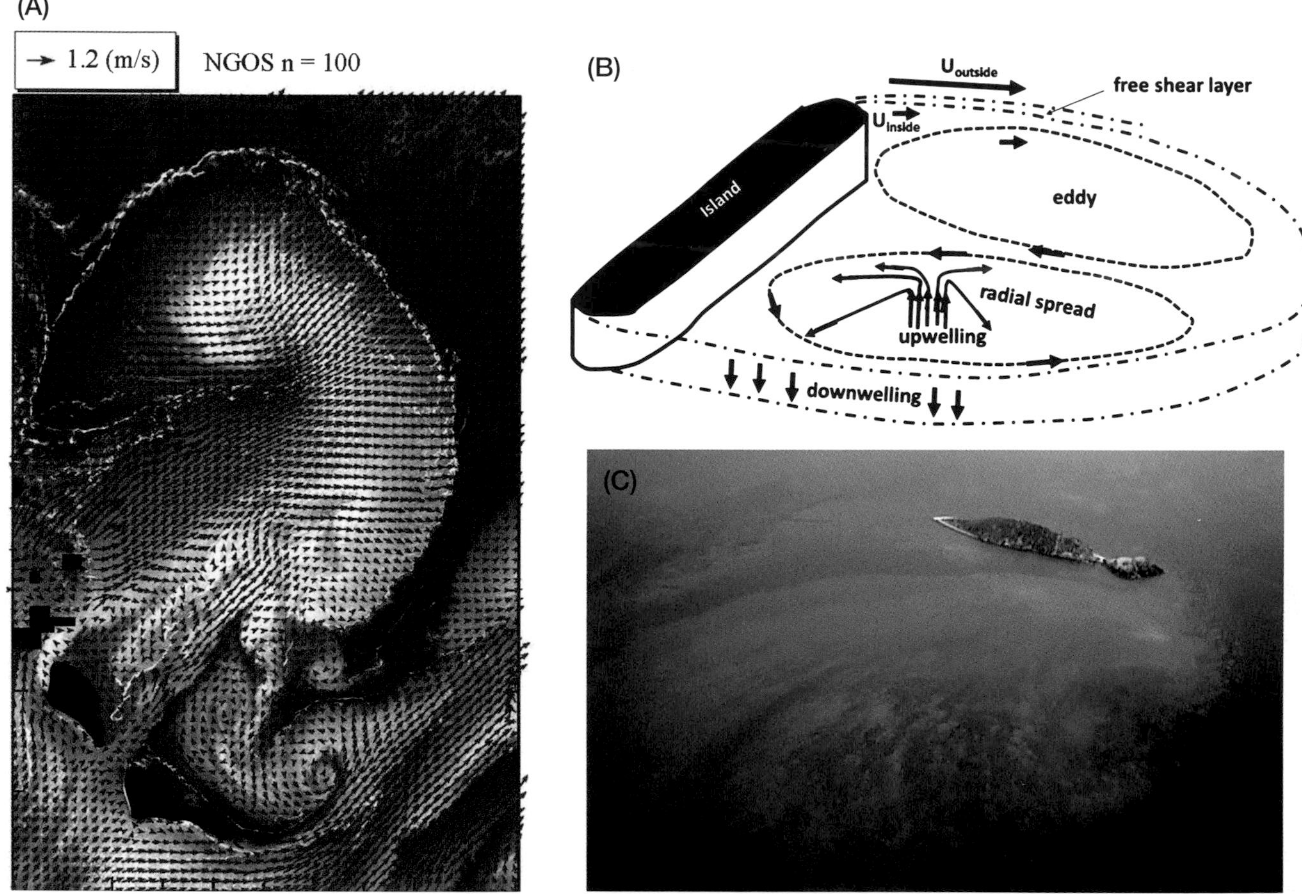

Fig. 4.4. (A) Satellite-derived instantaneous distribution of currents around Hull and Beverlac Islands in the southern GBR. **(B)** Sketch of the internal circulation and around an eddy in the lee of an island or a reef. **(C)** The eddy in the lee of Rattray Island at flood tide.

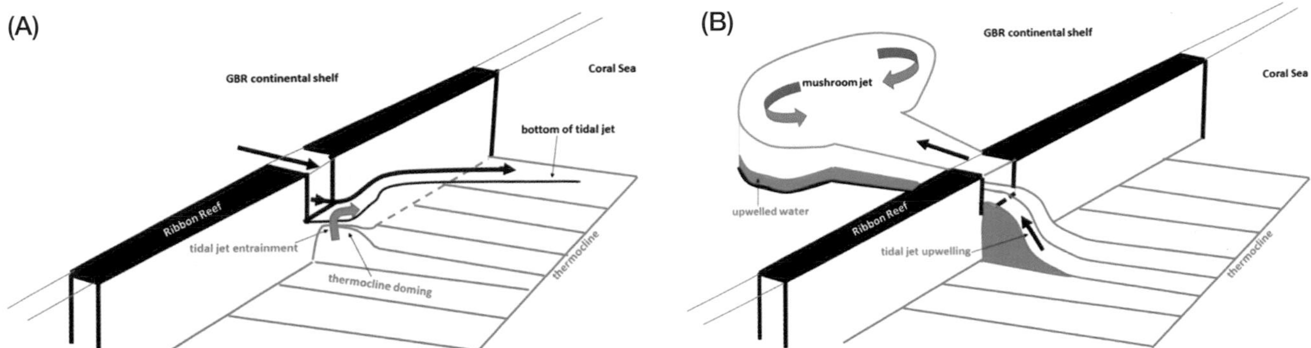

Fig. 4.5. The tidal jets and the internal upwelling and convergence zones near the Ribbon Reefs at **(A)** falling tide and **(B)** rising tide.

around, instead of through, the reef matrix (Fig. 4.6). 'Sticky water' aptly describes waters where dispersal is restricted through a combination of these factors. This process influences the transport of particles and the likelihood they will disperse a great distance from the source. For example, particles can be retained in a close mosaic of reefs. In contrast, the 'sticky water effect' is minimal for isolated reefs in strong mainstream currents where particles are quickly dispersed. This is of great relevance in estimating levels of connectivity among reefs.

Shelf waves and upwelling

The transport by the wind is downwind at the surface, but this is not the case through the whole water column. Transport of particles in the wind affected layer (Eckman layer) deviates to the left in

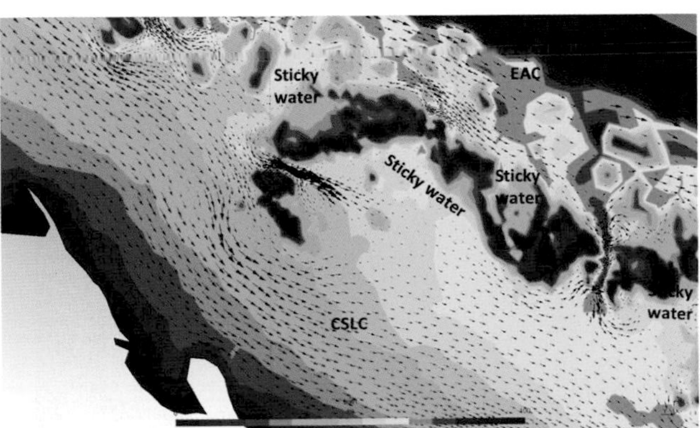

Fig. 4.6. The SLIM model-predicted net currents around Cape Upstart and Old Reef, illustrating how a dense reef matrix deflects an incoming current sideway around the reef mosaic. The colour bar at the bottom indicates the depth (dark blue to red = 0 to 100 m).

the southern hemisphere with depth and more to the right in the northern hemisphere as a result of the rotation of the Earth. On average, therefore, particles will be transport shoreward in the Eckman layer with a wind from the south and offshore with a wind from the north. When surface waters are transported away from the shore, a sea level low is created, which is replaced by cool water from beneath.

The three-dimensional nature of the water column on the shelf slope usually varies by season. There is always a surface well-mixed layer but its thickness can vary by 70 m from season to season (Fig. 4.7A). The changing wind generates low-frequency reversing currents on the shelf slope. In turn, this and the tides generate internal waves (Fig. 4.7B); the internal waves upwell deep, nutrient-rich water when the mixed layer is the thinnest (i.e. when the thermocline is closest to the surface). Upwelled water can then spill over the shelf and form a nutrient-rich, colder, bottom-tagging water layer over the GBR shelf, and enrich the outer GBR with nutrients. The cool water will often not reach the surface; it typically slides over the shelf as a high density water mass that remains close to the substratum. In some cases, these intrusions can make it most of the way across the shelf. Although data on the periodicity of upwelling are sparse, upwelling signals are most common from November to March when winds from the north are frequent and when the oceanic surface mixed layer is the shallowest for the year (Fig. 4.7A). This, combined with the influence of major currents, are the major mechanisms for generating upwelling of nutrient-rich waters. Upwelling in the clear waters of the GBR appears to an oxymoron because tropical waters are generally thought to be oligotrophic. The mixing of upwelled nutrients with planktonic assemblages can at times be detected by satellite as clear chlorophyll signatures (Fig. 4.7C) all along the length of the GBR.

Wind and associated wind-driven swell often have a great influence on circulation patterns in lagoons because waters of lagoons are often shallow (1–4 m deep).

Convergences

Convergences are caused by a variety of physical features that include eddies, jets, wind, fronts, internal waves and freshwater plumes. Convergence zones accumulate plankton and floating objects such as flotsam, drift algae and aggregations of cells (e.g. *Trichodesmium*); (see Chapter 16). The convergence zones of windrows that form parallel to the direction of the wind are Langmuir cells that aggregate flotsam and organisms such as plankton, coral eggs and jellyfishes. Tidally induced fronts are generated through differences in the density of waters masses, differences in depth (i.e. water will travel relatively faster in deeper water creating a shear zone or convergence) and the edge of eddies. When lagoons heat up, sea water becomes less dense and on the outgoing tide a thermal front forms where the warm water meets relatively cool inter-reefal waters. The differences in temperature can range from 0.2°C to 1.5°C and a convergence will form at the front. These and other tidal fronts generally dissipate with a change of the tide.

Reef-generated internal waves

Island internal waves result when there is a density difference with depth, and on the GBR this is

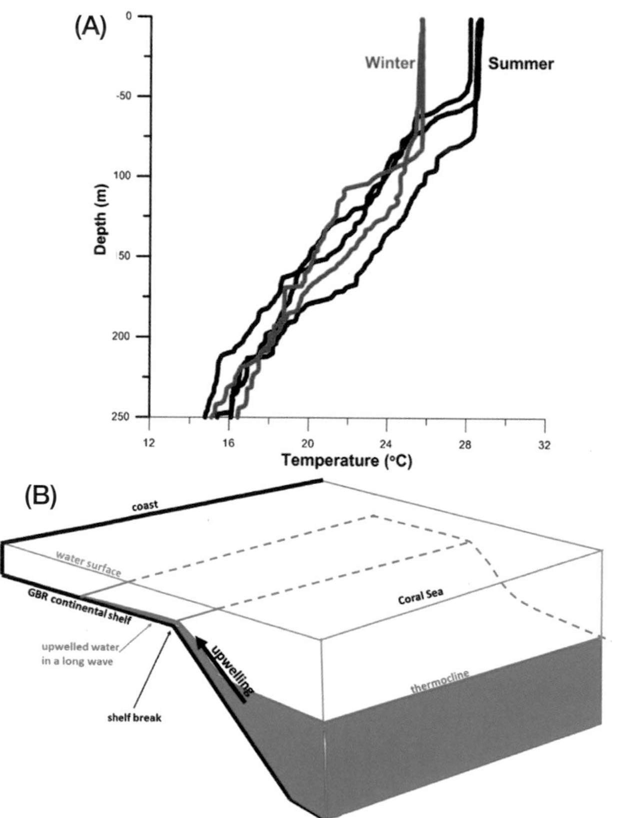

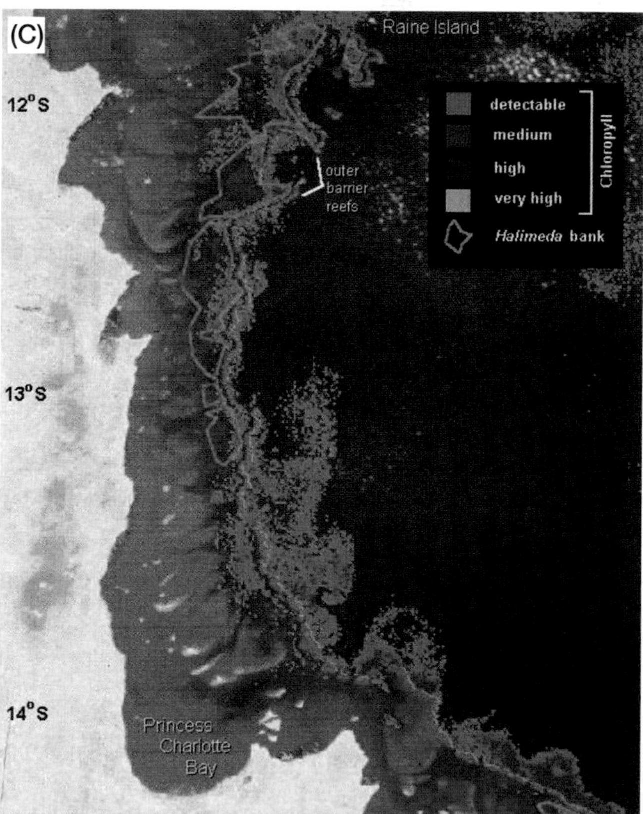

Fig. 4.7. **(A)** Waters of the shelf slope of the GBR are generally stratified. Water temperatures at over 200 m are often ~10°C lower that that at the surface. The examples we present here show a well-mixed water column (i.e. similar temperature from the surface to typically 50–100 m depth). **(B)** Upwelling by low-frequency internal waves on the GBR continental slope. **(C)** Satellite view of near-surface chlorophyll *a* along the GBR (Map: NOAA).

usually restricted to reefs on the outer shelf and in the Coral Sea, such as Myrmidon Reef and Raine Island. The reversing tidal currents deform the thermocline, initiating an internal wave; as the tide changes, the internal wave does not have time to smoothly adjust itself to the changes in pressure from the tides, so very large amplitude (typically 70 m) internal waves can result. They propagate upwards along the island slopes in a thin layer, typically only a few tens of metres thick, and can form an internal tidal bore (i.e. an internal shock wave). They also radiate seaward a packet of internal waves. Parallel convergences form at the surface over the rear of each wave. The largest internal waves in the world (amplitude of 270 m) have been found around isolated oceanic coral reefs.

The deformation of the thermocline upward over a reef combined with channelling by local topography will result in an upwelling that will bathe benthic organisms in cool water and this can constitute both a positive effect (e.g. a nutrient supply that enhances productivity) and a negative

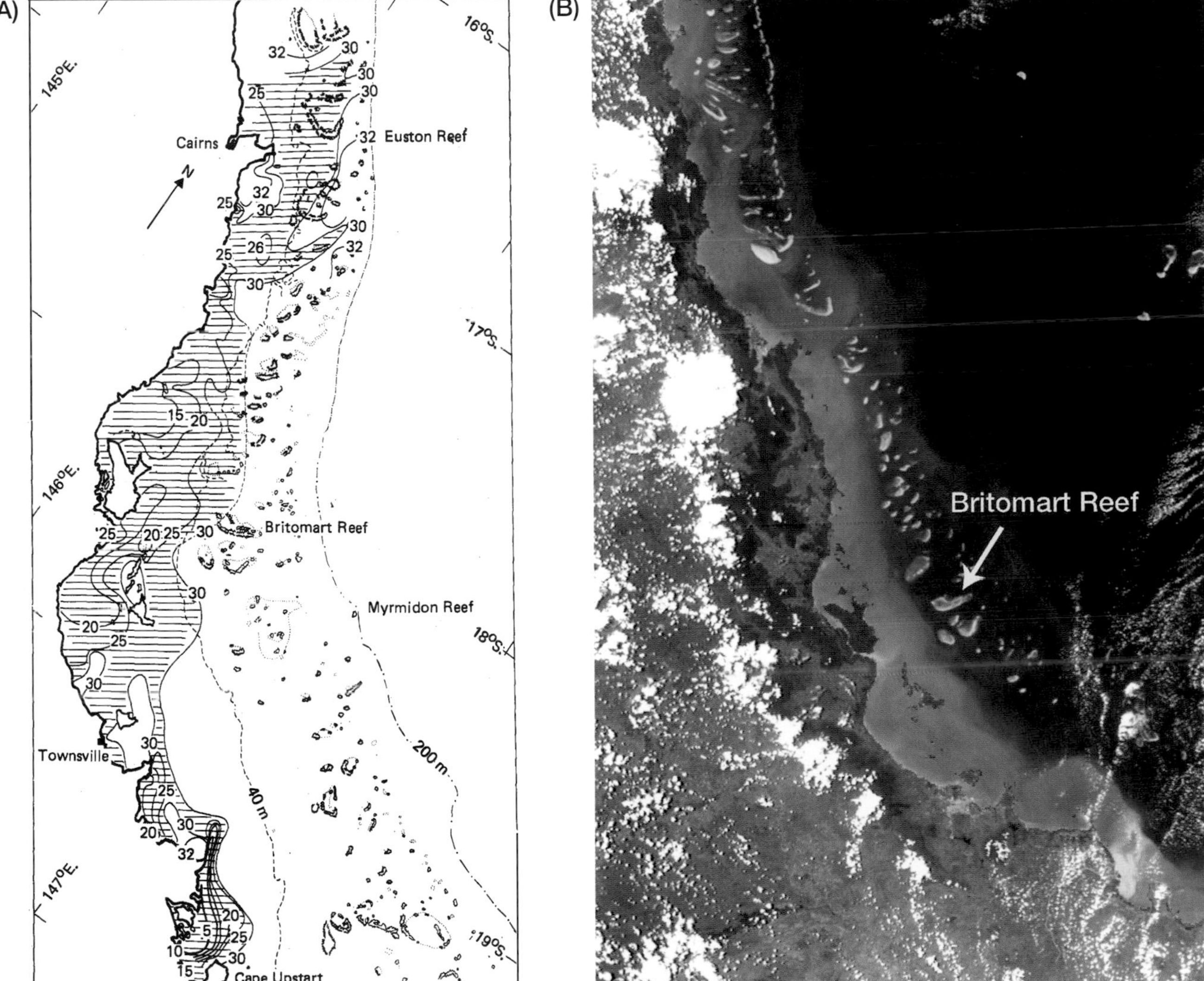

Fig. 4.8. **(A)** Surface salinity on 26–27 January 1981, during the peak of the Burdekin River flood (Source: Modified from Wolanski E, van Senden D (1983) Mixing of Burdekin River flood waters in the Great Barrier Reef. *Australian Journal of Marine and Freshwater Research* **34**, 49–63). **(B)** Satellite image of the turbid Burdekin River plume in February 2007. The Burdekin River plume moved longshore northwards, formed a surface coastal boundary current, and merged with the plume of the Herbert River (west of Britomart Reef), and remain distinct until Cairns where it turned seawards. (Image: MODIS sensor, NASA).

effect (e.g. thermal shocks potentially stressing organisms).

Buoyancy-driven flows

The flow of GBR rivers is stochastic with respect to time of year due to the timing of rain events. For example, large rain events coincide with cyclones or major weather fronts from the south in some years, while in other years significant rainfall is rare or absent. The Burdekin River has a huge catchment area of 128 860 km^2. The river is often dry, but over short periods of time it can be Australia's largest river, in terms of volume. In some flood events, up to 1690 million m^3 per day have been recorded. These huge volumes of fresh water, and associated sediment and nutrients from the land, are transported in a thin surface plume that can extend northwards for over 200 km (Fig. 4.8A). River plumes set up a three-dimensional structure that will have some influence on local flow and the transport of particles in the vicinity of plumes. These plumes are highly stratified in salinity, with fresher water on top, and the buoyancy, together with the Coriolis effect, make the plume drift northwards at velocity of 0.1–0.3 m/s even against a prevailing southward mean oceanic current. By turbulent mixing, the width of the plume grows as the water moves northwards away from the river mouth. Plumes are significant turbidity signatures that can be viewed from space (Fig. 4.8B); this sediment quickly rains out of the plume to deposit on the bottom (Fig. 4.9A). The 'marine snow' flocs from these plumes are very sticky and readily smother coral polyps and can cause mortality of juvenile corals.

During the dry season, strong winds entrain muddy sediment in suspension in embayments, such as Trinity Bay in the Cairns region of the GBR. When the wind calms, this water is negatively buoyant because of the suspended sediment load. The density-driven flows makes this water cascade downwards into deeper water as a bottom-tagging nepheloid layer a few metres thick, carrying the mud towards coastal and inner shelf reefs that it will help degrade (Fig. 4.9B).

The predictable and the unpredictable in GBR oceanography

The predictable

There is a predictable seasonality of oceanography on the GBR. During the period from December to March incursions of cool water are upwelled onto the shelf and these in turn can generate a pulse in primary productivity that can be observed as a chlorophyll signature from space (Fig. 4.7C). Because the incursions are of high density water, the water column becomes stratified as cool water

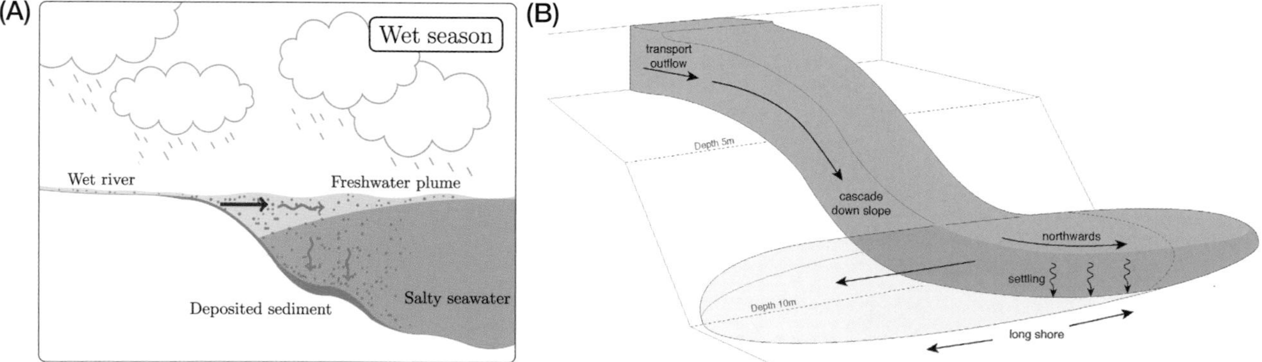

Fig. 4.9. (A) Sketch of the Burdekin River plume: the plume is thin and is at the surface; the suspended fine sediment falls out of suspension and deposits on the bottom (Source: Modified from Delandmeter *et al.* 2015). **(B)** Density currents generated by muddy water escaping from Trinity Bay off Cairns, cascading down to deeper water and being entrained by the prevailing currents on the shelf. (Source: Modified from Brinkman R, Wolanski E, Spagnol S (2004) Field and model studies of the nepheloid layer in coastal waters of the Great Barrier reef, Australia. In *Shallow Flows.* (Eds GH Jirka and WSJ Uijttewaal) pp. 225–229. Balkema Publishers, Leiden, Netherlands).

'tongues' slide onto the shelf. December to April is also the wet season from Tully north and the time that cyclones result on major 'dumps' of rain in drainages of the wet and dry tropics (i.e. Cape York to Fraser Island). These freshwater incursions increase stratification and reduce salinity in coastal waters (Figs 4.8, 4.9A).

Although winds can affect GBR waters at any time of the year, the dry season is characterised by strong trade winds (especially from the SE) that alter transport and resuspend sediments from the bottom. Superimposed on seasonal changes are other processes such as ENSO that influence the strength of the SEC on longer timescales.

The unpredictable: El Niño southern oscillation (ENSO)

The intensity of the SEC and resultant EAC and NQCC on the east coast of Australia, and the LC on the west coast, is projected by the Southern Oscillation Index (SOI). La Niña is sometimes referred to as 'super-normal conditions' as the westward flow of water from Peru is exaggerated. This results in high and positive values of SOI. The SEC is the largest during long periods of steady SOI and smallest during periods of rapidly variable SOI. During El Niño conditions (negative values of SOI), the upwelling fails off Peru. There is a 'decadal oscillation' that corresponds to periods of time when SOI is more regularly in the positive than the negative and visa-versa; although in the last 10 years this periodicity has been questionable. Variation in SOI has a great impact on biology: the failure of upwelling off Peru correlates with movements of whales and pelagic fishes and death of seabirds that can no longer find prey. Galapagos iguanas that rely on cool upwelled waters to support biomass of their algal food die *en masse* as the algae disappear. The number of lobsters that recruit to the west coast of Australia varies with strength of the LC. When flow is weak during El Niño, recruitment of puerulus larvae to limestone reefs is poor.

Coral bleaching on the GBR and other parts of the world have been recorded during La Niña and El Niño, but the biggest bleaching events have been recorded during El Niño conditions (1998, 2002, 2016, 2017). Patches of water that generally exceed 31°C stagnate on the GBR (Fig. 4.3C) and, if they persistent, will lead to the breakdown of the symbiosis and the loss of the symbiotic algae of corals (zooxanthellae), and mass mortality in many cases.

Some ecological consequences of the oceanography of the GBR

Connectivity

The fish, corals and other invertebrates spawn eggs and sperm at particular times of the year. The eggs are fertilised and the larvae that develop will drift with the water. The level to which reefs are self-seeded in these larvae (i.e. larvae that return to their natal reefs) and to which reefs are connected (i.e. the exchange of larvae between reefs) partly depends on the water currents during the drift period. Since these currents vary with the wind and the influence of the SEC – itself controlled by large scale oceanic features within the Coral Sea – oceanography predicts that the level of self-seeding and connectivity may vary seasonally and inter-annually from reef to reef. In areas where the reefs are scattered at relatively low density on the shelf, a reef is usually both self-seeded and connected with other reefs located 'upstream'. In areas where the reefs are scattered at high density, the currents are steered around the dense reef matrix instead of flowing through the reef matrix, because of the sticky water effect; reefs within the dense matrix are highly self-seeded and little connected with reefs outside. Recent evidence indicates that self-seeding by recruit fishes can be as high as 30% to 60% and it is likely that the behavioural abilities of presettlement fishes, as well as current, influence connectivity. Fishes, in their planktonic dispersal phase, greatly increase in mobility during development. Depending on the size of larvae, they can swim at speeds of typically 5–10 cm/s, and exceptionally 35–50 cm/sec, and they can orientate to reefs. Magnetic and sun compass senses can be used, as well as cueing into the sounds and smell of reefs (even the reef from which they were spawned). These abilities can greatly reduce dispersal and may explain why a high proportion of larvae may return to their natal reef.

Freshwater input and cyclones

Freshwater input has a great influence on the biology of inshore waters. In many parts of the world, it has been demonstrated that catch and recruitment rates of prawns and fishes vary with freshwater input: recruitment rates usually go up with input of fresh water. Changes in freshwater runoff may happen through anthropogenic alteration of catchments and through changes in rainfall that relate to global warming (see Box 4.2). There is also a concern that an increase in runoff and nutrients could affect inner shelf reefs through phase shifts (i.e. from coral to algae) and improve the survival of crown-of-thorns starfish (COTS) larvae.

The wet season is the same as the cyclone season and river floods often result from cyclones. Cyclones can thus be very destructive because of the huge input of sediment and freshwater input to the GBR and also of wave height on reefs. Physical destruction can be considerable and the swathe of damage is asymmetrical with respect to the position of the eye of the cyclone. Cyclones rotate clockwise in the Southern Hemisphere and anticlockwise in the Northern Hemisphere as typhoons (Asia) or hurricanes (America). The impact of cyclones on the GBR is greater on the southern side of the eye due to the greater fetch from the open ocean and the larger wave height. The destruction of reef habitat and death of organisms can be substantial. Storm swell can blast large chunks of coral (including large *Porites*) onto the reef flat. Damage will usually only be to the windward side of reefs, but life on coastal fringing reefs can be all but obliterated. Great changes in habitat type (e.g. from a species-rich assemblage of live coral to coral rubble) will have a great influence on local species diversity of most taxa (see Chapter 5).

Further reading

Andrews JC, Gentien P (1982) Upwelling as a source of nutrients for the Great Barrier Reef ecosystems: a solution to Darwin's question. *Marine Ecology Progress Series* **8**, 257–269. doi:10.3354/meps008257

Box 4.2. Consequences of global warming on GBR oceanography

Global warming will cause changes in the physical and chemical oceanography of the GBR, as well as related biological change. There is great speculation on exactly what will happen, but predictions include the following: global warming is likely to change the weather, and in doing so will alter within year patterns of wind, rainfall and the location and temperature of warm water patches. Furthermore, the intensity of events and inter-annual fluctuations are likely to vary due to the likely increased frequency of the El Niño–La Niña. This will increase the variability of the net currents in the GBR. Thus global warming is likely to cause increases in the frequency and duration of bleaching events on the GBR through warm water intrusions and stagnation. There is some evidence for that in that major bleaching events have occurred in 1998, 2002, 2010, 2016 and 2017. Storm frequency and intensity may also increase and this will have a direct physical effect on reefs. Increased fresh water and nutrient input from cyclones would facilitate phase shifts from coral-dominated to algal-dominated inshore reefs and the nature of the pelagic environment would change as would its suitability to the survival of larvae. For example, the survival of COTS larvae may increase, causing greater frequency of COTS outbreaks. Other forms of plankton may win or lose with major changes in the abundance and species richness of the plankton. An increase in atmospheric CO_2 will lower the pH of the sea, thus making it more acidic by 0.3–0.5 of a pH unit by 2100. This is critical for corals, shellfish and some plankton (e.g. coccolithophores) because a reduction in pH affects carbonate concentration ions, making it more difficult to precipitate calcium carbonate. Altered pH can also have unexpected effects such as altering the predator escape responses in small fishes. A consequence of warming seas is that they are likely to change patterns of connectivity because temperature has a great influence on the time that larvae spend in the plankton.

Andutta F, Kingsford M, Wolanski E (2012) 'Sticky water' enables the retention of larvae in a reef mosaic. *Estuarine, Coastal and Shelf Science* **101**, 54–63. doi:10.1016/j.ecss.2012.02.013

Brinkman R, Wolanski E, Deleersnijder E, McAllister F, Skirving W (2002) Oceanic inflow from the Coral Sea into the Great Barrier Reef. *Estuarine, Coastal and Shelf Science* **54**, 655–668. doi:10.1006/ecss.2001.0850

Burgess SC, Kingsford MJ, Black KP (2007) Influence of tidal eddies and wind on the distribution of presettlement fishes around One Tree Island, Great Barrier Reef. *Marine Ecology Progress Series* **341**, 233–242. doi:10.3354/meps341233

Burrage D, Steinberg CR, Skirvig WJ, Kleypas JA (1996) Mesoscale circulation features of the Great Barrier Reef region inferred from NOAA satellite imagery. *Remote Sensing of Environment* **56**, 21–41. doi:10.1016/0034-4257(95)00226-X

Delandmeter P, Lewis S, Lambrechts J, Deleersnijder E, Legat V, Wolanski E (2015) The transport and fate of riverine fine sediment exported to a semi-open system. *Estuarine, Coastal and Shelf Science* **167**, 336–346. doi:10.1016/j.ecss.2015.10.011

Furnas MJ (2003) *Catchments and Corals: Terrestrial Runoff to the Great Barrier Reef.* Australian Institute of Marine Science, Townsville.

Glynn W (1988) El Nino-Southern Oscillation 1982–1983: nearshore population, and ecosystem responses. *Annual Review of Ecology and Systematics* **19**, 309–346. doi:10.1146/annurev.es.19.110188.001521

Grimes CB, Kingsford MJ (1996) How do estuarine and riverine plumes of different sizes influence fish larvae: do they enhance recruitment? *Marine and Freshwater Research* **47**, 191–208. doi:10.1071/MF9960191

Kingsford MJ, Wolanski E, Choat JH (1991) Influence of tidally induced fronts and Langmuir circulations on distribution and movements of presettlement fish around a coral reef. *Marine Biology* **109**, 167–180. doi:10.1007/BF01320244

Mann KH, Lazier JRN (1991) *Dynamics of Marine Ecosystems: Biological-physical Interactions in the Ocean.* Blackwell, Oxford, UK.

Munday PL, Leis JM, Lough JM, Paris CB, Kingsford MJ, Berumen ML, *et al.* (2009) Climate change and coral reef connectivity. *Coral Reefs* **28**, 379–395. doi:10.1007/s00338-008-0461-9

Nof D, Pichevin T, Sprintall J (2002) Teddies and the origin of the Leeuwin Current. *Journal of Physical Oceanography* **32**, 2571–2588. doi:10.1175/1520-0485-32.9.2571

Suthers I, Taggart CT, Kelley D, Rissik D, Middleton JH (2004) Entrainment and advection in an island's tidal wake, as revealed by light attenuance, zooplankton, and ichthyoplankton. *Limnology and Oceanography* **49**(1), 283–296. doi:10.4319/lo.2004.49.1.0283

Thomas CJ, Lambrechts J, Wolanski E, Traag VA, Blondel VD, Deleersnijder E, *et al.* (2014) Numerical modelling and graph theory tools to study ecological connectivity in the Great Barrier Reef. *Ecological Modelling* **272**, 160–174. doi:10.1016/j.ecolmodel.2013.10.002

Weeks SJ, Bakun A, Steinberg CR, Brinkman R, Hoegh-Guldberg O (2010) The Capricorn Eddy: a prominent driver of the ecology and future of the southern Great Barrier Reef. *Coral Reefs* **29**, 975–985. doi:10.1007/s00338-010-0644-z

Wernberg T, Bennett S, Babcock RC, de Bettignies T, Cure K, Depczynski M, *et al.* (2016) Climate-driven regime shift of a temperate marine ecosystem. *Science* **353**, 169–172. doi:10.1126/science.aad8745

Wolanski E (2001) *Oceanographic Processes of Coral Reefs: Physical and Biological Links in the Great Barrier Reef.* CRC Press, Boca Raton FL, USA.

Wolanski E, Andutta F, Deleersnijder E, Li Y, Thomas CJ (2017) The Gulf of Carpentaria heated Torres Strait and the northern Great Barrier Reef during the 2016 mass coral bleaching. *Estuarine, Coastal and Shelf Science* **194**, 172–181. doi:10.1016/j.ecss.2017.06.018

Wolanski E, Colin P, Naithani J, Deleersnijder E, Golbuu Y (2004) Large amplitude, leaky, island-generated, internal waves around Palau, Micronesia. *Estuarine, Coastal and Shelf Science* **60**, 705–716. doi:10.1016/j.ecss.2004.03.009

Wolanski E, Kingsford MJ (2014) Oceanographic and behavioural assumptions in models of the fate of coral and coral reef fish larvae. *Journal of the Royal Society, Interface* **11**, 20140209. doi:10.1098/rsif.2014.0209

5

Coral reef habitats and their influence on reef assemblages

M. J. Kingsford, C. Syms, M. Srinivasan and G. P. Jones

Coral reefs as habitats

Coral reefs often characterise the immediate subtidal environment in tropical regions. Although any hard substratum can support corals, more generally the hard base (that may be 1–2 km thick on some reefs) consists mostly of dead corals. This base is generated by the death and assimilation of the carbonates fixed by coral polyps into the hard substratum and a veneer of live corals grows on this base (Chapter 9). The Great Barrier Reef (GBR) is characterised by a mosaic of reefs of differing morphologies (Chapter 2) that stretch on a north–south axis for over 2000 km and in places up to 300 km wide. The architecture of individual reefs is largely based on a sculptured rock base and habitat-forming coral of different species that can vary depending on exposure to the predominant winds (i.e. windward and leeward sides of reefs). A key to understanding patterns of the diversity in species richness and diversity of functional groups of organisms lies in their habitat preferences and what is driving habitat heterogeneity.

Recent perturbations on the GBR and other parts of the world have caused great changes to reefal habitats. These perturbations include: climate change and related coral bleaching and mass mortality; cyclones; riverine runoff and the input of nutrients to oligotrophic ecosystems; outbreaks of crown-of-thorns starfish (COTS); and overfishing (see Chapter 11). All of these factors have the potential to affect reef architecture and the organisms associated with habitats. Accordingly, understanding habitat dynamics is critical to assessing the health of reefs.

In this chapter we will discuss the role of coral reefs as habitats for other organisms and their importance within the broader tropical subtidal landscape (or maybe, more appropriately, seascape). We will not restrict our discussion to particular taxa, but rather use occasional examples to illustrate general points about coral reef habitats and habitat associations. Although some authors have lumped the entire range of environmental factors such as temperature, water quality and so forth into the definition of an organism's 'habitat', we instead will use a more literal definition of habitat from the Latin *habitare* 'the place where something lives'. So 'habitat' under this definition becomes a spatial concept, and reflects an emphasis on habitat as being what an organism perceives within its range of movement and response. There is some circularity in the discussion in that habitats include physical attributes such as dead carbonate structures and habitat-forming organisms that add to the heterogeneity and combine to form the reef architecture. We will start by considering organisms in the context of their contributions to habitat structure, how they can influence or simply respond to temporal and spatial variability in habitat heterogeneity. Some emphasis is given to the consequences of changes in habitats because this can be a major driver in determining biodiversity on the GBR.

Habitat formers, determiners and responders

An important distinction between temperate rocky reefs and coral reefs is that living organisms (e.g. hermatypic corals and octocorals) generate the primary hard superficial benthic structure in coral reefs. Although temperate macroalgae such as giant kelps, such as *Macrocystis pyrifera*, also generate habitat structure by extending physical complexity into the water column, corals contribute to complexity, whether dead or alive. Corals can be considered as **habitat formers** (also referred to as framework organisms). This biological structuring has several important consequences for the ecology of coral reefs. Coral growth forms vary greatly from one species to another, and even within the same species under different environmental conditions. Growth forms range from featureless encrusting forms, to massive corals with large physical mass but little small-scale structure, through to branching forms with complex structure and interstices that can provide shelter to a large number of species, and individuals within a species (Fig. 5.1). Consequently, the coral species composition and environmental conditions at a site can, to a large extent, determine the habitat structure. Corals also exist in a range of assemblage types. Monospecific stands of corals will also generate habitats with a very similar structure across their extent, whereas mixed-species stands will generate more variable habitats. In addition, not all hard bottom in a coral reef is covered by corals, so further habitat heterogeneity is generated by rock, rubble and sand substrata, and other habitat formers such as soft corals.

Fig. 5.1. Hard corals are important habitat formers (or framework organisms). Coral growth structures range from featureless encrusting forms, to massive corals with large physical mass, in this case over 3 m high **(A)**, through to tabular **(B)** and branching **(C)** forms. Even within the branching forms, the degree of branching may vary, generating a range of different potential shelter sites. This structure remains intact even after death, until being eroded away by biological and physical processes. Eroded corals then become carbonate sand, which in turn provides habitat for a new range of species. (Photos: Mike Kingsford.)

Because corals are living organisms, they are subject to a range of ecological processes that alter their ability to survive and grow at a site. Corals may be outcompeted by other species, their reproductive success is temporally variable, and they may be stressed by poor environmental conditions.

Corals are also the prey for other organisms. COTS and *Drupella* whelks can kill entire colonies of coral and, in the case of COTS, entire assemblages across large swathes of reefs. Following death, the physical structure provided by the corals quickly erodes and this reduces the complexity of the physical habitat. Species such as COTS are **habitat determiners**. Their ecological role alters the physical structure of the habitat. In this way, a feedback loop of ecological interactions occurs. Habitat determiners are attracted to habitats of a certain type, but they in turn can alter that habitat type and may move on. Habitat determiners, however, need not necessarily be destructive, but can help in maintaining the health of reefs. For example, urchins, such as *Diadema*, graze algae, and surgeonfishes, such as *Ctenochaetus striatus*, remove detritus and algae from the reef top and by doing so both species help maintain space that is suitable for coral recruitment.

Coral reefs are important because of the organisms or **habitat responders** that live on them. Although fishes are the most obvious occupants of coral reefs, a vast array of organisms from a wide range of phyla including crustaceans, molluscs, worms and so forth make up tropical reef biodiversity. Associations between organisms and the coral reef habitat are complex. Some organisms, such as coral gobies (*Gobiodon*, *Paragobiodon*), are very specialised in their habitat association and will be found only in a few species of hard corals (Fig. 5.2A, B). They are also obligate specialists – they never occur on a reef if their preferred coral is not found there. Most coral reef habitat responders are not obligate specialists, however. Many fishes, such as the lemon damsel *Pomacentrus moluccensis*, are commonly associated with particular habitats such

Fig. 5.2. Habitat specialisation occurs at a range of spatial scales. Obligate specialists may associate with a specific range of hard branching coral species such as the goby *Gobiodon histro* with coral *Acropora nasauta* **(A)** and *Gobiodon citrinus* also with *A. nausata* **(B)**. They are never found in the absence of these coral species. Facultative specialists such as the damselfish *Pomacentrus moluccensis* **(C)** are found on a range of hard coral species and growth forms **(D)**, and may even be found in areas devoid of hard corals. Generalists such as the wrasse *Cheilinus digramma* **(E)** are not typically associated with any particular coral type, but are generally found in areas that contain a range of different structural habitats **(F)** (Photos: A,B, P. Munday; C,E, Great Barrier Reef Marine Park Authority; D,F, Mike Kingsford).

as branching hard corals (Fig. 5.2C, D), but they are not restricted to particular species of coral, and indeed may be found in habitats that are devoid of live coral. Other fish species may be found in reef habitats irrespective of live coral, such as many wrasses (e.g. *Cheilinus digramma*, Fig. 5.2E, F). Measuring the predictability of the composition of habitat-responding communities, given habitat availability, is fraught with difficulties. Various studies have reported a difference of anywhere between 5% and 70% coral cover is required to detect changes in abundance of habitat responders such as fishes. Coral cover alone does not sufficiently explain patterns of variability in all habitat responders because other factors such as competitive interactions play a role. It is clear, however, that organisms associated with specific types of corals ('specialists') are more vulnerable to perturbations than those that have broad habitats preferences ('generalists').

Spatial variability in coral reef habitats

It is important to realise that coral reefs are not simply homogenous areas of corals on which organisms live. Spatial variability occurs at a range of scales (Fig. 5.3). At the finest scale, relatively homogenous areas, such as single coral heads, rubble patches or algal covered rock, form **habitat patches** at scales from centimetres to metres. These habitat patches may indeed form the entire habitat of extreme specialists. Coral gobies, for example, are restricted to a limited species range of live coral heads and they will not move from a single coral head. For many organisms, though, the habitat patch is not necessarily the smallest habitat unit. In a patch of staghorn coral, for example, some species may live at the base of the coral whereas others may live near the top. In this way, habitat patches can also be subdivided in a range of **habitat responders**. The microhabitat need not be completely contained within a single habitat patch. Some species use microhabitats at the transition of different patches. Physical, temporal and biological variability generates heterogeneity in the environment at larger scales of tens to hundreds of metres. On coral reefs this is often termed a **habitat zone** – that is, a collection or **mosaic** of different habitat patches – and is usually characterised by a combination of biotic, physical and physiographic factors.

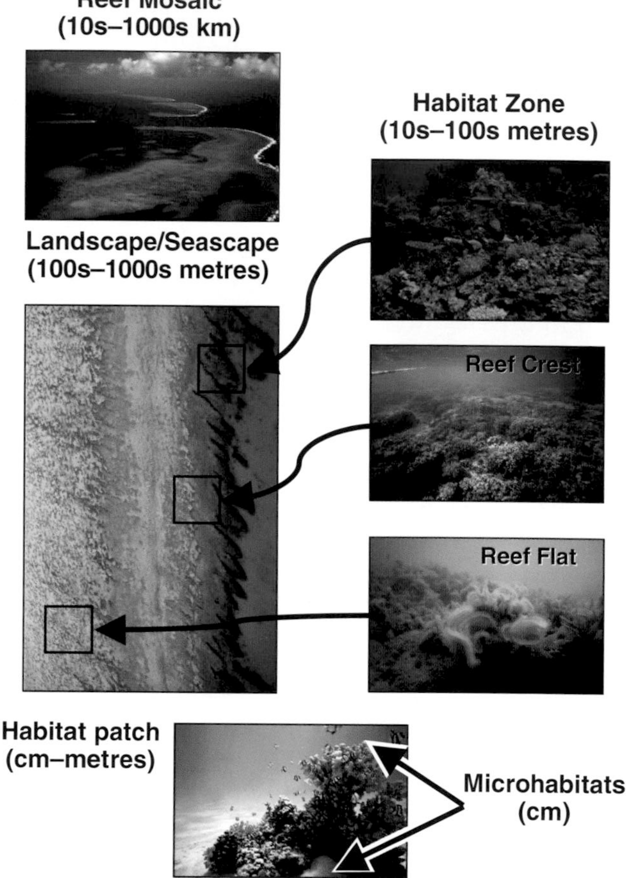

Fig. 5.3. Habitat heterogeneity occurs at a range of spatial scales. Individual coral stands, rubble areas and sandy areas, among others, form habitat patches. Within patches, however, the physical structure may not be homogenous so finer scale microhabitats such as the base versus the top of a coral head may exist. These habitat patches themselves form mosaics that typically occur in physiographic zones with characteristic physical conditions such as aspect to wave exposure. Depth is an important factor in zonation and even, with the same aspect over a vertical range of a few metres, reef tops and reef crests may have a very different appearance and species composition than reef slopes. These habitat zones are part of a wider landscape of different coral and non-coral habitat types at the scale of hundreds of metres, and generate heterogeneity at the scale of a single reef. The reefs themselves form a landscape or seascape that varies at scales of ten of kilometres across the shelf, and hundreds to thousands of kilometres along the GBR (Photos: Great Barrier Reef Marine Park Authority).

The windward and leeward sides of reefs have a different architecture that influences patterns of biodiversity. On the GBR, the windward side (normally the seaward side of reefs facing the continental shelf break) typically has a shallow reef crest, which may also be present or absent on the leeward side of reefs. This is a shallow zone in which coral growth rates are high, but the corals are also subjected to wave action and currents (Fig. 5.3). The reef crest will consist of stands of different coral species, interspersed with rock covered with encrusting algae. The reef crest sharply breaks to a reef slope, which is generally steeper, with a stronger depth gradient on seaward sides of reefs that are exposed to the prevailing winds, and hence have much more coral growth with well-developed reef crests and slopes (but also subject to more wave disturbance). This reef slope is not just a continuous band of corals, however. On exposed reefs in particular, the reef slope is punctuated by a surprisingly regular 'spur and groove' seascape in which coral rubble accumulates as a result of storm events, wave disturbance and rips.

On the leeward side of the reef crest, a shallow reef-flat zone often containing rubble, and fleshy macroalgae such as *Padina* or *Sargassum* species might occur. This zone is in very shallow water, exposed to wave action, with little vertical physical structure and heterogeneity. On reefs with lagoons, the reef-flat zone will give way to a lagoonal habitat zone, which is relatively sheltered from wave disturbance. The lagoon may often be very coral-rich in clear lagoons, but possibly devoid of corals in turbid lagoons. In the centre of the lagoon, fine sediments occur, providing yet another set of microhabitats. Windward and leeward seascapes of reefs are good predictors of where some organisms will be found. For example, large *Porites* colonies (metres wide and high) are rare on the exposed sides of reefs, but are often common on the leeward sides of reefs and in lagoons. On the leeward side of a typical reef, the lagoon will give way to a back-reef habitat zone. The back-reef is often rich in corals because visibility is good; not as good as the exposed side, but wave disturbance is much lower. This collection of habitat zones forms a seascape within which biodiversity can vary greatly on a single coral reef. While there is considerable small-scale spatial variability in habitats, at the spatial scale of hundreds of metres to kilometres the physical structure of the reef is relatively unchanging and predictable at ecological timescales and the spatial distribution of habitat responders across a reef landscape becomes very predictable.

Clearly the integration of habitat patches into zones in the reef landscape generates an enormous range of different types of places for different types of organisms to live, even taxonomically related species. And, although these organisms can be characterised as 'coral reef' species, they may not actually be associated with corals themselves. However, the reefs themselves form part of a larger seascape of multiple reefs that constitute the GBR (Fig. 5.3).

At larger spatial scales, geological, oceanographic, geographic, environmental and biogeographic patterns exert additional effects on the individual coral reefs that make up the GBR (see Chapter 2). **Cross shelf** variability at scales of tens of kilometres are associated with large differences in freshwater influence, turbidity, wave exposure and nutrient load that can vary with level of runoff near the mainland and upwelling near the shelf break (see Chapter 4). At spatial scales of hundreds to thousands of kilometres, the biogeography of reefs varies with latitude and the variation in mainstream currents and seawater temperature that goes with it. Other seascape morphologies may vary, for example, from the ribbon reefs of the northern GBR to reefal mazes of the Pompey complex on the southern GBR. The complexities of larval transport can also vary and influence the species available to form, influence or simply occupy habitats. Spatial scale, therefore, has profound effects on the habitat formers, determiners and responders that live on the reefs.

Temporal changes in coral reef habitats

Coral reefs, and associated habitats, change through time. Coral reefs are subject to a wide range of natural and anthropogenic perturbations, which can alter their structure and their heterogeneity (Fig. 5.4). At large spatial scales, the Pacific Decadal Oscillation can generate ocean-wide changes in temperature and primary productivity regimes. At slightly shorter temporal scales, El Niño Southern Oscillations (ENSO) may generate ocean-wide fluctuations in temperature, wind strength and oceanographic upwelling over time periods of 3–7 years. These fluctuations may in turn contribute to Pacific-wide disturbances such as coral bleaching during warm water events, and cyclones. The fear is that warm water events are increasing in frequency and that major bleaching events, as Hughes and colleagues documented in 2015–16 on the GBR, will become more common and give reefs no time to recover habitat-forming organisms.

Biological disturbances may also operate at large scales, such as clearances caused by urchins and parrotfish (Fig. 5.4A, B). COTS outbreaks can kill corals over vast areas of reef (Fig. 5.4C, D). Large-scale perturbations typically take a long time to recover and, if they occur together (e.g. coral bleaching and COTS outbreaks), their combined effect may further increase recovery time. Physical conditions will also facilitate outbreaks of biological disturbances. For example, particular combinations of suitable oceanographic conditions have the potential to increase survival rates of COTS through increased larval nutrition and condition and dispersal to suitable habitat.

Conversely, some environmental conditions may favour survival of pathogens, which may affect habitat determiners such as *Diadema* urchins. The interaction between physical and biological processes is important because of the different scales of each. Physical processes, which usually occur at a large scale, may effectively force biological processes to occur at a similar scale. This spatial autocorrelation (i.e. where the abundance or ecological interactions of organisms are more similar at locations closer to each other) has been termed the Moran effect. The consequence of this is that minor biological effects that would appear to exert only local effects can, if generated by physical forcing, occur over a much larger spatial extent than would be predicted. This was very clear in the Caribbean where a pathogen wiped out *Diadema* at spatial scales of hundreds of kilometres, which in

Fig. 5.4. Different scales of coral disturbance. Parrotfish **(A)** can generate local damage to hard corals **(B)** but rarely at large scales. Crown-of-thorns starfish outbreaks **(C)** can generate widespread damage to hard corals **(D)**, leaving dead corals that eventually erode to algal-covered rubble. Cyclones **(E)** can generate widespread damage to large areas of multiple reefs **(F)** (Photos: A, F, Mike Kingsford; B, C, Australian Institute of Marine science; D, E, Great Barrier Reef Marine Park Authority).

turn increased the cover of algae and reduced the recruitment and abundance of corals that resulted in fundamental changes in habitats, called a 'phase shift'.

Small-scale disturbances may also have important and immediate consequences for habitat structure. Storm damage can range from localised damage to individual coral heads, up to complete coral removal across 100 m swathes within windward sides of reefs, and across several reefs (Fig. 5.4E, F). Cyclones generally cross the GBR form east to west and generally leave a swathe of habitat damage on a scale of tens of kilometres (Fig. 5.5). Periodically, however, cyclones may take a track from north to south, and these events can cause broader scale destruction. Cyclone Hamish, for example, tracked along the outer GBR for over 1000 km in 2008 and caused widespread damage (Fig. 5.5A). These types of disturbances are not necessarily detrimental to a reef's overall health and regeneration can be relatively quick (<7 years) after a disturbance (Fig. 5.5B, C). Removal of a coral colony by storm damage frees up space for recolonisation by other species. It also generates heterogeneity in the substratum, so habitat responders that prefer bare rock or rubble can occupy the space that is cleared by the disturbance. In this way, local disturbances within a coral reef can actually increase species diversity. In low disturbance regimes, competitively dominant species, which may be fast growing in disturbed habitats, can form large monospecific stands. The diversity of corals in these stands, and habitat responders will be low. In contrast, highly disturbed areas will have a low coral diversity, and the substratum will consist primarily of encrusting algal and low-profile corals. These organisms provide little three-dimensional reef architecture, and consequently species diversity in highly disturbed habitats will also be low. However, at intermediate levels of disturbance, the abundant fast growing corals (e.g. acroporids) are reduced due to their higher susceptibility to disturbance, which in turn frees up space for other coral species and creates more microhabitats. At intermediate disturbance

Fig. 5.5. **(A)** Cyclone tracks over the Great Barrier Reef 2007 to 2017; the destructive event Cyclone Hamish is indicated (CH). **(B)** An area of reef at 12 m deep on the windward side of One Tree Island immediately after cyclone Hamish. **(C)** The same area in 2017 (Satellite tracks: John Nott using NOAA data; Photos: B, C, Mike Kingsford).

levels, coral diversity increases. Competition/disturbance mechanisms, therefore, generate a range of habitat types, which enables multiple species with different habitat preferences to occupy the same area.

Although natural disturbance is a normal part of coral reef ecology, and an important process that generates habitat variability, anthropogenic disturbances also exert effects at a range of spatial and temporal scales. Boat anchors may generate local coral damage. Artisanal fishing practices, such as netting, may generate local coral damage in reef systems. Commercial fish extraction practices such as cyanide and explosives, not common on the GBR but widespread in other parts of the world, may generate more widespread physical damage. Extraction of certain fish species and their related ecological functions on reefs, may itself have biologically mediated effects. For example, extraction of herbivorous fishes such as parrotfishes (Scaridae) may lead to increased growth of macroalgae, which may in turn lead to lower coral cover and, importantly, reduce the ability of corals to recruit onto bare space. At larger scales, increased sedimentation and eutrophication may result from land-based human practices such as deforestation, agriculture and sewage outfalls. The resulting changes in water quality are usually long lived, and may alter the trophic regime, light levels and increase stress to corals via sedimentation. These stresses may in turn lead to increased susceptibility of habitat-forming corals to environmental fluctuations such as temperature changes (see Chapter 11).

Natural and anthropogenic temporal changes cause habitat heterogeneity by perturbing the reef. Multiple disturbances and the frequency of events, therefore, are important in determining what the reef mosaic will look like. Moderate disturbances, which fragment large coral colonies will reduce the amount of physical structure but, because many corals can grow from fragments, this may not result in much loss of coral cover. Increased levels of physical disturbance may remove and kill entire colonies, and thus provide bare space. This space will undergo a natural succession of community states in which the space is colonised by encrusting coralline algae, which then provide a suitable substratum for coral settlement. Rapidly growing forms of corals, such as branching and plating corals, then occupy this space but may be supplanted over time by more wave-resistant forms. Extreme physical disturbances or widespread coral death may result in a reef being covered by coral rubble. Whether the rubble forms a substratum for coral or algal communities may be dictated by the numbers of herbivores on the reef. Exclusion of herbivores can result in algal takeovers of areas previously occupied by corals and fleshy and tufting algae such as *Padina* can occupy space following COTS outbreaks or bleaching. It is important to realise that coral reefs are not static, stable, unchanging ecosystems. The abundances of coral reef organisms change at temporal scales ranging from days to years to multiple decades, and at spatial scales ranging from a single coral head to entire reefs in which their preferred habitat occurs. Species that are not as habitat specific will be found on a wider range of reefs. Pelagic species may occur around coral reefs simply because they form the only physical structure in the ocean, which in turn may provide shelter and alter water currents that, in combination may attract prey and provide suitable spawning sites. If alternative structures are available, these species may occupy them irrespective of whether they were generated by, or covered by corals. These different temporal and spatial scales of habitat variability combine synergistically to provide an enormous range of different types of potential habitat for a huge diversity of organisms on the GBR.

Habitat structure and the distribution, diversity and dynamics of coral reef assemblages

Distribution and abundance

The distribution and abundance of most reef-associated organisms are closely linked to the habitats formed in different reef zones. We see major spatial changes in the structure of reef responder assemblages, with most species exhibiting a strong preference for either fore-reef or back-reef habitats, or for reef flat, reef crest or reef slope habitats. Most species also exhibit a preferred depth range, with those in shallow water having narrower depth ranges than those in deeper water. Within their typical habitats, most species have a preference for different microhabitats, where they spend most of their time either foraging or sheltering. This may include habitat-forming coral species, other benthic organisms, algal covered turf areas, bare rock, rubble or sand, caves or gutters. Habitat structure typically varies across the reef and down the reef slope, and the distribution and abundance of species can be determined by the availability of their preferred microhabitats. This is especially true for coral-associated species that depend on corals for food, shelter or living space. The abundance of corallivorous butterflyfishes, for example, is often directly correlated with live coral cover. In extreme cases, a single coral-dwelling fish (e.g. *Gobiodon* species) or coral-dwelling crab (e.g. *Trapezia* species) may be associated with a single coral species (e.g. *Acropora*), with most coral colonies above a certain size occupied by a pair of adults. In this case, the distribution and abundance of both the habitat former and responder are closely linked. You cannot have one without the other.

Patterns of distribution and abundance of reef organisms can be shaped at different life history stages. Many small species select their preferred habitats when they settle onto the reef after their brief pelagic larval stage. Hence, the adult distribution pattern can be already evident at the time of 'recruitment' as juveniles join the adult population. Others may settle across a range of habitat types, but exhibit better chances of survival in preferred habitats, especially those affording better shelter when they are small and vulnerable. Many larger species change their preferred habitats as they grow, and progressively move from one habitat to another. In extreme cases, this may involve recruiting into adjacent habitats, such as seagrass or mangroves, and migrating onto coral reefs as adults. Others may settle onto shallow reef flat areas and move down the reef slope as they grow.

Diversity and community structure

The species richness or diversity of reef-associated assemblages can depend on the strength of interactions among species, such as competition, predation and mutualism. It also depends on individual species responses to habitat structure and how habitat structure can mediate these important ecological interactions. High fish species diversity is often associated with reefs that have high topographic complexity, with lots of caves, overhangs, shelter holes and gutters that support a range of species with different habitat and shelter requirements (Fig. 5.6A). The relationship between fish diversity and coral cover may be more complex, with low fish diversity at low coral cover (where only non-coral associated species thrive) and at high coral cover (where only coral-specialists thrive) (Fig. 5.6B). Fish diversity can be significantly higher at intermediate coral cover, because it supports generalist fish species and those that are specialised on live and dead coral substrata. Because most species are specialised to some degree, experimental work has shown that fish diversity increases with coral diversity (Fig. 5.6C). All else being equal, a greater proportion of fish species tend to be associated with branching and corymbose corals, rather than massive or encrusting corals, presumably because these provide superior shelter from predators (Fig. 5.6D). Other typical patterns include higher fish diversity where there is greater spatial heterogeneity or a greater mix of a range of substratum types, including different coral species, dead corals, rubble and sand (Fig. 5.6E). Over time, habitats that alternate between high and low coral cover, through cycles of disturbance and recovery, may actually promote species diversity in responder communities, compared with habitats that persist in one stable state (Fig. 5.6F). Fish diversity tends to decline along a depth gradient, because shallow water, coral-associated species cannot survive beyond the depth limits of corals (Fig. 5.6G). The more pristine outer GBR

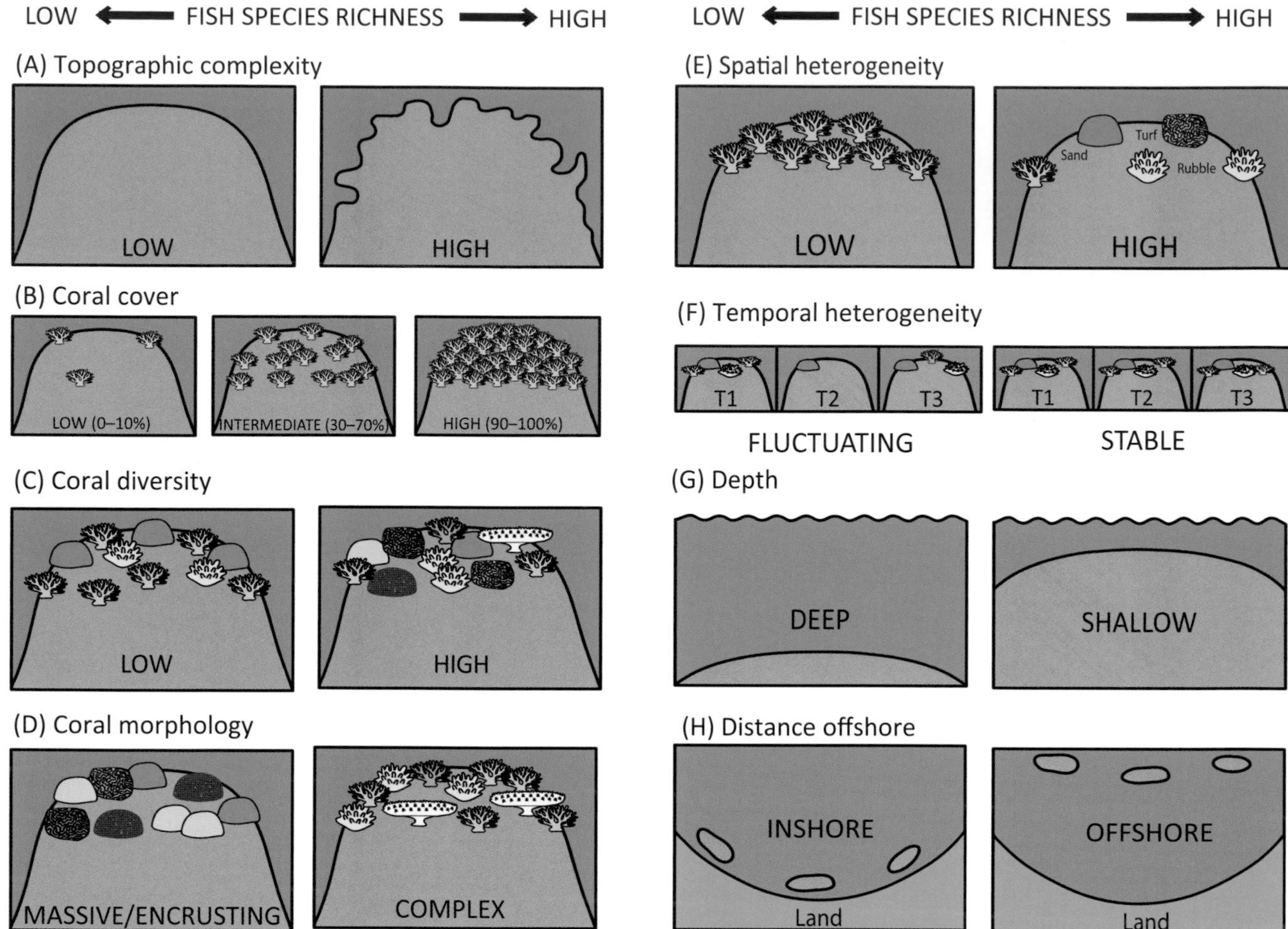

Fig. 5.6. Features of the habitat known to have an influence on coral reef fish species richness. Fish species richness is higher with: **(A)** higher topographic complexity; at **(B)** intermediate coral cover; **(C)** with higher coral species diversity; **(D)** with habitat dominated by branching and corymbose corals, rather than encrusting or massive corals; **(E)** with great spatial heterogeneity of microhabitats, including live and dead coral substrata; **(F)** with greater temporal variability in habitat composition; **(G)** in shallow water, compared with deep reefs; and **(H)** on offshore reefs, compared with coastal reefs (Drawings: Maya Srinivasan).

reefs tend to support higher fish diversity than coastal reefs, which may be associated with some of the above factors (Fig. 5.6H).

Disturbance and habitat degradation

Given the close association between reef-associated organisms and their habitat, it is not surprising that many coral-associated species are threatened by the long-term degradation of coral reef habitats. While they may be adapted to and cope well with moderate levels of natural disturbances such as cyclones, biodiversity will be threatened if such events become more intense and frequent. All disturbances – including cyclones, COTS outbreaks, heat stress and coral bleaching, and coastal run-off – have negative effects on coral communities, which, in turn, have flow on effects to other reef-associated responder communities. These disturbances differ in their immediate effects on coral cover, coral diversity and topographic complexity, and so may affect different components of the reef biodiversity. However, when extreme, all these natural and human impacts can drive assemblages towards a reduced diversity of species that can persist on dead reefs of low complexity. When corals bleach or are killed by COTS, the loss of coral-feeding and obligate coral-dwelling fishes can be immediate. During bleaching events, some colourful fish species are exposed to increased predation, because they are no longer camouflaged on the white reef-scape. Many fish species persist on bleached coral until the moment the coral actually dies (Chapter 12). Longer term effects of coral loss can occur, even if adult fishes can survive, because dead reefs no longer attract juvenile recruitment. In the longer term, as the topographic complexity of dead reefs erode, even larger predatory fishes no longer find suitable shelter, and their numbers decline. Some herbivorous fishes increase in numbers as coral cover declines and algal food supplies increase. However, many herbivores are specialised feeders on clean filamentous algae and cannot survive on other macro-algae or sediment-laden turfs that can proliferate on highly impacted reefs.

Conservation and management

Clearly protecting most reef organisms requires taking into account their habitat requirements and providing effective habitat protection. In establishing the marine protected area program for the GBR, care went into distributing no-take reserves ('green zones') along all axes of the main environmental gradients such as latitude and shelf position. Identifying bioregions and choosing to protect representative reefs within these regions maximised the number of species falling within the green zones. Other approaches to selecting sites for reserves could further increase species representation, including prioritising unique habitats and/or highly complex habitats that are unusually diverse.

It must be remembered that green zones primarily protect exploited species from fishing or collecting. They cannot themselves protect habitats from extrinsic disturbances such as cyclones and increasing temperatures, although they are effective in protecting habitats from destructive fishing practices and coral disease associated with discarded fishing gear. Most small reef fish on the GBR are not targeted by fishers, and so we do not expect their numbers to increase in reserves, but we do want to maintain their numbers at natural levels. To effectively achieve this, reserve implementation needs to be integrated with other management actions aimed at protecting reef health, including controls on coastal development, damaging farming practices and other efforts to improve water quality. Although green zones are not immune to extrinsic disturbances, there is some evidence to suggest that reserves recover more quickly following these events.

Fig. 5.7. A coral trout *Plectropomus leopardus* sheltering under a tabulate coral (Photo: Mike Kingsford).

It is well known that commercially and recreationally important reef fishes such as coral trout (*Plectropomus* species – Fig. 5.7) increase to three to four times the abundance and biomass in coastal green zones. Studies on larval dispersal show that reserves make a large contribution to juvenile recruitment in fished areas. However, coral trout really are 'coral' trout, and when coral cover declines as a result of cyclones or freshwater plumes, numbers decline in both fished and unfished areas. In this situation, the reserves that by chance escape habitat damage become critically important as the main sources of reproduction and larval supply to drive population recovery. At this stage, we do not know if there are particular reefs that are naturally resilient to a range of different disturbances, but if there are, protecting them will be a key conservation measure on the coral reefs of the future.

Wrap up

To summarise: living organisms on coral reefs contribute reef architecture, and hence ecological processes such as competition, predation and biological disturbance interact with physical processes such as water movement and storms to determine the physical structure of habitats. Accordingly, these processes, can exert very strong effects on those organisms that respond to the habitat-forming organisms. Temporal and spatial variability in habitat structure generates a range of opportunities for habitat-responding organisms to coexist at a range of spatial scales. Recognising spatial scales of habitats means that their management can be a relatively simple process, as long as big enough areas of space are controlled and protected. This diversity of places to live, in combination with the vast array of organisms that are available to live in these places, undoubtedly maintains biodiversity on the GBR.

Further reading

Connell JH (1978) Diversity in tropical rain forests and coral reefs. *Science* **199**, 1302–1310. doi:10.1126/science.199.4335.1302

De'ath G, Fabricius KE, Sweatman H, Puotinen M (2012) The 27-year decline of coral cover on the Great Barrier Reef and its causes. *Proceedings of the National Academy of Sciences of the United States of America* **109**, 17995–17999. doi:10.1073/pnas.1208909109.

Done TJ (1992) Effects of tropical cyclone waves on ecological and geomorphological structures on the Great Barrier Reef. *Continental Shelf Research* **12**, 859–872. doi:10.1016/0278-4343(92)90048-O

Graham NAJ, Jennings S, MacNeil M, Mouillot D, Wilson SK (2015) Predicting climate-driven regime shifts versus rebound potential in coral reefs. *Nature* **518**, 94–97. doi:10.1038/nature14140

Halford A, Cheal AJ, Ryan D, Williams DM (2004) Resilience to large-scale disturbance in coral and fish assemblages on the Great Barrier Reef. *Ecology* **85**, 1892–1905. doi:10.1890/03-4017

Hughes TP, Rodrigues MJ, Bellwood DR, Ceccarelli D, Hoegh-Guldberg O, McCook L, *et al.* (2007) Phase shifts, herbivory, and the resilience of coral reefs to climate change. *Current Biology* **17**, 360–365. doi:10.1016/j.cub.2006.12.049

Jones GP, Andrew NL (1993) Temperate reefs and the scope of seascape ecology. In *Proceedings of the Second International Temperate Reef Symposium*. Wellington, New Zealand. (Eds CN Battershill, DR Schiel, GP Jones, RG Creese and AB MacDiarmid) pp. 63–76. NIWA Marine, Wellington, New Zealand.

Jones GP, Syms C (1998) Disturbance, habitat structure and the ecology of fishes on coral reefs. *Australian Journal of Ecology* **23**, 287–297. doi:10.1111/j.1442-9993.1998.tb00733.x

Jones GP, McCormick MI, Srinivasan M, Eagle JV (2004) Coral decline threatens fish biodiversity in marine reserves. *Proceedings of the National Academy of Sciences of the United States of America* **101**, 8251–8253. doi:10.1073/pnas.0401277101

Messmer V, Jones GP, Munday PL, Brooks A, Holbrook SJ, Schmitt RJ (2011) Habitat biodiversity as a determinant of fish community structure on coral reefs. *Ecology* **92**, 2285–2298. doi:10.1890/11-0037.1.

Williamson DH, Ceccarelli DM, Evans DM, Jones GP, Russ GR (2014) Habitat dynamics, marine reserve status and the decline and recovery of coral reef fish communities. *Ecology and Evolution* **4**, 337–354. doi:10.1002/ece3.934.

6

Seabed environments, habitats and biological assemblages

C. R. Pitcher, P. J. Doherty and T. J. Anderson

The modern Great Barrier Reef (GBR) rises from a shallow (0–100 m) continental shelf of ~240 000 km^2, between the north-east Australian coast and the Coral Sea. Geologically, this area includes the part of the GBR that extends north of the Marine Park into the Torres Strait. Shallow water coral reefs are widely recognised as the iconic habitat of the GBR region, but they comprise only about 5% of the habitat area. This chapter focuses on the diversity of benthic habitats and biota of the deeper shelf seabed that represents the other 95% of the area.

Only a small fraction of this deeper seabed area had been examined in detail before 2003. The recent understanding presented here draws on new information from extensive sampling during 2003–2006 by the GBR Seabed Biodiversity Project, which mapped the distribution of seabed habitats (including mud, sand and gravel flats, algae and seagrass beds, sponge and gorgonian gardens, hard and shoal grounds) and their associated biological diversity. The project identified more than 5000 taxa, from which taxonomic research has described additional new species, and conducted assessments that contributed to conservation and management of the Marine Park.

The current GBR sea bed has alternated between terrestrial and marine environments, due to sea level changes arising from repeated glacial periods over geological history (Chapters 2 and 3). Today's sea bed habitats and biodiversity still partly reflect this history and are broadly influenced by coastal processes (Chapter 13) on one side and oceanic processes (Chapter 4) on the other, with multiple sub-regional patterns due to differences in other environmental drivers. Overlaid on these broad patterns, the majority residual variation in the biota results from numerous biological and stochastic processes.

Geophysical context and processes

Offshore from Townsville in the central GBR (~18.2–19.2°S), where many early studies were done, the continental shelf is ~120 km wide. On the outer 40–50 km of the shelf, an open matrix of coral reefs rises from 50–70 m depth. Between the coast and the reef matrix, a 75 km wide stretch of water known as the GBR Lagoon is almost devoid of reefs, possibly due to the action of cyclones, which are particularly frequent in the central GBR. The disturbing effects of cyclones reach the sea bed over much of the shelf, transporting fine particles north and inshore, leaving behind only a thin veneer of coarse particles on the mid- and outer-shelf.

North of 18°S, the shelf narrows (to ~50 km wide) and becomes shallower (typically 30 m, rarely >50 m). The outer shelf becomes delimited by a nearly continuous barrier of ribbon-like reefs, with steep descents into the Coral Sea. Platform reefs are common behind the Ribbon Reefs, often quite densely packed across the mid- and outer-shelf, and the GBR Lagoon becomes narrower (30 km to 20 km).

South of 19°S, to 22°S, the shelf broadens to as much as 250 km and generally becomes much deeper with substantial areas >70 m. Two lines of closely spaced reefs separated by narrow channels form the Pompey Reef Complex (PRC) built on a shallow (30–50 m) relic limestone platform that dominates along much of the outer shelf in this region (Chapter 2 and 3). The Lagoon also broadens to >100 km and deepens (50–70 m) to the southeast, becoming the Capricorn Channel with depths in excess of 100 m at its mouth.

At the southern end of the GBR (23.0–24.5°S), the Capricorn–Bunker Reefs rise from 30–40 m depth on the outer half of an 80 km wide sandy shelf.

Coastal influences

Along much of the coast, rivers export terrigenous sediments to inshore seas. After heavy rainfall, turbid flood plumes carry suspended solids and nutrients into the GBR Lagoon, affecting water quality inshore and across the shelf for up to several months. Over the last 8000 years, this process has resulted in a ~15 km wide inshore deposit of muddy sediments, the Holocene Wedge, which tapers from the coast, particularly between 12°S and 21°S.

Elsewhere, the inshore sediments are largely silica sands – especially in the southern GBR, where the shelf is extensively covered by silica sands transported northwards from the Great Sandy Region.

On exposed coasts, to depths of ~20 m, wave action from trade winds regularly re-suspends sea bed sediments, creating unstable habitat, redistributing finer particles northwards to less exposed areas and raising turbidity.

Oceanic influences

In the Coral Sea, the westerly South Equatorial Current bifurcates at the continental margin between 14°S and 16°S, producing a northward boundary current flowing into the Gulf of Papua as the Hiri Current, and a southward boundary flow as the East Australia Current (EAC). The open reef matrix of the deeper central GBR shelf allows episodic inflows of water from the EAC following relaxation of the trade winds. These upwellings of cool nutrient-rich water onto and across the shelf are most common during the Austral summer. Marginal upwelling is also observed in the Capricorn region.

Oceanic tides exert strong influences on parts of the shelf. The topography of the southern GBR creates a tidal node, which causes large amplitude tides and extreme tidal currents in the Broad Sound and Shoalwater Bay region. The same processes also cause very strong currents through the narrow passages between reefs in the PRC, which otherwise limit the exchange of oceanic water onto the shelf.

In the far northern GBR, the out-of-phase tides of the Coral and Timor Seas also cause extreme tidal currents, with similar effects. The ribbon barrier reefs in the north also impede the exchange of oceanic water onto the shallow shelf; nevertheless, tidal jets flowing into passages between some ribbon reefs can pump nutrient-rich Coral Sea water onto localised areas of the outer shelf.

In these areas of strong tidal influences, the currents re-suspend sediments, progressively depositing them in less energetic areas, and the strong vertical mixing stimulates elevated productivity.

Biophysical relationships and assemblage patterns

The Seabed Biodiversity Project examined the relationships between the major environmental drivers (including those outlined earlier) and the biology observed by video and collected in epibenthic sled and scientific trawl samples at almost 1400 sites. These relationships were used to produce integrated landscape maps of seabed assemblages that are needed for management in the Marine Park. Such maps were produced by predicting species distributions and assemblage patterns in

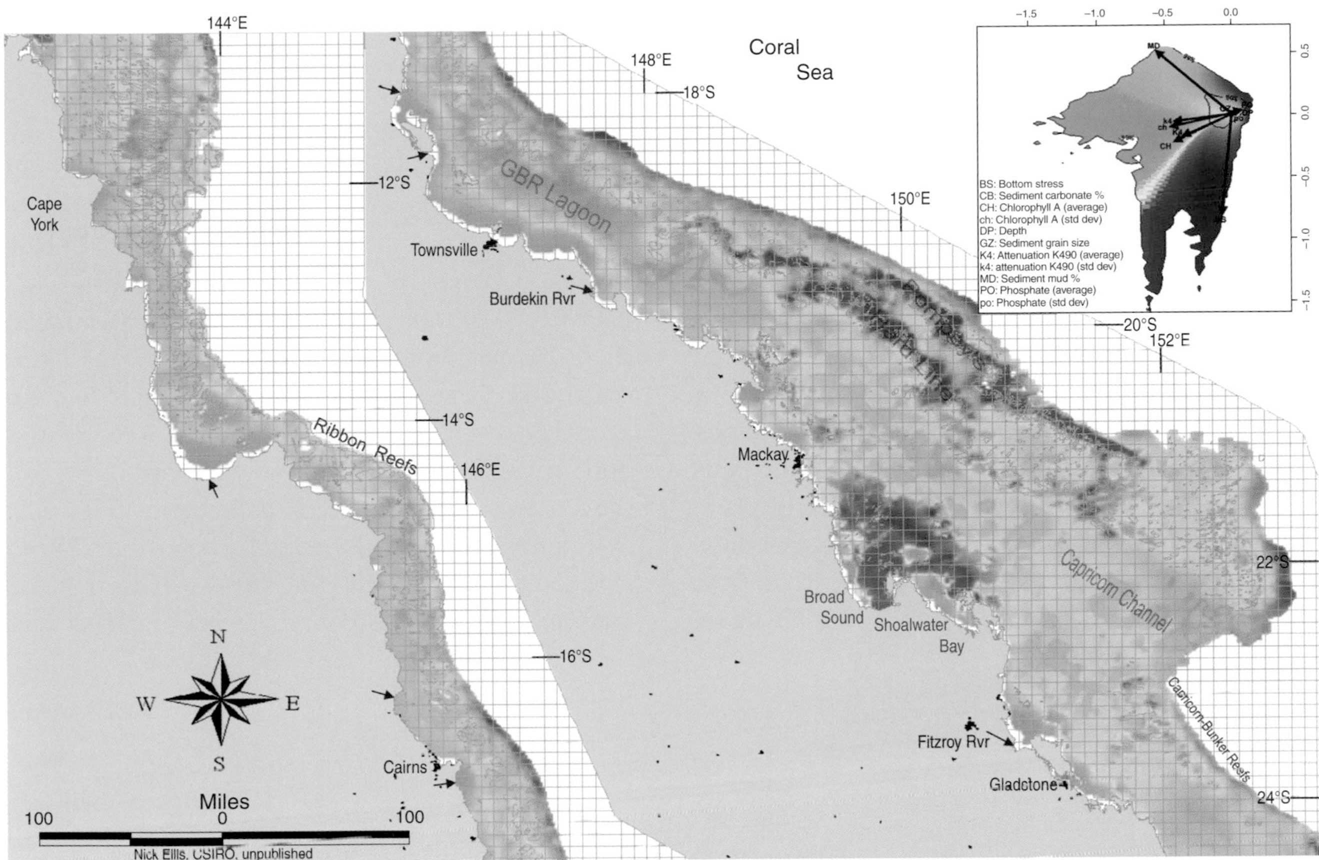

Fig. 6.1. Biophysical map of the GBR continental shelf. Inset: colour key showing distribution of >170 000 seabed 0.01° grid cells on the first two principal dimensions of environmental variation, representing 65% of the total, with similar colours indicating similar environments; contours show 99th and 50th percentiles of grid cell density; labelled arrows indicate direction of major physical factors (Image: N. Ellis, CSIRO).

unsampled areas, on the basis of modelling their sampled abundance with 28 environmental variables thematically mapped for the GBR. Although this surrogate approach has limitations, it is the only feasible option given the vast size of the GBR and the relatively sparsely distributed sampling (sites averaged ~12 km apart) over the shelf sea bed.

The biophysical modelling identified sediment grain size (particularly the percentage of mud), seabed irradiance, bed-stress (the force of water currents on the seabed), depth, temperature, salinity, chlorophyll/turbidity and some nutrients as the major drivers associated with the distribution patterns of seabed habitats and assemblages. While the correlations between the 28 environmental variables in the GBR are complex and multidimensional, they have been simplified to just two dimensions (Fig. 6.1), after rescaling them in proportion to the importance of their associations with biological patterns. The most common environments comprising 50% of the seabed, (coloured light grey in the centre of the biplot) are mostly coarse carbonate sands, often with sparse algae and benthos; the corresponding map shows the distribution of such areas. Less common and more extreme environments are shown in higher intensity colours and the arrows indicate the axes of the major environmental drivers, with the length indicating their association with biological composition.

Major biological patterns and important physical factors

The primary factor influencing biological assemblages is mud (MD vector, Fig. 6.1). The terrestrial muddy inshore areas along much of the length of the GBR (Fig. 6.4A) are often also very turbid (K4) and/or have high levels of chlorophyll (CH; green areas Fig. 6.1). These inshore (green–blue) areas tend to be devoid of visible biological habitat attached to the sea bed (Fig. 6.2 (white), Fig. 6.4A), but typically are ~60–80% bioturbated (Fig. 6.2 (grey), Fig. 6.4B) indicating the presence of burrowing animals such as worms, shrimps and bivalves. The mobile fauna has low diversity and low biomass (Fig. 6.3), but still includes many small fishes, prawns, crabs, mantis shrimps, gastropods and heart urchins – many of which feed on the deposits. A few sessile filter feeders, such as sea pens, survive in these soft sediments. Further offshore, where the turbidity is lower, the muddy sediments (light-blue areas, Fig. 6.1) are of carbonate, indicating their biological origin. Typically, these habitats are ~40% bioturbated (Fig. 6.2) and have a fauna similar to the inshore muds, but with some stalked and cryptic sponges and surface-dwelling bryozoans (lace corals) (Fig. 6.3). In the north, carbonate muddy habitats extend across most of the shelf (Fig. 6.1) given the protection of the ribbon reefs. In the deeper Capricorn Channel, the carbonate muddy habitats are among the most barren (Fig. 6.2), with ~20% bioturbation, and have the lowest diversity and biomass (Fig. 6.3) comprising a few crabs and fishes. Typically, with greater distance across the shelf, the sediment comprises more sand and gravel (Fig. 6.4C, D) of biogenic carbonate and may support a variety of biohabitats (Fig. 6.2). While these can include bioturbated sand (Fig. 6.4C), more commonly the surface biota include starfish, crabs, gastropods, fish, algae, and

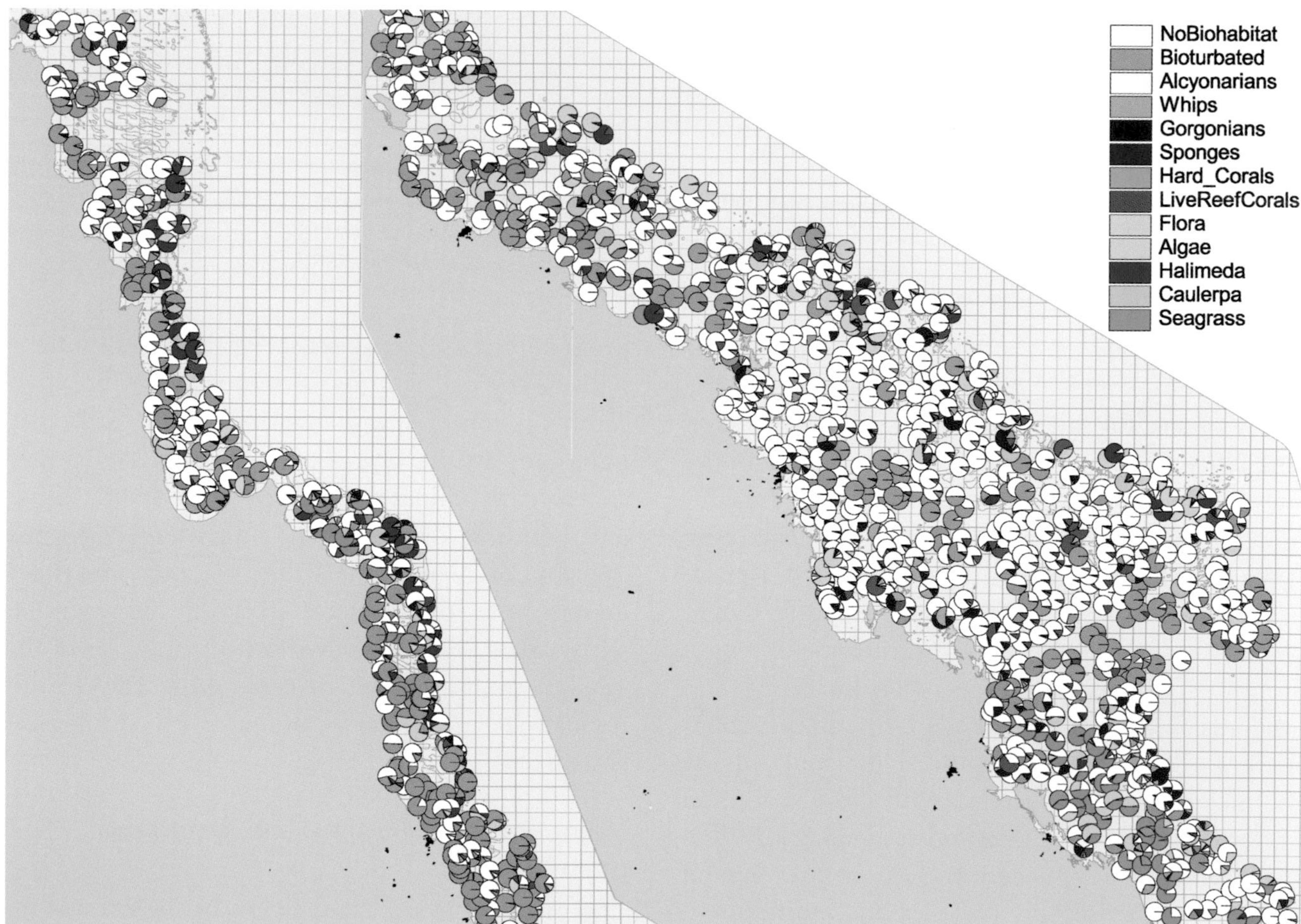

Fig. 6.2. Map of the distribution of broad biological seabed habitat types observed during towed video camera transects. Latitudes and longitudes are as per Fig. 6.1 (Image: CSIRO).

filter-feeding alcyonarians and sponges – sometimes in great abundance (Fig. 6.3).

At the opposite extreme to mud, bed stress (BS) is one of the strongest biophysical forces (red–orange areas, Fig. 6.1). The inshore vicinity of Broad Sound and Shoalwater Bay (orange) has the largest tidal range in the GBR and is accompanied by extreme currents (lower left orange branch of key) that scour sediments away, exposing rubble, stones and rock. Offshore, these tidal forces also cause extreme currents of clearer water (lower right red branch of key) to surge through the narrow channels of the PRC (red), again scouring the sea bed to the limestone base (Fig. 6.4E). Strong bed stress also occurs in the far northern GBR and Torres Strait, and in some local areas such as the Whitsunday Passage. In many of these areas, bare sea bed is often interspersed with gardens of colourful filter-feeding sea fans, whips and sponges (Figs 6.2, 6.4F, G) that benefit from the stable hard sea bed and elevated productivity. In some offshore passages, encrusting bryozoans form extensive biogenic rubble habitat (Fig. 6.4H). The fauna in high bed stress environments has very high diversity and moderate biomass (Fig. 6.3). It includes many species of bryozoans, sponges and gorgonians, as well as fishes, crustaceans, bivalves, starfish, urchins and corals. The scoured sediments are typically deposited in ripples, waves and dunes (Fig. 6.4I) on the fringes of these high stress (red) areas. In offshore sandy areas with medium currents, crinoid feather stars may be extremely abundant on the seabed (Fig. 6.4J). Also, relic coralline outcrops and shoals with a rich sessile biota,

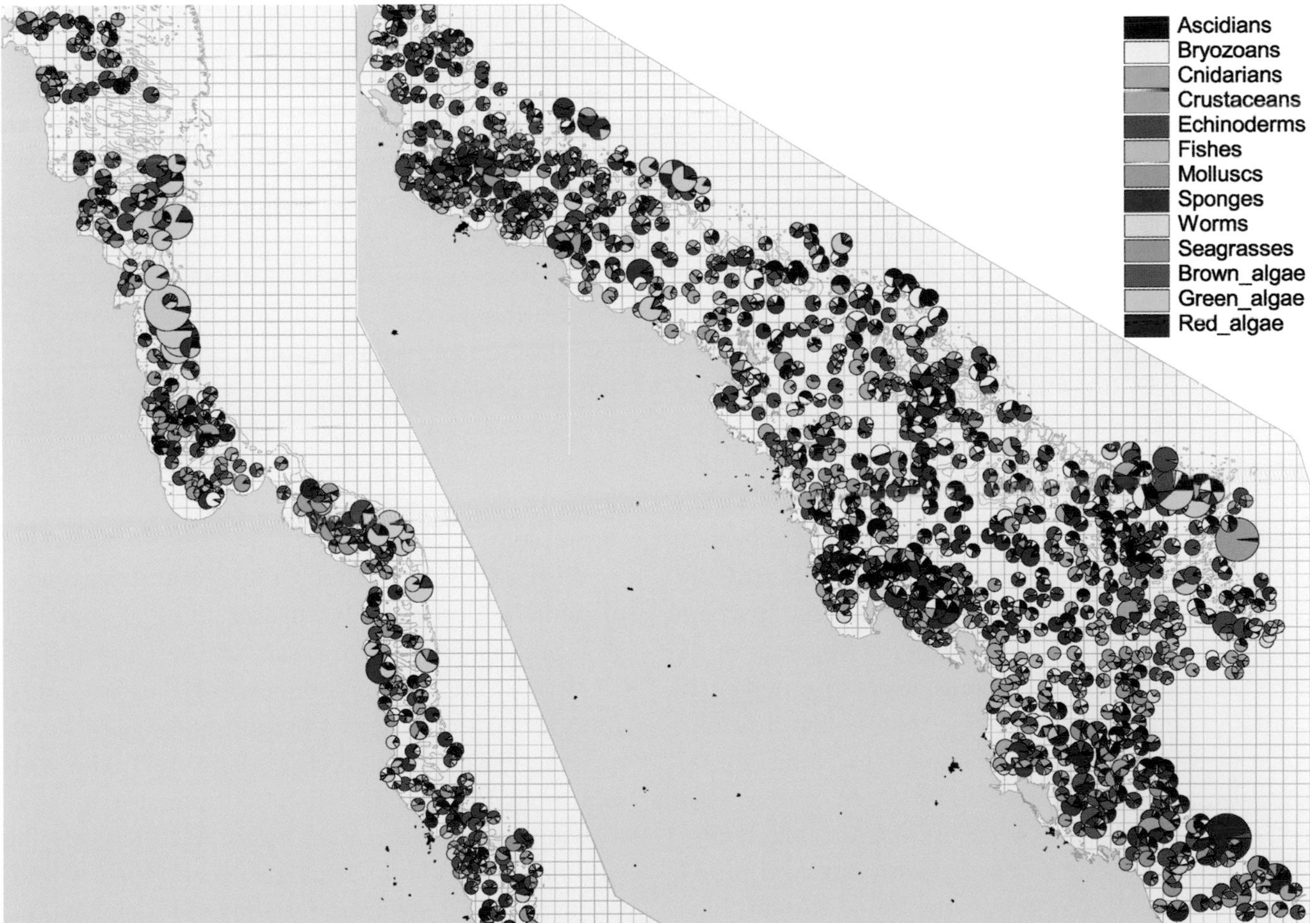

Fig. 6.3. Map of the distribution of biomass of the major biological groups sampled by an epibenthic sled. Latitudes and longitudes are as per Fig. 6.1 (Image: CSIRO).

including living hard corals, may occur in deep areas between emergent coral reefs providing substantial habitat for the development of mesophotic communities (e.g. Fig. 6.4K, see also Chapter 7).

The deeper clear waters near the outer edge of the continental shelf may be influenced by upwelled nutrients (e.g. PO, indigo areas, Fig. 6.1). Despite the depth, algae (including crustose coralline algae (Fig. 6.4L) that are adapted to low-light conditions) may be prolific in these clear nutrient-rich waters (Fig. 6.2) and contribute to habitats of moderate diversity and biomass (Figs 6.3, 6.4M). Areas where upwellings intrude onto the shelf can also be seen in Fig. 6.1 (faint indigo), particularly offshore from Townsville. Here, along the outer margins of the inshore turbid/muddy areas, where the water is clear enough to allow sufficient light to reach the seabed, a >200 km long band of mixed algae (including *Caulerpa*, Fig. 6.4N) and patchy seagrass (primarily *Halophila spinulosa*) proliferates (Fig. 6.2). Similarly, around the Turtle and Howick Islands in the central northern GBR (~14.5°S), meadows of dense *H. spinulosa* occur (Fig. 6.2). Algae and *H. spinulosa* habitats also occur over much of the shelf in the Capricorn region (Figs 6.1, 6.2, 6.4O). On the shallow shelf just inside the outer barrier ribbon reefs in the northern GBR, the signature of nutrients pumped in by tidal jets can be seen faintly in Fig. 6.1. These have encouraged the development of extensive banks of *Halimeda* algae (Figs 6.2, 6.4P) up to 15 m thick, comprised of the deposited carbonate skeletons of these algae. Recent high-resolution mapping has shown that the extent of these banks in the far northern GBR and their complex three-dimensional structure had previously been underestimated. *Halimeda* also occurs elsewhere (Fig. 6.2), but has not formed such banks. These varied marine plant communities form substantial areas of habitat for other biota, including numerous species of green, brown and red algae, small fishes, gastropods, sea slugs, crustaceans, starfish, sea cucumbers, urchins and corals (Fig. 6.3), which may attain moderate to high biomasses and, particularly in the case of mixed algal–seagrass meadows, very high species diversity. The moderately diverse biota of *Halimeda* banks includes numerous species of green algae, small fishes, gastropods, urchins, corals, sponges and alcyonarians.

Recent analyses indicate that these bio-physical relationships of the far northern GBR (north of ~12°S) overlap with south-eastern Torres Strait, affirming that the GBR does not end at Cape York. Similarly, those of the southern GBR (south of ~24°S) overlap into the adjacent Hervey Bay.

General patterns of abundance and diversity of the major phyla

The GBR Seabed Biodiversity Project collected samples of sea bed biota representing more than a dozen major phyla, and quantified them in terms of biomass, abundance, frequency of occurrence and numbers of species. Our understanding of these biota is dependent on the devices used to sample them. The Seabed Project used an epibenthic sled to sample sessile biota and slow moving invertebrates (Fig. 6.3) living on the sea bed or in the top few centimetres of the sediment, and a small research trawl to sample the more mobile fauna living just above the sea bed, such as fishes and crustaceans. However, animals living down in the sediments such as small worms, crustaceans and molluscs (infauna), were poorly sampled – they usually are collected by a grab or core. Similarly, biota living in rugged habitat or large mobile species may not have been well observed by video (tow- or baited camera). Hence, any census is incomplete.

Within these limitations, the sled and trawl confirmed the dominant biomasses of algae, particularly green algae, in the areas outlined above (Figs 6.2, 6.3). After algae, sponges are the next most abundant group: encrusting and massive morphotypes may reach high biomasses in the higher current areas, stalked and cryptic types are sparsely distributed in sedimentary areas. Ascidians are the next most abundant sessile group and have a similar pattern of distribution, followed by cnidarians, which tend to be more restricted to the higher current areas and harder substrata, as are bryozoans.

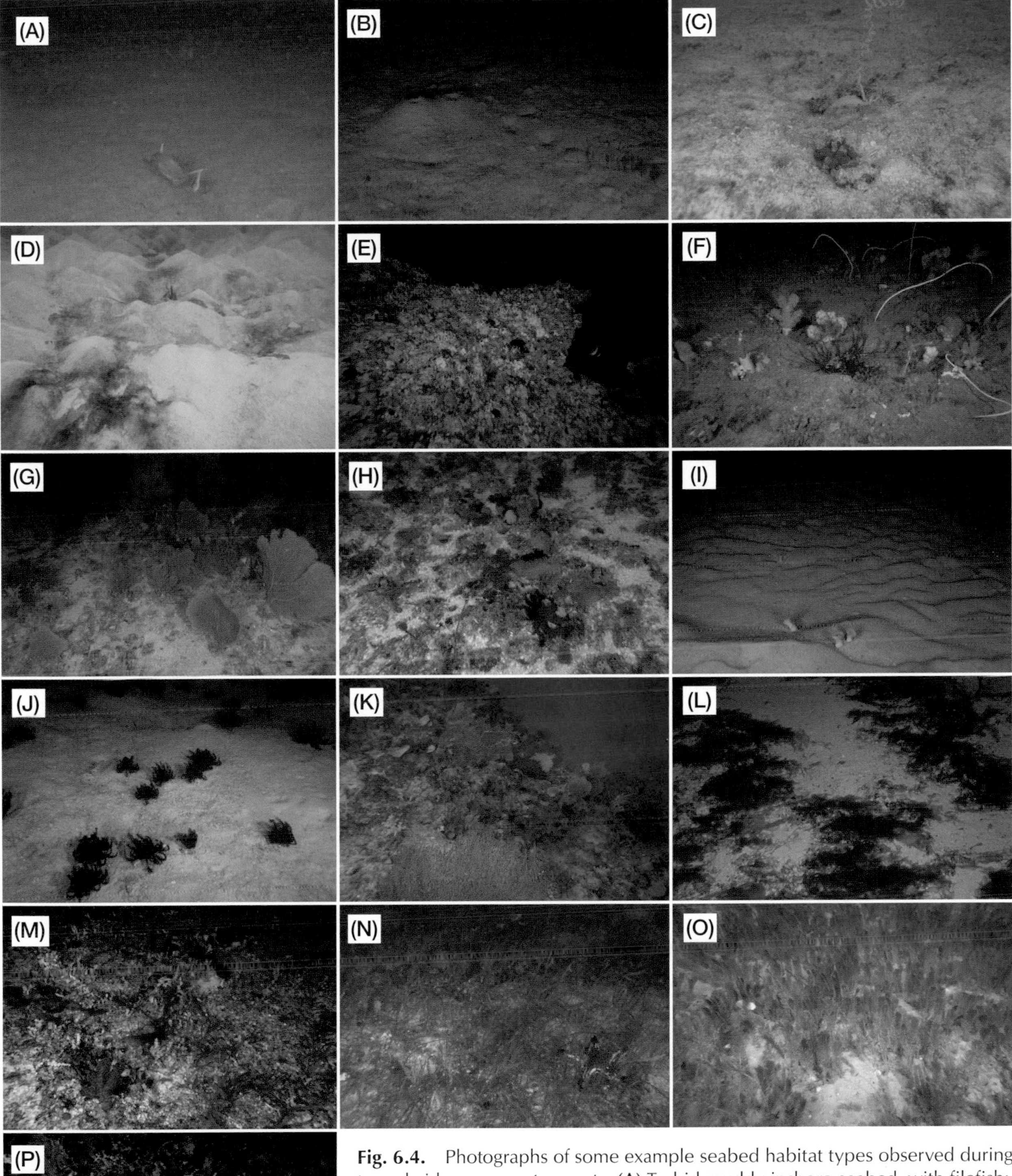

Fig. 6.4. Photographs of some example seabed habitat types observed during towed video camera transects. **(A)** Turbid muddy inshore seabed, with filefish; **(B)** bioturbated silty inner shelf sea bed; **(C)** coarse outer shelf sediment with sparse biota; **(D)** large bioturbation mounds in offshore sand; **(E)** scoured rocky seabed in extreme current area; **(F)** soft corals in strong current channel; **(G)** gorgonian garden on hard ground; **(H)** bryozoan rubble in strong current channel; **(I)** rippled sand in inshore high current area; **(J)** crinoids on sand in offshore strong current area; **(K)** shoal ground in deep water; **(L)** *Ulva* growing on patches of coralline algae at shelf edge; **(M)** diverse algae and coral near shelf edge; **(N)** dense algal bed (*Caulerpa*); **(O)** seagrass (*Halophila spinulosa*) bed; **(P)** *Halimeda* algae bank (Photos: CSIRO).

Echinoderms are the most abundant and widespread of the mobile invertebrates, with an overall biomass between that of sponges and ascidians. Molluscs, although widespread in softer sediments, appear to be about half as abundant. Fishes, better sampled by trawl, are next in abundance – inshore and muddy areas in particular tend to have high relative proportions of fishes. Crustaceans are much less abundant. These were followed by worms, elasmobranchs and minor phyla, none of which were well sampled by either device.

The ordering of these groups by frequency of species occurrence differed from that of biomass. Fish species occurred most frequently, followed by crustaceans, molluscs, echinoderms, sponges, corals, algae and ascidians. In terms of numbers of species, sponges were the richest with more than 1100 taxa, followed by molluscs (>1000), fishes (>850), crustaceans (almost 600), echinoderms (>500), algae (>400), corals (almost 400), bryozoans (>300) and ascidians (>300). These statistics indicate a very high diversity for the GBR sea bed, yet the true diversity is much greater, given the selectivity of the sampling devices used and the fact that some phyla were not fully sorted. Ongoing taxonomic investigations of these samples, all of which are lodged with the Queensland Museum, will continue to reveal this diversity.

Human influences on seabed habitat and assemblages

Terrestrial runoff is estimated to carry substantially higher loads of sediments, nutrients and contaminants after >150 years of human activities in GBR catchments and to have serious implications for coastal coral reefs (see Chapters 11, 13). Consequently, ambitious targets for water quality improvement have been set as part of the Reef 2050 Long-term Sustainability Plan. Although coastal processes do influence the composition of shelf sea bed biota, as outlined earlier, it is unclear whether anthropogenic increases in turbidity and sedimentation have caused changes to sea bed benthos in recent decades.

Prawn trawling is a widespread activity on the seabed in the GBR. Previous research showed that trawling can have direct impacts, particularly on biota that are easily removed and/or slow to recover, but stressed the importance of assessing these results in the context of the exposure of vulnerable biota to trawl effort. The Seabed Project completed this risk assessment, contributing to legislative requirements that all fishing in the Marine Park be sustainable. Trawlers now use devices to reduce the incidental catch in their nets of non-target marine life (by-catch). These devices are effective for elasmobranchs and some fishes, which is beneficial because the distributions of these groups have the greatest overlap with trawl effort. The risk assessments have indicated that the majority of vulnerable sessile fauna have distributions that overlap little with trawl effort. Combined with recent management that has reduced trawl effort substantially, this means that these fauna are unlikely to be at significant ongoing risk from trawling.

The Marine Park Zoning Plan (Chapter 14) contributes to conserving benthic biodiversity by protecting 20–100% of all sea bed habitats and species populations assessed, and the 2004 rezoning increased levels of protection by ~30%. These management actions provide greater assurance for the future of the diversity of habitats and biota on the seabed in the GBR.

Further reading

Benthuysen JA, Tonin H, Brinkman R, Herzfeld M, Steinberg C (2016) Intrusive upwelling in the Central Great Barrier Reef. *Journal of Geophysical Research. Oceans* **121**, 8395–8416. doi:10.1002/2016JC012294

Fabricius KE, Logan M, Weeks S, Brodie J (2014) The effects of river run-off on water clarity across the central Great Barrier Reef. *Marine Pollution Bulletin* **84**, 191–200. doi:10.1016/j.marpolbul.2014.05.012

Haywood MDE, Pitcher CR, Ellis N, Wassenberg TJ, Smith G, Forcey K, *et al.* (2008) Mapping and characterisation of the inter-reefal benthic assemblages of the Torres Strait. *Continental Shelf Research* **28**, 2304–2316. doi:10.1016/j.csr.2008.03.039

Hurrey LP, Pitcher CR, Lovelock CE, Schmidt S (2013) Macroalgal species richness and assemblage composition of the Great Barrier Reef seabed. *Marine Ecology Progress Series* **492**, 69–83. doi:10.3354/meps10366

Larcombe P, Carter RM (2004) Cyclone pumping, sediment partitioning and the development of the Great Barrier Reef shelf system: a review. *Quaternary Science Reviews* **23**, 107–135. doi:10.1016/j.quascirev.2003.10.003

Mazor T, Pitcher CR, Ellis N, Rochester W, Jennings S, Hiddink JG, *et al.* (2017) Trawl exposure and protection of seabed fauna at large spatial scales. *Diversity & Distributions* **23**, 1280–1291. doi.org/10.1111/ddi.12622.

McNeil MA, Webster JM, Beaman RJ, Graham TL (2016) New constraints on the spatial distribution and morphology of the *Halimeda bioherms* of the Great Barrier Reef, Australia. *Coral Reefs* **35**, 1343–1355. doi:10.1007/s00338-016-1492-2

Pears RJ, Morison A, Jebreen EJ, Dunning M, Pitcher CR, Houlden B, *et al.* (2012) 'Ecological risk assessment of the East Coast Otter Trawl Fishery in the Great Barrier Reef Marine Park'. Technical report. Great Barrier Reef Marine Park Authority, Townsville, <http://hdl.handle.net/11017/1147>

Pitcher CR, Doherty P, Arnold P, Hooper J, Gribble N, Bartlett C, *et al.* (2007) 'Seabed biodiversity on the continental shelf of the Great Barrier Reef World Heritage Area'. CRC Reef Research Final Report. AIMS/CSIRO/QM/QDPI, Brisbane, <http://fish.gov.au/reports/Documents/Pitcher_et_al_2007a_GBR_Seabed_Biodiversity_Final_Report.pdf>

Pitcher CR, Haywood M, Hooper J, Coles R, Bartlett C, Browne M, *et al.* (2007) 'Mapping and Characterisation of Key Biotic and Physical Attributes of the Torres Strait Ecosystem'. CSIRO/QM/QDPI Task Final Report to CRC Torres Strait. CSIRO Marine and Atmospheric Research, Brisbane. <http://www.cmar.csiro.au/e-print/open/2007/pitchercr_a.pdf>

Pitcher CR (2014) Quantitative indicators of environmental sustainability risk for a tropical shelf trawl fishery. *Fisheries Research* **151**, 136–147. doi:10.1016/j.fishres.2013.10.024.

Pitcher CR, Doherty P, Venables W, Browne M, De'ath G (2007) Indicators of protection levels for seabed habitats, species and assemblages on the continental shelf of the Great Barrier Reef World Heritage Area. CSIRO/AIMS report to Marine and Tropical Sciences Research Facility. CSIRO Marine and Atmospheric Research, Brisbane. <http://www.cmar.csiro.au/e-print/open/2007/pitchercr_x.pdf>

Pitcher CR, Ellis N, Venables W, Wassenberg TJ, Burridge CY, Smith GP, *et al.* (2016) Effects of trawling on sessile megabenthos in the Great Barrier Reef, and evaluation of the efficacy of management strategies. *ICES Journal of Marine Science* **73**, i115–i126. http://icesjms.oxfordjournals.org/content/73/suppl_1/i115.

Pitcher CR, Lawton P, Ellis N, Smith SJ, Incze LS, Wei C-L, *et al.* (2012) Exploring the role of environmental variables in shaping patterns of seabed biodiversity composition in regional-scale ecosystems. *Journal of Applied Ecology* **49**, 670–679. doi:10.1111/j.1365-2664.2012.02148.x

Schiller A, Herzfeld M, Brinkman R, Rizwi F, Andrewartha J (2015) Cross-shelf exchanges between the Coral Sea and the Great Barrier Reef lagoon determined from a regional-scale numerical model. *Continental Shelf Research* **109**, 150–163. doi:10.1016/j.csr.2015.09.011

Sutcliffe PR, Hooper JNA, Pitcher CR (2010) The most common sponges on the Great Barrier Reef seabed, Australia, include species new to science (Phylum Porifera). *Zootaxa* **2616**, 1–30.

Sutcliffe PR, Klein CJ, Pitcher CR, Possingham HP (2015) The effectiveness of marine reserve systems constructed using different surrogates of biodiversity. *Conservation Biology* **29**, 657–667. doi/10.1111/cobi.12506/abstract>

Sutcliffe PR, Mellin C, Pitcher CR, Possingham HP, Caley MJ (2014) Regional-scale patterns and predictors of species richness and abundance across twelve major tropical inter-reef taxa. *Ecography* **37**, 162–171. doi:10.1111/j.1600-0587.2013.00102.x

Sutcliffe PR, Pitcher CR, Caley MJ, Possingham HP (2012) Biological surrogacy in tropical seabed assemblages fails. *Ecological Applications* **22**, 1762–1771. doi:10.1890/11-0990.1

Woolsey E, Byrne M, Webster JM, Williams S, Pizarro O, Thornborough K, *et al.* (2013) *Ophiopsila pantherina* beds on subaqueous dunes off the Great Barrier Reef. In *Echinoderms in a Changing World.* (Ed. C. Johnson) pp. 175–179. Taylor & Francis Group, London, UK.

7

The Great Barrier Reef outer-shelf

T. C. L. Bridge, J. M. Webster, T. L. Sih and P. Bongaerts

Introduction

The outer-shelf of the Great Barrier Reef (GBR) can be broadly defined as the region of the continental shelf between the outer-most emergent reefs and the upper limit of the continental slope at ~200 m depth. The outer-shelf is exposed to clear, oligotrophic water from the Coral Sea, and is subjected to prevailing wind and wave energy from the ocean – all factors that strongly influence the ecological communities that occur there. The combination of its offshore location and exposure to wind and swell make research on the outer-shelf logistically challenging; consequently, this region remains poorly known compared with the GBR lagoon and inner-shelf. However, the last decade has seen a rapid increase in research focused on both the geology and biology of the outer-shelf, particularly in depths >30 m. This research has revealed a diverse range of physical habitats and ecological communities.

As recently as the 2004 GBR re-zoning, the majority of the outer-shelf of the GBR was considered to be 'non-reef' habitat, with the exception of isolated deep-water 'shoals' occurring seaward of the outer-shelf emergent reefs in the central GBR. However, recent multibeam bathymetry sea floor mapping and scientific coring has confirmed that the outer-shelf actually supports an extensive series of submerged reefs, which unlike the 'emergent' reefs, do not break the sea surface. These reefs had originally developed as shallow-water fringing or barrier reefs when sea levels were up to 120 m lower than the present. However, most of these submerged reefs have 'drowned', or were unable to keep pace growing with rapidly rising sea level and associated environmental changes following the end of the last ice age, ~20 000 years ago. These submerged reefs now represent an extensive habitat for the development of mesophotic coral ecosystems – reef communities that occur in the middle to lower photic zone, generally in depths of 30–150 m.

The existence of submerged reefs on the seaward shelf-edge of the GBR was recognised as early as 1925, but were not afforded any further interest until decades later. In the 1980s, scientists described parallel lines of submerged reefs at several sites along the GBR shelf-edge and suggested they represented fossil coral reefs that grew at lower sea levels. Recognition of the importance of the submerged shelf-edge reefs for understanding the effects of sea-level fluctuations on the growth and long-term dynamics of the GBR motivated greater research interest in the mid-late 2000s. This included a major scientific drilling expedition by the International Ocean Discovery Program (IODP) that recovered sediment cores through the submerged reefs, finally confirming their age and origin. Although the primary aims of these projects were geological, they also provided an opportunity to document the biodiversity of the GBR outer-shelf sea bed habitats, which until that time were virtually unknown.

The advent of SCUBA diving in the 1960s greatly increased scientific knowledge of coral reef ecosystems, including on the GBR. However, logistical and safety concerns associated with SCUBA diving at depth meant that few studies examined reefs at depths >30 m. One notable exception was a series of submersible dives conducted in 1984, which explored outer-reef slopes to >200 m depth at Myrmidon, Bowl and Ribbon No. 5 Reefs. This pioneering work provided the first description of mesophotic coral reefs on the GBR and described several different ecological communities, including huge stands of high coral cover (predominantly *Leptoseris* and *Pachyseris*) to depths of >90 m at Myrmidon Reef. Following this 1984 expedition, mesophotic coral reefs on the GBR received little further research attention until 2007, when an expedition by the RV *Southern Surveyor* conducted detailed investigations of four sites along an 800 km stretch of the GBR outer-shelf. The expedition collected high-resolution multibeam bathymetry, dredge samples and detailed benthic images collected by a new underwater robot known as autonomous underwater vehicle (AUV). This combination of tools enabled the first systematic investigation of the benthic (sea floor) habitats and communities associated with the GBR outer-shelf. The last decade has seen an exponential increase in research interest on mesophotic coral ecosystems using both robotics and advanced diving technology such as closed-circuit rebreathers, which have enabled scientists to dive to depths of up to 150 m. The GBR now represents one of the best-studied mesophotic reef systems globally, although mesophotic depths still remain poorly studied compared with the shallow-water reefs.

Geomorphology of the GBR shelf-edge

The morphology of the continental margin of north-eastern Australia changes markedly with latitude, which in turn strongly influences both the type and extent of habitats found along the GBR shelf-edge (Fig. 7.1). In the northern GBR, north of latitude ~16°S, the continental shelf is relatively narrow, in places only ~50 km wide. The shelf edge is rimmed with long, linear ribbon reefs before plunging steeply into the deep water of the basin adjacent to the GBR. South of latitude ~16°S in the central GBR, the continental shelf widens to >200 km and the outer-shelf area slopes more gently, enabling the development of extensive series of submerged reefs to 140 m. South of latitude ~23°S in the southern GBR, the shelf narrows again in the vicinity of the Capricorn-Bunker group

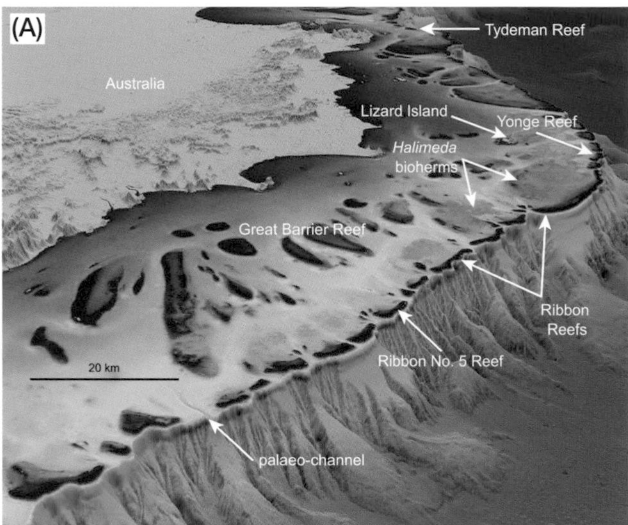

Fig. 7.1. Latitudinal change in the morphology of the Great Barrier Reef outer-shelf. **(A)** In the northern GBR, the outer-shelf is steep and narrow. **(B)** In the south, the outer-shelf becomes wider and more gently sloping, with more prominent development of submerged reefs (Source: Adapted from Robin Beaman, www.deepreef.org, CC BY 4.0).

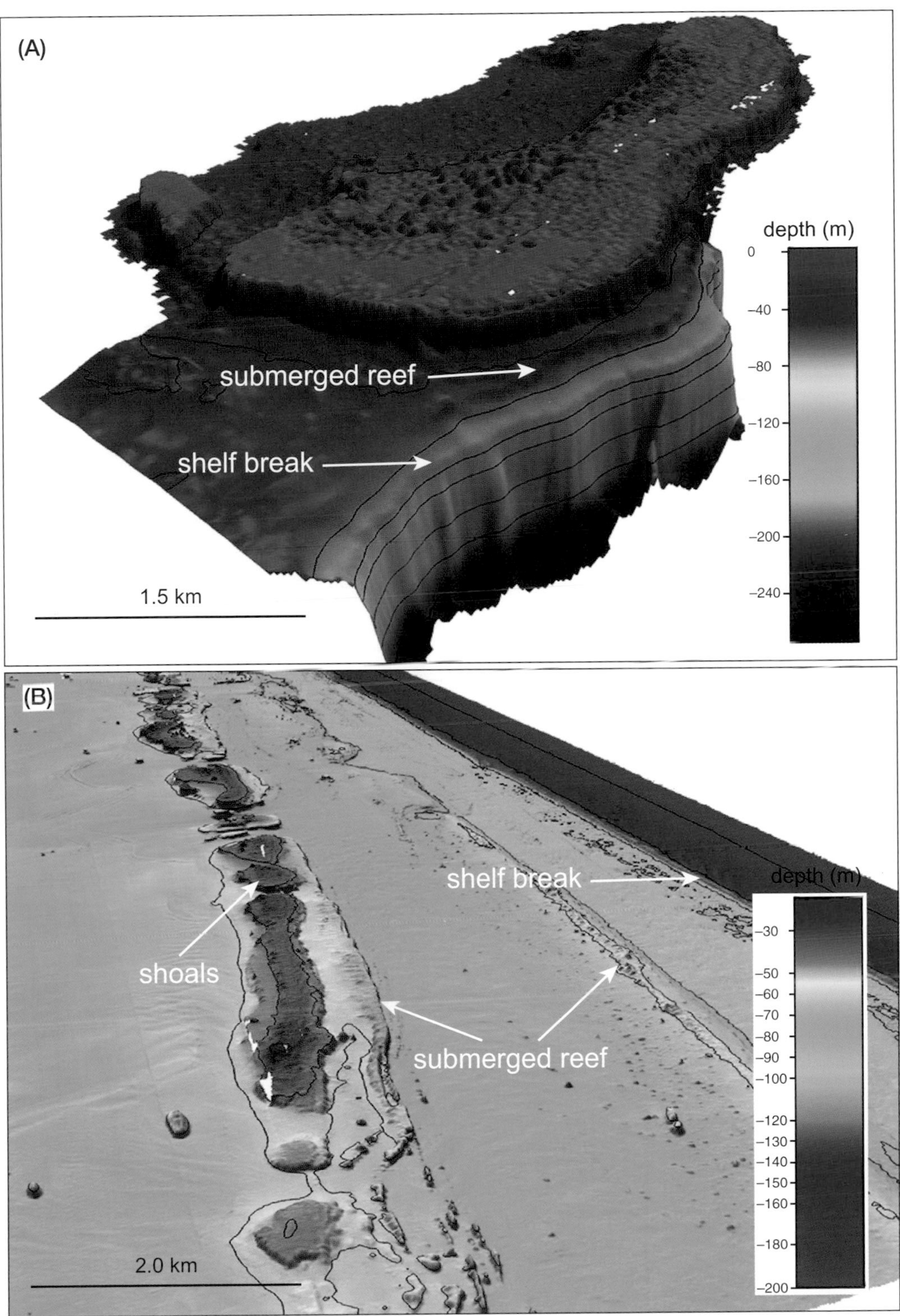

Fig. 7.2. Comparison of reef morphology between the northern and central GBR. **(A)** Outer-shelf at No. 5 Ribbon Reef, northern GBR (14°S), showing steeply sloping reef front, narrow submerged reef at ~45 m and steeply sloping wall below the shelf break at ~70 m. **(B)** Gently sloping outer-shelf at Hydrographers Passage, central GBR (21°S), showing parallel lines of submerged reefs in depths of ~15–150 m (Source: Adapted from Robin Beaman, www.deepreef.org, CC BY 4.0).

of reefs. The outer-shelf here also has a relatively gentle gradient but with less-well-developed submerged reefs compared with the central GBR outer-shelf.

The 'ribbon reefs' in the northern GBR may be up 28 km in length, and are separated by narrow passages less than 1 km across. The ribbon reefs form a true 'barrier' between the GBR lagoon and the offshore Coral Sea. Seaward of the ribbon reefs, water depths of 500 m are reached within a few hundred metres distance from the reef front (Fig. 7.2A). The steep shelf-edge profile in this region limits the space available for submerged reefs seaward of the emergent ribbon reefs. The reef fronts generally consist of steep or vertical walls to ~30–40 m, adjoining a rubble slope. In places, narrow, submerged reefs occur at depths of ~45 m, followed by a second rubble slope that extends to the shelf break at ~75 m. At this point, a narrow 'brow' of submerged reef and low pinnacles forms the junction between the rubble slopes of the shelf-edge and the near-vertical wall below the shelf-break.

Directly landward of the ribbon reefs, the outer-shelf is characterised by thick bioherms (or mound-like) accumulations of the green calcareous algae *Halimeda* (Fig. 7.3). Recent studies have revealed that the northern GBR bioherms are three times larger than original estimates (from ~2000 km^2 to ~6000 km^2), and their areal extent is roughly equivalent to that of the northern GBR shallow-water reefs. The *Halimeda* bioherms form a complex system of honeycomb-like surfaces with circular ring shapes ~300 m in diameter. Several large ancient river channels up to 100 m deep formed during periods of lower sea level cross the shelf in the region. These channels snake across the GBR lagoon and exit through the inter-reef passages between the ribbon reefs before connecting to the continental slope and basin via a complex system of submarine canyons (Fig. 7.1A). The benthic communities associated with these deep-water (>300 m) submarine canyons are poorly studied, but based on knowledge of similar canyon systems elsewhere in the world, these habitats likely represent hot-spots for deep-sea biodiversity.

South of latitude 16°S in the central GBR, the outer-shelf emergent reefs shift from 'ribbon' to 'platform' reefs, and are set back from the shelf-edge. A series of linear submerged reefs lie parallel to the shelf-break in depths of ~40 to ~140 m (Fig. 7.2B). The 'shoals' referred to in the GBR bioregionalisation maps of the early 2000s actually refer to the topographic high points of these submerged reefs, some of which may rise to within 5–10 m of the surface and provide an ideal habitat for shallow-water coral reefs (Fig. 7.4). However, most of these submerged reefs occur in depths of 40 m or below. At 17°30′S, the upper slope is marked by a massive submarine landslide that delivered

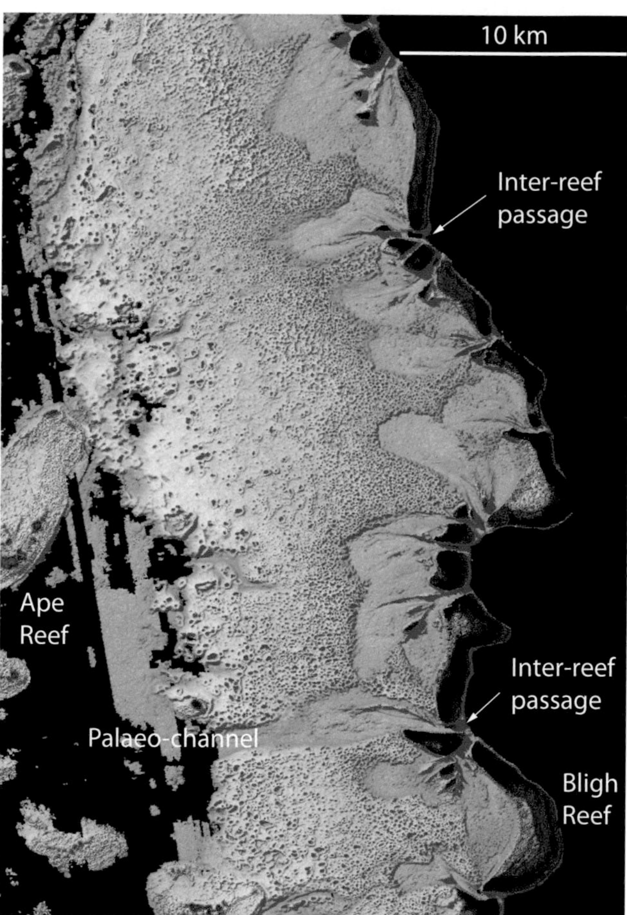

Fig. 7.3. Close up bathymetric image showing the shelf edge region in the northern GBR Cape York region. Here the complex morphology of *Halimeda* bioherms can be observed grading from reticulate to donut shaped forms from east to west. Palaeo-channels can also be observed crossing the shelf and the bioherms and exiting through the incised inter-reef passages between the Ribbon Reefs (Source: From McNeil *et al.* 2016).

~32 km^3 of debris into the adjacent basin. The tops of these blocks, called the Gloria Knolls, now form a deep-water habitat (~1100 m) for a cold-water coral community. The central GBR continental slope between latitudes 18° and 19°S, is also characterised by numerous smaller submarine landslides in relatively shallow depths of <200 m. These landslides may be related to the location of the ancient

Fig. 7.4. Examples of benthic habitats on the GBR outer-shelf: **(A)** high coral cover on the top of a submerged shoal near Cairns, 14 m depth (Photo: D. Kline); **(B)** robust corals on the exposed outer-reef crest at No. 5 Ribbon Reef (Photo T. Bridge); **(C)** lower reef slope at 40 m depth, Great Detached Reef (Photo: P. Bongaerts); **(D)** huge, monospecific stand of *Acropora tenella* at Great Detached Reef, 40 m depth (Photo: P. Bongaerts); **(E)** closed-circuit rebreather diver at the bottom of the Starkey River palaeo-river channel at 90 m. The steep canyon wall is clearly visible to the left of the diver (Photo: S. Mitchell, www.mesophotic.org, CC BY 4.0); **(F)** Live *Halimeda* growing on a *Halimeda* bioherm, northern GBR (Photo: E. Kennedy).

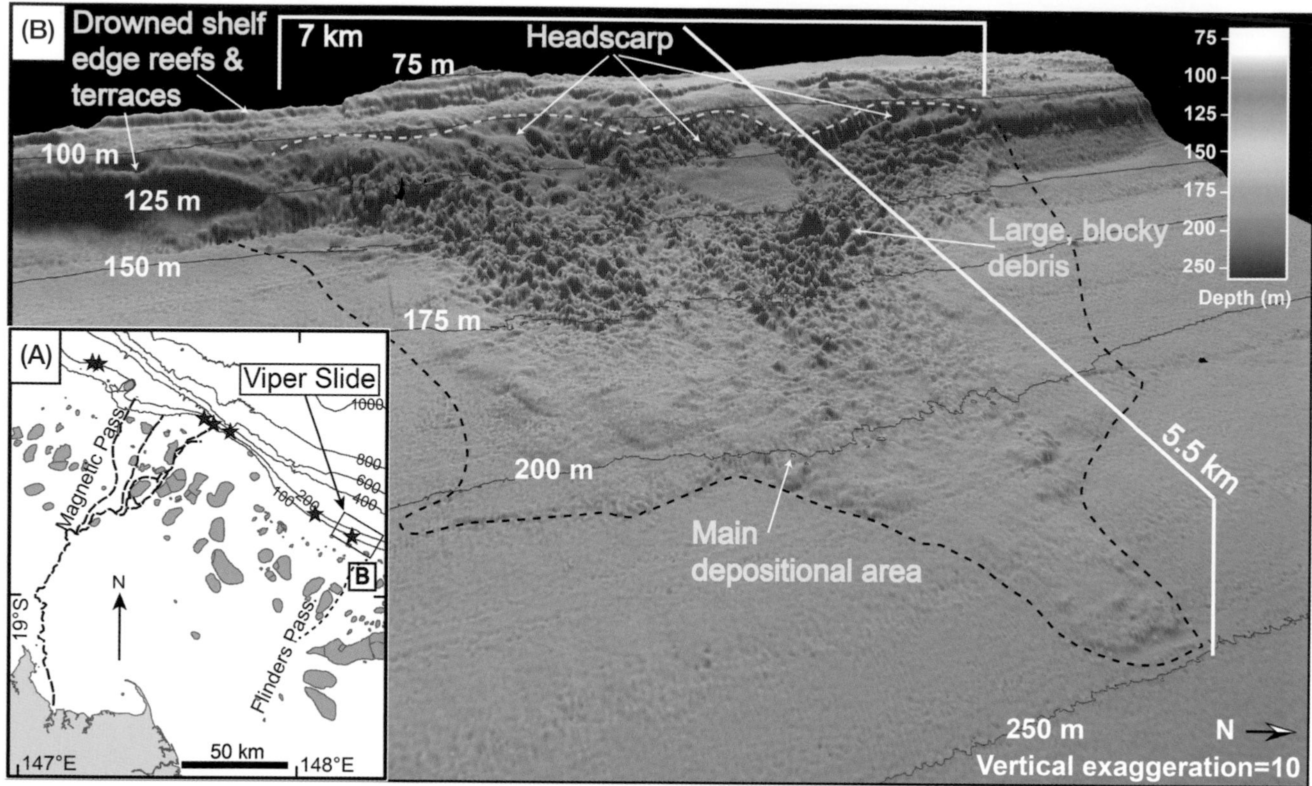

Fig. 7.5. **(A)** Map showing the distribution shelf-edge and upper-slope submarine landslides along the central GBR margin (modified from Webster *et al.* 2016). **(B)** High-resolution 3D bathymetry showing the geomorphology of the Viper Slide that cuts the shelf-edge and the resulting debris field.

river deltas formed during lower sea levels (Fig. 7.5). Recent investigations of the largest of these underwater landslide (18 km^2, 0.025 km^3) confirm that a 7 km section of the shelf failed catastrophically ~14 000–20 000 years ago, significantly altering the geomorphology and benthic habitats of the shelf-break and upper slope in this region. Further south of latitude 22°S, the shelf in Capricorn-Bunker region is marked by a prominent paleo-channel that can be traced across almost the entire continental shelf. A well-developed submerged reef at 50–60 m depth, along with clear terraces at 90 m, 100 m and 110 m, occur along the shelf-edge.

Habitats and biodiversity

Strong cross-shelf gradients in environmental conditions such as turbidity, salinity and water clarity are a characteristic feature of the GBR ecosystem, and are reflected in corresponding cross-shelf patterns in marine biodiversity. The inner- and mid-shelf areas are strongly influenced by freshwater inputs and sediment runoff from the mainland. The importance of factors such as sedimentation and nutrient levels on the composition and dynamics of ecological communities on the GBR is widely reported. However, the influence of sedimentation and water quality on modern habitats and communities across the outer-shelf is significantly lower than further inshore. The dominant broad-scale oceanographic feature in the region is the South Equatorial Current (SEC), which flows westwards across the South-West Pacific and brings clear, low-nutrient oceanic water to the GBR outer-shelf. The SEC interacts with the Australian continent along the outer-shelf of the central GBR at latitude ~17–20°S, and splits into the southward-flowing East Australian Current (EAC) and the northward-flowing Hiri Current. The outer-shelf is fully exposed to swell from the open ocean, and wave energy is therefore a key factor influencing ecological communities. In addition to the prevailing south-easterly trade winds, the GBR region also experiences regular tropical cyclones from November to April. Upwelling of cold, nutrient-rich waters also occurs on the outer-shelf, and in some places extends onto the continental shelf. Upwelling can strongly influence the biodiversity occurring on the GBR outer-shelf. For example, the presence of the *Halimeda* bioherms in the northern GBR is due to upwelling caused by tidal currents flowing through the narrow passages between the ribbon reefs. Recent interest in documenting habitats and communities of the GBR outer-shelf have revealed the presence of a wide range of habitats, including submerged reefs, mesophotic coral reefs, sand plains and *Halimeda* meadows (Fig. 7.6). These meadows, which occur on the outer-shelf of the central GBR, differ from the *Halimeda* bioherms further north because they do not form raised biogenic structures. Nonetheless, live *Halimeda* covers most of the sea floor in between submerged reef pinnacles in this region.

The outer-shelf 'shoals' are ubiquitous features of the GBR outer-shelf from around Port Douglas (latitude ~16°S) to at least as far south as Hydrographers Passage (21°S), and probably even further. Isolated shoals with very high coral cover have been identified and surveyed as far as latitude 24°S, between the Capricorn-Bunker and Lady Elliot Islands. In the central GBR, the outer-shelf shoals form a long, linear submerged barrier system with very consistent morphology characterised by spur-and-groove type features. These spurs and grooves are typically several metres high, and reflect their high-energy environment. The outer-shelf shoals are generally around 14 m deep at their shallowest point, although some isolated patches may rise close to the surface.

Benthic communities associated with submerged shoals have been examined offshore from Cairns, Townsville and Hydrographers Passage. These shoals generally exhibit similar ecological communities over this relatively large latitudinal range spanning some 600 km of the outer-shelf. The tops of the shoals are characterised by hard corals typically associated with high-energy environments, including various species of *Acropora* (*A. robusta*, *A. hyacinthus* and *A. clathrata*), *Isopora* and *Pocillopora* (Fig. 7.6A). Deeper sections below 20–25 m are often dominated by octocorals from the family Xeniidae, particularly *Cespitularia* and *Efflatounaria*, which may form large, monospecific stands (Fig. 7.6B). Shoals south of the Capricorn-Bunkers exhibit a much high cover of branching *Acropora*. The clear waters of the outer-shelf provide good conditions for coral growth and some shoals exhibit very high coral cover.

Seaward of the shoals lies the linear 'drowned' reefs and terraces, which formed in shallow depths during periods of lower sea level but now support extensive mesophotic coral ecosystems. The benthic communities occupying these deeper reefs can be broadly divided into 'upper mesophotic' (down to 60 m) and 'lower mesophotic' (60–150 m) communities. Upper mesophotic communities vary considerably in composition among sites, although are generally dominated by phototrophic taxa. Common sessile macrobenthic taxa include hard corals, soft corals, sponges and macroalgae (Fig. 7.6C, D). Lower mesophotic communities are dominated by heterotrophic octocorals, as well as sponges, black corals (Antipatharia) and a few species of hard corals adapted to low-light habitats (Fig. 7.6E, F).

Hard corals are probably the best-studied group on mesophotic reefs in the GBR. Approximately 200 coral species have been recorded from the mesophotic (>30 m depth), around 45% of all known coral species in the region. Species recorded represent a wide range of taxonomic groups, including a relatively high proportion of species from genera such as *Acropora*, which are typically considered characteristic of shallow water. Many species are now known to occur at much greater depths than previously thought; however, only a few appear capable of occurring at lower mesophotic depths. Only 29 of the 230 species recorded below 30 m are recorded from depths >60 m. Of the 29 species recorded, only a relatively small

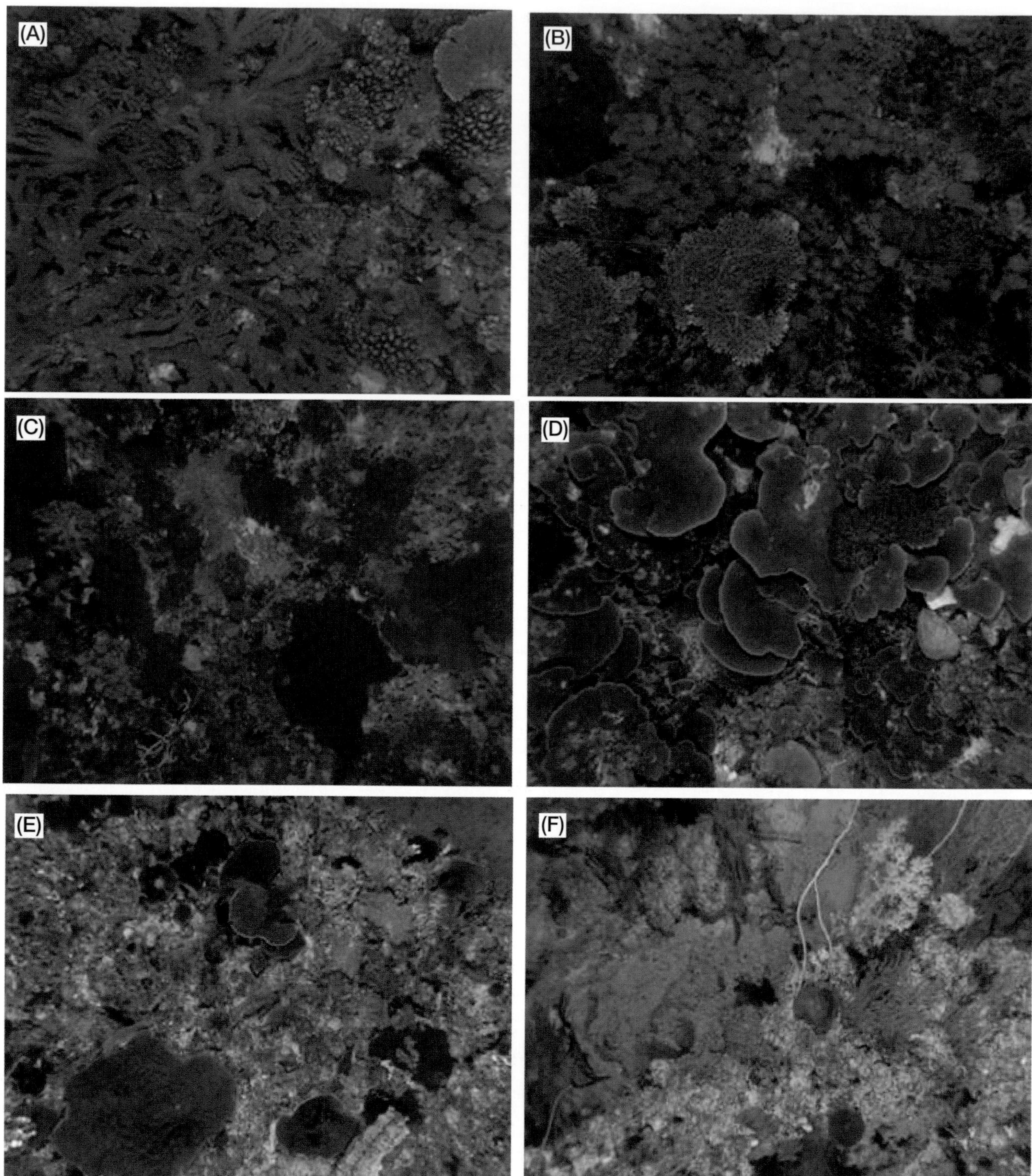

Fig. 7.6. Examples of ecological communities on the GBR outer-shelf from AUV images: **(A)** *Acropora*-dominated coral community on the top of shoals, 22 m, Hydrographers Passage; **(B)** benthic community comprising zooxanthellate Scleractinia and Octocorallia on the leeward side of shoals, 42 m, Hydrographers Passage; **(C)** zooxanthellate Scleractinia and Octocorallia and autotrophic sponge *Carteriospongia*; 60 m, Hydrographers Passage; **(D)** *Montipora*-dominated community, 55 m, Viper Reef; **(E)** *Leptoseris*-dominated lower mesophotic coral community, 85 m, Viper Reef; **(F)** Heterotrophic octocoral and *Leptoseris* community, 100 m, Viper Reef (Photos: IMOS AUV facility operated by the ACFR, University of Sydney, CC BY 4.0).

proportion commonly occur in the lower mesophotic, including *Echinophyllia aspera* and a few species of *Leptoseris*, which are common at mesophotic depths across the Indo-Pacific. The deepest known occurrence of a phototrophic (light-dependent) coral on the GBR is a colony of *Leptoseris* from 125 m at Yonge Reef near Lizard Island. To date, most coral biodiversity research has focused on steeply sloping lower reef slopes, rather than on submerged reefs, which are difficult to access using SCUBA. The number of recorded species will no doubt increase further with additional research effort, particularly in poorly studied habitats such as submerged reefs.

Octocorals are probably the most abundant habitat-forming benthic taxon on the central GBR outer-shelf. Xeniid octocorals including *Cespitularia* and *Efflatournaria* are abundant on the outer-shelf shoals. Numerous zooxanthellate taxa extend into upper mesophotic depths, including the *Cespitularia, Anthelia* and the Alcyoniids *Sinularia* and *Sarcophyton*. Azooxanthellate octocorals are also abundant on the outer-shelf below the depths affected by storm waves, but are rare in wave-exposed habitats. Azooxanthellate octocorals are found throughout the mesophotic and become increasingly dominant with depth to the point where they are the most abundant sessile benthic taxon in the lower mesophotic zone. A diverse assemblage of azooxanthellate octocorals, including many taxa that appear restricted to mesophotic depths, occupies submerged reefs along the shelf-break. At least five genera collected on the 2007 Southern Surveyor expedition (*Callogorgia, Heliania, Paracis, Pteronisis* and *Pterostenella*) represented new records for the GBR, highlighting the incomplete information available regarding the biodiversity of mesophotic reefs on GBR outer-shelf.

Other conspicuous taxa common within the mesophotic include black corals, sponges and algae, although research on the ecology and biodiversity of these groups has been minimal. The green calcareous algae *Halimeda* is common at mesophotic depths at many locations on the GBR outer-shelf. In addition to the extensive *Halimeda* bioherms in depths of 20–40 m in the northern GBR, *Halimeda* bioherms up to 20 m in height occur in depths of 80–100 m to the north of Myrmidon Reef in the central GBR. Vast 'meadows' of *Halimeda* also occur at depths of 50–80 m on inter-reef terraces seaward of Viper Reef, and likely also other areas of the central GBR. Since *Halimeda* is scarce in similar habitats in other regions (e.g. Noggin Pass), the existence of these meadows may be due to localised phenomena such as upwelling. Other conspicuous macroalgal taxa, particularly *Caulerpa*, appear relatively abundant in some regions. However, the diversity of algae on the GBR outer-shelf remains poorly described.

A diverse fish fauna occupies the outer-shelf of the GBR, with unique assemblages occupying different depths. The fish fauna from shallow-water outer-shelf reefs is well described, and is generally more diverse in terms of the number of species of key reef fish families than inner- and mid-shelf reefs. Likewise, diversity of fish communities on inter-reef habitats on the continental shelf increases with distance across the shelf. Fish communities associated with the mesophotic outer-shelf and continental slope are less studied, but recent research has revealed that mesophotic and sub-mesophotic fish faunas form distinct communities stratified by depth. Deep (>100 m) outer-shelf habitats support a unique array of deep-specialist snapper and grouper species, including members of the predominantly deep-dwelling genera *Pristipomoides, Etelis, Wattsia* and *Gymnocranius* (Fig. 7.7). Even depth-generalist genera such as *Lutjanus, Lethrinus* and the subfamily of groupers *Epinephelinae* contain deep-specialist species that occur exclusively in the lower mesophotic, such as the comet grouper (*Epinephelus morrhua*) and several species of bar cod. Other species exhibit wide depth ranges (e.g. the coral trout species *Plectropomus laevis* and *P. leopardus*), which occur from the shallows to depths of at least 100 m. Numerous highly mobile species also occur at mesophotic depths on the outer-shelf, and there is evidence that some species of jacks (Family Carangidae) and sharks (Family Carcharhinidae) may use deep habitats

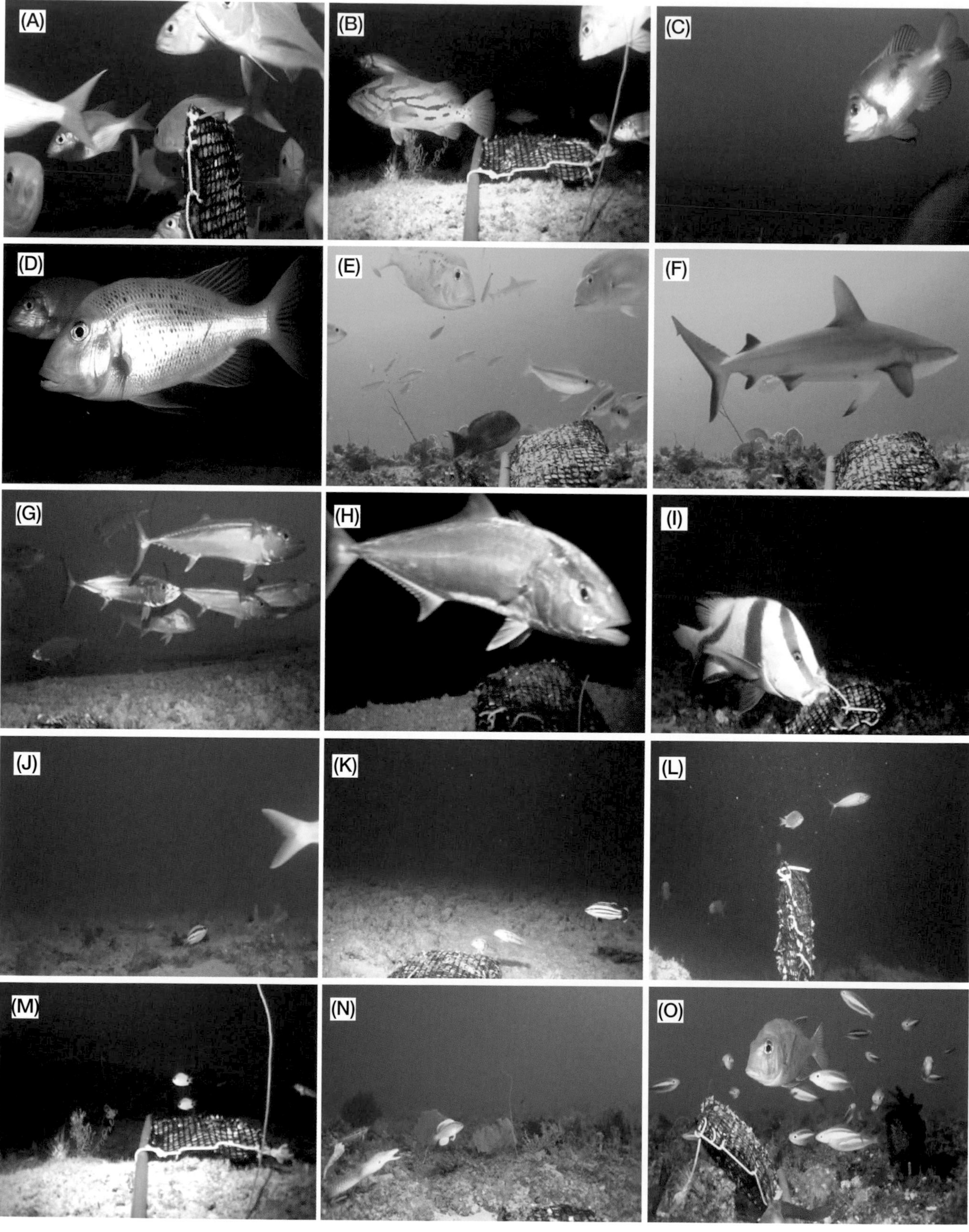
(A)
(B)
(C)
(D)
(E)
(F)
(G)
(H)
(I)
(J)
(K)
(L)
(M)
(N)
(O)

more than previously thought. Amberjacks (*Seriola dumerili, S. rivoliana*), dogtooth tuna (*Gymnosarda unicolor*) and other large, mobile predators commonly occur along steep slopes and drop-offs, and may benefit from nutrient upwelling along the GBR shelf-break congregating fish biomass around the relatively narrow area along the shelf-edge. Many fish species exhibit ontogenetic movements between different habitats, including deep reefs, at different stages of their life histories. For example, the red emperor *Lutjanus sebae*, uses deep *Halimeda* banks for spawning and shallower inshore habitats for nursery and intermediate life history stages. However, the importance of deeper habitats for specific life history stages of many species remains unknown.

Many mesophotic snappers, jacks and groupers are important fishery-target species, and deep reefs may provide important spatial refuges for these taxa. Understanding the importance of deep habitats for fish biodiversity and their vulnerability to anthropogenic and environmental disturbances is an important research priority, particularly with respect to the importance of these habitats for threatened, vulnerable and endangered inhabitants such as scalloped hammerhead (*Sphyrna lewini*), silvertip shark (*Carcharhinus albimarginatus*), tiger shark (*Galeocerdo cuvier*) and sandbar shark (*Carcharhinus plumbeus*).

Biological and ecological information on most fish species occurring on the deeper outer-shelf habitats are lacking due to the difficulties of sampling at depth, minimal sampling effort or the difficulties of sampling small-bodied reef fishes without SCUBA. Surveys using baited remote underwater video stations on the outer-shelf of the central GBR have revealed that the GBR outer-shelf is occupied by poorly known or newly described deep-water representatives of common shallow-water reef families including damselfishes (e.g. *Chromis okamurai, C. mirationis, C. circumaurea*), wrasses (e.g. *Bodianus bennetti*), breams (*Nemipterus* and *Pentapodus* spp.) and eels (e.g. *Gymnothorax berndti, G. elegans* and *G. intesi*). This research documented numerous species that are almost certainly undescribed, as well as several species not previously recorded from the GBR. Some of these species were relatively common in underwater video (e.g. the pink-banded fairy wrasse, *Cirrhilabrus roseafascia*, and a potential new species, *Selenanthias* sp.), highlighting knowledge gaps regarding the GBR outer-shelf reef fish fauna. Increased sampling effort on deeper reefs would no doubt yield many new species and new species records for the GBR. In particular, the use of closed-circuit rebreather diving technology would substantially increase knowledge of smaller and cryptobenthic species that would not be detected using scientific fishing, trawling or submarine surveys.

Most of ecological research on the GBR outer-shelf to date has focused on mesophotic coral reef systems characterised by hard substratum. However, extensive soft-bottom communities also occur on the GBR outer-shelf, which are not as visually striking as reefs but nonetheless harbour unique communities. For example, dense aggregations of the bioluminescent brittle-star *Ophiopsila pantherina* (mean density of 418 adults/m^2) were observed on large sand dunes 2–6 m in height at Hydrographers Passage in the central GBR. *Ophiopsila pantherina* is a filter-feeder that forms aggregations on the lee side of dunes, where it extends four arms into the water column to feed while using its remaining arm to anchor itself into the sand. Numerous other species of asteroids and other echinoderms, including holothurians, have been observed at mesophotic depths on the GBR outer-shelf. Furthermore, other studies of the outer shelf sediments show

◄

Fig. 7.7. Examples of mesophotic fishes captured on Baited Remote Underwater Video Stations: **(A)** *Pristipomoides* spp.; **(B)** *Epinephelus morrhua*; **(C)** *Wattsia mossambica*; **(D)** *Gymnocranius euanus*; **(E)** *Gymnocranius grandoculis* and *Variola louti*; **(F)** *Carcharhinus amblyrhynchos*; **(G)** *Gymnosarda unicolor*; **(H)** *Seriola dumerili*; **(I)** *Lutjanus sebae*; **(J)** *Chromis okamurai*; **(K)** *Bodianus* sp.; **(L)** *Chromis circumaurea*; **(M)** *Chromis* sp.; **(N)** *Cirrhilabrus roseafascia* and *Gymnothorax elegans*; **(O)** *Pentapodus aureofasciatus*. (Images by Tiffany Sih, www.mesophotic.org, CC BY 4.0.)

live, well-preserved and relict larger benthic foraminifera, representing a range of shallow and deep assemblages depending on downslope transport and mixing.

Conservation of outer-shelf habitats

The outer-shelf lies far offshore and is generally not subjected to human pressures to the same degree as habitats further inshore. Nonetheless, numerous disturbances may affect outer-shelf habitats and should therefore be considered in conservation policies. The GBR zoning plan implemented in 2004 aimed to achieve representation of the many different habitats in the GBR Marine Park by spreading the distribution of no-take areas across and along the GBR. This approach resulted in many outer-shelf habitats meeting the minimum target of at least 20% of each habitat type being included in a no-take area – even though many outer-shelf habitats were poorly known at the time of the rezoning. This outcome demonstrates that effective representation of habitats in protected areas is achievable even when ecological knowledge of an area is limited, and provides a blueprint for future marine planning exercises. Nonetheless, different habitats along the outer-shelf remain threatened by processes that no-take areas cannot protect against, particularly the effects of climate change. For example, Severe Tropical Cyclone Yasi caused extensive damage to submerged shoals around Cairns and Townsville, and cyclone damage was observed down to depths of 65–70 m at Myrmidon Reef. After 4 years, recovery was well underway on the shoals, aided by the abundance of fast-growing taxa such as tabular *Acropora*. Severe tropical cyclones could be expected to cause loss of corals to greater depths than weaker storms, therefore more intense storms associated with climate change could lead to storm damage extending into deeper waters. Likewise, although deeper corals are often less affected by heat stress-related bleaching than those in shallower waters, severe heat stress events could nonetheless cause coral loss at mesophotic depths. Consequently, the future trajectory of GBR outer-shelf reefs will depend on global actions to reduce atmospheric carbon dioxide in addition to local-scale conservation actions.

Further reading

Bridge TC, Fabricius KE, Bongaerts P, Wallace CC, Muir PR, Done TJ, Webster JM (2012) Diversity of Scleractinia and Octocorallia in the mesophotic zone of the Great Barrier Reef, Australia. *Coral Reefs* **31**, 179–189.

Englebert N, Bongaerts P, Muir PR, Hay KB, Pichon M, Hoegh-Guldberg O (2017) Lower Mesophotic coral communities (60–125 m depth) of the northern Great Barrier Reef and Coral Sea. *PLoS One* **12**(2), e0170336. doi:10.1371/journal.pone.0170336.

Hopley D, Smithers SG, Parnell K (2007) *The Geomorphology of the Great Barrier Reef: Development, Diversity and Change.* Cambridge University Press, Cambridge, UK.

McNeil MA, Webster JM, Beaman RJ, Graham TL (2016) New constraints on the spatial distribution and morphology of the *Halimeda* bioherms of the Great Barrier Reef, Australia. *Coral Reefs* **35**(4), 1343–1355. doi:10.1007/s00338-016-1492-2

Sih TL, Cappo M, Kingsford M (2017) Deep-reef fish assemblages of the Great Barrier Reef shelf-break (Australia). *Scientific Reports* **7**(1), 10886. doi:10.1038/s41598-017-11452-1

Webster JM, George N, Beaman R, Puga-Bernabeu A, Hinestrosa G, Abbey E, Daniell JJ (2016) Submarine landslides on the Great Barrier Reef shelf edge and upper slope: a mechanism for generating tsunamis on the north-east Australian coast? *Marine Geology* **371**, 120–129.

Webster JM, *et al.* (2018) Response of the Great Barrier Reef to sea-level and environmental changes over the past 30,000 years. *Nature Geoscience* **11**, 426–432. https://doi.org/10.1038/s41561-018-0127-3

8

Primary production, nutrient recycling and energy flow through coral reef ecosystems

O. Hoegh-Guldberg and S. Dove

Corals reefs provide a spectacular contrast to the oceanic ecosystems that surround them. Although coral reefs are ablaze with thousands of species living within a highly productive ecosystem, the surrounding tropical and subtropical oceans are usually devoid of particles and inorganic nutrients, and have low rates of primary production as a result. Whereas coral reefs are often likened to the 'rainforests of the sea', the surrounding waters are often referred to as 'nutrient deserts' due to the low quantities of inorganic nutrients such as ammonium and phosphate ions. How a marine 'rainforest' exists in an oceanic 'desert' was one of the grand puzzles that faced the early explorers, including Charles Darwin who first noted these differences in his seminal book on coral reefs in 1842.

The various drivers behind the flow of energy and nutrients through coral reefs are explored in this chapter. After investigating the physical and chemical factors that define corals and the reefs that they build, we describe some of the fundamental biological relationships that lead to the capture of energy and uptake of nutrients by coral reefs, and how these essential requirements of life flow through these magnificent ecosystems. As you will see, although primary production is similar to that of other shallow marine ecosystems, shallow coral reefs have a high proportion of pathways that involve the recycling of nutrients between closely coupled primary producers and consumers. We will also see that the highly productive nature of shallow coral reefs is deceptive, with the rapid and efficient recycling of nutrients and energy through coral reefs, while the overall net growth of coral reefs is low. This observation lies at the heart of the puzzle as to how highly productive and diverse coral reefs can exist in the 'nutrient deserts' of tropical and subtropical oceans.

Environmental factors driving the development of shallow water tropical coral reefs

Coral reefs grow in the shallow, sunlit waters of tropical and subtropical oceans, and are built from the activities of range of calcifying organisms. Globally, shallow tropical coral reefs occupy 284 300 km^2, which is less than 0.1% of the global ocean floor – yet may provide habitat to as many as one in every four species in the ocean (see Section 3). Coral reefs line shallow coastal areas between 30°N and 30°S of the Equator. In Australia, coral reefs are found on the relatively shallow continental shelves, from northern NSW to the most northern regions of Queensland on the east coast, from Rottnest Island in the south-west to the north-west coast of Australia, and along the western edge of the Gulf of Carpentaria. The structure of coral reefs varies tremendously throughout this range. Shallow coral reefs range from poorly developed reefs that fringe continental islands and the Australian coastline to extensive carbonate barrier reefs (see Chapter 2). At latitudes from 22° to 30°S, extensive communities of corals form dense communities, but fail to accumulate limestone and are hence referred to as non-carbonate reef systems (i.e. reef erosion exceeds calcification, see Chapter 2). Similar trends are present when the depth at which growth extends beyond ~40 m. At this point, just like at higher latitudes, conditions such as light become marginal for coral reefs. Coral reefs, however, can also exist in the deep sea to depths of 6000 m, where light does not penetrate and where water temperatures can be as low as 1°C. The Scleractinian corals that deposit the calcium carbonate framework of these deep sea reefs can be thousands of years old (some colonies >2500 years old), are taxonomically highly diverse (>3300 species identified to date, and counting) and obtain all their energy heterotrophically by trapping tiny organic particles. Interestingly, despite very slow rates of coral calcification, deep sea coral reefs are nonetheless carbonate, presumably because physical erosion is very low and is limited by a lack of wave action, and bioerosion, like calcification, at these depths is not solar powered (as it is in the shallows). The slow rate of coral calcification, however, makes these deep sea coral reefs highly vulnerable to the introduction of physical erosion from human activities such as trawling.

Reef-building corals (order Scleractinia, class Anthozoa, phylum Cnidaria, see Chapter 22) are central to the existence of coral reefs. Reef-building corals are often referred to as the framework builders of coral reefs. They are not, however, the only calcifiers on coral reefs. Other organisms such as molluscs, foraminifera and calcareous algae contribute substantial quantities of calcium carbonate to reef structures and sediment. Calcareous algae (see Chapter 17) also play a particularly important role in reef frameworks by coating and 'glueing' the coral framework together in some instances, and by building thick edifices of their own in others. The framework provided by corals and other marine calcifiers forms the three-dimensional (3D) structure within which hundreds of thousands of species of animal, plant, fungi and bacteria live.

Several authors have explored the underlying conditions for where shallow tropical coral reefs thrive globally. Explanations have frequently focused on the fact that coral reefs form in warm seas, leading to the 'rule of thumb' that coral reefs are limited to waters that do not decrease below 18°C in the winter. This principle is often incorrectly perceived as the ultimate limit to the development of coral reefs. This conclusion, however, ignores the many other variables that also change at higher latitudes. Although reefs are adapted to their local temperature regimes, the amount of light as well as the concentration of carbonate ions (important for reef-building activities), are also important for shallow reef development. An exhaustive study of the environmental factors associated with coral reefs using data from close to 1000 coral reef locations found that light availability and the concentration of carbonate ions are as important as temperature in defining where carbonate coral reefs are found (Table 8.1).

The conditions that are associated with the distribution of shallow coral reefs vary across spatial and temporal scales. Variability across the year can be substantial, while diurnal variability in temperature is usually small (except in areas such as shallow intertidal reef crests). The seasonal variability of conditions becomes important on coral reefs at high latitudes, where extremes in both winter (that can be too cold and dark) and summer (too hot and bright) can cause stress on coral reef organisms. At these sites in Australia, interannual variability such as that associated with the El Niño cycle along the east coast of Australia can play a big influence on coral reefs through relatively warmer or colder years and extremes. In coming decades, due to rising background sea temperatures, most years will exceed the tolerance of symbiotic organisms, giving rise to mass coral bleaching and associated mortality events. The effect of these warmer than normal conditions can drive extraordinary damage

Table 8.1. Environmental factors identified by Kleypas *et al.* (1999) that are associated with more than 1000 shallow reef locations worldwide

Variable	Minimum	Maximum	Mean	Standard deviation
Temperature (°C); based on NOAA AVHRR-based sea temperature records				
Average	21.0	29.5	27.6	1.1
Minimum	16.0	28.2	24.8	1.8
Maximum	24.7	34.4	30.2	0.6
Salinity (ppt)				
Minimum	23.3	40.0	34.3	1.2
Maximum	31.2	41.8	35.3	0.9
Nutrients (μmol/L)				
NO_3	0	3.34	0.25	0.28
PO_4	0	0.54	0.13	0.08
Aragonite saturation (Ω-arag)				
Average	3.28	4.06	3.83	0.09
Maximum Depth of light penetration (m) calculated from the monthly average depth at which average light decreased below the perceived minimum for reef development of 250 μmol/m²/s				
Average	–9	-81	–53	13.5
Minimum	–7	–72	–40	13.5
Maximum	–10	–91	–65	13.4

to the Great Barrier Reef such as the loss of 50% of its corals in the summer of the years 2016 and 2017 (see Chapter 12).

The global distribution of carbonate and non-carbonate coral reefs is strongly (and perhaps not surprisingly) correlated with the concentration of carbonate ions, which is ultimately determined by ocean temperature, salinity and other factors such as the concentration of atmospheric carbon dioxide. The concentration of carbonate ions is highest at the equator and decreases at higher latitudes due to the effect of cooler sea temperature on the solubility of CO_2 (see Chapter 12). Shallow coral reefs do not exist at carbonate concentrations below 200 μmol/kg, and this has significance in terms of the problem of ocean acidification for coral reefs as discussed in Chapter 12.

Light is also major determinant of where shallow coral reefs are found. Because of the dependence of primary production and calcification on light, shallow coral reefs are limited to clear tropical and subtropical waters where depths are generally less than 100 m. Both light quantity and quality (wavelength) are important, driving the primary steps of photosynthesis within the symbionts inside the tissues of corals, and many other photosynthetic organisms. Shallow coral reefs form only where the average daily irradiance is no less than 10% of tropical surface irradiances (~250 μmol/m/s).

Several variables affect the light available for coral reefs. Light enters the outer atmosphere of the Earth (Fig. 8.1) and is selectively filtered such that some wavelengths (ultraviolet, infrared) are largely removed by the ozone layer and water vapour (e.g. clouds). The penetration of photosynthetically active radiation (PAR, wavelengths from 400–700 nm) is also reduced by dust and clouds. Then, at the surface of the ocean, more light is reflected away from the surface, with the amount reflected decreasing as the height of the waves increases. The light that enters the ocean is absorbed by water molecules, or is scattered and absorbed by dissolved compounds, plankton and suspended sediments. These interactions are wavelength dependent such that the spectral breadth and intensity of this radiation decreases with depth. In a uniform water column, the intensity of light decreases exponentially as described by Beer's Law (Fig. 8.1).

Different water columns vary with respect to the amounts of dissolved substances and particles they contain. Inshore or coastal waters that receive freshwater from rivers and land runoff often have large quantities of sediments, tannins and phytoplankton (often referred to as 'Gelbstoff' or 'yellow substance'). These waters have spectra that are green-yellow shifted and light attenuation coefficients (k, Fig. 8.1) that may range up to 0.5/m, which means that it gets very dim at quite shallow depths. Waters that are offshore, or are located away from rivers, have far less scattering and absorption. These waters may have light attenuation coefficients as low 0.01/m. The effect of these differences in light at depth is substantial, causing the depth limits of corals in typical inshore regions to be as shallow as 5 m, compared with offshore sites where the depth limits may be in excess of 50 m. Light at depth in offshore sites is blue shifted due to a relatively small influence of Gelbstoff substances and the greater relative influence of the water itself as a source of scattering and absorption. One of the key reasons why coral reefs are not found near major rivers is because the light environment deteriorates due to extensive sediments blocking light transmission through the water column, in addition to the lack of stable clean surfaces for corals to settle and grow as a result of high rates of sedimentation.

The light environment of coral reefs also varies in time. Seasonal changes in solar flux are greater at higher latitudes, while hourly light environments at any single point on a reef vary according to the angle of the sun. Coral reefs growing close to

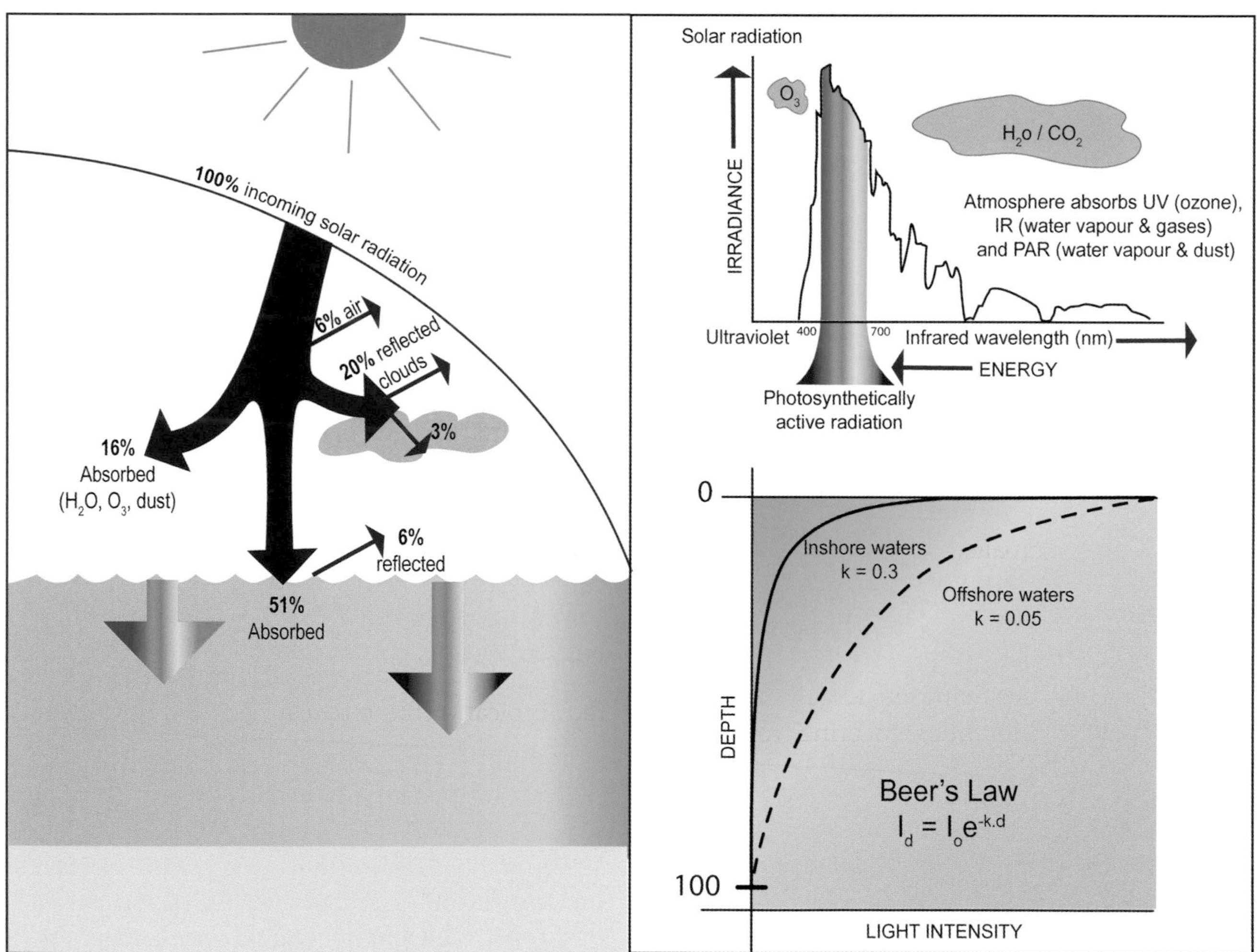

Fig. 8.1. Depiction of the pathways for solar radiation entering the ocean from the outer atmosphere. Radiation interacts with atmosphere components such as clouds, dust and specific gases such as carbon dioxide and water vapour. Light is reflected away from the surface as some crosses the surface interface of the ocean. Once it has entered the water, it is scattered and absorbed by water and components such as suspended sediments. Beer's Law relates the passage of light through the water column (z, metres) to the amount of suspended material in the water column. I_o, intensity of light at the surface, while I_d is the intensity at depth, d. The attenuation coefficient, k, is a measure of the extent to which light is absorbed and scattered by the water column and its internal constituents (Figure: D. Kleine and O. Hoegh-Guldberg).

high rocky islands, or patches of coral reef growing in crevasses, experience light environments that can fluctuate considerably. Tidal variations in depth can also significantly influence the short-term light environment as will variations in cloud cover over periods of minutes. At even finer time scales, effects such as the focusing of light through the lens effect of wave surfaces ('sun flecks') can increase the intensity of light over very short (millisecond) timescales.

Processes that underpin the primary productivity of coral reefs

The capture of the sun's energy is the initial step in the energy and nutrient cycles of biological systems on Earth. Only chemosynthetic organisms, which derive energy from the geothermal sources such as reduced compounds such as sulphide, stand apart from the overwhelming majority of organisms that are dependent ultimately on solar energy trapped by photosynthetic organisms. The myriad of photosynthetic organisms on coral reefs provide the basis for the vigorous energy and nutrient cycles that typify coral reefs.

Many organisms are photosynthetic on coral reefs. Prominent among these are cyanobacteria (Cyanophyta), microalgae (phytoplankton), macroalgae (green, red and brown seaweeds) and some dinoflagellates. The reef habitats they occupy are also diverse, with some of the most significant primary production occurring on largely uncharismatic 'mossy' communities covering rocks and sediments. Despite the large range of different organisms and primary producer communities, the capture of carbon dioxide occurs via photosynthetic processes that use the green pigment chlorophyll to trap solar energy.

In the balanced equation for photosynthesis, six molecules of carbon dioxide are incorporated while six molecules of water split in order to produce one molecule of sugar (glucose) and six molecules of oxygen.

$$6CO_2 + 6H_2O + Energy_{sunlight} \rightarrow C_6H_{12}O_6 + 6O_2$$

Respiration, on the other hand, acts in the reverse direction and oxidises organic molecules (here glucose) to produce carbon dioxide, water and energy.

$$C_6H_{12}O_6 + 6O_2 \rightarrow 6CO_2 + 6H_2O + Energy_{metabolic}$$

There are also two parts to the photosynthetic process. One part is referred to as the 'light reactions', which is where light energy is trapped by chlorophyll and is converted in the chemical energy of ATP and other molecules. The other part involves the 'dark reactions' (where the chemical energy that is trapped during the light reactions is used to 'fix' CO_2 into organic molecules). The dark reactions start with the fixation of CO_2 by the abundant enzyme ribulose bisphosphate carboxylase/oxygenase (or Rubisco) and involve the set of reactions comprising the Calvin-Benson Cycle. Whereas the light reactions are powered by the sun's energy, the dark reactions don't need to be, as long as the appropriate levels of ATP and other reduced molecules are available to power the enzymatically catalysed reactions involved.

Chlorophyll is the central pigment involved in the transduction of light into chemical energy. However, there is a range of 'accessory pigments' that are also important to the light reactions of photosynthesis. Accessory pigments interact with light in a variety of ways and consequently add colours to the tissues of different organisms from blue-green (typical of cyanobacteria or Cyanophyta) to red (typical of the red seaweeds or Rhodophyta). As light increases, so does the rate of gross photosynthesis (P_G). All organisms (whether photosynthetic or not) respire and release the embedded energy in carbon–carbon bonds within organic molecules. In photosynthetic organisms, the rate of respiration (R, usually measured in the dark when no photosynthesis is occurring) is subtracted from P_G to calculate the net rates of photosynthesis (P_N). This is essentially a measure of the rate at which organic carbon molecules (and the associated energy) accumulates during photosynthesis, over and above that consumed during respiration.

When measured per square metre of coral reef, values of P_N can be used as a measure of the net primary productivity. This is an important number as it defines the rates at which energy is being added to an ecosystem such as a coral reef, or for that matter, a sugarcane field or forest.

The relationship between P_N and light has a characteristically shaped curve (Fig. 8.2A). In the dark, P_N is negative and equals the rate of respiration (R). As light increases, however, P_N also increases until it equals zero. This is called the compensation irradiance (I_C), which is where organic carbon (energy) produced by photosynthesis exactly balances the consumption of organic carbon by respiration. No net accumulation of organic carbon (or energy) occurs at this point. I_C is also the compensation point at which the flux of oxygen into the organism just balances the rate at which oxygen is consumed by respiration. The reverse is true of carbon dioxide, which travels in the opposite direction to oxygen. The net rate of photosynthesis continues to increase as the light levels increase, with a net accumulation of organic carbon and production of oxygen.

The relationship between photosynthesis and light is linear at first, with a characteristic slope ('α') that is a measure of the efficiency of photosynthesis. Essentially α is a measure of the rate at which photosynthetic activity increases with an increase in light (quanta). α varies according to the type of organism, the different light-trapping processes and their acclimation state (Fig. 8.2A). As light increases, however, the slope begins to decline to zero, at which point the photosynthesis versus irradiance curve has reached a maximum value (referred to as $P_{N\ max}$). Further increases in light do not result in any further increases in photosynthetic production due to the saturation of the dark reactions of photosynthesis (i.e. the capacity of these reactions to convert energy and carbon dioxide into organic carbon molecules has been exceeded). At very high light levels, P_N may decline below $P_{N\ max}$ due to an often detrimental process called **photoinhibition**. This occurs when high light levels result in the light reactions capturing too much energy relative to the needs of the dark reactions. Much of the excess energy drives the conversion of oxygen into highly reactive 'active oxygen'. Oxygen receives the excitation energy (essentially becoming supercharged) and, if these resulting supercharged molecules of active oxygen are not neutralised, serious damage to cells and tissues can occur. Various antioxidant enzyme pathways exist for deactivating these super-changed molecules, yet these defense strategies can get overwhelmed in situations of high light and can damage to the dark reactions. Damage from active oxygen is an important step in the mechanisms driving 'coral bleaching' (see later).

The characteristics of photosynthetic versus irradiance curves, while being driven by light, can change to some extent through the process of 'photosynthetic acclimation' (often mistakenly referred to as 'photoadaptation', which wrongly implies that evolutionary 'adaptation' is involved). To optimise light capture under low light while reducing the risk of photoinhibition, photosynthetic organisms actively manipulate the efficiency of light capture (*a*) by adding or subtracting chlorophyll and other pigments to, or from, the photosynthetic components responsible for light capture. This ability to process captured light (that reduces the relative risk of photoinhibition) is high in organisms that grow in high light habitats, and low in organisms that grow in low light ones. Organisms can change their ability to process captured light by increasing the capacity of their dark reactions (Calvin-Benson cycle) to increase $P_{N\ max}$. These changes define many of the differences between shade and light-acclimated photosynthetic organisms (Fig. 8.2B).

Over day–night cycles, the photosynthetic activity of reef communities fluctuates between periods of net consumption (i.e. at night, Fig. 8.2C) and periods of net production of organic molecules (i.e. during the day, Fig. 8.2C). On the pathway between these two extremes, photosynthetic reef organisms pass through periods (usually a few hours after sunrise or before sunset) in which the net fluxes of carbon dioxide, oxygen and energy are zero (I_C, as described earlier). As the sun rises, the rate of

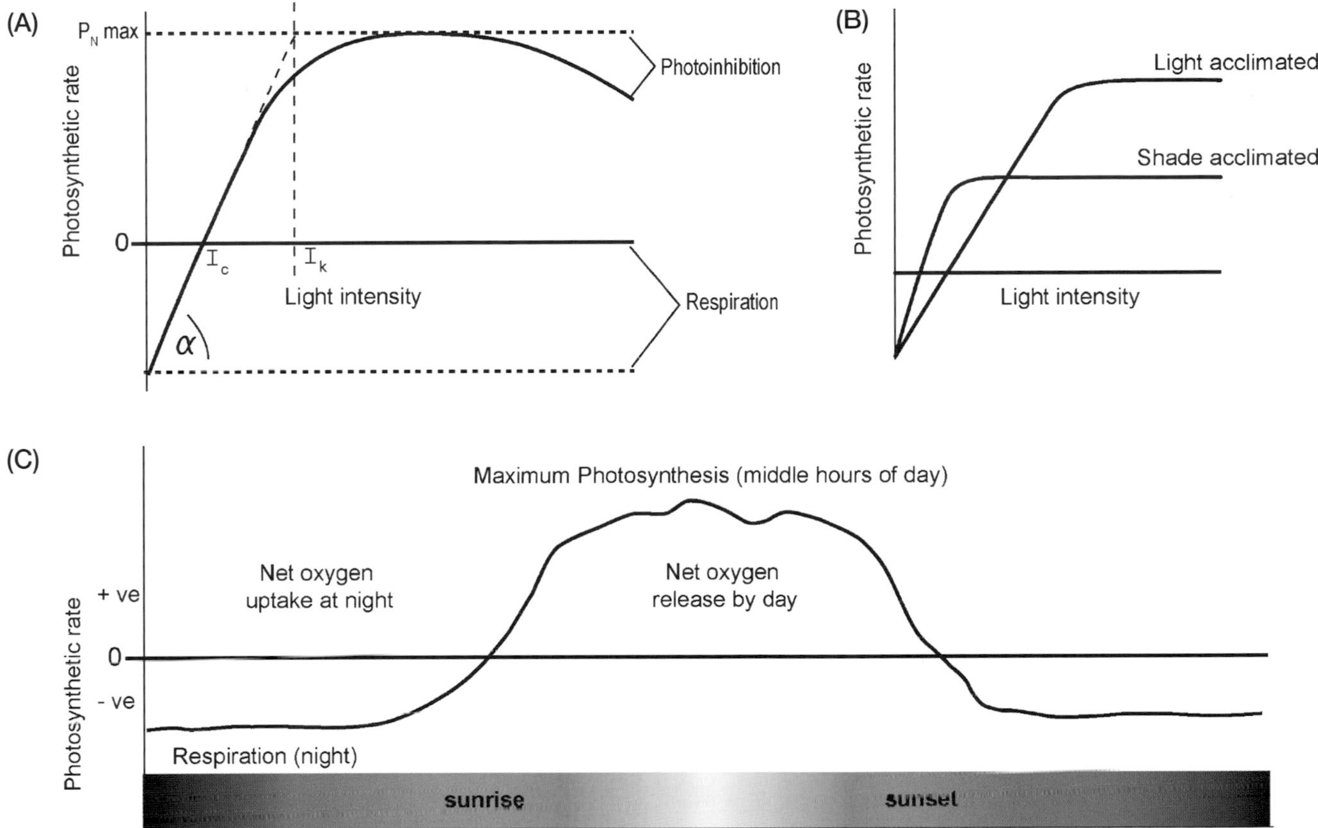

Fig. 8.2. Typical characteristics of photosynthetic processes occurring in marine organisms. **(A)** Typical photosynthetic curve showing how the rate of photosynthesis increases with light intensity until it matches the respiratory rate at the compensation irradiance (I_C) and eventually saturates at the net maximum rate of photosynthesis ($P_{N\ max}$). At this point, the maximal gross photosynthetic rate ($P_{G\ max}$) can be calculated from the sum of $P_{N\ max}$ and the respiration rate (R). The saturation irradiance (I_k) is a measure of the irradiance at which the photosynthetic rate is maximised. The efficiency with which light is converted into photosynthetic production is measured by the photosynthetic efficiency (α). If light levels continue to increase, the net rate of photosynthesis may decrease due to photoinhibition. **(B)** Organisms may acclimate or adjust physiologically to different light environments. In high light acclimated organisms, maximum rates of photosynthesis will be higher, while photosynthetic efficiencies will be lower. The opposite is true in organisms that have acclimated to living in low light conditions. **(C)** A typical pattern over time associated with the activity of a photosynthetic organism on a reef (net photosynthetic rate per hour). At night time, respiration dominates. At dawn, photosynthetic activity begins to increase as light levels rise until they achieve maximum levels of both light and photosynthetic activity at midday (Figure: D. Kleine and O. Hoegh-Guldberg).

photosynthetic activity becomes positive, increasing towards maximum values in the middle of the day. At this point, P_G approaches $P_{G\ max}$ that may be several times greater than the absolute value of respiration. The daily photosynthesis to respiration ratio (P:R ratio) can be derived from the integrated photosynthetic activity over a day ($P_{G\ 24\ h}$) divided by the respiration that occurred over the same period ($R_{24\ h}$). This ratio is used by many physiological ecologists to examine how dependent an organism is on energy derived directly from light. P:R ratios can be calculated for short (hourly) or long (daily, yearly) periods. Organisms that have P:R ratios that are greater than 1.0 are referred to as net autotrophic ('self feeding'; also referred to as phototrophic or 'light feeding') while organisms that have ratios of less than 1.0 are referred to as net heterotrophic ('feeding on others'). Organisms on coral reefs have a wide range of P:R ratios. Photosynthesis to respiration ratios can be applied to communities as well, giving important insight into how much organic carbon or energy is entering a particular patch of land or seascape. In the latter case, these measurements are easily incorporated into measurements of the net accumulation of organic carbon per unit time (primary production). These types of measurements provide an important basis from which to answer key questions such as whether coral reefs are a net sink or source of carbon dioxide.

Productivity varies across the planet, with some of the highest values being found in tropical rainforests (700–800 g C/m^2/year). The highest values for ecosystems in the ocean range from 200 to 500 g C/m^2/year and are associated with nutrient-rich coastal areas where light levels are high and where the upwelling of high concentrations of inorganic nutrients occurs (e.g. off the west coasts of tropical South America and Africa). The ocean conditions in which shallow coral reefs occur generally have primary production rates of only 0.1–1.0 g C/m^2/year. Shallow coral reefs, on the other hand, show average rates of primary production that range from 3–100 g C/m^2/year (averaged over all components including sand and rocky substrata) with some components (e.g. algal turfs, macroalgae, symbiotic corals) having rates that match even those of the highest values for primary production in terrestrial systems.

There is no single value for the primary production of all shallow coral reefs. Primary productivity varies according to whether a reef is inshore or offshore, or whether it is located at a high or low latitude site (Fig. 8.3). Values for primary productivity also vary according to the location within a reef system. The well flushed conditions of the fore reef have the highest primary production values (up to 1500 g C/m^2/year) while the often poorly circulated and more variable conditions (in terms of temperature and light) of back-reef areas can have lower production values (although still up to 300 g C/m^2/year). Primary productivity also varies with the component of reef ecosystem considered. Benthic microalgae and corals (and other symbiotic invertebrates) can have primary productivities of up to 2000 g C/m^2/year.

In addition to sunlight and carbon dioxide, photosynthetic organisms require inorganic nutrients, particularly inorganic nitrogen (usually in the form of ammonium or nitrate ions) and phosphorus (as phosphate ions). They obtain these compounds from the breakdown (waste) products of animals that have consumed other organisms, or by the breakdown of debris generated by predators (e.g. fish) or scavengers (crabs, starfish) and use them to

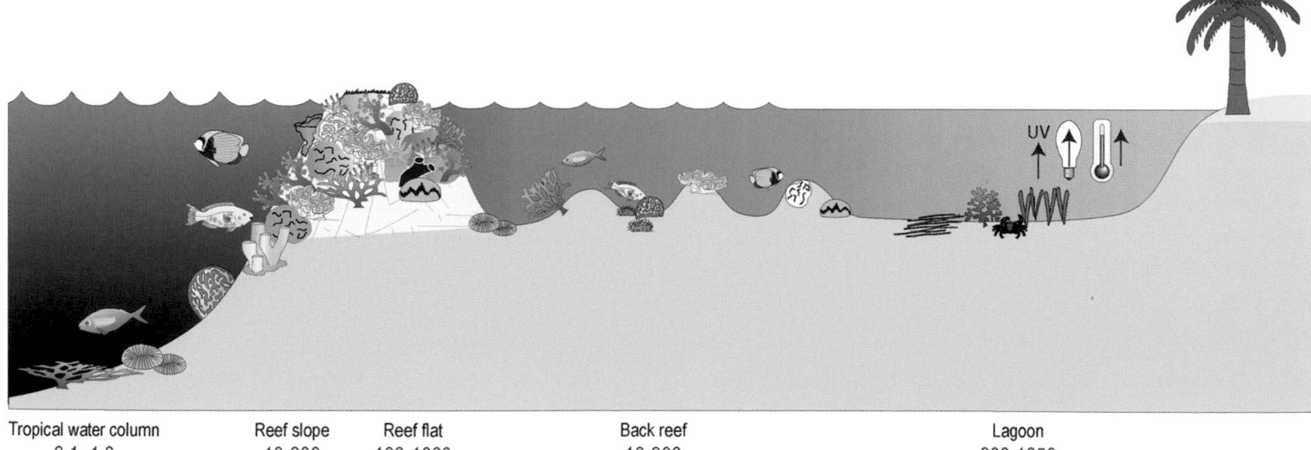

Fig. 8.3. Variation in primary productivity (red numbers: units = g C/m^2/year) and reef accretion (calcium carbonate) from the tropical water column to a typical coral reef coastline. Open ocean conditions are usually constant, although productivity is low due to a paucity of nutrients. Productivity and reef accretion is significant on the reef slope and crest due to more favourable conditions (e.g. optimal temperatures, light and water circulation). Going shorewards, conditions eventually become more extreme due to the ponding of water behind the reef crest (Figure D. Kleine and O. Hoegh-Guldberg).

build new organic molecules. A range of organisms are involved in these natural cycles, with different types of bacteria playing dominant roles in degradation. Huge populations of bacteria inhabit the sediments associated with coral reefs, which provide the ideal microenvironments for processing these compounds. One of the best known of these process pathways is the nitrogen cycle (Fig. 8.4), in which organic material settles on the surface of the sediments and is quickly buried by scavengers such as worms, molluscs and crustaceans.

In the upper few millimetres of the sediment, where oxygen levels are relatively high, a diverse range of microorganisms (e.g. bacteria such as *Vibrio*, as well as many actinomycetes and fungi) use proteinases to strip amine groups off proteins and release ammonium ions through a process called 'ammonification'. Some ammonium ions escape from the sediments into the water column for use by photosynthetic organisms to build amino acids and proteins. Other ammonium ions continue down a pathway of 'nitrification', in which the ammonium is oxidised to produce nitrate. This occurs in the oxygen-rich upper sediment layers by a set of bacteria that must have oxygen to survive (Fig. 8.4). Some nitrate released by this process leaves the sediment and is used by photosynthetic marine organisms, while the rest stays in the sediments and undergoes further denitrification.

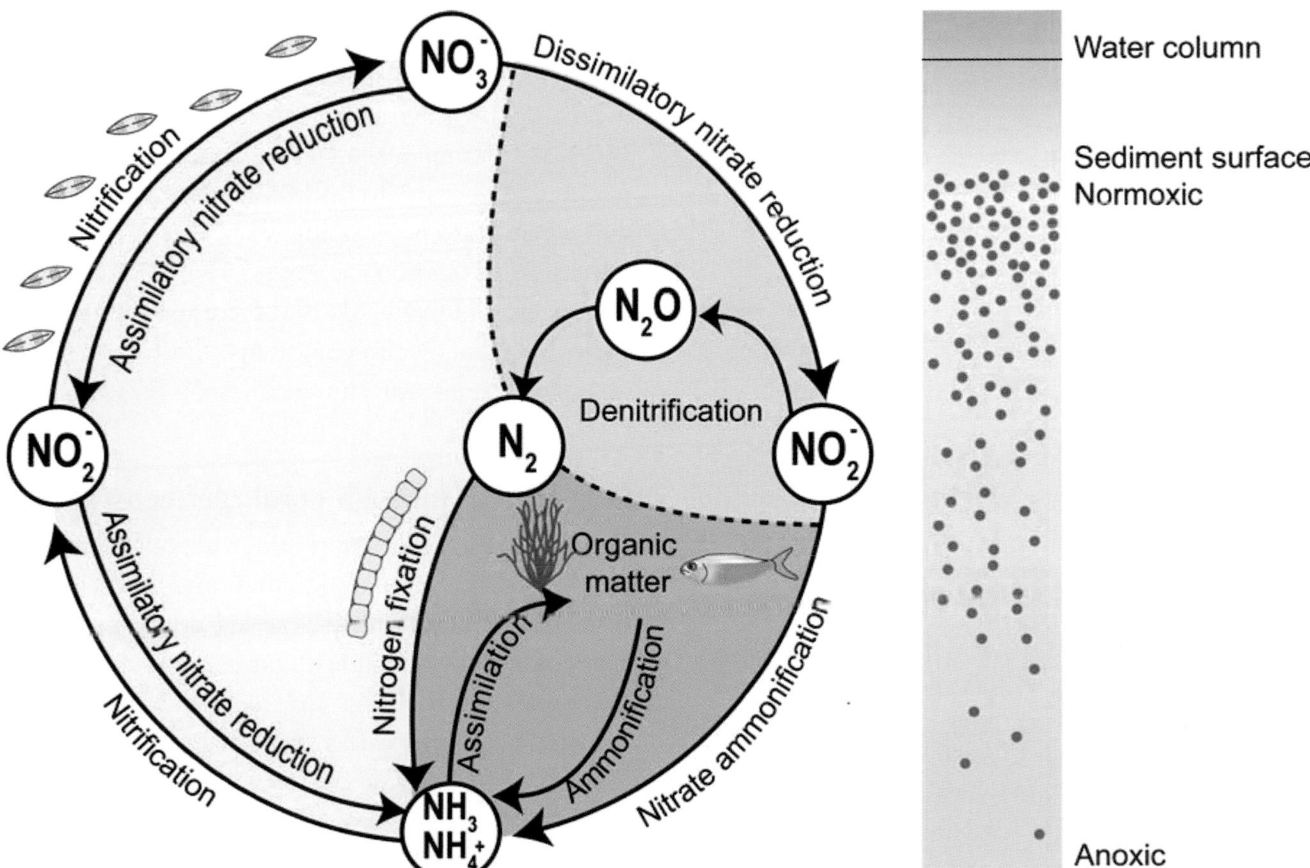

Fig. 8.4. The nitrogen cycle associated with coral reefs. The fixation of nitrogen into organic compounds occurs in the water column (above the sediments). Plants take up ammonium and may be eaten by herbivores, with the embedded ammonium being released and recycled, with the burial of some of this organic material. At this point, the ammonium ions released may participate in assimilatory nitrate reduction in the sediment layers that have relatively high oxygen levels. Nitrate may then be denitrified in the underlying sediments where oxygen concentrations are low. The colours associated with the left-hand diagram indicate when the majority of each part of the cycle occurs within the water and sediment profile (indicated in the right-hand diagram) (Figure: D. Kleine and O. Hoegh-Guldberg).

Fig. 8.5. Cyanobacteria are prominent members of the communities that cover most surfaces on coral reef, as well as forming large slicks in the waters surrounding coral reefs. Slicks such as the one shown here off the central portion of the Great Barrier Reef (showing the filamentous members of the common genus *Trichodesmium*) eventually sink to the bottom of the ocean and may contribute significantly to the nitrogen cycle of coral reefs (Photo: O. Hoegh-Guldberg).

The next step involves the conversion of nitrate to nitrite, which can only occur without oxygen, and consequently occurs deeper within the sediments, well below the first few millimetres of the upper oxygen-rich layers. Here, oxygen levels decrease to zero due to the lack of light and the abundance of metabolising organisms that use up oxygen and hence drive down the associated oxygen concentrations. The processes of ammonium oxidation and nitrate reduction are coupled. Aerobic nitrification sits adjacent to anaerobic denitrification (via the sediment gradient in oxygen availability), which is ideal for coupling the two processes together. The eventual outcome is that nitrate is reduced to nitrite, and nitrite is denitrified to nitrous oxide (NO) and/or nitrogen (N_2) gas, which is lost to the atmosphere.

Before leaving the topic of the nitrogen cycle, it is important to consider where inorganic nitrogen of coral reefs comes from. Cyanobacteria (so called 'blue-green algae') are important sources of inorganic nitrogen. One of their key characteristics is that they use an enzyme called nitrogenase to split the powerful triple bonds of atmospheric nitrogen gas (N_2) via a process called 'nitrogen fixation'. Nitrogen-fixing bacteria occur in a large number of habitats and ecosystems, including microbial mats and symbioses such as with sponges and legume plants. Recent work has revealed that nitrogen fixation is prolific in coral reefs and may play a critical role in supplying nitrogen to coral reef ecosystems. Nitrogen fixers are abundant on benthic surfaces (sediments, rocks) and in the water column of coral reefs in the form of surface blooms of the cyanobacterium *Trichodesmium* (commonly called 'sea sawdust', Fig. 8.5). Coral reefs are an exception to the general rule that high primary productivity must be matched by a large standing stock of inorganic nutrients (particularly N and P). Coral reef primary production is indeed high, yet the nutrient levels in the surrounding waters are exceeding low. What could be going on? To find the answer to this problem, we need to explore the ways that energy (contained in carbon compounds), and more particularly nutrients (N and P compounds), move through coral reef ecosystems after the initial stages of primary production.

Energy flow through coral reef ecosystems

Organic carbon that accumulates through primary production is eaten, degraded and recycled by other organisms, which are called 'consumers'. The mass of primary producers far exceeds that of primary consumers. This is due to inefficiencies in the transfer of energy and nutrients between trophic levels and to entropic considerations. The 'higher' plant and animal life of coral reefs forms a trophic pyramid, with a successive decrease in the mass of each successively higher trophic level (Fig. 8.6A) as discussed earlier. Although macroscopic organisms play visually prominent roles, microscopic life and processes on coral reefs are fundamental to trophic pyramids such as this through their mediation of energy and nutrient flows.

Herbivory

As discussed elsewhere (e.g. Chapters 6, 26, 28, 30), herbivory is an important process on coral reefs. An adequate level of herbivory is necessary for maintaining the balance between reef-building corals and macroalgae, and hence many of the characteristics of a coral reef. Any visit to an intact coral reef will yield countless observations of animals moving across the substrate and cropping marine plants, both small and large. Key herbivores include gastropod molluscs, chitons, echinoids, crabs and fish (Fig. 8.6). These organisms may be active at different times of the day and light levels (e.g. dawn, dusk, night) and tidal heights. Many gastropods (e.g. abalone) and chitons set out from their daytime hiding places to graze on benthic microalgae at night, while many grazing fish (e.g. scarids, siganids and acanthurids) are active only by day and will wait until the tide floods the reef crest under the cover of dusk or dawn before moving in to graze on these rich and highly productive reef areas. Grazing activity is often evident from scrape marks made by the tough beaks of grazing fish (Fig. 8.6B), or by the presence of feeding scars from grazing molluscs (Fig. 8.6C). Some grazing activity can be very subtle. Recent studies have revealed that porcellanid crabs (see Chapter 25), for example, are significant grazers on coral reefs, cropping the benthic microalgae from the substrate using their chelae that are like tiny scissors. In this case, there is little visual evidence of the grazing activities of these organisms despite their importance in terms of the mass of material removed each day from a coral reef.

Predation

Undisturbed coral reefs are visually exciting for many reasons, one of which is the presence of predators of all sizes and shapes (see Chapters 24–26, 30). Predation is a dominant driving process, from the tiniest predators such as the chaetognaths and fish larvae that eat copepods, to sharks that eat large organisms such as rays, turtles, groupers and even other shark species. In coral reef ecosystems, sharks are often referred to as apex predators (Fig. 8.6A). As with herbivores, some predators are nocturnal and have evolved to hunt in almost no light at all. Others prey on diurnal grazers and time their visits to the reef with the tidal cycle and particular light levels.

Particle feeding

Many of the organisms that live on coral reefs feed on particles such as plankton or small pieces of debris suspended in the water column, or deposited on submerged surfaces. Particle feeders are part of an important energy and nutrient loop (see next section). Particle feeders range from the largest (whale sharks and manta rays) to the smallest organisms (copepods) of reef-associated species. They also include sponges, which use special cells called choanocytes to drive water through small channels and trap particles. Many simple (e.g. hydroids, anemones) and advanced invertebrates (e.g. polychaetes, bivalve molluscs, ascidians) use cilia to drive water past their feeding structures and use sticky mucus to trap them before ingesting them. Other organisms such as crinoids use tube feet to pick particles out of the water while orienting feeding structures into the currents. Polychaete worms and sea cucumbers have specialised feeding structures that allow them to feed on particles deposited on sediment surfaces. Some fish actively pick particles out of the water column, while others sift sediments using modified gill rakers to find small pieces of detritus and debris for ingesting.

The assemblage of particle feeders along the reef crest has been termed the 'wall of mouths'. The 'wall' plays a crucial role in the accumulation of energy and nutrients on coral reefs. Few large zooplankton from the open ocean survive contact when they float by the cloud of particle feeding fishes at the edges of coral reefs (Fig. 8.6D). The amount eaten (and hence the energy acquired by the reef ecosystem) is highly significant, as shown when feeding rates are calculated per unit time per metre of reef-ocean interface (in one study, ~0.5 kg/m/day wet weight zooplankton, mostly larvaceans and copepods, entered the reef economy). It appears

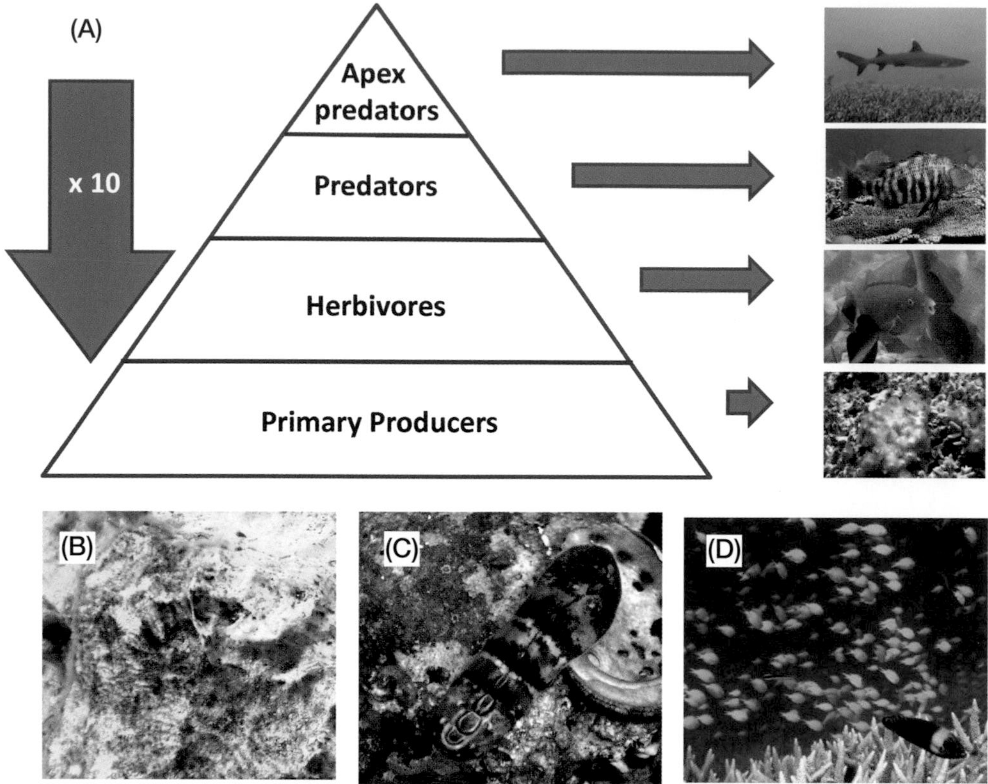

Fig. 8.6. Trophic interactions on coral reefs. **(A)** Trophic pyramid showing the flow of energy from primary producers such as the benthic algae to apex predators such as the White-tip reef shark shown. **(B)** Herbivores such as parrotfish leave characteristic scrape marks with their teeth as they feed. **(C)** Other invertebrates such as chitons graze the surfaces in between coral colonies and may be significant as cryptic reef grazers. **(D)** Another important input of energy to the reef system occurs through the large numbers of particle feeders that line the seaward edges of coral reefs forming the so-called 'wall of mouths' (Diagram: D. Kleine and O. Hoegh-Guldberg; Photos: O. Hoegh-Guldberg).

that the 'wall of mouths' represents yet another important part of the puzzle for how coral reefs maintain themselves in the nutrient-poor (oligotrophic) conditions of tropical seas.

Plant animal symbiosis: the rise of the holobiont

One of the hallmarks of coral reefs is the high number of mutualistic symbiotic relationships across a large range of organisms. These relationships number in the many thousands and involve all sorts of interactions, from those between gobies and burrowing shrimps to the intracellular symbioses between sponges and bacteria. One of the central hypotheses surrounding coral reefs is that the large proportion of mutualistic symbioses have arisen due to the low nutrient conditions, which embody the advantage of the close association of primary producer and consumer. One of the big advantages of these close associations is that the inorganic nutrients required by the primary producer are obtained directly from the animal consumer. This avoids the dilution that would otherwise happen if the nutrients and organic matter were to enter the water column. There is no better example of the ultimate close association than the mutualistic symbiosis of reef-building corals and their intracellular dinoflagellate symbioses (see Fig. 8.7 and Box 8.1). The close coupling of coral and *Symbiodinium* spp. has been in existence for at least 220 million years and is highly successful and is largely responsible for the huge reserves of limestone found in the upper layers of the Earth's crust. As pointed out elsewhere (see Chapters 2, 3 and 9), the limestone structures generated by corals

Box 8.1. The mutualistic endosymbiosis of corals and dinoflagellates

Reef-building corals and invertebrates from at least five invertebrate phyla form close associations with dinoflagellate protists from the genus *Symbiodinium*. Often referred to as zooxanthellae (a loose, non-taxonomic term), these single-celled plant-like organisms live within the endodermal cells (gastroderm) of reef-building corals. Here, they can photosynthesise like other phototrophs, but instead of retaining the organic carbon that they make, *Symbiodinium* can release up to 95% of this organic carbon to the host. This energy is used by the coral to grow, reproduce and produce copious amounts of calcium carbonate, which forms the framework of coral reefs. Importantly, the energy can also be released from corals as mucus, which forms an important source of food and energy for other reef dwellers. Only corals that have a symbiosis with *Symbiodinium* are able to calcify at the high rates that are typical of shallow reef-building corals and that are able to build the large 3D structures associated with shallow carbonate coral reefs. In return for this copious energy, *Symbiodinium* receives inorganic nutrients from the waste metabolism of the animal host (Fig. 8.7). Given the shortage of inorganic nutrients such as ammonium and phosphate ions in tropical and subtropical water columns, the provision of these nutrients is critical to the high rates of photosynthesis and energy production of *Symbiodinium*. Because there are benefits for both partners in this symbiotic relationship, and the fact that one cell lives inside the cells of another, this relationship is referred to as a **mutualistic endosymbiosis**. The tight recycling of energy and nutrients between the primary producer and consumer avoids the problem of the low concentrations of these materials in typical tropical seas. This is thought to be one of the key reasons why coral reefs have been able to prosper in the otherwise nutrient 'deserts' of tropical seas.

and other organisms provide the habitat for over a million species of plant, animal, fungi and bacteria worldwide.

The establishment of a mutualistic relationship between corals and symbiotic dinoflagellates centres around the maintenance of sublit, nutrient poor environments that are typically associated with shallow coral reef ecosystems. It is often also premised on the assumption that all symbiotic dinoflagellates are photoautotrophs (organisms that live entirely off organic carbon obtained from fixing CO_2), as opposed to mixotrophs, which are organisms that can supplement their photosynthetically fixed carbon with carbon externally acquired from either dissolved organic material (DOM) or particulate organic material (POM). As a mixotroph, the holobiont with symbiotic dinoflagellates shift from being a mutualistic symbiont to being a more parasitic symbiont under changing environmental conditions. The latter is consistent with the definition of a parasite as an organism living in or on another living organism, obtaining part or all of its organic nutriment from it, usually to the detriment of its host, but not its long-term viability.

The autotrophic, mixotrophic or heterotrophic status of a symbiont associating with a heterotrophic host is often dependent on the availability of food for the holobiont (symbiont + host). Using an example from another system (orchids), for example, the shift from autotrophic to heterotrophic orchids is linked to whether the orchid associates with a saprophytic or ectomycorrhizal fungi. The establishment of a parasitic relationship between orchids and fungi is made possible by an increased supply of organic carbon from the fungi's alternative food source. That is, instead of the saprophytic fungi seeking limited carbon in detritus, ectomycorrhizal fungi have unlimited organic carbon on tap from large canopy trees. The parasitic relationship becomes stable over time because it is not overly detrimental to the host fungi. In this relationship, the mixotrophic (parasitic) orchid associates with either saprophytic or ectomycorrhizal fungi. Similarly, the establishment of a mixotrophic symbiont within a heterotrophic coral host is, therefore, conceptually viable in reef ecosystems that are subject to variable nutrient fluxes associated with, for example, upwelling events, and/or river run-off. Interestingly, mixotrophy occurs among most, if not all, extant genera of dinoflagellates, inclusive of *Symbiodinium* (symbiotic dinoflagellates) with demonstrations, to date, of heterotrophic activity by *Symbiodinium* associated with gastropods, anemones and scleractinian corals.

Symbiotic dinoflagellates, even from the same clade, associate with a myriad of hosts, not just scleractinian corals living in nutrient poor environments. Likewise, symbiotic scleractinian corals exist in a range of nutrient and light environments.

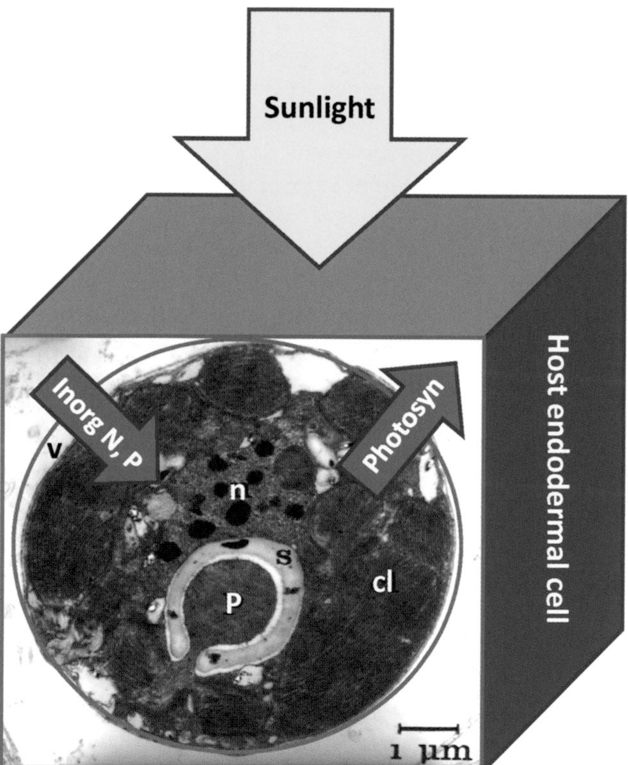

Fig. 8.7. Illustration of the relationship between *Symbiodinium* (transmission electron microscope, TEM, scale = 1 µm) and the host endodermal cells of reef-building corals. Sunlight illuminates the transparent host cells, which drives the photosynthesis of *Symbiodinium*. *Symbiodinium* passes copious amounts of the resulting photosynthetic products (labelled 'photosyn') to the host cell, which in turn provides *Symbiodinium* access to inorganic nutrients such as ammonium and phosphate that arise from host catabolism (labelled 'Inorg N, P'). Parts of the *Symbiodinium* cell are show: n, nucleus; s, starch cap; cl, chloroplast; p, pyrenoid; v, vacuole space between host vacuole membrane and outer plasmalemma of the enclosed *Symbiodinium* cell. (TEM image: O. Hoegh-Guldberg).

It is likely that a subset of the symbiotic dinoflagellates that associate with corals are facultative parasites, consuming their hosts to survive when photosynthesis is inhibited such that the ability of the symbionts to pass organic compounds to their hosts is reduced or absent. Symbiosis is not necessarily mutualism, and a coral that retains symbiotic dinoflagellates following a significant environmental disturbance is not necessarily fitter (growing/calcifying faster or more fecund) than a coral that incurs significant bleaching. A shift to a state with no symbiotic dinoflagellates, or a shift to a facultative parasite, can be equally detrimental to the ability of reef-building corals to calcify at a rate that offsets the rates of erosion in shallow tropical zones.

A growing number of studies has revealed a large number of other organisms that live in intimate contact with reef-building corals. Among these are bacteria that fix nitrogen to small plants that live inside the skeletons and are bioeroders, as well as many other acclimated residents of which the function is not clear. First coined by the evolutionary biologist, Lynn Margulis, the term 'holobiont' is now being regularly used to describe the collection of closely associated organisms. Margulis proposed that evolution might act on the holobiont as a collection of organisms in addition to acting directly on the organisms involved. Many studies use molecular techniques to probe the identity of the communities that associate with corals. Using these techniques, researchers have begun to identify a large and growing list of close associates of corals. Most of these are bacteria with proposed roles in maintaining the health of corals or acquiring key nutrients, and hence there is some interest in terms of studies aimed at trying to understand how current and future conditions may influence coral reefs.

Studies of food webs on coral reefs have identified a major role for the mucus generated by reef-building corals. Mucus is considered to be relatively cheap to produce due to the abundant photosynthetic energy available for corals in shallow habitats (hence the concept of 'junk carbon'). It is primarily produced to prevent the surfaces of corals from being colonised by fouling organisms and may have a role in protecting corals from excessive light (PAR and ultraviolet radiation, UVR). It tends to be sloughed off corals at the end of the day (generally after extensive photosynthesis has occurred). Corals on the intertidal reef flat at Heron Island (southern Great Barrier Reef) exude up to 4.8 L of mucus per square metre of reef area per day and, of that, up to 80% dissolves in the reef water. Although the dissolved component stimulates a burst of metabolic activity in the sediments where it is largely metabolised, the remaining particulate proportion is eaten by fish and other particle feeders on the reef crest. This transfer of energy is sizeable and relatively unique when compared with other marine food webs.

Coral reefs as the 'Beggars Banquet'

This chapter explored the production and flow of energy through coral reefs, which are highly productive ecosystems that prosper in the otherwise nutrient-poor waters of the tropics. While tropical oceans that surround most coral reefs have a primary productivity close to zero, coral reefs have some of the highest levels of primary productivity in the ocean. This productivity is a manifestation of the efficient photosynthetic processes and recycling that occurs within warm and sunlit tropical seas. In addition to energy, we also examined one of the key nutrient cycles of coral reefs – that of nitrogen – observing that nitrogen is regenerated by nitrogen fixation and that it cycles between the different organisms within the food web of coral reefs along with the energy of organic carbon bonds. The 'wall of mouths' of small fish and other particle feeders that forage at the interface between coral reefs and the open ocean play an important role in the acquisition of energy by coral reefs. As well, the efficiencies of mutualistic symbioses such as those seen between corals and symbiotic dinoflagellates have huge benefits to a wide range of organisms in the dilute nutrient conditions of tropical seas.

There is unlikely to be a single factor, however, that can explain why shallow coral reefs are so productive. The answer appears to involve a range of characteristics and mechanisms that generate and recycle nutrients. There is one important take-home message though: it is erroneous to conclude that much of the energy generated by coral reefs can be harvested as a net product of the system. The tight recycling of nutrients and energy means that the majority of primary production generated by coral reefs is rapidly recycled back into the ecosystem by the many pathways elucidated in this chapter. This may help us understand why fisheries on coral reefs are so susceptible to over-exploitation. In this regard, coral reefs have been likened to a 'beggar's banquet' where the table is set for a feast that looks at first glance to be generous and abundant. However, closer inspection reveals that very little can be eaten or taken away from the table, without the abundance on the table being reduced. This situation appears to be fundamentally different from that seen in other marine ecosystems such as kelp forests, where the rate of primary production can be large seasonally with substantial amounts of energy and organic carbon exported out of the ecosystem without the kelp forest disappearing. Another way of understanding this is to compare the measures of P_G and P_N on a community basis. The rate of gross photosynthesis of the community is large in both coral reefs and in kelp forests, but $P_{N\ community}$ of coral reefs is far less than $P_{N\ community}$ of kelp forests. These differences strike at the heart of the unique nature of coral reefs and may drive other emergent features such as the sensitivity of coral reefs to small changes in environment that surrounds them. These perspectives are important in the context of how we manage coral reefs and other ecosystems.

Further reading

Anthony KRN, Hoegh-Guldberg O (2003) Variation in coral photosynthesis, respiration and growth characteristics in contrasting light microhabitats: an analogue to plants in forest gaps and understoreys? *Functional Ecology* **17**, 246–259. doi:10.1046/j.1365-2435.2003.00731.x

Benson A, Muscatine L (1974) Wax in coral mucus – energy transfer from corals to reef fishes. *Limnology and Oceanography* **19**, 810–814. doi:10.4319/lo.1974.19.5.0810

Darwin CR (1842) *The Structure and Distribution of Coral Reefs.* Smith Elder and Company, London, UK.

Dove SG, Kline DI, Pantos O, Angly FE, Tyson GW, Hoegh-Guldberg O (2013) Future reef decalcification under a business-as-usual CO_2 emission scenario. *Proceedings of the National Academy of Sciences of the United States of America* **110**, 15342–15347. doi:10.1073/pnas.1302701110

Hamner WM, Jones MS, Carleton JH, Hauri IR, Williams D (1988) Zooplankton, planktivorous fish, and water currents on a windward reef face: Great Barrier Reef, Australia. *Bulletin of Marine Science* **42**, 459–479.

Hatcher BG (1988) Coral reef primary productivity: a beggar's banquet. *Trends in Ecology & Evolution* **3**, 106–111. doi:10.1016/0169-5347(88)90117-6

Hoegh-Guldberg O (1999) Coral bleaching, climate change and the future of the world's coral reefs. *Marine and Freshwater Research* **50**, 839–866. doi:10.1071/MF99078

Hughes TP, Baird AH, Bellwood DR, Card M, Connolly SR, Folke C, *et al.* (2003) Climate change, human impacts, and the resilience of coral reefs. *Science* **301**, 929–933. doi:10.1126/science.1085046

Kleypas JA, McManus J, Menez L (1999) Using environmental data to define reef habitat: where do we draw the line? *American Zoologist* **39**, 146–159. doi:10.1093/icb/39.1.146

Lesser MP, Mazel CH, Gorbunov MY, Falkowski PG (2004) Discovery of symbiotic nitrogen-fixing cyanobacteria in corals. *Science* **305**, 997–1000. doi:10.1126/science.1099128

Murray JM, Wheeler A, Freiwald A, Cairns S (2009) *Cold-Water Corals.* Cambridge University Press, New York, USA.

Muscatine L (1990) The role of symbiotic algae in carbon and energy flux in reef corals. In *Coral Reefs.* (Ed. Z Dubinsky) pp. 75–84. Elsevier, Amsterdam, Netherlands.

Wild C, Huettel M, Klueter A, Kremb SG, Rasheed M, Jørgensen BB (2004) Coral mucus functions as an energy carrier and particle trap in the reef ecosystem. *Nature* **428**, 66–70. doi:10.1038/nature02344

9

Calcification, erosion and the establishment of the framework of coral reefs

P. A. Hutchings, O. Hoegh-Guldberg and S. Dove

Much like a city represents an equilibrium between building construction and demolition, coral reefs are the net result of processes that form calcium carbonate (calcification) and those that take it away (physical and biological erosion). The resulting reef framework is quintessential to coral reefs, forming the habitat for tens of thousands of reef species. This chapter examines the major forces that control this equilibrium state, examining the physical, chemical and biological forces that are involved. It also investigates the role that humans have in influencing these processes, highlighting the large-scale effects of both local (eutrophication, land runoff) and global (global warming, ocean acidification) factors on reef structures. Some of these issues are also discussed further in Chapter 11 and 12.

Calcification

Calcification is highest in the warm, sunlit waters of the tropics and subtropics where the concentrations of carbonate ions are at their highest levels. As discussed in Chapter 2, these three factors are considered to be the major determinants for where shallow coral reefs grow, with waters becoming too cold, dim and low in crucial ions as one goes towards higher latitudes. The association of high rates of calcification with low latitude environments appears to have held for hundreds of millions of years, resulting in huge deposits of calcium carbonate within the Earth's geological structure in equatorial regions.

The organisms that calcify on shallow coral reefs are diverse and include a large number of phyla such as cnidarians, molluscs, crustaceans and foraminifera, as well as green and red coralline algae (see also Chapters 7 and 17). Many of the organisms that calcify at high rates are also symbiotic with dinoflagellates such as *Symbiodinium* (see Chapter 8). This association with organisms having high photosynthetic capacities is considered to be indicative of the high-energy requirements of the calcification process. These organisms take up calcium and carbonate ion from the super-saturated concentrations that are typical of tropical and subtropical waters, depositing either calcite or aragonite (two forms of calcium carbonate crystals). The form of calcium carbonate deposited depends mostly on the organism involved. Corals, for example, deposit aragonite while red coralline algae deposit Mg-calcite. The techniques for measuring calcification and de-calcification are outlined in Boxes 9.1 and 9.2, respectively.

Box 9.1. How to measure calcification

Accurate measurements of coral calcification are critical to any in-depth understanding of the rate that it plays in reef processes. There are several methods that have been used over the years to measure the rate at which calcium carbonate is deposited by corals and other calcifying organisms. Several these methods are outlined below, with some comments on their ease of use and efficacy.

Dye to measure linear extension

Corals build their skeletons by incrementally adding layers of calcium carbonate. If the thickness of added calcium carbonate is known, then it is possible to obtain a relative rate of calcification over time. This measure can be converted to an absolute measure of calcification when the thickness is multiplied by the area and the density of calcium carbonate. Many studies have used the fluorescent dye Alizarin, which binds effectively to proteinaceous elements of the coral skeleton. To undertake measurements with marker dyes, corals are incubated in non-toxic levels of the dye for several hours before being placed back in the field. Following exposure to the dye, which essentially marks the beginning of the period over which skeletal growth will be measured, corals are allowed to grow under field conditions before being harvested several months to years later. Back at the laboratory, the corals are killed and their skeletons cut into thin sections to show the annual banding patterns (exposed using X-ray photography). By measuring the distance between the skeletal surface (the site of the latest deposition of skeleton) and the fluorescent band (representing the beginning point), the linear extension of skeleton can be calculated as a function of time.

Using radioisotopes to follow the deposition of calcium atoms

Calcium is the cation that is deposited along with carbonate ions as the coral forms a skeleton. If $^{45}Ca^{2+}$ (a radioactive isotope of Ca^{2+}) is mixed into the water surrounding a coral, it immediately gets taken up by the coral as calcification occurs. By measuring the ratio of radioactive atoms to non-radioactive atoms of Ca^{2+}, the rate at which all Ca^{2+} atoms are deposited can be calculated. This technique is very precise and can yield an accurate measure of the rate of calcification of corals. The downside of this technique, however, is that it must be done in the laboratory, and is potentially hazardous if the right precautions are not taken. The conditions in the laboratory can be very different to the field and hence may not accurately depict calcification as it might occur on a coral reef. This method, like the alkalinity anomaly technique (next), is typically used to determine calcification rates over short period of time, rather than integrated rates from longer periods of over 24 h, months or years. These methods are essential if you would like to determine the difference between day and night rates of calcification, which also means that they are highly sensitive to the exact conditions applied.

Using changes in alkalinity to follow the deposition of carbonate ions

The alkalinity of a solution is lowered by two equivalents per mole of carbonate ion precipitated into calcium carbonate, meaning that the rate of calcification can be determined from the change in alkalinity of solution of a known volume over a set timeframe. When the technique incorporates corrections for certain nutrients that can also influence alkalinity, the technique is very precise and is ideal for determining rates of calcification over hours, and differences between night and day rates of calcification. However, as previously discussed, this technique is only really useful for short-term measurements.

Buoyant weight method

By measuring the difference in density between calcium carbonate and sea water, it is possible to convert the weight of an object that has been measured in sea water into an absolute measure of calcium carbonate deposition. Assumptions made in applying this technique are that the only negatively buoyant part of the coral colony is its calcium carbonate skeleton, which is largely true for the tissues of corals and many other organisms. The technique is typically used for determining calcification rates over timescales of months or longer and does not require the researcher to kill the coral or other organism. Conversion of buoyant weights into grams of calcium carbonate, however, requires precise knowledge of the density of the newly deposited material, which may change as a result of environmental variation. Many treatments including acidification tend to reduce the density of the skeleton rather than changing linear extension or volume.

Box 9.2. How are rates of bioerosion measured?

Although a piece of dead coral substratum can be cut open and the amount of calcium carbonate that has been removed calculated, this does not give any indications of the rate of its erosion. One solution for measuring bioerosion is to use experimental blocks of freshly killed coral that show no sign of boring and to lay them out on the reef (Fig. 9.6C) for fixed time periods. Because the original dimensions and density of the blocks are known, losses and gains can be calculated and assigned to the various organisms. Obviously, replicates need to be collected for each time period in order to determine the variation within a site before considering variation between sites. Thin sectioning of the blocks allows the distribution and density of the various microborers to be determined. Part of the block can be dissolved in order to extract the borers that can then be counted and identified to species level. Although net rates of grazing can be determined from changes to the dimensions of the block (see Fig. 9.6D), potential grazers in the region need to be identified, their densities calculated and the amount of calcium carbonate in their faecal pellets measured. In the case of scarids (parrotfish), the depth and dimensions of the feeding scars (Fig. 9.4A) can be measured and for echinoids, their population density can be measured and faecal pellets collected over a 24 h period to estimate the amount of calcium carbonate they contain in order to calculate rates of grazing. Increasingly, studies are using automatic video stations to measure the visitation and bite rate of some types of grazers such as fish. Knowing rates of calcification in the area, a balance sheet of losses and gains for the area can be constructed (i.e. a 'carbonate balance'). However, it must be noted that these rates may vary considerably within a reef, depending on amount of suitable substrate, and hence provide only an indication of rates of bioerosion. Also, we know that rates vary as the substrate age; consequently, analysis of bioerosion in geological cores requires different techniques.

Many marine organisms use calcium carbonate to create skeletons that either provide rigid support structures and/or protective shells or cases. These functions are likely to be crucial in the highly dynamic coral reef ecosystems, where competition and predation can be high, and physical stresses (e.g. storms) may be periodically extreme. Organisms also calcify using a range of different mechanisms. Most mechanisms, however, require the pH at the site of calcification to be raised above that of typical sea water or cellular pH to ensure that the majority of available dissolved inorganic carbon (DIC) is in the form of carbonate ions (CO_3^{2-}). Consequently, the rate of net calcification is greatest above a pH of 9 when the rate of precipitation is significantly greater than the rate of dissolution. In scleractinian corals, ~30% of the DIC that is used for calcification is extracted from sea water, with ~70% being derived from cellular respiration. In the absence of an effective mechanism to remove protons from the site of calcification (see Fig. 9.1), the mix of respired carbon dioxide with sea water introduced into this semi-confined space acidifies the fluid, is likely to significantly reduce the net rates of calcification. However, Ca^{2+}/H^+ ATPase pumps that are powered by a large number of mitochondria located in the calicoblastic ectoderm of the coral remove H^+ (protons) from the site of calcification and increase the pH and the ease with which precipitation of calcium carbonate can occur. This is facilitated by the photosynthetic activities of the endosymbiotic dinoflagellates by day, and potentially inhibited at night by respiratory processes that increase acidity by night (due to increasing amounts of carbon dioxide entering the tissues as a result of respiration). Photosynthesis not only provides carbohydrates for glycolysis by day, but also supersaturates the tissues adjacent to the site of calcification with oxygen facilitating mitochondrial oxidative phosphorylation and the production of ample ATP. By contrast, these deep tissues are depleted in oxygen at night time, favouring anaerobic as opposed to aerobic respiration. This potentially inhibits the activities of the Ca^{2+}/H^+ ATPase pumps, and increases the proton gradient which must be counteracted by increasing cellular acidosis. The existence of these mechanisms is supported by observations of light-enhanced, but

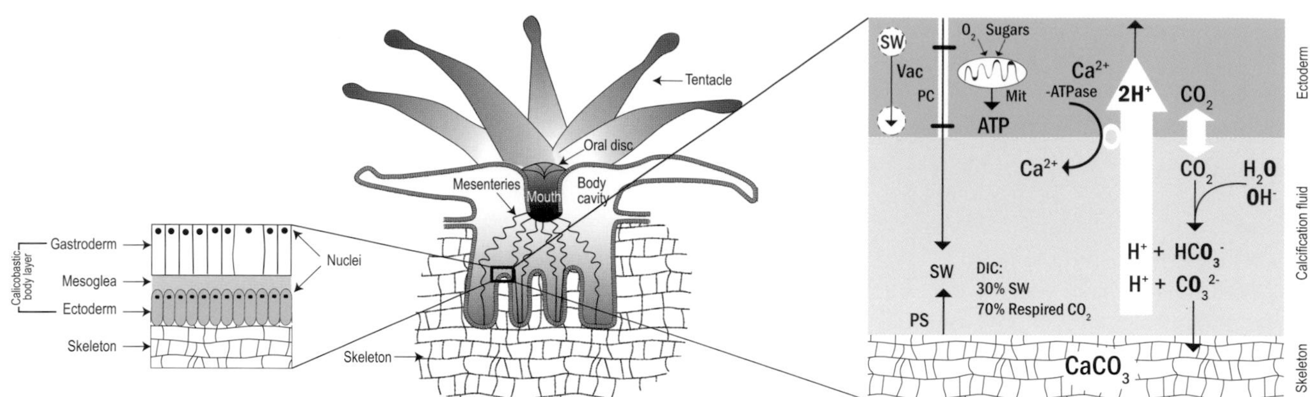

Fig. 9.1. Calcification in reef-building corals. Ca^{2+}-ATPase exchanges, powered by ATP produced from numerous mitochondria (Mit) located in the ectoderm of the calicoblastic body layer of a coral, remove protons (H^+) and introduce calcium (Ca^{2+}) into the calcification fluid that baths the skeleton. Dissolved inorganic carbon (DIC, e.g. CO_2, HCO_3^-, CO_3^{2-}) is introduced into this fluid via the diffusion of respired carbon dioxide and/or via the introduction of sea water (SW) into this space. SW can be introduced via vacuoles (Vac) that phagocytosis SW present in the gut (body cavity), via para-cellular (PC) transport through the intercellular spaces between epithelial cells, or via diffusion through a porous skeleton (PS) that may in some parts be unprotected by tissue and directly exposed to the environment (Image: D. Kleine, Dove and Hoegh-Guldberg, inspired by Cohen and McConnaughey (2003).

dark-suppressed rates of calcification in corals that host dinoflagellates, and by experiments that demonstrate that a supply of oxygen can increase night-time rates of calcification.

Nucleation of calcium carbonate requires bulk energy to concentrate calcium and carbonate ions within confined spaces and is assisted by the production of specialised proteins that are often referred to as 'skeletal matrix proteins'. These particular proteins tend to be highly anionic (covered in negative charges) and contain regions that are associated with enzymes such as carbonic anhydrase (that catalyses the rapid conversion of carbon dioxide to bicarbonate and protons, a reaction that occurs rather slowly in the absence of a catalyst, even under an optimal pH environment).

There are several ways calcium carbonate deposition is measured, with many being complementary as well as being informative with regards to different aspects of calcification. These are outlined in Box 9.1. The rates of calcification on coral reefs can be extremely high in equatorial or low latitude areas. On a more regional scale, the deposition of calcium carbonate varies with the presence or absence of rivers, where high nutrients and sedimentation may slow the deposition of calcium carbonate by reducing light and smothering. In this respect, inshore coral reefs on the GBR do not deposit calcium carbonate as fast as those reefs offshore. Again, this is a consequence of changes in factors such as light, temperature and nutrients. At the scale of a reef, calcium carbonate deposition can be quite dynamic and will vary between the slope, crest and back-reef areas as discussed already in Chapter 8. Coral reefs are also dynamic in geological timeframes, with the shape of deposited calcium carbonate varying over time in response to prevailing winds and currents. These aspects of reef construction are discussed in Chapter 2.

The skeletons of calcifying organisms build up over time, with the consolidation of skeletal material leading to the solid component of the framework of coral reefs. Calcifiers on coral reefs can have different morphologies and roles, such as the massive and branching structures. These structures are generated by corals and are essentially glued together into a consolidated framework by other organisms (Fig. 9.2A). The rate of calcification and erosion tend to substantially exceed the net rate of growth of carbonate coral reefs. Estimates of calcification suggest that rates vary from 1 to 2 m per century while rates of reef growth are ~1–2 m per millennium. Based on these rough figures, this would suggest that rates of calcification are between three and ten times higher than the rate at which calcium carbonate is removed by physical and biological erosion. As we will see later in this chapter, the balance between the two forces (calcification versus erosion) is critical to understanding the impacts of global change, such as ocean acidification.

Physical and biological erosion

The removal of calcium carbonate from coral reefs (erosion) is a key process on coral reefs that involves several elements including dissolution, physical breakage and the activities of several so-called bioeroders. These elements are intertwined and it is difficult to separate them. They are a feature of recent, as well as fossil, coral reefs where similar types of borers occur.

Wave action erodes the reef slowly over time by physical action and chemical dissolution of the reef calcium carbonate substratum. During storms, however, this rate will increase through wave action, with large boulders sometimes being dislodged such that they roll down the reef slope crushing and removing any coral colonies in their path (Fig. 9.2B, C). These forces can have significant

Fig. 9.2. Wave action is a key process on coral reefs and underpins the physical erosion of reefs. **(A)** Section through the reef crest at Heron I. (L, living corals; C, consolidated framework of dead corals stuck together by calcareous red algae); **(B)** reef crest at the Low Isles; **(C)** reef crest at Heron Island, showing the reef break where physical eroding forces are maximum; **(D)** Heron Island, spur and groove (Photos: O. Hoegh-Guldberg).

impacts on the shape of coral reefs (e.g. spur and groove formations (e.g. on Wistari Reef, Fig. 9.2D). Depending on the wave energy and the types of corals, the impacts of storms can be substantial, with branching corals most affected especially in shallow water. After a cyclone, for example, where wave action may extend many metres below the surface, the damage can appear as if the living veneer of the reef has just been peeled off and shed.

Biological erosion consists of the loss of reef substratum by boring and grazing. A suite of organisms including polychaetes, molluscs, sponges, barnacles, sipunculans and various microorganisms such as bacteria, fungi and algae bore into coral substrata. Endolithic algae colonise the skeletons of corals and specialise in living in very low light levels (Fig. 9.3A). Turf algae (Fig. 9.2B) often grow over dead skeletons providing food for grazing herbivorous fish (Fig. 9.3C, D) as well as invertebrate grazers such as chitons (Fig. 9.3E), echinoids (Figs 9.3F, G; 28.8A–F) and gastropods (Fig. 9.3H). These organisms physically bite or scrape the substratum to collect the algae (Fig. 9.3A) and with it they take particles of the substratum that has become honeycombed by the action of the borers (Fig. 9.4B). The calcium carbonate matrix, together with the algae, passes through the gut of the grazers and is ground up, separating the algae from the calcium carbonate. Cellulose enzymes break down the plant cells, the nutrients are then absorbed, and then the calcium carbonate is defecated as a fine powder. Recent studies have confirmed that most parrotfish are microphages that target cyanobacteria and other protein-rich autotrophic microorganisms that live on (epilithic) or within (endolithic) calcareous substrata. This contradicts the traditional view that parrotfish are major consumers of macroscopic algae. Swimming behind schools of large schools of parrotfish (scarids, Fig. 9.4C) one often sees the water column becoming cloudy as this fine powder is ejected (Fig. 9.4D). Similarly, the faecal pellets of grazing echinoids consist largely of compacted finely ground calcium carbonate. The lagoonal sediments, especially those offshore, are largely composed of these products of bioerosion and physical erosion – along with mollusc shells, carapaces of crustaceans, foraminifera tests and sponge spicules. Only sediments adjacent to the coast or large islands have a component of terrestrially derived sediments.

Storm activity will dislodge coral colonies both live and dead (Fig. 9.2B), which have often been weakened by borers attacking the base and branches of corals. These coral colonies and broken off branches of the staghorn corals (*Acropora*) can be washed down to the bottom of the reef slope or thrown up onto sand cays. This band of coral rubble, which is often well developed on the windward side of cays, forms an important coral reef habitat for a wide range of organisms. Coral rubble is itself subjected to further bioerosion, although often it develops a protective coat of coralline algae that provides some protection from the colonisation of the substratum by endolithic algae. Such surfaces lack endolithic algae and are therefore not grazed significantly by parrotfish and echinoids. Dead coral substrata adjacent to river mouths, where large plumes of sediment-loaded water flow out onto the reef during the wet season (see Chapter 13), tend to be covered in a thick layer of silt that again protects the substratum from endolithic algal colonisation. Live coral colonies have mechanisms to eject the sediment as it settles on the coral polyps, whereas it can accumulate on dead substratum.

Although live coral colonies are typically free of borers, the substratum is rapidly colonised by borers once a colony dies. Because all borers have pelagic larvae, it is difficult for them to settle on the living veneer of a coral colony without being eaten by the coral polyps. In cases where borers have settled on living colonies, it is presumed that larvae have settled on damaged polyps allowing them to metamorphose and rapidly bore into the substratum, before being eaten by neighbouring polyps. Once coral substratum becomes available for colonisation by borers, a distinct succession occurs, with the early settlers being bacteria, fungi and endolithic algae that appear to condition the substratum and facilitate the next suite of colonisers, primarily polychaete worms, and later sponges, sipunculans and

Fig. 9.3. **(A)** Endolithic algae inhabit the skeletons of corals, living among the crystals and over time weakening the skeleton (Photo: O. Hoegh-Guldberg). **(B)** Dead coral substratum covered by turf algae. (Photo: O. Hoegh-Guldberg). **(C)** The parrotfish *Scarus* sp. with well-developed jaws about to take a lump of dead coral substratum full of endolithic algae. (Photo: O. Hoegh-Guldberg) *D*, Jaws of *Bolbometopon muricatum* on the outer barrier near Lizard Island (Photo: D. Bellwood). **(E)** Close-up of the intertidal chiton, *Acanthopleura gemmata* from One Tree Island, nestled onto its home scar (Photo: B. Kelaher). **(F)** The grazing echinoid *Echinometra mathaei*, oral surface showing Aristotle's lantern partially protruding from the mouth that it uses to scrape off the surface of the coral (Photo: A. Miskelly). **(G)** Diagram of Aristotle's lantern. (Source: Illustration after Anderson DT (1996) *Atlas of Invertebrate Anatomy.* UNSW Press). **(H)** *Monodonta labio* (Trochidae) feeding. (Photo: K. Gowlett-Holmes).

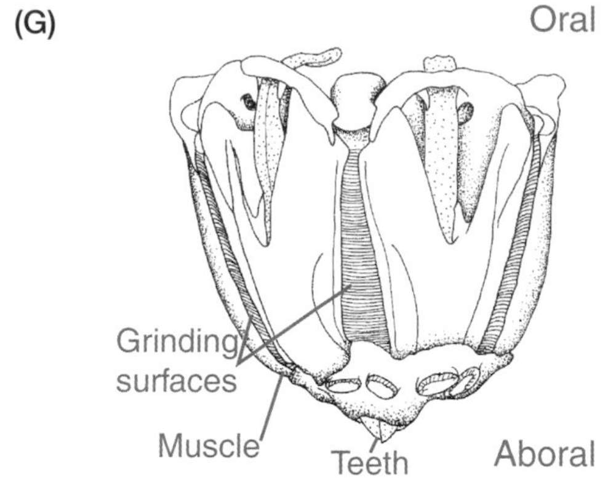

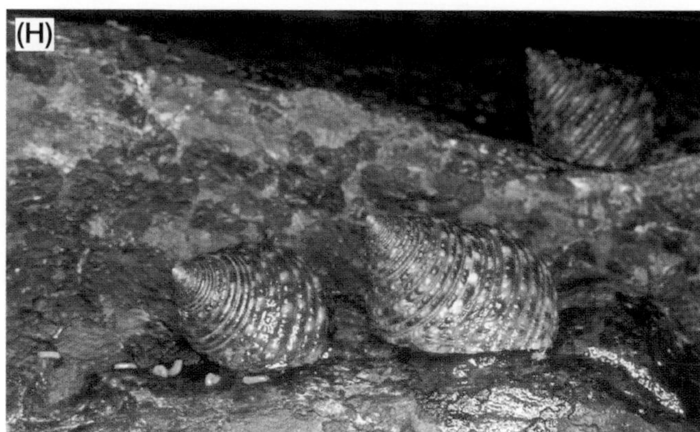

Fig. 9.3. (Continued)

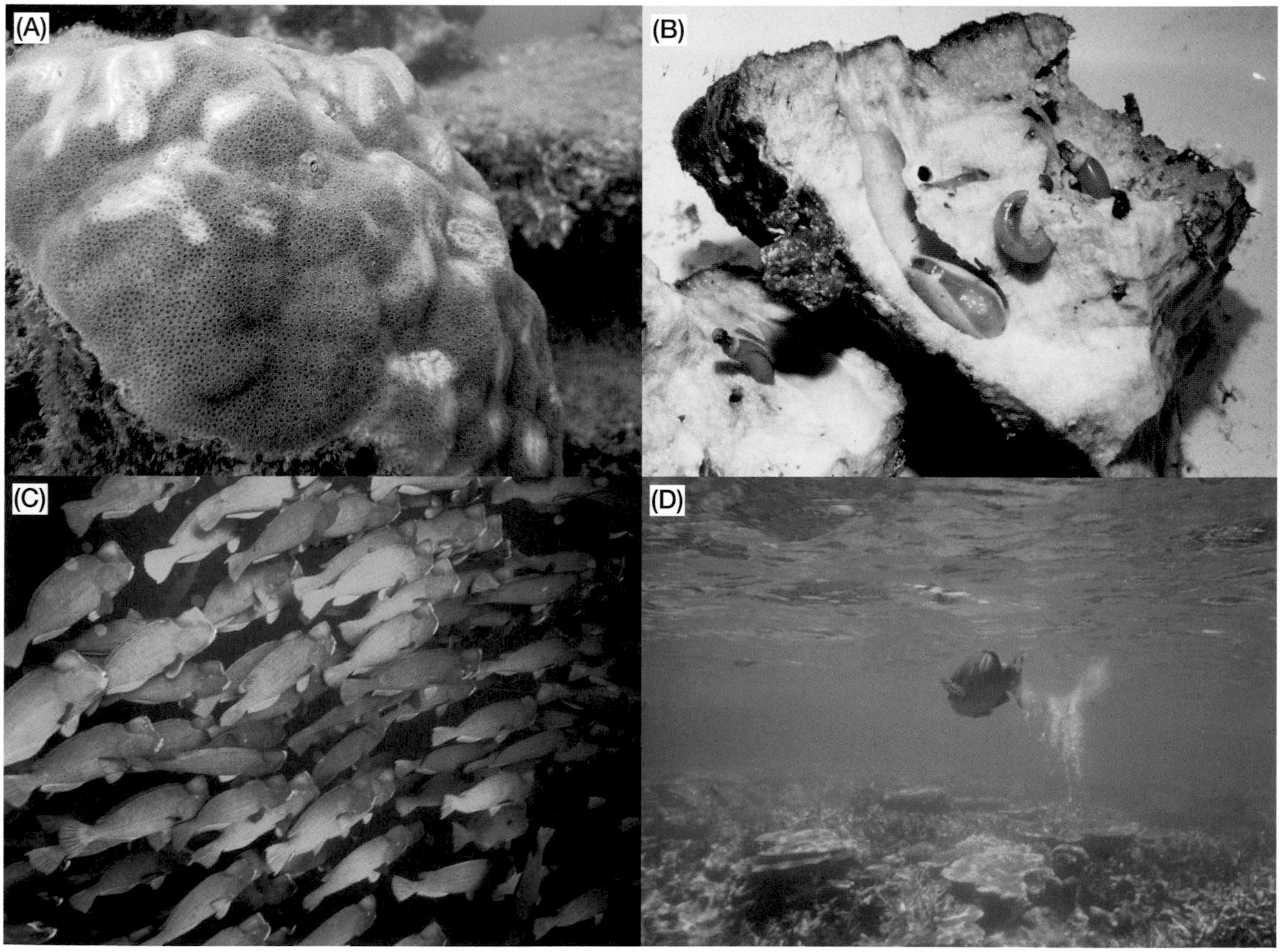

Fig. 9.4. **(A)** Multiple bite marks of a scarid (parrotfish) and a boring barnacle (centre) embedded in *Porites lutea* (Photo: O. Hoegh-Guldberg). **(B)** *In situ* dead coral habitat split open to reveal boring sipunculans and bivalves, burrow of boring bivalve (Photo: P. Hutchings). **(C)** School of *Bolbometopon muricatum* at Osprey Reef, Coral Sea (Photo: P. Hutchings). **(D)** Defecation by parrotfish: fine sediment produced by the grinding of the ingested coral fragments (Photo: D. Bellwood).

molluscs including bivalves and boring barnacles. Many of these early colonisers are short lived and their burrows tend to provide homes for non-boring organisms once they die. In contrast, the sipunculans, sponges, bivalves and some of the larger polychaete borers are long-lived (for many years) (Fig. 9.4A, B). All these organisms bore by either physically eroding the substratum or chemically dissolving it, or by a combination of these methods.

Rates of boring decrease once the borers are established, and subsequent rates are just sufficient to allow those organisms to grow. This is particularly true for sponge colonies. Borers must obviously retain a link to the outside of the substratum in order to obtain their food, for respiration and for discharging their gametes. Once established in the substratum, they are effectively entombed, often in flask-shaped burrows (Fig. 9.4B). They cannot leave their habitat, although some species of molluscs, primarily *Conus* spp., move over such substrate actively searching for particular boring species of polychaetes and sipunculans, inserting their proboscis into the burrow and then proceeding to extract and eat the worm.

Experimental studies have shown that recruitment of boring organisms is seasonal, with maximum recruitment occurring during the summer months. The recruitment of other types of borers, however, can occur throughout the year. This means that within weeks of substratum becoming available, it is already being colonised by borers. Recruitment also varies between years and this is presumably related to the availability of larvae, the supply of which will be influenced by weather patterns at the time of spawning. Net rates of bioerosion (i.e. losses due to grazing and boring plus gains from accretion from coralline algae and encrusting organisms, plus physical and chemical erosion) vary between sites on an individual reef as well as between reefs (Fig. 9.5, see Box 9.2 as to how to measure rates of bioerosion), and differences occur between oceans. Factors such as water quality and sediment load influence not only the rates, but the agents responsible for grazing and boring. As the substrate becomes more honeycombed, chemical erosion may be stimulated leading to greater risks of corals being dislodged during storms, as has already mentioned. However, if the substrate becomes heavily bored by sponges this may increase their flexibility and allow them to better withstand wave action.

On the GBR, various groups of scarids (parrotfish) are important grazers of microphages. Scarids can be divided into three distinct functional groups depending on the osteology and muscle development on the oral and pharyngeal jaws. Although all scarids have well-developed scraping plates or a beak (Fig. 9.3D), the actual development of these plates and associated muscles determines what they can feed on. One type, the 'croppers', remove only algae and associated epiphytic material, whereas 'scrapers' and 'excavators' remove pieces of substrate together with the algae, leaving distinctive feeding scars (Fig. 9.4A). The only difference between scrapers and excavators is the depth to which they can bite. These species are able to break down the calcium carbonate using their pharyngeal jaws. These last two categories are feeding on the surface layers of dead coral substratum containing endolithic algae (Fig. 9.3A). The large double header *Bolbometopon muricatum* (Fig. 9.4C) feeds almost exclusively on live coral, often on the faster growing species such as *Pocillopora* and the tabulate acroporids. On the GBR, many species of scarids occur and their distribution and functional roles such as grazing, erosion, coral predation and sediment reworking vary across the reef. Inner shelf reefs support large numbers of scarids, although their biomass is low, and they exhibit high rates of grazing and sediment reworking. In contrast, the outer-shelf reefs have much lower densities of scarids but with much higher biomass. In these outer reefs, scarids are responsible for higher rates of erosion through grazing and predation on corals. Mid-shelf reefs have intermediate values. In areas of overfishing, loss of scarids and other herbivorous species of fish, such as surgeonfishes, rabbitfishes and drummers, can have significant impacts on the reef, as the algae are not being removed by grazing leading to intense competition

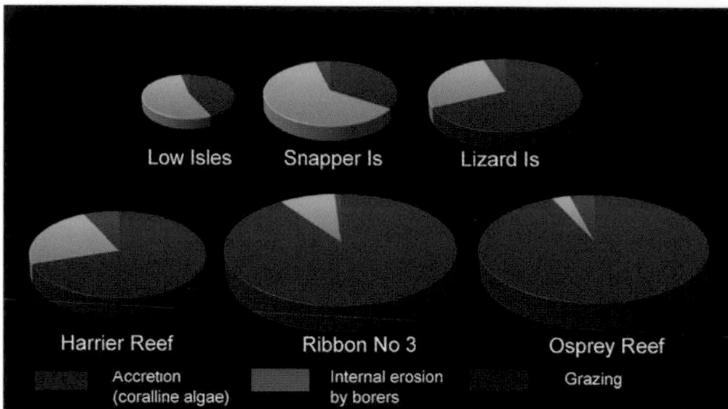

Fig. 9.5. Variations in rates of grazing, accretion and internal erosion by borers across the GBR from the Daintree River out into the Coral Sea along a gradient of increasing water clarity determined from experimental blocks as illustrated in Fig. 9.6C (Source: after Hutchings *et al.* 2005). Loss (grazing and boring) + gain (accretion) = net rates of bioerosion (Diagram: K. Attwood).

for space between corals and benthic macroalgae. This can lead to a shift from a coral dominated environment to one dominated by macroalgae.

In the Caribbean and on some French Polynesian reefs, echinoids are important grazers, especially *Diadema setosum* (Fig. 9.6A), *D. savignyi* and *Echinometra mathaei* (Fig. 9.3F, G). These species are also relatively uncommon on the GBR. These echinoids graze on the algal covered reefal substratum using their Aristotle's lantern, a complex series of calcareous plates that scrape the surface (Figs 9.3G, 28.8C). The densities of these species are influenced by water quality and high levels of nutrients encouraging algal growth. A study in Papeete, Tahiti, showed that overfishing has reduced fish populations that graze on juveniles of echinoids to low levels, which has allowed high densities of *Echinometra mathaei* (Figs 9.3F, 9.6) to develop, which thrive on the excessive algal growth that is being driven by high levels of nutrients in the water column. These high levels of nutrients are being washed down from nearby rivers where untreated sewage is being discharged. Excessive algal growth restricts coral recruitment and, over time, if water quality is not improved, rates of grazing and associated boring will far exceed rates of calcification and this is already leading to substantial loss of reef framework on this reef (Fig. 9.7). This will have massive flow-on effects, including loss of protection from storm activity on nearby low lying areas, loss of coral reefs leading to reduced fish landings, and the loss of tourism.

Species of *Echinostrephus* (Fig. 9.6B) on the GBR can often be seen nestling in their home depression that they have eroded in the reef substratum. They feed on plankton and suspended matter and basically stay in these depressions. The chiton, *Acanthopleura gemmata,* is common on the GBR in the intertidal zone (Fig. 9.3E). It forms deep depressions to which it returns after foraging on algae during low tides, mostly at night. It has been suggested that feeding at low tide may decrease the predation risk from fishes and sharks that move over the area as the tide rises. These chitons use their radulae to scrape the surface of the boulders, collecting the algae attached to the substratum. Examination of the extruded pellets of these chitons allows an estimate of the rate of sediment production to be calculated; in this restricted environment, rates are high.

A recent study of bioerosion from the inshore to outer reefs of the northern GBR and Coral Sea discovered that sediment flowing down the Daintree River heavily influenced rates and agents of bioerosion. Experimental substrata at sites at the mouth of the river and associated Low Isles, after short periods of exposure are covered in a thick layer of silt that restricts development of the turf algae and hence levels of grazing are low. The silt settles out from the turbid water column. In contrast, at sites out in the Coral Sea where water is clear (Fig. 9.4D), substrata are heavily bored with epilithic algae, encouraging high rates of grazing by scarids. Boring communities vary between inshore and offshore sites, with deposit-feeding polychaete species dominant at inshore sites, with filter and surface deposit feeders at offshore sites. Boring sponges are most abundant at inshore sites and boring bivalves at offshore sites. Net rates of erosion vary between sites and the relative importance of the components of erosion change markedly along the cross-shelf transect, with data on the distribution, abundance and species composition of scarids having

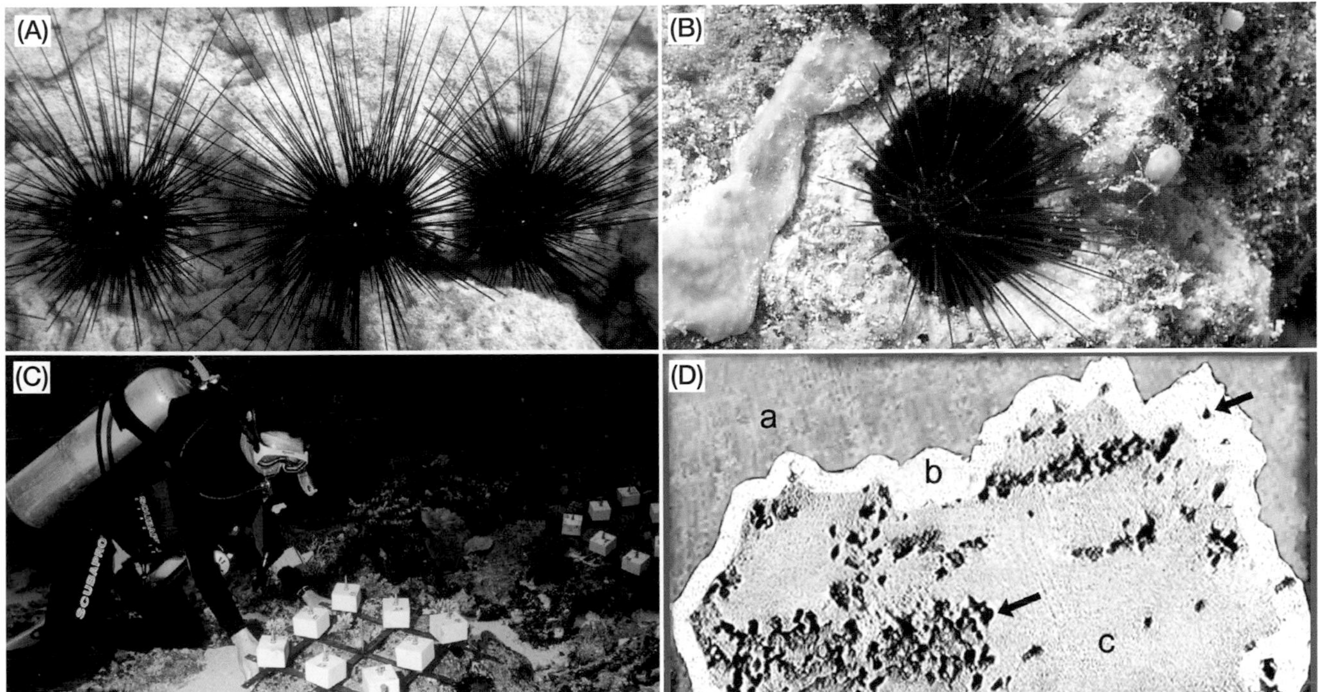

Fig. 9.6. **(A)** *Diadema setosum* a grazing echinoid linked to major erosion of western Indian Ocean reefs (Photo: O. Hoegh-Guldberg). **(B)** *Echinostrephus* sp., sitting in its home scar that it has created (Photo: O. Hoegh-Guldberg). **(C)** Experimental study of bioerosion at Osprey Reef, Coral Sea, two replicate grids with newly laid coral blocks to be exposed for varying lengths of time (Photo: J. Johnson). **(D)** Diagrammatic representation of coral block illustrating how the various components of bioerosion (i.e. grazing, accretion and boring) are determined from a series of sections through each block. Knowing the density of the coral block, these measurements can then be scaled up to rates per square metre and then net rates of bioerosion calculated: a, original block; b, accretion; c, block remaining after grazing and boring (Image: K. Attwood).

interesting trends across similar inshore to the Coral Sea transect (Fig. 9.5).

Tipping point: human influences on calcification and erosion

As human populations have expanded in the coastal areas of tropical and subtropical oceans, their influence on the environment that surrounds coral reefs has increased dramatically. Changes to the nutrient and sediment concentration of the waters surrounding corals have impacted the growth and calcification of a wide range of coral reef organisms. In the last two decades, these local impacts have been joined by global factors such as global warming and ocean acidification (see Chapters 11 and 12). Together, local and global factors have decreased the growth and calcification of reefs while at the same time probably increasing the rate of dissolution and/or bioerosion. These changes are complex, interactive, and have far-reaching consequences for both natural ecosystems and the human societies that depend on them.

Fig. 9.7. Experimental blocks after 6 months showing extensive grazing by *Echinometra mathaei* at Faaa, Tahiti. (Photo: M. Peyrot-Clausade).

The impact of coral bleaching, COTS outbreaks (see Chapters 5, 11 and 28) and a wide array of other factors have decreased the proportion of reefs covered by living corals, which is essential for the maintenance of carbonate reef substrates against the carbonate eroding activities of grazers and bioeroders. The loss of corals has in turn provided an increase in supply of suitable substrata for bioerosion, with rates remaining high or declining to prebleaching levels, depending on other factors such as water quality and supply of coral recruits. Clearly, if the growth and survival of coral reefs is to continue to decline under the rapid changes in global climate that are projected for this century, then there will be an increasing proportion of reefs that will be no longer growing and will be in net erosion. How fast a reef matrix can disappear is probably dependent on several factors. Some studies have suggested that accumulated calcium carbonate structures typical of many reefs can disappear quite quickly (see Fig. 9.7).

The additional problem of ocean acidification has arisen from the build-up of carbon dioxide in the atmosphere, as explored later in Chapter 12. In this particular case, roughly 30% of the carbon dioxide that has entered the atmosphere has been absorbed by the ocean. In the ocean, carbon dioxide reacts with water to create a dilute acid called carbonic acid. This acid releases protons that combine with carbonate, converting it to bicarbonate. The net effect is that the carbonate ion concentration has been declining (26% decrease since 1870) and will decline further as carbon dioxide builds up in the atmosphere (see Chapter 12). Decreasing carbonate ion concentrations will decrease the ease with which calcification can occur and will increase the tendency for calcium carbonate crystals to dissolve. This has many people concerned about whether projected increases in carbon dioxide in the atmosphere will tip the balance of coral reefs away from the accumulation of calcium carbonate and towards the erosion of this important resource.

The implications from the fact that reefs are increasingly eroding are serious and may involve a loss of the three-dimensional structure of reefs and the many services that are provided. The structure produced and maintained by corals is important as habitat for many thousands of species worldwide as well as being the 'front line' coastal defence throughout the world. Coral reefs dissipate over 97% of the energy of waves generated by storms and hence are critically important to protecting other ecosystems as well as human communities and infrastructure. The prospect of reef barriers disappearing as they erode in a warm and acidic sea may mean increased exposure of other ecosystems such as mangroves and seagrasses, which generally shelter behind the reef crests from the full force of ocean waves. These changes in wave energy, especially when combined with sea level rise, could also have dire implications for the extensive human infrastructure that often lines tropical coastlines.

Can carbonate coral reefs survive in a rapidly changing world?

In summary, the factors controlling rates and agents of bioerosion are complex and interrelated, as are those controlling reef growth. Superimposed on these factors are location and regional factors. Anthropogenic impacts seriously modify rates and agents of bioerosion, and commonly these are cumulative. For example, a reef can recover from a single mass coral bleaching and mortality event (Chapter 12), given time. However, if such events become more regular and occur with other stresses such as (say) a COTS outbreak or in a circumstance where over-fishing has occurred, reefs are unlikely to recover before being hit by the next heat stress (bleaching) event. At the same time, rates of bioerosion may also increase as dead coral substrate becomes increasingly available. Together, these changes may result in a rapid loss of reef structure, with devastating consequences for ecosystem services such as coastal protection or fisheries habitat.

Further reading

Allemand D, Tambutté É, Zoccola D, Tambutté S (2011) Coral calcification, cells to reefs. In *Coral Reefs: An Ecosystem in Transition*. (Eds Z Dubinsky, N Stambler) pp. 119–150.

Springer, Dordrecht, Netherlands. doi:10.1007/978-94-007-0114-4_9.

Chazottes V, Hutchings P, Osorno A (2017) Impact of an experimental eutrophication on the processes of bioerosion on the reef: One Tree Island, Great Barrier Reef, Australia. *Marine Pollution Bulletin* **118**, 125–130. doi:10.1016/j.marpolbul.2017.02.047

Clements KD, German DP, Piché J, Tribollet A, Choat JH (2016) Integrating ecological roles and trophic diversification on coral reefs: multiple lines of evidence identify parrotfishes as microphages. *Biological Journal of the Linnean Society of London*. doi:10.1111/bij.12914.

Cohen AL, McConnaughey TA (2003) Geochemical perspectives on coral mineralization. *Reviews in Mineralogy and Geochemistry* **54**, 151–187.

Gattuso J-P, Allemand D, Frankignoulle M (1999) Photosynthesis and calcification at cellular, organismal and community levels in coral reefs: a review on interactions and control by carbonate chemistry. *American Zoologist* **3c9**, 160–183. doi:10.1093/icb/39.1.160

Hoegh-Guldberg O, Mumby PJ, Hooten AJ, Steneck RS, Greenfield P, Gomez E, *et al.* (2007) Coral reefs under rapid climate change and ocean acidification. *Science* **318**, 1737–1742. doi:10.1126/science.1152509

Hutchings PA (2011) Bioerosion. In *Encyclopedia of Modern Coral Reefs–Structure, Form and Processes*. (Ed. D Hopley) pp. 139–156. Springer-Verlag, Berlin, Germany.

Hutchings PA, Peyrot-Clausade M, Osnorno A (2005) Influence of land runoff on rates and agents of bioerosion of coral substrates. *Marine Pollution Bulletin* **51**, 438–447. doi:10.1016/j.marpolbul.2004.10.044

Kleypas JA, Langdon C (2006) Coral reefs and changing seawater chemistry. *Coastal and Estuarine Studies* **61**(73), 73–110.

Kleypas JA, McManus J, Menez L (1999) Using environmental data to define reef habitat: where do we draw the line? *American Zoologist* **39**, 146–159. doi:10.1093/icb/39.1.146

Peyrot-Clausade M, Chabanet P, Conand C, Fontaine MF, Letourneur Y, Harmelin-Vivien M (2000) Sea urchin and fish bioerosion on La Reunion and Moorea Reefs. *Bulletin of Marine Science* **66**, 477–485.

Raven J, Caldeira K, Elderfield H, Hoegh-Guldberg O, Liss P, Riebesell U, *et al.* (2005) 'Ocean acidification due to increasing atmospheric carbon dioxide'. Special Report, Royal Society, London, UK.

Schönberg CHL, Fang JKH, Carreiro-Silva M, Tribollet A, Wisshak M (2017) Bioerosion: the other ocean acidification problem. *ICES Journal of Marine Science* **74**, 895–925. doi:10.1093/icesjms/fsw254

Tapanila L, Hutchings PA (2012) Trace fossils as indicators of sedimentary environments. In *Part V– Marine Carbonate Systems. Developments in Sedimentology*. Vol. 24, pp. 751–775. Elsevier, Amsterdam, Netherlands.

SECTION 2

Factors affecting the reef

10

Fisheries of the Great Barrier Reef

A. Chin, D. Cameron and R. Saunders

Introduction

The fisheries of the Great Barrier Reef World Heritage Area (GBRWHA) are diverse and dynamic. They reflect changing environmental, social, economic and regulatory conditions, as well as the stock status of fisheries resources. People have been fishing in the GBR for tens of thousands of years, with Australian Aboriginal and Torres Strait Islander peoples catching a wide variety of fish and shellfish for subsistence and cultural use (Smith 1987; Savage 2003; Cadet-James *et al.* 2017). In the 1800s, the arrival of Europeans in Australia, and later arrival of Pacific Islanders, brought new fishers and ways of fishing to North Queensland (Bolton 1970). Since the 1950s and 60s, fishing in the GBR has evolved into specific fisheries sectors, and fishing and fisheries continue to be important activities and industries that are recognised as some of the main uses of the 'multi-use' Great Barrier Reef Marine Park (GBRMP). Generally, GBR fisheries can be divided into four main sectors: (1) commercial fisheries; (2) recreational fisheries; (3) charter fisheries; and (4) Indigenous fisheries.

All these fisheries have unique traits including different impacts, benefits and values, ways of fishing, and trends in catch and effort. For some coastal communities, fishing is an integral part of their identity, lifestyle and/or livelihood. For Indigenous fishers, fishing may provide important sources of food, and the act of fishing maintains traditional knowledge and culture (Johannes and MacFarlane 1991; Cadet-James *et al.* 2017). It is also important to recognise that fisheries are complex social-ecological systems, where changes in social structures and resilience, management and policy, economics and the environment all affect the way fishing occurs, as well as the benefits it delivers and the impacts it has on other reef users and the environment (e.g. Tobin *et al.* 2010a; Sutton and Tobin 2012). These complex issues are important considerations for successful fisheries management.

The fisheries of the GBR have changed dramatically over the last two decades. Since the mid-1990s significant fishery management reforms have been introduced in the Queensland net, line and trawl fisheries. In 2004, the rezoning of the GBRMP increased spatial restrictions to fishing including increasing the area of 'no-take' marine reserves (Green Zones) from 3% to 33% of the GBRMP. While this initiative focused on biodiversity conservation, it had varying impacts on spatial access for fisheries. During the 2000s, new input and output controls (Box 10.1) were introduced that reduced the number of fishing licences and introduced harvest and effort quotas. Controls were tightened in the coral reef line fishery, including the introduction of total allowable commercial catch (TACC) limits and revised recreational 'in-possession' limits, that is, the total number of fish a person may take or possess at any one time.

As of December 2013, a A$9 million 'buyout' scheme had removed 69 large size mesh nets from

Box 10.1. Input and output controls

Fisheries management generally comprise arrangements that limit the amount and type of fishing effort (input), or the amount of species removed, including consideration of discards (output). **Input controls** include: limited entry licensing or permits to limit the number of fishers; limits on the size and type of fishing gear and vessels used; and limits on fishing effort and spatial and temporal access to the fishery. **Output controls** include: tools such as sex, size and in-possession limits; and catch and discard quotas, which limit the amount of fish that can be taken by a fishery sector or fishery.

the Queensland East Coast Inshore Fin Fish Fishery, removing significant potential fishing effort from the GBR (GBRMPA 2014). Additional commercial fishing effort was also removed in 2015 when the Queensland Government introduced new 'Net Free Fishing Zones', which removed commercial net fishing from specific areas around the cities of Cairns, Mackay and Rockhampton (DAF 2017). These changes in fisheries management arrangements, the GBRMP Zoning Plan and concomitant economic, social and environmental factors need to be considered holistically to understand how and why commercial fishing effort and harvest have declined in the GBR (Hughes *et al.* 2016) (Fig. 10.1).

Although commercial fishing activity has decreased, the number of recreational vessel registrations in Queensland has steadily increased (Fig. 10.1). Vessel registrations are used often as a proxy for recreational fishing effort, but these data have two major drawbacks: (1) they do not include recreational fishing that does not include fishing from a vessel; and (2) they include vessels and water craft that are not used for fishing. More recently recreational fishing participation seems to be declining (Taylor *et al.* 2012), but there needs to be improved information about recreational fishing effort (see 'Recreational Fisheries').

It should also be recognised that the status of fisheries resources of the GBR have changed overtime along with the fisheries. After concerns raised by commercial fishers about overfishing of black teatfish (sea cucumbers), the take of this species was closed in 1999 (GBRMPA 2014; Eriksson and Byrne 2015). Concerns about snapper (*Chrysophrys auratus*), which occurs in the southern GBR being overfished were documented in 2009 (Campbell *et al.* 2009), with no current signs of recovery, while

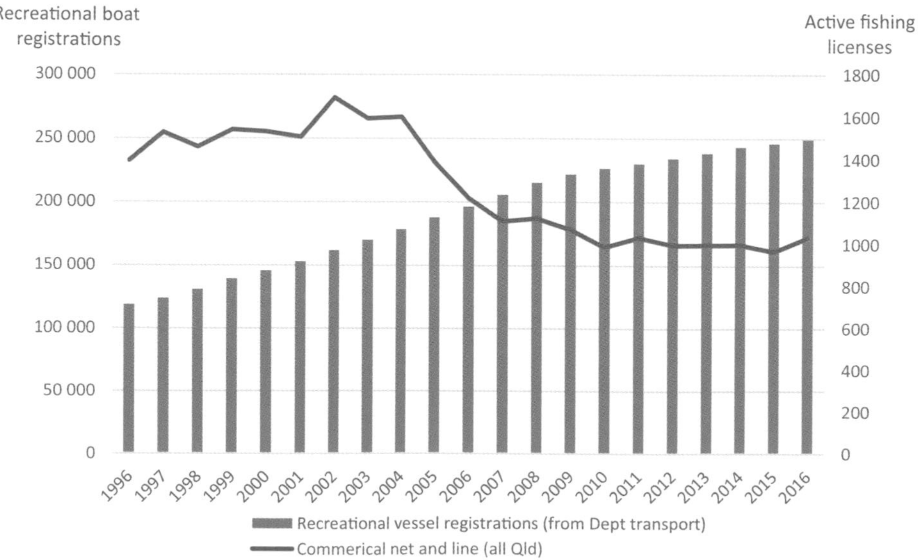

Fig. 10.1. The number of active commercial fishing licences (licences reporting catches) in the Great Barrier Reef has significantly reduced, while recreational boat registrations have consistently increased.

the sympatric pearl perch (*Glaucosoma scapulares*) was classified as transitional depleting in 2015 (DAF 2016). Similarly, average commercial catch rates for the spawning aggregation fishery targeting Spanish mackerel have declined by 90.5%, with the commercial extinction and loss of 70% of exploited spawning aggregation sites (Buckley *et al.* 2017).

Although the GBR has not experienced the extent of fishery declines evident in other tropical reef systems (e.g. Wilkinson 2008), significant issues remain regarding the performance, sustainability and future of GBR fisheries. While these concerns differ between fisheries sectors, some of the main concerns include:

- sustainability of fish stocks (although most stocks are believed to be sustainably fished, there are concerns about the stock status and fishing related trends for some species)
- insufficient monitoring and assessment to determine the status of many fished species
- incidental by-catch of species of conservation interest
- insufficient information regarding sustainability of other by-catch and discards in commercial and recreational fisheries
- illegal fishing and poaching
- impacts from external pressures including climate change, habitat degradation and loss, and pollution
- the need for improved management and recognition of Indigenous fisheries
- resource sharing, especially conflicts between commercial and recreational fishers
- latent effort (unused effort units, symbols and licences) that if activated, could rapidly increase fishing effort.

Overall, fishing in the GBR is rapidly changing due to political, social, economic, biological and other environmental factors. This chapter aims to provide an introduction and overview of the main fishery sectors and their management, and the status of the main targeted fish species.

Commercial fisheries

Commercial fisheries are fisheries that catch species for commercial sale. Before 1979, commercial fishing in Queensland was open access where the number of fishing licences issued was essentially unrestricted. Since then, the number of commercial fishing licences has been capped and gradually reduced to manage excessive fishing effort. The main commercial fisheries species include: prawns and crabs; coral reef fishes such as groupers, snappers and emperors; Spanish mackerel; and coastal fishes including smaller mackerels, barramundi, salmon and sharks. However, there are also harvest fisheries that selectively target marine aquarium fishes and coral, sea cucumbers and tropical rock lobster (crayfish). Fisheries for pearl oyster shell and trochus, which used to be important, are virtually non-existent due to lack of commercial demand and viability.

Commercial fisheries provide important seafood to coastal Queensland communities. Most Queenslanders prefer wild-caught seafood over imported or aquaculture product, and this has significant community and economic benefits (GBRMPA 2014). In the 2011–2012 period, commercial fisheries in the GBR were estimated to contribute A$160 million to the national economy, and to have driven ~975 jobs (GBRMPA 2014).

In Queensland, fishers can only legally sell their catch if they have a 'commercial fisher licence', which identifies them as commercial fishers. They also need a 'commercial fishing boat licence' to operate a vessel to commercially catch fish. Commercial fishers also have a series of 'symbols' attached to their fishing boat licence: these determine the fisheries in which the fisher is authorised to operate. These symbols denote the species that can be taken, the fishing gear that can be used, and the geographical area the fisher can fish (Business Queensland 2017). Collectively, these licences place some control over the amount of fishing effort, because there are only a limited number of licences and symbols available. Currently, no new licences are being issued, and a person who wants to use a commercial fishing boat licence must obtain a licence from an existing fisher (Business Queensland 2017). All commercial fishers are required to record their daily retained catch in a fisheries logbook. They are also required to report and record interactions with protected species such as marine turtles, cetaceans, dugongs, sea snakes, some sharks and sawfishes, and several fish species including Queensland grouper, barramundi cod and humphead Maori wrasse. These species must be immediately released if captured (DAF 2015).

The GBRMP Zoning Plan and marine parks legislation also restrict commercial fishing activities in the marine park. Large-scale commercial fishing such as purse seine fishing and longline fishing are not allowed due to their ecological risk. Since 2000, the Australian Government has required that all Commonwealth (federal) and state managed fisheries that export product, or interact with threatened or migratory species or cetaceans (whales and dolphins) **in Commonwealth waters** must be assessed and accredited under the *Environment Protection and Biodiversity Conservation Act 1999* (EPBC Act).

These approvals are subject to conditions designed to improve sustainability of the fishery. Other legal arrangements and policies (such as CITES – the Convention on International Trade in Endangered Species) can also affect GBR fisheries by requiring changes and/or assessments of fishery catches and practices.

The otter trawl, net and line fisheries are the GBR's largest commercial fisheries in terms of number of operators and harvest levels. In 2016, the combined catch from these fisheries was 6675 tonnes, which represents a significant reduction in catch levels since the 1990s, with declines in fishing activity driven by a range of factors including changes in fishing regulations and marine park management, fuels costs and market factors (Thébaud *et al.* 2014).

Every year, the status of a subset of key commercial fish stocks in Queensland are assessed according to protocols described in the Status of Australian Fish Stocks (SAFS) process (Flood *et al.* 2014). This process has been adopted by fisheries agencies in all jurisdictions to assess and report on the status of key exploited fish stocks in a consistent and easy to understand report. In Queensland, ~80 stocks are assessed over each 2-year period.

In the 2016 process, 65 of Queensland's key fish stocks were assessed. Of these, 41 were classified as sustainable, 16 were undefined due to lack of data and five stocks had negligible catch. Six stocks with sustainability concerns included snapper, mangrove jack (Gulf of Carpentaria), pearl perch, king threadfin (Gulf of Carpentaria), saucer scallops and barramundi (southern Gulf of Carpentaria).

Trawl fisheries

The trawl fisheries of the GBR target prawns, scallops, Moreton Bay bugs and squid. This fishery generates the most revenue of the commercial fisheries, with an estimated valuation (price paid to fishers at point of sale) of A$110 million each year (GBRMPA 2014). There are two types of trawlers: otter trawlers and beam trawlers. Otter trawlers are named for the 'otter boards', which are wooden or steel boards that keep the trawl net open as it moves along or just above the sea bed (see Flood *et al.* 2014, for details of different fishing gear). Almost all the trawlers operating in the GBR are otter trawlers, with 177 active otter trawlers operating in 2016 landing 3636 tonnes of catch. As with other commercial fisheries, the number of active licences (licences reporting catches) have declined (Fig. 10.2) as a result of management interventions as well as economic and environmental factors (GBRMPA 2014).

The harvest rates of prawns appear to be sustainable, with the Status of Key Australian Fish Stocks Report (Flood *et al.* 2014) stating that harvest of banana prawn (*Penaeus merguiensis*), eastern king prawn (*Melicertus plebejus*) and endeavour prawn (*Metapenaeus endeavouri*) is classified as sustainable. Analyses of trawl fishery catch data from Queensland and New South Wales indicates that catches of eastern king prawns in the southern GBRMP may have been sustained in recent years because of reduced fishing effort and catches in northern New South Wales. Harvest of tiger prawns (*P. esculentus, P. semisulcatus*) before 2000 may have exceeded maximum sustainable yield, but stocks appear to have recovered and the current harvest is now considered sustainable (Flood *et al.* 2014).

Recent assessments conclude that the east coast saucer scallop stock (*Amusium balloti*) appears to have collapsed. The 2016 harvest rates were the lowest recorded in 39 years and spawning stock biomass is estimated as being between 5% and 6% of 1977 levels (Yang *et al.* 2016). The fishery is in a state of recruitment overfishing that will require significant management measures to rebuild it, including the option of closing the entire fishery (Yang *et al.* 2016). The causes of the collapse are likely to be complex, and could include a combination of overfishing and environmental factors (Yang *et al.* 2016). A range of management measures are being implemented to reduce fishing pressure on the stock (as discussed later).

Trawl fisheries are managed with input and output controls. Trawl fishing effort is currently capped with a limited number of licences and there is a total effort unit cap for the GBR component of the fishery. However, trawl effort could significantly increase before the GBR effort cap is reached (GBRMPA 2014). Vessels are limited to 20 m length, and net sizes and configurations are also regulated. All nets must be fitted with turtle excluder devices (TEDs) and by-catch reduction devices (BRDs). The sustainability of by-catch of some deep water sharks and rays (Pears *et al.* 2012) and sea snakes in the trawl fishery in the GBR is of ongoing concern (Courtney *et al.* 2010).

Trawlers can operate only in General Use Zones of the GBRMP, and there are additional areas closed to trawling under fisheries legislation for fisheries management purposes. Trawl vessels are fitted with electronic vessel monitoring systems (VMS), which track vessel positions by satellite to monitor compliance with closed areas and provide improved spatial data to manage the fishery. As of 3 January 2017, six scallop replenishment areas have been permanently closed to trawling, and the harvest, retention or possession of scallops in other areas of the East Coast Otter Trawl Fishery has been banned between 1 May and 31 October (Business Queensland 2017). A major research effort to validate the status of the saucer scallop stock is currently in progress to provide information about effectiveness of current management to halt further collapse and to rebuild the fishery.

Line fisheries

Line fisheries in the GBR include the Coral Reef Fin Fish Fishery (CRFFF), which targets coral reef fin fish (coral trout, cods, emperors and tropical snappers), the Spanish mackerel fishery, and the hook and line component of the East Coast Inshore Fin Fish Fishery (which is predominantly a net fishery that targets many species including smaller mackerels, sharks, barramundi and tropical salmons; GBRMPA 2014). The coral reef line fishery contributes approximately A$31 million per year to the

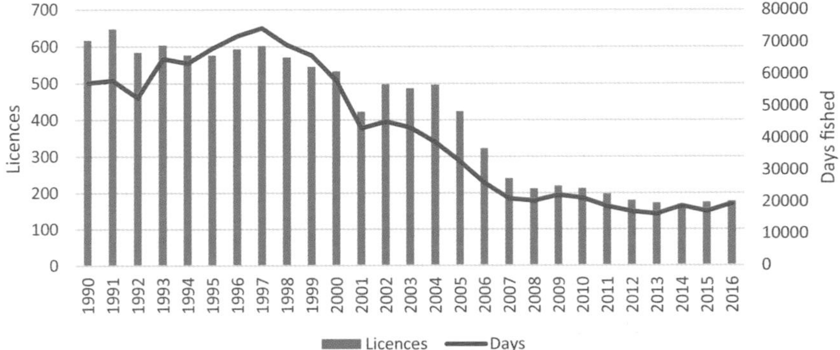

Fig. 10.2. Fisheries, marine park management changes and environmental and market forces have contributed to a significant reduction in active trawling licences and trawling effort in the Great Barrier Reef since 2004.

national economy, and also includes a significant export market, with live coral trout flown to markets in Asia (Thébaud *et al.* 2014).

The line fishery mainly uses baited hooks and lines, including trolling to catch fishes, with 280 active licences in 2016 that landed 1902 tonnes. As with other fisheries, catches have declined from historic high levels (Fig. 10.3) due to management interventions, economic factors and environmental impacts. In particular, cyclones can have a significant effect on coral trout fish behaviour, which makes them more difficult to catch for up to a year after the cyclone. (Tobin *et al.* 2010b; Leigh *et al.* 2014). This causes significant hardship on the fishery, and can change fishing patterns and create conflict as fishers move away from their usual fishing areas into new locations (Tobin *et al.* 2010b; GBRMPA 2014). After severe Cyclones Hamish (2009) and Yasi (2011) affected the central and southern GBR, fishing effort shifted northwards away from affected areas (GBRMPA 2014).

A stock assessment has been completed for common coral trout (*Plectropomus leopardus*), (Leigh *et al.* 2014) the species upon which the commercial fishery largely depends. The stock assessment suggests that the common coral trout population in areas open to fishing is at 60% of 1962 levels (Leigh *et al.* 2014), and catch levels are below estimated maximum sustainable yield (Flood *et al.* 2014). Several other coral trout species (*Plectropomus* spp.; *Variola* spp.) are also taken in the fishery, although to lesser amounts. The 2014 assessment for this complex of trout species is that the current catch is sustainable (Flood *et al.* 2014), but it should be noted that individual stock assessments have not been undertaken for these species. New monitoring programs were initiated in 2017 to provide better information for stock assessments.

Spanish mackerel (*Scomberomorus commerson*) is another key species in line fisheries. The current level of take is assessed to be sustainable, but harvest levels are close to sustainability reference points, and there is evidence that fished spawning aggregations have decreased in area and duration (Flood *et al.* 2014). Potential inflation of catches from fishing spawning aggregations adds considerable uncertainty to the stock assessment, which concludes that 2009 biomass could be at 51% or 39% of virgin (unfished) biomass, depending on which catch data are used (Campbell *et al.* 2012). Furthermore, analysis of historical catches shows that Spanish mackerel catch rates have declined by 90.5% between 1934 and 2011, and also shows the commercial extinction and loss of 70% of mackerel spawning aggregations (Buckley *et al.* 2017). Given the importance of the fishery, the shifted baseline from historical stock levels, and the uncertainty in current stock assessments, this fishery requires government attention to ensure its sustainability.

Many other fin fish species are also targeted in GBR line fisheries. A significant number of these are classified as sustainable stocks (DAF 2016), but many are also classified as 'undefined' because existing data are insufficient to confidently classify the stock. The annual stock status assessment

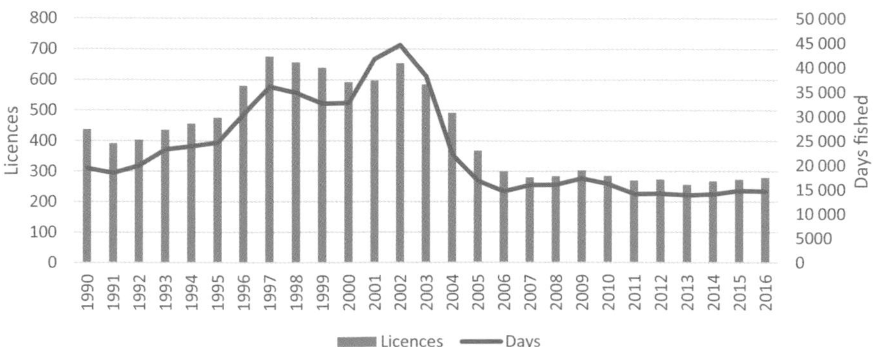

Fig. 10.3. Effort in Great Barrier Reef line fisheries has declined due to a combination of fisheries and Marine Park management changes, social and economic changes, market forces and environmental impacts.

process has classified pearl perch (*Glaucosoma scapulare*) as depleting since 2015, and snapper (*Chrysophrys auratus*) have been classified as overfished since 2010. Both species have recently been subject to formal stock assessments to assist fishery managers to determine the most appropriate management approach to rebuild the stocks.

Commercial fishers need licences and relevant symbols to operate in commercial line fisheries in the GBRMP. They are allowed to fish in General Use (light blue), Habitat Protection (dark blue) and Conservation Park (yellow) Zones of the GBRMP with trolling for pelagic species also allowed in Buffer Zones. All species taken in the commercial line fisheries are managed by size limits with some species also subjected to spawning closures. There are total allowable commercial catch quotas for coral reef fin fishes, Spanish mackerel, grey mackerel, spotted mackerel and sharks.

Net fisheries

Net fisheries in the GBR use mesh nets to target a wide range of species, except coral reef fin fish species for which it illegal to catch in nets. It is important to recognise that these fisheries do not operate in coral reef habitats, although some nets may be set near coastal inshore fringing reefs. This fishery mainly targets barramundi (*Lates calcarifer*), grey mackerel (*Scomberomorus semifasciatus*), sharks and threadfin salmon. The fishery is estimated to contribute over A$19 million to the economy in 2012. This fishery is also more widely dispersed across coastal communities than trawl or line fisheries, with many small-scale operators working out of small towns and communities, and selling product directly to local buyers.

As with other fisheries, fishing effort has declined over time (Fig. 10.4). Reported catches peaked in 2003 at ~2851 tonnes, but have decreased to a lower level ranging from 1907 tonnes (2011) to 1137 tonnes (2016). Barramundi are one of the key target species in this fishery. The status of barramundi is complex, with four separate stocks identified: Princess Charlotte Bay, North East Queensland, Mackay, and the Central East Coast (Flood *et al.* 2014). In some areas, captive-bred barramundi fingerlings introduced into freshwater dams can escape during flood events and mix with wild stocks. This issue, combined with difficulties estimating total recreational take and strong links between barramundi recruitment and environmental conditions, create uncertainty in estimates of barramundi biomass. The current status assessment suggests that harvest levels are sustainable (Flood *et al.* 2014), but the complexity of the stock and importance of the species to commercial and recreational fishers suggest that ongoing attention is needed to monitor sustainability of this iconic species.

There are two stocks of grey mackerel in the GBR that lie north and south of latitude 20°S (Lemos *et al.* 2014). The stocks are currently assessed as sustainable, but there is considerable uncertainty in estimates derived from recent

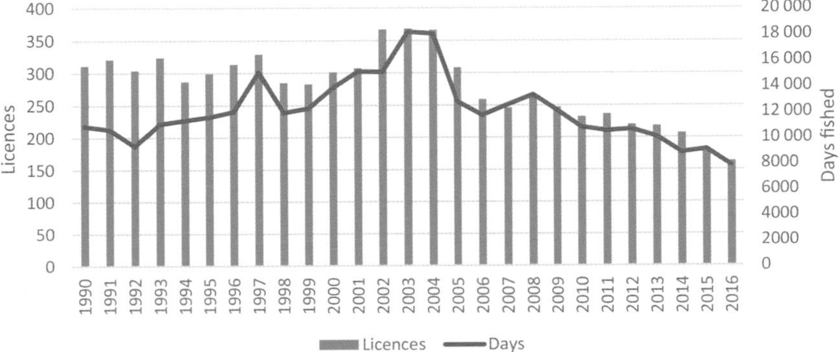

Fig. 10.4. Fishing effort in net fisheries has declined due to a combination of factors including changes in fisheries and Marine Park management changes, social and economic factors, market forces, and environmental impacts that affect the profitability of the fishery.

assessments. This has prompted recommendations for a cautious management approach such as devising separate TACC quotas for the two stocks, and increasing the minimum legal size limit (Lemos *et al.* 2014).

The status of sharks in the fishery is difficult to determine due to a lack of long-term data and uncertainty over the accuracy of species identification in existing datasets. A recent stock assessment suggested that harvest of the complex of key species, including the Australian blacktip shark (*Carcharhinus tilstoni*), common blacktip (*C. limbatus*) and spot tail shark (*C. sorrah*) are sustainable (Leigh 2016). However, the data available are limited and the assessment is compromised by the lack of species-specific catch rates and information about discards (Cortés 2016). Furthermore, other studies suggest that the Australian blacktip shark and pigeye shark (*C. amboinensis*) are potentially at risk of depletion (Harry *et al.* 2016). It is clear that further information is needed to determine the status of sharks in Queensland fisheries, and to develop clear management objectives for the fishery.

Two species of threadfin salmon are taken in the net fishery: the blue salmon (*Eleutheronema tetradactylum*) and the king threadfin salmon (*Polydactylus macrochir*). It is likely that both species comprise separate stocks along the Queensland east coast (Moore *et al.* 2011a,b). There is no stock assessment for blue threadfin, and the last stock assessment for king threadfin was completed in 2002, which suggested a maximum constant yield of 62 tonnes, a level exceeded every year since 1988 (Whybird *et al.* 2016). However, catch rates in the commercial fishery have increased since 1997, and catches have also increased in recreational fisheries despite decreasing effort. Catches may also be linked to rainfall events. These trends provide some evidence that catches are sustainable (Whybird *et al.* 2016). However, king salmon is also highly susceptible to fishing and has relatively low productivity, which has led to its assessment as potentially the teleost species at highest risk in the fishery (Tobin *et al.* 2010c). It is also taken in recreational fisheries for which there are limited data (see 'Recreational fisheries'). Given the species' complex population structures and ecology and the potential risk of overfishing, more information is needed to fully understand their population dynamics and status.

Input controls for the GBR net fisheries include a limit on the number of fishing licences, boat size, the length, depth and mesh size of nets, and several types of area closures including Marine Park 'green zones' and Conservation Park 'yellow zones', Dugong Protection Areas, and since 2016, Net Free Zones brought in by the Queensland Government (GBRMPA 2014; DAF 2017). There is also an annual closed season for barramundi between 1 November and 1 February. Output controls are presently in the form of size limits for key species, and a 1.5 m maximum size limit for all sharks. There is a TACC for grey mackerel (250 tonnes) and sharks (480 tonnes in the GBR, 120 tonnes in South-East Queensland). Net fishers are also generally required to be within 100 m of their offshore set nets to respond to incidental entanglement of species of conservation interest (Queensland Department of Primary Industries and Fisheries 2009).

Other Great Barrier Reef fisheries

Various other fisheries exist in the GBR. There is a large mud crab (*Scylla* spp.) fishery operating in mangroves and coastal areas, which landed 894 tonnes across the Queensland east coast in 2016. Catches peaked between 2011 and 2014, but have since returned to 2005 levels. There are also dive-based harvest fisheries that target sea cucumber, marine aquarium fish and live coral, and tropical rock lobster. These fisheries tend to be very selective and have little by-catch associated with them. However, concerns remain for some species such as sea cucumber, which are easily overfished, and indeed have historically been fished to collapse in the GBR (Eriksson and Byrne 2015).

Current issues in commercial fisheries

Commercial fisheries in Queensland face numerous challenges. The impacts of environmental events including a changing climate, changes in regulations, markets and the economics of the fishery, create a challenging operating environment for commercial fishers. Despite fishery improvements in several fisheries over the last two decades, concerns remain over the existing and potential impacts of the fishery on target stocks, as well as the GBR ecosystem. There are concerns regarding 'latent effort', which is where existing licences are currently not used, but if activated, could increase fishing catch or effort beyond sustainable levels. Large increases in effort can deplete target stocks, increase risks to by-catch species and cause disruption of economic performance of fisheries. In terms of sustainability, pink snapper and scallop are overfished, and there are concerns over pearl perch and sea cucumber (Eriksson and Byrne 2015). There are also concerns about fishing impacts (combined between commercial and recreational fisheries) on spawning aggregations of grey mackerel, black jewfish, golden snapper and barred grunter (*Pomadasys kakaan*) (GBRMPA 2014). Lastly, a large number of other species have yet to be assessed and are currently listed as 'uncertain' or 'undefined' (GBRMPA 2014).

Although the valuable commercial fisheries logbook program has been operational since 1988, concerns remain about the limited spatial precision of data reporting, and the lack of validation of data from this program. Historically, the complexity of some fisheries has meant that much data are pooled together as species groups such as 'tropical shark' or 'other reef species', which makes trends in individual species and populations very difficult to identify, and creates data uncertainty when trying to assess current stocks and evaluate risks. The lack of independent data for contemporary fisheries in the GBRMP across several fisheries also means that the quality of some stock assessments are suboptimal and any stock concerns are not identified in a timely manner. These issues are evidenced by the large number of species assessed as 'uncertain' or 'undefined'. Additionally, while the lack of detailed data creates uncertainty, the lack of a fisheries decision-making system in the form of harvest strategies undermines the ability of fisheries managers and the fishing industry to identify and respond to fisheries challenges in a timely manner. The development of harvest strategies has been identified as a key improvement needed to manage these fisheries (see 'The future of fisheries in the Great Barrier Reef').

The independent Commercial Fishery Observer Program, which provided data to independently validate fishery logbooks, ceased in 2012. This program enabled more detailed examination of the fisheries data and important observations of by-catch (especially in the trawl and net fisheries), including all discards (GBRMPA 2014; MRAG Asia Pacific 2014). Discards are a particular concern because data on discards are not recorded in logbooks and discard mortality of important target species (e.g. undersized and oversized fishes, female crabs) and non-target species is likely to be important for stock status determination and ecological risk assessments. Fishery interactions and incidental death of species of conservation interest (SOCI) continue to be a key environmental issue. SOCI include marine turtles, dolphins, crocodiles, whales, dugongs, sea snakes and sawfishes. Given the conservation status and restricted spatial range of stocks of some of these species, annual mortality of even a few individuals is likely to have significant impact on the viability of some populations (GBRMPA 2014). Interactions with SOCI species are recorded in a separate logbook and data reported to the Commonwealth Department of the Environment and Energy. However, the quality of the data rely on the willingness and capacity of individual fishers to record interactions correctly. The sustainability of by-catch of some deep water sharks and rays (Pears *et al.* 2012) and sea snakes in the trawl fishery in the GBR is of ongoing concern (Courtney *et al.* 2010). The Queensland Sustainable Fishing Strategy also commits to developing risk assessments for SOCI species that are intended to be incorporated in harvest strategies, but this process will take several years and the potential management outcomes are unknown.

Recreational fisheries

Recreational fishing is one of the main uses of the GBR, with ~65% of residents along the adjacent coast fishing in the region (Tobin *et al.* 2014). A 2013 recreational fishing survey found that 642 000 residents went fishing between August 2012 and September 2013, representing 15% of the Queensland population over 5 years old (Webley *et al.* 2015). In the GBR region, these fishers caught a wide range of target species, with the main species including coral trouts and cods (Serranidae), redthroat emperor (*Lethrinus miniatus*), tropical snappers (Lutjanidae), sweetlips (Haemulidae), threadfins, barramundi and bream (*Acanthopargus spp*). Many recreational fishers also took mud crabs. Line fishing with baited hooks was the most commonly used gear type (80% of fishers), followed by fishing with pots (13%). Recreational fishers regularly release fish alive, with almost half of the fish caught later released mostly due to size restrictions, with the highest release rates being for species such as barramundi, snapper, sharks, cod and grouper (Lynch *et al.* 2010; Taylor *et al.* 2012; Webley *et al.* 2015).

Queensland Government recreational fishing surveys indicate that the number of recreational fishers has declined since 2000 in spite of increasing population size. The causes of observed decline in recreational fishing participation is unknown, but could include a lack of time, loss of interest and perceived poor fishing quality, as well as increased competition for recreational time from other activities (Taylor *et al.* 2012). Recreational fishing participation is also declining in other states. There is also evidence that catch rates are declining, which could reflect reduced availability or catchability of fishes, or reduced accessibility to some target species (GBRMPA 2014).

State-wide and GBR-specific data about recreational fishing come from telephone surveys of Queensland residents, and voluntary participation in a 12-month fishing diary program. These surveys provide invaluable data about recreational fishing. However, given the length of time between surveys and the limited number of diarists whose data are used to extrapolate to total catch estimates, improvements in the survey are warranted. Additionally, the telephone survey and associated resident fishing diary program do not collect data on interstate recreational fishers visiting Queensland. In locations such as Lucinda in the central GBR region, fishing effort can increase by 500% in the winter months (Szczecinski 2012). In Lucinda, visiting seasonal fishers target javelin fish (*Pomadasys spp*), which has a higher harvest rate in the recreational fishery than the commercial fishery. This example highlights the need for improved data collection about seasonal recreational fishing.

There are no limits on the number of recreational fishers who can fish in the GBR or how frequently they fish, and there is no requirement for a recreational fishing licence. Marine National Park 'green' zones, Preservation 'pink' zones, and Scientific Research zones in the GBRMP prohibit all recreational fishing, while additional special management areas in some Conservation Park 'yellow' zones further restrict spearfishing. The 2004 zoning plan makes ~65% of the GBRMP available for recreational fishing. Temporal/seasonal fishery closures include two 5-day closures for coral reef fin fish in October and November each year to protect spawning fishes, and an annual 3-month closed season from 1 November to 1 February for barramundi during their peak spawning period. Output controls include size and in-possession limits for most targeted fish species. Nevertheless, more data on recreational fishing catch, effort, and the social and economic dimensions of recreational fishing are needed to adequately understand status and trends in the fishery to inform management. Specifically, these data will be vital in developing resource allocation policies and informing robust stock assessments (State of Queensland 2017).

Indigenous fisheries

There are over 70 Aboriginal groups with traditional connections to the GBR, as well as some Torres Strait Islander groups who have similar connections in the Far Northern GBR (GBRMPA 2014).

Indigenous peoples were the GBR's first fishers and fisheries managers, with connections to the GBR dating back 60 000 years (Savage 2003; GBRMPA 2014). Fishing is a very significant activity, with some communities relying heavily on marine resources for subsistence use. For example, seafood consumption in the Torres Strait is among the highest in the world (Johannes and MacFarlane 1991). However, fishing is also very important culturally. The act of fishing strengthens and maintains community networks, maintains traditional knowledge, and continues traditional practices and lore (Henry and Lyle 2003). Similarly, as Traditional Owners of 'sea country', many communities retain a strong sense of custodianship and connection to their fisheries resources (Henry and Lyle 2003). Indeed, to Indigenous Australians, natural resource management and cultural heritage and practice are a single entity where the use and management of natural resources such as fish are inseparable from cultural responsibility and practice (George *et al.* 2004).

A wide variety of fishing gear is used including mesh nets, fishing lines, hand spears, spear guns and harpoons (Smith 1987). Some communities also use fish traps (Henry and Lyle 2003; Nursey-Bray and Rist 2009), some being traditional stone traps with high cultural heritage value. A wide range of species are caught including archer fish, bream, barramundi, mullet, garfish, javelin fish, some stingrays and sharks, crabs, prawns, and numerous shellfish (Smith 1987; Henry and Lyle 2003; Cadet-James *et al.* 2017). Mullet and barramundi are especially important and are caught with nets (barramundi) and spears (mullet) (Smith 1987).

There is a considerable lack of data regarding the status, significance, values, trends and practices in Indigenous fisheries across Australia, including the GBR. It should be noted that the Indigenous information presented in this chapter stem almost entirely from two studies, one of which (Smith 1987) was limited to two communities. Given the diversity of Indigenous communities, cultures, practices and sea country connections, these data provide only a very limited 'snapshot' of the Indigenous fisheries of the GBR. The lack of information on Indigenous fisheries is concordant with calls for better recognition of Indigenous fisheries as a distinct fisheries sector that should be explicitly considered in fisheries management (Calogeras and Indigenous Reference Group 2012).

Under the *Native Title Act 1993*, Traditional Owners have legislated rights to continue traditional and cultural practices on their land and sea country, including the use of fisheries resources. The key factors determining what activities constitute traditional use are that: (1) the activity is carried out by a Traditional Owner of the land or sea country where the activity is taking place; and (2) that the intent of the activity and use of marine resources are for traditional purposes. As such, a Traditional Owner has the cultural and legal right to use modern fishing equipment to collect fish or harvest resources for traditional use. Under these legal arrangements, Traditional Owners fishing for traditional use are exempted from recreational catch and size limits, and have fishing access to all areas within their sea country.

While cultural practice and lore take precedence, many Traditional Owners acknowledge that the GBR is under increasing pressure and have voluntarily limited their fishing and hunting activities. Some Traditional Owner groups have also developed Traditional Use of Marine Resource Agreements (TUMRAs) that are a legally recognised agreements between the Traditional Owner group and the Queensland and Australian Governments about what activities and harvest can take place in their sea country (GBRMPA 2014). As of 2014, 14 Traditional Owner groups were engaged in formal management arrangements covering ~13% of the GBR (GBRMPA 2014).

The future of fisheries in the Great Barrier Reef

GBR fisheries face numerous challenges in the coming years, and several issues affect all fisheries sectors. Climate change can, and will likely, affect the relative abundance and distribution of several key target species (Welch *et al.* 2014). The ability for fisheries sectors, particularly the commercial sector, to adapt to the effects of a changing climate will be critical to their future viability.

Effective compliance and enforcement are important for all fisheries sectors. There are concerns over non-compliance with zoning requirements in coral reef line fisheries, and with commercial net requirements in the net fishery (GBRMPA 2014). Recreational fishers account for the most frequently reported fishing offences, with an increasing number of reported offences between 2011–12 and 2012–13. Although these increases are likely to be due to increased surveillance and enforcement effort, concerns about non-compliance remain (GBRMPA 2014).

Another key issue in GBR fisheries concerns resource allocation, or the sharing of fish between fishing sectors. This has historically been a high-conflict issue that remains unresolved, and hence developing an equitable and applicable resource allocation policy is a high priority for fisheries managers (State of Queensland 2017). The unresolved debate and conflict regarding resource allocation is exacerbated by insufficient data about the different fisheries sectors (State of Queensland 2017), especially for recreational and Indigenous fishing. Recreational fishing surveys recently commenced at boat ramps will provide validated data complementing that from the important State-wide recreational telephone surveys and fishing diary program. Despite fisheries management and monitoring capacity having been significantly reduced in recent times, the ongoing development of a fisheries monitoring and research plan is a very positive initiative.

In 2014, the Queensland Government commissioned an independent review of Queensland fisheries. This comprehensive review (MRAG Asia Pacific 2014) recommended a suite of improvements to fisheries management including:

- changing governance arrangements to give Fisheries Queensland decision making powers for technical/administrative matters
- developing a clear policy framework with clear operating principles and objectives
- developing harvest strategies for the most important fish stocks
- securing rights to sustainable catches
- improved stakeholder engagement
- a framework for resource sharing
- improved compliance
- improved data collection
- sufficient resources to manage and monitor fisheries.

The Green Paper on fisheries management reform in Queensland (Queensland Department of Primary Industries and Fisheries 2016) outlined challenges for fisheries management and proposed areas of significant reform. Feedback from the Green Paper confirmed strong support from all sectors to implement significant reforms in the way Queensland's fisheries resources are managed.

In response to this feedback, the Queensland Government recently released the Queensland Sustainable Fisheries Strategy 2017–2017. The strategy includes 33 actions across 10 key areas of reform. Some of the key actions include:

- additional monitoring and research programs
- improvements in data validation
- uptake of new monitoring technologies
- stakeholder working groups
- a clear resource allocation policy
- harvest strategies for key fisheries
- additional compliance officers
- satellite tracking of all commercial fishing vessels
- structural adjustment to improve sustainability and profitability of key fisheries.

There is also increased investment in stock assessments and a comprehensive ecological risk assessment program. The program is being supported by an expert panel with expertise in fisheries management, social and economic science, threatened species and stock assessment.

Through these reforms the Sustainable Fishing Strategy aims to tackle many of the issues raised in the review, and is a positive step in reforming Queensland's dynamic and diverse fisheries. Nevertheless, implementing these changes will require adequate resourcing, political will, and strong stakeholder engagement and support. Thus, although the Strategy provides a framework for improving the sustainability of Queensland's fisheries, it needs to be fully implemented to realise its potential.

References

Bolton GC (1970) *A Thousand Miles Away: A History of North Queensland to 1920.* Australian National University Press, Canberra.

Buckley SM, Thurstan RH, Tobin A, Pandolfi JM (2017) Historical spatial reconstruction of a spawning-aggregation fishery. *Conservation Biology* **31**, 1322–1332. doi:10.1111/cobi.12940

Business Queensland (2017) *Commercial Fishing in Queensland.* Vol. 2017. Queensland Government, Brisbane.

Cadet-James Y, James RA, McGinty S, McGregor R (2017) *Gugu Badhun: People of the Valley of Lagoons.* Aboriginal Studies Press, Canberra.

Calogeras C, and Indigenous Reference Group (2012) 'Second Fisheries Research and Development Corporation (FRDC) Indigenous Research Development and Extension (RD&E) Forum'. Report. Fisheries Research and Development Corporation, Canberra.

Campbell AB, O'Neill MF, Sumpton W, Kirkwood J, Wesche S (2009) 'Stock assessment summary of the Queensland snapper fishery (Australia) and management strategies for improving sustainability'. The State of Queensland, Department of Employment, Economic Development and Innovation, Brisbane.

Cortés E (2016) 'Desk Review of Queensland shark stock assessment for Fisheries Queensland'. Working paper for the Queensland Department of Agriculture and Fisheries. Self published. Panama City, Florida, USA.

Campbell A, O'Neill M, Staunton-Smith J, Atfield J, Kirkwood J (2012) 'Stock assessment of the Australian East Coast Spanish mackerel (*Scomberomorus commerson*) fishery'. Department of Agriculture, Fisheries and Forestry, Brisbane.

Courtney, A., Schemel, B., Wallace, R., Campbell, M., Mayer, D., *et al.* (2010) 'Reducing the impact of Queensland's trawl fisheries on protected sea snakes: report to the Fisheries Research and Development Corporation, Project 2005/053'. The State of Queensland/Fisheries Research and Development Corporation, Brisbane.

DAF (2015) *Protected and No-Take Species.* Vol. 2017. Queensland Department of Agriculture and Fisheries, Brisbane.

DAF (2016) *Summary of Stock Status for Queensland Species 2016.* Vol. 2017. Queensland Department of Agriculture and Fisheries, Brisbane.

DAF (2017) Net Free Fishing Zones. Vol. 2017. Queensland Department of Agriculture and Fisheries.

Eriksson H, Byrne M (2015) The sea cucumber fishery in Australia's Great Barrier Reef Marine Park follows global patterns of serial exploitation. *Fish and Fisheries* **16**, 329–341. doi:10.1111/faf.12059

Flood M, Stobutzki I, Andrews J, Ashby C, Begg G, *et al.* (2014) 'Status of key Australian fish stocks reports 2014'. Fisheries Research and Development Corporation, Canberra.

GBRMPA (2014) 'Great Barrier Reef Outlook Report 2014'. Great Barrier Reef Marine Park Authority, Townsville.

George M, Innes J, Ross H (2004) 'Managing sea country together: key issues for developing co-operative management for the Great Barrier Reef World Heritage Area. CRC Reef Research Centre Technical Report No. 50.' CRC Reef Research Centre, Townsville.

Harry AV, Saunders RJ, Smart JJ, Yates PM, Simpfendorfer CA, *et al.* (2016) Assessment of a data-limited, multi-species shark fishery in the Great Barrier Reef Marine Park and south-east Queensland. *Fisheries Research* **177**, 104–115. doi:10.1016/j.fishres.2015.12.008

Henry GW, Lyle JM (Eds) (2003) 'The National Recreational and Indigenous Fishing Survey July 2003'. Australian Government Department of Agriculture, Fisheries and Forestry, Canberra.

Hughes TP, Cameron DS, Chin A, Connolly SR, Day JC, *et al.* (2016) A critique of claims for negative impacts of Marine Protected Areas on fisheries. *Ecological Applications* **26**, 637–641. doi:10.1890/15-0457

Johannes RE, MacFarlane JW (1991) *Traditional Fishing in the Torres Strait Island'.* CSIRO Division of Fisheries, Marine Laboratories, Brisbane.

Leigh G (2016) 'Stock assessment of whaler and hammerhead sharks (Carcharhinidae and Sphyrnidae) in Queensland: Technical Report'. Queensland Department of Agriculture and Fisheries, Brisbane.

Leigh G, Campbell A, Lunow C, O'Neill M (2014) 'Stock assessment of the Queensland east coast common coral trout (*Plectropomus leopardus*) fishery'. Queensland Government, Brisbane.

Lemos R, Wang Y, O'Neill M, Leigh G, Helmke S (2014) 'East Queensland Grey Mackerel Stock Assessment'. Queensland Department of Agriculture, Fisheries and Forestry, Brisbane.

Lynch AMJ, Sutton SG, Simpfendorfer CA (2010) Implications of recreational fishing for elasmobranch conservation in the Great Barrier Reef Marine Park. *Aquatic Conservation* **20**, 312–318. doi:10.1002/aqc.1056

Moore BR, Stapley J, Allsop Q, Newman SJ, Ballagh A, *et al.* (2011a) Stock structure of blue threadfin *Eleutheronema tet-*

radactylum across northern Australia, as indicated by parasites. *Journal of Fish Biology* **78**, 923–936. doi:10.1111/j.1095-8649.2011.02917.x

Moore BR, Welch DJ, Simpfendorfer CA (2011b) Spatial patterns in the demography of a large estuarine teleost: king threadfin, *Polydactylus macrochir*. *Marine and Freshwater Research* **62**, 937–951. doi:10.1071/MF11034

MRAG Asia Pacific (2014) 'Taking stock: modernising fisheries management in Queensland'. MRAG Asia Pacific, Brisbane.

Nursey-Bray M, Rist P (2009) Co-management and protected area management: Achieving effective management of a contested site, lessons from the Great Barrier Reef World Heritage Area (GBRWHA). *Marine Policy* **33**, 118–127. doi:10.1016/j.marpol.2008.05.002

Pears RJ, Morison AK, Jebreen EJ, Dunning MC, Pitcher CR, *et al.* (2012) 'Ecological risk assessment of the East Coast Otter Trawl Fishery in the Great Barrier Reef Marine Park: Technical report'. Great Barrier Reef Marine Park Authority, Townsville.

Queensland Department of Primary Industries and Fisheries (2009) 'Guidelines for commercial operators in the East Coast Inshore Fin Fish Fishery, PR09_4406'. Department of Employment, Economic Development and Innovation, Brisbane.

Queensland Department of Primary Industries and Fisheries (2016) 'Green paper on fisheries management reform in Queensland'. Queensland Government, Brisbane.

Savage H (2003) 'Indigenous connections with the Great Barrier Reef'. Great Barrier Reef Marine Park Authority, Townsville.

Smith AJ (1987) 'An ethnobiological study of the usage of marine resources by two Aboriginal communities on the east coast of Cape York Peninsula, Australia'. James Cook University of North Queensland, Townsville.

State of Queensland (2017) 'Queensland Sustainable Fishing Strategy 2017–2027'. Queensland Department of Agriculture and Fisheries, Brisbane.

Sutton SG, Tobin RC (2012) Social resilience and commercial fishers' responses to management changes in the Great Barrier Reef Marine Park. *Ecology and Society* **17**, 6. doi:10.5751/es-04966-170306

Szczecinski N (2012) Catch susceptibility and life history of barred javelin (*Pomadasys kaakan*) in north eastern Queensland, Australia. James Cook University, Townsville.

Taylor S, Webley J, McInnes K (2012) 'Statewide recreational fishing survey'. State of Queensland, Department of Agriculture, Fisheries and Forestry, Brisbane.

Thébaud O, Innes J, Norman-López A, Slade S, Cameron D, *et al.* (2014) Micro-economic drivers of profitability in an ITQ-managed fishery: an analysis of the Queensland Coral Reef Fin Fish Fishery. *Marine Policy* **43**, 200–207. doi:10.1016/j.marpol.2013.06.001

Tobin A, Schlaff A, Tobin R, Penny A, Ayling A, *et al.* (2010a) 'Adapting to change: minimising uncertainty about the effects of rapidly-changing environmental conditions on the Queensland Coral Reef Fin Fish Fishery'. Final Report to the Fisheries Research and Development Corporation. Project 2008/103. Fishing and Fisheries Research Centre Technical Report No. 11, James Cook University, Townsville.

Tobin A, Schlaff A, Tobin R, Penny A, Ayling A, *et al.* (2010b) 'Adapting to change: minimising uncertainty about the effects of rapidly-changing environmental conditions on the Queensland Coral Reef Fin Fish Fishery. Final Report to the Fisheries Research & Development Corporation, Project 2008/103, Fishing and Fisheries Research Centre Technical Report No. 11'. James Cook University, Townsville.

Tobin AJ, Simpendorfer CA, Mapleston A, Currey L, Harry AJ, *et al.* (2010c) 'A quantitative ecological risk assessment of sharks and fin fish of Great Barrier Reef World Heritage Area inshore waters: a tool for fisheries and marine park managers: identifying species at risk and potential mitigation strategies'. Marine and Tropical Sciences Research Facility, Cairns.

Tobin R, Bohensky E, Curnock M, Goldberg J, Gooch M, *et al.* (2014) 'The Social and Economic Long Term Monitoring Program (SELTMP) 2014, Recreation in the Great Barrier Reef. Report to the National Environmental Research Program'. Reef and Rainforest Research Centre Limited, Cairns.

Webley J, McInnes K, Teixeira D, Lawson A, Quinn R (2015) 'Statewide recreational fishing survey 2013–14'. Department of Agriculture and Fisheries, Brisbane.

Welch DJ, Saunders T, Robins J, Harry A, Johnson J, *et al.* (2014) 'Implications of climate change on fisheries resources of northern Australia. Part 1: Vulnerability assessment and adaptation options: FRDC Project 2010/565'. James Cook University, Townsville.

Whybird O, Newman S, Saunders T (2016) King Threadfin. In *Status of Key Australian Fish Stocks*. Vol. 2017. Fisheries Research and Development Corporation, Canberra, <http://fish.gov.au/>.

Wilkinson C (2008) 'Status of coral reefs of the world: 2008'. Global Coral Reef Monitoring Network and Reef and Rainforest Research Centre, Townsville.

Yang W, Wortmann J, Robins J, Courtney A, O'Neill M, *et al.* (2016) 'Quantitative assessment of the Queensland saucer scallop (*Amusium balloti*) fishery, 2016'. Queensland Department of Agriculture and Fisheries, Brisbane.

11

Disturbances and pressures to coral reefs

M. S. Pratchett and M. O. Hoogenboom

Disturbances play an important role in structuring coral reef ecosystems and associated species assemblages. It is apparent, for example, that the most productive and biodiverse reef communities occur where coral reefs experience high turbulence and periodic disturbances. Although disturbances will, by definition, kill and remove some established reef organisms, this contributes to habitat heterogeneity and biodiversity and will often trigger rapid renewal and redevelopment of species assemblages and habitats. If, however, disturbances occur too often or are too severe, populations and communities will not be able to recover and reassemble completely in the period between successive disturbances, leading to progressive loss of reef species and systematic degradation of reef ecosystems (Chapter 12). Degradation of coral reefs may be further compounded by chronic pressures that undermine the physiological performance of coral reef organisms (Box 11.1). By constraining growth, reproduction and/ or population replenishment, chronic pressures reduce the capacity of reef species to recover effectively following periodic disturbances. In some cases, fundamental changes in the structure of habitats and communities may preclude recovery of formerly dominant species, forever altering the composition and function of reef ecosystems.

Degradation of coral reef ecosystems is commonly measured in terms of declines in the size and abundance of reef-building scleractinian corals, or more specifically the percentage of consolidated substrates that are occupied by scleractinian corals. Such changes in coral cover are readily apparent and easy to measure (Fig. 11.1), but scleractinian corals also have a major influence on the biodiversity, development, structure and productivity of reef systems. Scleractinian corals are the major habitat-forming organisms on coral reefs and the majority of species of reef-associated fishes are reliant on scleractinian corals at some stage of their lives (Chapter 5). Extensive coral loss leads declines in abundance for more than 60% of species of reef fishes, and many reef-associated fishes simply do not occur on coral reefs with <10% coral cover. Declines in the abundance and diversity of both habitat-forming corals and coral-associated species not only impacts on the aesthetics of reef ecosystems (Fig. 11.2), but also undermines reef productivity and biodiversity (Chapter 8). That is to say, there are important feedbacks between loss of corals and declines in abundance of functionally important reef-associated organisms, which further exacerbate the rate and extent of coral reef degradation. Most notably, widespread coral loss in the absence of sufficient grazing activity by herbivorous fishes and invertebrates may result in coral reef habitats becoming overgrown by seaweed, which will in turn inhibit recruitment and recovery by corals.

Coral reefs are among the most threatened natural ecosystems, owing to a long-history of anthropogenic degradation and exploitation, as well as their high susceptibility to emerging effects of global climate change. Globally, at least 19% of coral reefs have already been effectively destroyed, meaning that coral cover is so low (<5%) that these systems no longer effectively function as coral reefs, or the reefs have been physically destroyed by mining or land reclamation. A further 15% of reefs are threatened with similar fate within coming decades. Most of the worst affected reef systems are located in eastern Africa, southern Asia and the Caribbean, which are characterised by having large human populations living adjacent to coral reef environments. However, even relatively isolated reefs that are far removed from most direct anthropogenic pressures increasingly exhibit signs of sustained habitat degradation or marked declines in abundance of specific species, especially apex predators or reef invertebrates that are highly susceptible to exploitation.

Australia's Great Barrier Reef (GBR) is a very large and diverse coral reef system (see Chapters 2 and 5). Accordingly, the disturbances and pressures that affect coral reef ecosystems and associated species assemblages vary enormously along the length and breadth of the GBR. Reefs closest to the mainland, and especially those located near large rivers or urban centres, are particularly exposed to chronic pressures caused by land-based sources of pollutants and direct anthropogenic pressures. Degradation of many inshore reefs preceded scientific exploration and monitoring and coincided with intensification of agriculture and extensive land clearing along the Queensland coast in the early 1900s, which greatly increased the sediment and nutrient output from major rivers. Historical photographs of inshore reef systems exposed at low tide (e.g. Stone Island in 1890 by W. Saville-Kent, Fig. 12.1) clearly show vibrant coral growth in areas where there are now extensive mud flats. For reefs located further offshore, and reefs located north of Cooktown (where very few people live and there has been limited change in land use in catchments), reef degradation is much less apparent. In general, the abundance and diversity of contemporary coral assemblages within comparable reef habitats increases with increasing distance offshore, though these cross-shelf gradients are less pronounced in the far northern section of the GBR. Nonetheless, opportunistic and systematic monitoring of offshore coral reefs along the length of the GBR does reveal sustained declines in mean coral cover since the 1960s (Fig. 11.2). There are, of course, some reefs on the GBR that even today have extraordinarily high coral cover, but these are much less prevalent than they were in previous decades.

This chapter describes some of the major disturbances and pressures that are contributing to sustained and ongoing declines in the health and

Box 11.1. Acute disturbances versus chronic pressures

Documented declines in coral cover and associated degradation of reef ecosystems are commonly attributed to acute (or pulse) disturbances, which cause temporary, but sometimes extreme and widespread, coral loss. Examples include severe tropical storms, mass coral bleaching (heat stress), and outbreaks of coral predators or coral disease. The immediate consequences of such disturbances are readily apparent and easily quantified, though researchers tend to attribute all recent coral mortality to the most recent and apparent disturbance, neglecting to consider persistent background levels of coral mortality. Acute disturbances can also cause fundamental shifts in habitat structure, which have long-term consequences.

Chronic pressures are persistent changes in conditions or actions that impose almost constant, and often widespread, pressure on reefs systems and species. Examples include declining water quality due to excessive inputs of sediment, nutrients and pollutants, as well as the more insidious effects global climate change. Chronic pressures on reef ecosystems do not necessarily lead to significant increases in rates of coral mortality, but may nonetheless contribute to declines in coral cover by constraining coral growth, reproduction and recruitment. The contribution of chronic pressures to sustained and ongoing declines in coral cover is difficult to discern. However, chronic pressures may just as important as acute disturbances (if not more so) in structuring coral assemblages and reef ecosystems.

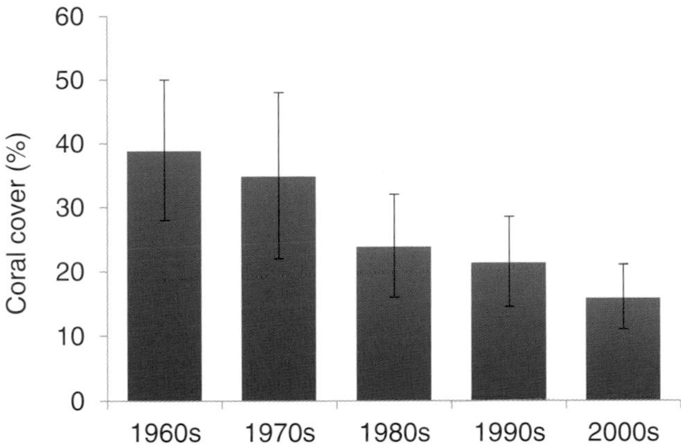

Fig. 11.1. Sustained declines in mean coral cover on Australia's GBR since the 1960s, based on the compilation of routine monitoring data as well as *ad hoc* survey data from published scientific studies.

status of coral reefs, with an explicit focus on the GBR. Many of these disturbances are not new, and coral reef assemblages have previously exhibited remarkable resilience, recovering and reassembling in the aftermath of devastating disturbances. However, there is now unequivocal evidence that the cumulative effects of different disturbances and pressures (many of which are linked to anthropogenic activities) are exceeding the resilience of coral reef systems. Most importantly, increasing ocean temperatures caused by sustained global warming has emerged as the foremost threat to coral reef ecosystems throughout the world. Other more direct anthropogenic disturbances and pressures are also increasing in their spatial extent and intensity, making coral reef species and systems ever more vulnerable to the perennial and natural disturbances that contributed to the formation and structure of modern coral reefs. In addition, there is an increasing array of different ways that anthropogenic activities are impacting on coral reef organisms and ecosystems, which are becoming apparent only due the scale and intensity of the impacts (e.g. increased shipping and associated increases in noise pollution) or are yet to even be recognised or understood.

Major (acute) disturbances

The sustained and ongoing degradation of coral-dominated reefs on the GBR has been the topic of considerable scientific interest and discussion for more than a decade. There is no question that condition of the GBR, already significantly altered by development of the Queensland coastline and modification of adjacent catchments since European settlement, has further declined in recent decades. However, the specific timing and cause(s) of contemporary reef degradation are widely debated. For the most part, sustained declines in coral cover

Fig. 11.2. Marked differences in **live coral** cover (high in photo **A**, low in **B**) are not only due to differences in the recent incidence of major disturbances, but may reflect persistent differences in underlying processes that structure reef systems (Photos: M. S. Pratchett).

since the 1960s are attributed to the cumulative impact of successive outbreaks of crown-of-thorns starfish (COTS). More recently, the extensive depletion of corals by ongoing outbreaks of COTS has been superimposed by unprecedented levels of coral bleaching and subsequent coral mortality. There have also been several other acute disturbances that have certainly contributed coral loss, such as very intense storms or cyclones and associated flood events. However, the relative contributions of these different disturbances to sustained declines in coral cover are difficult to discern. This difficulty arises because it is not just the overall level of coral mortality during one particular disturbance that is important, but the effects of disturbance of one type may affect susceptibility to disturbances of other types in the future, as well as constrain subsequent coral replenishment and recovery. Therefore, it can be erroneous to ascribe distinct episodes of coral loss exclusively to one type of disturbance: disturbances rarely occur in isolation and corals experience relatively high levels of background mortality even in the absence of major disturbances. The specific occurrence and spatial extent of major disturbances are described here, adding to descriptions of the specific effects of different disturbances on assemblages of reefs species in Chapter 5. However, we consider that these disturbances, along with a myriad of chronic pressures (discussed later), each have important contributions to the sustained coral loss and ongoing reef degradation on the GBR and need to be managed accordingly.

Outbreaks of crown-of-thorns starfish

Crown-of-thorns starfish (*Acanthaster* spp.) are one of the largest and most efficient predators on scleractinian corals. Each adult starfish is capable of consuming 5–11 m^2 of live coral each year, depending on their size. COTS occur on reefs throughout the Indo-Pacific, though there are distinct species located in the Red Sea (*Acanthaster* sp. A *nomen nudum*), southern Indian Ocean (*Acanthaster mauritiensis*), the northern Indian Ocean (*Acanthaster planci*) and the Pacific Ocean (*Acanthaster* cf. *solaris*), respectively. Fortunately, *Acanthaster* spp. are normally very rare, but during population outbreaks (Fig. 11.3) their densities may exceed 150 000 starfish/km^2, which equates to more than 10 000 starfish in an area the size of a typical rugby pitch. The combined feeding by high densities of large COTS inevitably leads to rapid and extensive depletion of coral prey. In the last decade, there have been new or renewed outbreaks of *Acanthaster* spp. at many locations throughout the Indo-Pacific, including the Red Sea, Maldives, Indonesia, Guam, the GBR, Vanuatu, Fiji and French Polynesia, contributing significantly to coral depletion and reef degradation.

The first documented outbreak of COTS on the GBR was detected in 1962, although there are anecdotal reports that outbreaks may have also occurred at other times throughout the 1900s. Since the 1960s, there have been three additional waves of outbreaks, starting in ~1979, 1993 and 2010. The initiation and spread of outbreaks of *A.* cf. *solaris* on the GBR have been fairly consistent in all four recorded outbreaks, initiating on midshelf reefs in the north-central region between Lizard Island (14.6°S) and Cairns (17°S). During the first documented waves of outbreaks (in 1962 and 1979), high densities of COTS were first detected (or at least reported) on reefs close to Cairns (e.g. Green Island), although increases in densities of *A.* cf. *solaris* may have occurred even earlier on reefs to the north. In 1993 and 2010, outbreaks were well advanced on reefs near Lizard Island before they were detected off Cairns. The exact timing and location where outbreaks start is critical for understanding and identifying potential causes or triggers of outbreaks on the GBR, suggesting that this area should be monitored intensively whenever initiation of a new wave of outbreaks seems possible.

Periodic outbreaks of COTS are at least partly attributable to their unique life-history characteristics. Most notably, large female *A.* cf. *solaris* are capable of spawning more than one hundred million eggs each year, though their reproductive potential is conditional upon their recent feeding history and, especially, access to preferred prey corals such as *Acropora*. Moreover, fertilisation rates, as well as developmental rates and survivorship of larvae, are subject to the vagaries of local environmental conditions. It is possible, therefore, that outbreaks will periodically arise through the effects of random environmental variation on reproductive success and/or larval survival. It is also apparent that initial outbreaks of COTS may result from a gradual accumulation in starfish numbers over multiple successive recruitment events, rather than a single mass recruitment event. Therefore, outbreaks may arise independently of any sudden or substantial changes in key demographic rates, such that any factor(s) responsible for the initial onset of outbreaks are likely to be very subtle and difficult to detect.

Although *Acanthaster* spp. are intrinsically predisposed to major population fluctuations, there are two factors that may contribute to increasing incidence or severity of outbreaks. First, nutrient enrichment (through runoff of sediments and nutrients from heavily modified catchments) may stimulate phytoplankton blooms that provide increased food for larval starfish, leading to increased rates of larval development, higher survivorship and ultimately more COTS settling on reefs. Also, overfishing of major predators and

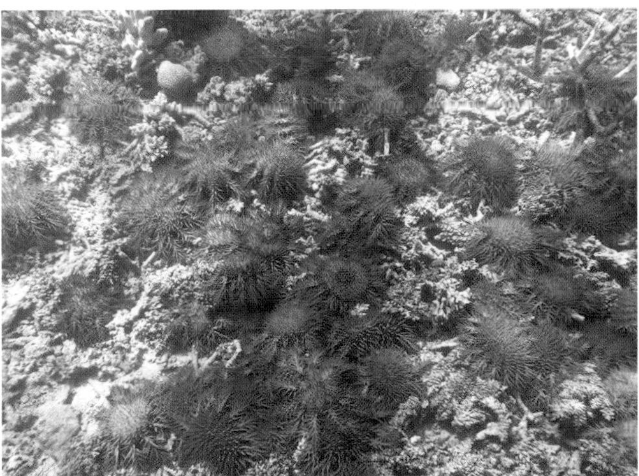

Fig. 11.3. High densities of crown-of-thorns starfish (*A.* cf. *solaris*) persisting on reefs in the northern GBR in 2014–15 even after localised depletion of coral prey (Photo: M. S. Pratchett).

subsequent disruption of the trophic structure within coral reef ecosystems may have reduced rates of predation on adult or juvenile COTS, allowing for higher densities and larger sizes of starfish. Neither nutrient runoff nor overfishing effectively account for outbreaks of *Acanthaster* spp., and redressing these issues is unlikely to entirely prevent outbreaks from occurring. For example, outbreaks of *A. mauritiensis* have recently occurred in remote, unoccupied atolls of the Chagos Archipelago, where there is no agriculture and negligible fisheries for coral reef species. Nevertheless, reductions in land-based inputs of sediment and nutrients and increasing constraints of fisheries exploitation will contribute to increasing the resilience of reef ecosystems (see Box 11.2), even if they do not directly prevent outbreaks of *Acanthaster* spp.

Currently, the most effective way to manage outbreaks of *Acanthaster* spp. is to directly kill or remove individual starfish from reef environments. The efficiency and effectiveness of direct controls was increased significantly by the recent development of a single-shot injection method, whereby starfish can be injected *in* situ with a single 10 mL dose of bile salts solution or other readily accessible acids (e.g. household vinegar). Direct culling of adult crown-of-starfish has since proven to be very effective in protecting limited reef areas, and might be scalable with appropriate investment. However, the best opportunity to prevent future outbreaks may be to concentrate control efforts in areas where outbreaks are known to initiate, thereby preventing reef-wide outbreaks from becoming established. This will require unequivocal evidence of an impending outbreaks and/or timely and appropriate investment in control activities well before the outbreaks become apparent.

Mass coral bleaching

Mass coral bleaching is the most obvious manifestation of changing environmental conditions on coral reefs, due to global climate change. When corals bleach, they lose their rich brown colouration due to declines in concentrations or pigmentation of photosynthetic symbionts (*Symbiodinium* spp.) within their tissues. Bleaching directly undermines the capacity of corals to obtain nutrition, ultimately resulting in the death of the colony if bleaching is too severe or too prolonged (see Chapter 12). Simultaneous bleaching of large numbers of corals of different species, which can occur at very large (near global) scales, is symptomatic of extremely adverse environmental conditions and is most often caused by elevated temperatures. Sustained increases in baseline ocean temperatures are occurring as a direct result of increasing carbon emissions, making it ever more likely that seasonal and inter-annual fluctuations in ocean temperatures will exceed the thermal tolerances and bleaching thresholds of scleractinian corals.

Like most disturbances, mass coral bleaching has selective impacts on different coral species. Whereas some corals are extremely resistant to increasing temperatures, rarely showing any visible signs of coral bleaching, there are other corals for which virtually all colonies turn pale very quickly during extreme temperature anomalies. For the most part, it is branching and erect corals, such as coral in the families Acroporidae and Pocilloporidae, that are the first to bleach and bleach most severely (Fig. 11.4). Given strong taxonomic differences in susceptibility to bleaching, recurrent mass bleaching is expected to drastically alter the composition of coral assemblages. However, the future state of coral communities will depend not only upon the differential bleaching susceptibilities of coral taxa, but also upon their capacity for recovery between successive bleaching events. Fortunately, the corals that experience greatest rates of bleaching and mortality are also rapid colonisers in the aftermath of major disturbances, such that future composition of coral assemblages subject to increasing incidence of coral bleaching is far from certain.

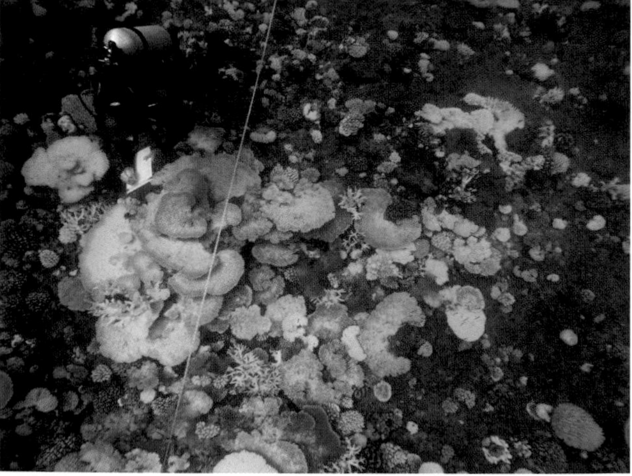

Fig. 11.4. One of the authors (M. Pratchett) surveying the extent of the bleaching at Chinamans Reef in March 2016 (Photo: C. A. Thompson).

On the GBR major instances of coral bleaching have been recorded at regular intervals extending back to 1980, when conspicuous bleaching of common corals (mostly, *Acropora* and *Montipora*) was noted at several reefs between Townsville and Cairns. Elevated temperatures and associated bleaching may have also contributed to dramatic alterations in the composition of coral assemblages on inshore reefs during a pronounced warm period that ended in the 1940s. However, the worst mass-bleaching events on the GBR, coinciding with the highest temperatures recorded in the modern era, occurred in 1998, 2002, 2016, and 2017. In 2016, severe coral bleaching was recorded along a 1000 km tract of reefs in the northern GBR, causing unprecedented levels of coral mortality across reefs that have been otherwise largely spared from any major disturbances for many years. The situation could have also been much worse, if it were not for high cloud cover, wind and rain associated with Ex Tropical Cyclone Winston that is likely to have prevented widespread bleaching in the southern GBR. Given sustained and ongoing increases in ocean temperatures, if corals are unable to acclimatise or adapt, coral bleaching events will become even more frequent and more severe in coming years (see Chapter 12).

Pervasive (chronic) pressures

Research on the causes of contemporary coral loss and reef degradation both in Australia and around the world has been preoccupied with readily apparent effects of major and acute disturbance

Box 11.2. Resilience thresholds and community transitions

In the context of coral reefs, resilience refers to the capacity of reefs to resist disturbances or to return to their original condition, both in terms of abundance and composition of assemblages of reef species, following a disturbance. In many cases, the loss of resilience is not apparent until reef assemblages fail to recover in the aftermath of major disturbances, and persistent shifts in the community composition have already occurred. In Jamaica, for example, formerly coral-dominated habitats were rapidly overgrown by fleshy seaweeds following the die-off of the herbivorous urchins (*Diadema antillarum*) in the 1980s. This seemingly rapid transition was the consequence of numerous factors, including the serial depletion of fishes from higher to lower trophic groups by fishing, which eroded the resilience of this system over preceding decades.

Cumulative effects of different and increasing pressures on coral reef ecosystems effectively reduces their resilience threshold, making undesirable community transitions much more likely to occur. While it is difficult, if not impossible, to stop some of the major disturbances (e.g. cyclones and mass coral bleaching), minimising anthropogenic pressures is a tractable management option that will increase resilience and ensure coral reef systems can withstand more moderate disturbance regimes.

events as described earlier. This is, however, somewhat at odds with the established literature on anthropogenic pressures to coral reef ecosystems, which almost invariably list overfishing, eutrophication and sedimentation as the foremost pressures on coral reef ecosystems. This in part reflects broad recognition of the emerging importance of environmental change in structuring and threatening coral reef ecosystems, but perennial and emerging anthropogenic pressures remain a major contributor to the degradation of coral reef ecosystems. Anthropogenic pressures are also increasing in extent, severity and diversity. There are also important interactions and synergies between acute disturbances and chronic pressures (Box 11.2) that must be considered to manage coral reef ecosystems effectively.

Fishing and harvesting

Exploitation of marine species, mostly for food, is one of the most pervasive and direct anthropogenic impacts across all marine systems. Even relatively light fishing pressure can have significant consequences for long-lived, slow-growing or highly vulnerable reef species. Many reef sharks, for example, are now threatened due to widespread exploitation to meet marked increases in demand (mostly in Asia) for shark-fin since the 1980s. More generally, coral reef fisheries are fully or overexploited in many tropical island nations due to critical reliance on coastal fisheries for food and livelihoods, combined with burgeoning human populations and increased use of destructive or industrialised fishing methods. Even in areas with relative moderate human densities and vast coral reef resources, projected increases in fisheries demand over coming decades cannot be sustainably met from coral reef ecosystems.

The major fisheries on the GBR include: (1) commercial hook and line fisheries as well as charter fishing operators that target coral trout and other carnivorous reef fishes; (2) commercial hand collection fisheries for lobster, sea cucumbers (bêche-de-mer) and aquarium specimens; (3) commercial trawl fisheries that operate in inter-reef areas; and (4) highly diverse recreational fishing activities. These fisheries are extensively regulated through a combination of catch and effort controls, as well as the designation of marine protected areas (MPAs) that restrict the spatial extent and distribution of different fishing activities (Chapter 10). Previous unsustainable fisheries that decimated populations of heavily harvested or highly vulnerable species (e.g. trumpet snails and marine turtles) have been abolished. However, there are current concerns about the ecosystem impacts of recreational fisheries, as well as sustained harvesting of reef sharks.

Recreational fishers are individually restricted in terms of both fishing effort (e.g. limits on the number of lines and hooks per person) and catch controls (e.g. protected species, bag limits and size limits). However, there is no practicable instrument (e.g. recreational fishing licences) to effectively constrain overall fishing effort.

Given increasing awareness of the functional or ecological importance of specific reef species, there are emerging conflicts between the need to balance increasing demand for fishes against the pressing conservation needs for protecting imperilled coastal ecosystems. Large predators, such as reef sharks, have an important role in structuring the trophodynamics (or the flow of energy through trophic pathways, Chapter 8) of reef ecosystems and are also extremely vulnerable to fishing. There are suggestions that functionally intact predatory assemblages ultimately benefit reef corals, though the various trophic linkages involved are extremely complex and poorly understood. Densities of reef sharks are already severely suppressed in areas of the GBR that are open to fishing, and there are concerns that this might compromise the health and resilience of these systems. Other reef fishes have much more direct and apparent effects on the function and condition of reef ecosystems. Herbivorous fishes, for example, limit the growth and abundance of seaweeds, which is shown to be critically important in facilitating recovery of coral assemblages in the aftermath of major disturbances. Fortunately, there is not yet any significant demand or established fisheries for herbivorous fishes on the GBR. However, herbivorous fishes are the mainstay of coral reef fisheries in many other countries, where larger carnivorous reef fishes have been already over-exploited. Australia should therefore, capitalise on current opportunities to legislate fisheries protection for herbivorous fishes. For more details regarding fisheries on the GBR, see Chapter 10.

Altered seawater chemistry

The increasing concentration of carbon dioxide (CO_2) in the atmosphere is not only contributing to global warming, but is also causing increasing concentrations of CO_2 in the ocean. As a result, the acidity of the oceans, and the levels of different forms of inorganic carbon present in sea water, are changing (Chapters 9 and 12). These changes can be beneficial for some organisms, including algae and seagrass for which rates of photosynthesis (carbon fixation) can increase when there is more inorganic carbon present in sea water. For marine organisms that build limestone (calcium carbonate) skeletons and shells, however, changing sea water chemistry, and specifically declines in the amount of dissolved aragonite, can impede calcification. Slower calcification can translate into slower growth, although corals can build skeletons that have a lower density of limestone in order to maintain the same growth rate using less material. Increased seawater CO_2 can also have negative effects on coral reproduction, including suppression of gamete fertilisation and larval metabolism, lowered settlement and metamorphosis, and smaller sized juveniles. Nevertheless, responses of different species to increased seawater CO_2 concentrations in experiments have been highly variable, and the overall effects of altered seawater chemistry on coral assemblages and ecosystem processes are poorly understood.

Predicting the effects of ocean acidification on coral assemblages and reef ecosystem functioning is challenging because not only are the direct effects of increased CO_2 highly variable among species, but there are also indirect effects that can alter assemblage structure. For example, the outcome of competition between pairs of coral species can be altered under experimental conditions with increased CO_2. Understanding how such changes will affect coral community structure requires assessment of whether competitive interactions will become more or less frequent in a high CO_2 world. If coral cover declines then the frequency of competitive interactions is likely to decrease, but if the availability of substratum suitable for coral larval settlement declines then competition could become more frequent. The behaviour of fishes that interact with coral colonies can be altered when seawater CO_2 concentrations increase, meaning that both the positive (e.g. nutrient provision) and negative (e.g. predation) effects of fish on corals might also change in a high CO_2 world. Finally, large stands of some species of seaweeds, which are generally considered to have negative effects on coral populations due to competition and recruitment inhibition, could potentially buffer effects of increased CO_2 on corals because the high photosynthesis rates of these algae draws inorganic carbon from sea water and this could restore normal seawater chemistry within the immediate vicinity of the algae. For the GBR, as with other reef regions globally, complex interactions within and among species, together with interactions between CO_2 concentrations and other environmental variables, will determine how ocean acidification influences species assemblages in the coming decades.

Particulate pollutants

Development and industry along the coastal fringe, and within catchment basins of coastal rivers, can lead to the input of several particulate pollutants into coral reef waters. Terrestrial sediment is the most widely studied of these particulates (see Chapter 13). However, other particulate contaminants can also make their way into the marine environment from land. The world's two largest exporters of coal, Indonesia and Australia, are home to extensive coral reefs. Coal exports are transported by ship, meaning that coal is stored and loaded at coastal port facilities, and transported along shipping lanes that are close to coral reefs and other sensitive coastal ecosystems. On the GBR, port expansions are projected to result in a substantial increase in coal shipments over the next decade. Recent research has revealed that small coal particles can cause mortality of coral tissues (Fig. 11.5), and also slow the growth rates of fish and seagrass. Similarly, small coal particles impede fertilisation of coral gametes, and can lower rates of embryo survivorship, whereas juvenile corals that have metamorphosed were not strongly affected. Such effects seem to be due primarily to physical processes, such as abrasion of tissues by coal particles or adhesion of particles to tissues, rather than due to chemical effects of the leaching of toxins from coal.

Plastic debris is a contaminant of emerging interest in coastal environments worldwide. Millions of tonnes of plastic waste enter the ocean on an annual basis, and this waste can be widely dispersed due to winds and currents. On the GBR, plastic debris levels are generally low compared with areas in South-East Asia where tens of thousands of plastic

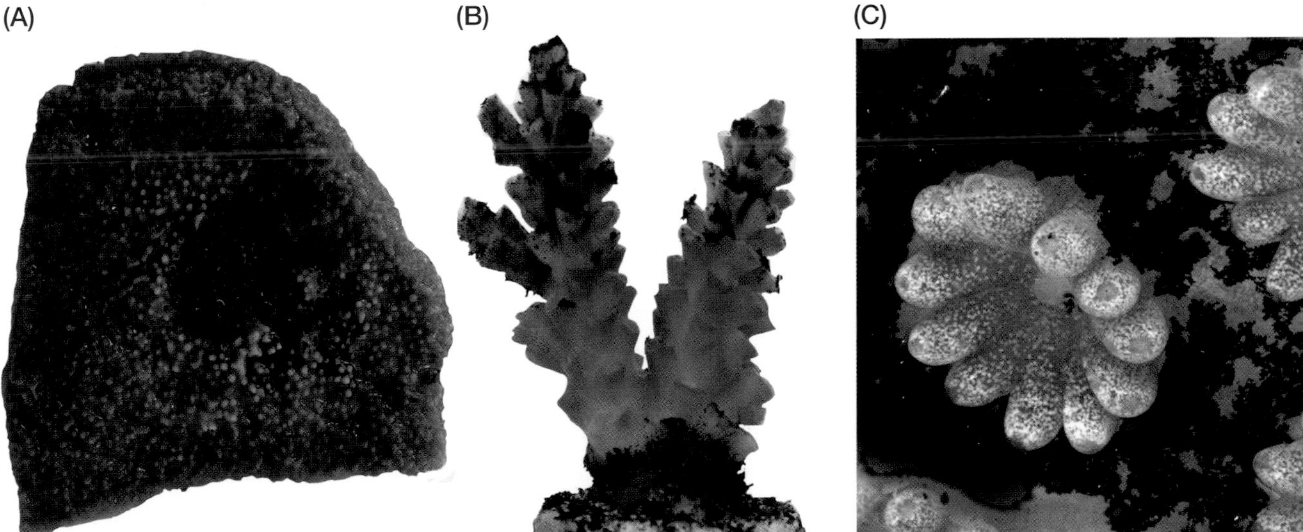

Fig. 11.5. Small coal particles: **(A)** smothering coral tissue causing tissue paling; **(B)** adhered to coral skeleton after loss of coral tissue on branch tips; and **(C)** trapped on the tissue surface between coral polyps soon after deposition onto the tissue (Photos: K. L. E. Berry).

items can be found within 500–1000 m of beachfront. Nevertheless, plastic debris does accumulate on GBR coastal beaches, with accumulation influenced by various processes including local tides and currents, and the orientation of beaches relative to prevailing winds. Over time, large plastic items become coated in a film of microbes and sessile invertebrates causing them to sink to the sea floor. On coral reefs, sunken plastic debris can become entangled around coral colonies, leading to smothering and tissue abrasion. These injuries to tissues can also increase their susceptibility to pathogens leading to an increased incidence of coral diseases. Eventually, plastic items break down into tiny pieces called 'microplastics' (usually defined as plastic fragments <2 mm diameter). On the GBR, microplastics have been detected in relatively low concentrations in water and plankton samples, and have also been found in beach sands and river sediments. Small microplastics can be similar in size to the normal prey of planktivorous fishes and corals. Research on whether microplastic ingestion has negative impacts on coral and fish health is in progress.

Dissolved pollutants

There is a long list of materials, including heavy metals, herbicides, pesticides and nutrients, that can cause harm to stony corals. Heavy metals, including copper, lead, zinc and nickel, can impede fertilisation of coral gametes and can also prevent coral larvae from metamorphosing from their free-swimming to reef-attached forms. These effects can lower coral recruitment success and slow rates of reef recovery after acute disturbances. Herbicides affect corals by inhibiting the photosynthesis by the symbionts present within coral tissue. Herbicides have negative effects on both adult and juvenile corals, with effects on juveniles most pronounced for species that contain symbionts within their larvae. Pesticides, can also reduce the survivorship of coral larvae, although effects of pesticides appear to be less severe, and less consistent, than the effects of herbicides.

Increased levels of nutrients, including nitrogen and phosphorus, can occur in coral reef waters due to terrestrial runoff. At very high concentrations, nutrients can have direct negative effects on tissues, particularly when nitrogen concentrations are high relative to phosphorus concentrations or vice versa. Nutrients can also have negative effects on corals by destabilising the symbiosis between corals and *Symbiodinium*. For example, when nitrogen availability is high, this can lead to symbionts retaining energy from photosynthesis and using it for symbiont cell division instead of transferring that energy to the coral host. Nutrients can also be detrimental to coral populations by promoting the growth of macroalgae that compete with corals for space on the reef.

Actions required to redress reef degradation

Climate change, and specifically ocean warming, is now the foremost threat to coral reefs throughout the world. Urgent action is, therefore, required to minimise carbon emissions and reduce longer term and devastating effects of climate change on coral reef ecosystems. A critical step in this process is to limit the extraction and export of coal, and rapidly transition to sustainable energy production. This will have the added benefit of reducing the risk and occurrence of pollutants associated with coal transport and exports. Most importantly, however, reductions in global carbon emissions will prevent extreme changes in environmental conditions and reduce rates of change to which species must acclimate and adapt in order to persist. Local management must also be focused on minimising all other disturbances and pressures that exacerbate vulnerability to climate change, or undermine the adaptive capacity of coral assemblages.

Reef restoration

Many coral reefs across the GBR are currently in very poor condition, owing to unprecedented bleaching in the northern sections in 2016 and 2017, as well devastating effects of Cyclone Debbie that passed over the central GBR and Whitsundays in March 2017. This has renewed calls for active intervention and restoration of severely degraded habitats. Many restoration projects have been conducted on coral reefs around the world, with the aim of restoring damaged or degraded reef habitats. Most of these projects simply involve transplanting coral species from relatively unaffected locations or habitats to damaged or degraded habitats, sometimes with fragmentation and *ex situ* grow out between collection and redeployment. There are inherent constraints on the types of corals that are used in restoration projects, such that restored habitats rarely match the species composition and diversity of the pre-disturbance coral assemblages. Effective coral restoration programs are also incredibly costly and very labour-intensive, which greatly constrains the area of reef habitat that can be practicably and feasibly restored. As long as there is a reasonable (even if slightly suppressed) natural supply of larvae, and the environmental conditions and state of the substrate will not overly constrain the settlement and subsequent survival of new corals, then replanting of corals (especially those harvested from other nearby habitats) is somewhat futile. There are also few instances where habitat modification (e.g. active removal of established seaweeds) will have any practicable and lasting benefits for the recovery of coral assemblages.

The management and conservation focus for the GBR, and coral reefs globally, must be firmly on reducing the disturbances that cause acute coral mortality, as well as anthropogenic pressures, that impact on coral health, replenishment and resilience. Only when or where disturbances and pressures can be effectively contained is it justified to invest in large-scale restoration of reef habitats and coral assemblages. Even then, careful consideration needs to be given to specific local limits or existing constraints on coral settlement, growth or survival, with focus on restoring, rather than necessarily replacing or supplementing, natural recovery processes. In some instances, there is significant appetite and necessary investment to accelerate recovery and restoration in small reef areas (e.g. immediately adjacent to established tourism infrastructure). These small-scale restoration programs are potentially viable and effective, but will require recurrent investment following each successive disturbance. Other projects that aim to engineer increased resistance and resilience among local coral populations and communities will contribute to increasing understanding of coral physiology and adaptation, but there are still questions about the scalability and practicability of actually restoring coral assemblages over large areas (e.g. along the 2200 km length of the GBR) in the face of environmental change and enhanced disturbance regimes. Moreover, corals that are resistant to specific pressures and disturbances (e.g. ocean warming) will still be vulnerable to other recurrent disturbances (e.g. outbreaks of coral predators), requiring comprehensive consideration and management of all disturbances and pressures.

Further reading

Bell JD, Ganachaud A, Gehrke PC, Griffiths SP, Hobday AJ, Hoegh-Guldberg O, *et al.* (2013) Mixed responses of tropical Pacific fisheries and aquaculture to climate change. *Nature Climate Change* **3**(6), 591–599. doi:10.1038/nclimate1838

Berry KLE, Hoogenboom MO, Flores F, Negri AP (2016) Simulated coal spill causes mortality and growth inhibition in tropical marine organisms. *Scientific Reports* **6**, 25894. doi:10.1038/srep25894

Birkeland C (Ed.) (1997) *The Life and Death of Coral Reefs*. Chapman and Hall, New York, USA.

Hughes TP, Day JC, Brodie J (2015) Securing the future of the Great Barrier Reef. *Nature Climate Change* **5**(6), 508–511. doi:10.1038/nclimate2604

Hughes TP, Kerry JT, Alvarez-Noriega M, Alvarez-Romero JC, Anderson KD, Baird AH, *et al.* (2017) Global warming and recurrent mass bleaching of corals. *Nature* **543**, 373–377. doi:10.1038/nature21707

Negri A, Vollhardt C, Humphrey C, Heyward A, Jones R, Eaglesham G, *et al.* (2005) Effects of the herbicide diuron on the early life history stages of coral. *Marine Pollution Bulletin* **51**(1–4), 370–383. doi:10.1016/j.marpolbul.2004.10.053

Pratchett MS, Caballes C, Rivera-Posada JA, Sweatman HPA (2014) Limits to understanding and managing outbreaks of crown-of-thorns starfish (*Acanthaster* spp.). *Oceanography and Marine Biology - an Annual Review* **52**, 133–200. doi:10.1201/b17143-4

Reichelt-Brushett AJ, Harrison PL (2005) The effect of selected trace metals on the fertilisation success of several scleractinian coral species. *Coral Reefs* **24**(4), 524–534. doi:10.1007/s00338-005-0013-5

12
Coral reefs in a changing world

O. Hoegh-Guldberg

Introduction

Coral reefs are extraordinary in terms of their breathtaking beauty and biodiversity, as well as having enormous value in terms of the ecological goods and services that they provide to hundreds of millions of people worldwide. In terms of the GBR, the direct economic benefits were recently valued at A$6 billion per year, representing employment for 69 000 people. It should be noted, that many of these economic estimates of the value of coral reefs to local economies are likely to be conservative, given that it is difficult to calculate the complete economic value on many of the contributions coral reefs make to people and their communities (e.g. what is the dollar value of clean beach sands, or the spiritual and cultural services that they provide?).

Despite their importance to humans, however, coral reefs are being heavily impacted by rapidly expanding human activities. These human impacts have steadily grown over time and are tied to changes such as rapidly increasing coastal populations, rising *per capita* consumption, pollution and unsustainable resource extraction. These activities have expanded dramatically over the last 50 years and human activities are now affecting every corner of the planet. Understanding how humans are interacting with coral reefs at global, as well as local, scales is becoming increasingly important. This is especially urgent as recent estimates suggest that the current rates of environmental change in the ocean exceed any seen in the last 65 million years at least. How we navigate these extraordinary times is very much the defining challenge for humankind over the coming decades and century.

In this chapter, we explore the changes that are occurring in the world's oceans, giving context to a discussion of how the world's most biologically diverse marine ecosystem, coral reefs, will survive over the coming decades and centuries of change. This discussion will review the trajectory of reef-building corals on the GBR, especially the loss of 50% of its corals over recent years (especially with respect to the thermal stress events of 2015–2017). These changes are extraordinary in the context of the long-term stability of regions such as the GBR over thousands of years. As will be developed here, the drivers of change on the GBR are complex and various, and are not a simple consequence of a single challenge.

Global change and the ocean

The ocean has a profound influence on our planet, interacting with the atmosphere, cryosphere, land and biosphere, as well as contributing to human welfare through the provision of resources, cultural, economic and livelihood benefits. The ocean also contributes through the regulation of atmospheric gas content, and the distribution of heat and water across the Earth, and hence the weather. In addition to these services, the ocean is an important source of protein for humans, providing 20% of the protein needs of 3 billion people across the planet. Coral reefs alone are estimated to provide significant amounts of food and livelihoods to 500 million people across the tropics and subtropics.

The original interactions of humans with ocean resources appear to have left few lasting signs of exploitation. During this time, humans exploited marine life at local levels, steadily improving boating and fishing technologies throughout the 10 000 years of the Holocene. Evidence of the impact of humans on coastal areas are seen in the changing composition and size distribution of shellfish in middens, tending to smaller sizes over time as resources were exploited. Over time, more challenging and substantial prey items were exploited through improvements in technology, until the advent of the Industrial Revolution where the energy revolution enabled fishing efforts to expand dramatically. From the 1950s onwards, fishing began to exploit most coastal and ocean resources. Powerful industrialised technologies such as globally operating fishing fleets together with refrigeration at sea collapsed the biomass of most fisheries by 80% within 15 years after the start of exploitation. Importantly, fishing increasingly not only impacted the organisms fished, but also began to affect entire ecosystems. These types of changes also impacted coral reefs, with the loss of fish stocks driving ecological transitions away from coral-dominated ecosystems. Targeting herbivorous fish in some regions, for example, decreased the herbivore controls on the organisms (i.e. seaweeds) that compete with corals for space, helping drive coral reefs towards seaweed dominated states.

There is little doubt that over-harvesting of fisheries is a major global challenge, for cold water as well as warm water coral reefs, and for many other ecosystems. In today's world, where we have over 7 billion people aspiring to consume at high levels, the global pressure on marine resources has become unsustainable. In addition to the rise of international fishing fleets, artisanal or local fishing has also expanded along coastal areas. Again, population growth and consumption patterns are major drivers of ecosystem change. Given that the population issue is often not articulated, yet is critically important to all of this, the population driver is often referred to as the 'elephant in the room'. Although we do not explicitly explore it here, most of the issues that we are discussing in this chapter can be traced back to the rapid increase in global population. The challenge will be how we develop the sustainable use of marine resources such as coral reefs while at the same time, assisting humanity in tackling poverty and disadvantage.

Human impacts on coral reefs involve a range of activities and influences. Changes to coastal land use (e.g. converting land from forests to agricultural use) has had a profound impact on the amount of sediment and nutrients flowing down river catchments and into coastal waters. In the case of Australia, the amount of sediment flowing down rivers and into the GBR region increased by 20-fold for some river catchments a few decades after the introduction of European farming methods (e.g. deforestation, hard-hoofed cattle) in the mid-19th century. As discussed in Chapters 8 and 13, coral reefs are sensitive to changes in the sediment and nutrient content of coastal waters. The paired photographs from the Queensland coastline (Fig. 12.1), taken roughly 100 years apart, illustrate the point that changes have been occurring to coral reefs within the GBR region for some time. In addition to sediments and nutrients, pollutants such as heavy metals, pesticides and antifouling compounds have also had a toll on inshore coral reefs and other ecosystems such as mangroves and sea grass beds within the GBR region.

Although sediment restricts light and smothers reef-building corals, inorganic nutrients such as ammonium and phosphate can trigger phytoplankton blooms that can improve the survival of ecologically influential species such as the crown-of-thorns starfish (COTS, *Acanthaster cf. solaris*), as discussed in Chapters 8 and 11. Individual COTS produce hundreds of millions of larvae each year, which suffer high mortality rates due to the normally low phytoplankton content (food) of

Fig. 12.1. Coastal reef fringing Stone Island near Bowen on the Great Barrier Reef, photographed by (A) Saville-Kent in 1893 and (B) A. Elliot in 1994 (Source: D. Wachenfeld, GBRMPA).

the normally clear coral reef waters. Consequently, even small changes in the availability of phytoplankton that lead to a fraction of a per cent in terms of added COTS survivorship can have a dramatic impact on the number of COTS in a particular region. Evidence for the role of elevated nutrients in COTS outbreaks has come from several sources including the observation that COTS outbreaks tend to a few years after major flood events, which boost local phytoplankton productivity (see Chapter 11). It should be recognised, however, that alternative hypotheses, such as the over-fishing of key predators such as the giant triton snail (*Charonia tritonis*) leading to outbreaks, have received some support from the scientific community over the years.

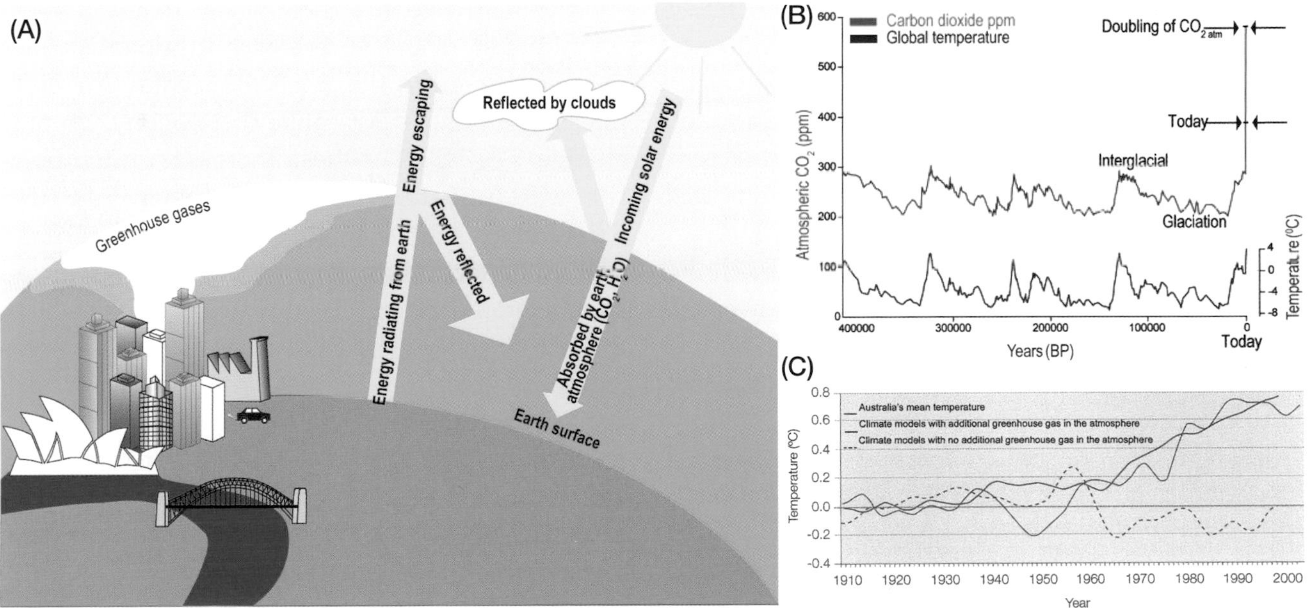

Fig. 12.2. (A) Key steps in the enhanced greenhouse effect (Figure: D. Kleine and O. Hoegh-Guldberg). (B) Global temperature and concentrations of carbon dioxide stretching back 400 000 years. (Source: Data derived from Petit *et al.* (1999). (C) Data illustrating projections of temperature when greenhouse gases are (upper black line) and are not (lower dotted black line) included in the calculations as compared with the mean temperature of Australia over the past 90 years (Source: Adapted from Karoly and Braganza 2005).

In the early 1980s, a new phenomenon began to affect coral reefs. Termed mass coral bleaching and mortality, sizeable sections of reefs in the number of parts of the world began to turn white and, in some cases, die. Investigation revealed that the symbiotic relationship between corals and their dinoflagellate symbionts (*Symbiodinium*) had broken down, with the brown symbionts leaving the tissues of their coral hosts *en masse*. Although this was a curiosity at the time, coral bleaching is now considered to be the most serious threat to coral reefs such as the GBR. In order to put this important problem into context, it is necessary to explore global climate change and its effect on the oceans.

The enhanced greenhouse effect

Radiation from the sun heats terrestrial, oceanic and atmospheric components of the Earth. About 30% of incoming energy is eventually reradiated back into space, with the remaining 70% driving an average global temperature of the Earth of around +14°C. The trapping of heat by the atmosphere is largely responsible for life on planet Earth. If there was no greenhouse effect, the average Earth's temperature would be –18°C. The term 'greenhouse effect' is used because of the similarities between the trapping of heat in the Earth's atmosphere to that seen in garden greenhouses. In the latter case, visible radiation from the sun passes through the glass panels of the greenhouse, warming interior surfaces that in turn re-radiate long wavelength infrared radiation. Infrared wavelengths, however, do not pass through the glass as efficiently as the incoming radiation, so heat begins to accumulate in the greenhouse.

In a similar way, solar energy enters the Earth's atmosphere and warms the air, rocks, ocean and other components. It too re-radiates infrared radiation, which is trapped by the atmosphere (Fig. 12.2A). Most (75%) of this trapping is done by the so-called greenhouse gases, principally carbon dioxide, water vapour and ozone, methane, nitrous oxide and chlorofluorocarbons (CFCs). Although these components are present only in trace amounts in the atmosphere, they have a huge impact on the heat budget of the Earth.

Life would not be possible without the warming effect of the atmosphere, so it is actually incorrect to say that the greenhouse effect is itself a problem for life on Earth. The problem that we currently face is associated with a rapid increase in the concentration of the greenhouse gases, a phenomenon referred to as the 'enhanced greenhouse effect'. The greater the concentration of greenhouse gases, the greater the amount of heat retained by the Earth. The increase is due to the burning of fossil fuels and other activities such as deforestation and agriculture, all of which are activities that release carbon dioxide (CO_2) and other greenhouse gases into the atmosphere. It is now unequivocal that these changes are forcing a major and relatively unprecedented change to the heat budget, climate and consequently biological systems of the Earth. The best way to familiarise yourself with the evidence is to refer to scientific consensus documents such as the latest reports from the Intergovernmental Panel on Climate Change (IPCC 2014; see Box 12.1).

The instrumental records of many nations unanimously show an increase in global surface temperature over the past 100 years. An early question in the study of anthropogenic climate change was whether these increases in temperature were 'unnatural' or whether they were part of a natural cycle? The answer has come from analysis of long-term records of temperatures deduced from isotopic ratios. In chemical reactions, isotopes of any atom (e.g. ^{16}O versus ^{18}O, or ^{12}C versus ^{13}C) participate in chemical reactions in which specific isotopic ratios are determined by temperature. In coral skeletons, the date when the chemical reaction that produced a particular monthly growth ring of calcium carbonate can also be precisely determined, by either counting the number of growth bands back from the living surface or by dating using other isotopic methods. Consequently, the isotopic ratio of oxygen converted into a measure of the sea temperature in which the coral was precipitating calcium carbonate can be precisely determined for a particular time. As some corals such as *Porites* may live for hundreds of years, with their fossil skeletons persisting for many thousands of years, these records provide an enormously valuable archive of past sea temperatures. Similar techniques have been used on tree rings, lake sediments and ice cores to derive long-term perspectives on how the Earth's temperature (and other key factors) have varied over thousands to tens of millions of years. As illustrated in Fig. 12.2B, current planetary temperatures are unprecedented in at least the last 420 000 years. Moreover, there is a direct link between the warming potential of greenhouse gases being added to the atmosphere and the average surface temperature of the planet. The latter demonstrates that models that do not include greenhouse gas drivers perform poorly in reconstructing past changes, and hence in projecting future changes (Fig. 12.2C).

Box 12.1. The Intergovernmental Panel on Climate Change (IPCC)

The United Nations Environment Program along with the World Meteorological Organisation formed the Intergovernmental Panel on Climate Change (IPCC) in 1988, in order to examine the evidence for the enhanced greenhouse effect and its influence on issues such as the climate, ecology and socio-economics of our planet. Since its formation, the IPCC has produced five assessment reports (1992, 1996, 2001, 2007 and 2014) that represent the global consensus of thousands of experts on climate change. There are three core working groups. The first is focused on reviewing the physical evidence and climate projections (Working Group 1: The Physical Science Basis) while the second is focused on the impacts of climate change, as well as the steps that human communities and governments might take in order to adapt to the consequences (Working Group 2: Climate Change Impacts, Adaptation and Vulnerability). Working Group 3 concentrates on ways that humans might reduce the rate at which greenhouse gases are building up in the atmosphere ('mitigation'). These working groups are formed several years before each report through nominations of the most qualified leading experts in each area by the world's nations. The scale of the collaboration and consensus associated with the IPCC reporting mechanism is unparalleled in history, given that each IPCC assessment is the consensus of thousands of scientists, which is publicly reviewed by a huge array of experts from science, economics, public administration, local to international governments, non-government agencies, industries and many others. Generally, trying to achieve consensus leads invariably to a more conservative assessment of the scientific evidence. Despite this, the fifth assessment report (WGI), which was released in 2013 stated:

> *Warming of the climate system is unequivocal, and since the 1950s, many of the observed changes are unprecedented over decades to millennia. The atmosphere and ocean have warmed, the amounts of snow and ice have diminished, sea level has risen, and the concentrations of greenhouse gases have increased.*

In addition to assessing whether or not changes been detected, and if these changes can be attributed to climate change, the IPCC also makes projections on how conditions are likely to change. This is based on a wide array of expert inputs, including the output of increasingly sophisticated global circulation models (GCM), which enable governments and many others to make informed decisions about the risks and opportunities that might arise from policy decisions and related actions. Recent changes to several physical and chemical variables as summarised in Fig. 12.3. Note, these changes are planetary in scale, with changes such as the loss of summer sea ice having fundamental ramifications for the entire global climate system through the increased absorption of heat as white reflective ice is replaced by dark heat absorbing ocean waters. Recent changes in ocean temperature, ocean acidity, oxygen, currents and storm systems are already driving some of the most fundamental changes in the way our planet works. The latest IPCC report also provides the summary map shown in Fig. 12.4. Not surprisingly, major changes have already occurred in the distribution and abundance, and health, of ocean ecosystems from the Equator to the polar regions. For more information on these changes, the reader is directed to the latest assessment report of the IPCC (www.ipcc.ch).

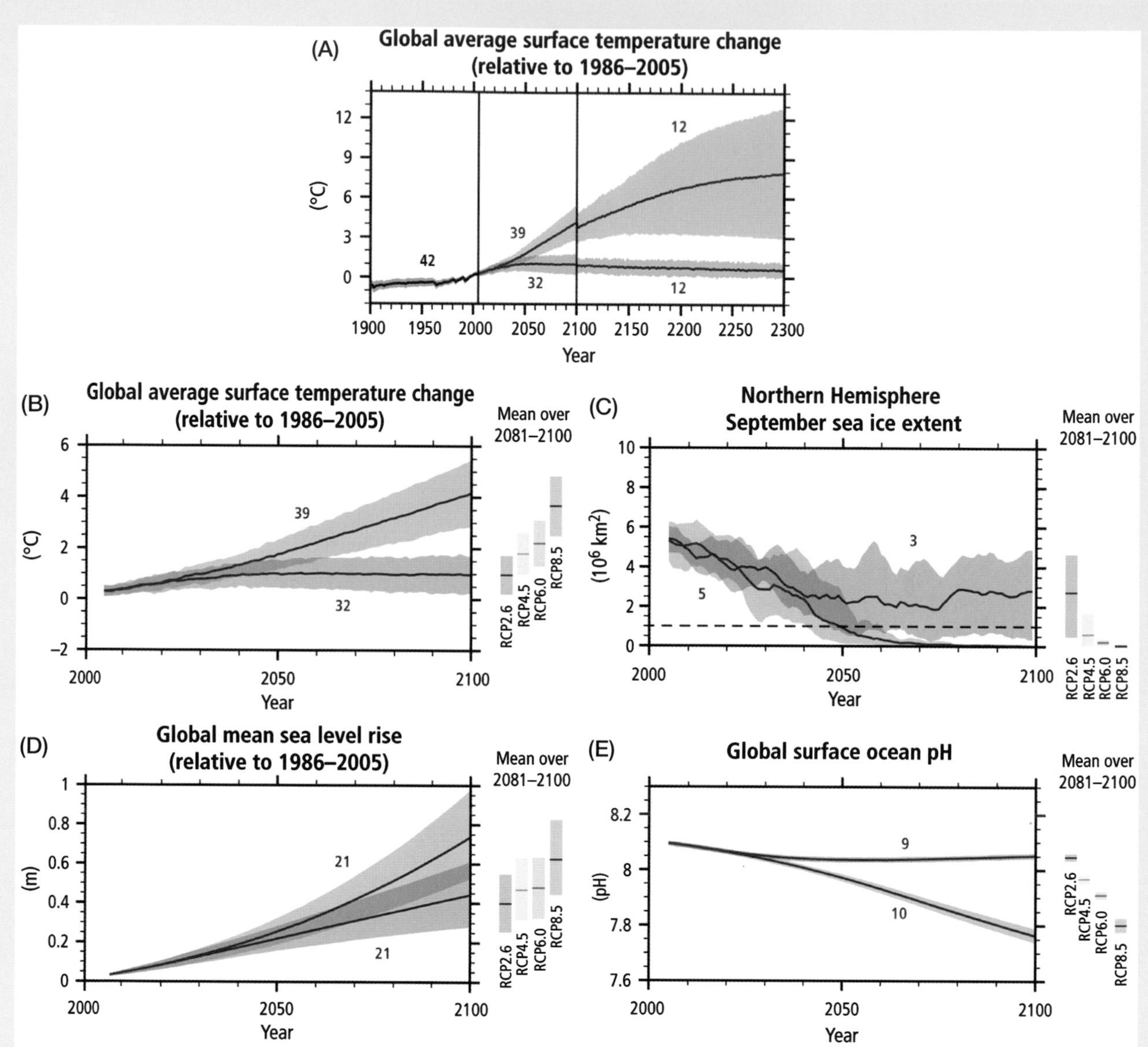

Fig. 12.3. Summary of recent and projected trends in key global variables. In each case, business-as-usual (RCP8.5, red lines and orange shading) are compared with scenarios (RCP2.6, blue lines and pale blue shading) where deep action to reduce greenhouse gas emissions is taken. **(A)** Global average surface temperature relative to 1986–2005 over the next three centuries; **(B)** similar but for the period up to 2100; **(C)** extent of summer sea ice in the Arctic; **(D)** global mean sea level rise; and **(E)** changes to global surface pH is the result of ocean acidification. Numbers indicate the number of independent CMIP5 models (cmip.llnl.gov/cmip5/) used to project changes. (Source: Reproduced with permission from IPCC (2014), p. 59).

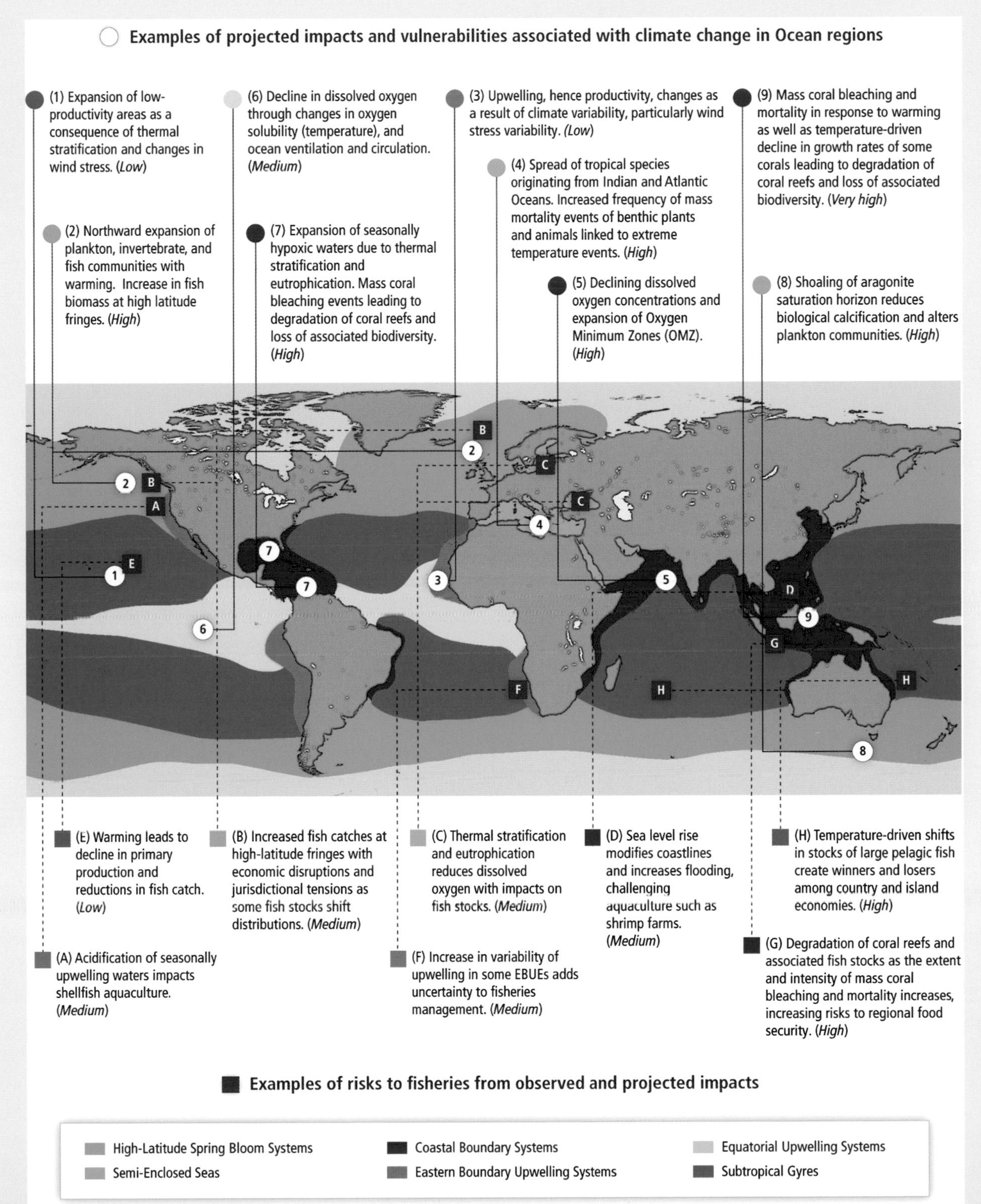

Fig. 12.4. Examples of projected impacts and vulnerabilities associated with climate change ocean (Source: Reproduced with permission from Hoegh-Guldberg *et al.* 2014, p. 1700).

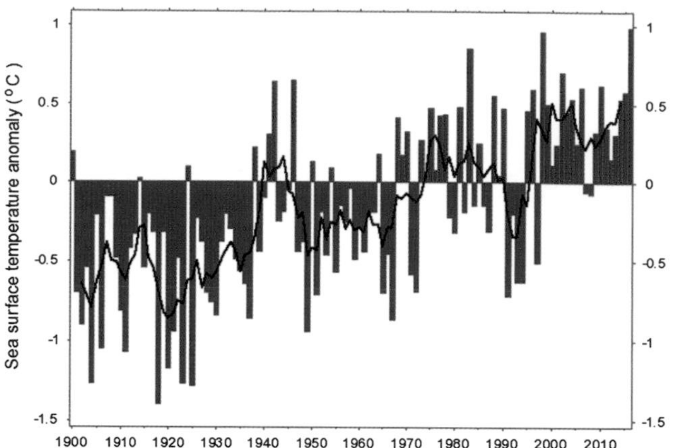

Fig. 12.5. April sea surface temperature anomaly for the Great Barrier Reef (1900–2016), relative to the average period 1961–1990. Black curve is the 5-year running mean. (Source: Bureau of Meteorology, Canberra. CC BY 3.0).

Climate change and the Great Barrier Reef

Australia's tropical waters have warmed significantly over the past 150 years, with much of the increase being seen over the past 50 years. In April 2016, the average sea surface temperature over the GBR was 1°C higher than the 1961–1990 average and probably ~2°C warmer on the pre-industrial period (Fig. 12.5). Rates of warming have been accompanied by decreasing pH due to ocean acidification, as well as changes to local weather patterns and extremes. In the latter case, there is growing evidence that extreme weather events (e.g. droughts, floods, very large cyclones) are increasing, while the capacity of many ecosystems such as coral reefs to recover has been decreasing. These changes are tipping the balance in favour of the decline of reef-building corals. For example, scientists from the Australian Institute of Marine Science (AIMS) recently reported that coral abundance in their extensive long-term surveys had decreased by 50% since the early 1980s. More recently, it had almost recovered to 1980 levels when the extraordinary mortality events of 2016–2017 occurred, dragging coral abundance down even lower on at least half of the GBR.

Mass coral bleaching and disease

Increasing summer sea temperatures have brought corals on reefs closer to their thermal limits, with the result that warmer than average years (arising due to natural variability) now push corals beyond their upper thermal thresholds. Corals respond to heat stress by losing their normal brown colour ('bleaching'). Coral bleaching occurs when the symbiosis between coral and their dinoflagellate symbionts (Fig. 12.6A, B) disintegrates, with the rapid movement of the brown symbionts out of the otherwise translucent tissues of the coral (Fig. 12.6C, D). As a result, corals change rapidly from brown pigmented appearance to pale or white. Bleaching is a generic response that occurs in response to a wide array of stresses, including reduced salinity, high or low irradiance, some toxins such as cyanide and many herbicides, microbial infection and high or low temperatures.

Coral bleaching has been known for over 70 years from reports of individual colonies or small patches of reefs were observed to have bleached. More recently, however, 'mass' coral bleaching events have affected coral reefs over hundreds, and sometimes tens of thousands, of square kilometres at a time. Mass coral bleaching events, on the other hand, have only been reported since 1979. Work done during the 1980s and 1990s revealed that mass bleaching events are triggered by warmer than normal conditions and can be predicted using sea-surface temperature anomalies measured by satellites. Light is an important co-factor. Corals that are shaded tend not to bleach as severely as those under normal irradiances. Corals also differ in their susceptibility, with some corals such as *Porites* and *Favia* being more tolerant of thermal stress than *Acropora*, *Stylophora* and *Pocillopora*. Differences in sensitivity probably relate to host characteristics such as tissue thickness and pigmentation, and possibly genotype of the symbiotic dinoflagellates within coral tissues. Coral bleaching is also affected by water motion, with corals in still, warm and sunlit conditions showing the greatest impact of thermal stress. The latter is consistent with the first observations of the association of coral bleaching with doldrums conditions typical of El Niño years in the eastern Pacific.

Mass coral bleaching has affected almost every coral reef worldwide since the 1980s and it occurs

Fig. 12.6. **(A)** A reef-building coral (*Acropora* sp.) with normal populations of dinoflagellate symbionts; **(B)** the key dinoflagellate symbiont of corals, *Symbiodinium*; **(C)** coral bleached due to rising environmental stress; **(D)** bleached coral reef in January 2006 in Keppel Island on the southern Great Barrier Reef (Photos: O. Hoegh-Guldberg).

on the GBR approximately every 3–5 years at the present time (late 20th and early 21st century). Mass coral bleaching has occurred most often in El Niño years (e.g. 1987–1988; 1997–1998, 2016–2017) and less often in the cooler non-El Niño or La Niña years (1988–1989; 1999–2000). Bleached corals tend to recover their dinoflagellate symbiotic populations in the months following an event if the stress involved is mild and short lived. But mortality (up to 100% of corals over large areas of coral reef) occurs following intense and long-lasting stress. This was seen in many parts of the world in 1998. Approximately 16% of corals that were surveyed before the global cycle of bleaching seen in the warm period had died by the end of 1998. This particular figure is an average and conceals the fact that in some oceans, such as the Western Indian Ocean, up to 46% of corals may have died. Similar global events have occurred in 2010 and 2016–2017.

Coral reefs in Australia have bleached repeatedly over the past 30 years. Mass bleaching events occurred in Australia in 1983, 1987, 1991, 1998, 2002, 2006 and in 2011, with large sections of the GBR and other reefs (e.g. Western Australia) bleaching in each case. Up until 2016, mortality rates on the GBR were relatively low (~10%) when compared with coral reefs elsewhere, mainly due to elevated temperatures being mild in comparison with other regions and times. This was not the case for some other Australian coral reefs. For example, very warm water sat above Scott Reef in the north-west waters of Australia for several months in 1998, resulting in extensive bleaching of corals down to 30 m and the loss of 95% of reef-building corals. Recent reports indicate that recovery of these reefs has occurred to some extent despite recruitment to these remote reefs being relatively low.

Coral disease, driven by pathogenic bacteria, is on the rise and may be connected to warmer than normal conditions and bleaching. Although coral disease affects less than 5% of the population, the incidence of diseases such as 'white syndrome' and other diseases (Fig. 12.7) are on the increase. Coral disease is currently not considered a major threat to coral reefs in Australia and many parts of the Pacific, but recent experiences in the Caribbean, where coral disease decimated populations of *Acropora* corals, suggest that understanding and monitoring coral disease is important.

The good fortune of the GBR in avoiding mass mortality ran out recently when exceptionally warm water drove very high rates of coral bleaching and mortality in sequential events over the period 2016–2017 (Fig. 12.8). The strong relationship between temperature and the mass mortality of corals enabled prediction of successive annual events in 2016–2017 (Hughes *et al.* 2017) almost 20 years ago (Hoegh-Guldberg 1999). Extending

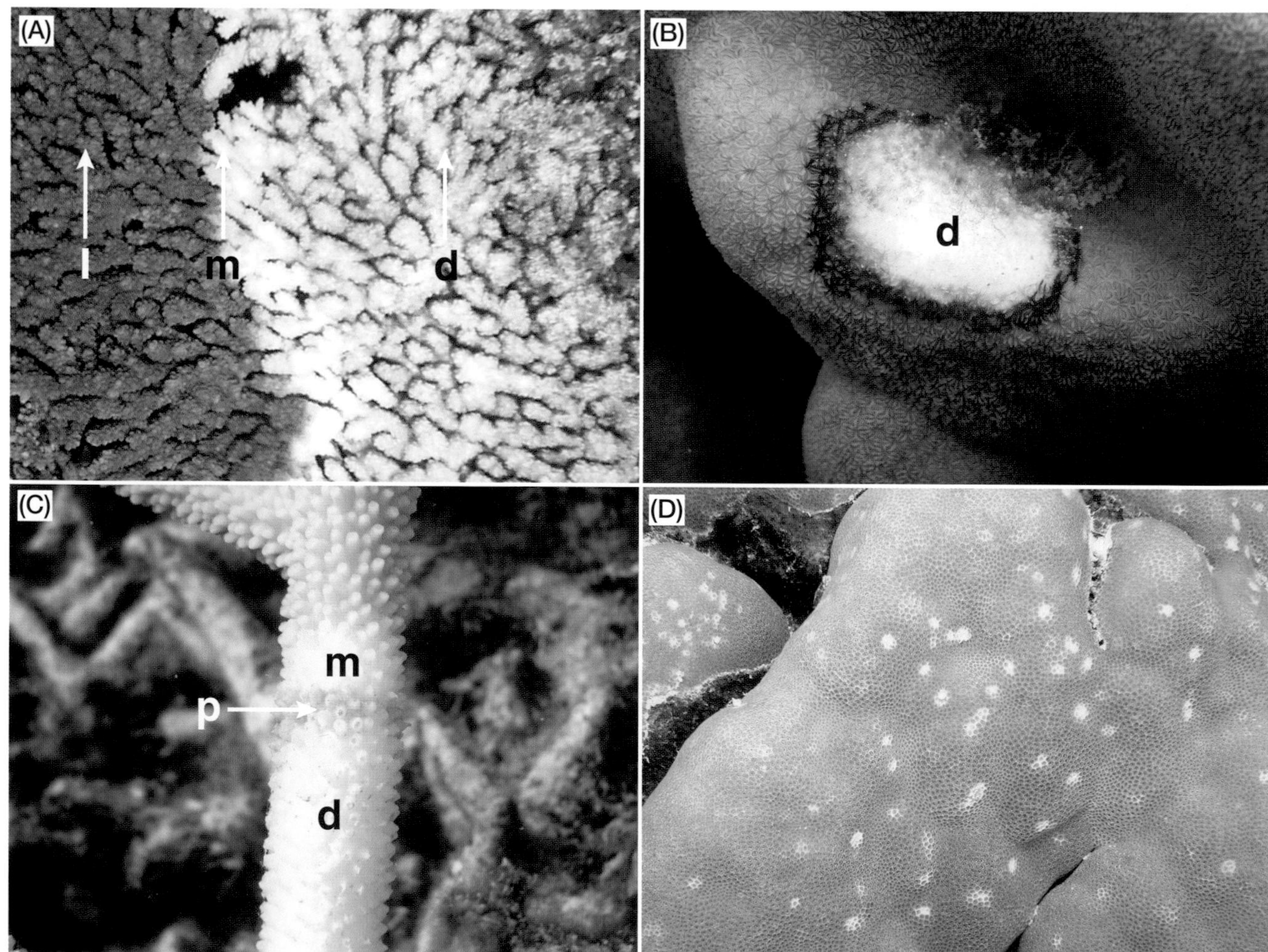

Fig. 12.7. Examples of diseases reported on Great Barrier Reef corals: **(A)** white syndrome on tabulate *Acropora* (Photo: G. Roff); **(B)** black band disease affecting *Pavona* sp. (Photo: G. Roff); **(C)** brown band disease on a branching *Acropora* (photo: O. Hoegh-Guldberg); and **(D)** white spot syndrome on *Porites* (Photo: O. Hoegh-Guldberg). Key: i, living tissue; m, advancing margin of disease; d, dead exposed coral skeleton; p, concentrations of ciliates.

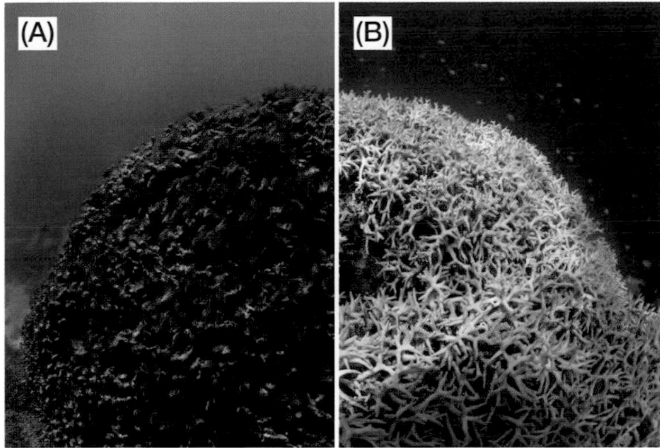

Fig. 12.8. (A) Bleached coral in March 2016 and **(B)** later in May 2016 after extensive coral bleaching and mortality at Lizard Island on the Great Barrier Reef (Source: The Ocean Agency). Note the disappearance of fish once the bleached corals in the left had died. Losing coral has very significant ramifications for the many species that are dependent on coral dominated reefs.

these projections, it is almost certain that back-to-back bleaching events will be common place in 5–10 years, driving coral cover within reef systems to very low levels. These changes and model verification reveal that an increase of 2°C over pre-industrial temperatures will lead to catastrophic annual mass coral bleaching mortality with little or no time between events such that coral populations will have time to cover. This will destine ecosystems such as the GBR to enter states that do not have significant coral cover or (eventually) the 3-dimensional reef structure that is essential for tens of thousands of species of fish and invertebrates. Such widespread habitat loss is almost certainly going to increase the rate of extinction within the borders of the world's largest Marine Park. For this reason, adopting and exceeding the goals of the Paris Agreement (Box 12.2) becomes exceedingly important if we are to retain any semblance of the today's coral reefs.

Ocean acidification

Approximately 30% of the CO_2 that is emitted by human activities is absorbed by the ocean. Once CO_2 enters the ocean, it reacts with water to form carbonic acid (Eqn 1), which disassociates, producing a bicarbonate (HCO_3^-) ion and a proton (H^+, Eqn 2). Protons are released then react with carbonate (CO_3^{2-}) ions to form additional bicarbonate ions (Eqn 3). At the alkaline pH, the flux of CO_2 into seawater leads to a decrease in the concentration of carbonate ions.

(1) $CO_2 + H_2O \rightarrow H_2CO_3$

(2) $H_2CO_3 \rightarrow HCO_3^- + H^+$

(3) $CO_3^{2-} + H^+ \rightarrow HCO_3^-$

The pH of open oceans has already decreased by 0.1 pH unit since the pre-industrial period, which sounds small until one considers that the pH scale is logarithmic and that the change is equivalent to a 30% increase in acidity (i.e. concentration of protons). This change in pH is accompanied by a 30% decrease in carbonate ions. Changes in seawater chemistry are highest at the polar regions due to lower sea water temperatures and hence higher solubility of CO_2. Current rates of ocean acidification are unprecedented both in the total amount of change and the rate at which it has been occurring. According to IPCC (AR5, 2014), 'The current rate and magnitude of ocean acidification are at least 10 times faster than any event within the last 65 Ma (*high confidence*) or even 300 Ma of Earth history (*medium confidence*).'

These changes are already having dramatic effects on marine life throughout the ocean. A recent study that synthesised the results and conclusions of hundreds of studies and revealed risks to the survival, calcification, growth, development and abundance of a broad range of taxonomic groups (i.e. from algae to fish) with considerable evidence of sensitivities for particular types of organisms. Organisms with shells and skeletons made out of calcium carbonate, for example, are particularly at risk, as are the early life history stages of a large number of coral reef organisms. Studies have even revealed that neural systems of fish are severely disrupted under projected ocean acidification. Interestingly, there were examples of

Box 12.2. The Paris Agreement

The United Nations Framework Convention on Climate Change (UNFCCC) is one of three conventions that were adopted at the Rio Earth Summit (1992) and which entered into force on 21 March 1994. The other two were the UN Convention on Biological Diversity (UNCBD) and the UN Convention to Combat Desertification (UNCCD). The ultimate goal of the UNFCCC and related instruments is to develop international policy and action that ultimately leads to the '... stabilisation of greenhouse gas concentrations in the atmosphere at a level that would prevent dangerous anthropogenic interference with the climate system ...'.

Since that time, 197 countries have ratified the convention, making them 'Parties to the Convention', which aims to prevent 'dangerous' human interference with Earth's climate system.

One of the early milestones of the UNFCCC was to establish the Kyoto protocol, which is an international agreement that committed the UNFCCC Parties to binding emission reduction targets. It was adopted in Kyoto in Japan on 11 December 1997, and entered into force on 16 February 2005. Initial actions on the six of most important greenhouse gases: CO_2, CH_4, N_2O, SF_6, hydrofluorocarbons (HFCs) and perfluorocarbons (PFCs) were taken under the first commitment of the Kyoto Protocol, which lasted from 2008 to 2012. A second commitment period of the Kyoto Protocol was established by the Doha Amendment in 2012, which commits the international community to a series of actions out to 2020. Further negotiations were held as part of the annual UNFCCC meetings to agree on actions to be taken after the second commitment period finishes in 2020. This resulted in the adoption of the Paris Agreement in December, 2015, which is a separate UNFCCC instrument to that of the Kyoto Protocol, which it replaces.

The Paris Agreement attracted signatures from 197 countries and came into force after being ratified by the required (at least) 55 UNFCCC parties that account for 55% of global greenhouse gas emissions. It is the world's first comprehensive climate agreement, with goals that match the expert consensus of the IPCC (described in Box 12.1). As we have already discussed, coral reefs such as the GBR will be destroyed if average global surface temperature increases to 2°C or more. This is also consistent with the sensitivity of a large number of other human and natural systems. Negotiations that occurred during the Paris Conference of the Parties (COP21) used the Fifth Assessment Report of the IPCC (AR5 IPCC 2014) to set the overall targets, as outlined in the text of the agreement:

> *'Emphasising with serious concern the urgent need to address the significant gap between the aggregate effect of Parties' mitigation pledges in terms of global annual emissions of greenhouse gases by 2020 and aggregate emission pathways consistent with holding the increase in the global average temperature to well below 2°C above pre-industrial levels and pursuing efforts to limit the temperature increase to 1.5°C above pre-industrial levels ...'*

Overall, the Paris Agreement has a unique structure. Previous instruments such as the Kyoto Protocol depended on top-down commitments to emission reduction targets that were set internationally through discussion and which were handed down to nations. Instead, the Paris Agreement depends on bottom-up 'Intended Nationally Determined Contributions' (INDC) for reducing greenhouse gas emissions that are proportionate to each country's opportunity and ability to support reductions. Every 5 years, the Conference of the Parties plans to meet and review INDC commitments and to adjust them if necessary. Importantly, the agreement adopts the principle of 'progression' such that any change has to be more ambitious than previous set of INDC. In this way, countries in the Paris Agreement are only able to increase their ambitions for greenhouse gas emission reductions rather than decrease them.

Mitigation of the underlying causes of rapid climate change is an urgent international priority. Instruments such as the Paris Agreement are centrally important in providing the policy instruments that are needed to reduce greenhouse gas emissions and stabilise the climate of our planet. If the Paris Agreement is successfully implemented, we will still see an additional 0.5°C increase in average global surface temperature, before planetary climate (e.g. temperature, storms, acidity) stabilises again. Although this will mean a further loss of corals of up to 70–90% of today's corals, it will mean that coral reefs will not be eliminated completely, as is likely to happen if we do not stay well below 2°C above the pre-industrial period.

taxa that did not show the same sensitivity to changes in pH and carbonate concentrations.

The effect of ocean acidification on calcification is one of the most important issues for coral reefs. The calcification of a wide range of marine organisms from microalgae (coccolithophores), molluscs (e.g. clams, pteropods) and corals, is strongly dependent on the concentration of carbonate ions in sea water. Corals were among the first organisms identified as having major problems with the rising concentration of atmospheric CO_2 and the decrease in the concentration of carbonate ions. These studies have shown consistently that the net calcification of coral reef communities effectively approaches zero at carbonate ion concentrations of 200 µmol/kg or less. Significantly, carbonate concentrations of 200 µmol/kg occur when atmospheric concentrations of CO_2 increases beyond 450 ppm. Given that coral reefs represent a balance between calcification and erosion (Chapter 9), atmospheric CO_2 concentrations need to remain well below 450–500 ppm if net reef accretion (calcification minus decalcification) is to remain positive against the forces of physical and biological erosion. Given that there is growing evidence that erosion (particularly bioerosion) is likely to increase under atmospheric CO_2, these thresholds become even more important (see Chapter 9).

Other impacts: rising sea levels, intensifying storms and changing local climates

The environment surrounding coral reefs is changing in additional ways to that of increasing ocean warming and acidification. As the ocean is warming, for example, it is also expanding. Land-based ice, such as that found in Greenland and Antarctica, is also melting and adding to the volume of the ocean. As a result of these two drivers (thermal expansion and the melting of land-based icesheet), mean sea level has risen by 20 cm over the past 100 years and is currently increasing at an average rate of 3.2 mm per year. Sea level is expected to have risen by as much as 1 m by the end of the century and will continue to increase by several metres over the next few centuries if immediate action is not taken to slow the rate of climate change. Coastal ecosystems such as mangroves, seagrass beds and saltmarsh are expected to shift towards the coastline as sea level rises. The ability of these important ecosystems to adapt to these changes will depend on the extent to which coastal development (cities, towns and agriculture) reduces the movement of these ecosystems to these new locations. It will also depend on other factors such as the delivery of sediment to coastal mangrove systems so that they are able to migrate to higher locations as sea level increases.

These changes are causing concern given the enormous human population and associated infrastructure (e.g. cities, ports) that is located in low-lying coastal areas of the world. Some low-lying countries such as Tuvalu, Kiribati and Bangladesh are already being affected by spring tides and storm surges that reach higher than ever before and now inundate towns, coastal agriculture and other infrastructure on a regular basis. The overall impact of sea level will also depend on how storm systems change, with effects of larger storms amplifying the effects of sea level rise. Although the number of storms has not increased, warmer oceans are adding to the energy of the storm systems that form, such that they are becoming more intense over time. There are some compelling facts about large storms in Australia. For example, there have been four Category-5 cyclones on the GBR since 2005, for example, with the next previous Category-5 cyclone occurring in 1918. Coupled with more fragile corals and reef structures due to ocean acidification, more intense storms place additional risks on the ability of coral reefs to persist as coral-dominated, carbonate structures. In some regions, such as the Caribbean, reef structures are being 'flattened' over time, with a huge loss of habitat and species that are dependent on the three-dimensional structure of coral reefs for habitat.

The volume of precipitation events is also increasing as a result of a warmer atmosphere holding ~6% more water for every 1°C of increase in average global surface temperature. Again, while it does not sound like a large change, 6% more water becomes voluminous if delivered in concentrated storm events, escalating the risks and size of flood events. When risks associated with sea level rise are added to these other risk factors, impacts become cumulative and potentially overwhelming for landscapes and watersheds. Other factors such as longer drought events (e.g. long-term drying trend along the eastern seaboard) these changes may lead to the loss of vegetation in catchments, causing further destabilisation of sediments and associated problems for coral reefs.

Escape clauses: rapid evolution, migrating out of trouble or hiding in deep water?

The robust and constant relationship between elevated sea temperature and mass coral bleaching and mortality is the basis for the success of projections for how coral communities are likely to change as ocean warming occurs. On the other hand, the accuracy of these projections should decay over time if reef-building corals were to become more tolerant over time due to acclimatisation (phenotypic change) or evolution (genetic change). Highly successful satellite predictions of mass coral bleaching and mortality, however, have used the same reference period (1985–1993 for over 30 years). Equally, the rapid loss of reef-building corals from tropical and subtropical regions should be slowing (and reversing) if the tolerance of corals to temperature and other changes was changing through acclimatisation and/or evolution. Lastly, the required rate of genetic adaptation needed is between 0.1–0.2°C per decade: a rate no doubt difficult for corals to match given their relatively long generation times (3–100 years or more) and hence slow rates of evolution. There are some ideas being explored as to how one might manipulate the population genetics by introducing heat-tolerant corals from warmer oceans into coral populations to help increase their thermal tolerance.

Another response by organisms facing increasingly hostile conditions is to relocate. This has been documented for a large number of marine plants and animals in response to ocean warming. Organisms such as tuna, for example, are tracking particular thermal environments as they change. Coral reef species are being reported at higher latitudes as temperatures increase, a phenomenon recently referred to as 'tropicalisation'. Insights from the fossil record also reveal that coral reefs were found at slightly higher latitudes than today during periods of warmer ocean temperature in the past. While these reports are fascinating, they do not represent sufficient support for the idea that entire coral reefs will move quickly and successfully to high latitudes as ocean warming continues. Observations are revealing that not all species are likely to survive at higher latitudes, leaving some interesting questions regarding the function of coral reefs if only a proportion reef species are present. Equally, moving to higher latitude will involve more than just changes to temperature. Questions remain as to how novel coral reef assemblages will be able to function with the lower light levels, reduced pH and carbonate ion concentrations, and different food web structures, of higher latitude environments.

As the climate has become more hostile for ecosystems in general, there has been increasing discussion of the role, and indeed existence, of potential refuges. In the case of coral reefs, refuges could provide some reef systems relief from the most severe effects of ocean warming, storms and acidification. As explored in Chapter 7, mesophotic reefs are deep water (40–100 m) reefs that often have significant populations of scleractinian corals yet may experience less heat stress and storm (cyclone) damage. If there are high levels of genetic connectivity between shallow and deep reefs, these mesophotic reefs could provide a source of corals for regenerating shallow reef systems as climate impacts occur on shallow water reef systems. Termed the 'deep sea refugia' hypothesis, researchers have found substantial differences in the rate of warming and acidification with depth, as well as examples of organisms that may span both the mesophotic and shallower reef regions. Other researchers found that mass coral bleaching and mortality do occur in mesophotic coral communities, and that cyclone damage can extend down to depths of 40–50 m. Researchers have also found that mesophotic coral species that use broadcast spawning are genetically connected to shallow water populations, thereby supporting the possibility of deep sea refugia playing a role in regenerating shallow water reefs. Counter to this, is the observation that mesophotic corals that brood their offspring are genetically separate from corals and shallow reef areas.

Implications for people in a coral reef poor world

As outlined at the beginning of this chapter, coral reefs are not only the most biologically diverse ocean ecosystem, but they are also vitally important for human communities throughout the tropics. Recent work done by the IPCC has identified the risks and vulnerabilities of people associated with tropical coastal regions in a warming and acidifying ocean. Ecosystems such as mangroves, coral reefs and seagrass are under threat from numerous climate change and non-climate change factors. Unfortunately, many of the services that the ecosystems provide are expected to diminish over time as a result. Understanding the nature of changes is important, however, if we are to assist people in finding solutions and helping them to adapt to the changes that they will increasingly face.

As has been outlined, coral reefs perform several important roles that are likely to be under increasing pressure. Coastal fisheries are likely to be reduced as coral abundance decreases and the three-dimensional structure of coral reefs declines, reducing food security and jeopardising livelihoods. Reduced structure and biodiversity will also degrade the attractiveness of coral reefs for tourism, reducing the opportunities for jobs and income. Similar challenges that face mangroves, seagrass and saltmarsh will further decrease the opportunities for food and livelihoods in tropical coastal regions. In helping the societies adapt, it will be important to seek alternative livelihoods such as sustainable aquaculture (e.g. seaweed cultivation) or careers that don't depend on coastal resources. Anticipating the loss of coastal protection from degrading coastal ecosystems will be vital in preparing people for greater exposure to stronger storms, waves and erosion.

The urgency of the need to act on climate change

We are already experiencing dangerous climate change despite the ambitions and goals of the UNFCCC. The once rare and now commonplace impact of intense storm systems in Australia, Caribbean, United States and many other places, plus catastrophic fires and floods give us insights into what the future might hold. Impacts of climate change on the ocean are no different, with the recent loss of coral reefs globally, including Australia's GBR, being of tremendous concern. And while developing ways of adapting to these changes, there is now an enormous urgency for us to deal with the root cause of the problem, which is the rise greenhouse gases such as CO_2 in the atmosphere. A key question here is: 'How much should we reduce our emissions of greenhouse gases, and over how long, in order to save places such as the GBR?'

In order to answer this, we need to consider the amount of CO_2 left to emit into the atmosphere before we exceed 2°C. Taking estimates from several different studies, we find that the number sits between 590 and 1370 Gt CO_2. Related to this, is how much CO_2 we release into the atmosphere on a yearly basis as a global society. This is ~40 Gt CO_2 per year. That allows us to calculate the number of years we have left before we must be at zero emissions of CO_2. Dividing the CO_2 budget remaining before we exceed 2°C (i.e. 590 to 1370 GT CO_2) by 40 Gt/year and rounding off, the calculation reveals that we have from 15 to 34 years before annual global emissions of greenhouse gases such as CO_2 must be zero.

Another interesting question concerns the impact on our 2°C budget, if we burn all of the proven (economically exploitable) fossil fuel reserves. In this case, the proven reserves of oil, gas and coal are somewhere around 3000 Gt-CO_2. If we divide this number (3000) by the remaining CO_2 budget, we get to see how much we will exceed the safe budget if we burn these proven reserves of fossil fuels. These numbers indicate that burning proven fossil fuel reserves over the next few decades will drive us to exceed the 2°C threshold by 200–300%. Another way of saying this, is that we need to leave 70–80% of fossil fuel reserves in the ground if we are to avoid a catastrophic outcome for people and ecosystems. It is important to note that there are additional, though less understood, unconventional fossil fuel resources. These contain fossil fuels that would deliver an estimated 15 000 Gt-CO_2 into the atmosphere, which would devastate the planet as we know it.

Naturally, the challenges involved in rapidly transitioning away from polluting technologies is a very large. Given that solving this problem is absolutely critical for life on our planet, it is a challenge from which we can no longer shy away. Already there are signs of rapid breakthroughs in renewable technologies such as solar, geothermal and wind energy generation that are making the burning of fossil fuels largely uneconomical. For this reason, the next few decades are likely to be some of the most exciting of all times. Hopefully, we will be able to prevent the total loss of ecosystem such as coral reefs such as the GBR. Only time will tell.

References and further reading

Burrows MT, Schoeman DS, Richardson AJ, Molinos JG, Hoffmann A, Buckley LB, *et al.* (2014) Geographical limits to species-range shifts are suggested by climate velocity. *Nature* **507**, 492–495. doi:10.1038/nature12976

Caddy M, Kamenev M (2016) Australia's most destructive cyclones: a timeline (updated). *Australian Geographic* website, <http://www.australiangeographic.com.au/topics/science-environment/2016/02/australias-most-destructive-cyclones-a-timeline>.

Costanza R, de Groot R, Sutton P, van der Ploeg S, Anderson SJ, Kubiszewski I, *et al.* (2014) Changes in the global value of ecosystem services. *Global Environmental Change* **26**, 152–158. doi:10.1016/j.gloenvcha.2014.04.002

Deloitte Access Economics (2017) *At What Price? The Economic, Social and Icon Value of the Great Barrier Reef.* Deloitte Access Economics, Brisbane, <https://www2.deloitte.com/content/dam/Deloitte/au/Documents/Economics/deloitte-au-economics-great-barrier-reef-230617.pdf>.

Gattuso J-P, Magnan A, Bille R, Cheung WWL, Howes EL, Joos F, *et al.* (2015) Contrasting futures for ocean and society from different anthropogenic CO_2 emissions scenarios. *Science* **349**, aac4722. doi:10.1126/science.aac4722.

Hoegh-Guldberg O (1999) Climate change, coral bleaching and the future of the world's coral reefs. *Marine and Freshwater Research* **50**, 839–866. doi:10.1071/MF99078

Hoegh-Guldberg O (2012) The adaptation of coral reefs to climate change: is the Red Queen being outpaced? *Scientia Marina* **76**, 403–408. doi:10.3989/scimar.03660.29A

Hoegh-Guldberg O (2014) Coral reefs in the Anthropocene: persistence or the end of the line? *Geological Society of London, Special Publications* **395**, 167–183. doi:10.1144/SP395.17

Hoegh-Guldberg O, Cai R, Poloczanska ES, Brewer PG, Sundby S, Hilmi K, *et al.* (2014) The ocean. In *Climate Change 2014: Impacts, Adaptation, and Vulnerability. Part B: Regional Aspects. Contribution of Working Group II to the Fifth Assessment Report of the Intergovernmental Panel of Climate Change.* (Eds VR Barros, CB Field, DJ Dokken, MD Mastrandrea, KJ Mach, TE Bilir, *et al.*) pp. 1655–1731. Cambridge University Press, Cambridge, UK.

Hoegh-Guldberg O, Thezar M, Boulos M, Guerraoui M, Harris A, Graham A, *et al.* (2015) *Reviving the Ocean Economy: the Case for Action – 2015.* Gland, Switzerland.

Hughes TP, Kerry JT, Álvarez-Noriega M, Álvarez-Romero JG, Anderson KD, Baird AH, *et al.* (2017) Global warming and recurrent mass bleaching of corals. *Nature* **543**, 373–377. doi:10.1038/nature21707

IPCC (2014) Summary for policymakers. In *Climate Change 2014: Synthesis Report. Contribution of Working Groups I, II and III to the Fifth Assessment Report of the Intergovernmental Panel on Climate Change.* (Eds Core Writing Team, R Pachauri and L Meyer) pp. 2–34. IPCC, Geneva, Switzerland. doi:10.1017/CBO9781107415324.

Jackson JBC (2001) Historical overfishing and the recent collapse of coastal ecosystems. *Science* **293**, 629–637. doi:10.1126/science.1059199.

Karoly DJ, Braganza K (2005) Attribution of recent temperature changes in the Australian region. *Journal of Climate* **18**, 457–464. doi:10.11.1175/JCLI-3265.1

Kroeker KJ, Kordas RL, Crim R, Hendriks LE, Ramajo L, Singh GS, *et al.* (2013) Impacts of ocean acidification on marine organisms: quantifying sensitivities and interaction with warming. *Global Change Biology* **19**, 1884–1896.

Lovelock CE, Cahoon DR, Friess DA, Guntenspergen GR, Krauss KW, Reef R, *et al.* (2015) The vulnerability of Indo-Pacific mangrove forests to sea-level rise. *Nature* **526**, 559–563. doi:10.1038/nature15538

McGlade C, Elkins P (2015) The geographical distribution of fossil fuels unused when limiting global warming to 2°C. *Nature* **517**, 187–190. doi:10.1038/nature14016

Myers RA, Worm B (2003) Rapid worldwide depletion of predatory fish communities. *Nature* **423**, 280–283. doi:10.1038/nature01610

Meinshausen N, Hare W, Raper SCB, Frieler K, Knutti R, Frame DJ, *et al.* (2009) Greenhouse-gas emission targets for limiting global warming to 2°C. *Nature* **458**, 1158–1162.

Petit JR, Jouzel J, Raynaud D, Barkov NI, Barnola JM (1999) Climate and atmospheric history of the past 420 000 years from the Vostok ice core, Antarctica. *Nature* **399**, 429–436.

Poloczanska ES, Burrows MT, Brown CJ, García Molinos J, Halpern BS, Hoegh-Guldberg O, *et al.* (2016) Responses of marine organisms to climate change across oceans. *Frontiers in Marine Science* **3**, 62. doi:10.3389/fmars.2016.00062

Pörtner HO, Karl DM, Boyd PW, Cheung WWL, Lluch-Cota SE, Nojiri Y, *et al.* (2014) Ocean systems. In *Climate Change 2014: Impacts, Adaptation, and Vulnerability. Part A: Global and Sectoral Aspects. Contribution of Working Group II to the Fifth Assessment Report of the Intergovernmental Panel on Climate Change.* (Eds CB Field, VR Barros, DJ Dokken, KJ Mach, MD Mastrandrea, TE Bilir, *et al.*) pp. 411–484. Cambridge University Press, Cambridge, UK, <https://www.ipcc.ch/pdf/assessment-report/ar5/wg2/WGIIAR5-Chap6_FINAL.pdf>.

Saunders MI, Leon JX, Callaghan DP, Roelfsema CM, Hamylton S, Brown CJ, *et al.* (2014) Interdependency of tropical marine ecosystems in response to climate change. *Nature Climate Change* **4**, 724–729. doi:10.1038/nclimate2274

UNFCCC (2015) *Adoption of the Paris Agreement.* United Nations, New York, USA, <https://unfccc.int/resource/docs/2015/cop21/eng/l09r01.pdf>.

Walsh K, White CJ, McInnes K, Holmes J, Schuster S, Richter H, *et al.* (2016) Natural hazards in Australia: storms, wind and hail. *Climatic Change* **139**, 55–67. doi:10.1007/s10584-016-1737-7

13

Terrestrial runoff to the Great Barrier Reef and the implications for its long term ecological status

J. Brodie and K. Fabricius

The Great Barrier Reef (GBR) is a large marine ecosystem adjacent to the north-east Australian coast, and is protected both as a Marine Park and through its World Heritage status. The land adjacent to the GBR forms the GBR Catchment Area (GBRCA) (Fig. 13.1) from which over 35 rivers and streams discharge into the GBR. As the GBRCA has been developed for agricultural, industrial, mining and residential use over the last 170 years, most of its rivers carry increasing loads of nutrients, sediments, pesticides and other pollutants into the GBR (Chapters 8 and 11).

River discharges of suspended sediments, estimated from monitoring and modelling studies, have increased by approximately five times, nitrogen exports by two times and phosphorus exports by three times. Pesticides, comprised of more than 40 individual active ingredients, are also being discharged to the GBR, with many at potentially ecologically relevant concentrations. The sources of these pollutants are the principal land uses on the GBRCA: sediments and nutrients from beef grazing, primarily through increased soil erosion; sediments, nutrients and pesticides (currently mainly herbicides) from cropping (sugarcane, banana and other horticulture, cotton and grain crops) associated with soil erosion, fertiliser and pesticide use; and a diverse array of other pollutants (e.g. heavy metals, persistent organic pollutants, plastics) and nutrients from urban sewage effluents and stormwater runoff.

Pollutants are transported into the GBR lagoon primarily during large river flow events following monsoonal/cyclonic rainfall. As the rivers flood, a proportion of the pollutant load is trapped on floodplains, especially in riparian vegetation and wetlands. However, since GBRCA rivers are relatively short and fast flowing, there is limited time for trapping, and most rivers discharge a high proportion of their fine particulate and dissolved pollutant load to the ocean.

Rivers can form plumes that transport pollutants hundreds of kilometres throughout the GBR lagoon and exposing the embedded ecosystems to terrestrial pollutants (Fig. 13.2). Within the plume, pollutant concentrations decline with distance from the river mouth via dilution, sedimentation and biological uptake. Most suspended sediments and particulate nutrients are initially deposited close to the river mouth (1–10 km) (Fig. 13.2) but part of the fine fraction then continues to be dispersed and transported via wind resuspension and

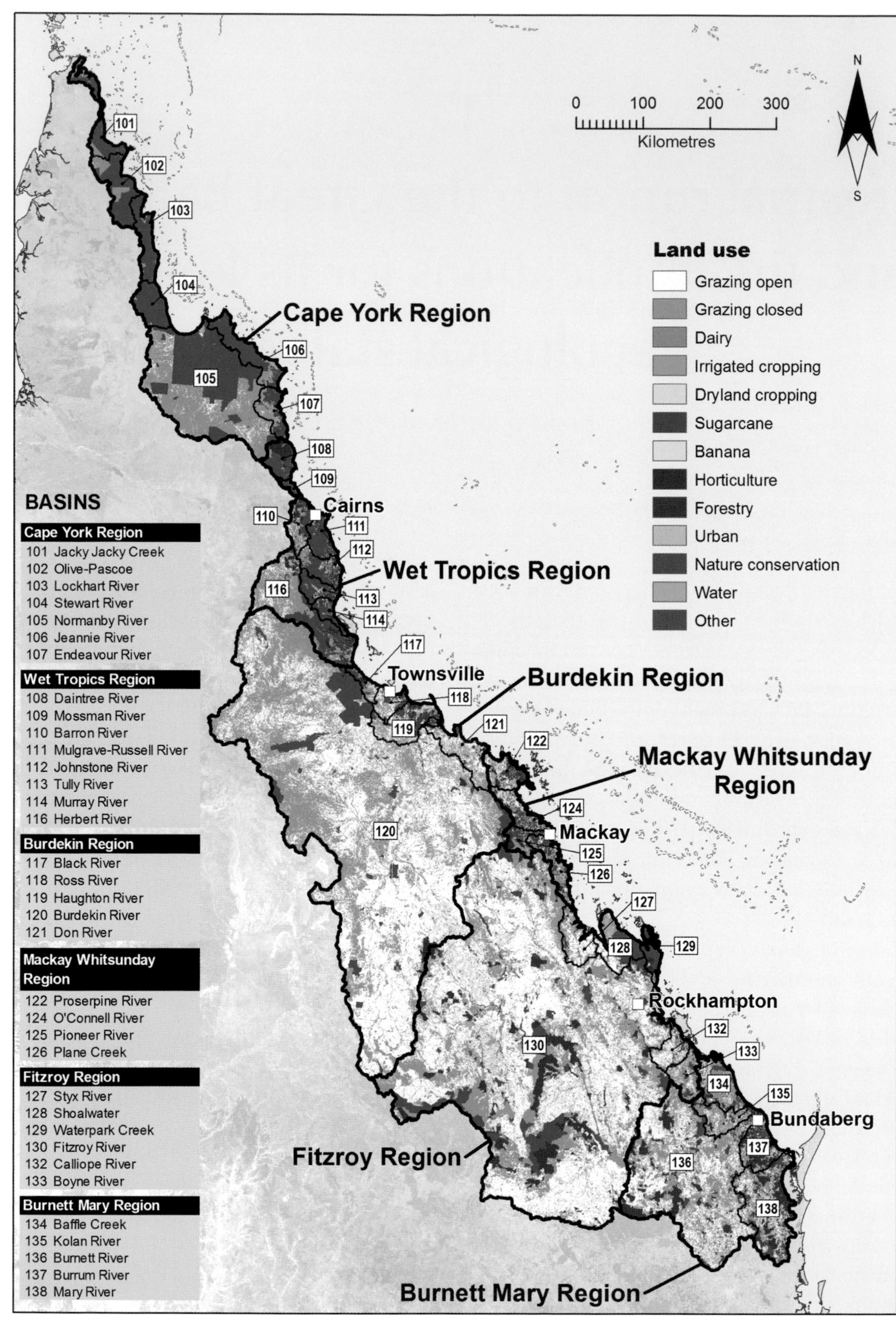
0 100 200 300
Kilometres
N
S
Land use
Grazing open
Grazing closed
Dairy
Irrigated cropping
Dryland cropping
Sugarcane
Banana
Horticulture
Forestry
Urban
Nature conservation
Water
Other
Cape York Region
Wet Tropics Region
Burdekin Region
Mackay Whitsunday Region
Fitzroy Region
Burnett Mary Region
Cairns
Townsville
Mackay
Rockhampton
Bundaberg
BASINS
Cape York Region
101 Jacky Jacky Creek
102 Olive-Pascoe
103 Lockhart River
104 Stewart River
105 Normanby River
106 Jeannie River
107 Endeavour River
Wet Tropics Region
108 Daintree River
109 Mossman River
110 Barron River
111 Mulgrave-Russell River
112 Johnstone River
113 Tully River
114 Murray River
116 Herbert River
Burdekin Region
117 Black River
118 Ross River
119 Haughton River
120 Burdekin River
121 Don River
Mackay Whitsunday Region
122 Proserpine River
124 O'Connell River
125 Pioneer River
126 Plane Creek
Fitzroy Region
127 Styx River
128 Shoalwater
129 Waterpark Creek
130 Fitzroy River
132 Calliope River
133 Boyne River
Burnett Mary Region
134 Baffle Creek
135 Kolan River
136 Burnett River
137 Burrum River
138 Mary River

Fig. 13.1. The Great Barrier Reef, its catchment area and land uses (Data source: Auslig Australia PL).

Fig. 13.2. Burdekin River plume following high discharge conditions in 2011 (Satellite image: NASA).

longshore currents further along the coast. Dissolved nutrients and herbicides are transported in the plume, with the bioavailable fraction of the nutrients stimulating phytoplankton blooms (Figs 13.2, 13.3).

The extent of exposure of ecosystems, such as coral reefs and seagrass beds, to the various land-sourced pollutants has been estimated through modelling and monitoring. The understanding of the effects of land-sourced pollutants on coral reef, seagrass and planktonic ecosystems, and specifically the GBR, have improved greatly over the last 30 years. Five different types of water quality issues can be distinguished when determining the effects of terrestrial runoff on coral reefs. These are: turbidity-related light attenuation (largely through resuspension of benthic fine sediment); enrichment with particulate organic matter (POM); dissolved inorganic nutrients leading to trophic shifts; sedimentation that can prevent reef recovery; and exposure to pesticides. A brief overview over these five types of water quality issues is given here.

Turbidity from suspended sediments and phytoplankton reduce light available for photosynthesis at deeper depths. Turbidity in the central and southern coastal, inshore and midshelf waters remains elevated for 4–8 months after river floods subside. Light requirements of seagrasses are relatively clearly known, and light availability determines their lower depth limit. For corals, the ability to handle low and variable light levels varies between species (Chapters 5, 7 and 8). Reduced light availability, especially during winter months, leads to the loss of sensitive species, and communities composed of fast growing high-light adapted taxa are being replaced by low-light tolerant taxa, including filter feeders. Turbidity also alters the larval settlement behaviour, feeding and abundances of some coral reef-associated fishes. Several types of herbivorous fishes avoid areas of high turbidity, which may contribute to the observed increases in macroalgal abundances in such areas.

Enrichment with POM also leads to shifts in reef communities. POM facilitates feeding and growth of some coral species, especially in high-flow environments. However, heterotrophic filter feeders will benefit from POM even more than corals do, hence ecosystems shift from communities composed of coral species that can grow at low food concentrations, to structurally simpler, more heterotrophic communities.

Dissolved inorganic nutrients are generally quickly removed from the water column through biological uptake, fuelling the productivity and high densities of bacteria, phytoplankton and fleshy benthic algae. Excessive nutrient concentrations can also help establishing macroalgal dominance on the seafloor, especially in areas where grazers (fish and/or invertebrates) are depleted. Healthy grazer populations may prevent algal proliferation and thus reduce the manifestation of undesirable effects of nutrient enrichment. Dissolved inorganic nutrients also promote the organic enrichment of benthos, sediments and suspended particulate matter, and they stimulate phytoplankton blooms. Large phytoplankton are food for the planktonic larvae of the crown-of-thorns starfish (COTS; *Acanthaster cf. solaris*), a major coral predator. COTS outbreaks have been a principal cause of coral mortality on the GBR (and throughout the Indo-Pacific coral province) over the last 60 years. Experiments suggest that the successful development of their larvae is food limited: their survivorship increases steeply with increasing availability of suitable food. The location and timing of high nutrient concentrations in the many Indo-Pacific locations correlate well with the location and times of primary outbreaks of COTS. In the GBR, mean summer chlorophyll concentrations in the central and southern GBR lagoon are about twice as high as in the far northern part of the GBR, and are elevated during large flood events. Increased nutrient availability due to high loads of river nutrients is therefore likely to be a major contributor to the high frequency of COTS population outbreaks that are now observed on the GBR.

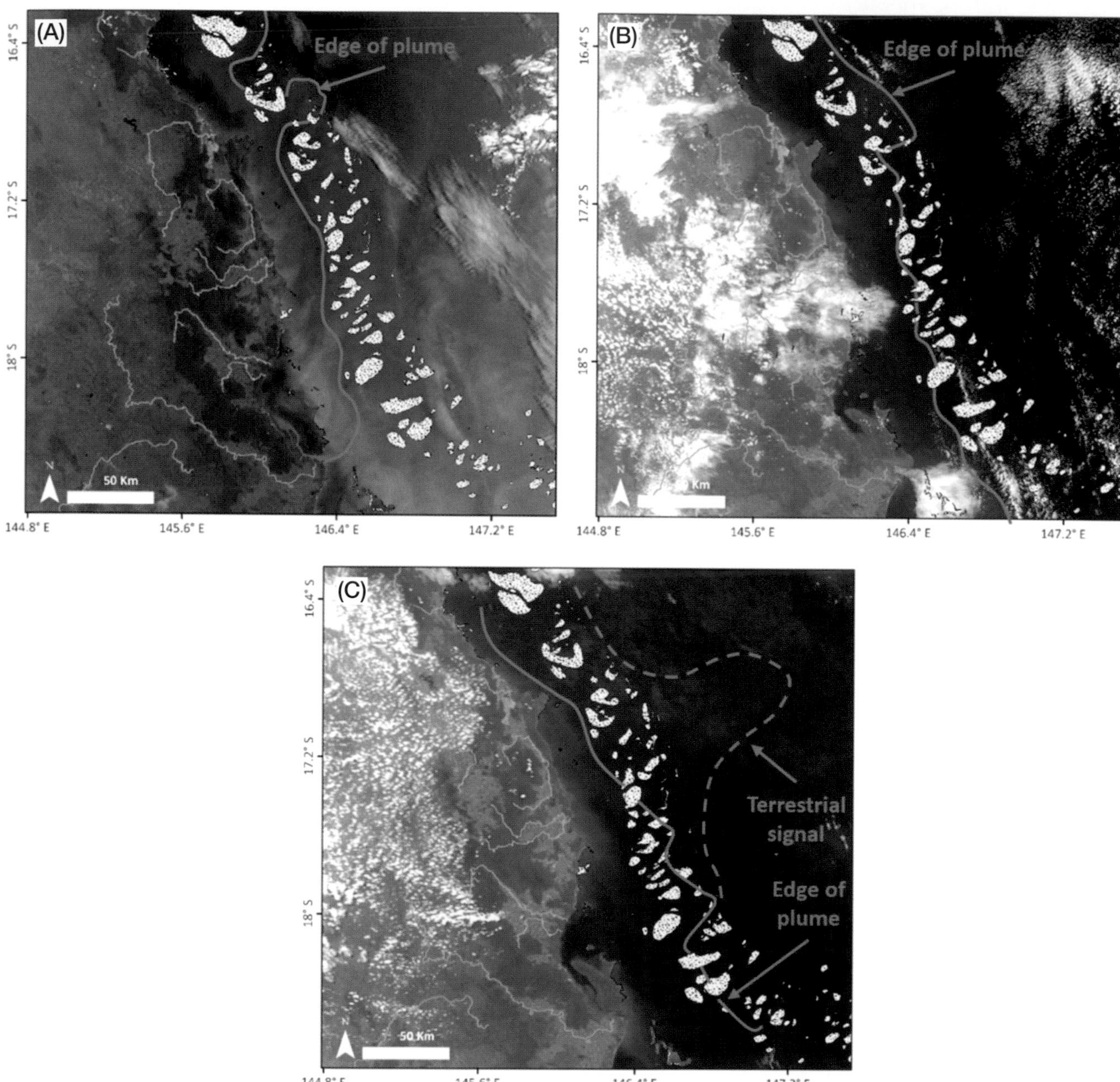

Fig. 13.3. Progression (**A–C**) of a multiple river plume in the Wet Tropics (9, 11, 13 February 2007, respectively) extending from the coast to beyond the outer reef. The lines show the outer edge of the plume made visible due to coloured dissolved organic matter and phytoplankton. Images (**A–C**) show the transformation from a plume dominated by terrestrial particulate matter into a plume dominated by a dissolved nutrient driven phytoplankton bloom. A proportion of the contained nutrients in the plume may be seen 'escaping' to the Coral Sea in image C (Image: NASA; figure prepared by Dr Caroline Petus, James Cook University).

Sedimentation also represents a severe disturbance for some inshore coral reefs. Sediments can smother corals and other reef organisms, especially small colonies and those with a flat growth form. Even more importantly, sedimentation also greatly reduces recruitment and the survival of early life stages in corals: settlement rates are near-zero on sediment-covered surfaces, and sedimentation tolerance in coral recruits is at least one order of magnitude lower than for adult colonies. Fine silts and sediments that are enriched with organic matter are more damaging for corals than inorganic carbonate sediment, because they facilitate bacterial growth, the formation of muddy marine snow and anoxia, and reduce seawater pH in the boundary layer. Many other groups of organisms, including crustose coralline algae, are also sensitive to sedimentation by organically enriched sediments, with negative consequences for reef biodiversity and ecological functions.

Herbicide residues (especially diuron, atrazine, simazine, ametryn, hexazinone and tebuthiuron) are commonly being detected in river waters and in river plumes. Residues of some pesticides are also being detected, presently in quite low concentrations, in the GBR during non-flood (dry season) periods. Herbicides can disrupt photosynthesis in marine plants including mangroves, seagrasses, coral zooxanthellae and crustose coralline algae. In addition, residues of various commonly used and detected pesticide types (herbicides, insecticides, fungicides) can directly affect the physiology of corals, especially their reproductive, larval and juvenile stages.

Overall, reduced recruitment success in corals, together with the promotion of macroalgae and increased frequencies of *A. cf. solander* outbreaks, are arguably the three most significant direct effects of terrestrial runoff on coral reefs on the GBR. For any specific location, the type and severity of response to terrestrial runoff depends on its hydrodynamic, geomorphological and biological properties, and whether they are predominantly exposed to turbidity, POM, dissolved inorganic nutrients, sedimentation or pesticides. Reefs that are surrounded by a shallow sea floor, corals located in poorly flushed bays or lagoons, deeper reef slopes, and frequently disturbed reefs are more likely to experience changes even at low levels of pollution, in particular when populations of herbivores are low. In contrast, well flushed shallow reef crests surrounded by deep sea floors are likely to be most resistant, especially when inhabited by healthy populations of herbivores that protect against overgrowth by sediment-trapping macroalgae. Severe exposure to terrestrial runoff leads to reduced reef calcification, increased bioerosion, shallower photosynthetic compensation points (see Chapter 9), changed coral community structure, and greatly reduced species richness. Hence, reef ecosystems increasingly simplify with increasing exposure to terrestrial runoff, compromising their ability to maintain essential ecosystem functions and to recover from disturbances, including the presently increasing level of human induced disturbances.

Pollutant discharges to the GBR from agricultural (and to a lesser extent urban) development on the GBRCA are most elevated in the Wet Tropics, Burdekin, Keppels and Whitsundays coastal waters. Here, coral diversity is ~50% reduced and the biomass of macroalgae is higher than expected for reefs at comparable latitudes and distances off the coast. In contrast, the high coral diversity and recruitment potential of inshore reefs in the remote and still relatively unpolluted region north of Princess Charlotte Bay is in stark contrast to the state of most inshore reefs in the central and southern regions of the GBR (Fig. 13.4). The increased frequency of outbreaks of COTS is also affecting offshore reefs along the entire central and southern section of the GBR essentially from Lizard Island to Mackay.

Syntheses of current scientific knowledge of water quality and the GBR have been published in 2001, 2008, 2013 and most recently in 2017. The 2017 Scientific Consensus Statement draws the following conclusions:

'Key Great Barrier Reef ecosystems continue to be in poor condition. This is largely due to the collective impact of land runoff associated with past and ongoing catchment development, coastal development activities, extreme weather events and climate change impacts such as the 2016 and 2017 coral bleaching events.

Current initiatives will not meet the water quality targets. To accelerate the change in on-ground management, improvements to governance, program design, delivery and evaluation systems are urgently needed. This will require greater incorporation of social and economic factors, better targeting and prioritisation, exploration of alternative management options and increased support and resources.'

Management responses have also been developed by the Australian and Queensland Governments through the GBR Reef Water Quality Protection Plan in 2003, with subsequent updates in 2009, 2013 and 2017 (the latter known as the Reef 2050 Water Quality Improvement Plan 2017–2022), and the Reef 2050 Sustainability Plan (to be revised in 2018). Water quality guidelines for the GBR were set in 2010 and river pollutant load targets for the protection of the GBR have been set at a basin scale in 2017. Best management practices for agricultural and urban activities have been tested for their effectiveness in improving water quality through the Paddock to Reef Integrated Modelling and Monitoring Program. The success of the management actions in improving farm practice, reducing pollutant loads and improving GBR lagoon water quality are assessed through joint government 'Report Cards' each year. The Report Card for the period up to 2016 shows very little progress in either improved management practices on farms or reductions in pollutant loads. The continuation of the current level of management response has been forecast to not reach any of the water quality targets set in 2017 at both GBR-wide and individual basin scales over the next decade.

Yet there is considerable published material that a change in approach by the governments, by providing more funding and stronger use of the existing legislation, could be effective in attaining much greater reductions in pollutant loads than the current policy mix. This would include actions such as:

- using a mix of policy instruments that include both voluntary and regulatory approaches
- implementing staged regulations to meet Reef outcomes. Consideration should be given to the use of existing legislation (e.g. the *GBRMP Act 1975* and the *Environmental Protection and*

Fig. 13.4. **(A)** Inshore reefs in the Princess Charlotte Bay region, in a relatively unpolluted environment, contrast with **(B)** inshore reefs further south that are exposed to agricultural runoff (Photos: K. Fabricius, AIMS).

Biodiversity Conservation Act 1999, in conjunction with the relevant Queensland legislation) to regulate catchment activities that lead to damage to the GBR.

- significantly increase funding to support catchment and coastal management to the required levels identified to address the pollution issues for the GBR by 2025.

Assessment of successful watershed management programs that lead to improved end-of-system ecosystem status clearly show that strong regulatory components were critical parts of successful policy mixes globally, and there are no reasons to believe that effectiveness should follow different rules in the GBRCA.

However, terrestrial pollution is not the only stress facing the ecosystems of the GBR, and it is the combination of multiple stresses (from global climate change, ocean acidification, fishing and pollution) that will escalate threats to the long-term viability of the system. The GBR has lost 50% of its coral cover between 1985 and 2012, mostly attributable to storms, COTS outbreaks and mass coral bleaching. Although poor water quality does not cause acute mass coral mortality, it slows coral growth and recovery from acute disturbances, such as heat stress-related mass coral bleaching and mortality. The GBR still has a high inherent capacity to recover in periods without disturbances, and recovery of the southern offshore reefs from COTS predation and storms since 2012 was encouraging. However, further mass bleaching events in 2016 and 2017 and a new COTS outbreak wave has intensified the pressure on coral populations, and many of the corals that have survived these two bleaching events are now eaten by another population outbreak wave of COTS. This situation makes a renewed focus on improving water quality even more urgent, since good water quality plays an important role in facilitating coral recruitment and curbing outbreak frequencies of COTS. However, management of water quality issues alone, no matter how effective, are unlikely to prevent long-term degradation of the GBRWHA without concurrent rapid and drastic reductions of carbon dioxide emissions.

Further reading

Australian and Queensland Governments (2017) *Great Barrier Reef Report Card 2016: Reef Water Quality Protection Plan.* State of Queensland, Brisbane, <http://www.reefplan.qld.gov.au/measuring-success/report-cards/2016/>.

Brodie J, Pearson RG (2016) Ecosystem health of the Great Barrier Reef: time for effective management action based on evidence. *Estuarine, Coastal and Shelf Science* **183**, 438–451. doi:10.1016/j.ecss.2016.05.008

Brodie J, De'ath G, Devlin M, Furnas M, Wright M (2007) Spatial and temporal patterns of near-surface chlorophyll a in the Great Barrier Reef lagoon. *Marine and Freshwater Research* **58**, 342–353. doi:10.1071/MF06236

Brodie J, Devlin M, Lewis S (2017) Potential enhanced survivorship of crown of thorns starfish larvae due to near-annual nutrient enrichment during secondary outbreaks on the central mid-shelf of the Great Barrier Reef, Australia. *Diversity (Basel)* **9**, 17. doi:10.3390/d9010017

Brodie J, Baird M, Waterhouse J, Mongin M, Skerratt J, Robillot C, *et al.* (2017) 'Development of basin-specific ecologically relevant water quality targets for the Great Barrier Reef. TropWATER Report No. 17/38'. James Cook University for the State of Queensland, Brisbane, <http://www.reefplan.qld.gov.au/about/assets/gbr-water-quality-targets-june2017.pdf>.

Fabricius KE (2005) Effects of terrestrial runoff on the ecology of corals and coral reefs: review and synthesis. *Marine Pollution Bulletin* **50**, 125–146. doi:10.1016/j.marpolbul.2004.11.028

Fabricius KE, Okaji K, De'ath G (2010) Three lines of evidence to link outbreaks of the crown-of-thorns seastar *Acanthaster planci* to the release of larval food limitation. *Coral Reefs* **29**, 593–605. doi:10.1007/s00338-010-0628-z

Fabricius KE, Logan M, Weeks SJ, Lewis SE, Brodie J (2016) Changes in water clarity in response to river discharges on the Great Barrier Reef continental shelf: 2002–2013. *Estuarine, Coastal and Shelf Science* **173**, A1–A15. doi:10.1016/j.ecss.2016.03.001

Great Barrier Reef Marine Park Authority (2009) *Water Quality Guidelines for the Great Barrier Reef Marine Park.* Great Barrier Reef Marine Park Authority, Townsville.

Great Barrier Reef Water Science Taskforce and the Office of the Great Barrier Reef Department of Environment and Heritage Protection (2016) 'Clean water for a healthy reef. Great Barrier Reef Water Science Taskforce final report'. Queensland Government, Brisbane, <www.gbr.qld.gov.au/taskforce/final-report/>.

Hughes TP, Kerry JT, Álvarez-Noriega M, Álvarez-Romero JG, Anderson KD, Baird AH, *et al.* (2017) Global warming and recurrent mass bleaching of corals. *Nature* **543**, 373–377. doi:10.1038/nature21707

Kroon FJ, Kuhnert PM, Henderson BL, Wilkinson SN, Kinsey-Henderson A, Abbott B, *et al.* (2012) River loads of suspended solids, nitrogen, phosphorus and herbicides delivered to the Great Barrier Reef lagoon. *Marine Pollution Bulletin* **65**, 167–181. doi:10.1016/j.marpolbul.2011.10.018

Kroon FJ, Thorburn P, Schaffelke B, Whitten S (2016) Towards protecting the Great Barrier Reef from land-based pollution. *Global Change Biology* **22**, 1985–2002. doi:10.1111/gcb.13262

Lewis SE, Brodie JE, Bainbridge ZT, Rohde KW, Davis AM, Masters BL, *et al.* (2009) Herbicides: a new threat to the Great Barrier Reef. *Environmental Pollution* **157**, 2470–2484. doi:10.1016/j.envpol.2009.03.006

Pratchett MS, Caballes CF, Wilmes JC, Matthews S, Mellin C, Sweatman HP, *et al.* (2017) 30 Years of research on crown-of-thorns starfish (1986–2016): scientific advances and emerging opportunities. *Diversity (Basel)* **9**, 41. doi:10.3390/d9040041.

Waterhouse J, Schaffelke B, Bartley R, Eberhard E, Brodie J, Star M, *et al.* (2017) '2017 Scientific Consensus Statement: land use impacts on Great Barrier Reef water quality and ecosystem condition'. Summary. State of Queensland, Brisbane,<http://www.reefplan.qld.gov.au/about/assets/2017-scientific-consensus-statement-summary.pdf>.

Wenger AS, Fabricius KE, Jones GP, Brodie JE (2015) Effects of sedimentation, eutrophication, and chemical pollution on coral reef fishes. In *Ecology of Fishes on Coral Reefs*. (Ed. C Mora) pp. 145–153. Cambridge University Press, Cambridge, UK.

14

Planning and managing the Great Barrier Reef Marine Park

J. C. Day

Introduction

The *Great Barrier Reef Marine Park Act 1975* (GBRMP Act) was pioneering legislation for its time, protecting the Great Barrier Reef (GBR) while also allowing sustainable use, provided such use did not compromise the conservation of the GBR. This federal Act created a statutory agency, the Great Barrier Reef Marine Park Authority (GBRMPA), whose responsibilities included recommending which parts of the GBR Region should be declared as Marine Park (GBRMP). The Act initially defined only the outer boundary of an area (i.e. the GBR Region) within which sections of the GBRMP could be created, and specified how zoning plans needed to be developed. None of the area-specific management arrangements, including zoning, were in place at this time.

Between 1979 and 1988, various sections of the GBRMP were sequentially declared and then zoned. The first zoning plan for a small section of the GBRMP (Capricornia) was finalised in 1981, and introduced a spectrum of zone types ranging from a General Use Zone (the least restrictive zone, allowing most reasonable uses), through to a Preservation Zone (a small 'no-go' area, set aside as scientific reference area). While refined, these zones still exist today (see Table 14.1).

In 1981 the GBR was included on the UNESCO List of World Heritage. This followed an evaluation by the International Union for the Conservation of Nature (IUCN) that stated 'If only one coral reef in the world were to be chosen for the World Heritage List, the Great Barrier Reef is the site to be chosen' (IUCN 1981).

By 1988, four different zoning plans had been finalised, collectively covering almost all of the GBR Region. The location of the original no-take zones within these zoning plans reflected a primary focus on coral reef habitats, which were considered the most important habitats at that time.

The statutory zone objectives shown in Table 14.1 are set out in the current (2003) Zoning Plan, which is subordinate legislation under the Act. Within each zone, certain activities are allowed 'as-of-right' (i.e. no permit is required, but users must comply with any legislative requirements in force), some activities require a permit and some activities are prohibited (see Fig. 14.1). The Zoning Plan provides details on what, and where, specific activities are allowed, and lists those activities requiring a permit.

Over the decades, our understanding of the GBR, its uses and its management have greatly increased through considerable research and adaptive management. The legislation has been amended in the light of operational reviews, and today the primary object of the Act is '... to provide for the long term protection and conservation of

Table 14.1. Current GBR zones and zone objectives (as of 2018)

Zone name	Zone colour	Statutory objective(s) of the zone
General Use Zone	light blue	to provide for the conservation of areas of the Marine Park, while providing opportunities for reasonable use.
Habitat Protection Zone	dark blue	(a) to provide for the conservation of areas of the Marine Park through the protection and management of sensitive habitats, generally free from potentially damaging activities; and (b) subject to (a), to provide opportunities for reasonable use.
Conservation Park Zone	yellow	(a) to provide for the conservation of areas of the Marine Park; (b) subject to (a), to provide opportunities for reasonable use and enjoyment, including limited extractive use.
Buffer Zone	olive green	(a) to provide for the protection of the natural integrity and values of areas of the Marine Park, generally free from extractive activities; (b) subject to (a), to provide opportunities for: (i) certain activities, including the presentation of the values of the Marine Park, to be undertaken in relatively undisturbed areas; and (ii) trolling for pelagic species.
Scientific Research Zone	orange	(a) to provide for the protection of the natural integrity and values of areas of the Marine Park, generally free from extractive activities; and (b) subject to (a), to provide opportunities for scientific research to be undertaken in relatively undisturbed areas.
Marine National Park Zone	green	(a) to provide for the protection of the natural integrity and values of areas of the Marine Park, generally free from extractive activities; and (b) subject to (a), to provide opportunities for certain activities, including the presentation of the values of the Marine Park, to be undertaken in relatively undisturbed areas.
Preservation Zone	pink	to provide for the preservation of the natural integrity and values of areas of the Marine Park, generally undisturbed by human activities.
Commonwealth Islands Zone	cream	(a) to provide for the conservation of the natural integrity and values areas of the Marine Park above low water mark; and (b) to provide for use of the zone by the Commonwealth; and (c) subject to (a), to provide for facilities and uses consistent with the values of the area.

the environment, biodiversity and heritage values of the Great Barrier Reef Region'. This reflects national obligations with respect to protecting the Outstanding Universal Value of the GBR World Heritage Area, but the Act also refers to other objectives, including allowing ecologically sustainable use of the GBR Region. However, all the subsidiary objectives must be consistent with the primary objective. The GBRMP is therefore a comprehensive multiple-use marine park; this means that the entire GBRMP is protected, but zoned to allow a range of uses and most reasonable activities to occur in different zones, and with the stated intention to minimise impacts and conflicts.

Today the GBRMP covers 344 400 km^2 – an area bigger than Victoria and Tasmania combined, or about the same size as Italy, Malaysia or Japan. The GBRMP extends 2300 km along the Queensland coast, and includes most of the waters from low water mark on the mainland coast, to the outer (seaward) boundary of the GBRMP. It is complex jurisdictionally, with both the Australian (federal) and Queensland (state) Governments involved in the management of the waters and islands within its outer boundaries.

The federal GBRMP includes marine areas below the low water mark plus 70 islands or parts of islands that contain lighthouses or

ACTIVITIES GUIDE

(see relevant *Zoning Plans* and *Regulations* for details)

	General Use Zone	Habitat Protection Zone	Conservation Park Zone	Buffer Zone	Scientific Research Zone [3]	Marine National Park Zone	Preservation Zone
Aquaculture	Permit	Permit	Permit [1]	✗	✗	✗	✗
Bait netting	✓	✓	✓ [2]	✗	✗	✗	✗
Boating, diving, photography	✓	✓	✓	✓	✓ [3]	✓	✗
Crabbing (trapping)	✓	✓	✓ [4]	✗	✗	✗	✗
Harvest fishing for aquarium fish, coral and beachworm	Permit	Permit	Permit [1]	✗	✗	✗	✗
Harvest fishing for sea cucumber, trochus, tropical rock lobster	Permit	Permit	✗	✗	✗	✗	✗
Limited collecting	✓ [5]	✓ [5]	✓ [5]	✗	✗	✗	✗
Limited spearfishing (snorkel only)	✓	✓	✓ [1]	✗	✗	✗	✗
Line fishing	✓ [6]	✓ [6]	✓ [7]	✗	✗	✗	✗
Netting (other than bait netting)	✓	✓	✗	✗	✗	✗	✗
Research (other than limited impact research)	Permit	Permit	Permit	Permit	Permit	Permit	Permit
Shipping (other than in a designated shipping area)	✓	Permit	Permit	Permit	Permit	Permit	✗
Tourism programme	Permit	Permit	Permit	Permit	Permit	Permit	✗
Traditional use of marine resources	✓ [8]	✓ [8]	✓ [8]	✓ [8]	✓ [8]	✓ [8]	✗
Trawling	✓	✗	✗	✗	✗	✗	✗
Trolling	✓ [6]	✓ [6]	✓ [6]	✓ [6,9]	✗	✗	✗

PLEASE NOTE: This guide provides an introduction to Zoning in the Great Barrier Reef Marine Parks.

1. Restrictions apply to aquaculture, spearfishing and harvest fishing for aquarium fish, beachworm and coral in the Conservation Park Zone.
2. No take of bream, flathead or whiting by commercial bait netters.
3. Except for One Tree Island Reef (SR-23-2010) and Australian Institute of Marine Science (SR-19-2008) which are closed to public access and shown as orange, all other Scientific Research Zones are shown as green with an orange outline.
4. Limited to 4 catch apparatus per person (eg. crab pots, collapsible traps or dillies).
5. By hand or hand-held implement and generally no more than 5 of a species.
6. Maximum of 6 hooks attached to no more than 3 hand-held rods or handlines per person.
7. Limited to 1 hook attached to 1 hand-held rod or handline per person. Only 1 dory detached from a commercial fishing vessel.
8. Apart from traditional use of marine resources in accordance with s.211 of the *Native Title Act 1993*, an accredited Traditional Use of Marine Resources Agreement or permit is required.
9. Pelagic species only. Seasonal Closures apply to some Buffer Zones.

Detailed information is contained in the *Great Barrier Reef Marine Park Zoning Plan 2003* and *Regulations* and the *Marine Parks (Great Barrier Reef Coast) Zoning Plan 2004*.

- Permits are required for most other activities not listed above.
- Commonwealth owned islands in the Great Barrier Reef Marine Park are zoned "Commonwealth Islands Zone" - shown as cream.
- All Commonwealth Islands may not be shown.
- Special Management Areas may provide additional restrictions at some locations.
- The Zoning Plan does not affect the operation of s.211 of the *Native Title Act 1993*.

ACCESS TO ALL ZONES IS PERMITTED IN AN EMERGENCY.

Fig. 14.1. Activities guide for the Great Barrier Reef (Source: Courtesy of the Great Barrier Reef Marine Park Authority © Commonwealth of Australia (GBRMPA)).

defence training areas under federal jurisdiction. The federal GBRMP does not include the majority of the ~1050 islands within the park's outer boundary, which are under Queensland jurisdiction; about half these islands are declared national parks. The federal GBRMP also does not include 13 coastal exclusion areas around ports, nor any Queensland 'internal waters' (e.g. waters within long narrow channels or bays on the landward side of a defined baseline).

The GBRMPA, a Commonwealth (federal) statutory authority, is the principal adviser to the Australian Government for the planning and management of the GBRMP, and is part of the Australian Government's Environment and Energy portfolio. The Australian Government's primary legislation for environmental regulation, the *Environment Protection and Biodiversity Conservation Act 1999*, also provides for such aspects as the protection of World Heritage values, biodiversity conservation, and the protection of threatened and migratory species.

While the GBRMP is today relatively small compared with many more recent large-scale marine protected areas (MPAs), on a global scale the GBR:

- is one of the best known MPAs
- is one of the most methodically planned and comprehensively managed MPAs, particularly over such a large scale
- contains the largest systematically declared network of highly protected areas
- contains the greatest species diversity of any World Heritage Area on the planet, primarily due to its extensive latitudinal and cross-shelf diversity (Day 2016).

As the world's largest coral reef ecosystem, the GBR is a significant global resource. The GBR and its associated features contribute significantly to Australia's economy, with the direct and indirect value-added contribution estimated at A$6.4 billion in 2015–16. This includes A$5.7 billion from the tourism industry, A$346 million from recreational activities and A$162 million from commercial fishing (Deloitte Access Economics 2017). This economic activity generates ~64 000 jobs, mostly in the tourism industry, which brings around 2 million tourists to the GBR each year. In addition, it is estimated some 14 million recreational visits are made to the GBR each year, and some 69 000 recreational vessels are registered in the area adjoining the GBR. All these industries, and their flow-on activities, underpin a significant and growing proportion of Queensland's regional economy. They rely on the continued health of the GBR system for their long-term economic sustainability.

While coral reefs remain the major drawcard for the area, reefs comprise less than 10% of the overall extent of the GBRMP. The remaining non-reefal areas are seagrass beds, shoals, sandy or muddy seabeds (61%), continental shelf/slope (15%) and deep oceanic habitats (16%) up to 250 km offshore.

Because of the iconic status of the GBR, many people think that the entire GBRMP is a marine sanctuary or a marine national park, and therefore protected equally throughout. Many do not understand that the GBRMP is a multiple-use area, allowing a wide range of activities and uses including many extractive industries (but not mining nor drilling for oil), while still protecting one of the world's most diverse ecosystems. The multiple-use zoning system provides high levels of protection for specific areas, while allowing a variety of other uses to continue in certain zones. Such uses include shipping, dredging, aquaculture, tourism, boating, diving, military training, commercial fishing and recreational fishing.

Under the GBRMP Act, the Authority has power to perform any of its functions in cooperation with Queensland, with an authority of that state, or with a local governing body in that state. This is important given the adjoining Queensland marine parks, national parks and islands.

Management of the Great Barrier Reef Marine Park

The Field Management Program (i.e. management 'on the water') is a jointly funded cooperative partnership between the Australian and Queensland Governments. The program coordinates the day-to-day activities and field operations required for the management of the GBRMP, the adjoining Queensland marine parks and the GBR World Heritage Area (which includes all the islands and the intertidal waters). Field management is undertaken by several state and federal government agencies working under formal arrangements or in partnership with the GBRMPA. These government agencies include the Queensland Parks and Wildlife Service, the Queensland Boating and Fisheries Patrol, the Queensland Water Police, Australian Border Force, the Australian Federal Police, the Australian Maritime Safety Authority and the Australian Quarantine and Inspection Service. The GBRMPA also directly participates in field management activities such as compliance, monitoring and the assessment of permits.

In addition to the field management arrangements, Queensland Government agencies with responsibilities for policy coordination, environment, local government, maritime matters, catchments, land use and fisheries are actively involved in administration and management of issues pertaining to the health and operation of the GBR. To undertake its functions effectively, the GBRMPA maintains liaison and policy coordination arrangements with all of these agencies, both at the operational and strategic levels. A close working partnership between Queensland and the GBRMPA has evolved over 30 years, and today includes such aspects as complementary zoning, joint permits and joint field management. This partnership has ensured the effective management of the complex and inter-related mix of marine, coastal and island issues, and provides for integrated management of the GBR on a whole-of-ecosystem basis.

This fundamental working relationship between the federal and Queensland Governments and their various agencies is important for the effective management of the GBR, particularly as the two jurisdictions have differing interpretations of some aspects such as the location of 'low-water', which would otherwise require this boundary to be identified and mapped. Staff of the Authority also maintain strong partnerships with a wide range of agencies, stakeholders, councils, Traditional Owners, community members and researchers with an interest in the protection, ecologically sustainable use, understanding and enjoyment of the GBR.

Having good relationships with scientists also helps the managers access the best available information for decision making that is essential for scientifically based management of the GBRMP. Over the last 30 years there has been a huge growth in scientific understanding of ecological aspects of the GBR, including ecosystem processes, connectivity, and the relationships between species and the physical environment.

Some sections of the GBRMP that were zoned in the 1980s were subsequently re-zoned, and by the late 1990s, ~15 800 km^2 (~4.6%) of the GBRMP were in 'no-take' areas (Marine National Park Zones), with a further 450 km^2 (~0.13% of the GBRMP) set aside within small 'no-go' areas (Preservation Zones).

Why was the entire GBR rezoned in the early 2000s?

By the mid-late 1990s, substantial research and surveys of GBR biodiversity had been undertaken, and reporting obligations under the World Heritage Convention had been developed. GBRMPA commissioned an expert report to more fully describe the Outstanding Universal Value of the GBR World Heritage Area. This confirmed that the existing zoning did not adequately protect the range of biodiversity and particularly the range of non-coral reef habitats by then known to exist within the GBRMP. Furthermore, there were growing concerns following the 1998–99 mass coral bleaching and recurrence of crown-of-thorns starfish outbreaks that the zoning was inadequate to ensure that the entire ecosystem remained healthy, productive and resilient into the future.

Between May 1999 and December 2003, the GBRMPA undertook a comprehensive planning and consultative program to develop a new zoning plan for the GBRMP. The primary aim of the program was to better protect the range of biodiversity in the GBR, by increasing the extent of no-take areas (locally known as 'green zones'), ensuring they included 'representative' examples of all the different habitat types, otherwise known as bioregions (hence the name, the Representative Areas Program or RAP). A further aim was to maximise positive and minimise negative impacts on the users of the GBRMP. Scientific input, community involvement and agency innovation all contributed to achieving these aims.

During the rezoning program, all components of the existing zoning plans were open for comment and alteration. Given the previous and existing zoning plans had all been progressively developed over 17 years, some of the terms, management provisions, zone names and zone objectives differed slightly between various sections of the GBRMP. A new single zoning plan was therefore developed for the entire GBRMP enabling the planners to also take into account various important planning tasks, including:

- zoning 28 new coastal sections that were added to the GBRMP during 2000 and 2001
- standardising terminology to make it consistent throughout
- implementing simpler coordinate-based boundary descriptions for all zone boundaries.

The following five factors contributed to the success of the new zoning plan:

1. ***Use of the best available scientific knowledge and independent experts*** – Independent experts greatly assisted in the development of several 'products' that were important to the planning process, and were widely available for discussion early in the planning program, in particular:
   - ➤ GBR bioregionalisation. A fundamental foundation for the new zoning plan was a map of 30 reef and 40 non-reef 'bioregions' that was developed early in the RAP.
   - ➤ A comprehensive range of biological and physical information across the GBRMP was used to define the bioregions. Staff of the GBRMPA initially collated information from numerous scientists with expert knowledge of the GBR. The most appropriate datasets were then used in classification and regression tree analyses to spatially cluster areas of similar species composition. Several workshops, comprising reef and non-reef experts, then used all these data and analyses, plus their experience, to spatially describe the biodiversity of the GBR and develop the map of 70 bioregions.
   - ➤ It was decided to map diversity at the scale of tens to hundreds of kilometres because this was a scale over which habitats change markedly. It was also a scale at which most relevant information was available and it was a meaningful scale for subsequent planning and management. Areas of relative homogeneity were labelled 'bioregions' to facilitate communication with stakeholders.
   - ➤ A draft version of the bioregionalisation was made available for public comment, recognising that many local 'experts', including commercial and recreational fishers, and coastal residents have specialist knowledge about the GBR. This led to additional information and nine major refinements to the draft regionalisation.
   - ➤ Operating principles for developing the new network. External natural science and social-economic cultural advisory committees were used to develop 11 biophysical operational principles and four socio-economic, cultural and management feasibility operational principles. These principles clarified the planning 'rules' up front for all to see before any new zones were proposed. The biophysical principles included recommendations for minimum percentages of no-take areas for each

bioregion and each known habitat type, plus recommendations for representing cross-shelf and latitudinal diversity and for replication. Given the uncertainty about what amounts would be adequate for effective conservation, the recommendations were considered to be the minimum, in the context of global experience. The social, economic, cultural and management feasibility principles aimed to maximise complementarity of zoning with human uses and values.

2. ***Integrated approaches*** – A combination of expert opinion, stakeholder involvement and analytical approaches were used to identify options for possible zoning networks. The linking of science, technical support and community participation was an essential three-way dynamic in the planning process. The analytical tools applied included marine reserve design software, adapted and expanded for use in the RAP, and a suite of GIS-based spatial analysis tools. The analytical software enabled the GBRMPA to integrate several data layers representing biophysical, social and economic values, and enabled several zoning options to be assessed.
3. ***Effective leadership*** – This occurred at multiple levels including political leadership, leadership of GBRMPA and within the agency, and sectoral leadership, and these were all important factors in the success of RAP.
4. ***High levels of public participation*** – The extensive public participation program during the RAP included a huge amount of public engagement and resulting public input. As required by the relevant legislation, there were two formal phases of community engagement during the RAP: the first calling for input into the preparation of a new zoning plan, and the second providing the draft zoning plan for public comment. The resulting 31 690 public submissions (10 190 in the first formal phase; 21 500 in the second phase), many of which included maps, were unprecedented compared with previous planning programs in the GBR. They necessitated the development of new, fast and effective processes for analysing and recording the range of information that was received by the GBRMPA. A large number of the submissions included spatial information, including ~5800 maps in the second formal phase alone. All submissions, including this spatial information, were considered, coded and analysed, and the maps were digitised and/or scanned.

   In addition, there were many hundreds of formal and informal meetings and information sessions in over 90 centres along and beyond the GBR coast. This included meetings with local communities, commercial and recreational fishing organisations, Traditional Owners, tourism operators, and conservation groups. Meetings also occurred with representative organisations such as Sunfish, the Association of Marine Park Tourism Operators, World Wildlife Fund Australia, and all branches of the Queensland Seafood Industry Association within the GBR catchment area.

   Many modifications were made to the draft zoning plan as a result of the detailed information provided in submissions, from these meetings and other information that was received; however, in some locations there were limited options available to modify proposed no-take zones, particularly in inshore coastal areas and still achieve the minimum levels of protection recommended by the scientists.

5. ***Consequent socio-political support*** – The significant changes between the initial zoning, the draft zoning plan and the final zoning plan, as accepted by Parliament, can be seen in the *Review of the Great Barrier Reef Marine Park Act 1975 – Review Panel Report*, released in 2006 (available at http://www.environment.gov.au/marine/gbr/publications/review-great-barrier-reef-marine-park-act-1975-review-panel-report). Maps 9, 10 and 11 on pages 69–71 of that report

highlight the differences. Many of the aspects outlined in this chapter are illustrated in the Capricorn-Bunker case study shown in more detail in the same report (i.e. pp. 78–90 of the same webpage).

In accordance with the legislation, the zoning plan was submitted to both Houses of the Federal Parliament in December 2003. Following a statutory review period, the Minister announced that the new zoning plan would be implemented on 1 July 2004.

In November 2004, the Queensland Government 'mirrored' the new zoning in most of the adjoining state waters (i.e. intertidal waters and some other areas deemed to be state waters), so now there is complementary zoning for virtually all the state and federal waters within the entire GBR World Heritage Area. Table 14.2 indicates the area of each zone type in effect today (2017).

All five aspects were important, but the last three were absolutely critical – having perfect science would not have guaranteed the RAP outcome; the successful outcome did, however, depend heavily on effective leadership, high levels of public participation and the consequent social-political support.

Table 14.2 also shows how the eight different zone types equate to four of the six IUCN global categories of protected areas (noting there is not an exact correlation between the GBR zone objectives and the IUCN categories).

While the comprehensive zoning plan remains the cornerstone of the multiple-use management approach in the GBRMP, effective marine conservation requires a variety of other marine spatial planning tools that are applied as additional layers of management; these include:

- statutory plans of management dealing with uses in key areas
- special management area regulations
- site plans
- other spatial and temporal management provisions (e.g. shipping lanes, agreements with Traditional Owners, closures at fish-spawning times, other fisheries regulations, defence training areas).

Table 14.2. Total area of zone types within the Great Barrier Reef Marine Park (as at 2017)

Zone name	Area (km^2)	% of GBRMP	Equivalent IUCN category
Preservation*	710	<1	IA
Marine National Park*	114 530	33.3	II
Scientific Research*	155	<1	IA
Buffer	9880	2.9	IV
Conservation Park	5160	1.5	IV
Habitat Protection	97 250	28.2	VI
General Use	116 530	33.8	VI
Commonwealth Islands	185	<1	II
Total	**344 400**	**100**	

* Effectively no-take zones where extractive activities are generally not permitted.

These tools have been found to be more effective than zoning for managing specific industries or activities such as tourism, shipping, ports, traditional use of marine resources, defence and some fisheries regulations. When used in conjunction with the underlying zoning, they collectively comprise the integrated management approach for the GBR and help achieve ecological protection and other management objectives.

As a result of the 2003 zoning, an independent review of the Act was undertaken in 2006. One outcome of the review was a legislated requirement for a comprehensive 5-yearly report on the status of, and outlook for, the GBR in order to guide future management action. The statutory report includes a systematic assessment of all habitats and species, ecosystem processes, uses, heritage values, the factors affecting all the values, and an assessment of management effectiveness. The Outlook Report was well received by government and was 'a significant major innovation … fundamental to tracking the effectiveness/inadequacies and the needs for continuing management efforts to maintain the OUV of the GBRWHA' (R.A. Kenchington *pers. comm.*). An important focus of the 2009 report was establishing a framework and methodology that was repeatable in the subsequent 2014 report. The assessment methodology has proven to be effective and been well accepted by stakeholders and the target audience of government decision makers.

Conclusions

The approach that was undertaken in the RAP has been recognised as one of the most comprehensive advances in the protection of marine biodiversity and marine conservation within any MPA. It built upon the lessons from the first round of GBR zoning, consolidating the large-scale approach to managing the GBR for both conservation and sustainable use. The final outcome, including the high-level protection of representative examples of every habitat, and the increase in no-take zones to more than 33% (over 115 500 km^2), comprises the world's largest systematic network of highly protected no-take zones.

The zoning plan, along with the many other spatial and temporal management tools applied in the GBRMP (as outlined above), collectively make up the comprehensive management approach, which has been adaptively improved over the past four decades.

This comprehensive management approach, however, will not guarantee the future or sustainability of the GBR. Although the past management approaches in the GBR are considered by many to have been both effective and appropriate, today's pressures and cumulative impacts mean that these past approaches are not sufficient to retain the values for which the GBR has been long been acclaimed. The GBR is under pressure from a wide range of human uses and natural impacts, and the values of both the GBRMP and the GBR World Heritage Area are not static. Furthermore, the use of the GBRMP has escalated rapidly in the 40+ years since its establishment. Use patterns and technology are constantly changing and the marine environment itself is dynamic.

Solutions to tackle today's pressures are not simple, because those pressures are the accumulation of many additive and synergistic activities rather than one or two identifiable threats. However, if measures to adapt to pressures such as climate change are not taken, and if certain development decisions continue to remove options for long-term conservation, the GBR's future is unlikely to be secure.

A wide range of strategies must continue to be implemented to ensure the GBR is protected now and into the future. These strategies must:

- take into account the effects of climate change through local, national and global efforts
- greatly improve water quality through the Reef Water Quality Protection Plan
- increase compliance with the Zoning Plan and supporting Regulations
- ensure sustainable fisheries
- ensure sustainable coastal development.

Unless all these strategies are more effectively implemented (and that will require more resources than are currently being applied), then the health and resilience of the GBR is threatened. The Australian Government's *GBR Outlook Report 2014* (p. vi), concluded:

> *Even with the recent management initiatives to reduce threats and improve resilience, the overall outlook for the Great Barrier Reef is poor, has worsened since 2009 and is expected to further deteriorate in the future. Greater reductions of threats at all levels, Reef-wide, regional and local, are required to prevent the projected declines in the Great Barrier Reef and to improve its capacity to recover.*

Similarly, IUCN's 2017 assessment of the outlook for each of the world's natural world heritage properties, has assessed the outlook for the GBR World Heritage Area as being of 'Significant concern'.

Given these assessments and dire predictions, more must be done at all levels to mitigate the wide range of the threats identified in many scientific studies and agency assessment reports. This is also essential to ensure a sustainable future for the many users, industries and communities that are dependent on a healthy GBR.

Australia is a relatively rich country and previously had a reputation for being a global leader in marine conservation, so this is doable. It will, however, require far greater resources and some hard decisions to restore the values of the GBR and to fulfil Australia's global responsibilities. Both are needed for the sake of the Reef and future generations.

For more information about the Marine Park, such as zoning and zoning maps, visit the GBRMPA website (www.gbrmpa.gov.au). Copies of the zoning maps are available in major centres along the GBR coast.

Further reading

Australian National Audit Office (2015) *Regulation of Great Barrier Reef Marine Park Permits and Approvals*. Australian Government, Canberra, <https://www.anao.gov.au/work/performance-audit/regulation-great-barrier-reef-marine-park-permits-and-approvals>.

Brodie J, Waterhouse J (2012) A critical review of environmental management of the 'not so Great' Barrier Reef. *Estuarine, Coastal and Shelf Science* **104–105**, 1–22. doi:10.1016/j.ecss.2012.03.012

Commonwealth of Australia (2015) *Reef 2050 Long-term Sustainability Plan*. Commonwealth of Australia, Canberra.

Day JC (2015) Marine spatial planning (MSP): one of the fundamental tools to help achieve effective marine conservation in the Great Barrier Reef. In *Transboundary Marine Spatial Planning and International Law*. (Eds D Hassan, T Kuokkanen and N Soininen) pp. 103–131. Earthscan (Routledge), London, UK.

Day JC (2016) The Great Barrier Reef Marine Park – the grandfather of modern MPAs. In *Big, Bold and Blue: Lessons from Australia's Marine Protected Areas*. (Eds J Fitzsimmons and G Wescott) pp. 65–97. CSIRO Publishing, Melbourne.

Day JC (2017) Effective public participation is fundamental for marine conservation – lessons from a large-scale MPA. *Coastal Management* **45**(6), 470–486[+ Supplementary]. doi: 10.1080/08920753.2017.1373452

Day JC, Dobbs K (2013) Effective governance of a large and complex cross-jurisdictional marine protected area: Australia's Great Barrier Reef. *Marine Policy* **41**, 14–24. doi:10.1016/j.marpol.2012.12.020

Deloitte Access Economics (2017) *At what price? The Economic, Social and Icon Value of the Great Barrier Reef*. Deloitte Access Economics, Brisbane, <https://www2.deloitte.com/content/dam/Deloitte/au/Documents/Economics/deloitte-au-economics-great-barrier-reef-230617.pdf>.

Dobbs K, Day J, Skeat H, Baldwin J, Molloy F, McCook L *et al.* (2011) Developing a long-term outlook for the Great Barrier Reef, Australia: a framework for adaptive management reporting underpinning an ecosystem-based management approach. *Marine Policy* **35**(2), 233–240.

Emslie MJ, Logan M, Williamson DA, Ayling AM, MacNeil A, Ceccareli D, *et al.* (2015) Expectations and outcomes of reserve network performance following re-zoning of the Great Barrier Reef Marine Park. *Current Biology* **25**, 983–992. doi:10.1016/j.cub.2015.01.073

Fernandes L, Day J, Lewis A, Slegers S, Kerrigan B, Breen D, *et al.* (2005) Establishing representative No-Take Areas in the Great Barrier Reef: large-scale implementation of theory on Marine Protected Areas. *Conservation Biology* **19**(6), 1733–1744. doi:10.1111/j.1523-1739.2005.00302.x

GBRMPA (2004) *Great Barrier Reef Marine Park Zoning Plan 2003*. Great Barrier Reef Marine Park Authority, Townsville.

GBRMPA (2005) *Report on the Great Barrier Reef Marine Park Zoning Plan 2003*. Great Barrier Reef Marine Park Authority, Townsville.

GBRMPA (2012) *Great Barrier Reef Biodiversity Conservation Strategy 2012*. Great Barrier Reef Marine Park Authority, Townsville.

GBRMPA (2014) *Great Barrier Reef Outlook Report 2014*. Great Barrier Reef Marine Park Authority, Townsville.

Harrison HB, Williamson DH, Evans RD, Almany GR, Thorrold SR (2012) Larval export from marine reserves and the recruitment benefit for fish and fisheries. *Current Biology* **22**, 1023–1028. doi:10.1016/j.cub.2012.04.008

Hughes TP, Day JC, Brodie J (2015) Securing the future of the Great Barrier Reef. *Nature Climate Change* **5**, 508–511. doi:10.1038/nclimate2604

IUCN (1981) *World Heritage Nomination IUCN Technical Review: Great Barrier Reef*. IUCN, Gland, Switzerland, <http://whc.unesco.org/archive/advisory_body_evaluation/154.pdf>.

IUCN (2017) World Heritage Outlook – Great Barrier Reef. [Online]. IUCN, Gland, Switzerland, <http://www.worldheritageoutlook.iucn.org/explore-sites/wdpaid/2571>.

Johnson JE, Marshall PA (Eds) (2007) *Climate change and the Great Barrier Reef: a Vulnerability Assessment*. Great Barrier Reef Marine Park Authority and Australian Greenhouse Office, Townsville.

Kerrigan B, Breen D, De'ath G, Day JC, Fernandes L, Tobin R, K (2010) *Classifying the Biodiversity of the Great Barrier Reef World Heritage Area*. Research Publication No. 104 Great Barrier Reef Marine Park Authority, Townsville.

Lucas PHC, Webb T, Valentine PS, Marsh H (1997) *The Outstanding Universal Value of the Great Barrier Reef World Heritage Area*. Great Barrier Reef Marine Park Authority, Townsville.

Macintosh A, Bonyhady T, Wilkinson D (2010) Dealing with interests displaced by marine protected areas: a case study on the Great Barrier Reef Marine Park Structural Adjustment Package. *Ocean and Coastal Management* **53**, 581–588. doi:10.1016/j.ocecoaman.2010.06.012

McCook LJ, Ayling T, Cappo M, Choat JH, Evans RD, De Freitas DM, et al. (2010) Adaptive management of the Great Barrier Reef: a globally significant demonstration of the benefits of networks of marine reserves. *Proceedings of the National Academy of Sciences of the United States of America* **107**, 18278–18285. doi:10.1073/pnas.0909335107

Russ GR, Cheal AJ, Dolman AM, Emslie MJ, Evans RD, Miller I, et al. (2008) Rapid increase in fish numbers follows creation of world's largest marine reserve network. *Current Biology* **18**, 514–515. doi:10.1016/j.cub.2008.04.016

World Heritage Committee (2014) Decision: WHC 38 COM 7B.63, Great Barrier Reef (Australia) (N 154). World Heritage Committee, Doha, Qatar, <http://whc.unesco.org/en/decisions/6049>.

World Heritage Committee (2015) Recording of Committee discussion about the Great Barrier Reef, 1 July 2015. World Heritage Committee, Bonn, Germany, <http://whc.unesco.org/en/sessions/39com/records/?day=2015-07-01#trv_TzLFMKfE12876>.

SECTION 3

Overview of reef biodiversity and organisms

15

Biodiversity

P. A. Hutchings and M. J. Kingsford

Declining biodiversity and species abundance on coral reefs is a global concern. Patterns of biodiversity are indicative of the 'health' of reefs and this is particularly important to determine in the face of multiple perturbations. Furthermore, there is a need to understand the mechanisms that regulate the structure of assemblages so that their resilience to change can be determined and potentially improved. Biodiversity has generally referred to species richness (i.e. number of species), but other definitions are also common and in some cases are perhaps more relevant to the challenges of managing reefs. It is critical that biodiversity is defined carefully because it could refer to species richness, genetic diversity, habitat diversity, structural diversity, the diversity of functional groups (e.g. trophic groups or life history stages) or even life history traits (e.g. feeding type, growth forms, reproductive strategy, longevity). Some definitions can be a proxy for species richness. For example, species representation and abundance are known to vary among habitats. Habitat richness, therefore, can be especially relevant where the taxonomy of organisms in the various habitats is poorly known and there are concerns to protect species richness. A critical component of the zoning plan of the GBR (the Representative Areas Program, see Chapter 14) was partly based on the protection of different habitats and used these as surrogates to conserve species diversity because the biota on the GBR is poorly known apart from the corals and fish. In this chapter we focus on patterns of species richness at different spatial and temporal scales. We also note that descriptions of biodiversity and an understanding of processes influencing biodiversity (e.g. the impact of human activities) are critical for ecosystem management.

Coral reefs occur in a broad band around the equator, wherever suitable depths occur, between latitudes of 30°N and 30°S, although some coral reef development occurs further south and north where water currents ensure that for most of the year water temperatures do not drop below 18°C (see Chapters 2 and 8). For example, the gyre from the East Australian Current diverges eastwards south of the GBR and enables coral reefs to be developed on the east coast of Lord Howe Island at 31°30′S. This is the southernmost coral reef in the world. Species richness varies at scales that range from ocean basins and seas to microhabitats on reefs (Box 15.1 and Box 15.2). The richest reefs in the world in terms of number of species occur in the triangle (= 'Coral Triangle') that includes the northern tip of the Philippines, the eastern and western tips of Indonesia, and Papua New Guinea, although today many of these reefs have been damaged by human activities, including, climate change, dynamite fishing, overfishing and coastal developments.

An illustration of this diversity is the recent intensive survey of all the coral reef environments within an area of 295 km^2 in New Caledonia, which resulted in the collection of over 2700 species of molluscs. This is much larger than the number of species recorded from similar areas anywhere else in the world. Of these molluscs, 32% were found only at one site and 20% of species were represented by just a single individual, indicating that rare species make up a considerable proportion of the fauna. In addition, many of the species were undescribed. In the entire Indo-Pacific, in the well-known group of opisthobranchs (nudibranchs) it is estimated that at least 30% remain to be formally described. In other invertebrate groups much of the fauna remains even to be collected, let alone formally described.

Many factors are responsible for the high diversity of biota found on coral reefs (e.g. Box 15.3), but in part it is due to the diversity of habitats found within coral reefs – ranging from soft sediments (both vegetated and unvegetated), various reefal habitats, and pelagic and coastal habitats that include mangroves and saltmarsh. Habitats are usually characterised by particular habitat-forming organisms and variation in geomorphology. For example, *Porites cylindrica* is a habitat-forming coral with spur and groove geomorphology, creating a shallow habitat (see Chapter 5). Further, the presence of coral *Porites cylindrica*, which has a complex morphology, is a good predictor of local reef fish species richness and abundance.

For all groups of organisms associated with coral reefs, the highest number of genera occurs in the Coral Triangle, with numbers declining both eastwards and westwards as well as north and south of this area. For example, while a total of 70 coral genera occur in the Indo-Pacific, most of these occur in the triangle with numbers declining across the Pacific with 10 in Hawaii and only two on the Pacific coast of America (Fig. 15.1). The decline is less across the Indian Ocean, with 50 genera present on the eastern Indian Ocean reefs. The Indo-Pacific is much richer than the Caribbean where only 20 coral genera occur, with only *Acropora, Favia, Leptoseris, Madracis, Montastrea, Porites, Scolymia* and *Siderastrea* being shared between these two coral realms, and no species of corals occur in both realms. Similarly, as one moves north or south of this triangle, numbers of genera decline, with only 30 genera found at Lord Howe Island and 50 at Ningaloo on the West Australian coast, and numbers continuing to decline southwards down to two genera at Rottnest Island in the far south-west corner of Australia. If we now consider species of corals we find similar patterns. The genus *Acropora* represents the largest extant group of reef-building corals and over 370 species have been described, with several major biogeographic patterns observed (Fig. 15.2). First, there is a distinct separation

Box 15.1. Major sources of spatial and temporal variation in species richness

Spatial variation

Among tropical seas and oceans (e.g. Caribbean versus Pacific)
With distance from diversity hot-spots (e.g. Indonesia, Philippines triangle)
With increasing latitude (e.g. along the Great Barrier Reef)
Cross shelf (e.g. from mainland to the shelf break of the GBR)
Among seascapes (e.g. reefs versus inter-reefal areas)
Windward versus leeward sides of reefs
Among habitats within reefs
With depth on reefs, the shelf and down the continental slope

Temporal variation

Evolutional time (speciation, extinction)
Climate change (e.g. changes species ranges)
ENSO and Pacific decadal oscillations events (e.g. bleaching causing local extinctions)
Seasonal change (especially at higher latitudes)
Pulse events (e.g. storms)

Spatial and temporal variation

Anthropogenic effects (fisheries, eutrophication, pollutants, climate change, introductions)
Natural perturbations (cyclones, tsunami, crown-of-thorns starfish, productivity)

Box 15.2. How do we measure biodiversity?

Louiseau and Gaertner commented that species richness has been the most common method for determining biodiversity on reefs, but this can greatly underestimate the 'multicomponent' aspects of reef biodiversity. They stated, investigators have often chosen indices of diversity without making a clear statement on why a method has been chosen. Indices including: species richness, the 'evenness' of species representation, and weighted indices for common (Simpson's) or rare species (Shannon's) have been the most widely used. In many cases, however, these metrics are poorly suited to the challenges of monitoring coral reefs and establishing sensible baselines for the future. Multiple investigators have pointed out that functional information on the structure of assemblage is critical in determining reef health and resilience to change. Pratchett and colleagues found that coral loss greatly affects fish diversity and that 'response diversity', critical for maintaining ecosystem function, typically varies by group; corallivores may respond poorly while herbivores, omnivores and carnivores respond well. This type of knowledge assists in setting priorities for how diversity and responses to perturbations can be measured. Similar conclusions have been made for the invertebrates associated with coral reefs.

between Atlantic and Indo-Pacific fauna. Some species are widespread throughout the Indo-Pacific, but there is a distinctive Indian and Pacific fauna and components of this fauna extend into the Indo-Australian arc. Endemicity occurs in several areas including even the relatively young areas of the Red Sea and Arabian Gulf. Some of these patterns can be explained by looking at the fossil record of this group as well as the geological history of the area (see Chapter 22). Other studies are required on other invertebrate groups to see if this pattern is a universal one, but such groups need to be speciose, well known taxonomically, possess a fossil record and have detailed distribution records, which does somewhat limit the choice of taxa.

Biogeographic patterns of diversity that are similar to corals can be found among other well-known groups such as fish, echinoderms, molluscs and some decapod crustaceans. For example, the distribution of species of damselfish across the south Pacific (Fig. 15.3) also clearly shows a reduction in the number of species from 110 in the Coral Triangle, slowly declining across the Pacific until members of the family are no longer found. Patterns of species representation are odd for some groups. For example, some tropical taxa have 'anti-tropical' distributions, in which they are found on the GBR and at high latitude reefs near Japan (e.g. Labridae, *Bodianus perditio*).

In summary, the Great Barrier Reef occurs just south of the Coral Triangle and is home to much of this Indo-Pacific fauna. Distinct regional biogeographical patterns occur within the GBR. For most groups, the northern GBR is more diverse in terms of number of species than the southern GBR. For example, 960 species of fish have been recorded for

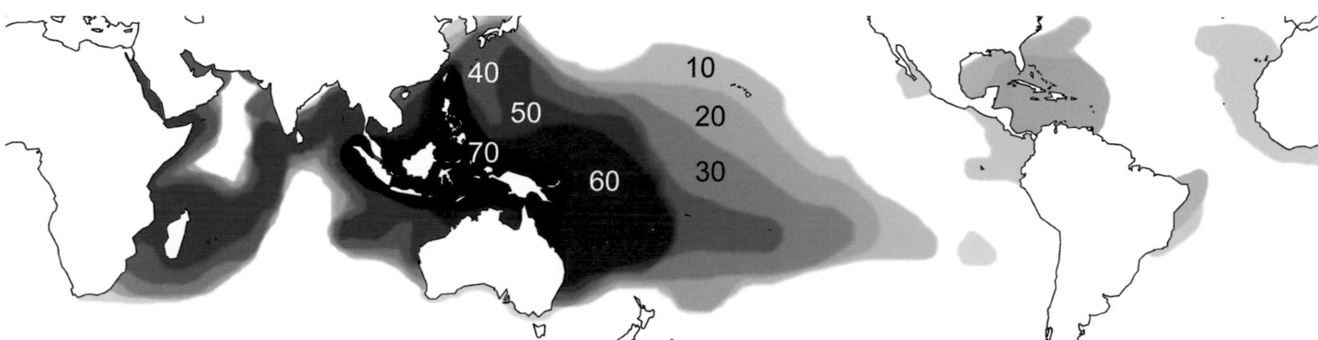

Fig. 15.1. Patterns of global reef-building coral genera worldwide. The greatest generic diversity is found in the Indo-Pacific around the 'Coral Triangle', with numbers tapering off to the east and west as well as north and south (Source: Australian Institute of Marine Science (drawings based on Veron 2000)).

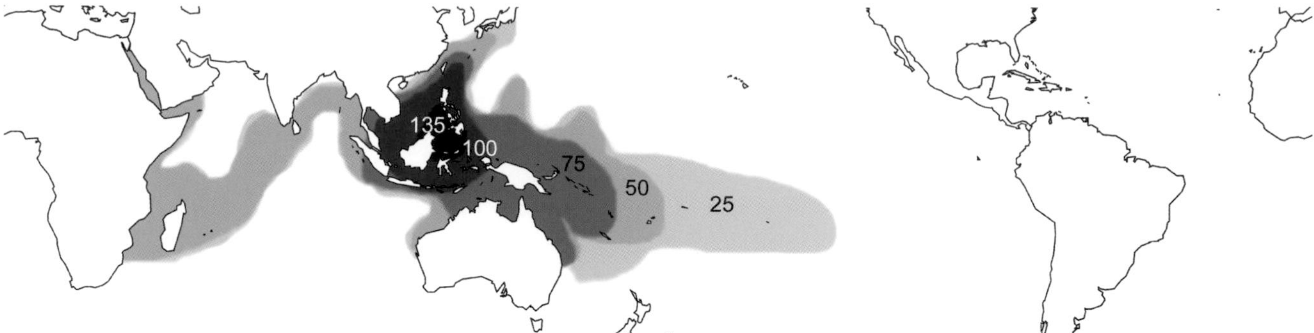

Fig. 15.2. Global diversity of *Acropora* species. The greatest diversity is found in the Indo-Pacific triangle (Source: Australian Institute of Marine Science (drawings based on Veron 2000)).

the Capricorn Group, whereas over 1500 species have been found in the Lizard Island area. While diversity of habitats may explain some of this increase, warmer sea temperatures play a potentially important role. Certainly some species occur throughout the GBR, whereas others are restricted to more northern regions. For example, the leopard coral trout (*Plectropomus leopardus*) and Chinese footballer (*Plectropomus laevis*) are found throughout the GBR whereas the square tail grouper (*Plectropomus areolatus*) is only found north of ~11°S (near Lizard Island).

In addition to latitudinal changes, the variation in diversity at spatial scales of tens to hundreds of kilometres is often greater across the shelf than with latitude. This has been well documented for fish on the GBR. Comparisons of fish faunas of inshore reefs with those on reefs in the GBR lagoon and with those on the outer barrier reveals significant differences both in the terms of abundance and species present (Fig. 15.4). Those of the inshore reefs that are subject to strong terrigenous influences are distinctively different to those on the mid and outer continental reefs. Inshore reefs tend to be depauperate in terms of species diversity compared to offshore reefs. These differences are apparent across each of the zones of these reefal systems, for example the fish communities of the outer reef slope differ depending where that outer reef slope is located. While the initial studies to document this cross diversity were carried out in the Townsville region, subsequent studies in other parts of the GBR confirmed that this is a widespread phenomenon. The complexity of habitats within each reefal system can easily mask this cross shelf variation; it was only when the distribution of species within a reef was carefully documented against habitat that these patterns became obvious. In the case of herbivorous fish, their distributions are related to the ways in which they feed.

Inshore reefs are characterised in terms of biomass by low levels of planktivores and herbivores, whereas outer reefs are dominated by these fishes. Mid shore reefs have intermediate biomasses of

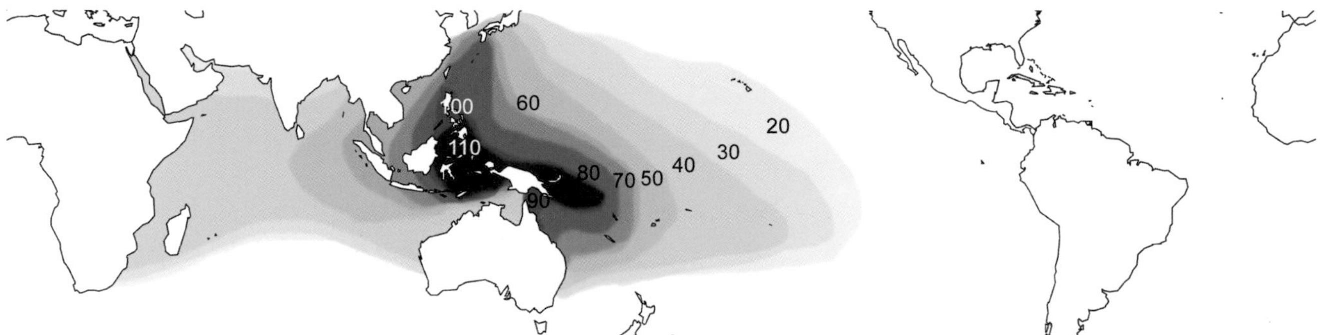

Fig. 15.3. Broad-scale patterns of damselfish species richness: distribution of damselfish (Pomacentridae) species across the indo-pacific. (Source: Data from Allen GR (1991) *Damselfishes of the World*. Mergus, Melle, Germany.)

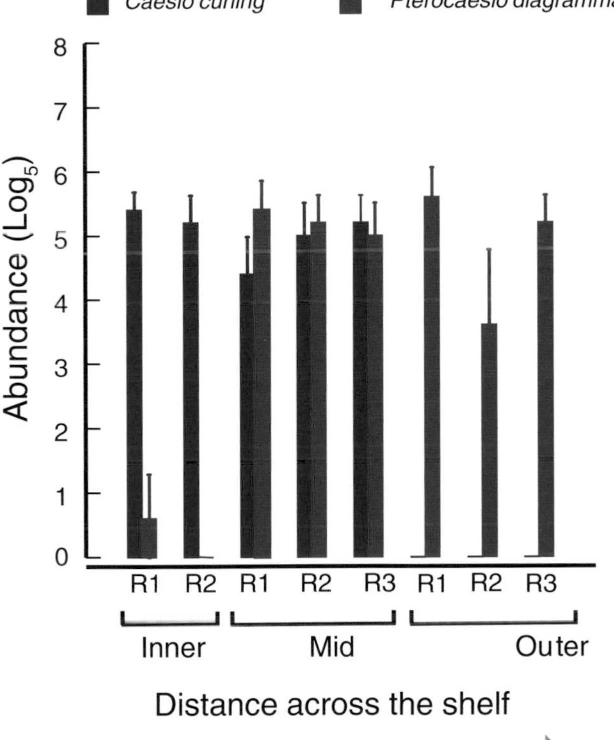

Fig. 15.4. Contrasting patterns of abundance of fusiliers (Caesionidae) across the Central Great Barrier Reef with counts at nearshore (Inner), mid-shelf (Mid) and outer shelf reefs (Outer). The data are based are log5 abundance from a 45 minute swim at five sites within each reef (mean +1SE) (Source: Based on Williams DM, Russ G, Doherty PJ (1986) Reef Fish: large scale distribution and recruitment. *Oceanus* **29**, 76–82). The photo shows *Pterocaesio diagramma* at an outer shelf reef (Image: MJ Kingsford).

herbivores and the highest levels of planktivores. It is generally thought that this relates to water clarity (i.e. turbidity) and food supply (e.g. availability of plankton) as well as patterns of larval dispersal (Chapter 4). Similarly, the distribution of coral species across the GBR as well as within a reefal system varies, with most species having well defined habitat requirements in terms of depth, water clarity and water movement.

In the following chapters, most authors have given some idea of the diversity of each group, where known. As will become clear, however, many of the specialists in the following chapters could give only partial estimates of the species richness of their groups because much of the diversity is currently undescribed. Even less is known about species distributional patterns along or across the GBR. This was recently highlighted by the sea bed biodiversity project (Chapter 6). In general, our knowledge tends to decline with decreasing size of the organism, with the meiofauna inhabiting the soft sediments most poorly known, along with the permanent members of the plankton.

In summary, it seems likely that for most invertebrate groups the number of species is higher in the northern GBR than in the southern, and distinct distribution patterns occur across the shelf as well as within a reef, depending on levels of exposure, depth and water clarity. Many invertebrates and most fish species sustain diversity by the input of pelagic larvae to reefs: this recruitment is a critical factor in determining viable populations. Events that modify the larval transport can affect connectivity and the raw material that influence biodiversity. It is also clear that perturbations influencing habitat-forming groups such as coral have a great influence on the species richness and abundance of invertebrates and fishes that rely on them for needs such as food and shelter. The resilience of 'functional groups of species' also have a great influence on the health of reefs and assessments of 'functional diversity' is relevant to many challenges facing reefs. As discussed in other chapters, coral reefs are dynamic environments that are affected

Box 15.3. Agents moulding patterns of diversity and endemism: size matters

Mantid shrimps give insight to the factors determining diversity and endemism in the Indo-west Pacific. Reaka and colleagues concluded that life histories, dispersal and speciation/extinction dynamics are the key factors influencing diversity. They found that body size is a reliable indicator of speciation and extinction potential and predicted that the same would be true for other groups. Where assemblages were dominated by small-sized taxa, speciation and extinction were likely to be high and there were very high levels of endemism (e.g. the Coral Triangle). In areas of high diversity, speciation would be more rapid than extinction. Away from centres of diversity, the larger body sizes and related fecundity increases the dispersal of shrimps and this slowed speciation. At more remote areas still, moving east into the Pacific and west into the mid-Indian Ocean, body size declined and extinction rates exceeded speciation rates resulting in very low species diversity.

Fig. 15.5. Mantid shrimp, *Gonodactylus* cf. *smithii*. (Photo: Roy Caldwell).

by multiple natural (e.g. cyclones) and anthropogenic impacts that influence patterns of diversity on the GBR. With climate change we will see habitat changes through coral bleaching. Furthermore, some species will expand their range south into cooler waters as flows from oceanographic features change and intensify (e.g. the East Australia Current), so altering patterns of larval dispersal. Key processes that need to be better understood include influences on functional diversity, speciation, endemism, coexistence, extinction, the vulnerability of taxa and the habitats in which they live as well as how physical forcing affects biodiversity in space and time.

Further reading

Biodiversity measures

Clarke KR, Warwick RM (1999) The taxonomic distinctness measure of biodiversity: weighting of step lengths between hierarchical levels. *Marine Ecology Progress Series* **184**, 21–29. doi:10.3354/meps184021

Krebs CJ (1989) *Ecological Methodology*. Harper & Row, New York.

Loiseau N, Gaertner JC (2015) Indices for assessing coral reef fish biodiversity: the need for a change in habits. *Ecology and Evolution* **5**, 4018–4027. doi:10.1002/ece3.1619

Biodiversity and ecosystem management

Frid CLJ, Paramor OAL, Scott CL (2006) Ecosystem-based management of fisheries: is science limiting? *ICES Journal of Marine Science* **63**, 1567–1572. doi:10.1016/j.icesjms.2006.03.028

Folke C, Carpenter S, Walker B, Scheffer M, Elmqvist T, Gunderson L, *et al.* (2004) Regime shifts, resilience and biodiversity in ecosystem management. *Annual Review of Ecology Evolution and Systematics* **35**, 557–581. doi:10.1146/annurev.ecolsys.35.021103.105711

Singh JS (2002) The biodiversity crisis: a multifaceted review. *Current Science* **82**(6), 638–647.

Effects of climate change

Hutchings PA, Ahyong S, Byrne M, Przeslawski R, Wörheide G (2007) Benthic invertebrates (excluding corals). GBR ecological vulnerability assessment. In *Climate Change and the Great Barrier Reef.* (Eds J Johnson and P Marshall) pp. 309–356. Great Barrier Reef Marine Park Authority and Australian Greenhouse Office, Townsville.

Przeslawski R, Ahyong S, Byrne M, Wörheide G, Hutchings P (2008) Beyond corals and fish: the effects of climate change on non-coral benthic invertebrates of tropical reefs. *Global Change Biology* **14**, 1–23.

Distribution of corals across the Indo-Pacific

Veron JEN (2000) *Corals of the World*. Australian Institute of Marine Sciences, Townsville.

Wallace CC (1999) *Staghorn Corals of the World*. CSIRO Publishing, Melbourne.

Origin of biodiversity in the region

Dornelas M, Connolly SR, Hughes TP (2006) Coral reef diversity refutes the neutral theory of biodiversity. *Nature* **440**, 80–82. doi:10.1038/nature05187

Reaka ML, Rodgers PJ, Kudla AU (2008) Patterns of biodiversity and endemism on Indo-West Pacific coral reefs. *Proceedings of the National Academy of Sciences of the United States of America* **105**, 11474–11481. doi:10.1073/pnas.0802594105

Wilson MEJ, Rosen BR (1998) Implications of paucity of corals in the Paleogene of SE Asia: plate tectonics or centre of origin? In *Biogeography and Geological Evolution of SE Asia*. (Eds R Hall and JD Holloway) pp. 165–195. Backhuys Publishers, Leiden, Netherlands.

Functional groups and indicator species

Bellwood DR, Hughes TP (2001) Regional scale assembly rules and biodiversity of coral reefs. *Science* **292**, 1532–1535. doi:10.1126/science.1058635

Bustos-Baez S, Frid CLJ (2003) Using indicator species to assess the state of macrobenthic communities. *Hydrobiologia* **496**(1–3), 299–309. doi:10.1023/A:1026169520547

Pratchett MS, Hoey AS, Wilson SK, Messmer V, Graham NAJ (2011) Changes in biodiversity and functioning of reef fish assemblages following coral bleaching and coral loss. *Diversity (Basel)* **3**, 424–452. doi:10.3390/d3030424

Stella J, Pratchett MS, Hutchings P, Jones GP (2011) Coral associated invertebrates, diversity, ecological importance and vulnerability to disturbance. *Oceanography and Marine Biology – An Annual Review* **49**, 43–104.

Regional patterns of fish distributions and functional groups

Bellwood DR, Hughes TP, Folke C, Nyström M (2004) Confronting the coral crisis. *Nature* **429**, 827–833. doi:10.1038/nature02691

Williams DM (1991) Patterns and processes in the distribution of coral reef fishes. In *The Ecology of Fishes on Coral Reefs*. (Ed. PF Sale) pp. 437–474. Academic Press, San Diego, CA, USA.

16

Plankton

M. J. Kingsford and A. D. McKinnon

The term 'plankton', derived from the Greek *'planktos'* = 'drifting', refers to the diverse communities of small organisms with apparently limited powers of locomotion relative to the water bodies in which they are found. Plankton is typically divided into phytoplankton (the single-cell 'plants of the sea'; Fig. 16.1) and zooplankton (Fig. 16.2). Although most planktonic organisms are too small to observe in detail without a microscope, neither 'small' nor 'limited locomotion' are strictly true: some gelatinous plankton such as siphonophores and pyrosomes are extremely large, and indeed colonies of the siphonophore *Praya* sp. can grow up to 40 m in length, making it the longest, if not the bulkiest, organism in the sea.

Many of the 'true jellyfishes', such as the Australian blubber (*Catostylus*) can swim at 4–6 m/s. Similarly, smaller planktonic organisms can be highly mobile; for instance the copepod *Oithona oculata* is 650 µm in length but capable of swimming at 20 mm/s, equivalent to 31 body lengths per second. These data can be compared to those for a FA18 Hornet fighter jet (17 m, Mach 1.8 (597 m/s), 35 body lengths per second). Regardless of these technicalities, the term plankton is widely applied to the diverse community of organisms inhabiting the water column, but not as highly mobile as fish and squid ('nekton').

For convenience, planktonic organisms are often divided into categories based on either size, trophic role or life history. The size-based approach divides plankton into seven categories, each of which includes several trophic roles (Table 16.1).

Plankton are often divided into autotrophs, which convert energy from light into chemical energy by photosynthesis, and heterotrophs, which either directly or indirectly depend on uptake of organic matter produced by autotrophs. The most important autotrophs of open waters are the phytoplankton, all of which contain chlorophyll, the pigment primarily responsible for photosynthesis. Measurement of the concentration of chlorophyll is widely used to estimate phytoplankton biomass, and can be quantified by satellite to facilitate broad-scale surveys (Box 16.1). Most other planktonic organisms (bacteria, ciliates, zooplankton etc.) are heterotrophs. In addition, many microorganisms are now recognised as mixotrophic, combining photosynthetic activity with heterotrophy.

Almost all marine organisms spend some part of their life history in the water column. The larval or pre-adult stages of organisms that either settle to the bottom ('benthos') or become strong swimmers ('nekton') as adults are defined as 'meroplankton', or temporary plankton, in contrast to those that spend their entire life history in the water column as permanent plankton, 'holoplankton'. It is common for the larval forms to look completely different from the adults and juveniles that are found on reefs and other tropical environments (Fig. 16.3).

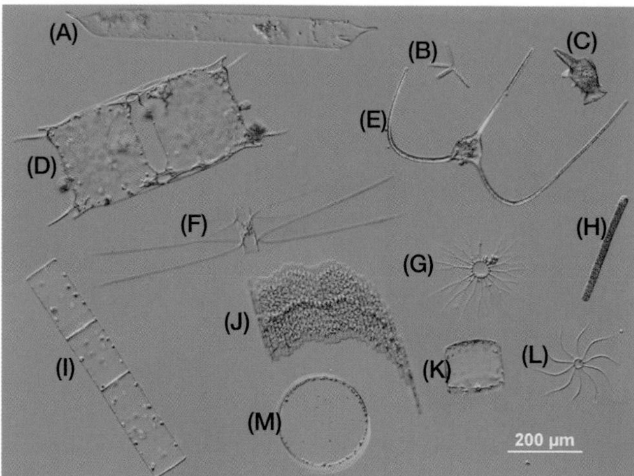

Fig. 16.1. Diversity of phytoplankton: **(A)** *Rhizosolenia* sp.; **(B)** *Thalassionema nitzschoides*; **(C)** *Dinophysis caudata*; **(D)** *Odontella* sp.; **(E)** *Ceratium* sp.; **(F)** *Chaetoceros* sp.; **(G)** *Bacteriastrum* sp. (intercalary disc); **(H)** Cyanobacterium; **(I)** *Pseudoguinardia* sp.; **(J)** *Bacillaria paxilifer*; **(K)** centric diatom in girdle view; **(L)** *Bacteriastrum* sp. (terminal disc); **(M)** *Thalassiosira* sp. (Image: Kirsten Heimann).

The composition of plankton samples varies greatly according to the method of capture. Samples concentrated from water samples are dissimilar to those collected by nets, and samples collected with fine mesh nets (100 µm aperture size or less) bear little resemblance to those collected with larger mesh sizes (e.g. 500 µm aperture size). Furthermore, nets generally turn soft-bodied plankton such as jellyfish to a bucket of amorphous mucus. An integrated understanding of the organisms making up the plankton therefore requires multiple methods.

In the revised version of this chapter we emphasise the ecology and role of planktonic organisms, rather than the morphology and taxonomy emphasised in the first edition. For identification of Great Barrier Reef (GBR) plankton, we refer the reader to the first edition for summary images and the

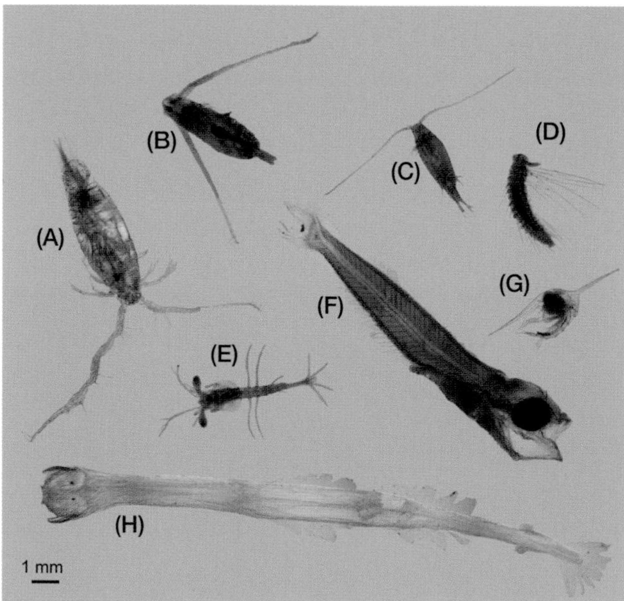

Fig. 16.2. Diversity of zooplankton: **(A)** large calanoid copepod (*Labidocera* sp.); **(B)** calanoid copepod (Eucalanidae sp.); **(C)** calanoid *Acartia* sp.; **(D)** polychaete worm larva; **(E)** prawn larva *Acetes* sp.; **(F)** fish larva; **(G)** crab larva (zoel stage); **(H)** arrow worm (Chaetognatha). (Image Kirsten Heimann)

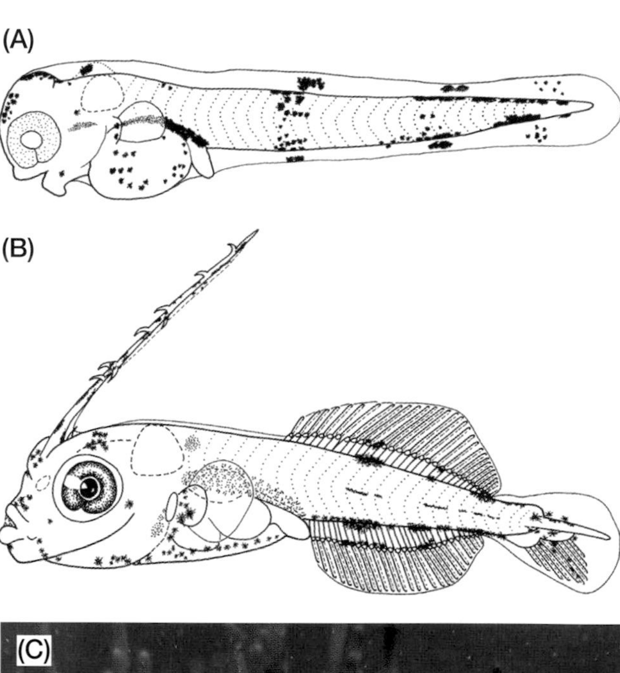

Fig. 16.3. Pre-settlement reef fish larvae *Oxymonacanthus* sp.: **(A)** at 3 mm total length without the dorsal spine found in juveniles and adults on reefs; **(B)** at 9 mm total length; and with an exaggerated dorsal spine compared with **(C)** larger reef-associated stages (Images: A, B, Jeff Leis; C, Mike Kingsford).

Table 16.1. Size classes of plankton

Size class	Size range	Representative organisms	Functional groupings
Femtoplankton	<0.2 µm	Viruses	Parasites
Picoplankton	0.2–2.0 µm	Archaea, bacteria, cyanobacteria (e.g. *Synechococcus*), Prochlorophytes (e.g. *Prochlorococcus*)	Primary producers, heterotrophs
Nanoplankton	2–20 µm	Cyanobacteria, diatoms, flagellates	Primary producers, grazers, predators
Microplankton	20–200 µm	Ciliates, coccolithophorids, diatoms, dinoflagellates, copepod juveniles	Primary producers, grazers, predators
Mesoplankton	0.2–20 mm	Copepods, larvaceans, chaetognaths, salps, *Trichodesmium* colonies	Primary producers, grazers, predators
Macroplankton	2–20 cm	Jellyfish, larval fish, euphausiids, mysids, pteropods, salps	Grazers, predators
Megaplankton	>20 cm	Jellyfish, siphonophores, colonial salps	Grazers, predators

Box 16.1. What is chlorophyll?

All plants contain the green pigment chlorophyll that is able to absorb energy from light to drive photosynthesis, by far the most common means by which living cells produce organic matter. In the ocean, the amount of chlorophyll is often used as a proxy of the biomass of microscopic phytoplankton, the main primary producers in open waters. Biomass is usually measured in units of carbon, but it is difficult to measure the biomass of phytoplankton directly as carbon, whereas it is easier to measure the concentration of chlorophyll using a spectrophotometer or a fluorometer. Moreover, it is possible to image the amount of chlorophyll in the surface ocean from satellite-based sensors, making synoptic pictures of chlorophyll concentration possible over large geographic areas (e.g. the GBR), provided that images of the surface ocean are not badly obscured by cloud cover. Chlorophyll concentration, in units of µg chlorophyll *a* per litre has become one of the most frequently used water quality variables, and has gained great acceptance as a measurement of phytoplankton biomass. However, it is important to realise that the ratio of carbon to chlorophyll can vary widely, depending on the physiological state of the phytoplankton community. Phytoplankton cells can 'green up' (i.e. add more pigment, under low light conditions).

Consequently, the carbon to chlorophyll ratio can vary by at least an order of magnitude, making the chlorophyll concentration in whole sea water rather a crude proxy of phytoplankton biomass. It is also important to note that the chlorophyll concentration is not a measure of primary production, though it is often used as such. Strictly speaking, production is increase in biomass over time.

Though it is sometimes reasonable to argue that high biomass is an indication of high production (e.g. Box 16.3), this is not always the case – clear waters such as those on the Great Barrier Reef can have very low phytoplankton biomass but high production as a result of rapid turnover of phytoplankton cells.

references and the online taxonomic resources listed at the end of this chapter.

Plankton and the Great Barrier Reef

At first impression, plankton do not seem to play a significant role in the waters of the GBR, which are characteristically crystal clear and apparently devoid of life. This impression is misleading, because most planktonic organisms in the waters of the GBR are either extremely small or transparent. Large jellyfishes are easily observed, but even clusters of tiny organisms can sometimes be seen with the naked eye. On calm summer nights, surface waters of the GBR can become bioluminescent, especially in the wakes of watercraft. This occurs because of the bioluminescent response of bacteria and microplankton including dinoflagellates, which generate light as a means of escaping their predators. One such species, *Noctiluca scintillans,* is appropriately called the 'sea sparkle'. Surface slicks of the cyanobacterium (also known as blue-green alga) *Trichodesmium* (Box 16.2, Fig. 8.5) and the spawn of corals can stretch for kilometres during the spring and summer months (looking like windrows of sawdust). Plankton can even be observed by satellite (Fig. 16.5).

Most plankton production is generated in the photic zone where sufficient light is available for photosynthesis. About 70% of all carbon fixed by primary producers on the GBR originates from phytoplankton (with the balance fixed by benthic organisms such as corals and macroalgae).

Phytoplankton account for ~50% of global primary production and, therefore, have a major role in cycling atmospheric carbon dioxide (CO_2). In tropical waters the photic zone may reach a depth of ~150 m due to the clarity of the water column. Below this depth, phytoplankton respiration will exceed the energy derived from photosynthesis. The pelagic environment can be a 'bottom-up ecosystem', where the biomass of plankton and in turn that of higher trophic levels (e.g. fishes, squid and whales) depends on concentrations of nutrients (especially dissolved forms of nitrogen and phosphorus) and trace elements (e.g. iron). This effect can be observed when 'new' nutrients are introduced into surface waters by riverine plumes or upwelling. Alternatively, it can be a 'top-down ecosystem' controlled by grazers, such as by

Box 16.2. *Trichodesmium* – 'sea sawdust'

Trichodesmium is a filamentous cyanobacterium that is abundant in the water column of tropical seas and sometimes aggregates at the surface (Fig. 16.4A). *Trichodesmium* will form visible blooms under ideal environmental conditions (usually calm). As well as occurring as single cells, *Trichodesmium* can form colonies 0.5–3 mm in diameter that are visible to the naked eye. Colonies are of two types: (1) 'tuft' colonies generated by the parallel alignment of filaments; or (2) 'puff' colonies with more radial alignment. Great drifts, usually of dead cells, are common on the surface of the ocean, a phenomenon that has been termed 'sea sawdust' by sailors (Fig. 16.4A).

Concentrations of dead cells are often in the convergence zones of tidal fronts, windrows and internal waves. *Trichodesmium* thrives in areas with low dissolved inorganic nitrogen because it is one of a select group of organisms able to fix atmospheric nitrogen ('diazotrophs'), giving it a competitive advantage over most primary producers that depend on dissolved forms of nitrogen such as ammonium and nitrate. Nitrogen is the limiting nutrient in GBR waters and is necessary for the production of molecules such as proteins essential for cell growth.

Trichodesmium is the major contributor to nitrogen fixation in the water column of GBR waters, which often exceeds 30 mg N/m/day. Hood *et al.* (2004) commented that concentrations of *Trichodesmium* develop in regions where the mixed layer is relatively thin (resulting in high mean light levels) and dissolved inorganic nitrogen (DIN) and concentrations and phytoplankton biomass are low for extended periods of time. The abundance of *Trichodesmium* is also limited by the availability of trace elements, such as iron, molybdenum, and vanadium, essential elements of the nitrogen-fixing machinery.

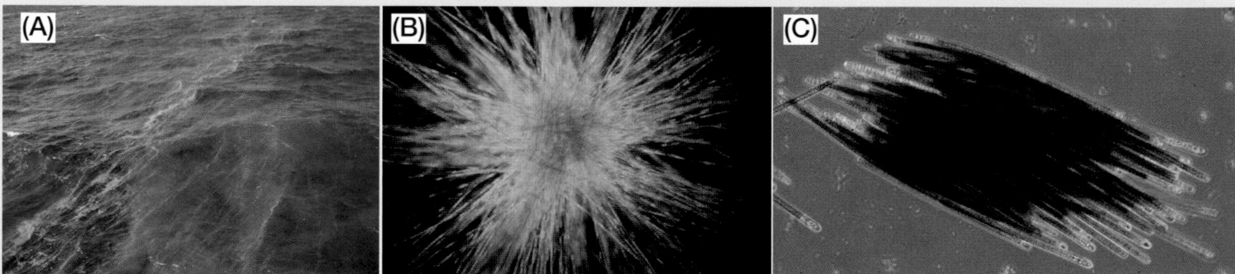

Fig. 16.4. **(A)** A raft of *Trichodesmium* cells aligning with windrows on the surface of GBR waters; **(B)** a 'puff' of *Trichodesmium*; **(C)** a 'tuft' of *Trichodesmium*. (Photos: A, Lachlan McKinna; B,C, Hans Paerl).

herbivorous copepods. Additionally, predators such as fish can remove herbivorous zooplankton, relieving grazing pressure on phytoplankton and resulting in an increase in phytoplankton biomass.

In contrast to high-latitude ecosystems, tropical systems generally have low variation in productivity and biomass, and pulses in production are event driven by storm-driven mixing, floods, cyclones and upwelling intrusions. Tropical systems are high turnover, low biomass systems and the waters are generally oligotrophic (i.e. low in nutrients) and clear. Most nutrients are present at such low concentrations that they approach the detection limits of analytical equipment. Phytoplankton production in GBR waters is primarily limited by the supply of nitrogen, most of which is fuelled by the remineralisation of organic matter via excretion and decomposition. The resultant dissolved ammonium is very rapidly recycled, with the pool of dissolved ammonium turning over in a matter of hours. Some seasonality can occur in GBR waters because of increased frequency of monsoonal events such as seasonal rains and upwelling events, both of which introduce 'new' nitrogen into the system as nitrate. In the 'dry tropics' (the central and southern GBR), the rains are not predictable, but a significant input of fresh water nevertheless results from cyclone/storm events that may stratify and slightly freshen the sea water and subsequently modify planktonic assemblages and processes.

Fig. 16.5. Satellite image showing *Trichodesmium* slicks (see also Fig. 8.5) within the convergences of an estuarine ebb-tide plume and in windrows beyond the plume. Puffs of clouds are seaward of the windrows and the plume, with some over the land, especially on the lower right side (Image: USGS Landsat-8 data processed by Norman Kuring, NASA Goddard Space Flight Centre).

Upwelling of cold, nutrient-rich, deep ocean water into the photic zone has a great influence on pelagic systems and is determined by currents, wind and topography. Regions of intense upwelling, such as the coast of Peru, the west coast of North America and South Africa, have productive waters and a wealth of consumers from copepods and krill to whales. As a result, these are the sites of some of the great fisheries of the world (e.g. Peruvian anchovy). Though of comparatively small scale, upwelling does occur in GBR waters, typically under wet season conditions with northerly winds (i.e. blowing southwards).

Of course, most planktonic organisms go about their business undetected by the casual observer. To observe femto- and picoplankton, it is necessary to employ technologies such as flow cytometry. Nano-, micro- and mesozooplankton can be sampled by specialised net systems and observed by light microscopy, but often not without disruption and destruction of their natural beauty.

Femtoplankton (<0.2 µm) and picoplankton (<2 µm)

Our understanding of the microbial ecology of GBR waters is very much in its infancy, because the necessary tools (flow cytometry and molecular biology) have only recently become available. Consequently, we are only beginning to appreciate the huge taxonomic diversity of these organisms, as well as their important role in many ecosystem processes. In GBR waters viruses occur in the order of 10^6 per millilitre and bacteria/Archaea in the order of 10^5 per millilitre. We know little of the role of viruses in GBR waters, but we do know that viruses are a major source of mortality of microbes, causing cell lysis that contributes to the pool of dissolved and particulate organic carbon: a process known as the 'viral shunt'. The microbial communities of inshore waters of the GBR are dominated

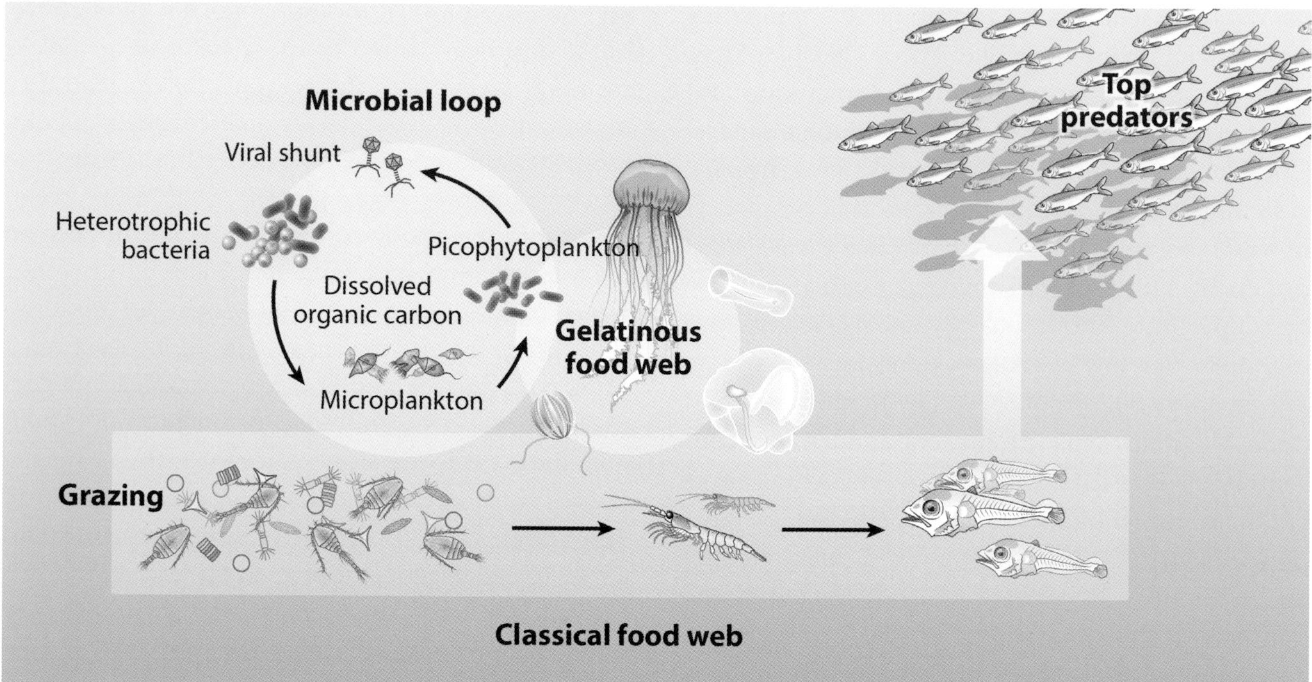

Fig. 16.6. Trophic pathways in the plankton, showing the interaction of the 'microbial loop' and the classical food chain.

by the bacterial orders Pelagibacterales and Synechococcales, and the Archaeal Class Marine Group II. Pelagibacterales account for one third of free-living bacteria in the surface waters of the ocean, and include the SAR11 clade, which are particularly small cells and probably the most numerous bacteria in the world. Synechococcales include the photosynthetic picoplankters *Synechococcus* (~1 µm in size), which is more important in GBR lagoon waters, and *Prochlorococcus* (0.5–0.7 µm in size), which is characteristic of more oceanic waters on the mid and outer shelf. Marine Group II Archaea are heterotrophs that reside in the photic zone, but little is known of their ecological role.

Heterotrophic bacteria are important components of marine ecosystems because of their ability to use dissolved organic matter (DOM) and make this available to higher trophic levels. In the waters of the GBR, DOM is derived not only from leakage from phytoplankton and cell lysis as a result of viral infection (viruses appear to be responsible for 50–100% of the mortality of heterotrophic bacteria), but also from coral mucus. Also, nitrogen-fixing bacterioplankton are an important source of nitrogen to GBR waters, which tend to be N-limited. These 'diazotroph' assemblages are dominated by the cyanobacterium *Trichodesmium erythraeum*, g-proteobacteria from the Gamma A clade, and d-proteobacterial phylotypes related to sulphate-reducing genera.

Picophytoplankton contribute ~80% of phytoplankton biomass in the waters of the central GBR, and about two-thirds of the productivity. Paradoxically, the number and biomass of picophytoplankton tends to be rather invariant, because their principal grazers are not much larger (2–20 µm), comprising hetero- or mixotrophic nanoplankton. Despite achieving maximal rates of photosynthesis these tiny primary producers can never outgrow their grazers because both have similar growth rates, imposed by size-related physiological constraints. This cycle, by which both autotrophic and heterotrophic picoplankton are either grazed by heterotrophic microbes or lysed by viruses, liberating DOM that is then recycled by heterotrophic bacteria, has become known as the 'microbial loop' (Fig. 16.6). This represents a short circuit that prevents the transfer of primary production to higher trophic levels via the 'classical food web', which consists of nanoplanktonic primary producers and their mesozooplankton grazers, which are of a large enough size to be available to tertiary predators such as larval fish. These two trophic pathways generally operate in parallel, though generally the microbial loop predominates in tropical waters such as those of the GBR. However, the two are linked by the entrapment of microbial material in marine snow aggregates that are also large enough to be used by fishes (Fig. 16.6).

Nanoplankton (2–20 μm)

This size fraction includes both phytoplankton (mostly diatoms and dinoflagellates) and zooplankton (flagellates, ciliates, radiolaria) components (Table 16.1). Common small diatom species such as *Leptocylindricus danicus* and *Cylindrotheca* spp. are capable of up to four doublings per day under the right conditions. When 'new' nitrogen is introduced into GBR waters by floods or upwelling, small diatoms such as these are able to outgrow their grazers, resulting in phytoplankton 'blooms' (see Box 16.3). Under 'normal' conditions, nanoplankton account for most of the one-third of non-picoplankton primary production, but this proportion can become more substantial under bloom conditions. Heterotrophic nanoflagellates are the most important grazers of picoplankton, and appear to be the primary source of mortality for picophytoplankton.

Microplankton (20–200 μm)

This size fraction includes both phytoplankton (mostly diatoms and dinoflagellates) and

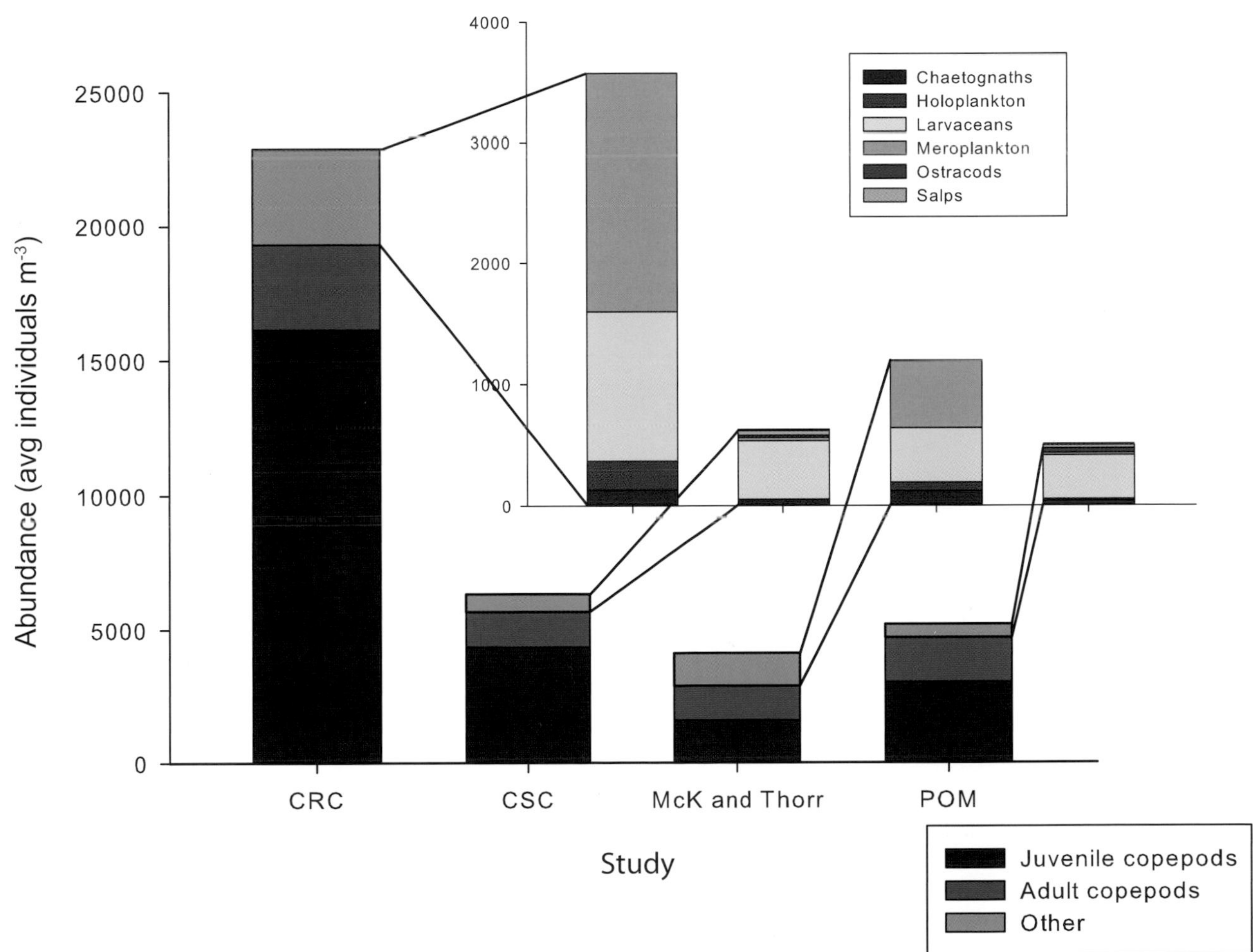

Fig. 16.7. Zooplankton community composition in GBR waters, based on four studies undertaken by AIMS: CRC (McKinnon *et al.* 2005), McK and Thorr (McKinnon and Thorrold 1993), POM (Alongi *et al.* 2015) and CSC (McKinnon unpublished). Mesh sizes were 73 μm in the CRC study, and either 100 μm or 150 μm in the other studies.

zooplankton (ciliates, radiolaria, copepod nauplii, meroplankton, etc.) components. The microzooplankton are the most important grazers of nanoplankton primary production, with heterotrophic and mixotrophic (i.e. heterotrophic but with the ability to photosynthesise) dinoflagellates being particularly important. Tintinnid ciliates with vase-shaped lorica (e.g. *Tintinnopsis* sp.) as well as naked ciliates such as *Strombidium* spp. are important grazers of nanoplankton. Naupliar stages of copepods are the major metazoan component of the microzooplankton, and are thought to be the most numerous life form on Earth.

Globally, microzooplankton consume 50–75% of primary production, and the few available unpublished measurements of microzooplankton grazing in GBR waters indicate grazing rates of the same order. Despite microzooplankton being the critical link between primary producers and the elements of the mesozooplankton that are important in the diets of larval fishes, there is a critical shortfall of information on the microzooplankton of the GBR.

Box 16.3. River plumes and plankton

Nearshore waters of the GBR are influenced by the input of freshwater from rivers, primarily in the wet season (December to March). The input of freshwater to coastal areas forms conspicuous plumes that can extend kilometres seawards, sometimes out to the reef itself, and hundreds of kilometres along the coast. Riverine waters bring with them increased concentrations of sediments, nutrients and pesticides from the catchments. The river plumes are initially sediment-rich, which limits light availability to phytoplankton and other primary producers, but once the sediment settles nutrients are rapidly taken up, causing 'blooms' that can be quite conspicuous. The phytoplankton responsible for these blooms are primarily diatoms that can double several times in day, both in number and in biomass. These rapid growth rates outstrip the ability of their grazers (e.g. copepods, which have generations times of 7–10 days) to control the accumulation of biomass, resulting in greatly elevated standing stocks of phytoplankton.

Fig. 16.8. Influence of the Burdekin River plume on the waters of the Great Barrier Reef. A sediment plume is obvious by the brown colour; the resultant phytoplankton bloom is indicated by green on the margins of the sediment plume and to the north-west of the plume (Image: NASA MODIS Aqua data processed by Lachlan McKinna, Go2Q Pty Ltd)

Mesozooplankton (200–2000 μm)

Copepods account for between 71% and 90% of the GBR mesozooplankton by number. Juvenile stages

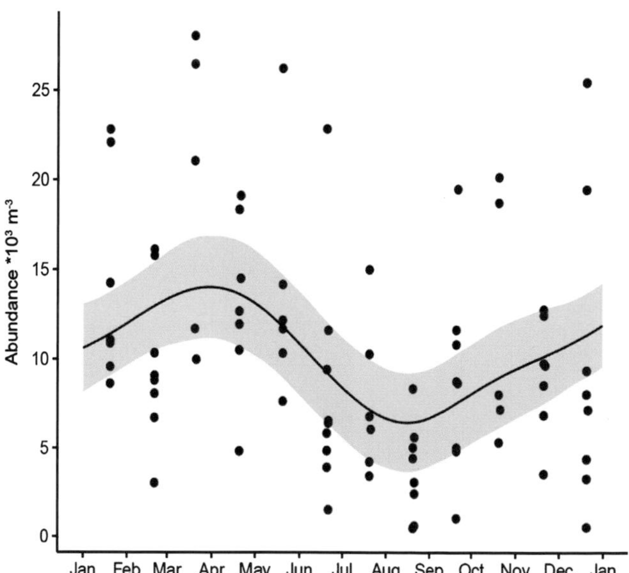

Fig. 16.9. Annualised trend of total >100 μm zooplankton sampled at the Yongala National Reference Station (IMOS). The trend line is a GAM fitted to the abundance data between 2009 and 2017 with any long-term trend factored out (though there was no noticeable change in abundance over this time period). The shaded area is the 95% confidence interval about the predicted value.

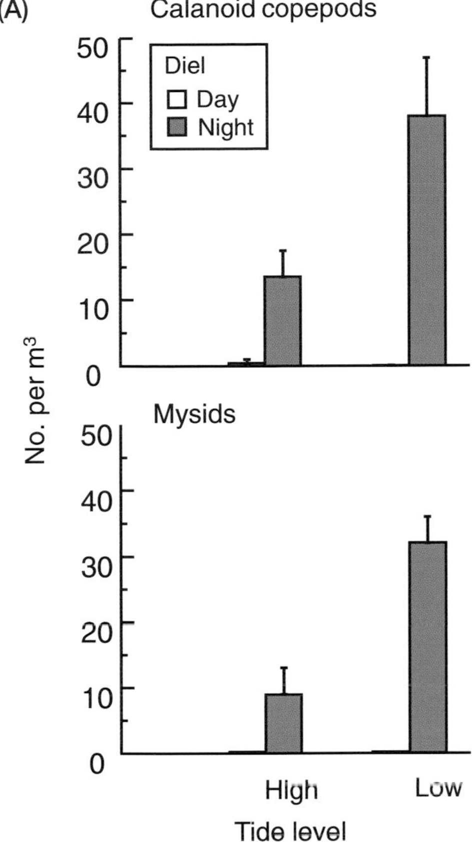

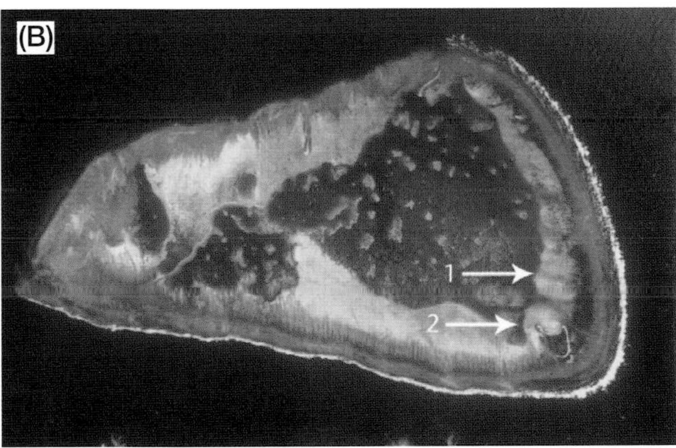

Fig. 16.10. **(A)** Variation in day–night (diel) patterns of plankton abundance within reef lagoons can be great. For example, in a ponding lagoon at One Tree Island, southern GBR, Drew and Kingsford (unpubl. data) found abundance of plankton was very low during the day, in contrast at night 'demersal plankton' rose from the substratum and was abundant in the water column. **(B)** This explanation was especially clear at night when the lagoon was ponding at low tide, so excluding outside waters (data in water column of 3–6 m: mean of two sites (1 and 2) in the lagoon and 4 times per site and n = 2 vertical hauls per time at each site).

of copepods, particularly nauplii, make up most of the copepod population (between 55% and 84%), depending on the mesh size of the nets used. The most abundant copepod taxa in GBR waters are species of the cyclopoid genus *Oithona* and of the calanoid family Paracalanidae (*Paracalanus, Bestiolina* and *Parvocalanus* species). Of the non-copepod zooplankton on the GBR, Larvacea are the most abundant holoplankton, followed by Chaetognatha (Fig. 16.7).

Estimates of the abundance of zooplankton are dependent upon the sampling strategy employed, especially the mesh size of the nets used. In general, there is an onshore–offshore gradient in zooplankton abundance, with abundances in excess of 50 000 organisms per m^3 occurring in sheltered inshore environments such as predominated in the CRC study (Fig. 16.7). High abundances can occur in other areas depending on events such as floods, upwelling, or advection as a result of the interaction of water currents and bathymetry. These effects tend to outweigh seasonality, though generally speaking both abundance and biomass tend to be greater in the wet season. At the Yongala National Reference Station, there was a maximum in abundance in April (after the conclusion of the wet season) and a minimum in late August (end of dry season; Fig. 16.9); a seasonal pattern also evident over a shorter time period (Fig. 16.7). In the GBR lagoon, diel changes (i.e. day–night changes) in zooplankton abundance and biomass tend to be very weak. The only differences in community composition relate to the occurrence of the larger component of the mesozooplankton: night-time samples tend to have bottom-associated taxa such as ostracod and mysids, as well as some of the larger calanoids (e.g. *Canthocalanus, Undinula*).

Mesozooplankton are important grazers of microplankton, including larger phytoplankton such as diatoms. Though microzooplankton are an important component of copepod diets, the role of detrital material such as 'marine snow' is also likely to be important in GBR waters but is yet to be quantified. Productivity by >150 µm mesozooplankton is ~0.18–3.39 mg C/m^3/day, depending on

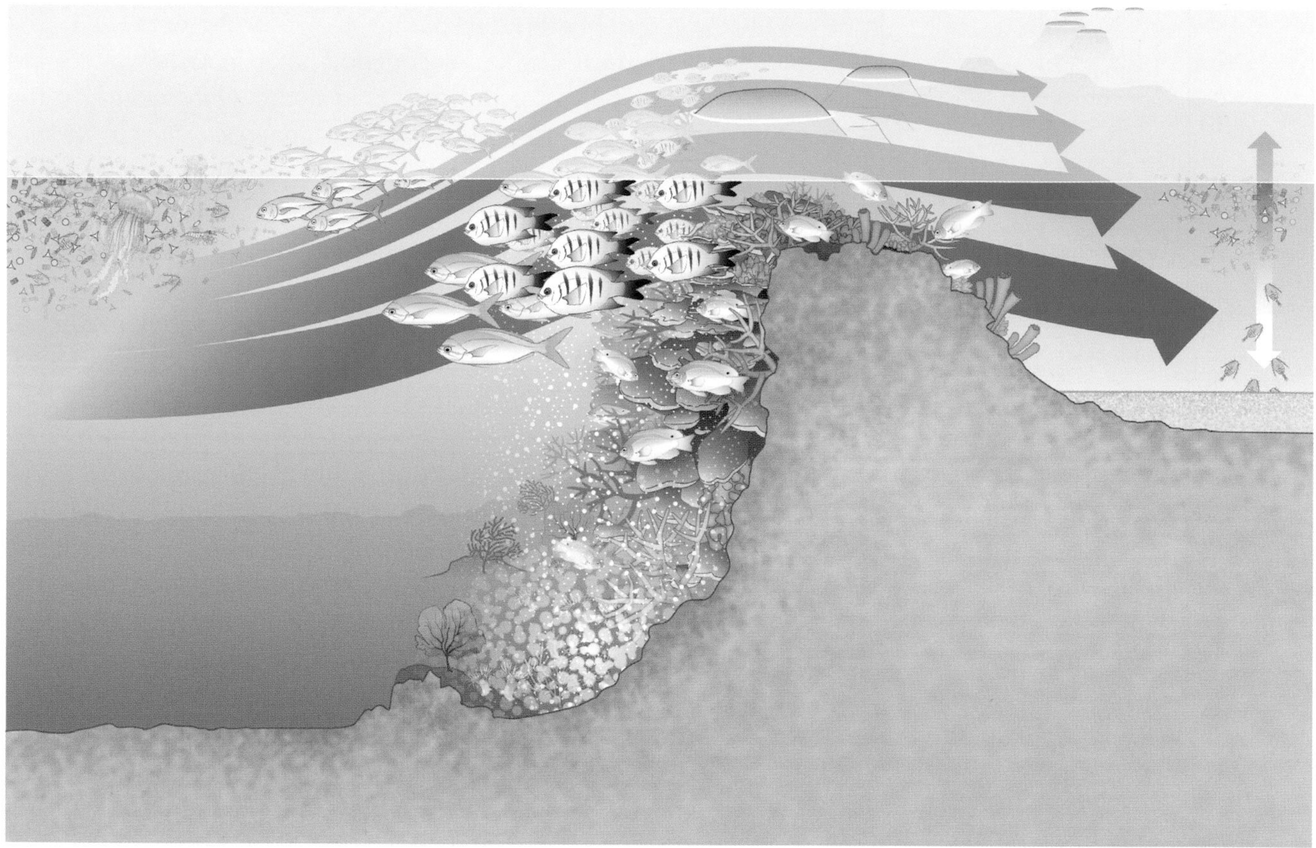

Fig. 16.11. Trophic linkages between plankton and coral reefs: 'the wall of mouths' (see also Chapter 8). Fish that feed on plankton (planktivores) crop the larger plankton as it flows past the reef, and other suspension feeders such as corals and sponges feed on the very small plankton. Plankton is also an important source of energy for the reef as the fall-out of the excreta planktivores reaches the bottom to drive production. Eventually the fish themselves become a source of food for reef predators. To the right, demersal zooplankton rises into the water column at night and becomes available to nocturnal planktivores such as cardinalfish.

the time of year and the model of growth applied. On this basis, it appears that this size fraction of zooplankton grazes ~8% of primary production per day in GBR waters.

Within coral reef lagoons, the emergence of demersal zooplankton contributes to markedly higher night-time abundances. During the day, plankton densities can be low but at night the demersal zooplankton (including copepods, mysids, amphipods and polychaete worms) rises into the water column (Fig. 16.10). This is when many corals and nocturnal fishes feed, and later provide nutrient-rich faeces that fall to the bottom, remineralise and provide a source of recycled nutrients to primary producers.

Macrozooplankton (>2 mm)

Tropical plankton are smaller than those of temperate or polar waters and so this fraction of the zooplankton is less conspicuous in GBR waters than in temperate waters. The macrozooplankton of the GBR are primarily gelatinous (Box 16.5). Mucous net feeders such as pelagic tunicates (salps and larvaceans) and pteropods are characteristic of the more oceanic waters of the GBR in the summer months. These animals are able to graze on the full spectrum of particulate matter, including the picoplankton. Hydrozoan and scyphozoan medusae are present all year round, but the abundance of nuisance species such as the box-jellyfish (*Chironex fleckeri*) and Irukandji jellyfish is strongly seasonal.

Comb jellies (Ctenophora) of the genera *Pleurobrachia* and *Beroe* are common in GBR plankton, but are too delicate to be adequately sampled by plankton nets. The macrozooplankton also contain larval and pre-settlement stages of fishes. Because macrozooplankton have better locomotion than the smaller plankton, their behaviour plays a more important role in determining their distribution. Larval fish, in particular, have been shown to swim long distances to settle on suitable coral reef substrate (Box 16.7).

Some plankton aggregate in convergences, and maintain their position for the purposes of reproduction (e.g. larvaceans and jellyfishes). Schools of mysids are an obvious feature of coral reef lagoons, and feed on detrital material in the epibenthos (i.e. just above the sediment). The larvae of prawns, crabs and fishes are capable of regulating their depth to take advantage of differential current flow. For example, lobster larvae (*Palinurus ornatus*) use surface currents for dispersal, then migrate to deeper water later in development for transport back towards the Australian coastline.

Marine snow

In situ observations indicate there is a huge abundance of large aggregates of particulate material ('marine snow' also known as 'sea snot') in the pelagic environment of the GBR, composed of aggregated coral mucus, gelatinous material, faeces, phytoplankton cells and bacterial flocs, often colonised by protozoan grazers. Mucus and gelatinous fragments are produced by a host of planktonic organisms that include larvaceans (Box 16.4), heteropods, salps and jellyfishes. The analogy with snow is that these particles are suspended in the water column but ultimately settle to the sea bed, where they can be an important source of food to benthic detrital feeders (Fig. 16.12).

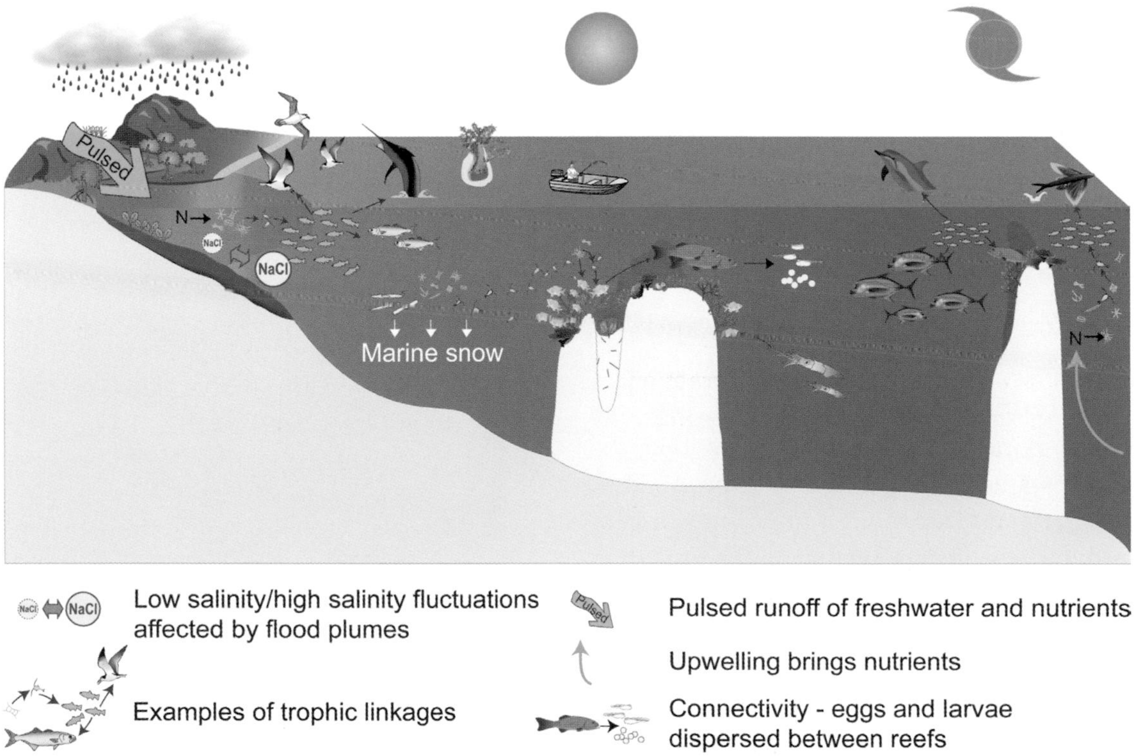

Fig. 16.12. Tropical pelagic food chain showing plankton as an important source of food for pelagic nekton and reef-associated species such as planktivorous invertebrates and fishes. Demersal zooplankton migrate from reefs and other substrata into the water column at night. Input of freshwater, upwelling and cyclones are factors that influence nutrient availability and growth of phytoplankton (Image: GBRMPA).

Box 16.4. Larvaceans

Larvaceans are common in waters of the GBR and other parts of the world. They are often abundant and are effective filter feeders on pico- and nanoplankton. Larvaceans are also a preferred prey of reef planktivores such as damselfishes.

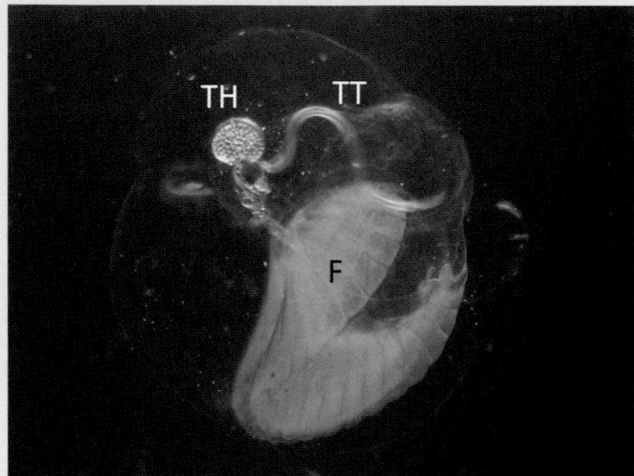

Fig. 16.13. *Oikopleura*, a common genus of Appendicularia (larvacean). The animal itself appears pale green in this image, and the major filters of the mucous house are pink. TH – Tadpole head, TT = Tadpole tail, F – filter the paddling motion of the tail generates a current so that water passes through the filters. There is also an escape hole so that the tadpole can abandon the house quickly in threatened (Image: Dr Fabien Lombard, Observatoire Océanographique de Villefranche, France).

Trophic linkages between plankton and coral reefs

The great diversity of organisms on coral reefs was considered a paradox by Darwin given that the surrounding waters appeared to be unproductive. We now understand that though standing stocks of phytoplankton are very low (normally less than 0.4 µg chlorophyll *a* per litre), they are turning over at a very rapid rate (sometimes up to four times per day). So, although there is little phytoplankton biomass (i.e. plankton is not obvious in the water column because the water is so clear) productivity can be quite high, and is rapidly transformed by trophic processes into grazer biomass or flocculant detritus.

Box 16.5. Gelatinous ('jelly') zooplankton – including deadly jellyfish

The largest types of plankton are gelatinous (see also Chapter 20) that are ecologically important as consumers of plankton, predators of fish larvae, and as a habitat on which some plankton can 'hitchhike'. When they die and fall to the bottom, they become a source of nutrition to benthic organisms. Also the venom of some is a threat to humans. Jelly plankton includes the true jellyfishes such as lion's mane, blubbers and box jellyfishes (e.g. stingers and Irukandji species). The hydroid jellyfish are diverse and abundant, but are generally very small; the largest, *Aequoria,* is only ~5 cm in diameter. Siphonophores such as the Portuguese-man-war (or 'blue-bottle') float on the surface as 'pleuston'; others are found beneath the surface where they clone and can form colonies that are metres in length. All of these jellyfish have stinging cells. In contrast, the 'comb jellyfish' (ctenophores), voracious though delicate predators, capture their prey with sticky cells. Salps and doliolids are relatives of the larvaceans (which live in gelatinous houses, Box 16.4) and can 'bloom' in great numbers stripping the water column of pico- and nanoplankton like vacuum cleaners. Traditional plankton nets often crush these delicate jelly plankton. Consequently, direct observations in the water and modern videography are generally the best way to study these animals.

Fig. 16.14. An aggregation of 'blubbers' (*Catostylus mosaicus*) on the Queensland coast. Each jellyfish would weigh over 3 kg and would have a bell of ~20–25 cm wide (Image: Mike Kingsford).

Plankton helps to support reefs, as tides and other currents sweep past and bring a supply of fresh plankton to the reef face (Fig. 16.11, Chapter 8). The rain of plankton on to the incurrent face of coral reefs encounters a 'wall of mouths' of planktivorous fishes that first capture that food source then remineralise it to subsidise nutrient pathways on the reefs themselves. Filter feeders such as sponges also have a role in sequestering very small plankton.

Physico-chemical drivers of pelagic processes

The movement of water has a great influence on organisms that largely drift. However, some directional control by these organisms can be obtained by varying their vertical position in water columns that have vertical stratification in terms of current speed and direction. Currents can influence the trajectories of plankton and can influence concentration. Some eddies are called 'phase eddies' that only occur at one phase of the tide (e.g. on ebb tides behind a reef), while other eddies (and larger gyres) persist over long time periods in mainstream currents. In addition to eddies, currents can interact with abrupt topography such as reefs and seamounts to alter the supply of plankton through upwelling.

Plankton often concentrate in convergences. In tropical waters it is common to see cyanobacteria (Fig. 16.5, Box 16.2), coral spawn, jellyfishes, spawning larvaceans and other plankton in convergence zones generated by physical phenomena that include: tidal and thermal fronts; tidal jets; the edge of eddies; windrows; and internal waves. As a result, therefore, convergences are often sites of intense biological activity and nekton (e.g. planktivorous fishes, piscivores) and birds are often attracted to the abundant food in these areas.

Oceanography can also influence the distribution of plankton with depth. Although the water column on the shelf of the GBR is usually well mixed (see Chapter 4), thermoclines and haloclines can stratify the water column into different water

Box 16.6. Crown-of-thorns starfish and larval transport

The crown-of-thorns starfish (COTS) is a predator of corals that undergoes large fluctuations in population size and periodically reaches densities of 'plague' proportion. When this occurs, the hard coral communities of the GBR can be virtually destroyed. COTS are amazingly fecund; large females are capable of producing up to 65 million eggs in a season. In the laboratory, COTS larval development takes 11–25 days when fed a phytoplankton diet, after which the last planktonic stage (late brachiolaria) settles to the substrate to become a juvenile starfish. In the field, larval COTS feed on particulate material in the water column and are transported by ocean currents. Recent research suggests that primary outbreaks originate from an 'initiation zone' located between Cairns and Lizard Island, and that the southward transport of water distributes larvae to the central section of the GBR.

It is virtually impossible to identify larval COTS in plankton samples by conventional means because many other larval forms of starfish look very similar. Recent research has applied molecular methods to detect not only the presence but the abundance of COTS larvae in plankton samples, making it possible to monitor the spread of COTS in the plankton, and the possible development of COTS outbreaks 2–3 years later, after the juvenile starfish have grown to adulthood. This technology can be developed and applied to all species of plankton, lessening the need for conventional taxonomic skills.

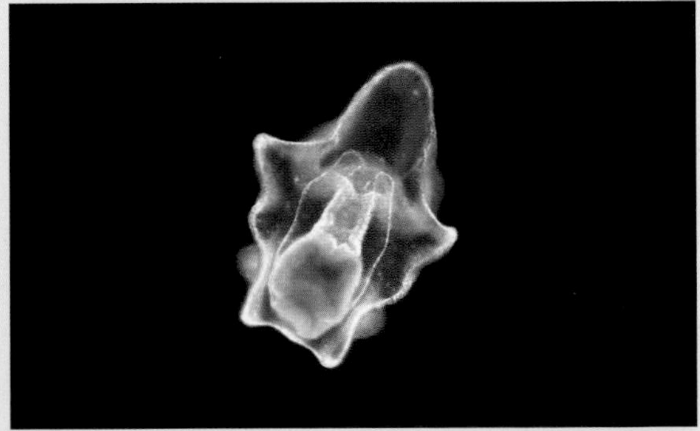

Fig. 16.15. Brachiolaria larvae of the crown-of-thorns starfish, reared in captivity (Image: Kennedy Wolfe).

masses. For example, at the edge of the continental shelf upwelling events are common and 10 m to 20 m deep 'wedges' of cool water move along the bottom over the shelf under lighter and warmer shelf waters. Different planktonic assemblages may be found above and below the thermocline in these water masses. In addition, some plankton will concentrate in and around the thermocline. Freshwater plumes will also cause vertical stratification and variation in abundance of plankton with depth.

During times of flood, river discharge brings elevated nutrient loads into the coastal zone of the GBR. Under these conditions, larger phytoplankton cells such as diatoms can outgrow their smaller cousins, resulting in phytoplankton blooms. These blooms are sometimes visible to the naked eye, but are very obvious features of satellite imagery that detects chlorophyll *a* (Box 16.3), the most common photosynthetic pigment found within phytoplankton.

Floods and upwelling stimulate an immediate increase in copepod egg production in the central GBR, but abundances of juvenile copepods peak ~6 weeks later because the generation time of these copepods is ~2 weeks, sustaining the effect of the flood for some time.

Under doldrum conditions (spring–summer), very obvious rafts of the cyanobacterium *Trichodesmium* form in surface waters of the GBR (see Fig. 8.5), colloquially referred to as 'sea sawdust' or 'spawn' (Box 16.2). In fact, these cells are very important in the economy of GBR waters because they are able to incorporate atmospheric nitrogen into their cells ('nitrogen fixation'), and nitrogen is a limiting nutrient for phytoplankton growth on the GBR. *Trichodesmium* is toxic to most grazers: only a few genera of small pelagic harpacticoid copepods are known to graze upon it directly.

Occasionally *Trichodesmium* slicks are blown shorewards where they decompose, generating unpleasant aromas and turning the water red as a result of the liberation of a pigment called phycoerythrin.

Contrary to popular belief, the waters of the GBR lagoon are actually quite productive – the average primary production rate is ~0.73 g C/m^2/day, at the high end of primary production rates measured in the Australasian region. However, this contrasts greatly with great upwelling areas of

Box 16.7. Larval fish are better swimmers that we thought

Reef fishes have a larval stage that remains in the plankton for days to months. 'Nemo' damselfishes (*Amphiprion* spp.) spend ~9 days as larvae, while surgeon fishes (Acanthuridae) may spend up to 3 months in the plankton. Initially they were thought to be hopeless swimmers that drifted in the ocean and their arrival and ultimate settlement on a reef was considered to be at the whim of currents. This view was partly based on larvae from the northern hemisphere limited swimming ability. Subsequent research has shown that, although for the first few days of life the 2.5 to 6 mm larvae of reef fishes could not swim well, the situation changed rapidly. Experiments have demonstrated the sustained swimming speeds of larger larvae are often greater than the current speeds measured around reefs. That was the first surprise. The second was that larval fishes have excellent abilities to orientate, even to the reef from which their parents spawned. Multiple experiments have shown that they can orientate using senses of smell, hearing and sun-compass orientation. Most recently fish larvae have even been found to detect magnetic fields that help with orientation. The tagging of eggs in a variety of female reef fish is now used to determine the proportion of fishes that disperse away from the reef of their parents and those that leave home and may disperse for tens to hundreds of kilometres.

Fig. 16.16. Soap fish larva *Diploprion* sp. with specialised larval fins; the larva was caught in a light trap (Image: Mike Kingsford).

the world (e.g. Peru) that have primary production of ~10 g C/m/day.

Plankton and climate change

Anthropogenically driven climate change is widely accepted to have elevated ocean temperatures and caused ocean acidification because of increased concentrations of dissolved CO_2. Elevated water temperatures have been shown to affect plankton abundance, distribution and phenology (the seasonal timing of events such as diapause). Most of what we know of the effects of increased temperature on the distribution and abundance of plankton comes from continuous plankton recorder surveys that have repeatedly surveyed the waters of the North Sea since the 1930s. These surveys have described substantial changes in planktonic assemblages in the North Atlantic, with the poleward movement of warm water species ('tropicalisation'). Similar surveys in the North Pacific have shown changes in phenology of large copepods that are important in the diet of seabirds. These copepods descend to depth to diapause over the winter, and warming seas have diminished the period of overlap between their occurrence in surface waters and the critical period of seabird breeding, impacting the survivorship of fledglings. In the waters of the GBR, such seasonal effects are much more subtle. The impact of global change on plankton of the GBR will be greatest due to changes in temperature, the availability of nutrients (e.g. upwelling and runoff) and changes in pH that can affect plankton with carbonate skeletons (such as coccolithophores and pteropods). On the GBR, it is expected that changes over the next 50 years are likely to vary by region (e.g. central section versus southern section). It is now known that warm waters of the southward-flowing East Australian Current (EAC) that bathes the outer reefs of the GBR, now make it almost 400 km further south than it did in the 1960s, greatly amplifying the tropicalisation effect by active transport of warm water plankton southward. Meroplankton in the EAC are now carried as far south as Tasmania, and changing currents will influence larval connectivity among GBR reefs and beyond (Box 16.6).

Further reading

Femtoplankton and picoplankton

Alongi DM, Patten NL, McKinnon D, Köstner N, Bourne DG, Brinkman R *et al.* (2015) Phytoplankton, bacterioplankton and virioplankton structure and function across the southern Great Barrier Reef shelf. *Journal of Marine Systems* **142**, 25–39. doi:10.1016/j.jmarsys.2014.09.010

Angly FE, Heath C, Morgan TC, Tonin H, Rich R, Schaffelke B, *et al.* (2016) Marine microbial communities of the Great Barrier Reef lagoon are influenced by riverine floodwaters and seasonal weather events. *PeerJ* **4**, e1511. doi:10.7717/peerj.1511

Crosbie ND, Furnas MJ (2001) Abundance, distribution and flow-cytometric characterization of picophytoprokaryote populations in central (17°S) and southern (20°S) shelf waters of the Great Barrier Reef. *Journal of Plankton Research* **23**, 809–828. doi:10.1093/plankt/23.8.809

Messer LF, Brown MV, Furnas MJ, Carney RL, McKinnon AD, Beymour JR, *et al.* (2017) Diversity and activity of diazotrophs in Great Barrier Reef surface waters. *Frontiers in Microbiology* **8**, 967. doi:10.3389/fmicb.2017.00967

Raleigh RH, Coles VJ, Capone DG (2004) Modeling the distribution of *Trichodesmium* and nitrogen fixation in the Atlantic Ocean. *Journal of Geophysical Research* **109**, C06006. doi:10.1029/2002JC001753

Nanoplankton and microplankton

Furnas M, Mitchell A, Skuza M, Brodie J (2005) In the other 90%: phytoplankton responses to enhanced nutrient availability in the Great Barrier Reef Lagoon. *Marine Pollution Bulletin* **51**, 253–265. doi:10.1016/j.marpolbul.2004.11.010

Schmoker C, Hernandez-Leon S, Calbet A (2013) Microzooplankton grazing in the oceans: impacts, data variability, knowledge gaps and future directions. *Journal of Plankton Research* **35**, 691–706. doi:10.1093/plankt/fbt023

Mesozooplankton

Doyle JR, McKinnon AD, Uthicke S (2017) Quantifying larvae of the coralivorous seastar *Acanthaster planci* on the Great Barrier Reef using qPCR. *Marine Biology* **164**, 176. doi:10.1007/s00227-017-3206-x

Mackas DL, Tsuda A (1999) Mesozooplankton in the eastern and western subarctic Pacific: community composition, seasonal life histories, and interannual variability. *Progress in Oceanography* **43**, 335–363. doi:10.1016/S0079-6611(99)00012-9

McKinnon A, Thorrold S (1993) Zooplankton community structure and copepod egg production in coastal waters of the central Great Barrier Reef lagoon. *Journal of Plankton Research* **15**, 1387–1411. doi:10.1093/plankt/15.12.1387

McKinnon AD, Doyle J, Duggan S, Logan M, Lønborg JC, Brinkman R (2015) Zooplankton growth, respiration and grazing on the Australian margins of the tropical Indian and Pacific oceans. *PLoS One* **10**, e0140012. doi:10.1371/journal.pone.0140012

McKinnon AD, Duggan S, De'ath G (2005) Mesozooplankton dynamics in nearshore waters of the Great Barrier Reef. *Estuarine, Coastal and Shelf Science* **63**, 497–511. doi:10.1016/j.ecss.2004.12.011

Others – studying plankton and plankton ecology

Hamner WM, Jones MS, Carleton JH, Hauri IR, Williams DM (1988) Zooplankton, planktivorous fish, and water currents on a windward reef face: Great Barrier Reef, Australia. *Bulletin of Marine Science* **42**, 459–479.

Hood RR, Coles VJ, Capone DG (2004) Modeling the distribution of *Trichodesmium* and nitrogen fixation in the Atlantic Ocean. *Journal of Geophysical Research* **109**, C06006.

Kingsford M, Leis J, Shanks A, Lindeman K, Morgan S, Pineda J (2002) Sensory environments, larval abilities and local self-recruitment. *Bulletin of Marine Science* **70**, 309–340.

McKinnon AD, Richardson AJ, Burford MA, Furnas MJ (2007) Vulnerability of Great Barrier Reef plankton to climate change. In *Climate Change and the Great Barrier Reef.* (Eds JE Johnson and PA Marshall). pp. 121–152. Great Barrier Reef Marine Park Authority, Townsville.

McKinnon AD, Williams A, Young JW, Ceccarelli D, Dunstan P, Brewin *et al.* (2014) Tropical marginal seas: priority regions for managing marine biodiversity and ecosystem function. *Annual Review of Marine Science* **6**, 415–437. doi:10.1146/annurev-marine-010213-135042

Pitt KA, Kingsford MJ, Rissik D, Koop K (2007) Jellyfish modify the response of planktonic assemblages to nutrient pulses. *Marine Ecology Progress Series* **351**, 1–13. doi:10.3354/meps07298

Suthers IM, Rissik R (2009) *Plankton: A Guide to Their Ecology and Monitoring for Water Quality.* CSIRO Publishing, Melbourne.

Identification

Boltovskoy D (1999) *South Atlantic Zooplankton.* Backhuys Publishers, Leiden, Netherlands.

Castellani C, Edwards M (2017) *Marine Plankton: A Practical Guide to Ecology, Methodology and Taxonomy.* Oxford University Press, Oxford, UK.

Dakin WJ, Colefax AN (1940) *The Plankton of the Australian Coastal Waters off New South Wales. Part I.* Australian Publishing Company, Sydney.

Graham LE, Wilcox LW (2000) *Algae.* Prentice Hall, Upper Saddle River, NJ, USA.

Larink O, Westheide W (2011) *Coastal Plankton: Photo Guide for European Seas.* Verlag Dr Friedrich Pfeil, Munich.

Leis JM, Carson-Ewart M (2000) *The Larvae of Indo-Pacific Coastal Fishes.* Brill, Leiden, Netherlands.

Online guides

Marine species identification portal: http://species-identification.org/index.php

Phytoplankton encyclopaedia project: https://www.eoas.ubc.ca/research/phytoplankton/

Australian marine zooplankton taxonomic guide and atlas: http://www.imas.utas.edu.au/zooplankton

Plankton portal: https://www.planktonportal.org

The National Reference Stations maintained by IMOS include one station in the waters of the GBR, near the site of the Yongala dive site south-east of Townsville. Physico-chemical and biological data, including phytoplankton pigment analyses and zooplankton counts, are available at the IODC portal: https://portal.aodn.org.au

Lynch TP, Morello EB, Evans K, Richardson AJ, Rochester W, et al. (2014) IMOS National Reference Stations: A Continental-Wide Physical, Chemical and Biological Coastal Observing System. *PLoS ONE* **9**(12), e113652. doi:10.1371/journal.pone.0113652

17

Macroalgae

G. Diaz-Pulido

Overview

Generalities

Macroalgae are often referred to as seaweeds, yet they are not actually 'weeds'. Rather, macroalgae is a collective term used for large algae that are macroscopic and generally grow in the sea. They differ from other plants, such as seagrasses and mangroves, in that macroalgae lack roots, leafy shoots, flowers and vascular tissues. In fact, algae are a polyphyletic group with diverse evolutionary histories. While still used, the term 'algae' is no longer recognised as a formal taxon, although it remains a useful name when referring to those protists that are photosynthetic.

The macroalgae of the Great Barrier Reef (GBR) are a very diverse and complex group of species and forms. Forms include crusts, foliose and filamentous thalli (thallus refers to the body of an alga), ranging from simple branching structures to complex forms with highly specialised structures. The specialised structures are adaptations for light capture, reproduction, support, flotation and/or substrate attachment. The size of coral reef macroalgae ranges from a few millimetres to plants of up to 3–4 m high (such as the brown alga *Sargassum*). However, in the more nutrient-rich temperate regions of the oceans, brown algal kelps may grow to over 50 m in length (e.g. giant bull kelp in California). These organisms are not found on coral reefs.

Tropical macroalgae occupy a variety of habitats, including shallow and deep sections of coral reefs, inter-reefal areas, sandy bottoms, seagrass beds, mangrove roots, rocky intertidal areas, or even within the skeletons of healthy and dead corals, shells and limestone material (endolithic algae). Macroalgae are the major food source for a variety of herbivores, are major reef formers and create habitat for invertebrates and vertebrates of economic interest. They also play critical roles in reef degradation, when coral-dominated reefs are replaced by rocky reefs covered in macroalgae.

Diversity of Great Barrier Reef macroalgae

Taxonomic diversity and classification

Macroalgae are taxonomically classified into four different Phyla: Rhodophyta (from the Greek *'rhodon'* meaning 'red rose' and *'phyton'* meaning 'plant': red algae); Ochrophyta (class Phaeophyceae, from the Greek *'phaios'* meaning 'brown': brown algae); Chlorophyta (from the Greek *'chloros'* meaning 'green': green algae); and Cyanobacteria (from the Greek *'cyanos'* meaning 'dark blue': blue-green algae). This systematic classification is largely based on the composition of pigments involved in photosynthesis.

There are ~880 species and varieties of macroalgae recorded for the GBR according to the Australian Marine Algal Name Index (AMANI) and recent surveys across the GBR shelf seabed. However, this account is preliminary and it is likely that the number will increase with future field exploration.

In fact, the macroalgal flora from the GBR, together with the northern Australian coast, is one of the lesser known floras on the Australian continent.

According to the AMANI database, the Rhodophyta are the most diverse phylum for the GBR, with 323 species contained in 131 genera, with *Laurencia* (27 species), *Polysiphonia* (19) and *Ceramium* (16) the most speciose. There are 111 species of Phaeophyceae, with more than 50% of the species belonging to *Sargassum* (47 species) and *Dictyota* (11). Thirty-two genera of brown algae have been recorded for the GBR. The Chlorophyta include 195 species in 51 genera, of which *Caulerpa* (36 species), *Halimeda* (23) and *Cladophora* (19) contain the highest number of species.

Functional group diversity

Besides traditional species classification, macroalgae can also be classified based on ecological terms following a functional form group approach. This approach takes into consideration key plant attributes and ecological characteristics, such as the form of the plant, size, plant toughness, photosynthetic ability and growth, grazing resistance, and so on. Functional group classification is helpful in understanding the distribution of algal communities and their responses to environmental factors, because morphologically and anatomically similar algae have similar responses to environmental pressures regardless of their taxonomic affinities. The functional approach is useful particularly when identification to species level is not possible and consequently has been widely used in ecological studies on coral reefs of the GBR.

The functional group approach includes three main algal categories (Table 17.1):

- **Algal turfs.** Assemblages or multispecies associations of minute filamentous algae and the early life history stages of larger macroalgae, with high productivity, fast growth and colonisation rates. Turfs are ubiquitous and have low biomass but dominate much of the reef framework's surface. Analogous to some grasslands in terrestrial environments, turfs owe their continued existence to herbivores that graze on them, thereby preventing overgrowth by fleshy macroalgae. The term 'epilithic algal community' or EAC is often used to refer collectively to the algal assemblage that grows on the substrate; usually this refers to an assemblage dominated by filamentous algal turfs.
- **Upright macroalgae.** Large algal forms, more rigid and anatomically more complex than algal turfs, abundant in zones of low herbivory

Table 17.1. Categories and functional groups of benthic macroalgae present on the Great Barrier Reef

Algal categories	Functional groups	Examples of common genera
Algal turfs (<10 mm height)	Microalgae	*Lyngbya, Calothrix*
	Filamentous	*Cladophora, Polysiphonia*
	Juvenile stages of macroalgae	
'Upright' macroalgae (>10 mm height)		
Fleshy	Foliose membranous	*Ulva, Anadyomene*
	Foliose globose	*Ventricaria, Dictyosphaeria*
	Foliose corticated	*Dictyota, Lobophora*
	Corticated	*Laurencia, Hypnea*
	Leathery	*Sargassum, Turbinaria*
Calcareous	Calcareous articulated	*Halimeda, Amphiroa*
Crustose algae	Calcareous crustose	*Porolithon, Peyssonnelia*
	Non-calcareous crustose	*Ralfsia, Cutleria*

such as the intertidal, reef flats, inter-reefal areas or inshore reefs where strong wave action, heavy predation, or water quality limit grazing. They often contain chemical compounds that deter grazing from fishes.

- **Crustose algae.** Calcareous plants that grow completely adhered to the substrate forming crusts, with slow growth rates in general, and are abundant in shallow reefs with high herbivory pressure. This group includes species from the families Peyssonneliaceae and Corallinaceae.

Spatial and temporal distribution

Biogeography

The algal flora of the GBR belongs to the Solanderian biogeographical province in terms of the benthic algal flora of Australia. This province is less diverse when compared with southern Australia (Flindersian province) and New South Wales (Peronian province). Endemism on the GBR is low, because most species are widely distributed in the Indo-East Pacific biogeographical region, and many GBR species are also thought to be present in the tropical Atlantic. However, recent studies using DNA markers have revealed that some of the species that look identical are actually distinct.

Spatial distribution

The distribution and abundance of macroalgae on coral reefs are determined by the resources they require (i.e. light, carbon dioxide, mineral nutrients and substrate), the effects of environmental factors (e.g. temperature, salinity and water movement), individual rates of recruitment, mortality and dispersal, and biological interactions such as competition and herbivory. Macroalgal communities of the GBR are highly variable, showing latitudinal, cross shelf and within reef variation in composition and abundance. Many GBR macroalgae are also highly seasonal.

Cross shelf distributions – offshore and inshore reefs

Offshore reefs usually have low abundance of fleshy macroalgae and high cover of algal turfs and crustose coralline algae (CCA) compared with inshore reefs. Some fleshy macroalgae, such as the green fleshy macroalgae *Caulerpa*, *Chlorodesmis*, *Halimeda*, and the reds *Laurencia*, *Galaxaura* and *Liagora* are common in offshore reefs but in low

Table 17.2. Example of zonation of benthic algae in the Great Barrier Reef

Zones	Crustose corallines	Common fleshy macroalgae		
		Reds	***Browns***	***Greens***
Upper reef slope and reef front	*Porolithon* *Neogoniolithon* *Lithophyllum*	*Predaea*, *Galaxaura*	*Lobophora*	*Chlorodesmis*
Reef crest	*Porolithon* *Neogoniolithon*	*Laurencia*		*Caulerpa racemosa* *Chlorodesmis*
Reef flat	*Porolithon*	*Acanthophora* *Laurencia* *Gelidiella* *Hypnea*	*Dictyota*, *Padina* *Sargassum* *Hydroclathrus* *Chnoospora*	*Caulerpa* *Chlorodesmis* *Halimeda* *Dictyosphaeria*
Lagoon			*Hydroclathrus*	Cyanobacteria *Halimeda* spp. *Caulerpa* spp.
>10 m deep and cryptic	*Lithothamnium* *Mesophyllum* *Neogoniolithon*	Turfs	*Lobophora* *Melanamansia*	*Rhipilia* *Halimeda* *Caulerpa*

abundance. CCA are abundant and diverse on offshore reefs and play significant roles in reef construction. Common taxa on offshore reefs include the CCA *Porolithon onkodes,* and species of *Neogoniolithon* and *Lithophyllum.* The cross-shelf distribution of the algal functional groups is predominantly affected by fish grazing and water quality (nutrient availability and sedimentation).

Inshore reefs usually have abundant and conspicuous stands of fleshy macroalgae. In particular, brown macroalgae of the order Fucales such as *Sargassum, Hormophysa, Turbinaria* and *Cystoseira,* form dense and highly productive beds of ~2 m height. Other fleshy brown macroalgae such as *Lobophora, Dictyota, Colpomenia, Chnoospora* and *Padina* and the red *Asparagopsis taxiformis,* may also be abundant in shallow inshore reefs. *Lobophora variegata* can be particularly abundant in inshore reefs, especially between branches of corals and after coral disturbance such as bleaching. Crustose coralline algae are common but are not abundant.

Within reef distribution

Algal zonation is quite clear in rocky intertidal coasts but is normally diffuse in subtidal reefs, where algal communities are distributed as a continuum along environmental gradients (e.g. depth). Several reef zones can be recognised in a cross-section of an offshore reef from shallow to deep areas (Table 17.2):

- **intertidal and beach rock** – diverse fleshy macroalgal communities, reduced grazing by large animals, intense solar radiation
- **reef lagoon** – limited macroalgal growth due to sandy bottom; however, the microphytobenthos community that grows on sand can be highly productive
- **reef flat (back-reef)** – diverse fleshy macroalgal communities, low grazing
- **reef crest** – abundant CCA and algal turfs, intense grazing and wave action
- **reef front and upper reef slope** – abundance and diversity of macroalgae decreases with increasing depth, algal communities dominated by turfs and CCA, poorly developed fleshy macroalgal populations
- **walls** – low algal cover and high coral cover, some upright calcareous macroalgae such as *Halimeda* can be locally abundant

Particular microhabitats such as crevices and the territories of damselfishes play important roles in locally increasing the diversity of fleshy and turf algae.

Inter-reefal areas and *Halimeda* beds. Macroalgal communities associated with seagrass beds, particularly in deep, soft-bottom inter-reefal (situated between reefs) areas and lagoonal areas (located between mainland and reef) along the GBR are quite abundant and rich in species. In contrast to seagrasses, most algae do not attach to sand, although several green macroalgae have adapted to such environments by developing special anchoring features. This is the case for green algae such as *Halimeda, Caulerpa* and *Udotea,* which are commonly found intermixed with seagrasses. Large red foliose macroalgae are also common in inter-reefal habitats. Macroalgae growing on leaves of seagrasses are called epiphytes and may play

Box 17.1. Inter-reefal areas and *Halimeda* beds

Seaweeds are abundant in the deep water, inter-reefal areas of the northern part of the GBR. Large mounds formed from the green calcareous alga *Halimeda* are estimated to cover up to 6167 km^2 in this region and may be up to 20 m high. These *Halimeda* meadows occur principally in the northern sections at depths between 20 m and 40 m, but there are also some in the central and southern sections of the GBR, where they have been found at depths down to 96 m. The GBR contains the most extensive actively calcifying *Halimeda* beds in the world, thus playing important role as calcium carbonate sinks. Tidal jets and localised upwelling events in the northern section of the GBR provide the nutrients needed to sustain extensive deep (30–45 m) meadows of *Halimeda*.

Box 17.2. Preservation

Macroalgae used for taxonomic purposes should be collected with the anchoring systems and preferably with reproductive structures. Macroalgae can be preserved in a solution of 4% formalin or in ethanol 70% (although decolouration may occur). However, dry herbarium specimens are easier to use and transport. White cardboard acid-free sheets are used for mounting specimens. Each herbarium sheet should contain basic information on the locality of collecting, date, depth, habitat, colour and the collector's name. Algal tissue can also be preserved in silica gel for molecular purposes. Permits are required: see Chapter 14.

important roles as food for invertebrates and vertebrates in seagrass meadows. *Halimeda* mounds grow in nutrient-rich upwelling water in between reefs that make up the outer barrier reef, particularly in the northern GBR (see Box 17.1).

Seasonality

The abundance, growth and reproduction of many GBR macroalgae are highly variable in time. Large seaweeds such as *Sargassum* are strongly seasonal, with peaks in biomass and reproduction generally during the summer and lowest biomass during the winter. Extensive blooms of fleshy brown macroalgae such as *Chnoospora* and *Hydroclathrus* are common on shallow reef flats during winter. Due to these strong seasonal changes, some authors argue that the seasonality of the GBR flora may be as strong as that from temperate zones.

Common genera and identification

Identification of coral reef macroalgae to the genus level is relatively easy, but the identification at the species level is more difficult and generally requires examination under a compound microscope (see Box 17.2 for preservation methods). The reproductive structures, internal tissues, cell organisation, and so on, are key features required for rigorous species identification, and molecular sequencing is generally used for accurate identification.

Red algae

- *Amphiroa:* heavily calcified, branches cylindrical to flattened, composed of smooth segments linked by very short non-calcareous joints. Pale pink to red, also called geniculated Corallinaceae. Common throughout the GBR (Fig. 17.1A).
- *Asparagopsis taxiformis:* plants soft, with creeping stems and upright fluffy or feathery tufts. Bright pink and common on inshore reefs (Fig. 17.1B).
- *Corallophila:* red filamentous alga with creeping axes and erect branches with pointed tips, usually found overgrowing corals (Fig. 17.1C).
- *Eucheuma:* similar to some species of *Hypnea* but the thalli are tougher and rubbery. Sometimes found between branches of branching *Acropora* corals (Fig. 17.1D).
- *Galaxaura:* lightly calcified, dichotomously branched, branches cylindrical or flattened, smooth or hairy. Pink to red, sometimes with chalky appearance.
- *Hypnea:* branches cylindrical and generally bearing numerous relatively short spine or tooth-like branchlets with pointed tips. Colour variable from pale brown to dark purple. Usually between branches of hard corals (Fig. 17.1E).
- *Jania:* heavily calcified, similar to *Amphiroa* but plants are smaller (few mm) and branches are predominantly cylindrical. Widespread in shallow and deep reefs (Fig. 17.1F).
- *Laurencia:* plants generally bushy, branches usually cylindrical with blunt tip branchlets. Colour variable, some with green branches and pink tips, others ranging from orange, red to pink. Common on reef flats (Fig. 17.1G).
- *Amansia:* red to reddish-brown, branches leaf-like and grouped in rosettes, with marginal teeth. Generally found on crevices and other low-light microenvironments (Fig. 17.1H).
- *Peyssonnelia:* encrusting, rounded to fan-shaped plants. Calcification on the lower side and fleshy surface. Colour variable, dark red, purple

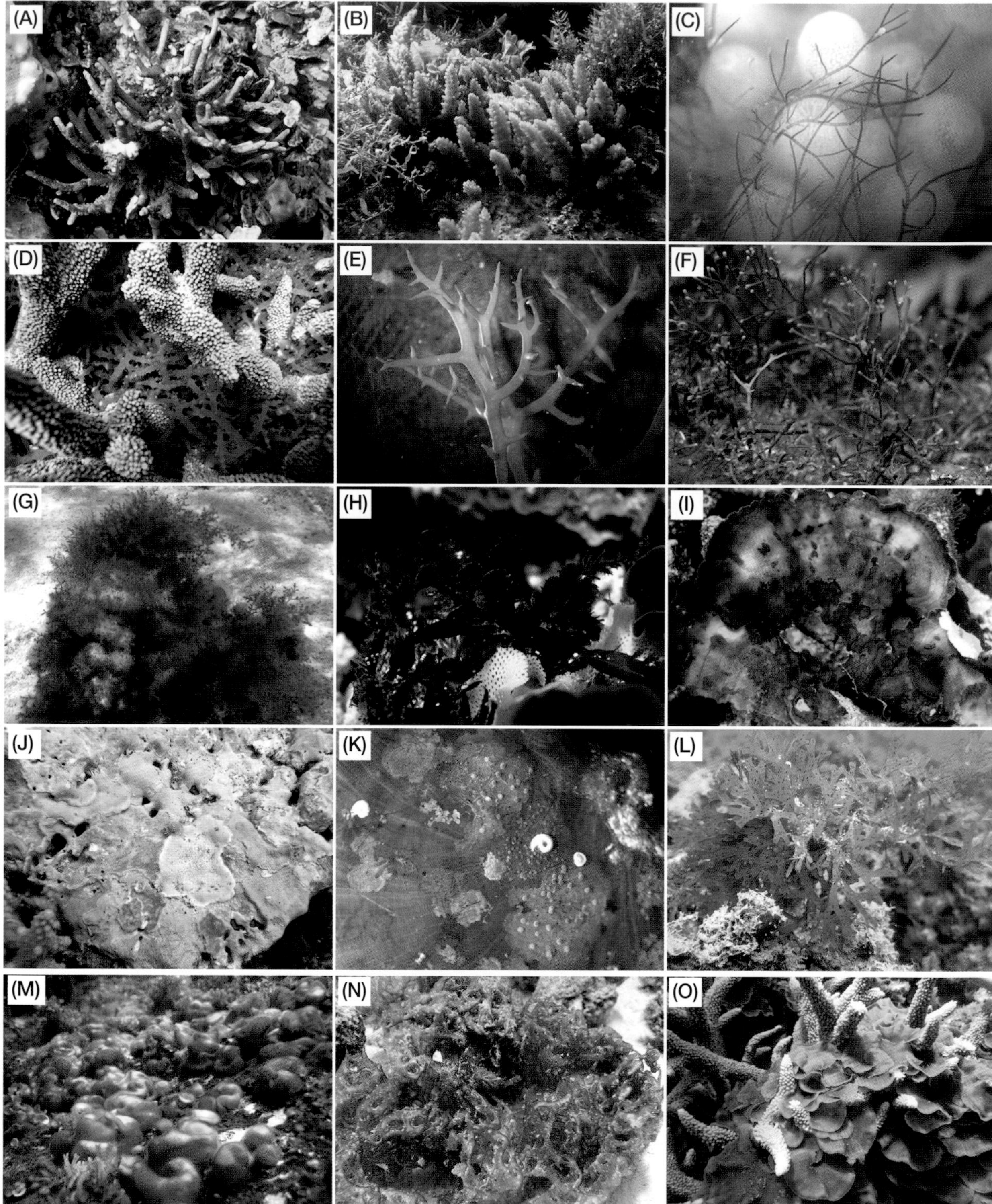

Fig. 17.1. Common genera and species of benthic algae from the Great Barrier Reef: **red algae (A)** *Amphiroa* sp.; **(B)** *Asparagopsis taxiformis*; **(C)** *Corallophila huysmansii* (= *Centroceras huysmansii*); **(D)** *Eucheuma denticulatum*; **(E)** *Hypnea pannosa*; **(F)** *Jania* sp.; **(G)** *Laurencia* cf. *intricata*; **(H)** *Amansia glomerata*; **(I)** *Peyssonnelia* sp.; **(J)** *Porolithon onkodes*; **(K)** crustose coralline algae epiphytic on *Lobophora variegata*; **brown algae (L)** *Dictyota* sp.; **(M)** *Colpomenia sinuosa*; **(N)** *Hydroclathrus clathratus*; **(O)** *Lobophora variegata* (Photos: G. Diaz-Pulido).

pink or red yellowish. Common on overhangs and cryptic microhabitats (Fig. 17.1I).

- *Porolithon:* encrusting, heavily calcified with chalky texture. Pink crusts of several mm thick. Common on the reef crest (Fig. 17.1J). Other crustose coralline algae are epiphytic (grow on the surface of other benthic algae) (Fig. 17.1K).
- *Polysiphonia:* filamentous, usually a few millimetres tall, examination under the microscope shows cylindrical polysiphonous branches with a 'banding' appearance, similar to *Sphacelaria*. Pink to red, red-brown or brown. Common and abundant on algal turfs.

Brown algae

- *Dictyota:* plants creeping or erect, flattened, strap or ribbon-like (without midrib as in *Dictyopteris*). Branching dichotomous (forked). Light brown and several species with blue-green iridescence. Common (Fig. 17.1L).
- *Chnoospora:* plants forming mats or cushion-like clumps, branches are slightly flattened and dichotomous. Light brown. Common on reef flats.
- *Colpomenia:* plants rounded or irregularly lobed with hollow interior. Light to pale golden brown. Common on inshore reefs (Fig. 17.1M).
- *Feldmannia (= Hincksia):* small (few mm) filamentous plants, fine, erect, uniseriate (one row), pale yellowish and translucent. Reproductive structures (sporangia) somewhat corn cob-like. Common on algal turfs.
- *Hormophysa:* similar to *Sargassum* but blades with internal oblong air bladders. Yellow to dark-brown. Common on inshore refs.
- *Hydroclathrus:* light to pale golden brown, net-like structure, perforated. Common on calm and sheltered waters (Fig. 17.1N).
- *Lobophora cf. variegata:* creeping, rounded to fan-shaped plants, sometimes encrusting. Pale to dark brown and usually with concentric bands and radiating lines. Common to locally very abundant, particularly between branches of corals (Fig. 17.1O).
- *Padina:* upright, sheet-like, fan-shaped plants, with concentric bands and also whitish bands due to carbonate deposition. Similar to *Lobophora* but has characteristic inrolled outer margin. Common on reefs flats (Fig. 17.2A).
- *Sargassum:* plants erect, leathery, some up to a couple of metres are the tallest seaweeds in the GBR. They typically have a stipe, leaf-like fronds with a midrib (central vein), and air vesicles or floats. Very abundant on inshore reefs (Fig. 17.2B, C).
- *Sphacelaria:* filamentous, small (few mm) turfing plants, branch cells (observed under compound microscope) arranged in regular transverse tiers, individual cells rectangular and elongated longitudinally. Distinctive dark brown cell at the tip of each filament. One of the most abundant taxa of algal turfs.
- *Turbinaria:* plants erect, tough, leathery, with closely placed top-shaped branches with spiny margin, each containing an embedded air bladder. Common on reef flats (Fig. 17.2D).

Green algae

- *Caulerpa:* all species have a creeping stolon attached by rhizoids and erect green branches or fronds. Fronds are very variable including leaf and feather-like, others have cylindrical to club-shaped, spherical-like branchlets. Common on sandy and reef bottoms from shallow and deep reef (Fig. 17.2E–G).
- *Chlorodesmis:* plants forming tight clumps or tufts of repeatedly forked filaments. Bright green and common on shallow reefs (Fig. 17.2H).
- *Dictyosphaeria:* plants spherical to irregularly lobed, light green, surface hard and tough composed of one layer of angular or polygonal cells, resembling a honeycomb. Hollow to solid inside depending on the species. Common (Fig. 17.2I).
- *Halimeda:* plants erect, lightly to heavily calcified, pale to dark green. Branches formed by calcified segments separated by deep

Fig. 17.2. Common genera and species of benthic algae from the Great Barrier Reef: **brown algae (A)** *Padina* sp.; **(B)** *Sargassum tenerrimum;* **(C)** *Sargassum* spp.; **(D)** *Turbinaria ornata;* **green algae (E)** *Caulerpa cupressoides;* **(F)** *Caulerpa* sp.; **(G)** *Caulerpa racemosa;* **(H)** *Chlorodesmis fastigiata;* **(I)** *Dictyosphaeria versluysii;* **(J)** *Halimeda* cf. *discoidea;* **(K)** *Halimeda* sp.; **(L)** green band of an endolithic algal community (including *Ostreobium* spp.) within the skeleton of *Porolithon onkodes;* **(M)** *Udotea* sp.; **(N)** algal turf overgrowing recently dead coral; **cyanobacteria (O)** cyanobacteria growing on dead coral; **(P)** filamentous cyanobacteria under compound microscope (Photos: G. Diaz-Pulido).

constrictions. Segments can be flattened (triangular to discoid or kidney-shaped) to cylindrical (Fig. 17.2J, K).

- *Ostreobium:* microscopic green filaments, cylindrical to inflated, usually within skeletons of healthy and dead corals and other carbonate substrates. Widespread on deep and shallow reefs (Fig. 17.2L).
- *Udotea:* upright calcified, stalked and fan-shaped plant, anchored to the substratum by a rhizoidal mass. Grey-green (Fig. 17.2M).
- *Ulva:* bright green, sheet-like or membranous blades. Uncommon on reefs. Species of *Enteromorpha* (now belong to the genus *Ulva*) are small (a few mm) and have the form of a hollow tube; common on algal turfs.
- *Ventricaria:* globose plants up to several centimetres in diameter. Glossy dark green with bright reflective glare. Usually epiphytised by pink crustose calcareous algae. Common throughout the reef.

Reproduction

Macroalgae reproduce either asexually or sexually. Asexual reproduction involves the release of spores (propagules) or by fragmentation (pieces of plant braking off to produce new individuals). In sexual reproduction, male and female gametes are released into the water, but there are some examples where female gametes are retained by the parent and the resulting embryo develops (at least temporarily) on the parent gametophyte.

Macroalgae have complex life cycles, of which there are three types in coral reef algae:

(1) **haplontic life cycle** (meiosis of a zygote occurs immediately after karyogamy)
(2) **diplontic life cycle** (the zygote divides mitotically to produce a multicellular diploid individual)
(3) **diplobiontic life cycle or alternation of generations** (the haploid and diploid phases are alternated, each phase consisting of one of two separate, free living organisms: a gametophyte, which is genetically haploid, and a sporophyte, which is genetically diploid).

Ecological roles of macroalgae

Contribution to primary production

A large proportion of the primary production (the creation of organic matter by plants from inorganic material such as CO_2 and sunlight during photosynthesis) in a coral reef comes from the contribution of benthic algae. Net primary production is variable, ranging from 148 to 500 g C/m^2/year for algal turfs, 146 to 1095 g C/m^2/year for fleshy macroalgae, and 73 to 475 g C/m^2/year for crustose coralline algae. Planktonic microalgae and algal symbionts of scleractinian corals also contribute to reef productivity but to a much lesser degree. The organic matter (carbon) produced by benthic macroalgae enters the reef food chain either by: (1) consumption by herbivorous fishes, crabs and sea urchins; (2) release of dissolved organic matter by the algae into the water column where it is consumed by bacteria that in turn may be consumed by a variety of filter feeders; or (3) export to adjacent ecosystems such as seagrass meadows, mangroves or to the sea floor by currents and tides. See Chapter 8 for further details on the energy flow through coral reefs.

Nitrogen fixation and nutrient cycling

Filamentous cyanobacteria such as *Calothrix* living in algal turf communities (Fig. 17.2O) fix significant amounts of atmospheric (inorganic) nitrogen into ammonia, which is then used by the cyanobacteria themselves to build organic matter. Because of the rapid growth rates of cyanobacteria and intense grazing on turf communities, the organic nitrogen is then distributed throughout the reef, contributing to reef nutrition (Chapter 8). Macroalgae take up, store and release nutrients, thereby contributing to nutrient cycling in coral reef ecosystems.

Construction

Many macroalgae make important contributions to the construction of the reef framework by depositing calcium carbonate ($CaCO_3$). CCA (e.g. *Porolithon* and *Pneophyllum*) are important framework builders and framework cementers in coral reefs, whereby they bind adjacent substrata and provide a calcified tissue barrier against erosion. This process is particularly important on the reef crest environments of the GBR. Crustose calcareous algae such as *Peyssonnelia* are also important in deeper areas at the edge of the continental platform in the southern GBR (80–120 m), where they form large algal frameworks of several metres high. Vertical growth of crusts of *Porolithon* is very slow (~1.5 mm/year) but since *Porolithon* may be locally abundant in reef crests, deposition of calcium carbonate within the tissues of CCA can be up to 10.3 kg $CaCO_3$/m^2/year in some parts of the GBR (e.g. Lizard Island). CCA deposit a form of calcium carbonate (high-magnesium calcite) that is more soluble than the aragonitic skeleton of corals; therefore CCA are perhaps the most sensitive group of calcifying organisms to the impacts of ocean acidification in the GBR.

Upright calcareous algae such as *Halimeda*, *Udotea*, *Amphiroa* and *Galaxaura* contribute to the production of marine sediments that fill in the spaces between corals. The white sand of beaches and reef lagoons is largely the eroded calcium carbonate skeletons of these algae. Calcium carbonate is deposited as aragonite in *Halimeda*, with an estimated production of around 2.2 kg $CaCO_3/m^2/$year. Calcification may be an adaptation to inhibit grazing (a defensive mechanism), resist wave shock and to provide mechanical support.

Facilitation of coral settlement

Some CCA induce settlement and metamorphosis of coral larvae and a range of other invertebrates in the GBR, thus playing a critical role in reef resilience. This interaction seems to be mediated by chemicals released by the alga and by the presence of some bacterial strains on the surface of the CCA.

Roles in reef degradation

Macroalgae play important roles in reef degradation, particularly in ecological phase shifts, where abundant reef-building corals are replaced by abundant fleshy macroalgae. Reductions in herbivory due to overfishing, and increases in nutrient inputs leading to eutrophication (e.g. sewage and fertiliser), have been suggested as causes of increased abundance of fleshy macroalgae leading to coral overgrowth and reef degradation. Coral bleaching, crown-of-thorns starfish outbreaks, extreme low tides, coral diseases, cyclones, and so on, result in coral mortality, providing an environment that is rapidly colonised by diverse algal communities (Fig. 17.2N). Early stages of the colonisation process are dominated by biofilms of cyanobacteria and diatoms, which are subsequently replaced by filamentous algal turfs. Fleshy macroalgae and CCA may overgrow algal turfs during later stages of the succession. Such disturbances, and particularly those due to climate change (e.g. bleaching), may lead to an overall increase in the total amount of macroalgae (see Chapters 11 and 12 for further details). Dominance by thick mats or larger, fleshy macroalgae may contribute to reef degradation by overgrowing corals, inhibiting coral settlement and recruitment, contributing to coral diseases (e.g. Fig. 17.2P), and thereby decreasing the aesthetic value of reefs.

Bioerosion

Endolithic algae that live within the skeletons of both healthy and dead corals, as well as other calcareous substrates, contribute to reef erosion and destruction. These algae are generally filamentous and microscopic but form a thin dark green band visible to the naked eye underneath the coral and crustose algal tissue (Fig. 17.2L). Some examples of carbonate-boring algae include the greens *Ostreobium* spp., cyanobacteria *Mastigocoleus testarum*, *Plectonema terebrans* and *Hyella* spp. and some red algae. Endolithic algae penetrate and dissolve the calcium carbonate, weakening the reef framework and thus hasten other erosive activities. Studies at One Tree Island on the GBR have shown rates of bioerosion by endolithic algae to range between 20 and 30 $g/m^2/$year. For more information on bioerosion see Chapter 9.

Further reading

Algae of Australia Series (2007) *Algae of Australia: Introduction*. Australian Biological Resources Study, Canberra and CSIRO Publishing, Melbourne.

Borowitzka LJ, Larkum AWD (1986) Reef algae. *Oceanus* **29**, 49–54.

Clayton MN, King RJ (Eds) (1990) *Biology of Marine Plants*. Longman Cheshire, Melbourne.

Cribb AB (Ed.) (1996) *Seaweeds of Queensland: A Naturalist's Guide*. The Queensland Naturalist's Club, Brisbane.

Diaz-Pulido G, McCook LJ (2002) The fate of bleached corals: patterns and dynamics of algal recruitment. *Marine Ecology Progress Series* **232**, 115–128. doi:10.3354/meps232115

Huisman JM (Ed.) (2000) *Marine Plants of Australia*. University of Western Australia Press, Perth, WA.

Hurrey LP, Pitcher CR, Lovelock CE, Schmidt S (2013) Macroalgal species richness and assemblage composition of the Great Barrier Reef seabed. *Marine Ecology Progress Series* **492**, 69–83. doi:10.3354/meps10366

Kraft GT (2009) *Algae of Australia: Marine Benthic Algae of Lord Howe Island and the Southern Great Barrier Reef, 2: Brown Algae*. Australian Biological Resources Study, Canberra and CSIRO Publishing, Melbourne.

Littler DS, Littler MM (Eds) (2003) *South Pacific Reef Plants*. Offshore Graphics, Washington DC, USA.

McCook LJ, Jompa J, Diaz-Pulido G (2001) Competition between corals and algae on coral reefs: a review of evidence and mechanisms. *Coral Reefs* **19**, 400–417. doi:10.1007/s003380000129

McNeil MA, Webster JM, Beaman RJ, Graham TL (2016) New constraints on the spatial distribution and morphology of the *Halimeda* bioherms of the Great Barrier Reef, Australia. *Coral Reefs* **35**, 1343–1355. doi:10.1007/s00338-016-1492-2

Price IR, Scott FJ (Eds) (1992) *The Turf Algal Flora of the Great Barrier Reef: Part I. Rhodophyta*. James Cook University of North Queensland, Townsville.

Womersley HBS (Ed.) (2003) *The Marine Benthic Flora of Southern Australia. Rhodophyta – Part III D.* State Herbarium of South Australia, Adelaide.

Online guides

Australian Marine Algal Name Index: http://www.anbg.gov.au/abrs/online-resources/amani/

International Database of Information on Algae: http://www.algaebase.org/

18

Mangroves and seagrasses

N. C. Duke and A. W. D. Larkum

Overview

Sandwiched between two of the world's iconic tropical ecosystems of coral reefs and rainforests, are two important coastal communities: mangroves and seagrasses. While corals flourish in shallow warm seas, and rainforests cover wetter upland regions, all are dependent on this unique association. Mangroves inhabit the sheltered intertidal margin part barely above mean sea level. Seagrasses occupy depths from intertidal to deeper habitats, depending on the clarity of the water column. Like coral reefs, each of these biota-structured ecosystems play an important role in coastal processes with highly developed linkages and connectivity between and among them. These relationships are vital to the survival of each. For example, while sediment-loving mangroves depend on shorelines sheltered by coral reef structures, they in turn protect sediment-sensitive corals from receiving unwanted materials flushed downstream from surrounding land catchments.

General description role in ecosystems

Mangrove and seagrass communities are characterised by a small number of widely distributed angiosperm taxa, having evolved mostly post-Cretaceous, within the last 60–100 million years. The relatively recent evolution of these habitats may help explain their comparatively low species diversity today, but this is arguably more related to the harsh environmental factors that characterise these communities. The relatively rich mangrove and seagrass floras of today are testament to their adaptive and evolutionary success for living in the intertidal zone and shallow waters. These highly specialised plants flourish in soft sediments, tapping rich estuarine nutrients with their distinctively vascular root systems. Mangroves further provide significant habitat and structure where their biomass accumulation, while readily seen as contiguous with adjacent rainforests, is also analogous to that created by coral reefs.

While important details of relevant phyletic origins remain lacking, ancestral mangrove and seagrass plant taxa are known to have reinvaded marine environments in multiple episodes from the diverse number of respective plant lineages dominated by angiosperms. Their evolution appears constrained and directed by key functional attributes essential for survival in saline and aqueous settings where isotonic extremes, desiccation and hydrologic exposure combine as uniquely harsh constraints on organisms living in tidal zones and estuaries. To achieve this, mangroves and seagrasses share several ecophysiological traits and have evolved mechanisms to cope with life at the land–sea interface, such as salt tolerance, translocation of gases to aerate their roots, and specialised reproductive strategies. For mangroves, this includes the significant development of viviparous propagules. Both plant communities perform important ecosystem services such as sediment stabilisation, nutrient processing, shoreline protection, providing habitat and nursery. Genetic diversity of mangroves and seagrasses can be high based on the number of plant families represented. But, despite this, species numbers are often quite low compared with adjacent communities of tropical rainforests and coral reefs. By contrast, the diversity of organisms that reside in, and use, mangroves and seagrasses can be quite high. Many fish and shellfish, including important commercial, recreational and artisanal fisheries, spend all or part of their life cycle in mangroves and/or seagrasses. Other fishery species offshore depend on the abundant supply of food resources coming from these highly productive shoreline ecosystems.

The land–sea interface is a dynamic environment where subtle natural changes in climate, sea level, sediment and nutrient inputs have dramatic consequences in the distribution and health of mangroves and seagrasses. Local human disturbance of mangroves and seagrasses include eutrophication, dredging/filling, overfishing and sedimentation. The combined pressures of human disturbances and global climate change have led to mangroves and seagrasses becoming 'endangered communities'. Furthermore, in a recent extreme weather event in the Gulf of Carpentaria, 7400 ha of mangroves were killed along 1000 km of shoreline (see Box 18.1). It is not known how seagrasses fared. These types of impacts are currently under further investigation. In other instances, small-scale restoration projects have demonstrated the extreme difficulty in scaling up mitigation efforts for effective, large-scale restoration. Urgent protective measures are needed to avoid the loss of mangroves and seagrasses, and the resulting environmental degradation of coastal ecosystems of the GBR coast.

Mangroves

Overview

Mangroves are a diverse group of predominantly tropical trees and shrubs growing in the upper half of the intertidal zone of coastal areas worldwide. They are well known for their morphological and physiological adaptations for life coping with salt, saturated soils and regular tidal inundation, notably with specialised attributes such as: exposed breathing roots above ground, extra stem support structures, salt-excreting leaves, low water potentials and high intracellular salt concentrations to maintain favourable water relations in saline environments, and viviparous water-dispersed propagules. Mangroves are often mistakenly thought of as a single entity. But, like coral reefs, mangroves are as functionally diverse and complex as the range of species, variants and morphotypes present at particular locations (Fig. 18.1). Also like coral reefs, they provide essential structure and habitat for a host of marine and intertidal species, including residents among their dense forests and complex roots and for visitors with each flooding tide (see Box 18.2 – green sea turtles eat mangrove fruits at high tides). Mangroves also have analogies with tropical rainforests, having comparable canopy habitat for birds, mammals and insects. This overlap is re-enforced by ancestral links between these plant habitats. But, despite the shared features, mangroves include specialist attributes and dedicated resident biota found nowhere else. Examples include specialist mangrove forms of the: robin, mistletoebird, mistletoes, grapsid crabs, molluscs, herbivorous insects and numerous floral visitors.

Taxonomy and functional morphology

Mangroves are not a genetic entity, but an ecological system (Tomlinson 2016). Mangrove vegetation includes a range of functional forms, including trees, shrubs, a palm and ground fern. These generally exceed 0.5 m in height; and normally grow above mean sea level in the intertidal zone of marine coastal environments, and estuarine margins. Mangrove plants do not come from a single genetic source, and the only plant families that are comprised exclusively of mangrove taxa are Avicenniaceae and Sonneratiaceae. Around the world, the total number of mangrove plants is around

Fig. 18.1. Some mangrove species found along the coastline of GBR catchments in Queensland: **(A)** exposed roots of milky mangrove, *Excoecaria agallocha*; **(B)** small tree of the grey mangrove, *Avicennia marina* – inset showing mature fruits; **(C)** leaves and mature fruits of river mangrove, *Aegiceras corniculatum*; **(D)** shrubs of myrtle mangrove, *Osbornia octodonta*; **(E)** knee roots of orange mangrove, *Bruguiera gymnorhiza* – inset showing leafy shoot with distinctive red flower buds; and **(F)** small tree of stilt mangrove, *Rhizophora stylosa*.

80 taxa from 21 families, consisting mostly of angiosperms. Bordering the GBR World Heritage Area, there are 44 taxa of 19 families (Table 18.1), representing more than 50% of the worlds' genetic variation in mangrove plants, and almost all of Australia's. Some species, such as *Avicennia marina, Rhizophora stylosa* and *Bruguiera gymnorhiza* (Fig. 18.1), are widespread in the Indo West Pacific region, while others, such as *Ceriops australis, Bruguiera exaristata, Diospyros littorea,* are more restricted to the Australasian region.

It is noteworthy that a rare and endangered species, *Bruguiera hainesii* was recently discovered for Australia and the GBR region by a community scientist in mangroves around Cairns (Cooper *et al.* 2016; see Box 18.3). This discovery highlights the need for dedicated botanical surveys to better characterise and describe the diversity and distribution of plants in GBR tidal wetland communities.

Mangrove species are commonly located on mainland islands, sand cays and reef atolls, never falling below 60% presence and often above 80%. This group (Fig. 18.1A–F) includes nine species: *Avicennia marina, Aegiceras corniculatum, Bruguiera gymnorhiza, Excoecaria agallocha, Lumnitzera racemosa, Osbornia octodonta, Pemphis acidula, Rhizophora stylosa* and *Sonneratia alba*. Only *R. stylosa* and *A. marina* were considered major dominants. All species of this group were generally described as less climatically sensitive. Their wide ecological amplitude makes them tolerant of a wide range of salinities from seaward exposure to periodic pulses of fresh water. Each species has further individual traits with, for example: *R. stylosa* confined to the front of mangroves, often fronting the sea and associated with moderate salinities but not extremes; *E. agallocha* are mostly associated with dry margins; while *O. octodonta* is characteristically common on sandy substrates.

Role of mangroves and the filters of the coast

Mangroves and tidal wetlands are essential to the sustainability of highly productive natural coastal environments. Mangroves have many well-acknowledged roles in coastal connectivity supporting enhancements in biodiversity and biomass not possible otherwise (Mumby *et al.* 2004). At another level, commercial advantages decry the importance of mangroves, where around 75% of the total seafood landed in Queensland comes from mangrove estuarine related species. These messages clearly indicate that healthy estuarine and nearshore marine ecosystems are biologically and commercially important – and these natural systems are intimately related, connected and dependent. So, where one is impacted, the effect will be felt more widely than might otherwise be expected. This is the case whether these ecosystems are viewed as sources of primary production with complex trophic linkages, as nursery and breeding sites, or as physical shelter and buffers from episodic severe flows and large waves.

These ecosystems and their linkages are seriously threatened. The GBR, for example, is threatened in this way by another insidious factor, in addition to global warming and coral bleaching. This immense and unique natural wonder is seriously threatened by smothering plumes of mud that greatly exceed prior natural flood levels. This corresponds with a century of large-scale land clearing and conversion of coastal forested wetlands into agricultural, port, urban and industrial developments. Coastal rivers have become little more than drains transporting eroded mud to settle in estuaries, in coastal shallows and on inshore reefs. Mangrove-lined estuaries have offered some respite and dampening of this effect, but in recent years this final bastion of coastal sediment filters are succumbing also to the increasing and unrelenting pressures of sea change populations expanding across coastal and estuarine regions of the GBR catchments (e.g. Jupiter *et al.* 2007).

Seagrasses

Australia is fortunate in having nearly half the 60 species of the world's seagrasses, 14 of which occur on Australian coral reefs. However, in reef waters in Australia, seagrasses may be inconspicuous because they are grazed heavily by invertebrates, fish, turtles and dugongs (Fig. 18.6). In protected

Table 18.1. The 44 mangroves of the GBR coastal catchments (updated from Duke 2006; 2016)

Also see Fig. 18.2. Those highlighted in bold are commonly found on islands along the GBRWHA.

Family	Taxa
Acanthaceae	*Acanthus ebracteatus* Vahl – two subspecies including: subsp. *ebracteatus*; and subsp. *ebarbatus* *Acanthus ilicifolius* L.
Arecaceae	*Nypa fruticans* Wurmb
Avicenniaceae	***Avicennia marina* (Forssk.) Vierh. - two varieties including: northern, var. *eucalyptifolia*; and southern, var. *australasica*** *Avicennia officinalis* L. (Northern Torres Strait only)
Bignoniaceae	*Dolichandrone spathacea* (L.f.) K.Schum.
Bombacaceae	*Camptostemon schultzii* Mast.
Caesalpiniaceae	*Cynometra iripa* Kostel
Combretaceae	*Lumnitzera littorea* (Jack) Voigt ***Lumnitzera racemosa* Willd.** *Lumnitzera* × *rosea* (Gaudich.) C.Presl ex Tomlinson, Bunt, Primack & Duke
Ebenaceae	*Diospyros littorea* (R.Br.) Kosterm.
Euphorbiaceae	***Excoecaria agallocha* L.**
Lecythidaceae	*Barringtonia racemosa* (L.) Spreng.
Lythraceae	***Pemphis acidula* J.R.Forst. & G.Forst.**
Meliaceae	*Xylocarpus granatum* J.Koenig. *Xylocarpus moluccensis* (Lam.) Roem.
Myrsinaceae	***Aegiceras corniculatum* (L.) Blanco**
Myrtaceae	***Osbornia octodonta* F.Muell.**
Plumbaginaceae	*Aegialitis annulata* R.Br.
Pteridaceae	*Acrostichum aureum* L. *Acrostichum speciosum* Willd.
Sterculiaceae	*Heritiera littoralis* Aiton
Sonneratiaceae	***Sonneratia alba* J.Sm. in A.Rees** *Sonneratia caseolaris* (L.) Engl. *Sonneratia* × *gulngai* N.C.Duke *Sonneratia lanceolata* Blume *Sonneratia ovata* Backer (Northern Torres Strait only)
Rhizophoraceae	*Bruguiera cylindrica* (L.) Blume *Bruguiera exaristata* Ding Hou ***Bruguiera gymnorhiza* (L.) Savigny ex Lam. & Poiret** *Bruguiera* × *hainesii* C.G. Rogers *Bruguiera parviflora* (Roxb.) Griff. *Bruguiera* × *rhynchopetala* (W.C.Ko) X.J.Ge & N.C.Duke *Bruguiera sexangula* (Lour.) Poir. *Ceriops australis* (C.T.White) E.Ballment, T.J.Sm. & J.A.Stoddart *Ceriops pseudodecandra* Sheue, Liu, Tsai & Yang *Ceriops tagal* (Perr.) C.B.Rob. *Rhizophora* × *annamalayana* Kathiresan *Rhizophora apiculata* Blume *Rhizophora* × *lamarckii* Montrouz. *Rhizophora mucronata* Lam. ***Rhizophora stylosa* Griff.**
Rubiaceae	*Scyphiphora hydrophylacea* C.F.Gaertn.

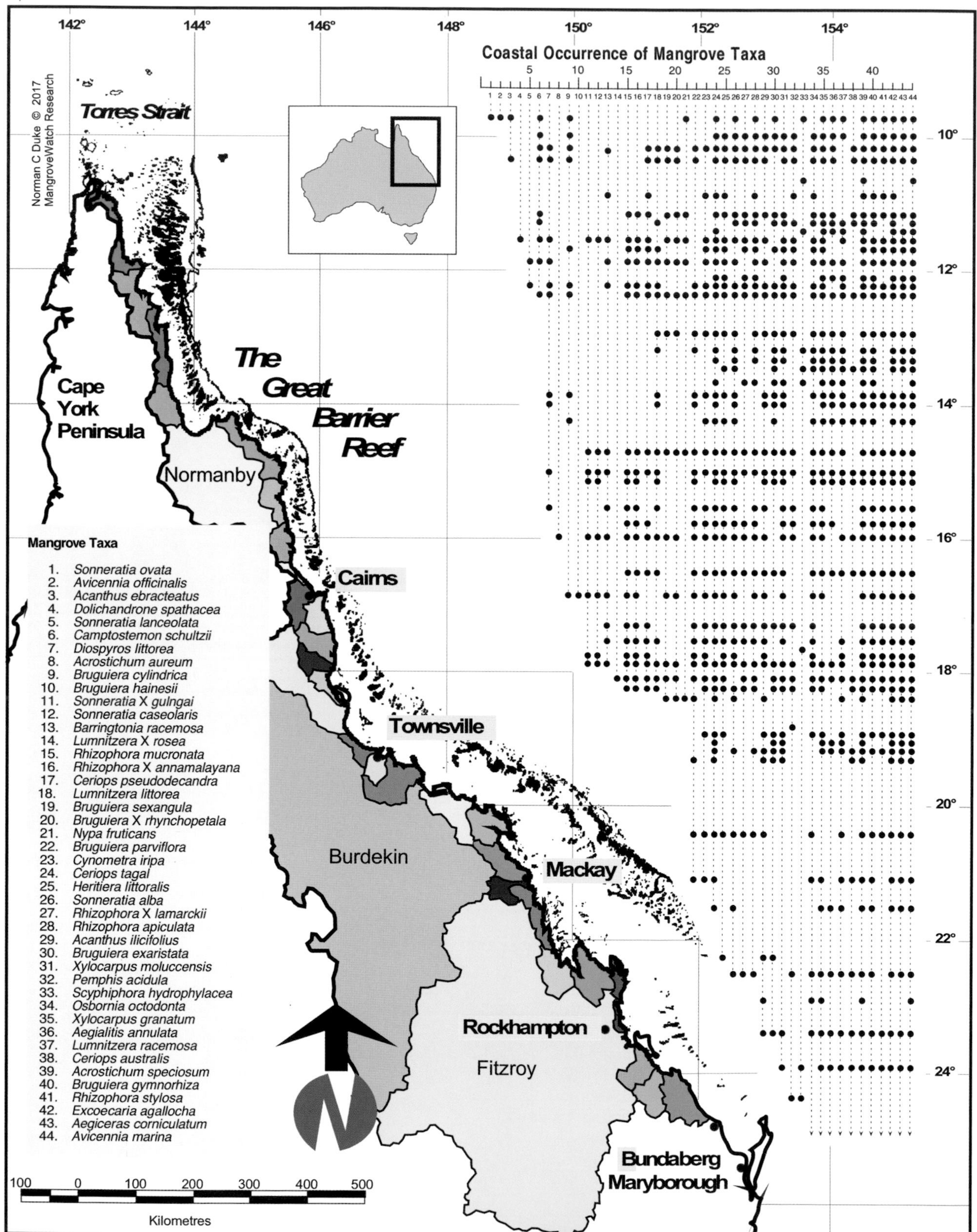

Fig. 18.2. Distribution of mangrove species along the coastline of GBR catchments in Queensland (Source: From Duke 2006, with updates in 2016).

Box 18.1. Unprecedented large-scale dieback of mangroves in Australia's Gulf of Carpentaria co-incident with coral bleaching on the GBR

Between late 2015 and early 2016, extensive areas of mangrove tidal wetland vegetation died along 1000 km of the shoreline of Australia's remote Gulf of Carpentaria (Fig. 18.3). The dieback was severe and wide-spread, affecting around 7400 ha, or 9%, of mangrove vegetation in the southern area from the Roper River estuary in the Northern Territory across to near Karumba in Queensland. Worryingly, early observations of this dieback were co-incident with periods of notably high temperatures along the nearby north-eastern coastline of Australia – the site of severe and widespread coral bleaching.

With no previous reports of such extensive, unexplained dieback of mangroves anywhere in the world before now, the extent, severity and timing of this occurrence of mangrove dieback is globally significant. In the region, there have been no severe storms or large human impacts such as a large oil spill, so the cause appears mostly to do with the extreme weather conditions associated with the unusually severe El Niño event. This is the first recorded instance of such a severe impact on mangroves that can be attributed to severe drought and hot weather conditions coupled with a 20 cm drop in sea level over several months. Further investigations into this unusual occurrence and its cause are underway.

Fig. 18.3. Contrasting images of live and dieback mangrove shorelines in the Gulf of Carpentaria taken in 2016. This dieback was coincident with the same El Niño event that caused severe coral bleaching in the GBR region during the summer of 2015–2016 (Photos: N. Duke).

Box 18.2. Fruiting mangroves and grazing green turtles

Green turtles seek out maturing mangrove propagules to include in their diets (Fig. 18.4). Several observations come as evidence and clues of this previously unknown behaviour. It has been noted for some time that turtles frequent mangrove-lined waterways at high tide. They are also known for occasionally becoming caught-up as tides recede in tree limbs and stranded on exposed mudflats. Firm evidence of feeding, however, has only recently been described. In 2000, Limpus and Limpus reported purposeful cropping of mature propagules of *Avicennia marina* in Shoalwater Bay. This has been further supported by observations of *A. marina* propagules within mature turtle guts interspaced with seagrass. The feeding behaviour to be interpreted from these observations is that some turtles feed on seagrass at low tide, and for a short time at high tide, they take in the nutritious mangrove propagules. This behaviour is dependent on periods of higher tides and the distinct seasonality of propagule availability. However, the extent to which turtles seek out this occasional food bonanza is seemingly shown by more recent observations in Great Sandy Straits of some turtles purposely stranding themselves on mudflats near mangroves (risking desiccation, predation and ignoring low tide feeding) to gather more of this previously unrecognised food source.

Fig. 18.4. Green turtle (*Chelonia mydas*) cropping grey mangrove (*Avicennia marina*) propagules from foreshore trees at high tide (Illustration: Fran Davies for Duke 2006, based on observations of Limpus and Limpus 2000).

areas they form dense, lush meadows that harbour a diversity of animal and algal life (Box 18.4).

Seagrasses are not a monophyletic group, being found in up to five families (depending on taxonomic definition) of monocotyledonous (grass-like) plants. However, they all have a creeping stem (rhizome) and underwater flowering and pollination – and are the only vascular plants that inhabit the sea.

Box 18.3. A newly discovered mangrove for Australia in the GBR region

A local citizen scientist has made an amazing discovery in a busy Cairns' waterside suburb bordering the GBR – finding the first recorded instance of a rare mangrove species never seen before in Australia (Fig. 18.5). The new species record had previously been found in numbers only in Singapore, on the Malay Peninsula, and a small number of isolated locations in New Guinea and the Solomon Islands.

Hidetoshi Kudo made the remarkable discovery of Haines Orange Mangrove, *Bruguiera hainesii*. The species is listed as rare and endangered on the IUCN Red List and, according to the records, less than 250 of the plants exist. It is significant that Mr Kudo found another 50 of the rare plants in the Cairns site. This is widely recognised as one of the most exciting recent day botanical discoveries for this country. In addition, coupled with the discovery was the significant range extension of another related mangrove species, *Bruguiera cylindrica*, by at least 150 km further south than previously known.

The mature plants were found in dense mangroves of a small estuary bordering a populated area surrounded by urban bikeways, managed parklands, controlled drainage channels, and a cleared telegraph line access corridor. This is an area of mangroves that was thought to have been well known, surveyed and explored. These discoveries highlight our lack of basic botanical knowledge about even something as obvious as these sizable tree species. There clearly remains the prospect of further unknown occurrences of mangrove and other plant species in the GBR region.

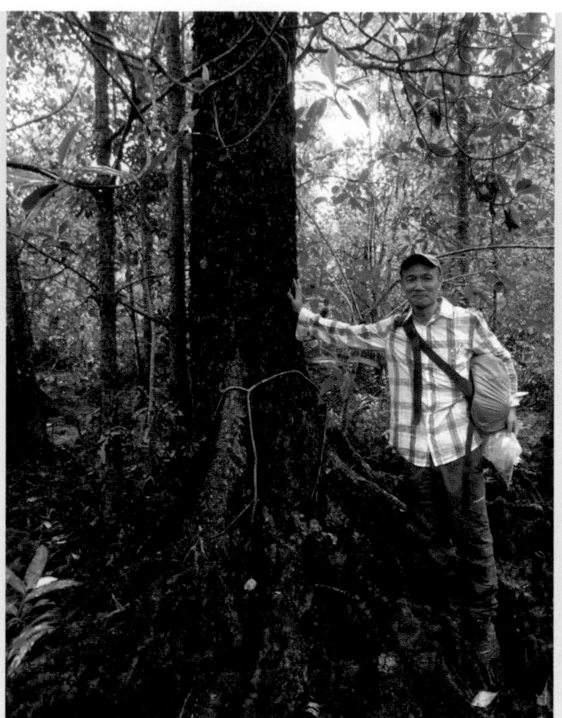

Fig. 18.5. The newly discovered mangrove species, *Bruguiera hainesii* with its discoverer, Hidetoshi Kudo, in the Cairns and GBR location (Photo: N. Duke).

Seagrasses grow mainly on soft substrates (silt and sand) in shallow, sheltered marine or estuarine situations. There are two exceptions to this generalisation: (1) *Thalassodendron ciliatum*, which grows directly on hard substrates (such as dead coral), as it has specialised roots, which allow firm attachment to these substrates; and (2) many species of *Halophila* that can grow at considerable depths, especially *H. capricorni*, *H. decipiens*, *H. spinulosa* and *H. tricostata*, which can grow down to 60 m. Thus these latter species of *Halophila* are often found growing on deep sediments outside coral reefs or in inter-reefal environments on the GBR.

In the main, seagrasses on coral reefs are found in lagoons or in shallow inter-reefal areas, where they support large populations of herbivores. They, therefore, form a very important source of primary production. In the GBR Zone, these same species of seagrass are often found growing on shallow, sheltered sediments in coastal situations, where there are no corals.

Box 18.4. Seagrass beds and macrograzers

The larger species of seagrass in the Great Barrier Reef Zone – *Zostera capricorni, Cymodocea serrulata, C. rotundata, Thalassia hemprichii, Enhalus acoroides, Thalassodendron ciliatum* – form dense beds near the coast and in sheltered bays offshore on continental islands. These beds are usually formed from several species of seagrass and present a wide diversity of habitats for infauna and inflora. Many epiphytes, epizoans and an extensive microflora occur on the seagrass fronds. Many animals and algae grow on or in the sediment between the seagrass plants, and organic substances and oxygen released into the sediments from the roots of seagrasses support a specialised habitat of microflora, protists and invertebrate grazers around the roots (Figs 18.7, 18.8).

It is therefore not surprising that macrograzers are found in or near seagrass beds. The dugong (*Dugong dugon*) has received the most attention in Australia, because this marine mammal is almost entirely dependent on seagrass beds, and the seagrasses themselves, for their food. Marine turtles also feed on seagrass beds, although this has been much better documented in the Caribbean that in Australian waters.

Fig. 18.6. A dugong feeding on a shallow bed of *Halodule uninervis* (Photo: A. Larkum).

Fig. 18.7. *Zostera muelleri* on the bed of Gladstone Harbour being studied by oxygen sensors (Photo: Ponlachart Chotikarn).

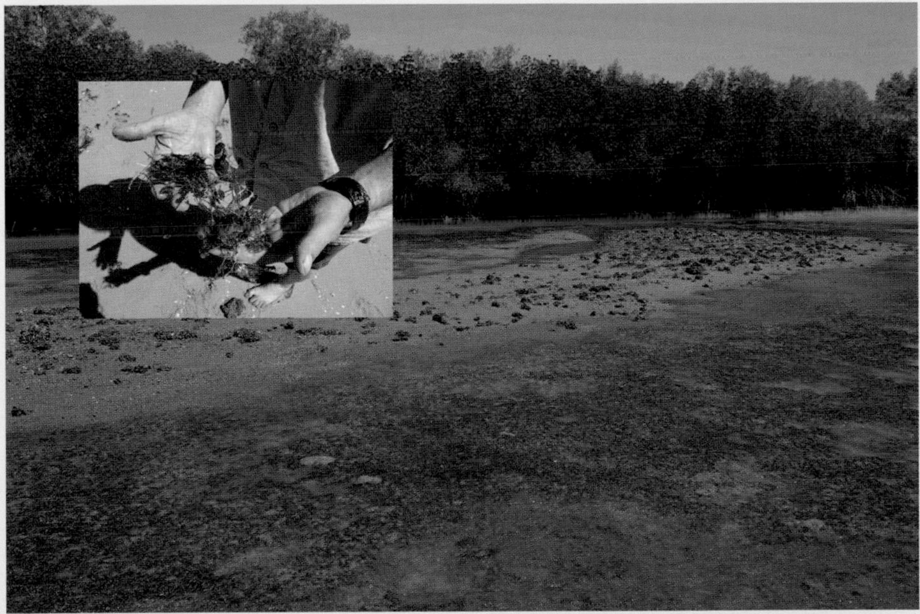

Fig. 18.8. Seagrass habitats in GBR waters are vulnerable to fouling by epibiota that flourish with increased nutrient loads from coastal runoff. Inset showing epibiota attached to seagrass leaves (Photos: N. Duke).

Only three species of seagrass, all in the genus *Halophila*, are present in the Bunker-Capricorn Groups at the southern end of the GBR (*H. capricorni*, *H. decipiens* and *H. spinulosa*). Further north, seagrasses are both more numerous in terms of species may be represented in a fairly small area and they are more conspicuous. Further information on seagrasses is provided in 'Further reading'.

Further reading

Cooper WE, Kudo H, Duke NC (2016) *Bruguiera hainesii* C.G.Rogers (Rhizophoraceae), an endangered species recently discovered in Australia. *Austrobaileya* **9**(4), 481–488.

Den Hartog C (1969) *Seagrasses of the World*. North Holland Publishing Company, Amsterdam, Netherlands.

Duke NC (2006) *Australia's Mangroves: the Authoritative Guide to Australia's Mangrove Plants*. The University of Queensland and Norman C Duke, Brisbane.

Duke NC (2014) Mangrove Coast. In *Encyclopedia of Marine Geosciences*. (Eds J Harff, M Meschede, S Petersen and J Thiede) pp. 412–422. Springer, Amsterdam, Netherlands, <http://link.springer.com/referenceworkentry/10.1007%2F978-94-007-6644-0_186-1>.

Duke NC (2016) *Mangrove Click! Australia: Expert ID for Australia's Mangrove Plants*. MangroveWatch Publication, Currumbin, <https://itunes.apple.com/us/app/mangrove-au/id1157235522?mt=8>.

Duke NC, Schmitt K (2016) Mangroves: unusual forests at the seas edge. In *Tropical Forest Handbook*. (Eds L Pancel and M Köhl) pp. 1693–1724. Springer-Verlag, Berlin, Germany.

Duke NC, Meynecke J-O, Dittmann S, Ellison AM, Anger K, Berger U, *et al.* (2007) A world without mangroves? *Science* **317**, 41–42. doi:10.1126/science.317.5834.41b

Duke NC, Kovacs JM, Griffiths AD, Preece L, Hill DJE, Oosterzee Pv, *et al.* (2017) Large-scale dieback of mangroves in Australia's Gulf of Carpentaria: a severe ecosystem response, coincidental with an unusually extreme weather event. *Marine and Freshwater Research* **68**, 1816–1829. http://dx.doi.org/10.1071/MF16322

Jupiter SD, Potts DC, Phinn SR, Duke NC (2007) Natural and anthropogenic changes to mangrove distributions in the Pioneer River Estuary (Queensland, Australia). *Wetlands Ecology and Management* **15**, 51–62. doi:10.1007/s11273-006-9011-9

Lanyon J (1986) *Seagrasses of the Great Barrier Reef*. Special Publication Series No.3. Great Barrier Reef Marine Park Authority, Townsville.

Larkum AWD, McComb AJ, Shepherd SA (Eds) (1989) *Biology of Seagrasses*. Elsevier, Amsterdam, Netherlands.

Larkum AWD, Orth R, Duarte CA (2006) *Seagrasses: Biology, Ecology and Conservation*. Springer Verlag, Berlin, Germany.

Limpus CJ, Limpus DJ (2000) Mangroves in the diet of *Chelonia mydas* in Queensland, Australia. *Marine Turtle Newsletter* **89**, 13–15.

Mumby PJ, Edwards AJ, Arias-Gonzalez JE, Lindeman KC, Blackwell PG, Gall A, *et al.* (2004) Mangroves enhance the biomass of coral reef fish communities in the Caribbean. *Nature* **427**, 533–536. doi:10.1038/nature02286

Rogers K, Boon P, Branigan S, Duke NC, Field CD, Fitzsimons *et al.* (2016) The state of legislation and policy protecting Australia's mangrove and salt marsh and their ecosystem services. *Marine Policy* **72**, 139–155. doi:10.1016/j.marpol.2016.06.025

Schmitt K, Duke NC (2016) Mangrove management, assessment and monitoring. In *Tropical Forest Handbook*. (Eds L Pancel and M Köhl) pp. 1725–1760. Springer-Verlag, Berlin.

Tomlinson PB (2016) *The Botany of Mangroves*. 2nd edition. Cambridge University Press, Cambridge, UK.

Waycott M, McMahon K, Mellors J, Calldine A, Kleine D (2004) *A Guide to Tropical Seagrasses of the Indo-West Pacific*. James Cook University, Townsville.

19

Sponges

J. N. A. Hooper

Overview

The Phylum Porifera is the most primitive of the multicellular animals, with a most ancient geological history. Porifera appeared early in the history of life on Earth, established in the Proterozoic, with the major class Demospongiae first appearing in the fossil record during the Precambrian Ediacaran age (~750 million years ago (Mya)). By the Middle Cambrian (~500 Mya) demosponges were thriving, and today represent ~85% of all living species. During the massive radiation of life forms in the Ordovician (490–460 Mya), and for the next 100 million years or so, sponges formed extensive barrier and fringing reefs around the ancient continents and were the primary reef builders in these ancient oceans. Sponges declined as the primary reef-builders at the end of the Devonian (350 Mya), and today they are not significant reef builders in shallow waters, unable to compete with the faster growing zooxanthella-bearing corals, although several hard-bodied (lithistid and hypercalcified) reef-building groups of sponges persist in modern day reefs. Although sponges are still a major component of modern-day coral reef ecosystems, they are often overlooked by the curious naturalist because they are frequently hidden among the more prominent corals on the reef, or live in the less frequently visited deeper waters surrounding reefs. Nevertheless, in terms of their species diversity, they outnumber both the hard and soft corals combined. In some habitats they provide pivotal ecological services, such as significant contributions to coral reef primary productivity, filtration of the waste products and toxins from other reef organisms, and major recycling of calcium carbonate and silicon back into the reef system through bioerosion and sequestration, respectively. Although sponges are vulnerable to habitat disturbance, such as the effects of trawling and smothering from sediments, they appear to be less sensitive to ocean warming and ocean acidification, particularly phototrophic and bioeroding species, such that sponges may be 'winners' from global environmental change.

Diversity

Sponges are predominantly marine, living from the intertidal to the abyssal zone, with a small number also found in freshwater habitats. Worldwide there are ~8900 species described in the literature that are considered to be 'valid', with about twice this number estimated for all oceans, lakes and rivers. Approximately 1600 species have been described among the Australian fauna so far, although an escalated collection effort over the past couple of decades, mainly from tropical and subtropical waters and spurred on by drug discovery from marine organisms, has discovered a fauna three times this size. This work leads to an estimated Australian fauna of at least 5000 species. So far only ~450 species are described in the literature for all Queensland waters, including the coast, the Great Barrier Reef (GBR) and Coral Sea island territories. Extensive surveys over the past two decades, however, reveal that more than 2500 sponge species actually live here, with the majority remaining undescribed (new species).

Among the better known sponge faunas on the GBR are, not surprisingly, those from the most frequently visited localities in the vicinity of the major marine research stations: Heron and One Tree Islands in the Capricorn-Bunker Group off Gladstone (387 species); Orpheus and Palm Islands off Townsville (107 species); Low Isles off Port Douglas (134 species); and Lizard and the Direction Islands NE of Cairns (212 species known so far). We also know of other species-rich 'hotspots' in more remote areas on the GBR that are less frequently visited (and hence not necessarily biased by collection effort), including the Swain Reefs (304 species) at the southern end of the GBR, the Ribbon Reefs (204 species) and the Howick Turtle Island group (210 species known so far) in the northern region of the GBR. In between these 'hotspots' is a variable mosaic of diversity and species richness, with the central region of the GBR generally less rich than either the northern or southern sectors. These observations also have some support from population genetics analysis of ribosomal DNA of a widely distributed calcareous sponge, *Leucetta chagosensis* (see Fig. 19.4C), which showed clear genetic divergence between northern and southern GBR populations, both of which are genetically more closely related to Indonesian populations than they are to each other, suggesting weak connectivity between northern and southern regions and indicative of significant exogenous larval recruitment and colonisation from regions outside the GBR.

Although we now know the GBR contains a highly diverse sponge fauna, and that sometimes sponges occur in dense local populations ('sponge gardens'), there still remains a significant challenge to place these faunas into an international context through the processes of rigorous taxonomy. Only by this strategy can we accurately determine which of these species are unique/endemic to the GBR (or to a particular reef system within the GBR), and which are truly widespread and distributed over large (international) spatial scales. Like many marine invertebrate taxa, several sponge species have long been perceived to be widely distributed, ranging from the Red Sea to the central western Pacific islands (reported as 5–15% of regional faunas). This notion of cosmopolitanism is now gradually diminishing with increased application of molecular techniques at population levels, with the outcome being that many so-called widely distributed morphospecies may consist of several sibling species with high genetic diversity that is not, or only barely, manifested at the morphological level across their wide geographic ranges. This problem is exacerbated by the high plasticity of growth form renowned among the phylum Porifera, challenging even the most experienced taxonomists to differentiate 'regional variants' of widespread morphospecies. Estimates of sponge diversity, therefore, based on morphospecies, may be grossly underestimated.

Distribution and abundance

Sponges may live in all types of coral reef habitats, but in reality they exhibit very patchy distributions, such that in one particular area they may form the dominant structural benthos, whereas in another adjacent area they may be practically absent. This is not unusual for marine invertebrates, where at small (local) spatial scales (i.e. encompassing different habitats within a single reef, up to groups of adjacent reefs tens of kilometres apart), spatial heterogeneity is common (in terms of both species diversity and abundance/biomass), and has been widely reported for sponges across all ocean basins. Many factors may significantly influence local sponge distributions. Terrestrial influences, such as freshwater input, turbidity, sedimentation, light penetration, nutrient levels, food particle size availability, and so forth have been found to explain differences between sponge faunas in the lagoon, closer to the land, and those living on the outer reefs. Geomorphological differences between reefs may also markedly influence the composition and distribution of their resident sponge faunas, including factors such as microhabitat availability, the nature and quality of the substrate (coralline versus non-coralline, soft versus hard), aspect of the sea bed, exposure to waves and currents, depth and other factors. In fact, adjacent reef systems (only tens of kilometres apart) have been reported to have as little as 15% similarity in their species compositions, with the presence or absence of particular niches (such as caves, a reef-flat, a lagoon, spurs and grooves) showing strong correlation with the presence or absence of particular species. Other factors that influence patchy sponge distributions include small scale random events, such as patterns and timing of arrival and survival of larvae and asexual propagules, effects of severe storm events on fragmentation and dispersal, and the history of (and changes to) current patterns and other barriers to larval/propagule dispersal. From our current understanding, sponges appear to have very limited sexual reproductive dispersal capabilities (an absence of any feeding pelagic larval stage), with short larval lives (with a reported maximum of 72 h in the water column before settlement). Oviparous species, such as the ubiquitous giant barrel sponge, *Xestospongia testudinaria* (see Fig. 19.7G) that broadcast eggs and sperm into the water, last only several days at most. Conversely, at smaller spatial scales at least, it is thought (also now with some molecular genetics support) that clonal dispersal and larval recruitment are predominant, and small-scale endemism appears to be common among sponges, possibly through genetic isolation of remnant populations of once widespread species.

At larger spatial scales (i.e. from biogeographic provinces to ocean basins), factors such as historical changes to physical barriers and current patterns, climate change impacts, presence or absence of carbonate platforms have had large scale influences on present day sponge faunas. Unlike some other marine invertebrate phyla, there are no apparent latitudinal gradients of sponge species richness from temperate to tropical waters (both have patchy mosaics of very rich faunas, on both sides of the Australian continent). The sponge fauna composition changes substantially, however, along the east coast of Australia, with subtropical–tropical faunal transition zones (or species turnover points) occurring in the vicinity of the Tweed River, Hervey Bay–Fraser Island, the Mackay–Townsville region, Cape Flattery north of Cooktown, and on the eastern side of Cape York.

Reef habitats

Once thought of as predominantly niche generalists, opportunistically scattered over reefs wherever larvae settled or propagules landed, sponges are increasingly recognised as being predominantly niche specialists with marked habitat preferences. Distinct species assemblages characterise particular habitats, although there are several ubiquitous species found vicariously throughout the reef. Some of the more prominent sponge habitats occurring on the GBR are described here.

Reef flats and rock pools

Of all the habitats, the reef flats and shallow parts of lagoons contain probably the most significant of the coral reef sponge faunas, at least in terms of their visibility and provision of ecological services. These are the phototrophic (or autotrophic) species, belonging primarily to two orders of Demospongiae, the Dictyoceratida and Haplosclerida, that derive most of their nutrition from the photosynthetic products of their resident symbiotic cyanobacteria, and in the process contribute significantly to overall net primary productivity of entire reef systems. Phototrophic species (which derive most of their energy from sunlight through photosynthetic symbionts, see Chapter 8) include representatives from several orders, with the common species on the GBR being *Phyllospongia papyracea*, *Carteriospongia foliascens*, *Strepsichordaia lendenfeldi*, *Haliclona cymaeformis*, *Cymbastela coralliophila* and *Lamellodysidea herbacea* (Fig. 19.1A–H). Some of these species can be found in very large populations, especially in the clear waters of outer

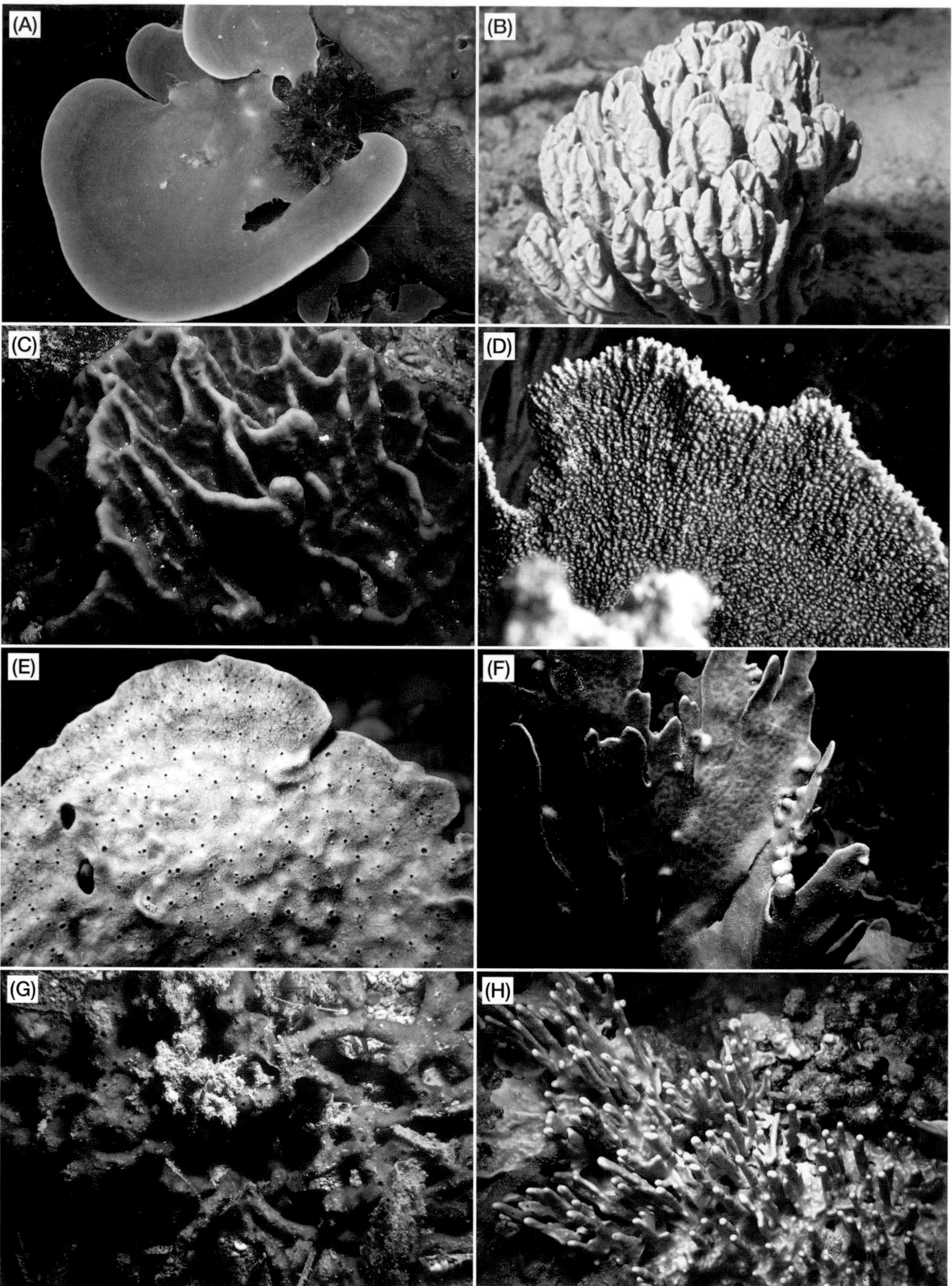

Fig. 19.1. Phototrophic sponge species living on coral reef flat and in coral rubble in shallow waters: **(A)** *Collospongia auris*; **(B)** *Lendenfeldia plicata*; **(C)** *Cymbastela coralliophila*; **(D)** *Carteriospongia foliascens*; **(E)** *Strepsichordaia lendenfeldi*; **(F)** *Phyllospongia papyracea*; **(G)** *Haliclona cymaeformis*; **(H)** *Lamellodysidea herbacea* (Photos: J. Hooper).

Box 19.1. Chemical defence

Sponges are among the most toxic of animals, with toxins derived from one or a combination of the sponge's own metabolic products, sequestered chemicals excreted by other reef organisms and filtered by sponges, and symbioses with a variety of microbial infloras. These toxins are thought to have a variety of biological functionality, including repelling predators, out-competing other sessile reef organisms for space, controlling parasites and microbes, chemical recognition between host and resident in-faunas, and bioerosion of the substrate. Many of these chemicals have also demonstrated toxicity against human pathogens, cancers, parasites and other therapeutic properties of interest to the pharmaceutical industry, and over the past few decades the majority of new chemical structures and new classes of chemical compounds have been isolated from sponges, including species from the GBR.

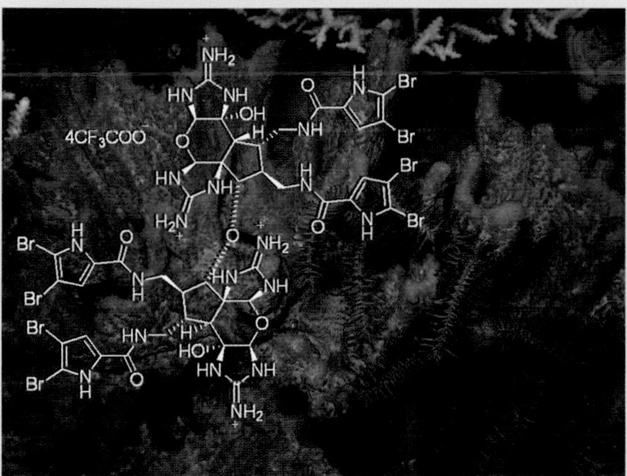

Fig. 19.2. 'Flabellazole a', a new P2x7 antagonist, isolated from *Stylissa flabelliformis* from the GBR. P2x7 antagonists may provide new treatment for inflammatory diseases.(Photo: J. Hooper; molecular structure T. Carroll, Natural Products Discovery Griffith University).

reefs. These phototrophic sponges are also unusual in having (probably truly) widespread geographic distributions, appearing to be very similar on both sides of the continent, with some also common in the Indo-Malay archipelago and south-western Pacific islands, although no molecular study of these widespread populations has yet been attempted to test their alleged conspecificity.

Another highly diverse fauna on the reef flat and in the lagoon shallows are the coral rubble, under-rubble and boulder sponges, living in crowded, encrusting and sciaphilic communities. These sponges range from thin crusts of no more than several millimetres in thickness, competing with each other for space and other resources using an arsenal of chemicals (Box 19.1 and Fig. 19.2), to massive slimy sponges that bind the coral rubble together and form the paving substrate. Examples of these include: *Myrmekioderma granulatum, Clathria aceratoobtusa, Leucetta microraphis, Aplysinella rhax, Neopetrosia chaliniformis, Gelliodes fibulata* and *Hyrtios erectus* (Fig. 19.3A–H, Fig. 19.4A).

Reef builders, caves and crevices

In prehistoric reefs, many hard-bodied sponges, such as 'lithistids', 'stromatoporoids', 'chaetetids' and 'sphinctozoans' (all now considered grades of construction in body plan, and not taxonomic clades), were the major structural components of reefs. In modern day reefs, hard-bodied sponges have only a minor structural role, although they are still considered to be of some importance in accreting coral skeletons in deeper waters where light is limiting and coral growth is rudimentary. Nevertheless, living representatives of these 'reef-building sponges' are still found on coral reefs and other deeper reefs, with their ancient body plans persisting over many millions of years. Some of these sponges (such as *Astrosclera willeyana*) have a solid calcitic skeleton (contributing to reef building, analogous to the skeletons of modern hermatypic corals), as well as discrete siliceous spicules within the soft tissues, virtually identical to those seen in their fossil ancestors from the Lower Cretaceous (160 Mya) and Triassic (250 Mya) respectively. Others have only a solid skeleton composed of aragonitic crystals (a form of calcium carbonate) with no free spicules (e.g. *Vaceletia crypta*, with a continuous fossil record from the Middle Triassic, 245 Mya to present). Others have a

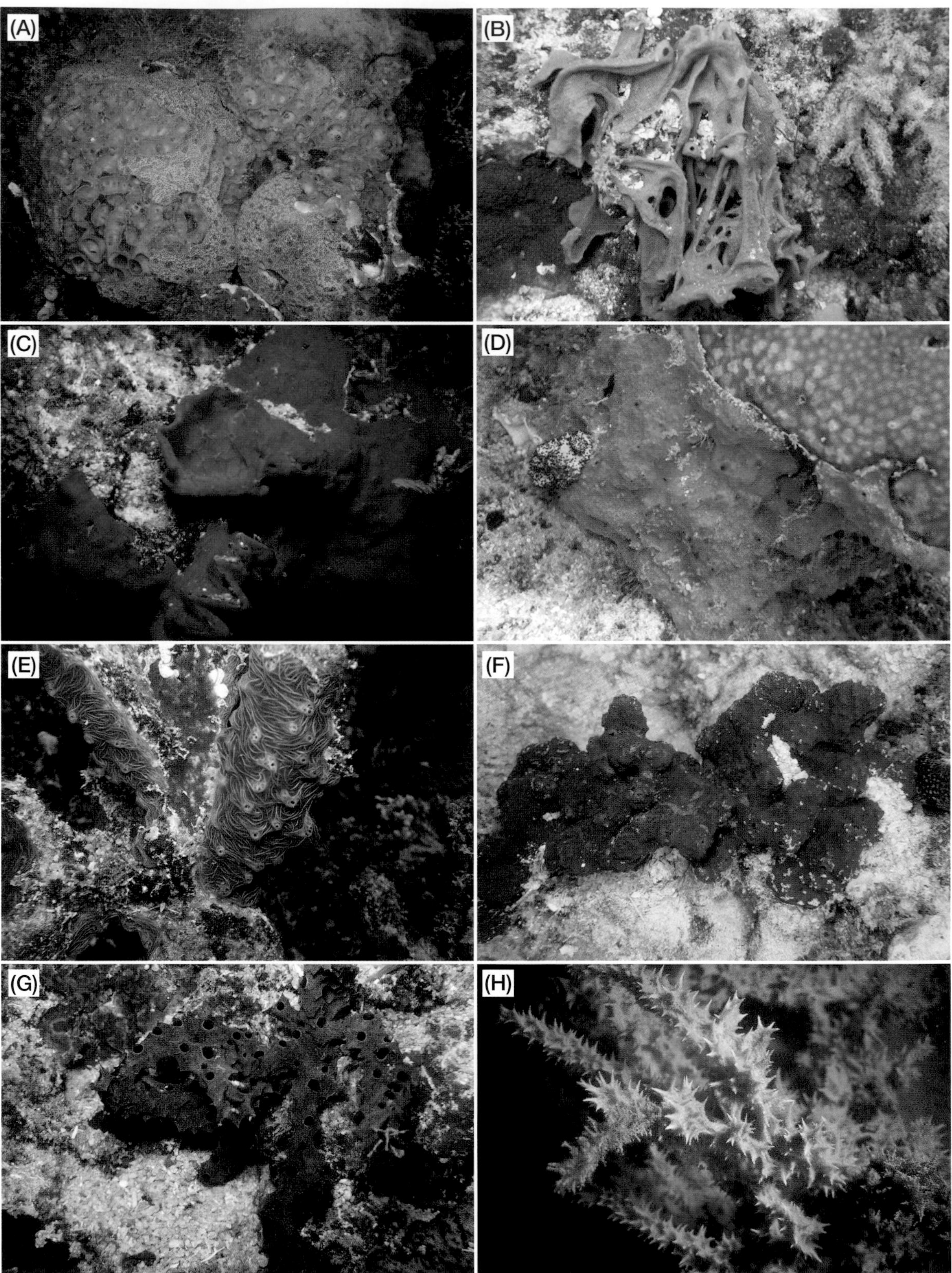

Fig. 19.3. Sponges of the reef flat and shallow lagoon, including the cryptic, coral rubble, and under-boulders communities: **(A)** *Myrmekioderma granulatum*; **(B)** *Leucetta microraphis*; **(C)** *Clathria* (*Microciona*) *aceratoobtusa*; **(D)** *Stelletta* sp.; **(E)** *Chalinula nematifera*; **(F)** *Aplysinella rhax*; **(G)** *Chelonaplysilla* sp.; **(H)** *Gelliodes fibulata* (Photos: J. Hooper).

Fig. 19.4. Sponges of the reef flat and shallow lagoon, including the cryptic, coral rubble, and under-boulder communities: **(A)** *Hyrtios erectus*; **(B)** *Suberea ianthelliformis*. Reef-builders and cave faunas: **(C)** *Leucetta chagosensis*; **(D)** *Levinella prolifera*; **(E)** *Ulosa spongia*; **(F)** *Soleneiscus radovani*; **(G)** *Astrosclera willeyana*; **(H)** *Petrosia (Strongylophora) strongylata* (Photos: J. Hooper).

solid skeleton of either linked or rigid calcareous spicules (e.g. *Plectroninia hindei*, with a body plan known from the Mid Miocene, 23 Mya) or a rigid basal mass of calcite, and some (the so called 'lithistid' rock sponges) have only siliceous skeletons composed of special spicules called desmas, ranging from entirely fused and forming a rock hard skeleton to loosely articulated rendering the body more flexible (e.g. *Theonella swinhoei*, with ancestors recorded from the Tertiary, 65 Mya). These so-called 'living fossil' sponges are usually found in shaded or dark habitats, such as in crevices, deep caves or under coral rubble, rarely in full light, and due to their hard coralline texture may be confused with corals by the novice. Similarly, caves and dark crevices are also home to a variety of soft-bodied demosponges, sometimes lacking pigments (e.g. *Petrosia (Stongylophora) strongylata*), sometimes colourful (e.g. *Ulosa spongia*), and especially the multitude of frequently brightly coloured calcareous sponges living on the reef (e.g. *Leucetta chagosensis, Levinella prolifera* and *Soleneiscus radovani*) (Fig. 19.4C–H).

Reef slope

Between the reef crest and the base of coral reefs, across the shelf of the GBR, occur large, predominantly heterotrophic sponges that feed on food particles and waste products filtering down from the coral reef above. On some reefs, particularly those closer to the coast, these faunas number several hundred species, which in some instances occur as large populations. Their morphologies are as diverse as the conditions they live under, such as: flexible whips, fingers and fans adapted for coping with high currents (*Ianthella basta, Axos flabelliformis, Clathria (Thalysias) cervicornis*); soft tubes, vases and other shapes that predominate in silty, turbid water where inhalant and exhalant pores are located on different surfaces to prevent smothering (e.g. *Echinochalina (Protophlitaspongia) isaaci* and *Fascaplysinopsis reticulata*); and several ubiquitous, amorphous, bulbous, massive, spherical (and other shaped) forms that appear nearly anywhere they can settle and survive (e.g. *Cinachyrella schulzei, Stylissa massa, Stylissa carteri, Acanthella cavernosa, Pipestela terpenensis, Pericharax heteroraphis,* and *Agelas axifera*) (Figs 19.5A–H, 19.6A–H).

Deeper lagoon and inter-reef

Sponges living on the sea bed in between the reefs are generally very different to those found on the coral reefs, living in high current, low light (mesophotic), turbid waters where they burrow into soft sandy and muddy sediments (e.g. *Oceanapia renieroides* and *Disyringa dissimilis*), in seagrass and *Halimeda* beds (e.g. *Oceanapia sagittaria*), or attached to hard objects on the sea floor (e.g. *Xestospongia testudinaria, Melophlus sarasinorum* and *Liosina paradoxa*) (Fig. 19.7A, C, G, H). Growth forms include elongate species with root-like tufts or bulbs for anchoring in soft sediments and long tubes to prevent smothering, flexible fans and whips in high current areas, and massive barrels and volcanoes attached to rock and coral outcrops. This fauna comprises a significant proportion of the 'benthos', providing important habitat for other marine species, ranging from aggregating fishes to numerous crustacean in-faunas. In a recent major survey across the length and breadth of the GBR, conducted under the auspices of the Cooperative Research Centre for the GBR (CRC Reef), over 1200 morphospecies of sponges were discovered from the inter-reef 'mesophotic' region, which is particularly susceptible to human impacts such as trawling (see also Chapter 11). Whereas several species living on the coral reef flats may be found on both sides of the Australian continent, these deeper water GBR lagoonal and inter-reef species appear to differ significantly in composition from the fauna found at similar latitudes on the west coast of the continent.

Reef bioeroders (sponges as 'parasites')

A special group of sponges are responsible for significant carbonate recycling on the reef, collectively termed excavating ('boring' or bioeroding) sponges (see also Chapter 9). They are responsible for extensive damage to hard and soft corals, as well as shellfish and other molluscs, and are among the most destructive internal bioeroding organisms of

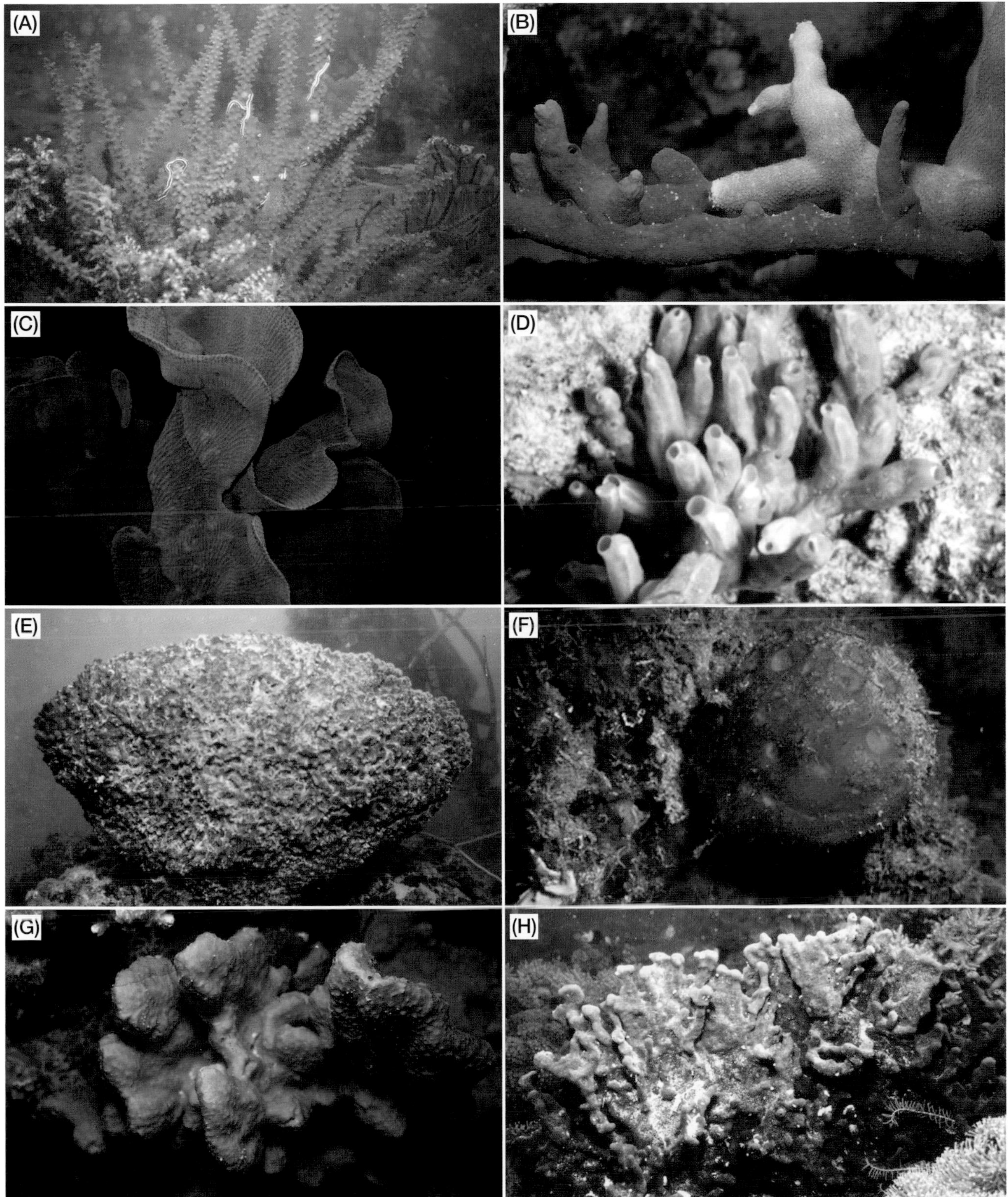

Fig. 19.5. Reef slope faunas: **(A)** *Axos flabelliformis*; **(B)** *Clathria (Thalysias) cervicornis*; **(C)** *Ianthella basta*; **(D)** *Echinochalina* (*Protophlitaspongia*) *isaaci*; **(E)** *Fascaplysinopsis reticulata*; **(F)** *Cinachyrella schulzei*; **(G)** *Stylissa massa*; **(H)** *Pipestela terpenensis*. (Photos: J. Hooper.)

Fig. 19.6. Reef slope faunas: **(A)** *Pericharax heteroraphis*; **(B)** *Agelas axifera*; **(C)** *Acanthella cavernosa*; **(D)** *Reniochalina* sp. (photo: Chris Ireland); **(E)** *Phycopsis fusiformis*; **(F)** *Callyspongia aerizusa*; **(G)** *Pipestela candelabra*; **(H)** *Stylissa carteri*. (Photos: *A–C, E–H*, J. Hooper.)

coral reefs in terms of their effects (such as weakening coral platforms and producing dead coral rubble). The rates of destruction by these organisms range up to 15 kg/m^2 per year. Much of the damage caused to corals during storms has been attributed to weakening of basal structures by bioerosion (see also Chapter 9). An excavating mode of existence has been independently acquired by several sponge orders. Some of these (e.g. *Terpios*) simply overgrow coral at rapid rates, periodically resulting in extensive tracts of coral bleaching and the destruction of large tracts of coral. Others burrow into dead coral, eventually occupying the entire original coral head, with breathing tubes (fistules) protruding (e.g. *Coelocarteria singaporensis* and *Siphonodictyon* sp.). The most significant of these are the 'clionaids' belonging to the families Clionaidae, Thoosidae and Spirastrellidae (e.g. *Cliona* sp., *Spheciospongia vagabunda* and *Cliona montiformis*) (Fig. 19.7B, E, F). Clionaids excavate chambers within the coral skeleton using a cellular process undertaken by special etching cells secreting acid phosphatase and lysosomal enzymes that dissolve organic matter and produce limestone chips that are physically liberated into the sea water via the sponge exhalant canal. Etching initially produces a cavity with sponge papillae protruding outside the coral (alpha stage), after which the external papillae fuse to produce a continuous sponge crust covering the coral (beta stage), eventually becoming massive and consuming the entire coral (gamma stage). It is estimated that between 20% and 40% of the sediment surrounding coral reefs is composed of liberated calcite chips produced by sponges bioeroding corals. Although clionaid sponges have the ability to invade living coral tissue and to survive direct contact with coral polyps, their ecological success may be largely due to their ability to undermine and erode the coral skeletal base, thus avoiding contact with the coral polyp defensive mucus and nematocysts.

Sponge body plans and classification

The Phylum Porifera is defined by their unique possession of chambers lined by a single layer of flagellated cells (choanocytes or collar cells) that actively beat to produce a unidirectional water current through the body, connected to the external water column by a system of differentiated inhalant and exhalant canals with external pores (ostia and oscula, respectively), together forming a highly efficient aquiferous system that maintains basic metabolism and contributes significantly to reef filtration (Fig. 19.8). Sponges have a cellular grade of construction without true tissues, with their highly mobile populations of cells capable of differentiating into other cell types (totipotency), thus conferring a plasticity to growth form. The outer and inner layers of the sponge individuals are formed by special T-shaped or flattened 'skin' cells (exopinacocytes and basipinacocytes) that lack a basement membrane (except in the class Homoscleromorpha). The middle layer (or mesohyl) is variable among the orders of sponges but always includes motile cells and usually some skeletal material. Sponge skeletons are essentially divided into the ectosome ('skin') and choanosome (body containing the choanocyte chambers). Adult sponges are generally sessile, attached to the seabed or other substrate for most of their lives (although some are capable of slow movement), and most have motile larvae that swim or crawl away from their parent. Body plans range from simple (asconoid and syconoid, found in a few calcarean sponges) through to complex (leuconoid, occurring in most sponges), produced by varying degrees of infolding of the body wall and complexity of water canals throughout the sponge. Adults are asymmetrical or radially symmetrical, and have evolved an amazing range of growth forms best described as highly irregular and sometimes completely plastic, frequently altered by prevailing external conditions (currents, turbidity, salinity, etc.). Sponges also have evolved an amazing array of colours, some linked to dietary carotenoid proteins and others with a photoprotection functionality.

The current classification of the Porifera is based primarily on features of the organic (collagen fibres and filaments) and inorganic skeletons (discrete

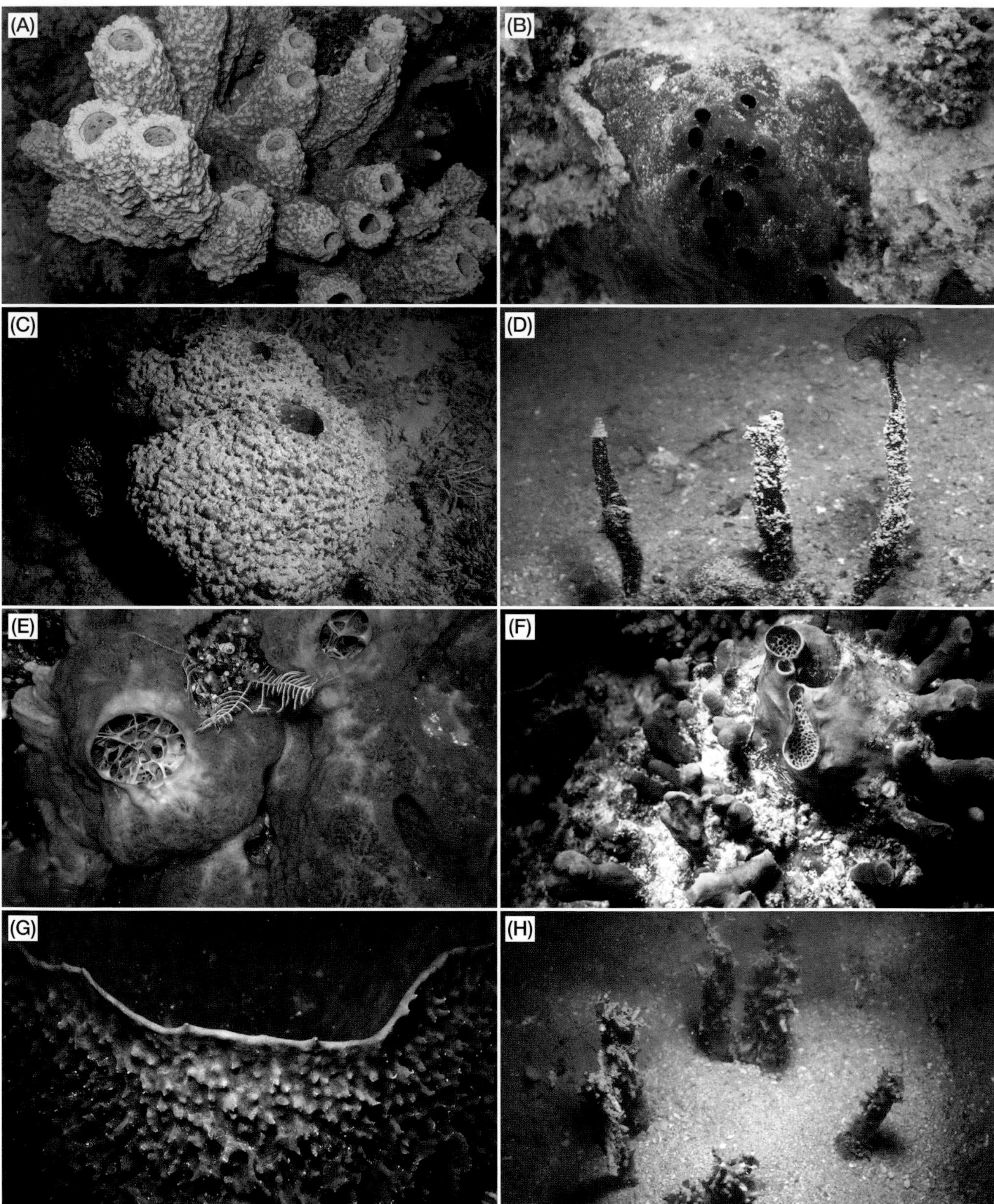

Fig. 19.7. Deeper lagoon and inter-reef faunas, reef bioeroders and soft sediment faunas: **(A)** *Liosina paradoxa*; **(B)** *Spheciospongia vagabunda*; **(C)** *Melophlus sarisinorum*; **(D)** *Oceanapia sagittaria*; **(E)** *Cliona* sp.; **(F)** *Coelocarteria singaporensis*; **(G)** *Xestospongia testudinaria*; **(H)** *Oceanapia renieroides* (Photos: J. Hooper).

and/or fused spicules composed of calcium carbonate or silicon dioxide), with some species also having a hypercalcified basal skeleton of solid limestone. The taxonomic scheme is primarily morphologically based, and as complex as the diversity of sponges – the study of sponge taxonomy is not for the faint-hearted. Applying taxonomic principles to sponges is made even more difficult by the occurrence of frequent character losses, modifications and apparently convergent features reappearing within the classification. No attempt is made here to provide more than a very basic summary, with a list of further reading provided. There are four distinct classes of living sponges (plus a fifth extinct one): **Calcarea**, having calcitic spicules with three or four rays, sometimes a solid basal calcitic skeleton, and a unique 'hollow' viviparous blastula larva; **Hexactinellida**, with tissues largely syncytial composed of multinucleated cells, supported by a skeleton of discrete and/or fused siliceous spicules, the larger ones three or six-rayed; **Demospongiae**, with siliceous spicules in many (but not all) species, and/or with a skeleton of organic fibres or fibrillar collagen supported by spicules with one, two or four rays divided into megasclere and microsclere categories; and **Homoscleromorpha**, with flagellated surface 'skin' cells, a basement membrane lining both the external and internal surfaces, a unique viviparous cinctoblastula larva, and a skeleton, if present, composed of four-rayed siliceous spicules called calthrops or reduced derivatives, without differentiation between megascleres or microscleres. Only Calcarea, Demospongiae and Homoscleromorpha have so far been recorded from the GBR, although Hexactinellida live in deeper waters on the continental slope and shelf adjacent to the GBR. An overview of the phylum, including a taxonomic revision and identification keys for ~25 orders, 127 families and 700 genera, has recently been undertaken but species-level identifications remain appallingly difficult, with few easily accessible taxonomic publications that would be useful to a non-specialist audience. Further useful reading is listed at the end of the chapter, including general reading on sponge biology, sponge cell biology, a web checklist of the published Australian sponge fauna (including Queensland species) with keys to genera, a web list of all published sponge species worldwide, and sponge higher classification. The recent escalation of the molecular study of sponges will certainly have a major impact on our current ideas of the phylogeny and classification of Porifera, and to this end a Sponge Barcoding Project (based on a systematic use of molecular tools) is well underway and is also available on the web.

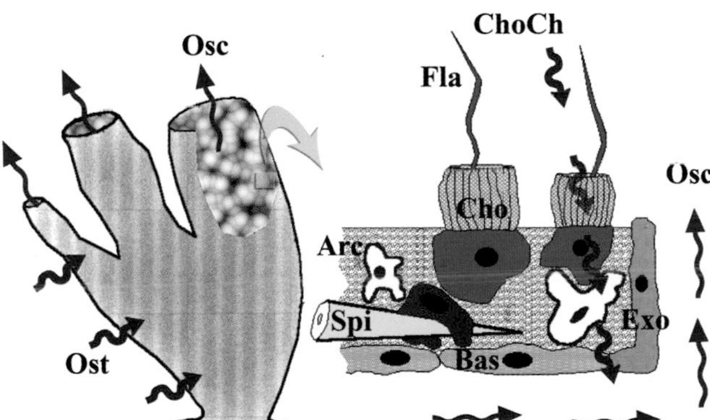

Fig. 19.8. Diagrammatic sponge morphology: Arc, totipotent phagocytotic cells (archaeocytes); Bas, basipinacocytes lining internal aquiferous system; Cho, choanocytres or collar cells; ChoCh, choanocyte chamber (lined by choanocyte cells); Exo, exopinacocytes (lining exterior surfaces); Fla, flagellum on choanocytes; Ost, inhalant pores (ostia); Osc, exhalant pores (oscula); Spi, spicules (siliceous or calcitic depending on class). Red arrows (inhalant water current with food particles, etc.); blue arrows (exhalant water current with waste products) (Source: Modified from UCMP Berkeley).

Reproduction and life history

Sponges use several reproductive strategies based around their characteristic cellular totipotency. Asexual reproduction involves the production of propagules such as buds and fragments containing a sufficient number of cells from which complete sponges can develop. A few euryhaline species of the genus *Mycale* produce gemmule-like bodies, but true gemmules are restricted to freshwater sponges of the order Haplosclerida. Most groups have considerable means of asexual propagation, such as fragmentation from storm events, which is thought to be an important mechanism for sponge recruitment, and all have extensive regenerative powers that appear to be vital for sustaining local populations. Sponges have sexes that are separate, or sequentially hermaphroditic, producing eggs and sperm at different times. Although there are no gonads or reproductive ducts, sexual reproduction involves the production of gametes by the choanocytes and totipotent archaeocytes, with fertilisation often (but not always) internal. Individuals release sperm externally via the exhalant current, whereas their oocytes reside in the incurrent aquiferous system to minimise self fertilisation. Sperm are engulfed by choanocytes, which become amoeboid, travelling to and transferring them to the oocytes. Cleavage leads to a solid steroblastula or hollow coeloblastula, with internally brooded, viviparous embryonic development in many cases, and larvae leaving the parent for dispersal. Other sponges are oviparous, with females shedding their eggs externally as zygotes or early embryo stages, rarely as unfertilised oocytes, although the details of embryology still remain unknown for most species. Other forms of development have also been recorded, such as elimination of the free-swimming larval stage and embryos brooded in the maternal sponge before being expelled as young adults. Eight different larval types are known, but few of these have been adequately investigated. Most embryos develop into free-swimming (lecithotrophic) or demersal crawling larvae, ciliated to a greater or lesser extent, 0.05–5.00 mm long, with a brief planktonic phase, short longevity (maximum of 72 h recorded so far), and, unlike most marine invertebrates, have no planktotrophic stage.

Release of propagules (gametes, zygotes or embryos) is asynchronous in viviparous species but may be highly synchronous in oviparous sponges, triggered by factors such as temperature and lunar cycle. A prominent member of the GBR sponge community, *Xestospongia testudinaria*, is oviparous and broadcasts eggs in spawning events that were synchronised among populations of the same species, with timing found to be correlated with the lunar cycle. Molecular studies of individuals in local sponge populations show that most have high levels of genetic variability, not high genetic relatedness as would be expected if asexual recruitment was predominant, with some evidence that both asexual and sexual propagules are important for population structure, whereby sponge fragments that disperse and reattach may contain incubated sexual propagules.

Larval settlement and metamorphosis is thought to be influenced by a variety of environmental stimuli (such as light, gravity, physical and chemical features of the substrate), with the former best studied to date. There are examples of both photonegative and photopositive responses among the phylum. Larval competence (the threshold and duration of larval maturity required for settlement) is not thought to be as important for sponges as for many invertebrates because the high cellular totipotency allows fragmented larvae to attach unselectively. Growth rates, regenerative abilities after damage, and longevity is still poorly understood, but what little is known to date demonstrates that these vary considerably across groups of sponges and the habitats they occupy (Box 19.2). Using various direct (C_{14}) and indirect measurements (e.g. growth rate extrapolation indices), some species are known to reproduce and die in less than 1 year (such as some soft-bodied *Haliclona* spp.), or are highly seasonal in their growth, biomass and ultimately survival (*Chondrilla australiensis*); some species live for many decades (*Aplysilla* sp.), to over 400–500 years (*Astrosclera willeyana*) and it is claimed that sponges belonging to the hexactinellid family Rosellidae living in Antarctica are among the oldest living animals on the planet, with individuals estimated to be over 1500 years old.

Feeding

Sponges filter sea water to eat, exchange gases and excrete waste products. Filtration is an active process involving choanocytes lining chambers. Each choanocyte has a central flagellum that actively beats to create a water current, surrounded by a collar of cilia that traps food particles such as plankton and bacteria, as well as detritus. A water current containing food enters the sponge through an osculum and is initially filtered through a series of sieve-like pores (diminishing in size), finally ending up at the collar cells. Food particles are actively carried across the cell wall, engulfed by archaeocytes, and are transferred throughout the mesohyl to other cells. Filtered water leaves the sponge via the exhalant canal system. Unlike most multicellular animals, digestion and excretion of waste products occurs within cells, not within any common body cavity. There may be 7000–18 000 choanocyte chambers per cubic millimetre of sponge, and each chamber may pump ~1200 times its own volume of water per day. Thus, a sponge is capable of pumping around 10 times its body volume each hour, making them the most efficient vacuum cleaners of the sea. In areas where sponge populations are abundant, they remove substantial amounts of both particulate organic matter (POM) from the water column (such as plankton and detritus), and dissolved organic matter (DOM) from benthic production (such as macroalgae and coral mucus), and as such they play an important role in benthic–pelagic coupling in coral reef systems. In addition to these ecosystem services such as nutrient and calcium recycling, sponges have also been shown to have a major role in global silica cycling.

Some sponges, particularly those growing on coral reef flats and shallow lagoons, also have a unique symbiosis with cyanobacteria, providing the sponge with nutrients derived from photosynthesis to supplement those obtained by the sponge from normal filter feeding activities (phototrophy or autotrophy). These extra nutrients greatly augment sponge growth rates and competitive ability in coral reef systems. There are also often huge populations of bacteria and Archaea living within sponge cells and/or within the sponge mesohyl (in some species with up to 40% of cells being prokaryotes, hence the term 'sponge hotels'), with which the sponge cells interact at various levels (from predation to commensalism). Furthermore, sponges with a high microbial abundance may pump between 50% and 90% less volume of water than those with a low microbial abundance, relying more on dissolved than particulate metabolism, respectively. In addition to nutrients acquired through filter feeding activities, sponges may

Box 19.2. Sponge farming

Libby Evans-Illidge, Carsten Wolff and Alan Duckworth, AIMS

Sponges have been used for cosmetic, bath or industrial applications since early Grecian times (one early reference is in Homer's *The Iliad*). However, modern supply, which is predominantly from wild harvest fisheries, is unable to meet a well-established global demand. This shortfall presents an opportunity for sponge production through in-sea aquaculture. Sponges grow easily from cuttings, on lines or mesh panels suspended in the water column. Research has shown that commercially viable sponge farming can be achieved within a sustainable environmental footprint using basic infrastructure, and this opportunity is currently being explored by remote coastal Aboriginal communities in the GBR region and elsewhere in the south Pacific (Fig. 19.9).

There is further opportunity to expand the species targeted for sponge farming to include those that elicit promising bioactive compounds. To date, many of these have not progressed in drug development due to the lack of reliable supply. Bulk wild collection is not suitable because the compounds are often produced in low yields, and their structural complexity makes them initially difficult to synthesise. Access to large quantities of sponges is necessary, and development of sponge aquaculture has provided a production option. Although this opportunity is potentially quite lucrative, it is also high risk and at best transient for any one chemical entity. Although aquaculture has already played an important role in supply for drug development research and proof of concept, history shows that industry prefers synthesis for global market supply, often based on a variation and simplification of the natural molecule.

Fig. 19.9. Sponge farming trials at Palm Island on the GBR. Sponge explants being grown in mesh panels (Photo: Australian Institute of Marine Science).

ingest a myriad of toxic chemicals excreted by other plants and animals from the coral reefs above, which they modify (sequester) and reuse for their own purposes. The combination of chemicals produced by normal sponge metabolism, those sequestered from the sea water, and those produced by or in combination with the resident microbial populations makes sponges among the most toxic of all life forms, and hence of great interest to the pharmaceutical industry (Box 19.1 and Fig. 19.2).

Predation and defence

Sponges are most unappetising by human standards due to a combination of high toxicity and a generally low ratio of soft tissue to mineral skeleton. The ability to digest and modify the waste products and chemicals produced by other organisms that live in, on or near sponges may at least partly account for their diverse, frequently novel and often highly toxic biochemistry. Nevertheless, sponges do have many recorded predators such as molluscs, echinoderms, fishes and turtles. On the GBR and in north-western Australia, sea cucumbers (*Synaptula* species) are frequently seen congregating on sponges (in particular branching and lamellate *Haliclona*, *Axos* and *Ianthella* species), feeding on the mucus exudate (Fig. 19.5A). Nudibranchs are also active feeders on sponge mucus and collagen, and sometimes these molluscs are quite specific as to the species or genus of sponge upon which they prey. Nudibranchs also ingest the sponge's toxic chemicals, concentrating, modifying and reusing (sequestering) them for their own chemical protection. The predators are often the same colour as the sponge, having ingested the sponge's characteristically brightly coloured carotenoid and other pigments (Fig. 19.3C). Other documented predators of sponges on the GBR include green and hawksbill turtles, many species of grazing fishes and asteroid, ophiuroid and holothurian echinoderms. There is now some good evidence to show that sponges also use their extensive arsenal of chemicals as both offensive and defensive weapons, such as repelling predators, deterring parasites and competing for space, and that the concentration, toxicity and/or secondary modification of particular compounds may vary seasonally and in response to predation intensity.

Threats and conservation

Despite there being ~8900 species of sponges now known worldwide, only 20 species are currently included on threatened species lists worldwide, and none from the Australasian faunas. Yet sponges attract increasing conservation interest as they become more dominant on reefs as corals decline in response to climate change, ocean acidification and habitat destruction.

Although the current conservation status of sponges does not appear to indicate they are particularly threatened, little information is available for most species – particularly when only less than half the estimated sponge fauna is currently described. Dredging and bottom trawling are obvious threats to sponges through habitat disturbance and detachment from the substrate, increased turbidity and suspended sediments leading to smothering. Experimental trawling on the GBR sea bed (lagoon) quantified the significant impact trawling and dredging had on large epibenthic invertebrates, including sponges. It demonstrated a correlation between increased trawl effort and the reduction in species' biomass, as well as a reduction in habitat complexity that key commercial species such as fish and crustaceans depend on for ecosystem services. Vulnerable sponge species included those with the largest biomass (such as *Xestospongia testudinaria*, *Ianthella basta* and *Callyspongia* sp.) and the most prevalent species across the GBR shelf (such as *Xenospongia patelliformis, Theonella xanthus, Spheciospongia vagabunda* and *Coscinoderma nardorus).* Similarly, although not yet quantified experimentally for the GBR fauna, reports from other coral reef systems have demonstrated catastrophic storm events can have major impacts on sponges, with the loss of up to half of local populations recorded. Resilience to storm impacts appears to be related to differences in the relative proportions of inorganic spicules and proteinaceous spongin that forms the sponges' skeletal fibres. Also, although most species potentially have regenerative powers, the success of fragments and larvae resettling is dependent on the extent of substrate loss, and the growth rates of the species, which where known are also generally slow.

The scale and impact of other drivers, such as microbial pathogens, land runoff and eutrophication, and ocean temperature and acidification are less well understood. Over the past few decades several species of sponges in the Caribbean and Mediterranean in particular, have suffered sponge necrosis syndrome whereby sponge tissues become discoloured, degrade, with eventual exposure of the skeleton and mortality. The potential primary causal agents for the disease are thought to be consortia of bacteria, cyanobacteria and fungi, with fungi predominant within necroses. Conversely, similar symptoms were also recorded in populations of *Ianthella basta* from the GBR and Torres Strait, from which there was no evidence that microbes were the primary cause of brown spot lesions and ensuing necrosis, and it is thought that disease-like symptoms in sponges may also be attributed to environmental stressors such as water temperature, coastal agricultural runoff, physical damage, predation and chemical interactions with other sessile invertebrates.

Notwithstanding these various threats, as ocean warming, increased CO_2 levels and ocean acidification are predicted to continue causing the decline of reef-building corals, there is growing evidence that sponge abundance on coral reefs has been increasing globally, shifting from coral-dominated faunas to macroalgae and sponge reefs. This increase is partially attributed to a decrease in the spatial competition with corals. However, perhaps more significantly, sponges are able to filter and digest DOM from the water (including mucus produced by macroalgae and corals), which is otherwise inaccessible to most other coral reef animals. This nutrition allows sponges to continually produce new choanocyte cells, shedding the old cells into the water column as POM and food for other detritus-feeding species. This phenomenon (a 'positive feedback loop' or the 'sponge loop'), contributes significantly to carbon flow in otherwise nutrient-poor waters characteristic of coral reefs, enhancing the sponges' growth at the expense of corals.

Current experimental evidence conducted on species from the GBR suggests that some sponges will be 'winners' in the face of rapid environmental change. The combined factors of increased ocean warming and acidification appears to benefit phototrophic species (reduced mortality, necrosis and bleaching), but produced temperature stress in heterotrophic species (increasing levels of tissue necrosis and bleaching, elevating respiration rates and decreasing photosynthetic rates). Increased ocean acidification alone had little adverse effect for either nutritional group, and it is thought that as sea water pH decreases (a result of increasing carbon dioxide), sponge bioerosion also increases significantly due to weakening of the corals' calcitic skeletons. It has been hypothesised that climate change may produce a shift in the composition of coral reef sponge communities favouring phototrophic and bioeroding species in particular.

Further reading

ABIF-Fauna (2004) Australian faunal directory (website). Department of Environment and Energy, Canberra, <https://biodiversity.org.au/afd/taxa/PORIFERA>.

Becerro MA, Uriz MJ, Maldonado M, Turon X (2012) Advances in sponge science: phylogeny, systematics, ecology. *Advances in Marine Biology* **61**, 1–432.

Becerro MA, Uriz MJ, Maldonado M, Turon X (2012) Advances in sponge science: physiology, chemical and microbial diversity, biotechnology. *Advances in Marine Biology* **62**, 1–355.

Berquist PR (1978) *Sponges*. Hutchinson, London, UK.

Hall KA, Hooper JNA (2014) SpongeMaps – an online community for sponge taxonomy. <www.spongemaps.org>.

Hooper JNA, Van Soest RWM (2002) *Systema Porifera: A Guide to the Classification of Sponges. Vols 1 and 2*. Kluwer Academic/Plenum Publishers, New York, USA.

Pitcher CR, Doherty P, Arnold P, Hooper JNA, Gribble N (2007) 'Seabed biodiversity on the Continental Shelf of the Great Barrier Reef World Heritage Area. AIMS/CSIRO/QM/QDPI CRC Reef Research Task Final Report'. CSIRO Marine and Atmospheric Research, Hobart, <http://era.daf.qld.gov.au/id/eprint/1704/1/CRC_GBR_Seabed_Biodiversity_Final_Report__Fri20July07c-sec.pdf>.

Simpson TL (1984) *The Cell Biology of Sponges*. Springer-Verlag, New York, USA.

Van Soest RWM, Boury-Esnault N, Vacelet J, Dohrmann M, Erpenbeck D, De Voogd NJ, *et al.* (2012) Global diversity of sponges (Porifera). *PLoS One* **7**(4), e35105. doi:10.1371/journal.pone.0035105

Van Soest RWM, Boury-Esnault N, Hooper JNA, Rützler K, de Voogd NJ, Alvarez B, *et al.* (2017) World Porifera database. <http://www.marinespecies.org/porifera>.

Wörheide G (Ed.) (2006) The sponge barcoding project (website). GeoBio-Center. Ludwig Maximilian University of Munich, Munich, Germany, <http://www.palaeontologie.geo.uni-muenchen.de/SBP/>.

20

Pelagic jellyfishes and comb jellies

L. Gershwin and M. J. Kingsford

Overview

The jellyfishes are conspicuous, but poorly understood, members of the Great Barrier Reef (GBR) fauna. Much attention has been given over the past 70 years to two groups in particular: the so-called 'box jellyfishes' and 'Irukandjis', both of which are highly dangerous and have killed humans. However, the dangerous species comprise only a small fraction of the jellyfishes that make their home in the GBR region.

Jellyfishes are among the most intriguing of animals, tantalising the child in all of us with their strange shapes, often bright colours, and sometimes flashing lights, and yet, even the milder stinging varieties seemingly represent a threat (Box 20.1).

'Jellyfishes' is a term that is used to loosely include gelatinous animals from three major phyla: the Phylum Cnidaria (stinging jellies), the Phylum Ctenophora (comb jellies), and the Phylum Chordata (salps and doliolids). Cnidarian jellies are close relatives of corals and anemones: they are characterised by having stinging cells called cnidocytes that contain organelles known as nematocysts used for prey capture and defence. The cnidarian jellies are grouped into four classes: the Hydrozoa (e.g. water jellies and siphonophores; Figs 20.1, 20.2); the Scyphozoa or true jellies (e.g. moon jellies, sea nettles, and blubber jellies; Figs 20.3, 20.4); the Cubozoa (e.g. box jellies, Irukandjis and jimbles; Fig. 20.5–20.7); and the Staurozoa (e.g. stalked jellies). Ctenophoran jellies – or ctenophores – do not have stinging cells, but have sticky cells called colloblasts on their tentacles to capture their prey (Fig. 20.8, 20.9). The ctenophores come in a wild array of shapes, and are grouped into those with tentacles (Class Tentaculata) and those without (Class Nuda). The third jellyfish-containing phylum, the Chordata, possesses neither stinging cells nor sticky cells. Larval chordates – including salps and humans – possess a notochord, which is the embryological precursor to the backbone; therefore, chordate jellies (pelagic tunicates, such as salps, pyrosomes, doliolids and appendicularians) are more closely related to us than to other types of jellyfish.

While appearing to be simple creatures consisting of no more than mucus and stinging cells, jellyfish have great potential to alter ecosystems through predation and uptake and excretion of nutrients. Cnidarian and ctenophoran jellies are carnivorous, preying on eggs, larvae and small animals (zooplankton), while salps and doliolids are mainly herbivorous filter feeders of tiny plant plankton (phytoplankton). 'Blooms' of all these jellies are found in the GBR and many other parts of the world. The medusa stage of cnidarian jellyfishes is the most obvious, and often the most voracious, stage of the life cycle; however, the polyp stage represents the 'seed bank' of the species, making most jellyfish species incredibly difficult, if not impossible, to eradicate.

Jellyfish introduced into exotic environments can cause great harm to the ecosystem (e.g. *Mnemiopsis* in European seas, and *Phyllorhiza* in the Gulf of Mexico and Mediterranean). So, too, extreme concentrations of native species can cause serious impacts on the local biota (e.g. blooms of *Chrysaora* in the Gulf of Alaska and *Nemopilema* in Asia, and *Chrysaora* and *Aequorea* in Namibia). Jellyfishes can also be an important food source for many species, such as fishes and turtles, and they are heavily fished in some parts of the world (e.g. China).

Evolution and diversity

Jellyfishes are an ancient group, found in the fossil record before the Cambrian explosion; credible examples of fossil jellyfishes from the famous Ediacaran fauna (originally identified in South Australia) date back ~585 million years, and appear relatively unchanged through the eons. Many

Fig. 20.1. Some hydrozoans from the Great Barrier Reef: **(A)** *Turritopsis lata*; **(B)** *Physalia utriculus*; **(C)** *Zygocanna* sp.; **(D)** *Olindias* sp.; **(E)** *Porpita* (Photos: L. Gershwin).

Box 20.1. Facts about jellyfishes

- There are well over 100 species of jellyfishes that make their home in the GBR region, and all but ~10 are regarded as harmless.
- Above average concentrations of jellyfishes are called 'blooms' or 'swarms'.
- Box jellyfish are found primarily close to the coast, but Irukandji jellyfish are found across the shelf of the GBR around reefs and islands as well as along the coast. At least 20 species of jellyfish are believed to cause Irukandji syndrome, ranging from as far north as the United Kingdom and Boston and as far south as Southern Australia.
- All jellyfishes are animals with no bones, no brain, no heart, and no lungs or gills; the only hard parts they have are the statoliths that are part of the balance systems (in the rhopalium) (Fig. 20.7)
- Box jellyfish and Irukandjis have well developed eyes (up to 24 of them) with lenses, retinas and corneas, just like our eyes, and we know experimentally that they can see light and dark with some eyes and images with others, impressive given they have no brain!

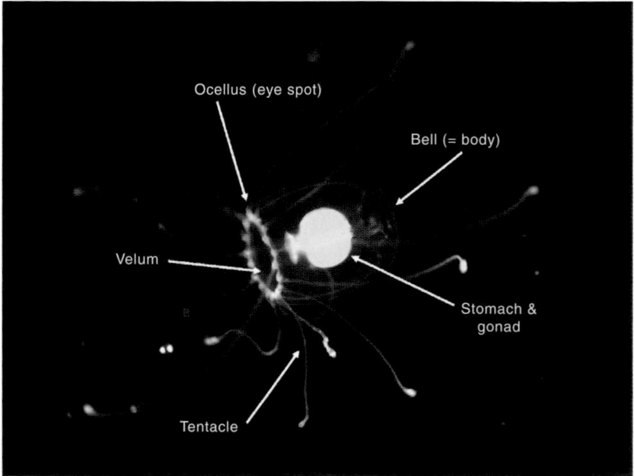

Fig. 20.2. Features of hydrozoan jellyfish (Photo: L. Gershwin).

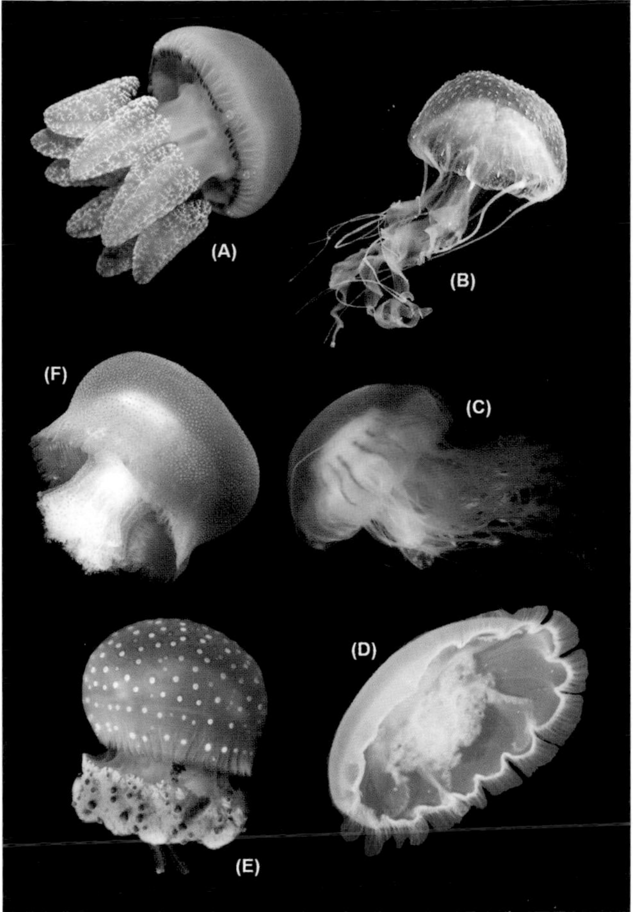

Fig. 20.3. Some scyphozoans from the Great Barrier Reef: **(A)** *Catostylus mosaicus*; **(B)** *Pelagia noctiluca*; **(C)** *Cyanea* sp.; **(D)** *Aurelia* sp.; **(E)** *Phyllorhiza* sp.; **(F)** *Bazinga rieki* (Photos: A—D, L. Gershwin; E, F. Denis Riek).

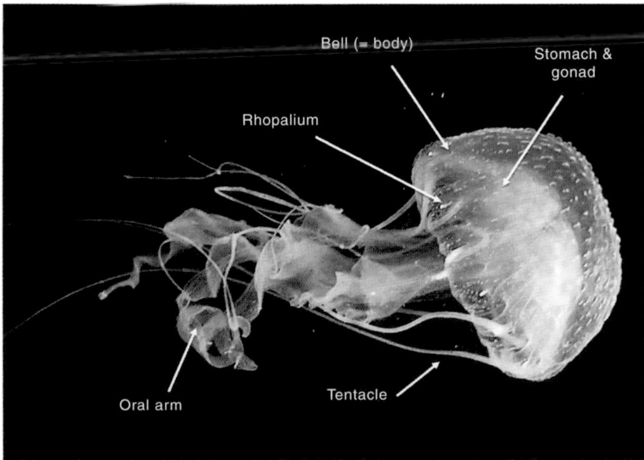

Fig. 20.4. Features of scyphozoan jellyfish (Photo: L. Gershwin).

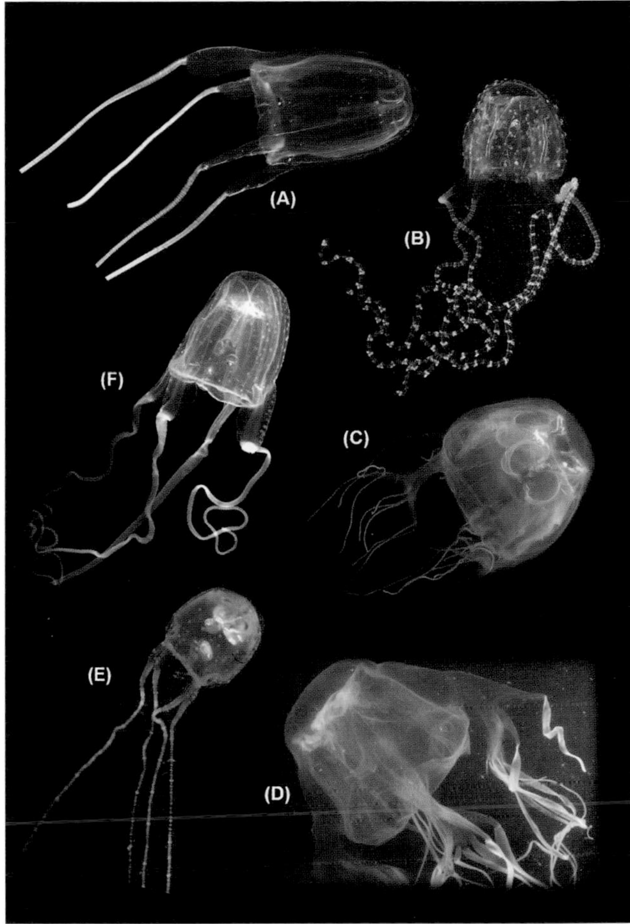

Fig. 20.5. Some cubozoans from the Great Barrier Reef: **(A)** *Malo kingi*; **(B)** *Carukia barnesi*; **(C)** *Chiropsella bronzie*; **(D)** *Chironex fleckeri*; **(E)** *Copula sivickisi*; **(F)** *Morbakka fenneri* (Photos: A—E, L. Gershwin; F, Merrick Ekins).

interpretable fossils exist from the Cambrian and scattered more recent periods. Even the box jellyfishes, which are considered 'the pinnacle of development' among jellyfishes, were fully developed by the Middle Pennsylvanian (~300 million years ago (Mya)). This long persistence of jellyfish through changing conditions and mass extinctions suggests that they are remarkably resilient and adaptable creatures.

Globally, probably a few thousand species of jellyfish exist. With few specialists, however, we lack clarity on the exact number. Currently, well over 100 jellyfish species are recorded from the GBR region. Although this may appear depauperate, many of these species are still awaiting formal classification. Most are tiny and inconspicuous hydromedusae, presenting no serious medical threat, giving a 'sea lice' sting at the very most. However, they may at times occur in such dense aggregations that visibility may become severely diminished and even minor stings become annoying if in large numbers.

Life history

The cnidarian jellyfishes typically have a complex (metagenic) life cycle wherein the pelagic medusa (sexual jellyfish stage) alternates with a sessile polyp (asexual hydra stage, Figs 20.10, 20.11). In most cases, the jellyfish are either male or female, and spawn freely into the water column, though fertilisation is sometimes internal (e.g. the cubozoan *Copula*, which mates following courtship, or the scyphozoan *Aurelia*, in which the females fertilise internally by ingesting free-floating sperm strands). The fertilised embryo grows into a 'planula larva' stage: a tiny ciliated capsule- or teardrop-shaped creature, which seeks a suitable place to settle and transform into a polyp. The polyp produces many clones through asexual reproduction, and may produce environmentally resistant cysts (podocysts). When the conditions are right, the benthic forms undergo a metamorphosis (budding or strobilation) to produce many tiny juvenile jellyfish; in scyphozoans these are a special flower-shaped stage called an ephyra (Fig. 20.10). The baby ephyrae and medusae are voracious feeders on a diversity of plankton and generally grow very fast. For most of the GBR species, life cycles are largely assumed through knowledge of closely related species from other regions (Figs 20.10, 20.11). In most cases, the appearance of the jellyfish stage is highly seasonal.

Siphonophores are highly modified pelagic hydrozoans: each is essentially a superorganism, or a colony of genetically identical, morphologically dissimilar clones, arranged along a common stem. So little is known about the reproduction of most siphonophores that we can only generalise from the few that are known. A fully grown siphonophore colony is asexual itself, but buds off short-lived medusa-like sexual gonozoids called eudoxids. Some species are monoecious (budding off both males and females), while others are dioecious (budding separate sexes).

The life cycle of most ctenophores is poorly known, but most are thought to be simultaneous hermaphrodites, bearing both sperm and eggs at the same time. Ctenophores lack a polyp stage, and instead develop directly from the fertilised egg into a larva. They often occur in dense aggregations, which may facilitate external fertilisation.

Salps have the most complex and intriguing life cycle of all the jellies, and are capable of budding into huge colonies. They are entirely pelagic (i.e.

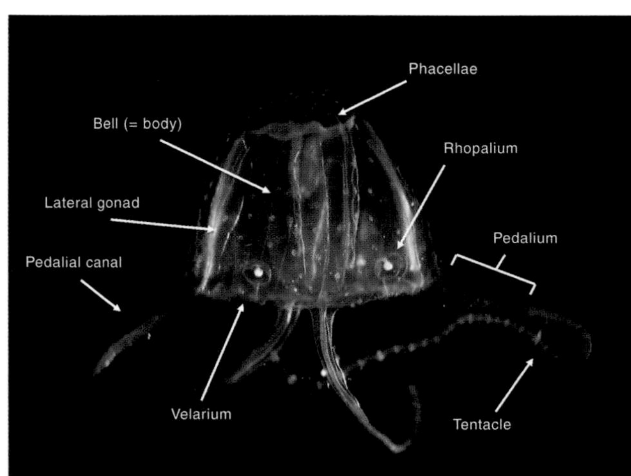

Fig. 20.6. Features of cubozoan jellyfish (Photo: L. Gershwin).

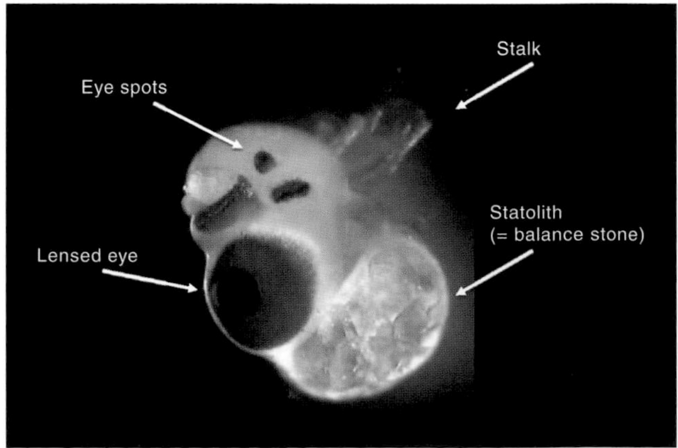

Fig. 20.7. Features of the cubozoan rhopalium (Photo: L. Gershwin).

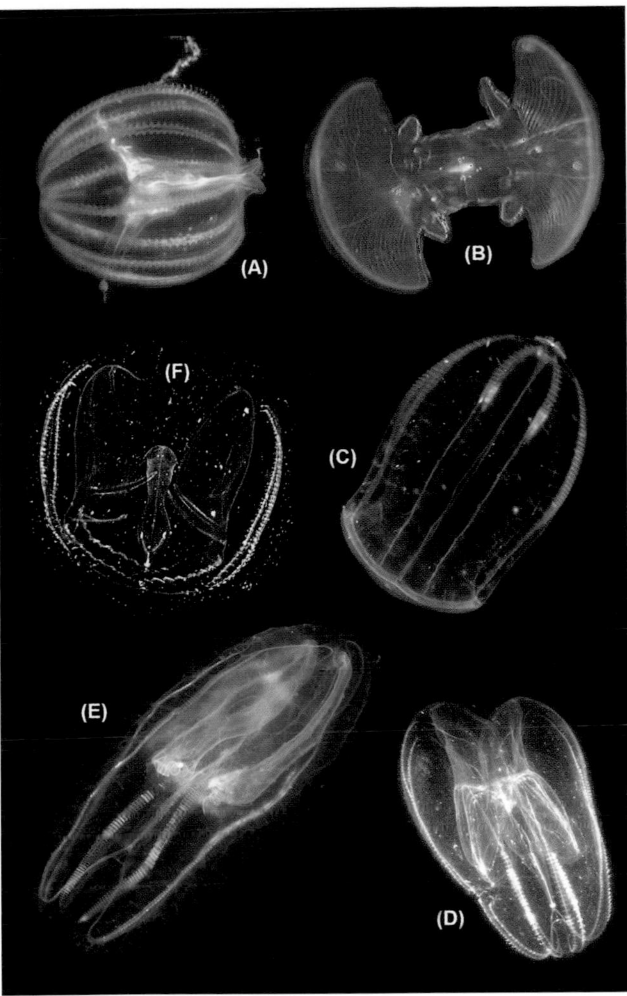

Fig. 20.8. Some ctenophores from the Great Barrier Reef: **(A)** *Pukia falcata*; **(B)** *Ocyropsis* sp.; **(C)** *Beroe* sp.; **(D)** *Bolinopsis* sp.; **(E)** *Leucothea filmersankeyi*; **(F)** *Ocyropsis* sp. (Photos: L. Gershwin).

lacking a benthic polyp). Most salps alternate between two main life stages: an asexual solitary stage, which buds off chains of clone-mates; and a sexual aggregate stage, in which the youngest zooids nearest the parent are female and the oldest ones at the end of the chain are male. The life history of other pelagic tunicates (pyrosomes, doliolids and appendicularians; see Chapters 16, 29) is progressively more complex.

Distribution

Because jellyfishes are pelagic (free-swimming), they occur throughout the GBR region, though some have restricted distributions and their numbers vary greatly with time of year. Box jellyfish, for example, are generally restricted to coastal and nearshore island localities from spring through to autumn. In contrast, stings from Irukandji jellyfishes have been reported from nearly every coastal and island/reef locality where recreational activities take place throughout the GBR (Box 20.2).

The larger, showier species such as lion's manes (*Cyanea* spp.), blubbers (*Catostylus* sp.), spotted blubbers (*Mastigias* spp.), and sugar-bowl jellies (*Netrostoma nuda*) are occasionally common throughout the GBR region and can even reach bloom proportions, where hectares of closely packed jellyfish can be found in some places. Curiously, two of the commonest jellies globally, the

Box 20.2. Irukandji syndrome

At least 20 species of jellyfish have been linked with a bizarre constellation of symptoms known as 'Irukandji syndrome'. Most cases of Irukandji syndrome are attributable to box jellyfish; however, it can also be caused by a few non-cubozoan species such as a rare Pacific form of the Portuguese-man-or-war (*Physalia*), a small hydrozoan (*Gonionemus*) found in the north-eastern US and Russia, a type of Lions Mane (*Cyanea*) found in the Irish Sea, and a large rhizostome found in China and Japan (*Nemopilema*). The initial sting is typically mild, and is followed after a 5–40 minute delay by any or all of the following: severe lower back pain, nausea and vomiting, difficulty breathing, profuse sweating, cramps, spasms, muscular restlessness, and a feeling of impending doom. Irukandji syndrome can be fatal. In the GBR region, Irukandji stings are commonest during periods of low wind, and anecdotally Irukandji jellyfishes are more prevalent on days with dense 'jelly button' or 'sea lice' blooms, especially if salps are present (see Chapter 16). The best prevention is a full-body lycra suit (or neoprene wet suit), which provides good protection for ~80% of the body's skin surface. If stung, douse the wound with vinegar to neutralise microscopic stinging cells on the skin and seek immediate medical advice.

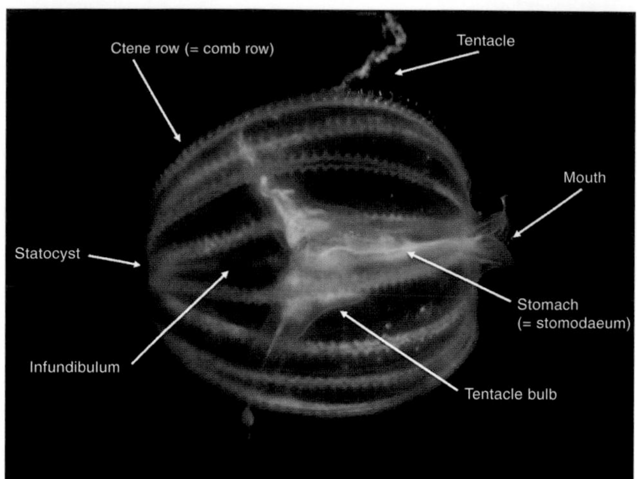

Fig. 20.9. Features of ctenophores (Photo: L. Gershwin).

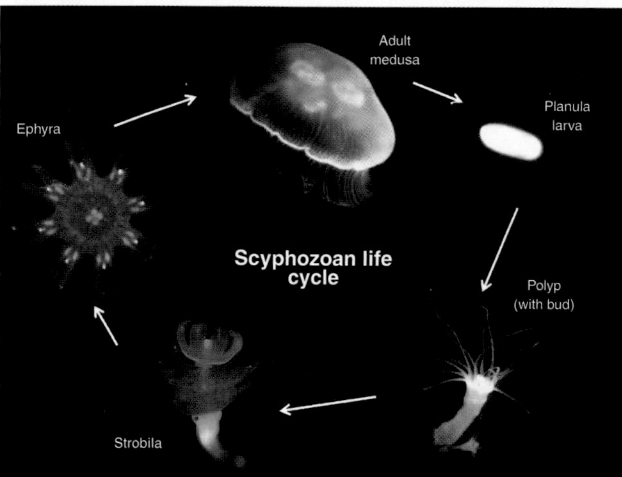

Fig. 20.10. Life history of a scyphozoan jellyfish, *Aurelia* sp. (stages from multiple taxa shown) (Photos: L. Gershwin).

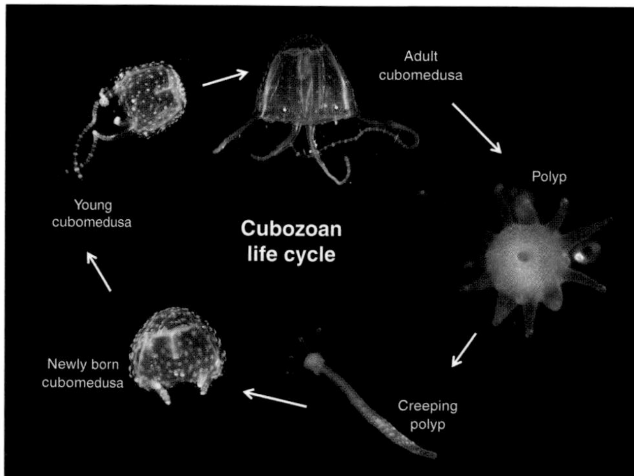

Fig. 20.11. Life history of a cubozoan jellyfish (stages from multiple taxa shown) (Photos: L. Gershwin).

moon jellyfish (*Aurelia* spp.) and the blue bottles (*Physalia* spp.) are only sporadically found in the GBR region, where they may occur in dense but short-lived blooms. *Physalia* have a bladder that sits above the water and therefore their abundance often increases during prolonged periods of onshore winds when they are blown from the open ocean. However, the majority of jellyfish species have only been reported from a few reefs or islands, but this is probably more due a paucity of studies than a reflection of their true abundance.

Predation and defence

Medusae and ctenophores eat a diversity of other species of plankton, including the eggs and larvae

Box 20.3. Stinging cells, sticky cells and mucous nets

Nematocysts are basically a capsule within a cell called a cnidocyte, with a spring-loaded harpoon (thread or tubule) coiled up inside, and an external hair trigger at one end (Fig. 20.12). The thread has rows of spines near the base, and usually a relatively smooth shaft along the rest of its length. When the hair trigger is activated, the thread is discharged inside-out (i.e. everting spines-first), with tremendous force, ~40 000 times the force of gravity. Nematocysts can easily penetrate crab carapaces, fish scales and human skin, delivering their subcutaneous dose of venom. Nematocysts discharge following a combination of mechanical and chemical stimulation.

All cnidarian jellyfishes have nematocysts, which they use for defence and prey capture. Instead of nematocysts, ctenophores have intriguing organelles called colloblasts on their tentacles which lack venom but ensnare prey with adhesive. A nematocyst might be thought of as a tiny harpoon with poison, while a colloblast is more like a rope covered in honey.

In contrast, salps and doliolids feed like a jet engine, sweeping tiny prey into the body and filtering the water over continually renewable mucus nets.

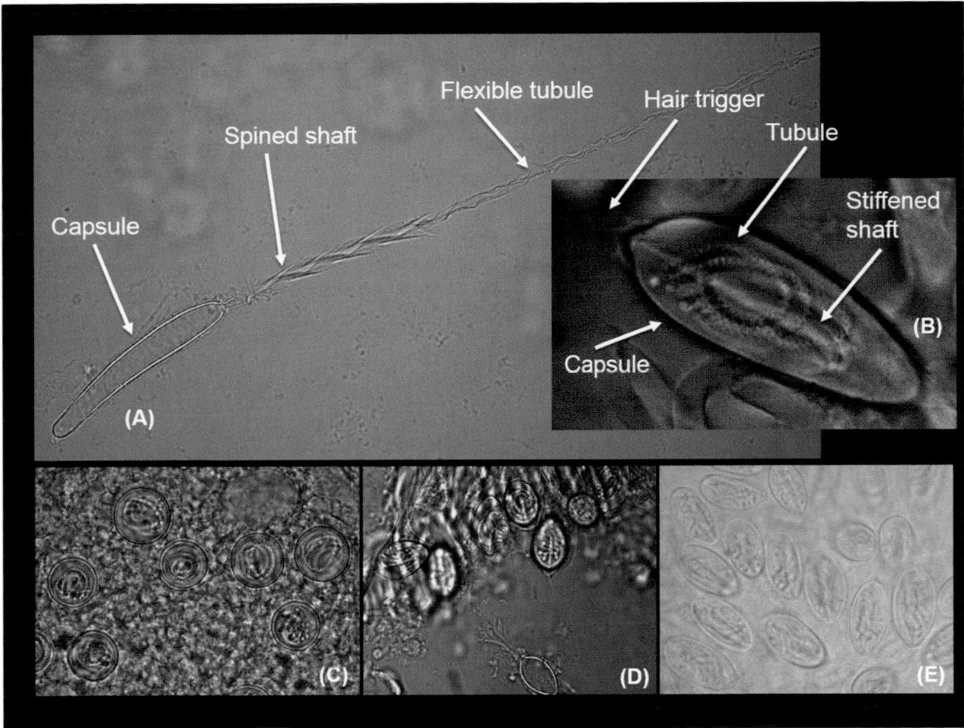

Fig. 20.12. The morphology of nematocysts (stinging cells). **(A)** discharged mastigophore from *Chironex*; **(B)** undischarged mastigophore from *Malo*; **(C)** undischarged isorhizas from *Malo*; **(D)** discharged stenotele (lower-centre) and undischarged stenoteles (upper) from *Carukia*; **(E)** undischarged euryteles from *Carybdea* (Photos. L. Gershwin).

of reef-associated organisms such as fishes; they use stinging cells or sticky cells to subdue their prey (Box 20.3). Jellyfish are typically carnivorous, preferring zooplankton prey rather than phytoplankton. Their own kind can also be on the menu: for example, *Cyanea* spp. (lions mane) eat *Aurelia* (moon jellies) and others, while the ctenophore *Beroe* preys on other ctenophores such as *Pleurobrachia*. Some species of rhizostomes (Class Scyphozoa) have symbiotic algae (zooxanthellae) in their tissues, similar to those of corals: for example, *Mastigias* (spotted blubbers) and *Cassiopea* (upside down jellies, which live on the bottom). These symbiotic jellyfishes are typically found in warm shallow waters, where they 'farm' their algal symbionts; these jellyfish may obtain a significant portion of their nutritional requirements from their algae.

Some jellyfish have definite feeding preferences. For example, box jellyfish (*Chironex fleckeri*) from the GBR region have been demonstrated to change their dietary habits as they grow, starting out preferring crustaceans such as prawns and switching to fish later in life. They may even hunt their prey through orientation assisted by their 24 eyes! While the huge appetites of jellyfish make them fearsome predators, their soft bodies also make them an easy meal. Turtles, crabs, seabirds, and many fishes, such as sunfish, triggerfish and leatherjackets, actively hunt jellyfishes.

Slippery slope to slime

The term 'slippery slope to slime' was coined in reference to the decline of coral reefs and the politics of their conservation. Despite contributing hundreds of billions of dollars to the global economy each year, coral reefs are rapidly degrading from communities rich with fish, turtles, sharks and diverse invertebrates to rubble, seaweed and slime. Meanwhile, debate over the primary cause of decline, such as protected areas that are too small and piecemeal, and lack of clear conservation goals, prevents meaningful action. A similar simplification of pelagic habitats is occurring in many locations around the world, with a similar lack of focus on solutions, while jellyfish blooms continue to cost marine industries huge financial losses.

Bloom dynamics, or the cycles and triggers of jellyfish blooms that may result in degraded pelagic ecosystems, is a hot topic of modern jellyfish research. Many jellyfish are tolerant of a wide range of environmental conditions and pelagic assemblages, and are capable of exploiting perturbed ecosystems, whether through pollution (especially eutrophication), overfishing, species introductions, climate change or coastal construction. Water quality conditions that threaten the survival of higher organisms may facilitate the development of jellyfish blooms. Numerous phase-shift examples exist around the world where jellyfish have essentially 'flipped' the local ecosystem, becoming the dominant predator and/or competitor for plankton (see, for example, Box 20.4, Box 20.5, Box 20.6).

Box 20.4. Phase shifts and alternative stable states

It seems that a stable food web based on microbes and jellyfish can co-exist alongside the more traditional food web that supports fish, marine mammals and seabirds. These alternative stable states can persist with no apparent impact, until some threshold or trigger causes a phase shift, with jellyfish taking over as the dominant predator. Once shifted, the simpler ecosystem can be very resilient against shifting back.

One good example is from the Benguela Current off Namibia in south-western Africa, where jellyfish biomass now exceeds that of once prolific finfish. This highly productive upwelling region historically supported large stocks of fish, as well as abundant marine mammals and seabirds, but heavy fishing pressure has reduced these stocks considerably. In particular, small pelagics (anchovies and sardines) were overfished, releasing jellyfish from the pressure of predation and competition. Since the 1990s, two jellyfish have dominated this ecosystem. Because of the predation and competition that jellyfish now exert on fish, this phase shift may be irreversible.

Jellyfish employ several simple, but powerful, mechanisms in their struggle to survive, which severally or collectively can have a devastating effect on other species. One is sometimes referred to as the 'jellyfish double whammy', where jellyfish eat the eggs and larvae of other species as well as the food sources that these larvae would eat (thus influencing larval survival). This combination of predation and competition enables jellyfish to potentially take over as top predator and keep other species down. Another mechanism is their opportunistic growth, where they can rapidly bloom, grow and reproduce in response to favourable conditions, or some stages can go dormant for years at a time in poor conditions. In particular,

Box 20.5. Comb jellies: alien invaders

Comb jellies are well known as invaders. For example, *Mnemiopsis leidyi*, a lobate ctenophore in the Class Tentaculata were inadvertently carried in ships' ballast water from the eastern United States to the Black Sea in the early 1980s. Without natural predators, its numbers had exploded to one billion tons of biomass by 1990. This one species single-handedly caused the collapse of an ecosystem and the sardine fisheries that went with it. This case has been well studied, and it is believed that the jellyfish ate the larvae of other species including sardines, as well as the food sources that the larvae would eat; this combination of predation and competition prevented other species from regaining viable populations. Ironically, numbers were thought to have dropped in recent years as a result of predation on *M. leidyi* by *Beroe*, another ctenophore that was also introduced in the system.

Elsewhere in the world, other ctenophore and medusa species have been considered pests, either through exotic species introductions, or through local perturbations that allowed the jellyfishes to alter existing ecosystems; perturbations include overfishing, pollution and changes in water temperature. Jellyfish can bloom rapidly when conditions are ideal or go for long periods of time without food if necessary. Their potential to quickly change pelagic assemblages is therefore great.

Box 20.6. Salps and their kin: plankton gluttons

Pelagic tunicates (salps, pyrosomes, doliolids, and appendicularians; also see Chapter 16) are marine herbivores with extremely high feeding rates and short generation times. These unique features allow them to rapidly exploit pulses of nutrients, thus reaching high population densities faster than most other planktonic species.

Salps and their kin feed by filtering phytoplankton and other small particles through a mucus net. Because of their sheer density, they can rapidly clear the water of food particles, outcompeting fish larvae and other grazers and consumers. The faecal pellets discharged by salps and dead bodies sink rapidly towards the seafloor, enhancing nutrient flux and aiding in carbon sequestration.

Although it has long been thought that salps and other gelatinous species are a trophic dead end, recent studies have shown that many species of vertebrates and invertebrates feed on them. However, it appears that in some cases where salps have consumed large quantities of toxic algae, vertebrates ingesting them are killed by the toxic overload.

they are capable of cloning in at least 13 different ways, and they can even 'degrow' in the absence of food.

Many of the impacts of human enterprise are creating oceans that favour jellyfish. For example, warming water can stimulate their growth and reproduction. The runoff of nutrients from land, creating eutrophic conditions, may be beneficial to symbiotic algae hosted by the jellyfish, as well as for helping to facilitate plankton blooms that in turn could support more jellyfish. However, empirical evidence for some of these ideas is lacking and jellyfish are also common in relatively unaltered pelagic ecosystems (e.g. polar regions). Fishing also has the potential to help jellyfishes by removing their predators and competitors, while pollution that negatively impacts other species may leave jellyfish relatively unaffected.

Further reading

Alldredge AL, Madin LP (1982) Pelagic tunicates: unique herbivores in the marine plankton. *Bioscience* **32**(8), 655–663. doi:10.2307/1308815

Arai MN (1997) *A Functional Biology of Scyphozoa*. Chapman & Hall, London, UK.

Ates RML (1988) Medusivorous fishes, a review. *Zoologische Mededelingen (Leiden)* **62**(3), 29–42.

Ates RML (1991) Predation on Cnidaria by vertebrates other than fishes. *Hydrobiologia* **216/217**, 305–307. doi:10.1007/BF00026479

Ates RML (2017) Benthic scavengers and predators of jellyfish, material for a review. *Plankton & Benthos Research* **12**(1), 71–77. doi:10.3800/pbr.12.71

Castellani C, Edwards M (2017) *Marine Plankton: A Practical Guide to Ecology, Methodology, and Taxonomy*. Oxford University Press, Oxford, UK.

Ennion J (2017) What exactly are jellyfish, anyway? *The Australian Geographer,* 30 March, <http://www.australiangeographic.com.au/topics/science-environment/2017/03/what-are-jellyfish>.

Gershwin L (2013) *Stung! On Jellyfish Blooms and the Future of the Ocean*. University of Chicago Press, Chicago IL, USA.

Gershwin L (2016) *Jellyfish: A Natural History*. Ivy Press, London, UK.

Gershwin L, Zeidler W, Davie PJF (2010) Ctenophora of Australia. *Memoirs of the Queensland Museum* **54**(3), 1–45.

Gershwin L, Richardson AJ, Winkel KD, Fenner PJ, Lippmann J, *et al.* (2013) Biology and ecology of Irukandji jellyfish (Cnidaria: Cubozoa). *Advances in Marine Biology* **66**, 1–85. doi:10.1016/B978-0-12-408096-6.00001-8

Harbison GR (1985) On the classification and evolution of the Ctenophora. In *The Origins and Relationships of Lower Invertebrates*. (Eds SC Morris, JD George, R Gibson and HM Platt) pp. 78–100. Oxford University Press, Oxford, UK.

Holstein TW, Tardent P (1984) An ultra high-speed analysis of exocytosis: nematocyst discharge. *Science* **223**(4638), 830–833. doi:10.1126/science.6695186

Kingsford MJ, Mooney CM (2014) The ecology of box jellyfishes (Cubozoa). In *Jellyfish Blooms*. (Eds KA Pitt, CH Lucas) pp. 267–302. Springer, Dordrecht, Netherlands.

Kingsford MJ, Pitt KA, Gillanders BM (2000) Management of jellyfish fisheries, with special reference to the order Rhizostomeae. *Oceanography and Marine Biology - an Annual Review* **38**, 85–156.

Kramp PL (1961) Synopsis of the medusae of the world. *Journal of the Marine Biological Association of the United Kingdom* **40**, 7–382. doi:10.1017/S0025315400007347

Lynam CP, Gibbons MJ, Axelsen BE, Sparks CAJ, Coetzee J, Heywood BG, Brierley AS (2006) Jellyfish overtake fish in a heavily fished ecosystem. *Current Biology* **16**(13), R492–R493. doi:10.1016/j.cub.2006.06.018

Lynam CP, Heath MR, Hay SJ, Brierley AS (2005) Evidence for impacts by jellyfish on North Sea herring recruitment. *Marine Ecology Progress Series* **298**, 157–167. doi:10.3354/meps298157

Mayer AG (1910) *Medusae of the World. Vols 1 and 2, the Hydromedusae; Vol. 3, the Scyphomedusae.* Carnegie Institution, Washington DC, USA.

Moller H (1984) Reduction of a larval herring population by a jellyfish predator. *Science* **224**, 621–622. doi:10.1126/science.224.4649.621

Pitt KA, Lucas CH (Eds) (2014) *Jellyfish Blooms.* Springer, Dordrecht, Netherlands.

Roux J-P, van der Lingen C, Gibbons MJ, Moroff NE, Shannon LJ, Smith ADM, *et al.* (2013) Jellyfication of marine ecosystems as a likely consequence of overfishing small pelagic fishes: lessons from the Benguela. *Bulletin of Marine Science* **89**(1), 249–284. doi:10.5343/bms.2011.1145

Totton AK (1965) *A Synopsis of the Siphonophora.* British Museum of Natural History, London, UK.

Williamson JA, Fenner PJ, Burnett JW, Rifkin JF (Eds) (1996) *Venomous and Poisonous Marine Animals: A Medical and Biological Handbook.* NSW University Press, Sydney.

Online resources – safety and information on jellyfish

An authoritative view of jellyfish taxonomy is provided in WORMS – the World Register of Marine Species: http://www.marinespecies.org/.

Australian Marine Stinger Advisory Services – a resource for marine stinger management: www.StingerAdvisor.com.

Curating Jellyfishes: preservation, storage, and photography: www.StingerAdvisor.com/curating.html.

The Jellyfish App: a smartphone app available on the IOS and Android platforms. See the App Store or Google Play for more info, or www.TheJellyfishApp.com.

Marine Stingers, a portal maintained by Surf Life Saving Queensland: www.marinestingers.com.

21

Hexacorals 1: sea anemones and anemone-like animals (Actiniaria, Zoantharia, Corallimorpharia, Ceriantharia and Antipatharia)

C. C. Wallace and A. L. Crowther

Overview

Because they have no skeleton and do not participate in the building of reefs, sea anemones and their kin (class Anthozoa; orders Actiniaria, Zoantharia, Corallimorpharia, Ceriantharia and Antipatharia) are not as intensely studied as their relatives, the hard corals (order Scleractinia in the same class). Nevertheless, they are prominent, and in some cases well known, members of coral reef communities. The largest sea anemones, reaching almost a metre in diameter, are hosts to anemone fish, crustaceans and various invertebrates that live among the anemones' tentacles in fascinating symbiotic relationships. Other anemones play guest rather than host and gain mobility by attaching to molluscs, crabs or shells housing hermit crabs. Zoanthids and corallimorphs can be abundant occupiers of space, sometimes when this is made available by bleaching or other damage to corals; they too may have symbiotic associations with many other reef organisms, thus securing space in the reef column, protection from predators and/or access to food and energy sources. Like hard and soft corals, the members of these groups have life cycles involving a short larval stage and a long polyp phase. The adult occurs as a single polyp in Actiniaria (sea anemones) and Ceriantharia (tube anemones) and usually, but not invariably, as a colony or clone of interconnected polyps in Corallimorpharia (jewel anemones) and Zoantharia (zoanthids). Antipatharia (black corals) are exclusively colonial. The major external characteristics of the polyp are a column, an oral disc bearing the mouth, siphonoglyph(s) (flagellated furrows that direct water currents into the gastrovascular cavity) and tentacles, and (in solitary forms) the pedal disc or bulb allowing attachment or anchorage. In colonial forms, polyps are joined by basal stolons or common coenenchyme. Common to all groups, main internal characters include the actinopharynx leading from the mouth and siphonoglyphs, and mesenteries that may be complete or incomplete (reaching or not reaching the actinopharynx at their upper limit) and on which the gonads develop (see Fig. 21.1 for examples of these characters in various orders).

Animals from these orders contribute greatly to the taxonomic and ecological diversity of the Great Barrier Reef (GBR) and to the dominant role of the Anthozoa in the coral reef environment. They are

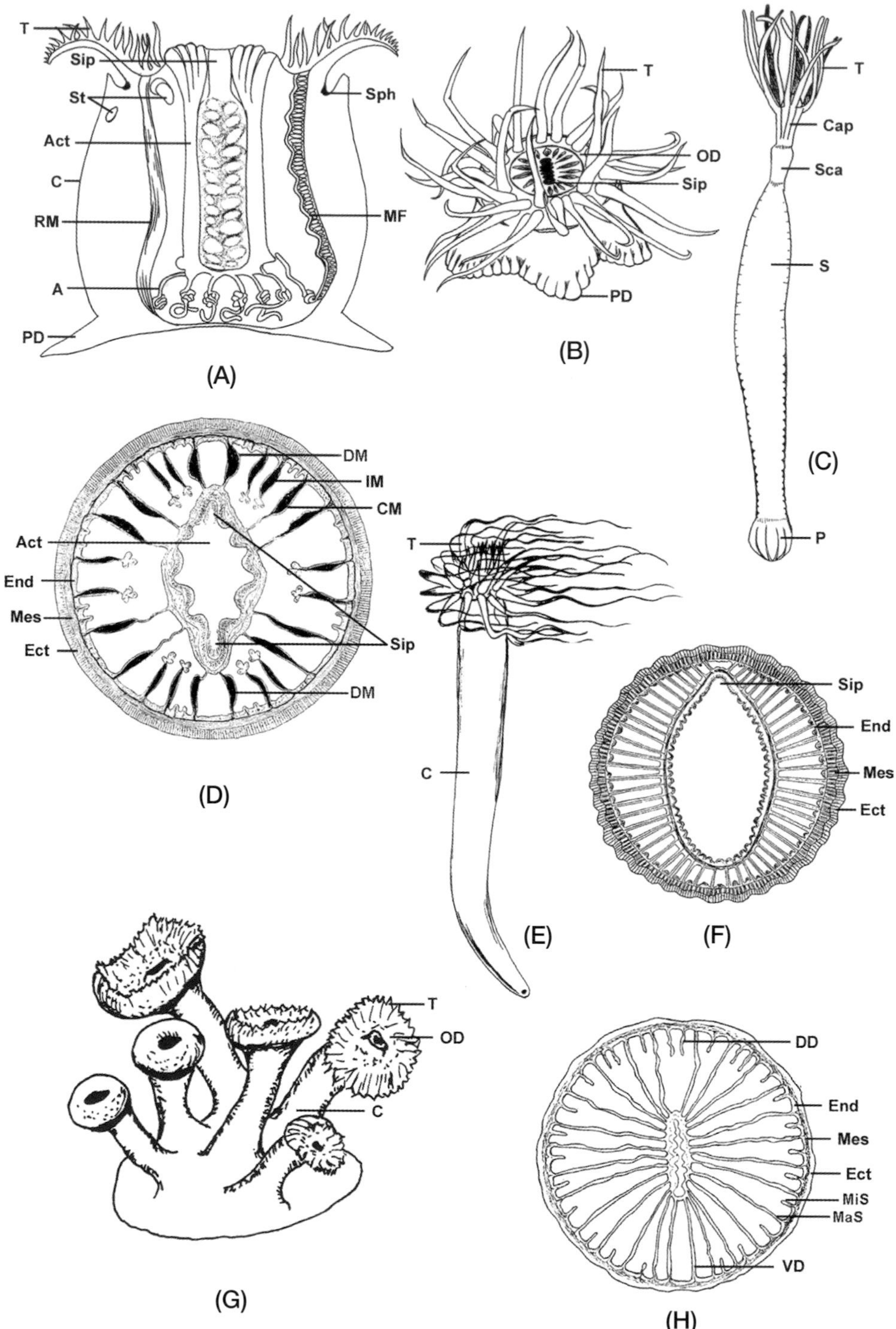

Fig. 21.1. Anatomical features. **Actiniaria: (A)** diagrammatic sea anemone internal features; **(B)** external features (*Anthopleura handi*); **(C)** external features, Edwardsiidae; **(D)** transverse section through actinopharynx region. **Ceriantharia: (E)** external features; **(F)** transverse section through actinopharynx region. **Zoantharia: (G)** external features; **(H)** transverse section through actinopharynx region. A, acontia; Act, actinopharynx; C, column; Cap, capitulum; CM, complete mesentery; DD, dorsal directives; DM, directive mesentery; Ect, ectoderm; End, endoderm; IM, incomplete mesentery; MaS, macrocneme; MF, mesenterial filament; Mes, mesoglea; MiS, microcneme; OD, oral disc; PD, pedal disc; P, physa; RM, retractor muscle; S, scapus; Sca, scapulus; Sip, siphonoglyph; Sph, sphincter; St, stomata; T, tentacles; VD, ventral directive. (Images redrawn from various sources by W. Napier.)

mostly carnivorous, feeding on tiny planktonic or benthic animals or particulate matter, and sometimes using symbiotic dinoflagellates in a similar manner to the hard corals. Their attractive radially symmetrical form and dramatic colouration, combined with relatively simple environmental requirements (at least in some cases) make this group very popular with the aquarium trade. Because of this, some are well known to hobbyists and their food, habitat requirements and even breeding and dividing patterns are recorded on numerous web pages, although scarcely documented in the scientific literature.

Whereas the classification and identification of Scleractinia and Alcyonaria are primarily based on skeletal elements (solid skeleton in hard corals and elaborate spinules in soft corals), other features of the polyp, both macro- and microscopic, must be used for these predominantly skeleton-free orders. Valid identification of smaller specimens often requires verifying details of external and internal anatomy under a dissecting microscope, as well as histological preparation of thin sections and tissue squashes for high power microscopic study. Fortunately for the field observer, many coral reef species have large size, bright colouration, unique habitat preferences or typical associates that allow for field identification with reasonable certainty.

Order Actiniaria (sea anemones)

Throughout the world, sea anemones occur in most marine environments, from rocky and soft intertidal to deep abyssal substrata, and in many sizes from more than 1 metre in diameter down to a millimetre or two. Large sea anemones can be seen in both tropical reefs and deep Antarctic waters, but the families represented differ, with many of the deep-sea species having strong gelatinous column walls or heavy external coatings and very large mouths, while tropical species tend to have soft bodies, smaller mouths and strong defence mechanisms. Not surprisingly, many of the tiny sea anemones remain undescribed or unrecorded for Australia and their presence in ecosystems may go unnoticed. Some sea anemones harbour symbiotic dinoflagellates and small-bodied species are sometimes used as indicators for laboratory study of heat tolerance and other characteristics in these organisms.

Sea anemones occupy many habitats and microhabitats on coral reefs. Some favour elevated sites on the reef surface, some extend from recesses within the framework and others burrow within the sandy floor. Sometimes closely related species have obvious differences in habitat preference; for example, the two large anemonefish hosts *Heteractis magnifica* and *H. crispa* favour exposed high points and protected crevices, respectively.

The 'swimming anemone' *Boloceroides mcmurrichi* lives unattached or very lightly attached, and is observed sporadically on reefs, sometimes in 'swarms' of large numbers of individuals. Many other anemones have the capability of changing location slowly by actions of the pedal disc. Most anemones are highly retractile and expandable, capable of greatly varying the column and tentacle length, closing the oral disc through the actions of a strong marginal sphincter, or even of disappearing from sight completely by withdrawing into the substratum.

Although all Cnidaria have nematocysts (stinging capsules) in their ectodermal tissues, sea anemones have a great variety of these and some coral reef inhabitants are extremely dangerous because of the toxic properties of their nematocysts. Anyone wishing to explore the taxonomy of anemones must master the various categories of these cellular features, as the types of nematocysts present are used in family determinations, and genera and species can have typical dimensions and proportional composition of these nematocysts in different body regions (see Fig. 21.2 for some common types). Both external and internal morphological features are needed for the identification of sea anemones. Much remains to be resolved in the classification of sea anemones, and relationships between and among genera and families are not always clear. This is an active research area and new developments can be seen as updates in the 'Hexacorallians of the World' database (see 'Further reading').

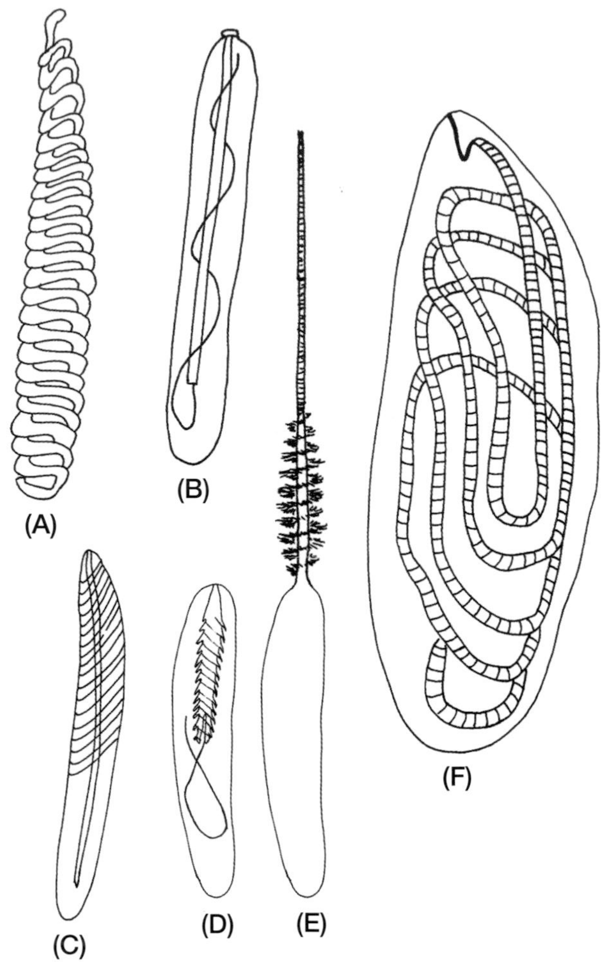

Fig. 21.2. Cnidae types in Actiniaria; **(A)** spirocyst; **(B)** basitrich; **(C)** microbasic b-mastigophore; **(D–E)** microbasic p-mastigophore (E fired); **(F)** holotrich (Images redrawn from various sources by W. Napier.)

External anatomy

The sea anemone column may have up to three recognisable regions (from the base: scapus, scapulus and capitulum: see Fig. 21.1) and above the column is a flat or domed oral disc bearing the mouth and tentacles in two to many cycles. The mouth is oval or slit-like and has one, two or more siphonoglyphs. The margin of the oral disc is often folded into a groove or fosse. Tentacles in Actiniaria are mostly hollow and simple but may be branched, sometimes elaborately so, or carry structures containing numerous nematocysts. Other projections such as verrucae (adhesive projections), acrorhagi (projections bearing concentrated nematocysts), and marginal spherules (projections with fewer nematocysts) may occur on the column, margin or fosse, respectively. A large group of sea anemones (known as 'acontiate' anemones) have nematocyst-bearing structures called acontia (fine hair-like projections from the base of the mesenteries), which can be extruded through various parts of the body, and sometimes through permanent holes in the column called cinclides.

Internal anatomy

The mouth leads to an actinopharynx, below which is the gastrovascular cavity. The body cavity is divided by paired mesenteries, some complete (reaching the actinopharynx) and others incomplete, in typical arrangements. Gonads (oocytes and testes) develop from endodermal interstitial cells that ripen within the mesenteries. The arrangement of gonads is used in the diagnostic process, with either hermaphroditic (both sexes occurring in each polyp) and gonochoric (sexes separate) states occurring in different species. As might be expected in soft-bodied animals without skeletons, anemones are equipped with a variety of muscles, which serve to maintain body shape, control tentacles and maintain the position of the animal on or in the substratum. Often the presence/absence, position, size and shape of the muscles (especially of the marginal sphincter muscle responsible for closing off the oral disc) are diagnostic.

Life history and ecology

All sea anemones are capable of sexual reproduction and many are also capable of asexual reproduction by division of the polyp and separation of the two resulting polyps. The bulb-tentacle anemone, *Entacmaea quadricolor,* and the magnificent sea anemone, *Heteractis magnifica,* are well known for forming huge clonal groups of individuals, occupying many square metres, by this method. Other species, most notably the aquarium pest, *Aiptasia pulchella,* produce numerous cloned individuals by a process known as pedal laceration, and the swimming anemone, *Boloceroides mcmurrichi,* is able to reproduce by shedding tentacles, which then bud a complete new anemone. Sexual reproduction involves the development of eggs and sperm in separate gonads, followed by release of the gametes into the water column for external fertilisation (in hermaphroditic and possibly some gonochoric species) or fertilisation of eggs *in situ* by sperm released from other individuals and brooding of larvae inside or close to the female parent (in gonochoric species).

Diversity

Of ~150 species of anemones known from Australian waters (including Antarctica), around 40 are known so far from the GBR (Figs 21.3–21.7, Box

Fig. 21.3. Some sea anemones of the Great Barrier Reef. **(A)** **Heteractis magnifica*; **(B)** **Heteractis crispa* (with *Amphiprion clarkii*); **(C)** **Heteractis aurora*; **(D)** **Heteractis malu*; **(E)** **Stichodactyla mertensii*; **(F)** **Cryptodendrum adhaesivum*; **(G)** **Entacmaea quadricolor*; **(H)** +*Telmatactis* sp.; **(I)** *Thalassianthus* sp.; **(J)** +*Megalactis griffithsi*; **(K)** *Actinodendron glomeratum*; **(L)** sand-burrowing anemone, Edwardsiidae. *, anemonefish host; +, dangerous because of stinging potential (Photos: *A, E, F, G*, A. Crowther; *B, C, D, I, J, L*, P. Muir; *H*, W. Napier; *K*, R. Steene).

Fig. 21.4. Merten's sea anemone *Stichodactyla mertensii* with the clownfish *Amphiprion clarkii* (Photo: P. Muir).

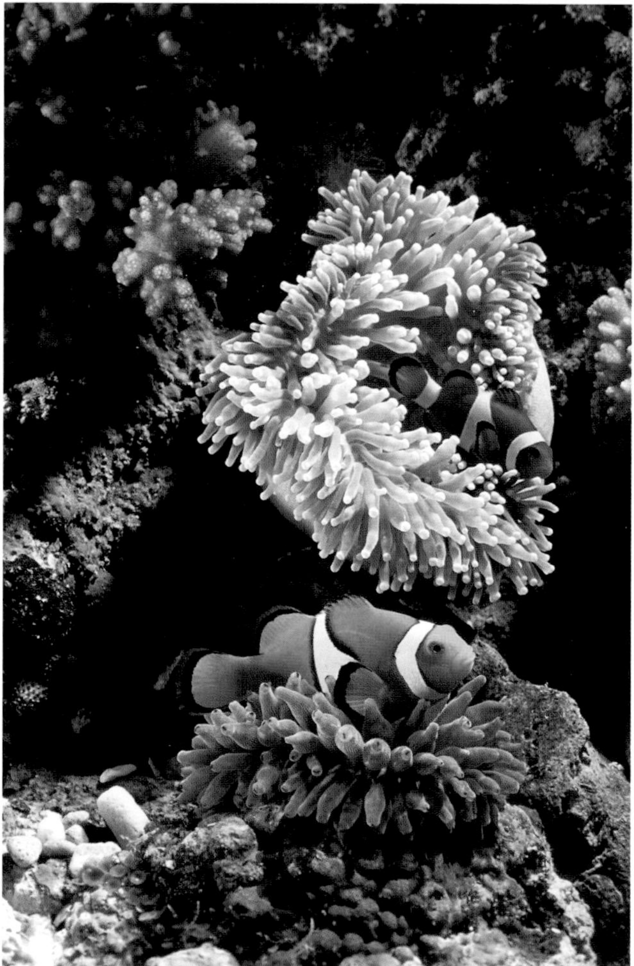

Fig. 21.5. Clown anemonefish in *Entacmaea quadricolor* in ReefHQ aquarium, Townsville (Photo: GBRMPA).

21.1). This is likely to be a gross underestimate, particularly in regard to smaller and cryptic species and sand-dwelling species, especially those from the family Edwardsiidae, which is under review worldwide at present. As with corals, many of the anemone species found on the GBR have broad ranges through the Indo-Pacific.

Preservation and study

For taxonomic research, sea anemones should ideally be examined alive, in both the field and

Fig. 21.6. *Calliactis polypus* is often seen in association with hermit crabs, and can be seen when the crab is on the move at night, feeding on the reef surface (Photo: P. Harrison).

Fig. 21.7. The 'boxer crab' (*Lybia* sp.) carries juveniles of the sea anemone *Triactis producta*, or other Aliciidae species, in its claws, presumably for protection from predators. Presumably, also, this is a dispersal mechanism for the anemone (Photo: R. Steene).

Box 21.1. Sea anemones as hosts and guests

Hosts

One of the best known and loved associations on coral reefs is that between large tropical sea anemones and the small, brightly coloured fish known as 'clownfish' or 'anemonefish'. Fish and sea anemones live together in a close association known as 'symbiosis'. It is also mutualism, because both species benefit from the arrangement and concessions are made by the partners to make it possible for the two to coexist. Ten species of sea anemone are known to play host to anemonefishes. These come from the families Stichodactylidae (species of *Heteractis* and *Stichodactyla*), Actiniidae (*Entacmaea quadricolor* and *Macrodactyla doreensis*) and Thalassianthidae (*Cryptodendrum adhaesivum*) (see Fig. 21.3 for some of these species). Like most sea anemones, these hosts have a well-developed defensive system based on nematocysts within the ectoderm of the tentacles. The fish swim around among the tentacles despite this hazardous feature and benefit from the protection from larger predators that would be harmed by the tentacles. The anemones themselves have also been found to benefit by protection from predators due to the defensive activities of the anemonefishes and several other ecological advantages have been documented.

At least 28 fish species are involved in anemone associations throughout the Indo-Pacific, all from the family Pomacentridae and most from the genus *Amphiprion* (see Figs 21.3B–E, 21.4, 21.5). Some anemonefish are quite host specific, but most can occur in many hosts. A hierarchy exists among the fish species themselves, which means that certain species are much more likely than others to successfully colonise and monopolise the available hosts. Other factors too, such as habitat and geography, play a role in the availability and likelihood of settlement in a host by a particular fish species. The fish species differ in the way they approach new hosts, but all have some mechanism for developing a tolerance to the sting of the anemone via a mucous surface coating developed following gradual contact with the tentacles.

Guests

There are also numerous associations of sea anemones with other host organisms, where again the anemone's stinging capabilities provide some protection for the host and the host gives the anemone access to resources. One thing that is difficult for anemones (at least after the larval phase) is moving around, and some of them avoid this problem by living in association with motile animals such as slow-moving molluscs or mollusc shells inhabited by hermit crabs.

laboratory, to note external morphological features using a microscope for small specimens. For preservation and histology, the animal should be relaxed before being fixed in 4% formaldehyde and then 75% alcohol. The most popular relaxant is magnesium chloride, added in very small amounts up to a concentration of ~1.8% over a period of 1–3 hours (depending on the size and response of the animal), until the animal is unresponsive to touch. Permits are required to collect (see Chapter 14).

Order Zoantharia ('Zoanthids')

These are mostly colonial forms (except for species of the large tropical free living zoanthid genus *Sphenopus*), in which the polyps lack a pedal disc and are joined by coenenchyme, which often contains a thick mesoglea (Fig. 21.8A–F). In one group of tropical zoanthids, including *Palythoa* and *Parapalythoa*, this coenenchyme incorporates sand and other material within a mesogleal matrix, so that the columns of the polyps are not visible. The number of valid zoanthid species known to be present on the GBR is around 13, but revisions and new studies on zoanthids from the Indo-Pacific (including genetic studies) may increase this number, and clarify the identity of species representatives of many genera. Identification of genera is often possible from external features, but even this needs to be confirmed by internal observation in many cases (e.g. the genus *Zoanthus* has a divided marginal sphincter). Confirmation of some identifications requires histological sectioning. Much progress is being made in zoanthid taxonomy and natural history, and many new publications on this group can be expected over the coming years.

Anatomy

The column wall is without special external structures as seen in Actiniaria. The oral disc has unbranched tentacles, which occur in an exocoelic

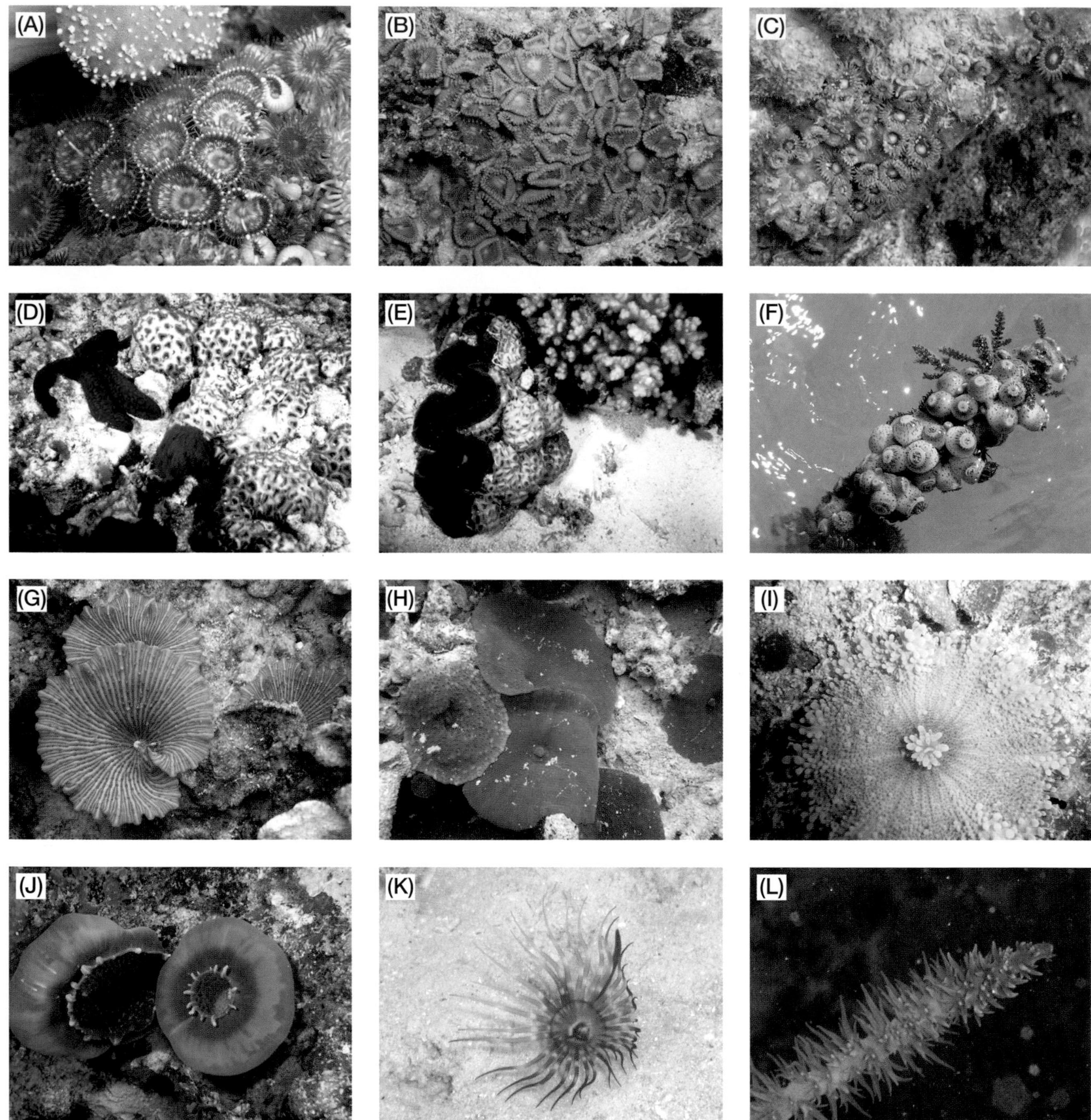

Fig. 21.8. Zoantharia: (A) *Protopalythoa* sp.; **(B)** *Palythoa heliodiscus*; **(C)** *Zoanthus* cf. *vietnamensis*; **(D)** *Palythoa tuberculosa* on reef flat; **(E)** *Palythoa tuberculosa* on giant clam; **(F)** *Acrozoanthus australiae* (note open polyps below water line); **Corallimorpharia: (G)** *Discosoma* sp.; **(H)** *Discosoma* sp.; **(I)** *Ricordia* sp.; **(J)** *Amplexidiscus fenestrafer*. **Ceriantharia: (K)** *Cerianthus* sp. or *Arachnanthus* sp. **Antipatharia: (L)** Black coral *Antipathes* sp. (Photos: *A, B, I, J, K, L*, P. Muir; *C, F, G, H*, A. Crowther; *D, E*, GBRMPA.)

and an endocoelic cycle (these cycles often appearing as a single marginal ring). The mouth is usually slit-like and has a single siphonoglyph, which is ventral. The arrangement of septa (akin to mesenteries of Actiniaria) of zoanthids is unique in living Anthozoa and new septa form only to either side of the ventral septal pair. In general, each pair of septa has one large septum (macrocneme, extending to the actinopharynx) and a small septum (microcneme), except for the dorsal directives, which are both microcnemes, and the ventral directives, which are both macrocnemes. This is called the 'brachycnemic' arrangement (as in Fig. 21.1H), but a few zoanthids also have another complete pair of macrocnemes on each side (the 'macrocnemic' arrangement).

Life history and ecology

Colonies of zoanthids may grow attached to the substratum by a broad colony base (e.g. in *Palythoa* species) and many are epizoic, often using specific animal species, worm tubes (e.g. *Acrozoanthus australis*), sponges, gorgonians and numerous other organisms as an attachment surface. Many species host symbiotic dinoflagellates (zooxanthellae) in the endodermal tissues. Sexual reproduction probably occurs in all species and follows similar patterns to those of corals and sea anemones. At least one species has been seen to spawn during the coral mass spawning event on the GBR. Zoanthids also form clones by division of colonies and large areas of reef flat may contain representatives of only a few genotypes because of this.

The intertidal reef flat often appears to be a preferred habitat for zoanthids, probably due to the rapid growth and cloning ability of the genera *Palythoa* and *Protopalythoa* that occur here and are particularly obvious a few months after damage to corals and other organisms, such as that caused by cyclones. Species of *Protopalythoa* and *Zoanthus* occur in calmer sublittoral and back-reef areas. Zoanthids can be very toxic, their tissues containing a toxin known as palytoxin. For this reason they are not as popular as corallimorphs for aquarium culture.

Order Corallimorpharia ('jewel anemones' or 'corallimorphs')

Usually grouped with sea anemones, this relatively small order of around 40 named species is sometimes regarded as 'skeleton-free corals', although the close relationship with Scleractinia is questioned by recent genetic studies. They occur as clones of polyps connected by stolons (Fig. 21.8G–J). The polyps have a broad, thin oral disc and short, featureless column. The mouth protrudes from the centre of the oral disc. There is a variable number of short tentacles, usually occurring throughout the oral disc, but sometimes in rows and sometimes there are specialised tentacles (bearing nematocyst batteries in rounded tips). The nematocyst composition (cnidome) is less complex than in Actiniaria, but some corallimorph nematocysts are very large. Corallimorphs are often very brightly coloured and a single species may occur in many colour morphs. There are probably around 20 species on the GBR, although further study may discover more.

Corallimorphs may be non-zooxanthellate (family Corallimorphidae) or zooxanthellate (Discosomatidae and Ricordeidae). Many are formidable predators. The large discosomatid *Amplexidiscus fenestrafer* has been observed to feed by enveloping small fish in its oral disc. Other discosomatids may be able to form monocultures by defending territory using nematocysts stored in the tips of the marginal tentacles. They are mostly subtidal and many avoid bright light. They are generally very tolerant of environmental conditions and are popular as aquarium subjects for this reason, as well as their attractive appearance. Histological sectioning of these disc-shaped animals is very difficult, and external features are regarded as very important for identification.

Order Ceriantharia ('tube anemones' or 'cerianthids')

These animals occur within soft sediments with only the extended tentacles and a little of the long, soft column visible (Fig. 21.8K). The mouth is usually hidden by a crown of short tentacles and this is surrounded by much longer tentacles that may extend in the direction of water flow. The individual produces and occupies a flexible tube that may be up to 1 metre long. They are identified by dissection and identification of internal features.

Cerianthids occur from the intertidal to the deep sea, and they are quite frequently encountered on reef lagoons and the off-reef sea floor. Most feed at night, retracting (at least partially) into their tubes by day. Associates from several phyla, including lophophorates, may occur in the tube.

Order Antipatharia ('black corals')

These are colonial anthozoans in which the polyps occur on the outside of an axial skeleton of hard, black, proteinaceous material that bears tiny thorn-like projections (Fig. 21.8I). The polyps are non-retractile and bear only six tentacles. Two main colonial structures occur: single whip-like forms that may be coiled but do not branch (genus *Cirripathes*), and bushy forms, in which a thick central trunk divides into increasingly smaller branches and eventually very thin branchlets (genus *Antipathes*). While *Cirripathes* are common on the GBR, the large tree-like forms mostly occur out of scuba diving range.

There is a long history of exploitation of black coral skeleton for jewellery. Colonies large enough to have formed the desired dense, shiny black skeleton at their bases are believed to be very old and usually occur far deeper than 20 m: these are sometimes found on Australian beaches after heavy weather. All international trade in black coral, even in the form of the final product, is subject to regulation under the Convention for International Trade in Endangered Species (CITES).

Further reading

Burnett WJ, Benzie JA, Beardmore JA, Ryland JS (1997) Zoanthids (Anthozoa, Hexacorallia) from the Great Barrier Reef and Torres Strait, Australia: systematics, evolution and a key to species. *Coral Reefs* **16**, 55–68. doi:10.1007/s003380050060

Carlgren O (1949) A survey of the Ptychodactiaria, Corallimorpharia and Actiniaria. *Kungliga Svenska Vetenskaps-Akademiens Handlingar* **1**, 1–121.

Erhardt H, Knop D (2005) *Corals Indo-Pacific Field Guide*. IKAN, Undterwasserarchiv, Frankfurt, Germany.

Fautin DG (2006) *Hexacorallians of the World*. The University of Kansas, Lawrence, KS, USA, <http://hercules.kgs.ku.edu/hexacoral/ anemone2/index.cfm>.

Fautin DG, Allen GR (1997) *Anemone Fishes and their Host Sea Anemones*. Western Australian Museum, Perth WA.

Fosså SA, Nielsen AJ (1998) *The Modern Coral Reef Aquarium. Vol. 2*. Birgit Schmettkamp Verlag, Bornheim, Germany.

Napier WR, Wallace CC (2018). *Acrozoanthus australiae* (family *Zoanthidae*). Website. Department of Environment and Energy, Canberra, <www.environment.gov.au/cgi-bin/species-bank/sbank-treatment2.pl?id=81832>.

Saville-Kent W (1893) *The Great Barrier Reef of Australia: Its Products and Potentialities*. W. H. Allen & Co., Waterloo Place, London, UK.

Scott A, Harrison PL (2005) Synchronous spawning of host sea anemones. *Coral Reefs* **24**, 208. doi:10.1007/s00338-005-0488-0

Sinniger F, Montoya-Burgos JIP, Chevaldonne´ P, Pawlowski J (2005) Phylogeny of the order Zoantharia (Anthozoa, Hexacorallia) based on the mitochondrial ribosomal genes. *Marine Biology* **147**, 1121–1128. doi:10.1007/s00227-005-0016-3

van der Land J (Ed.) (2008) *UNESCO-IOC Register of Marine Organisms (URMO)*. Website. Netherlands Centre for Biodiversity Naturalis, Leiden, Netherlands, <http://www.marinespecies.org/urmo>.

Wagner D, Luck DG, Toonen RJ (2012) The biology and ecology of black corals (Cnidaria: Anthozoa: Hexacorallia: Antipatharia). *Advances in Marine Biology* **63**, 67–132. doi:10.1016/B978-0-12-394282-1.00002-8

Wallace CC, Richards Z (2018) Species Bank entries for *Bolocerоides mcmurrichi, Cryptodendrum adhaesivum, Entacmaea quadricolor, Heteractis magnifica, Heterodactyla hemprichii, Macrodactyla doreensis, Stichodactyla haddoni* and *Triactis producta*. Website. Department of Environment and Energy, Canberra, <http://www.environment.gov.au/biodiversity/abrs/online-resources/species-bank/records.html>.

WoRMS (2004) *World Register of Marine Species: Hexacorallia*. Website. World Register of Marine Species, at Netherlands Centre for Biodiversity Naturalis, Leiden, Netherlands, <http://www.marinespecies.org/aphia.php?p=taxdetails&id=1340>.

22

Hexacorals 2: reef-building or hard corals (Scleractinia)

C. C. Wallace

Overview

The order Scleractinia, like other members of class Anthozoa, are Cnidaria that have a life cycle involving a short larval stage and a long polyp phase. The polyp lays down an external skeleton by the activity of specialised cells within the external tissue layer at its base: the 'calicoblastic ectoderm'. Most corals on coral reefs form colonies of interconnected polyps by growth and division from an original polyp. The presence of symbiotic dinoflagellates (*Symbiodinium* or 'zooxanthellae') in the tissues of these colonies enhances production of skeleton, and these rapid-growing corals are known as the 'hermatypic', 'zooxanthellate', or 'reef-building' corals. They have played the major role in building the present fabric of the Great Barrier Reef (GBR) over some 6000 to 8000 years and previous reef formations over many millennia, as the carbonate structures of the Queensland continental shelf and Coral Sea testify (see Chapters 2 and 3). Living corals also provide the three-dimensional structure in and around which the other creatures of the reef spend their lives. Their health is vital to the wellbeing of innumerable organisms and ultimately to the continued existence of the GBR itself.

More than 450 species of corals have been recorded from the waters of the GBR, eastern coastal Australia, and continental and Coral Sea islands. These complex invertebrate animals with single-celled plant symbionts are currently the subject of intense focus in relation to the impacts of global climate change and local pollution and exploitation processes, both on the GBR and throughout the world (see Chapter 12). Since the first edition of this book, new technologies – especially for molecular study of the coral genome and microbiome, analysis of large databases and remote exploration of the sea bed – have changed profoundly the way scientists are able to interpret species boundaries, evolutionary patterns and distributions of reef-building corals. While much of this research is still unfolding, this chapter will incorporate some findings and their implications.

As one very significant example, it has been discovered that a coral colony hosts not only zooxanthellae, but also a myriad of other microorganisms, known as the coral's 'microbiome', while the sum of host plus microorganisms is known as the 'holobiont'. All these elements can be genetically assessed and used to examine, in a far more detailed way than ever before, topics ranging from health and disease of the colony to the evolutionary pathway it has followed to reach its current status. It is important to also note that azooxanthellate corals (those without zooxanthellae) occur in great diversity right around Australia (more than 250 species recorded to date), at depths extending to beyond

1000 m (Cairns 2004). Information about them plays a very important role in the interpretation of evolution of reef-building corals.

Anatomy

Polyps

The basic coral polyp is simple in structure (Fig. 22.1). The column is cylindrical and without adornment, the tentacles hollow and, with few exceptions, simple, although one or two specialised larger tentacles may be developed. The tentacle number is based on a primary cycle of six (as compared with eight in the soft corals), with cycles being added between the preceding cycles. The oral disc bears a round or oval mouth without siphonoglyphs (the flagellated furrows found in anemones and zoanthids), and this leads to a pharynx and the paired mesenteries. In the process of replicating polyps to form a colony, division occurs either within the polyps or external to them. When division is incomplete, polyps may exist in continuous connection (as in the brain corals). In general, the coral colony can be thought of as the sum of its polyps, with processes such as feeding and gonad formation being performed in individual polyps but contributing to the general nutrition, growth and sexual propagation of the colony.

Skeleton

Because of the intimate relationship between soft body parts and exoskeleton in corals, the skeleton has long been regarded as a proxy for the entire animal. A vast terminology has developed to cover the various skeletal components and their configurations. These terms are used so often in descriptions of corals that it is very useful to master the basics of them (Fig. 22.2). The skeleton formed by

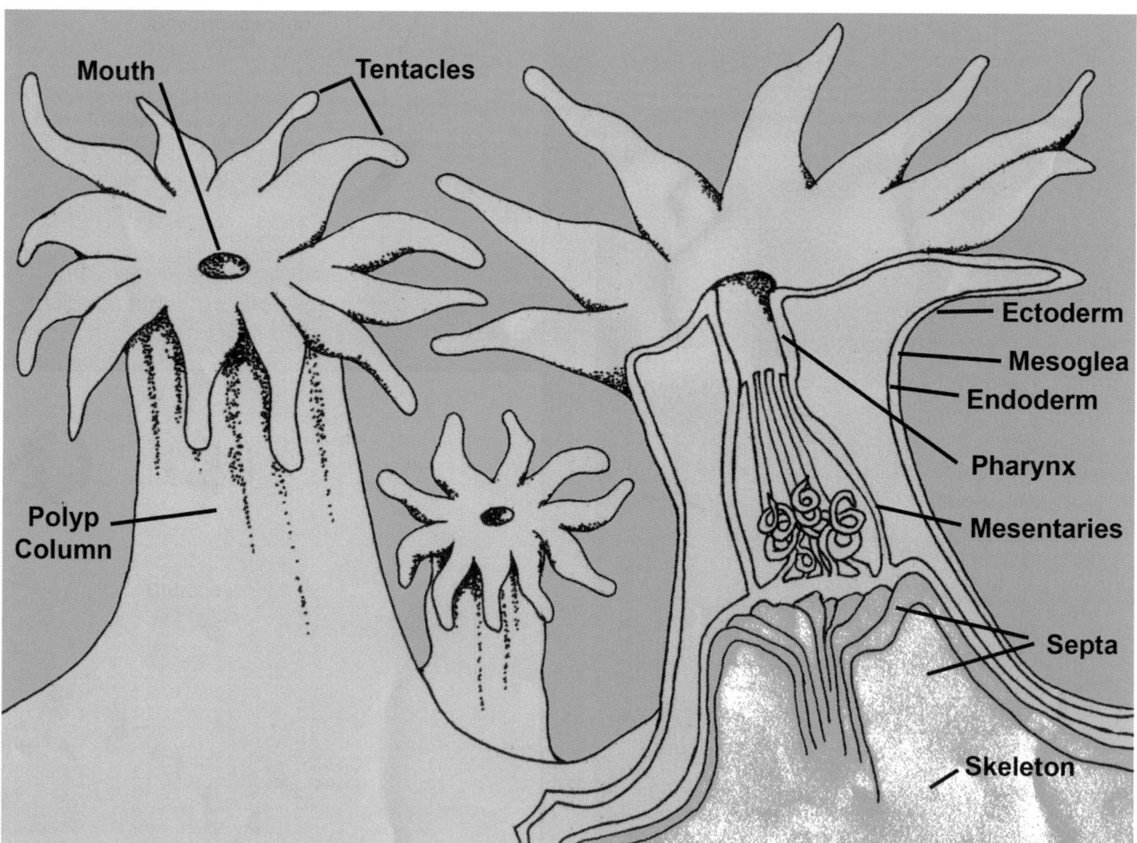

Fig. 22.1. Relationship of tissue to polyp in an idealised coral colony. Entire polyp on left, cut-away polyp, exposing skeleton, on right (Source: Adapted from Wallace and Aw 2000).

the colony or part of the colony is the **corallum** and that formed by the polyp is a **corallite**. This has dividing elements called **septa** (singular septum), which support the mesenteries and often a central element, the **columella**, which is also part of the support system of the polyp. Septa may extend outside the polyp as **costae** (singular costa). Interconnecting skeleton is referred to as the **coenosteum**. Supporting structures below and between the polyps are **dissepiments** and these represent layers

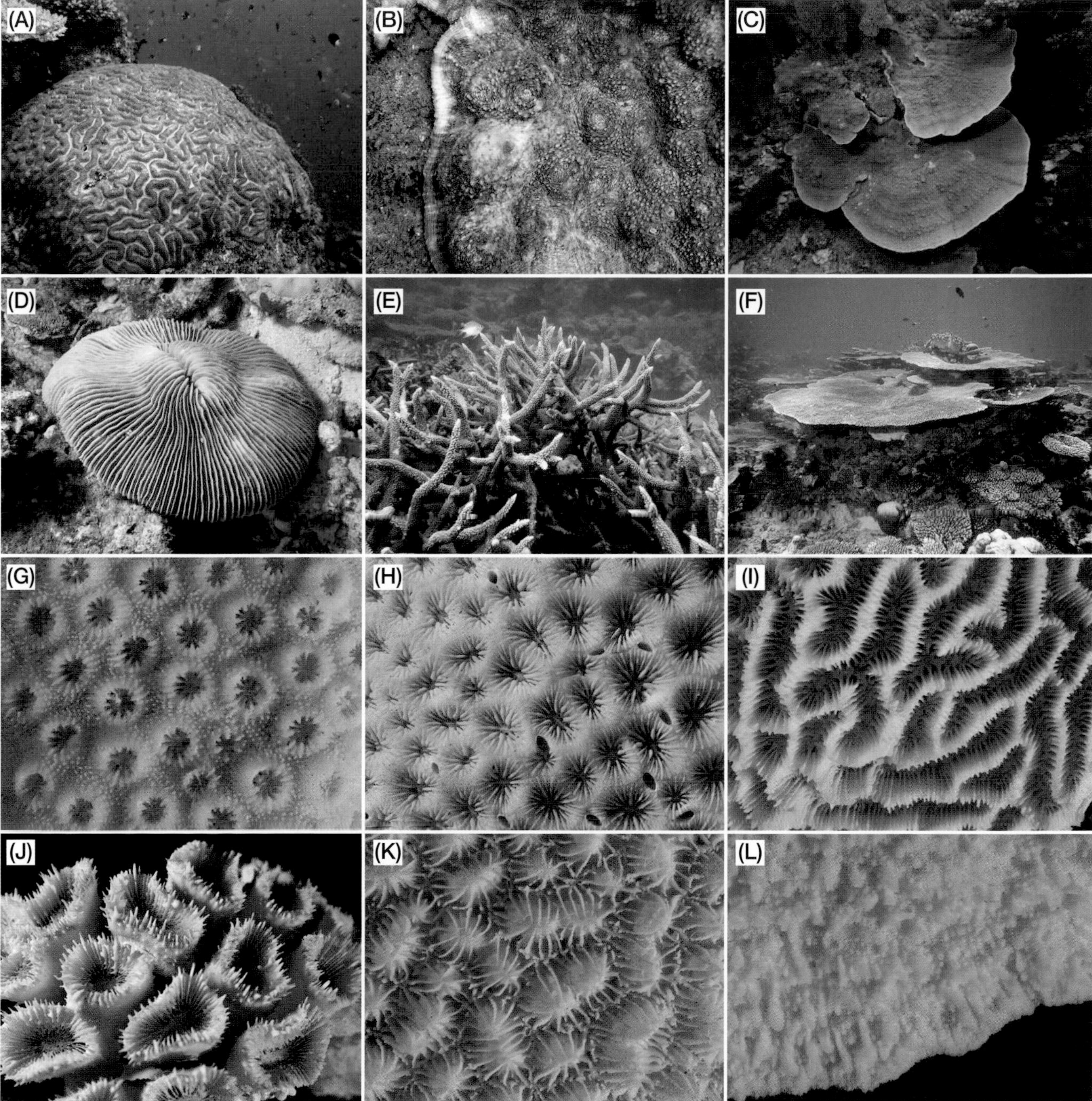

Fig. 22.2. Structures and terms used in the description of reef corals. Terms relating to overall colony shape: **(A)** massive; **(B)** encrusting; **(C)** plating; **(D)** solitary free-living; **(E)** branching arborescent; **(F)** branching tabular. Terms relating to forms of corallite arrangement in the corallum: **(G)** plocoid; **(H)** ceroid; **(I)** meandroid; **(J)** phaceloid; **(K)** hydnophoroid; **(L)** plocoid with coenosteum features (Photos: P. Muir).

of skeleton laid down as the polyps and their connecting tissues move upwards in the colony as they (and the colony) grow. All these elements may be ornamented in various ways and even the basic process by which they are laid down affects their appearance and forms the traditional basis of the classification of families.

Life history and ecology

All corals have the ability to reproduce sexually by producing male and female gametes for fertilisation and subsequent development of ciliated larvae known as planula larvae (or planulae), which settle on the reef surface, transform into individual polyps and begin to divide and lay down skeleton to form the colony. Individual polyps in a colony may all be male or female (the gonochoric or dioecious condition) or both sexes can be developed within each polyp (hermaphroditic or monoecious condition). There are two very different modes of production of larvae, usually characteristic of a genus or family. Sperm may be released into the water column by the polyps of one colony to find their way into the body cavity of other colonies, where eggs are fertilised and develop in or on the polyps (Figs 22.3A). This is the brooding condition, with corals having this life cycle being referred to as 'brooders'; until the 1980s, this was thought to be the only type of life cycle in corals (see Box 22.1). In an alternative life cycle, shown by corals referred to as 'broadcast spawners', eggs and sperm are released into the water column for external fertilisation and development (Figs 22.3B, 22.4). Because gametes from the same colony do not usually self-fertilise, this second mode requires many colonies of the same species to release gametes at the same time. Going beyond this, the spawning times of species in many families and genera may coincide or overlap to some extent during 'mass spawning' events at specific times of the year (see Box 22.1).

The larvae of brooding corals are released fully developed and ready to settle on the reef, whereas embryogenesis and development of larvae by broadcast-spawning corals requires a developmental period of up to 10 days in the sea before the planula is ready to settle. Dispersal of larvae outside the home reef is thus far more likely in broadcasters than brooders. Some corals are also subject to asexual reproduction by breakage and redistribution of branches or formation of polyp balls or polyp bailout and this may contribute considerably to population size and biomass.

Feeding and nutrition within a coral colony generally involve two sources: (1) acquisition of food by individual polyps, digestion in the polyp's gut and distribution of digestate through the colony via the coelenteric connections between polyps; and (2) photosynthetic activity of the symbiotic dinoflagellates that live in the polyp's tissues. Coral polyps are basically carnivorous, with their food source being to some extent determined by polyp, mouth and tentacle size. Food sources include planktonic organisms, incidental detritus and mucus with contained microorganisms. Corals themselves form the prey of some animals, notably the crown-of-thorns sea star *Acanthaster* cf. *solaris*, which sometimes occurs in 'outbreak' numbers, causing severe loss of living corals from large tracts of reef. Other predators include molluscs such as *Drupella* sp., which feed as aggregations, and numerous fish ranging from small butterflyfish, which feed on individual polyps, to large grazers such as parrotfish, which ingest whole chunks of coral (Cole *et al.* 2008).

Corals are also known to suffer from a wide range of infectious and non-infectious diseases. Although there is a long literature of documenting the impact of such diseases on coral populations around the world, microbiome studies have facilitated further understanding of pathogens and the stress factors that can be involved in outbreaks of some diseases. In the past, the GBR has been relatively less impacted by coral diseases than, in particular, the Caribbean. However, there is a strong and serious interest in understanding and monitoring for disease presence and impact in GBR waters (Willis *et al.* 2004).

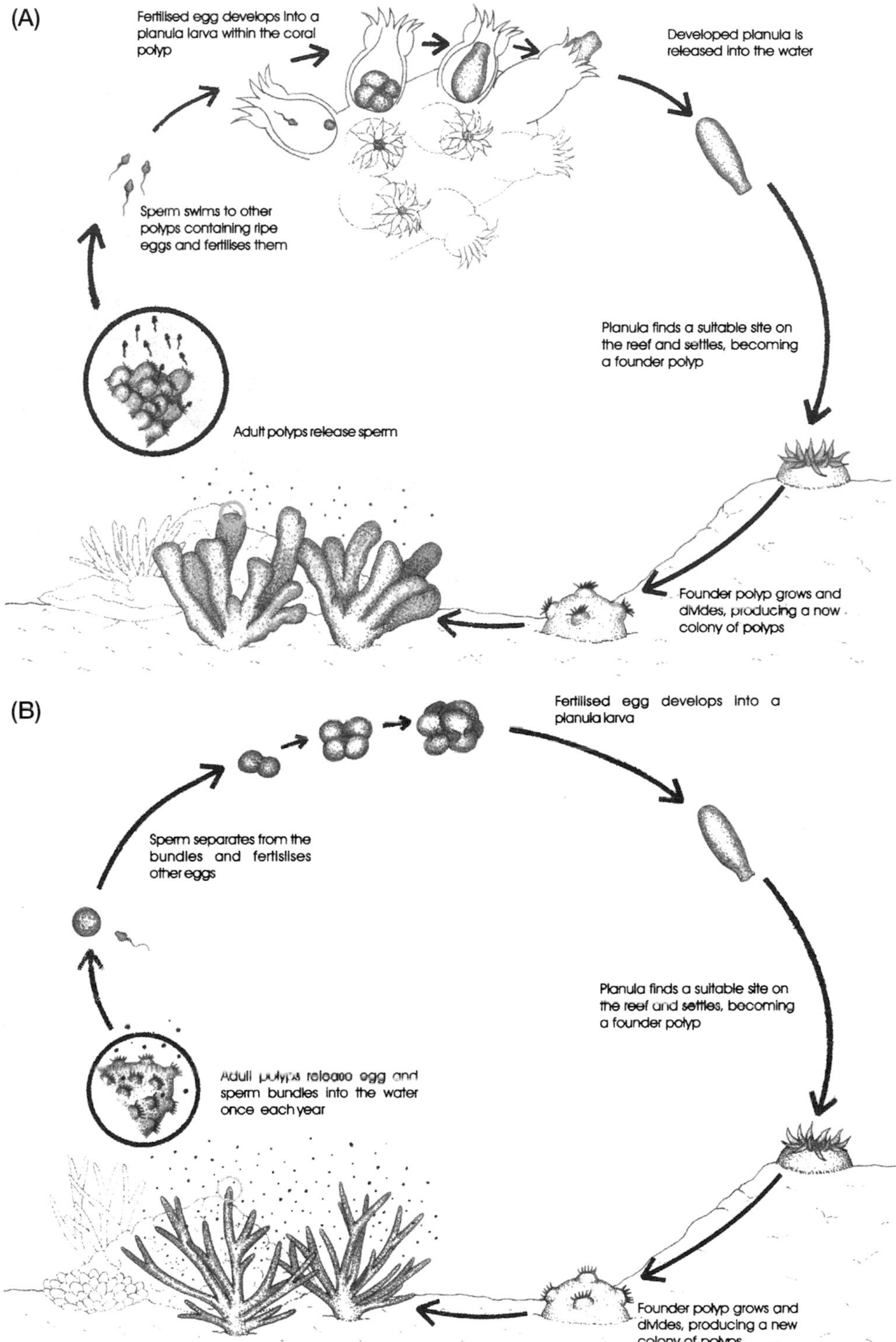
(A)
Fertilised egg develops into a planula larva within the coral polyp
Developed planula is released into the water
Sperm swims to other polyps containing ripe eggs and fertilises them
Planula finds a suitable site on the reef and settles, becoming a founder polyp
Adult polyps release sperm
Founder polyp grows and divides, producing a new colony of polyps
(B)
Fertilised egg develops into a planula larva
Sperm separates from the bundles and fertislises other eggs
Planula finds a suitable site on the reef and settles, becoming a founder polyp
Adult polyps release egg and sperm bundles into the water once each year
Founder polyp grows and divides, producing a new colony of polyps

Fig. 22.3. Reproductive cycles in corals: **(A)** cycle in brooding species; **(B)** cycle in hermaphroditic broadcast-spawning species. Other variations such as broadcast spawning in gonochoric species, occur in some corals (Image adapted from Wallace and Aw 2000).

Box 22.1. The coral mass spawning phenomenon

Corals from the majority of species on the GBR develop eggs and sperm once a year and release these for external fertilisation in multispecies spawning events. These occur at night during the week following the full moon in October to February, with November being the major spawning month. The discovery of this extraordinary phenomenon in Australia in the 1980s was one of the major advances in coral reef science of the 20th century. Mass spawning, as it has been dubbed, presents a paradox: on the one hand it ensures ample reproductive material for the cross-fertilisation required to ensure development of the next generation of each species, at the same time saturating the food supply of predators for a very short period; on the other, it also provides opportunities for mismatches, in which sperm could fertilise eggs of the wrong species to form hybrids or non-viable embryos. The discovery of mass spawning led to revision of ecological and evolutionary models and predictions and provided unprecedented opportunities for research, management planning and experimentation using gametes. Its impact has been worldwide and the phenomenon has now been reported from most countries bearing coral reefs.

Fig. 22.4. Coral colonies spawning during the mass coral spawning event on the Great Barrier Reef: **(A)** *Acropora tenuis* spawning egg/sperm bundles for external fertilisation (note gametes on the water surface); **(B)** female *Fungia* species spawning eggs (these will be externally fertilised by sperm released by a male); **(C)** *Galaxea fascicularis* spawning large egg/sperm bundles for external fertilisation (Photos: GBRMPA).

Species concepts

Although the skeleton has always provided the basis of coral species description, the ideal of a biological species concept for corals has always been assumed and vigorously examined whenever new data became available. For example, when gametes from known parents became available following the discovery of mass spawning, work immediately began on looking for breeding boundaries between species. These boundaries would theoretically apply at any stage of the reproductive process, such as timing of release of the eggs and sperm, activation of chemical or physical barriers to fertilisation or embryo development, then settlement of a planula to become a viable colony. Over some decades now, detailed laboratory crosses have been conducted on GBR corals, particularly those involved in mass spawning events, to test for compatibility between individuals and species. While strong breeding barriers between species were seen in many cases, these experiments also showed that hybrids were regularly formed, often leading to fully settled polyps. Hence there remains a real possibility that hybridisation plays a role in the diversification of coral communities.

Less frequently, actual hybrids have been observed on the reef. The three morphological species of *Acropora* (staghorn corals) traditionally recognised in the Caribbean have been demonstrated to be actually two species plus their first generation hybrid, with the hybrid being found only where both parent species are nearby. The greater diversity of species in Indo-Pacific waters makes it difficult to detect such a clear hybrid system as this on the GBR. Genetic studies frequently indicate more complex interrelationships involving multiple species (Richards *et al.* 2013). Some research has detected syngameons involving several well known morphologically defined species, which share genetic material from ancient sources as well as from ongoing hybridisations.

Paradoxically, it has also been found that unseen genetic boundaries can be present within apparently single species populations. Molecular analyses indicate that these may consist of two to many 'genetic lineages' or 'cryptic lineages' that can be recognised in various combinations in populations in different regional locations. Sometimes these have been regarded as cryptic species, but usually some level of genetic exchange can be detected among the lineages. Demonstrating these lineage groups and determining their geographic distribution is a slow process, but is proving productive in terms of indicating populations that may be rapidly evolving in response to environmental change. Typically, populations are studied along a gradient (e.g. the Kuroshio current in the NW Pacific) or across large parts of the Indo-Pacific. Sequencing and lineage-sorting programs are run on samples taken from the regions to detect the number and type of lineage groups present. These studies show that, although numerous cryptic lineages may occur together in tropical locations, often a single genotype dominates at the margins of distribution of the species. It is hypothesised that these margins or edge-zones might provide the opportunity for evolutionary innovation and the emergence of a new species from the marginal genotype. This is of particular interest in the light of global climate change.

Thus it is emerging that the traditional morphological definition of a species does not cover the range of ongoing evolutionary changes revealed by genetics. In contrast to this, fossil discoveries, some made since the previous edition of this book, show that many coral species living today have been in existence for up to 18–20 million years, without any noticeable change in skeletal morphology, micromorphology or microstructure (see Box 22.2). For example, colonies of 12 extant species of the coral genus *Acropora* (see later) were found in early Miocene deposits 17.9 million years old in central Borneo. The preservation of these fossils was such that they could be compared directly with freshly sampled modern colonies using electron microscope imaging and other standard comparisons. The resolution of this paradox over the coming years should make for very interesting developments in our understanding of coral species evolution, as well as the process of diversification of corals in the Indo-Pacific (Santodomingo *et al.* 2015).

Classification and identification

This section looks at some examples of the corals occurring on the GBR, and the roles they play in the ecology of the reef, based on their extraordinary variety of form. Guidance for identification can be found in numerous publications, internet databases and underwater identification tools. When accurate identification to species level is required, especially for those corals with tiny polyps, further microscopic inspection may be necessary. For this purpose, registered coral collections may be consulted in museums and some universities. Additionally, coral identification workshops are held by experts in association with ACRS meetings and elsewhere. This section comes with a caveat regarding future change (see Box 22.2).

Major coral families and their significance in coral reefs

While all corals contribute in some way to the ecology of the GBR, certain families and/or genera dominate by making a major contribution to the reef framework, coral cover, species diversity, age structure, viability, coral/organism interactions and many other aspects of the overall operation of the reef. These groups tend to be the focus of the most research and to provide much of the information relevant to understanding and managing coral reefs in the face of a multiplicity of demands and threats to their persistence as healthy ecosystems. Presented here is a discussion of some major families represented on the GBR, re-arranged into their

Box 22.2. Renaissance in coral taxonomy and nomenclature

By the end of the 20th century, a seemingly definitive and consistent description of the coral species of the world, including variability, distributions and life history information, was available, first in book form and later as internet resources. These developments presaged a new era in which important issues of the environment and its management could be examined, without the hindrance of unresolved nomenclature.

Unstoppable change, however, was afoot. A revolution in higher level classification of corals is being led by molecular biology (genetics), which is also being used to study boundaries and relationships between species, population level genetic diversity and even the identification of individuals and species. Early findings indicated that much of coral taxonomy was at odds with evolutionary patterns, as shown by phylogenetic analyses using various genetic markers. Expansion into genomic studies, in which large sequences from the coral genome could be analysed and compared, dramatically altered the way that the identity and relationships of coral species and higher level taxa could be assessed. At the same time, enhanced microscopic technology was allowing palaeontologists to distinguish more clearly than ever before small microscopic structures, such as tiny teeth on the septae or costae, referred to as 'micromorphology' of the coral, and internal arrangement of aragonite fibres within the skeleton, now referred to as 'microstructure'. Each group of scientists (molecular biologist and palaeontologist) found itself in need of confirmation of findings by the other, as well as new ground-truthing from field biologists.

So at the beginning of this century, practitioners from at least three streams of evidence – palaeontology, biology (life history and ecology) and molecular biology – began to collaborate and publish together on what has become known as 'integrative taxonomy' – describing and comparing lineages and taxa using a combination of molecular, environmental and palaeontological characters. Since the previous edition of this book, many significant integrative studies have proposed major changes in the phylogeny and higher level classification of corals. The blossoming of integrative taxonomy continues at pace, but some important published findings to date are as follows:

- A profound distinction has been found between Atlantic and Pacific ocean corals, with some shared genera and families having to be fully revised.
- Many previously well-established families are paraphyletic (mixed origins) and nomenclatural changes have been made to correct this.
- Modern corals have evolved along two major clades, currently named Robust and Complex.
- Azooxanthellate deep-water corals do not form a separate group, but are found throughout most family clades.
- Some individual species have been attributed to the wrong genus or family and these are being re-allocated.

Although very distinguished, well-accepted works describing these name changes have been published (see examples in 'Further reading'), some findings remain contested or have not yet been incorporated into general identification systems for the GBR. The World Register of Marine Species (WoRMS), particularly its facility (WoRMS taxon match (http://www.marinespecies.org/aphia.php?p = match), is a useful online service that aims to keep abreast of changes as they are published. These changes illustrate the way science progresses. At the grass-roots level of identifying species, genera and families on the reef using the available published guides, they currently add a little difficulty for the reef student or researcher. Most will be incorporated into the general nomenclatural system over time, without too much disruption to the work of caring for the natural environment of the GBR.

membership of the two modern clades, Complex and Robust (see Box 22.2).

Complex clade

Family Acroporidae

This family includes the two most diverse genera of living zooxanthellate corals, *Acropora* and *Montipora* and also *Isopora*, as well as three other extant genera (Fig. 22.5 A–F). These genera often dominate the outer reef flat and reef front, down to around 15 m, after which more species from the less prominent families can be found. *Acropora* (known as the staghorn coral genus), has a unique branching mode in which one type of corallite (the axial) forms the centre of the branch and buds off many more of a second kind (the radials) at its tip as it extends. This has led to numerous colony shape possibilities. Additionally, growth of the colony may be symmetrical and determinate (slowing down as a determined shape is reached) or asymmetrical and indeterminate, the colony growing to fill available space on the reef.

Acropora has over 125 species worldwide and over 70 on the GBR. As many as 40 *Acropora* species may occur in 500 m^2 of reef front. Probably because of its great abundance, *Acropora* suffers the most from storm and cyclone damage, coral diseases, grazing by predators such as the crown-of-thorns sea star *Acanthaster* cf. *solaris* and aggregations of *Drupella* gastropods, and coral bleaching due to elevated sea-surface temperatures and other factors (see Chapter 12). *Acropora*, *Isopora* and *Montipora* are usually abundant among recruiting corals in natural and experimental situations and some species also recruit asexually by growth of fragments after storm damage. The genus *Alveopora*, previously in family Poritidae, is now included in Acroporidae because it is genetically distant from family Poritidae but morphologically close to Acroporidae in its simple skeleton formed by a single kind of skeletal element, rather than two kinds as in the Poritidae. Species of *Isopora* and *Alveopora* are brooders, whereas all other Acroporidae are broadcast spawners.

Family Poritidae ('golf ball' corals, 'bommie' corals)

Colonies in this family are massive, encrusting, plating or branching (Fig. 22.5G–I). In all cases, corallite walls are formed by upward growth of fine trabecular elements, so that the septa can be seen as growing upwards from the corallite floor, rather than inwards from the wall (e.g. as in Acroporidae). This family contains one of the most remarkable zooxanthellate genera, *Porites*. Although species of *Porites* can exist in branching and plating forms, it is the massive species that provide examples of the longest-lived and largest solid coral forms. Growth in these massive colonies is usually in increments of 1–2 cm per year, with two growth seasons being recorded by a 'dark' and a 'light' (dense and less dense) band. By counting and measuring these bands, it is possible to calculate the age of the colony and witness the history of environmental conditions that the colony has encountered. Colonies of over 700 years old have been identified and these sometimes form the basis of the huge 'bommies' found within lagoons and just outside the reef slope on reefs of the GBR. Not surprisingly, species of *Porites* form the subject of numerous studies of coral growth, vulnerability to environmental conditions, effects of coastal runoff and many other aspects of coral reefs (see Lough 2010 and Chapter 12).

Some situations favour the development of *Porites*-dominated coral assemblages, such as calm, shallow offshore shoals or fringing reefs. *Goniopora* may also be present in these habitats. This genus is most notable for its long-columned polyps that are extended during the day. It is a popular aquarium subject and is apparently tolerant of high turbidity and low-light situations.

Family Agariciidae

This family has several genera, all colonial (Fig. 22.5L). One genus, *Leptoseris*, occurs almost exclusively in low light situations such as overhangs and crevices in the reef as well as at mesophotic depths (below 30 m), with colonies of some species of *Leptoseris* being recorded from below 100 m. Its flat plating form, seen also in some species of other agariciid genera *Pavona* and *Pachyseris*, is ideally suited to the capture of light in the low-light conditions found in this zone. Because of their location, these family members find refuge from the coral bleaching events that cause widespread damage to corals on shallower parts of the reef.

Robust clade

Family Merulinidae (includes genera previously in Faviidae, Pectiniidae and Trachyphlliidae)

The family Faviidae, previously one of the most ubiquitous and recognisable families to scientists, reef lovers and aquarium enthusiasts throughout the world, has now been subsumed as a subfamily (Faviinae) of the Atlantic-only family Mussidae, with all Indo-Pacific members joining the previously small family Merulinidae (Fig. 22.6F–J, 22.7J–L) (Huang *et al.* 2014). In this family, the genera are massive, encrusting, plating or less commonly branching, and corallites have numerous septa, well-developed columellae and walls formed by simple fan systems of calcium carbonate crystals (trabeculae). These corals contribute to the reef by

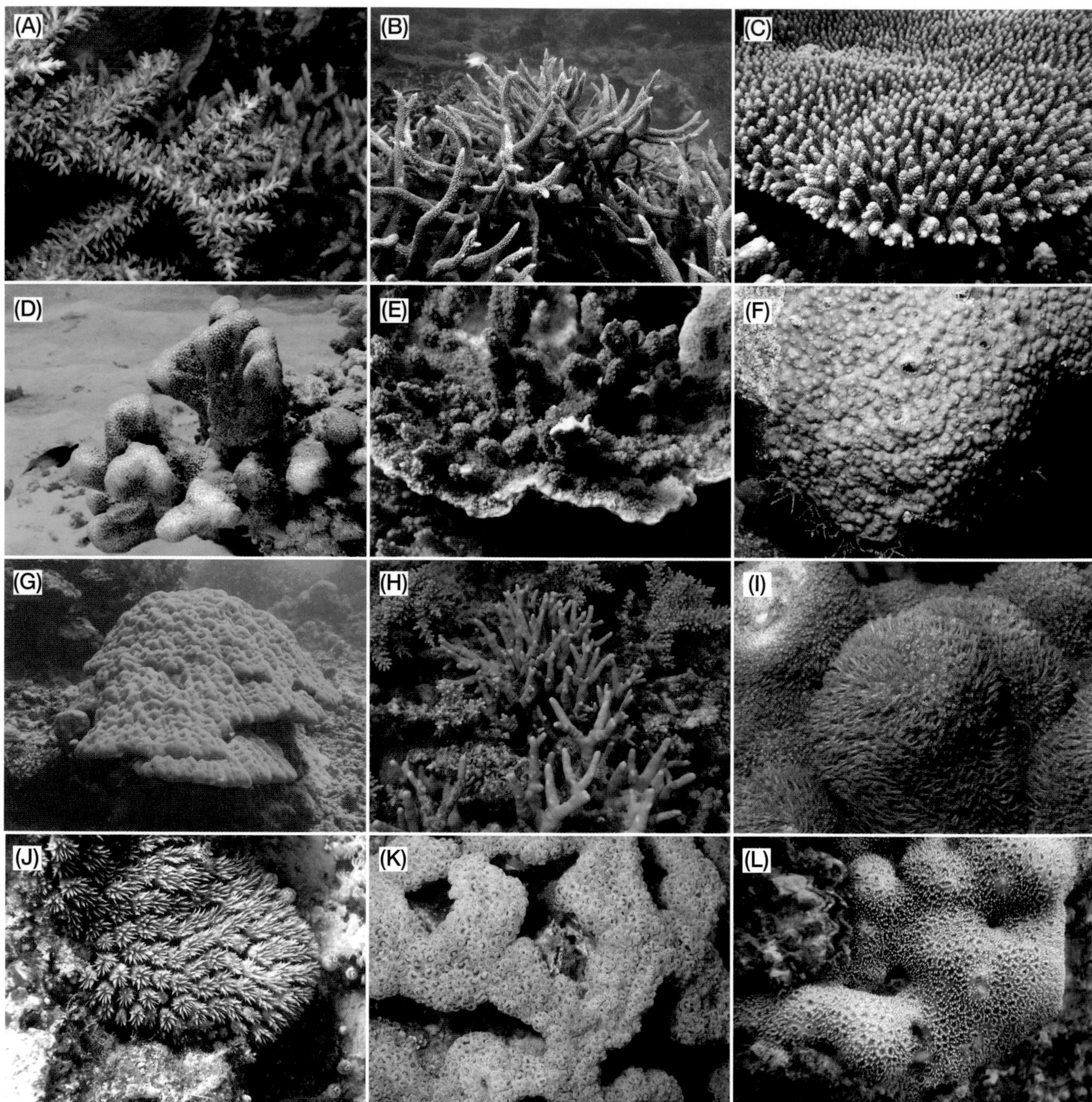

Fig. 22.5. Examples of genera and species from the Complex clade. Family Acroporidae: **(A)** *Acropora echinata;* **(B)** *Acropora muricata;* **(C)** *Acropora hyacinthus;* **(D)** *Isopora palifera;* **(E)** *Montipora hispida;* **(F)** *Astreopora gracilis.* Family Poritidae **(G)** *Porites lutea;* **(H)** *Porites cylindrica;* **(I)** *Goniopora lobata.* Family Euphylliidae: **(J)** *Galaxea astreata;* **(K)** *Euphyllia ancora.* Family Agariciidae **(L)** *Pavona maldivensis.* (Photos: P. Muir.)

their densely constructed skeletons and their wide-ranging environmental tolerances. Genera are distinguished by the type of polyp division (extra- or intratentacular) and the arrangement of corallites (either separated or in contact, or combined in valleys to form the meandroid form that leads to the name 'brain corals'). Species distinctions within genera are based on dimensions of the corallites,

Fig. 22.6. Examples of genera and species from the Complex clade **(A)** and the Robust clade (B–L). Family Dendrophylliidae: **(A)** *Turbinaria mesenterina. Family* Pocilloporidae: **(B)** *Pocillopora damicornis;* **(C)** *Pocillopora eydouxi;* **(D)** *Stylophora pistillata;* **(E)** *Seriatopora hystrix.* Family Merulinidae (includes Indo-Pacific Faviidae): **(F)** *Dipsastraea* (previously *Favia*) *maritima;* **(G)** *Favites halicora;* **(H)** *Coelastrea* (previously *Goniastrea*) *aspera;* **(I)** *Platygyra sinensis.* Family Lobophyliidae: **(J)** *Oxypora lacera.* Family Diploastreidae **(K)** *Diploastrea heliopora.* Family Fungiidae: **(L)** *Fungia fungites* (Photos: All Paul Muir except E, GBRMPA).

features of the septa, costae, coenosteum and columella and number, shape, ornamentation and elevation of the septa above the corallite.

Merulinids can be seen in most reef habitats. They are common on the intertidal reef flat and submerged shallow shoals, and also on fringing reefs along the coastline and continental islands. They can be the dominant coral on fringing reefs, such as the reefs of Moreton Bay at the port of Brisbane (Wallace *et al.* 2009) and on the exposed reef flats on the western side of Magnetic Island, off Townsville. Merulinids form a major component of the coral mass spawning events in eastern Australia, and exhibit several different strategies in the way they release their gametes; for example, gametes may be negatively rather than positively buoyant in some species. The genera *Goniastrea, Coelastrea* and *Leptastrea* appear to be among the most tolerant of all corals to heat stress and exposure to silty conditions. On reef flats, *Platygyra* and *Goniastrea* commonly develop into 'microatolls', with living coral surface confined to the vertical sides of the colony, surrounding a dead top, which sometimes gets colonised by algae or other corals. This morphology comes about when the top of the colony gets regularly exposed to the air and/or sunlight on low tide.

Family Pocilloporidae

In this family the genera have relatively small colonies, a branching mode of growth, simple, small corallites with reduced walls, only two septal cycles and always separated by coenosteum (Fig. 22.6B–E). All genera are brooders and, in some species of *Pocillopora*, planulae can be produced asexually by budding. An asexual form of reproduction known as 'polyp bail-out' is also seen in this group (Fordyce *et al.* 2017). *Pocillopora* is regarded as a 'weedy' genus, its species recruiting soon after catastrophic events and living short lives, with relatively regular monthly release of larvae throughout the year. Early observations of the breeding patterns of genera in this family led to the misconception that all corals release brooded larvae throughout the year, making larvae available for settlement on the reef at all times. This misconception influenced ecological thinking about reef corals until the 1980s, when mass broadcast spawning in corals was discovered (see Box 22.1). The best-known species are *Stylophora pistillata, Pocillopora damicornis* and *Seriatopora histrix*, which have been used for much published experimentation on the effects of water chemistry, light, temperature and other factors on corals. Because of their abundance, compact size and easy access in shallow waters, these corals are also marketed frequently as curios for home décor and as live aquarium subjects, with *Pocillopora* sp. often sold as 'brown-stem coral'.

Family Fungiidae ('mushroom corals')

Most genera and species in this family occur as individual polyps that grow very large instead of multiplying, and that remain free living (unattached to the reef) during adult life (Figs 22.6L, 22.7A–C). These corals contain zooxanthellae but are mostly regarded as non reef-building, or 'ahermatypic', because they generally do not contribute much to the reef structure (though there are notable exceptions when they and their dead predecessors occur in large mounds of thousands of individuals). Sexual reproduction often involves gonochoric gamete development (individuals of separate sexes), followed by simultaneous release of eggs or sperm by many individuals of the same species (see Fig. 22.4B). Polyps settle on the reef to form an attached juvenile coral that may bud to form a 'stack' of cloned individuals: as these grow, they break off the stack and lie free on the reef floor. Generally found in the subtidal parts of reefs, fungiids may occur in great abundance and diversity. The Fungiidae have been studied intensively for several decades, and their nomenclature has been refined, particularly by raising many subgenera of the type genus *Fungia* to the level of genus (Gittenberger *et al.* 2011).

Family Lobophyliidae (previously Indo-Pacific Mussidae ['spiky brain corals'])

Genera in this family have massive or encrusting colonies, with pronounced teeth on the septa and corallite walls formed by complex trabecular growth, which leads to very sturdy corallites, including some of the largest corallites to be seen in colonial corals (Fig. 22.7D–G). Polyp tissues are very fleshy and often very colourful and these corals contribute by their great bulk to the reef structure. They are also very obvious participants in mass spawning events on the GBR, expelling extensive bundles of large and brightly coloured eggs and sperm during the spawning night. The

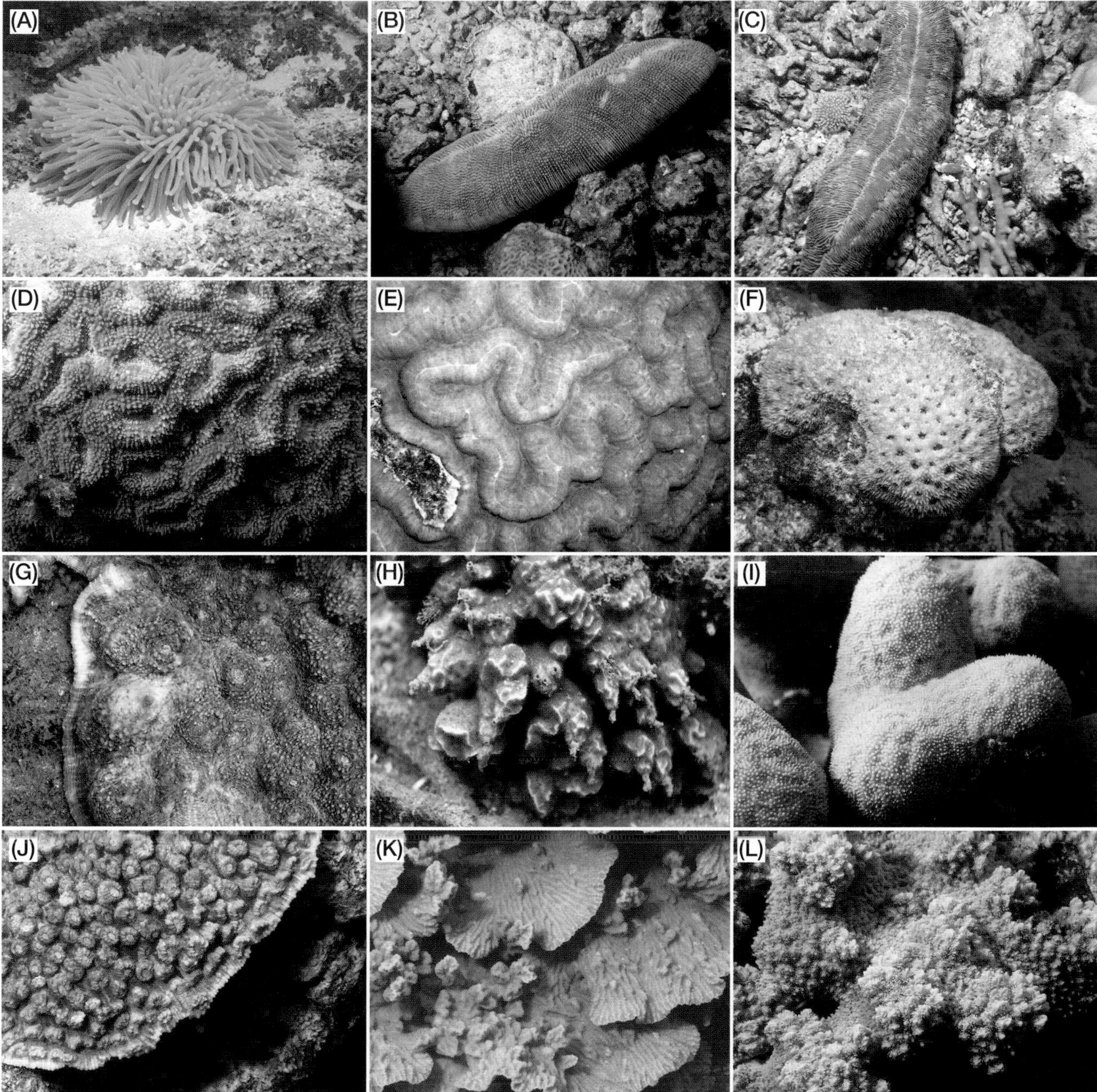

Fig. 22.7. Examples of genera and species from the Robust clade. Family Fungiidae: **(A)** *Heliofungia actiniformis*; **(B)** *Ctenactis echinata*; **(C)** *Herpolitha limax*. Family Lobophylliidae (previously Indo-Pacific Mussidae): **(D)** *Lobophyllia hemprichii*; **(E)** *Lobophyllia* (previously *Symphyllia*) *recta*; **(F)** *Acanthastrea echinata*; **(G)** *Echinophyllia aspera*. Family Psammocoridae (previously Siderastreidae): **(H)** *Psammocora haimeana*. Family Coscinaraeidae (previously Siderastreidae): **(I)** *Coscinaraea exesa*, Family Merulinidae: **(J)** *Mycedium elephantotus*; **(K)** *Merulina ampliata*; **(L)** *Hydnophora exesa* (Photos: P. Muir).

revolution in coral higher taxonomy began with the splitting of the family Mussidae into its Atlantic and Indo-Pacific components. The Indo-Pacific component came to be called Lobophyllidae when the Atlantic Ocean members of Mussidae were shown to be different from the Indo-Pacific members in both genetic and micromorphological features: the family name Mussidae stayed with its Atlantic type genus *Mussa*.

Conclusion

As pointed out in Chapter 9, scleractinian corals are the major habitat-forming organisms on the reef and ultimately the providers of a place to live for most of the other reef organisms. Since the first edition of this book, the doors have opened to new and revolutionary ways of interpreting corals, their relationships, evolution, classification, lifestyles and distributions. The number of workers, from researchers, reef managers and collaborators to people of specific expertise from around the world, willing to apply and enrich this information has never been greater. At the same time, risks to the continuation of healthy coral assemblages on the GBR and in its adjacent waters are greater than ever seen during the time of developing this knowledge and workforce. Hence there is, as never before, an urgency to find and apply solutions. With the rest of the scientific world, Australians have embraced the study of a 'new' habitat: the mesophotic depths (below 30 m) and are looking in detail at this habitat as a refuge of sorts from the impacts on shallower waters (see Chapter 7). Other studies are looking at the potential for the most resilient coral species, populations or individuals to be seeded onto damaged reefs to assist with their recovery. See Chapter 12 for a sober discussion of what it will take to turn around the degradation of the GBR and ensure its survival into the future. We have recently learned, from the study of fossil assemblages in neighbouring parts of the world (Santodomingo *et al.* 2015), that many coral species living on GBR and Coral Sea reefs are up to 18–20 million years old – an enormous length of time in which to track and accommodate changes in the underwater world. Thus the task ahead for us all is to do our best to understand some more, and to make more effort to become part of the solution.

Further reading

Anatomy, life history and ecology

Cole AJ, Pratchett MS, Jones GP (2008) Diversity and functional importance of coral-feeding fishes on tropical coral reefs. *Fish and Fisheries* **9**, 286–307. doi:10.1111/j.1467-2979.2008.00290.x

Fordyce AJ, Camp EF, Ainsworth TD (2017) Polyp bailout in *Pocillopora damicornis* following thermal stress. *F1000Research* **6**, 687. doi:10.12688/f1000research.11522.2.

Harrison PL, Wallace CC (1990) Reproduction, larval dispersal and settlement of scleractinian corals. In *Ecosystems of the World: Coral Reefs*. (Ed. Z Dubinsky) pp. 133–208. Elsevier, Amsterdam, Netherlands.

Okubo N, Mezaki T, Nozawa Y, Nakano Y, Lien Y-T, Fukami H, *et al.* (2013) Comparative embryology of eleven species of stony corals (Scleractinia). *PLoS ONE* **8**(12), e84115. doi:10.1371/journal.pone.0084115.

Willis BL, Page CA, Dinsdale EA (2004) Coral disease on the Great Barrier Reef. In *Coral Health and Disease*. (Eds E Rosenberg E and Y Loya) pp. 69–104. Springer, Berlin, Germany.

The microbiome

Ainsworth TD, Krause L, Bridge T, Torda G, Raina J-B, Zakrewski M *et al.* (2015) The coral core microbiome identifies rare bacterial taxa as ubiquitous endosymbionts. *International Society for Microbial Ecology Journal* **2015**, 1–14. DOI10.1038/ismej.2015.39

Klaus JS, Janse I, Heikoop JM, Sanford RA, Fouke BW (2007) Coral microbial communities, zooxanthellae and mucus along gradients of seawater depth and coastal pollution. *Environmental Microbiology* **9**, 1291–1305. doi:10.1111/j.1462-2920.2007.01249.x

Sweet MJ, Bulling MT (2017) On the importance of the microbiome and pathobiome in coral health and disease. *Frontiers in Marine Science* **4**, 9. doi:10.3389/fmars.2017.00009

Thompson JR, Rivera H, Closek CJ, Medina M (2015) Microbes in the coral holobiont: partners through evolution, development, and ecological interactions. *Frontiers in Cellular and Infection Microbiology* **4**, 176.

Species concepts, population biology and evolution

Barbeitos MS, Romano SL, Lasker HR (2010) Repeated loss of coloniality and symbiosis in scleractinian corals.

Proceedings of the National Academy of Sciences of the United States of America **107**, 11877–11882. doi:10.1073/pnas.0914380107

Kitahara MV, Lin M-F, Forêt S, Huttley G, Miller DJ, Chen CA (2014) The 'naked coral' hypothesis revisited – evidence for and against scleractinian monophyly. *PLoS One* **9**, e94774. doi:10.1371/journal.pone.0094774

Richards ZT, Miller DJ, Wallace CC (2013) Molecular phylogenetics of geographically restricted *Acropora* species: implications for threatened species conservation. *Molecular Phylogenetics and Evolution* **69**, 837–851. doi:10.1016/j.ympev.2013.06.020

Richards ZT, van Oppen MJH, Wallace CC, Willis BL, Miller DJ (2008) Some rare Indo-Pacific coral species are probable hybrids. *PLoS One* **3**(9), e3240. doi:10.1371/journal.pone.0003240

Romano S, Cairns SD (2000) Molecular phylogenetic hypotheses for the evolution of scleractinian corals. *Bulletin of Marine Science* **67**, 1043–1068.

Rosser NL (2015) Asynchronous spawning in sympatric populations of a hard coral reveals cryptic species and ancient genetic lineages. *Molecular Ecology* **24**, 5006–5019. doi:10.1111/mec.13372

Santodomingo N, Wallace CC, Johnson KG (2015) Fossils reveal a high diversity of the staghorn coral genera *Acropora* and *Isopora* (Scleractinia: Acroporidae) in the Neogene of Indonesia. *Zoological Journal of the Linnean Society* **175**, 677–763. doi:10.1111/zoj.12295

Schmidt-Roach S, Miller KJ, Lundgren P, Andreakis N (2014) With eyes wide open: a revision of species within and closely related to the *Pocillopora damicornis* species complex (Scleractinia; Pocilloporidae) using morphology and genetics. *Zoological Journal of the Linnean Society* **170**(1), 1–33. doi:10.1111/zoj.12092

Stolarski J, Roniewicz E (2001) Towards a new synthesis of evolutionary relationships and classification of Scleractinia. *Journal of Paleontology* **75**, 1090–1108. doi:10.1017/S0022336000017157

Stolarski J, Kitahara M, Miller DJ, Cairns SD, Mazur M, Meibom A (2011) The ancient evolutionary origins of Scleractinia revealed by azooxanthellate corals. *BMC Evolutionary Biology* **11**(316), 1–10.

Suzuki G, Keshavmurthy S, Hayashibara T, Wallace CC, Shirayama Y, Chen CA, *et al.* (2016) Genetic evidence of peripheral isolation and low diversity in marginal populations of the *Acropora hyacinthus* complex. *Coral Reefs* **35**, 1419–1432. doi:10.1007/s00338-016-1484-2

Voolstra CR, Li Y, Liew YJ, Baumgarten S, Zoccola D, Flot J-F, *et al.* (2017) Comparative analysis of the genomes of *Stylophora pistillata* and *Acropora digitifera* provides evidence for extensive differences between species of corals. *Scientific Reports* **7**, 17583. doi:10.1038/s41598-017-17484-x

Integrative taxonomy, nomenclature and phylogeny

Arrigoni R, Stefani F, Pichon M, Galli P, Benzoni F (2012) Molecular phylogeny of the Robust clade (Faviidae, Mussidae, Merulinidae, and Pectiniidae): an Indian Ocean Perspective. *Molecular Phylogenetics and Evolution* **65**, 183–193. doi:10.1016/j.ympev.2012.06.001

Arrigoni R, Vacherie B, Benzoni F, Stefani F, Karsenti E, Jaillon O, *et al.* (2017) A new sequence dataset of SSU rRNA gene for Scleractinia and its phylogenetic and ecological applications. *Molecular Ecology Resources* **17**, 1054–1071. doi:10.1111/1755-0998.12640

Budd A, Stolarski J (2009) Searching for new morphological characters in the systematics of scleractinian reef corals: comparison of septal teeth and granules between Atlantic and Pacific Mussidae. *Acta Zoologica* **90**, 142–165. doi:10.1111/j.1463-6395.2008.00345.x

Dai C-F, Horng S (2009). *Scleractinia fauna of Taiwan. 1 The Complex Group.* National Taiwan University, Taipei, Taiwan.

Dai C-F, Horng S (2009). *Scleractinia fauna of Taiwan. 2 The Robust Group.* National Taiwan University, Taipei, Taiwan.

Gittenberger A, Reijnen BT, Hoeksema BW (2011) A molecularly based phylogeny reconstruction of mushroom corals (Scleractinia: Fungiidae) with taxonomic consequences and evolutionary implications for life history traits. *Contributions to Zoology (Amsterdam, Netherlands)* **80**(2), 107–132.

Huang D, Arrigoni R, Benzoni F, Fukami H, Knowlton N, Smith ND, *et al.* (2016) Taxonomic classification of the reef coral family Lobophylliidae (Cnidaria: Anthozoa: Scleractinia). *Zoological Journal of the Linnean Society* **178**, 436–481. doi:10.1111/zoj.12391

Huang D, Benzoni F, Fukami H, Knowlton N, Smith ND, Budd AF (2014) Taxonomic classification of the reef coral families Merulinidae, Montastraeidae, and Diploastraeidae (Cnidaria: Anthozoa: Scleractinia). *Zoological Journal of the Linnean Society* **171**, 277–355. doi:10.1111/zoj.12140

Kitahara MV, Cairns SD, Stolarski J, Blair D, Miller DJ (2010) A comprehensive phylogenetic analysis of the Scleractinia (Cnidaria, Anthozoa) based on mitochondrial CO1 sequence data. *PLoS One* **5**, e11490. doi:10.1371/journal.pone.0011490

Identification guides

Cairns SD (2004) The azooxanthellate Scleractinia (Coelenterata: Anthozoa) of Australia. *Records of the Australian Museum* **56**, 259–329. doi:10.3853/j.0067-1975.56.2004.1434

Kelley R (2016) Indo-Pacific coral finder: the world's first searchable underwater ID smart guide to corals (website) 3rd edn. Russell Kelley, Townsville, <http://www.russellkelley.info/print/indo-pacific-coral-finder/>.

Veron JEN (2000). *Corals of the World*. Vols 1–3. Australian Institute of Marine Science, Melbourne.

Veron JEN, Stafford-Smith MG, Turak E, DeVantier LM (2017). Corals of the world (website), <www.coralsoftheworld.org/>.

Wallace C, Aw M (2000). *Staghorn corals: a 'Getting to know you and Identification' Guide.* OceanNEnvironment, Carlingford, Australia.

Wallace CC, Done BJ, Muir PR (2012) Revision and catalogue of worldwide staghorn corals *Acropora* and *Isopora* (Scleractinia: Acroporidae) in the Museum of Tropical Queensland. *Memoirs of the Queensland Museum* **57**, 1–255.

Wallace CC, Fellegara I, Muir PR, Harrison PL (2009) The scleractinian corals of Moreton Bay, eastern Australia: high latitude, marginal assemblages with increasing species richness. *Memoirs of the Queensland Museum* **54**, 1–118.

Climate change and human impacts

Carpenter KE, Abrar M, Aeby G, Aronson RB, Banks S, Bruckner A *et al.* (2008) One-third of reef-building corals face elevated extinction risk from climate change and local impacts. *Science* **321**, 560–563. doi:10.1126/science.1159196

Huang D (2012) Threatened reef corals of the world. *PLoS One* **7**(3), e34459. doi:10.1371/journal.pone.0034459

Keshavmurthy S, Fontana S, Mezaki T, del Caño González L, Chen CA (2014) Doors are closing on early development in corals facing climate change. *Scientific Reports* **4**, 5633. doi:10.1038/srep05633

Lough JM (2010) Climate records from corals. *Wiley Interdisciplinary Reviews: Climate Change* **1**, 318–331. doi:10.1002/wcc.39

Muir PR, Marshall PA, Abdulla A, Aguirre JD (2017) Species identity and depth predict bleaching severity in reef-building corals: shall the deep inherit the reef? *Proceedings. Biological Sciences* **284**, 20171551. doi:10.1098/rspb.2017.1551

Pollock FJ, Lamb JB, Field SN, Heron SF, Schaffelke B, Shadrawi G, *et al.* (2014) Sediment and turbidity associated with offshore dredging increase coral disease prevalence on nearby reefs. *PLoS One* **9**(7), e102498. doi:10.1371/journal.pone.0102498

23

Octocorals

P. Alderslade and K. Fabricius

Taxonomic overview

Octocorals (subclass Octocorallia, in the class Anthozoa, phylum Coelenterata) are exclusively marine sessile animals, with a mobile larval phase. The distinguishing characteristic of this subclass is that polyps always bear eight tentacles (hence octocoral), which are typically fringed by one or more rows of lateral processes called pinnules along both edges (Fig. 23.1). The popular term 'soft coral' points to the fact that most octocorals, in contrast to the related hard corals, have no massive solid skeleton. Instead, their colonies are supported by one or more of the following: tiny calcareous granules called sclerites, which can be fused but in most cases are separately embedded in the tissue; a hydroskeleton or a branched or unbranched internal axis that can be constructed in several ways. Some axes contain sclerites that are either free, or cemented together, or embedded in a proteinaceous matrix (called gorgonin). Some axes even have nodes of sclerites embedded in gorgonin alternating with internodes of cemented sclerites. Other types of axes contain no sclerites and instead consist of flexible gorgonin with or without various amounts of calcareous fibres, or the axis consists simply of solid calcium carbonate. In the latter, axes may also be articulated, consisting of nodes of gorgonin alternating with internodes of solid calcium carbonate.

The terminology used in the literature to refer to octocorals can be confusing. The term 'soft coral' is commonly used to refer only to those octocorals that have no massive skeleton or internal axis, and the term 'gorgonian' is commonly used when referring to octocorals (other than sea pens and blue coral) that arise from the substrate with the support of some form of internal axis. The terms 'sea fan' or 'sea whip' are often used synonymously with the term gorgonian, but gorgonians can also be bushy or tree-like. It was once thought that the different families and genera demonstrated a continuum from the simplest forms of soft coral to the most complex gorgonians, but evidence from DNA analyses indicates that many forms have evolved independently on several occasions. Perhaps reflecting this, the term 'soft coral' is now often used when referring to both soft corals and gorgonians, and although it is still useful to refer to those two groups separately, here 'soft coral' is mostly used inclusively.

Although octocorals occur in all seas and oceans and at most depths, they form a conspicuous and very diverse group of reef inhabiting organisms on Indo-West Pacific coral reefs. The number of octocoral species is estimated to be ~3000–4000 globally, but the number of shallow-water Indo-West Pacific species is unknown. Many species still await taxonomic description and many genera are in urgent need for revision. Currently, three orders of Octocorallia are distinguished: the Order Alcyonacea (soft corals); the Order Helioporacea (blue coral); and the Order Pennatulacea (sea pens).

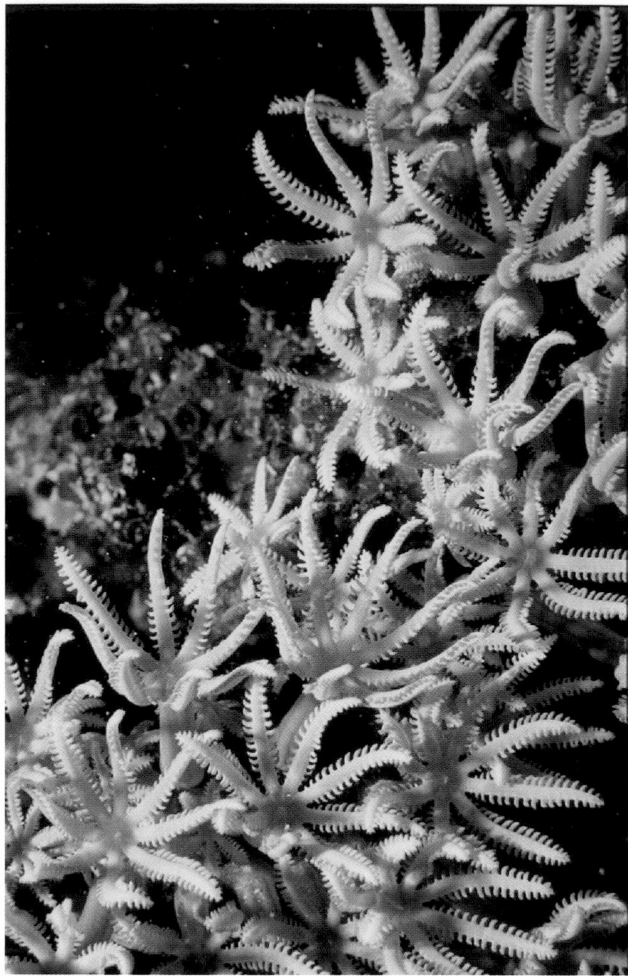

Fig. 23.1. Species of the Subclass Octocorallia are characterised by bearing polyps with eight tentacles, which in most cases are fringed by one or several rows of pinnules on both edges (Photo: K. Fabricius).

The majority of octocorals found in tropical and subtropical shallow water Indo-Pacific coral reefs are soft corals, with ~100 genera in 21 families described to date. In comparison, the sea pens show less diversity, with about nine genera in five families living in shallow Indo-West Pacific soft bottom habitats (often only emerging at night). Helioporacea is represented by a single species in the Indo-West Pacific. The much earlier separation into several different orders and suborders was abandoned a few decades ago when it was realised that there is no distinct morphological dividing line between soft corals and gorgonians. Recent phylogenetic analyses have confirmed those earlier classifications to be largely invalid, and significant changes to the current classification system are likely to occur within the next decade.

Few comprehensive taxonomic inventories of Indo-West Pacific octocorals exist (e.g. from the Great Barrier Reef (GBR), North-West Australia, New Guinea, New Caledonia, Micronesia, Japan, South-Eastern Africa and the Red Sea), and there are still major gaps in the understanding of octocoral diversity and biogeography. On the GBR, an estimated 289 morpho-species were represented in 993 specimens collected from 102 sites around Lizard and Heron Islands (northern and southern GBR) (M. Ekins, *pers. comm.*) between 2008–2010. In the same time period, detailed systematic taxonomic inventory of reefs in the Kimberley region in Western Australia yielded 206 species or morpho-species (M. Bryce, *pers. comm.*), and a study from Palau in Micronesia recorded 150 species (Fabricius *et al.* 2007).

Biology and ecology

Octocorals of the GBR include a diverse range of species with widely contrasting biological properties and ecological requirements. Life expectancy and growth rates of most soft corals are largely unknown. Some (especially among the family Xeniidae) are fast colonisers with a short life expectancy, but others appear to be quite slow growing, extending by only one to a few centimetres per year. Some large *Sinularia* colonies (family Alcyoniidae) are probably hundreds of years old, and large gorgonian colonies may also be many decades old. Octocoral colonies also commonly shrink, for example, when torn by storm waves or damaged by moving rubble or predation. As in other modular organisms the relationship between size and age is therefore weak.

Dispersal strategies include asexual propagation and sexual reproduction. Some taxa rapidly colonise substrata via the asexual generation of new colonies at the terminal ends of stolons ('runners') formed by the parental colony. Other forms of asexual propagation involve budding of miniature colonies that fall off the parental colony and settle nearby, and fragmentation, when a middle part of a colony dies yet the edges survive and reorganise into complete colonies. Pelagic larvae are typically the product of sexual reproduction. Some taxa are gonochoric (males and females are separate colonies), while others are hermaphroditic (producing both male and female gametes). Two modes of sexual reproduction exist. First, 'broadcasting' species release their gametes into the water column where fertilisation occurs. The pelagic larvae are then dispersed by currents over many kilometres until ready to settle days to weeks later. Second, in 'brooding' species, eggs are fertilised within the parental colony, and the resulting larvae develop on the colony surface until they are ready to detach and settle nearby soon after their release.

The food of octocorals consists of small suspended plankton particles filtered from the water column; they rely on water currents to carry the particles towards their polyps. Actively swimming zooplankton is not ingested because octocorals' stinging cells are only weakly developed and unable to paralyse large zooplankton items. Much of the food is therefore derived from phytoplankton, minute detrital particles, flagellates and very small zooplankton. More than half of the warm shallow-water Indo-West Pacific octocorals also contain endosymbiotic dinoflagellate algae (often called zooxanthellae) in their tissue. These algae provide photosynthetic carbon to their host, while the coral in return provides nutrients and shelter to the algae. This finely tuned symbiosis between animal and alga depends on the availability of light for photosynthesis. The azooxanthellate heterotrophic taxa are suspension feeders that strongly depend on currents to transport food particles towards the polyps; they are mostly found in high-flow environments. Heterotrophic taxa are easily visually distinguished by their bright yellow, orange, red, pink, purple or snow-white colouration, while their phototrophic relatives tend to have duller colours. Zooxanthellate phototrophic taxa include many of the 'true' soft corals (the Alcyoniina group), especially most genera within the species-rich families Nephtheidae, Alcyoniidae and all Xeniidae. In contrast, most sea fans, and also several of the soft corals (e.g. *Dendronephthya*) do not contain dinoflagellate endosymbionts.

As sessile organisms without a protective skeleton, octocorals would appear to be vulnerable to predation. However, with the exception of a few snails (e.g. the egg cowry *Ovula ovum*), a few species of butterfly fish that selectively feed on coral and octocoral polyps, and the occasional grazing by sea turtles, nudibranchs or by *Diadema* sea urchins, remarkably few organisms feed on octocorals, and overall feeding pressure appears low. Many species are protected against predation, fouling by algae or overgrowth from neighbouring organisms through feeding deterrent, toxic or allelopathic secondary metabolites. Many of these substances have been investigated for their bioactivity and pharmaceutical relevance.

Octocoral colonies, although not contributing to reef growth, provide shelter to a range of other reef-inhabiting organisms. For example, some species of brittle stars (Ophiuroidae), feather stars (Crinoidea), shrimps, ctenophores and fishes (gobies and pygmy sea horses) are exclusively found on the surface of specific octocoral colonies. Most of these associates use the octocoral colony as a perch or for shelter; however, a few of these associates appear to also feed on octocoral mucus.

After hard corals, octocorals are the second most common group of macrobenthic animals on the GBR. Mean octocoral cover of the GBR regions ranges from 3% to 35% on outer reef slopes, but cover can be as high as 70% in current-swept yet wave-protected environments such as channels between reefs or islands, and near zero on wave-exposed macro-algal dominated turbid and silty inshore reef crests.

Octocorals are highly diverse not only taxonomically, but also ecologically. The GBR is situated on a wide continental shelf and contains a remarkable range of marine habitat types for octocorals. They include wave-beaten outer barrier reef walls with steep drop-offs and oceanic water clarity, more protected mid-shelf reefs with sheltered lagoons and current-flushed flanks, extensive inter-reefal areas with soft bottom environments and outcrops of hard substratum, seagrass meadows and *Halimeda* mounds, and inshore coral reefs within the reach of terrestrial influences. Each of these habitats houses a distinct octocoral assemblage. For example, outer-shelf reefs contain diverse octocoral communities with high abundances of Xeniidae, while many Xeniidae and Nephtheidae are missing on turbid inshore reefs. Mid-shelf reefs have the highest species richness of all reefs on the GBR. Deep-water reef slopes and inter-reefal habitats are inhabited by azooxanthellate taxa such as many gorgonians and *Dendronephthya*, as well as ubiquitous and tolerant zooxanthellate taxa such as *Sinularia* and *Sarcophyton*.

Ecological surveys have shown that the taxonomic richness of octocorals is strongly related to water clarity and amounts of sediments deposited. Changes in taxonomic richness and community composition in octocorals have therefore been suggested to be suitable as indicators of past and recent disturbance by poor water quality on the GBR and other reef environments.

Species richness at a given site is affected by three factors. First, the biogeographic location determines the regional species pool present. On the GBR, the species richness in octocorals strongly attenuates with increasing latitudes: many more genera and species occur in the tropical far northern part than on the southern end of the GBR. Second, environmental conditions (especially turbidity, light availability and water currents) determine what cross-section of the local species pool occurs at that locality. Third, at any point in time, local and regional species richness also depends on disturbance history, specifically the nature and intensity of the disturbance, and the time since past disturbances have removed colonies. For octocorals, disturbances include storms with high wave energy (dislodging or damaging colonies), episodes of high water temperatures (causing coral bleaching), chronically reduced water clarity (reducing photosynthesis) and sedimentation (hampering larval settlement and smothering colonies). After a disturbance, fast colonisers re-establish if propagules from surviving colonies (locally or farther upstream) are available, whereas slow-colonising or slow-growing taxa will take much longer to return to their previous abundance. Chronic disturbance such as water pollution reduces diversity, because only persistent species can flourish. In order to understand regional and local biodiversity patterns, biogeographic settings, environmental requirements of taxa and disturbance histories need to be investigated simultaneously.

A brief guide to the identification of octocoral genera of the Great Barrier Reef

The identification of octocorals is based primarily on colony form, the nature, location and

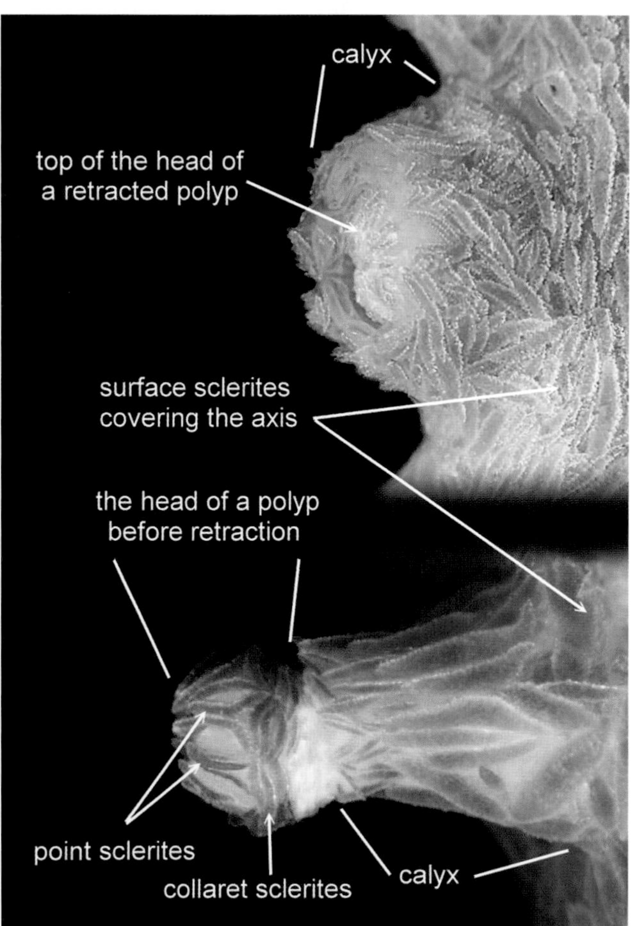

Fig. 23.2. Close-ups of two gorgonians showing calyces, retracted and non-retracted polyps and sclerite arrangements.

arrangement of the sclerites and the nature of the internal axis if one is present. (See Box 23.1 for collection and preservation methods.) Octocorals are modular animals that, except in rare occurrences, are constructed of several polyps united in a common tissue mass called the coenenchyme. Five types of polyps are found in octocorals: autozooids, siphonozooids, mesozooids, oozoids and acrozooids. The first three forms are found in soft corals, with mesozooids being rare, but all are found in sea pens. The first type, autozooids, are generally responsible for feeding and reproduction, and are present in all species. Autozooids always have eight tentacles that, except in a few cases, bear pinnules along each edge. The second type of polyp, the siphonozooids, are very small, and the tentacles are rudimentary or entirely absent. They are primarily for water circulation, but can also be involved in reproduction, and are found in several soft coral families. Taxa that have both polyp types are referred to as dimorphic (e.g. *Lobophytum*).

Autozooid polyps are described as being either contractile or retractile. By expelling water and using mainly longitudinal muscle fibres, a contractile polyp has the ability to fold the deflated tentacles over or into its mouth and shrink its body down to become substantially smaller. A retractile polyp has a similar ability, but it can also invaginate the upper part of the polyp, the head, into that part of the polyp that is located below the flat surface of the colony or into a short, projecting tube- or volcano-shaped extension of the colony surface, which is called a calyx (Fig. 23.2).

The growth form of a colony may be stolonate, membranous, encrusting, fleshy, erect, and either whip-like or branching, with or without an axis (Fig. 23.3). An axis that contains sclerites is considered as part of the coenenchyme, which is divided into an outer cortex, which includes the polyps,

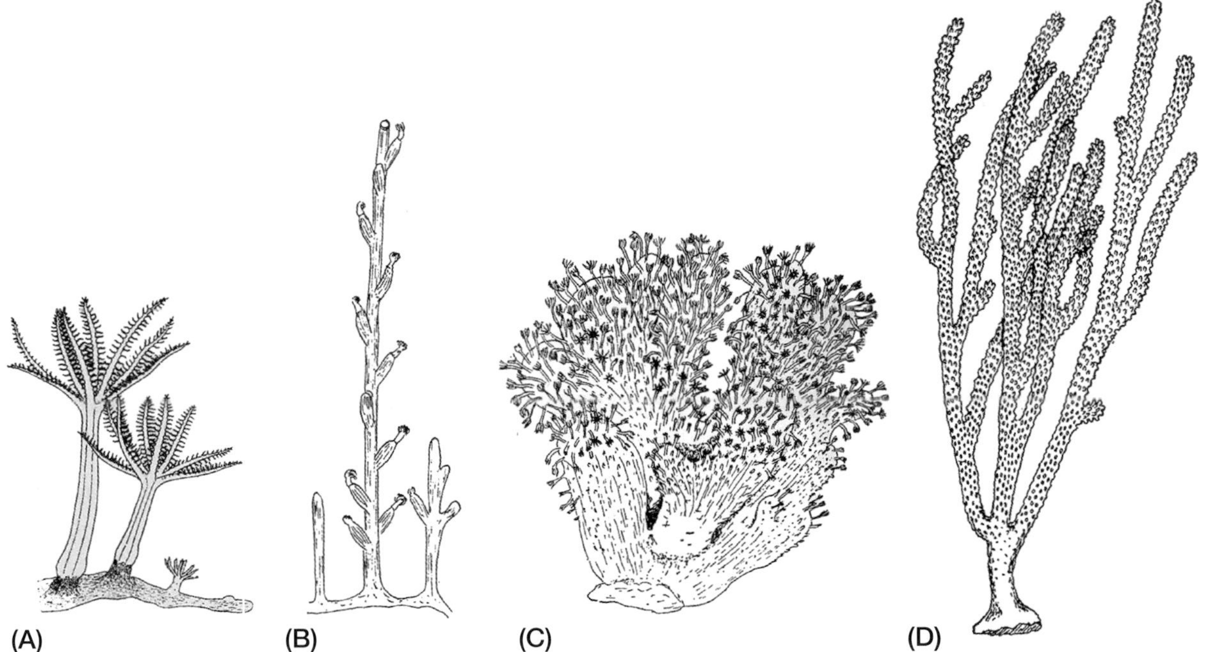

Fig. 23.3. The most commonly encountered octocorals growth forms: **(A)** single polyps connected by stolons as found in *Clavularia*; **(B)** tall axial polyps that bud off lateral polyps and are united at their base with stolons as found in *Carijoa*; **(C)** prototype 'soft coral' where polyps are embedded in a thick fleshy tissue mass called coenenchyme as found in *Klyxum*; **(D)** prototype 'gorgonian' with upright growth form and polyps arranged around a central axis as found in *Rumphella*. (Sources: Illustration A, Gohar HAF (1948) A description and some biological studies of a new alcyonarian species *"Clavularia hamra"* Gohar. *Publications of the Marine Biological Station Gahdaqa (Red Sea)* **6**, 1–33. Illustrations B-D, modified from Hyman 1940).

and an inner axial medulla. In gorgonians without sclerites in the axes, only the outer surrounding layer with the polyps comprises the coenenchyme.

Calcareous sclerites and axis structures are other important features used to identify octocorals. Sclerites are present in most species, ranging in size from ~0.02 mm to >10 mm. Depending on their shape, they have common names such as spindles, rods, capstans, platelets, thorn-scales, double heads, rooted-leaves, 8-radiates, double stars and needles, and their details will differ from one species to another (see figures accompanying the descriptions of the genera).

The brief descriptions of some of the more commonly encountered genera presented here provide a first glimpse at the diversity of features, shapes,

Box 23.1. Techniques used to collect, preserve, and investigate octocorals

When taxonomic identification is attempted, a photograph showing the entire colony and its growth form is advantageous, and close-ups of details of the surface and polyp structures are useful for reference. It is possible with some practice to identify most octocorals to genus level underwater or from photographs, but few can be identified to species level, because a compound microscope is needed to investigate sclerites in detail. Most soft corals have different sclerites in the upper polyp-bearing surface of the colony (lobes, branches), the interior of the polyp region, the surface and interior of the base, and within the polyps. How the sclerites are arranged within the polyps and polyp tentacles can also be informative. For a full diagnosis, until a satisfactory level of field confidence is achieved, it is therefore necessary to collect a sample with all of the main colony regions present. Note that a sampling permit is required for collection on the GBR (see Chapter 14) and, in general, sampling should be minimised wherever possible. If a sampling permit exists and sampling is considered essential, with very small colonies the whole specimen may have to be collected. With larger specimens this is neither necessary nor practical. In broad, thickly encrusting colonies it is sufficient to remove a segment shaped like a pie-slice, which includes upper lobes (branches or ridges, and polyps), the surface layer of the colony side down to the base, and the included interior portions. In gorgonians, one branch is sometimes sufficient to obtain surface, interior and polyp material for sclerite examination, but for species determination more material may have to be collected, because sclerites can vary between branches and from colony base to upper (younger) colony parts. The branching pattern is an extremely important diagnostic feature that may be captured by a good photograph to minimise sampling. Growth form is especially important in the family Ellisellidae because several genera only differ in this character.

Long-term storage of small to medium-sized specimens is best in 70% ethanol in fresh water, in which colonies will last indefinitely. It is not recommended to initially fix octocorals in dilute formalin, because it is virtually impossible to remove all traces before preserving in alcohol and the calcareous octocoral sclerites will eventually be eroded by the acids derived from the oxidation of formaldehyde to formic acid (even if buffered). This will make identification impossible and complicate DNA analyses. If dilute formalin is initially used to reduce shrinkage or prepare samples for histology, the specimens should be extensively soaked in running water before long-term storage and the alcohol changed regularly if there is to be any chance for all traces of formaldehyde to be removed. Numerous specimens reliably tagged, or individually placed in plastic bags with holes, can be kept together in a single large container suitable for alcohol storage. Larger gorgonian branches may have to be air-dried, preferably after fixing in alcohol, unless a container of sufficient size can be found, but air-dried samples are more difficult to work with, they are more prone to abrasion and breakage, and susceptible to mould and insect attack. A compromise then is to keep a portion of a large specimen in alcohol.

To examine sclerites, they must be freed from the coenenchyme in which they are embedded by the use of concentrated bleach (sodium hypochlorite), which dissolves the organic tissues and leaves the sclerites untouched. Unless the sclerites are large, a sample <2 mm^2 is sufficient for examination. The sample is cut from the relevant part of the colony with a scalpel under a dissecting microscope and placed into one or two drops of bleach on a microscope slide. Once the bubbles have ceased, the sclerites are spread out by stirring, a cover-slip is applied and the sample is investigated under a compound microscope. Preparations that have dried can be rehydrated with one or two drops of water.

Fig. 23.4. Pennatulacea, or sea pens, generally inhabit soft bottom habitats and are often hidden in the sand during the day: **(A)** *Cavernularia* sp.; **(B)** *Virgularia* sp. (Photos: K. Fabricius).

colours and forms found in octocorals of the GBR. Each genus is introduced by a brief text, together with a plate showing a representative underwater photograph, and the forms of sclerites typically found in that genus. In some cases, these brief descriptions will be insufficient to facilitate the reliable identification of a genus because the variability between species within genera can be high (a single photograph often does not capture the range of shapes and forms found within a genus). To prevent the premature naming of such genera, we list the names of similar looking genera in the text, even if they are not shown in this chapter. It is advised to refer to a more comprehensive field guide and taxonomic literature before an identification is attempted.

Fig. 23.5. The blue coral, *Heliopora coerulea* (family Helioporidae): **(A)** whole colony; **(B)** cross-section of a broken branch and close-up of the colony surface (Photos: K. Fabricius).

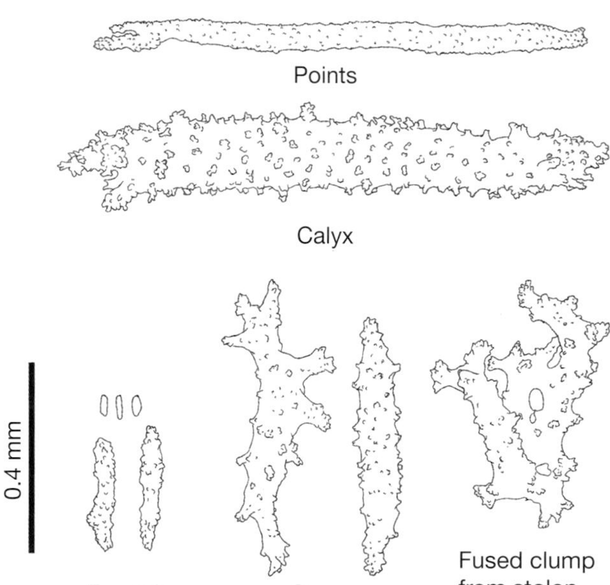

Fig. 23.6. *Clavularia* sp. (family Clavulariidae): **(A)** with expanded and retracted polyps, connected by stolons. Also note the white larvae, which are brooded on the colony surface for a few days after spawning; **(B)** some sclerites representative of this genus (Photo: K. Fabricius; illustration: P. Alderslade).

Order Pennatulacea

The sea pens, or Pennatulacea (Fig. 23.4), are a diverse group of octocorals, but only nine genera in five families have so far been recorded in shallow waters of the tropical and subtropical Indo-West Pacific. Because they inhabit soft bottom habitats and often are completely contracted during daylight and only emerge at night, they are infrequently encountered by divers and snorkelers. Pennatulacea are characterised by their large central primary or axial polyp (oozooid), which is usually supported by a proteinaceous axis. The lower part of the oozooid forms the colony 'foot' or penduncle that digs into sand or mud, anchoring the colony in soft substratum. The upper part, the rachis, reaches into the water column when expanded, and bears the autozooids and siphonozooids. In some shallow-water species, the emergent part somewhat resembles a robust feather (hence the name sea pen), while others can be shaped like cigars.

Order Helioporacea

Family Helioporidae

Heliopora coerulea, known as the 'blue coral' (Fig. 23.5), has a calcified branching skeleton with a brownish-blue, very smooth surface and blue core. Colonies can reach several metres in diameter. Polyps when expanded are <3 mm in diameter, and very fine. It is found in shallow-water areas in which the water temperature remains above 22°C all year round. At present this is the only known representative of this order and family on the GBR, but research elsewhere indicates a second and cryptic species exists (i.e. a species that looks virtually the same), so more than one species may exist on the GBR. Similar genera: none.

Order Alcyonacea

Family Clavulariidae

Genera in this family consist of cylindrical or bluntly conical polyps usually joined only at their bases by reticulating stolons, which may coalesce into thin membranous expansions. In some genera, tall cylindrical polyps develop long secondary polyps that resemble branches. In a few instances, polyps are also connected by extra, transverse, bar-like stolons above the basement layer. Nearly all of the species in this family have sclerites, and it is not unusual for these to be fused into clumps or tubes. Sclerites include smooth branched rods, and prickly or tuberculate 6–radiates, spindles and platelets.

Clavularia (Fig. 23.6). Individual polyps are one of the largest of any soft coral on the reef. They can grow up to 40 mm tall and are united basally by ribbon-like stolons. The polyp head is completely retractile into the lower part of the body, which is stiffened by spindle-shaped sclerites to form a calyx. Sclerites of the stolons are warty spindles, often fused, while those of the tentacles are short rods and small platelets. These corals can be brown, cream or greenish. On the GBR, *Clavularia* is restricted to latitudes north of 20°S. Similar genera: *Anthelia* (family Xeniidae).

Family Tubiporidae

Tubipora (organ-pipe coral) (Fig. 23.7) is the only genus in this family. Colonies consist of a large (sometimes >20 cm), solid skeleton of red, hard calcareous tubes formed from fused sclerites, connected at regular intervals by horizontal platforms. Each tube is formed and occupied by a single polyp, which is connected to other polyps by stolons located in canals within the horizontal plates. Stolons and non-extendable parts of the polyps consist of thin, soft tissue, the tentacles contain minute platelets.

It is commonly thought that only one species exists, *Tubipora musica* Linnaeus, 1758, and specimens are usually given this name. There are actually quite a few species, and several them occur on the GBR. They can usually be told apart by the different shapes and colours patterns of the polyp tentacles, which do not always bear visible pinnules. Unfortunately, the name *Tubipora musica* was based on numerous old descriptions of dry skeletons, almost certainly from several different species, and the location simply given as the Indian Ocean. Organ-pipe coral is uncommon but widely distributed. Similar genera: none.

Family Alcyoniidae

Members of this family can dominate some inshore octocoral communities. Growth forms are often massive or encrusting, with some colonies measuring several metres across. Most colonies have a bare basal section (the stalk or trunk), and an upper, polyp-bearing part that may be flat, undulating, or divided into lobes, ridges or short branches. Monomorphic and dimorphic forms are represented; that is, siphonozooid polyps are

Fig. 23.7. *Tubipora* sp. (organ-pipe coral, family Tubiporidae): **(A)** close-up of polyps; **(B)** a broken colony showing the skeletal structure; **(C)** some sclerites found in the polyp tentacles (Photos: A, B, K. Fabricius; C, P. Alderslade).

present among the autozooids in some genera. The interior of a colony may be compressible and jelly-like if the sclerite content is low, and rigid and solid if it is high. Sclerites include tuberculate or prickly spindles, clubs, 6- or 8-radiates, ovals and dumb-bells.

Sinularia (Fig. 23.8). Members of this genus form encrusting colonies, with the upper surface formed into knobs, ridges, simple or branched lobes. In fact, they represent the greatest diversity of colony size and shape of any soft coral genus, with colonies growing up to several metres in diameter. The polyps are retractile and monomorphic. The interior sclerites are large warty spindles; surface sclerites are mostly clubs. The colour may be brown, cream, yellow or green. *Sinularia* is a very species-rich genus that is widely distributed and found even at high latitudes and dark depths (despite the presence of zooxanthellae). Similar genera: *Lobophytum* (which has two types of polyps), *Klyxum*.

Sarcophyton (Fig. 23.9). Colonies have a distinct stalk and an upper polyp-bearing region that is dish-shaped. Although the margin can be relatively simple, it is usually undulate or folded. Members of

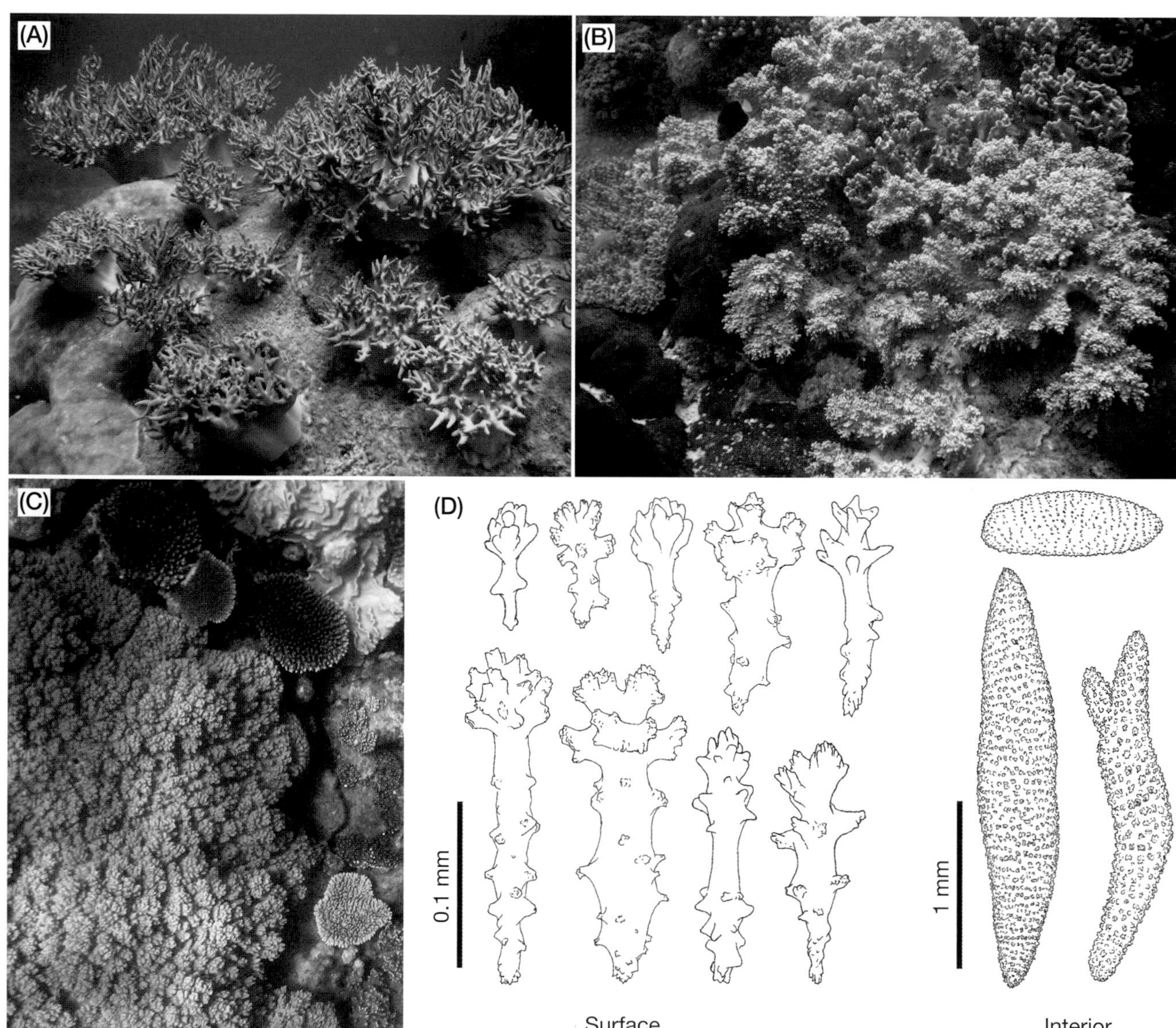

Fig. 23.8. *Sinularia* spp. (family Alcyoniidae): **(A–C)** some of the many and diverse growth forms found in this species-rich genus; **(D)** some sclerites representative of this genus (Photos: K. Fabricius; illustration: P. Alderslade).

this genus can grow up to 1 m in diameter. Polyps are abundant and dimorphic. The autozooids are retractile. The interior sclerites are spindle-shaped; surface ones include many clubs. The colour may be brownish, cream, yellow or green. The genus is very widely distributed. Similar genera: none.

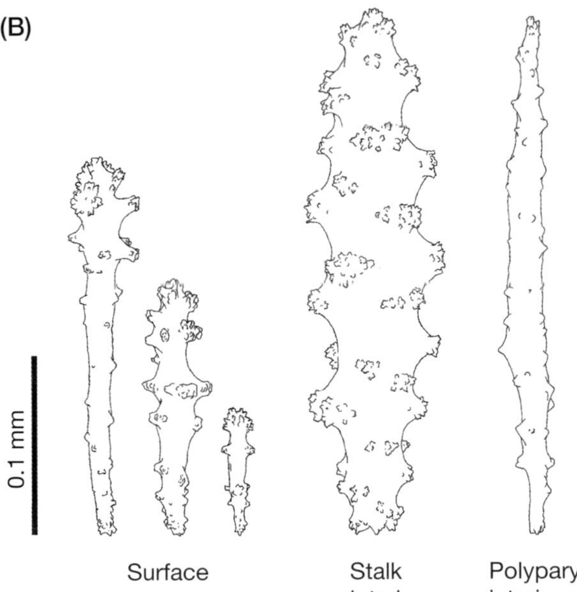

Fig. 23.9. *Sarcophyton* sp. (family Alcyoniidae): **(A)** colonies with expanded and retracted polyps; **(B)** some sclerites representative of this genus (Photo: K. Fabricius; illustration: P. Alderslade).

Lobophytum (Fig. 23.10). Colonies are encrusting, often large, with an upper surface that is lobate, digitate or ridged, and covered in abundant dimorphic polyps. Autozooids are retractile. Sclerites of the interior are spindle to barrel-shaped with tubercles often in girdles. Surface sclerites are generally poorly formed clubs. The colour may be brown, cream, yellow or green. Members of this genus are widely distributed, and abundant in high light environments. Similar genera: *Sinularia*, *Klyxum* (which both do not have dimorphic polyps).

Klyxum (Fig. 23.11). Colonies are soft and fleshy with partly subdivided lobes, and although some very large species exist they are usually relatively small. Polyps are highly contractile and although they may appear to be retracted they are not. The density of the sclerites is often very low, causing colonies to appear translucent. The characteristic coenenchymal sclerites are spindles that have prominent cone-shaped tubercles and they are often only present in the base. If the polyps have sclerites, they are minute discs and flattened rods with a granular surface. Granular rods may also occur in the interior of the lobes in small numbers. The colour may be creamish or pale brown often with darker polyps. This species is quite common in more turbid sheltered regions and infrequent elsewhere.

Family Nephtheidae

A large number of genera with a wide range of ecological characteristics are grouped in this family, some probably erroneously. Most have bushy, globe-shaped or tree-like growth forms, while a few are massive, or consist of finger-like lobes united by a common base. Many genera contain highly coloured species that are azooxanthellate. In most cases, the polyps, singly or in small clusters, are more or less restricted to the upper and outer twigs or branches. In a few cases, polyps grow directly on main branches. A small number of broad primary polyp canals extend longitudinally through the stem of arborescent colonies, subdividing into groups that extend up into the distal lobes and branches where many polyps can connect to

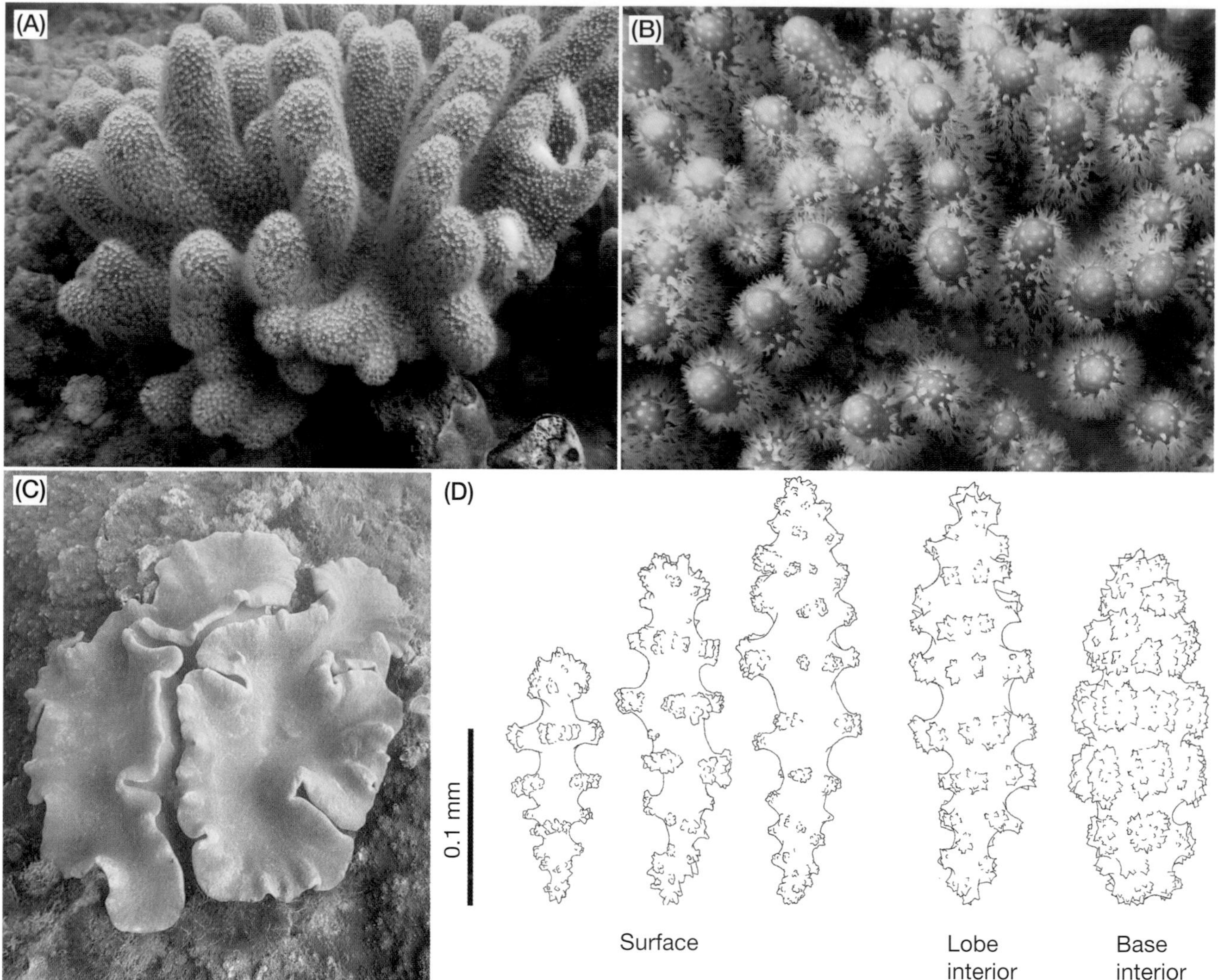

Fig. 23.10. *Lobophytum* sp. (family Alcyoniidae): **(A–C)** some of the growth forms found in this genus. Note in B the tiny polyps (pores) interspersed among the larger polyps, which are characteristic for this genus; **(D)** some sclerites representative of this genus (Photos: K. Fabricius; illustration: P. Alderslade).

each canal. The canal walls are generally thin, with few sclerites, permitting colonies to easily inflate with water, dramatically increasing their size ('hydroskeleton'). Although some genera are soft and floppy, others have a rough or distinctly prickly feel due to long, protective sclerites projecting beyond the polyp head. In the stem and branches, the sandpaper-like texture can be attributed to numerous, strongly sculptured, spiny sclerites in the surface layer. Sclerite forms include prickly needles, leafy clubs, irregular shaped spiky forms, and tuberculate and thorny spindles, often extensively ornamented along one side. Members of this family are most often found in clear-water habitats.

Litophyton (Fig. 23.12). Colonies are arborescent. Non-retractile polyps, supported by a bundle of sclerites, are arranged in lobes or catkins. Sclerites are irregular, spindle or caterpillar-like. The colour is generally brown to green. *Litophyton* is found mostly in clear waters, often growing in clusters of several to many colonies. Similar genera: *Stereonephthya*, *Chromonephthea*, *Lemnalia*. *Nephthea* is a junior synonym of *Litophyton*.

Dendronephthya (Fig. 23.13). Colonies are arborescent with the terminal twigs forming rounded, umbellate or tree-like groupings. Non-retractile polyps are arranged in small groups on the end of the twigs, and several in each group are protected

Fig. 23.11. *Klyxum* sp. (family Alcyoniidae): **(A–D)** colonies; **(E)** some of the sclerites representative of this genus (Photos: K. Fabricius; illustration: P. Alderslade).

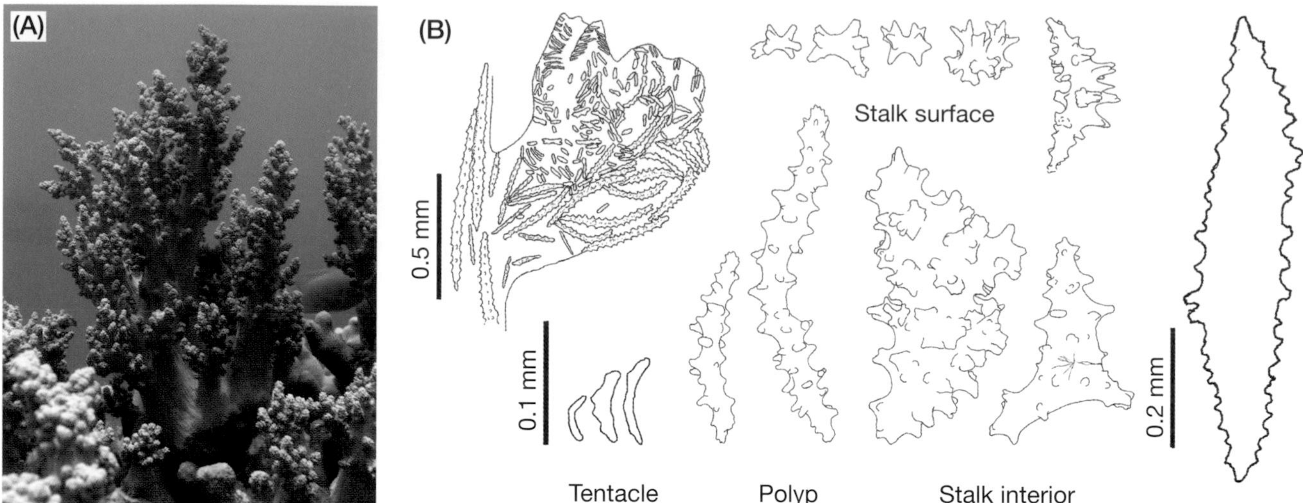

Fig. 23.12. *Litophyton* sp. (family Nephtheidae): **(A)** colony; **(B)** some sclerites representative of this genus (Photo: K. Fabricius; illustration: P. Alderslade).

by long supporting sclerites. Colonies are brightly coloured, azooxanthellate. The sclerites are similar to *Litophyton*. *Dendronephthya* are found in high currents, generally growing as individual colonies. Similar genera: *Scleronephthya*, *Chromonephthea*, *Litophyton* and *Stereonephthya*.

Lemnalia (Fig. 23.14). Colonies are arborescent, formed from bare stalks and branches and generally thin twigs. Non-retractile polyps are isolated on the twigs. The colour is cream to brownish. Interior sclerites are long, thin needles. Surface forms include capstans, crescents and brackets. *Lemnalia* are found in clear waters, growing as individual colonies, small clusters or large assemblages. Similar genera: *Litophyton* and some *Paralemnalia*.

Capnella (Fig. 23.15). Colonies are small, arborescent to lobed. The lobes are crowded with incurved polyps that are covered in club-like

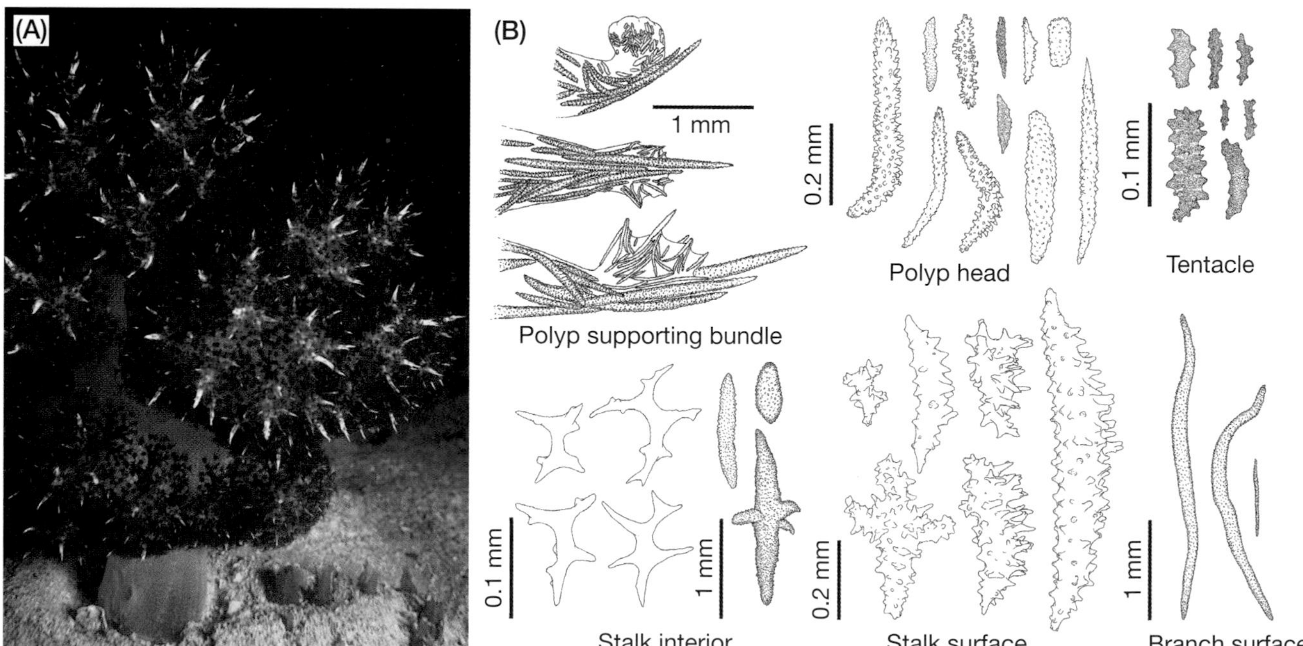

Fig. 23.13. *Dendronephthya* sp. (family Nephtheidae): **(A)** colony; **(B)** some sclerites representative of this genus (Photo: K. Fabricius; illustration: P. Alderslade).

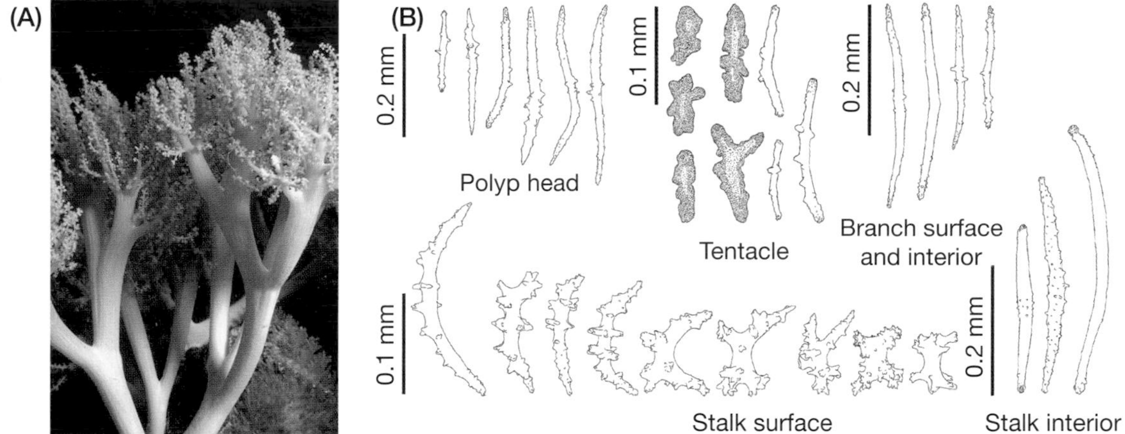

Fig. 23.14. *Lemnalia* sp. (family Nephtheidae): **(A)** colony; **(B)** some sclerites representative of this genus (Photo: F. Dipper; illustration: P. Alderslade).

Fig. 23.15. *Capnella* sp. (family Nephtheidae): **(A–B)** colonies; **(C)** some sclerites representative of this genus (Photos: K. Fabricius; illustration: P. Alderslade).

sclerites. Interior sclerites are often globular, while the surface ones are leafy to spiky capstans. The colour may be grey or beige. *Capnella* are found in a wide range of habitats, as individual colonies or in small groups of colonies.

Family Xeniidae

Members of this family often dominate offshore octocoral communities. Xeniid genera are mostly small and soft, and often quite slippery to touch. Colony growth forms can be lobate, or thin and membranous or have short, cylindrical stalks terminating in a domed polyp-bearing region, or short but arborescent. Polyps can be very small or quite tall and are usually not retractile. All species are zooxanthellate, often with opalescent pastel colours (white, pink, light iridescent blues and greens); some inshore species are dark brown. Some species have pulsating polyps, where the autozooids continually open and close the tentacle basket. Members of one genus have siphonozooids at least during some periods of their lives. Not all of the species in this family have sclerites. Where sclerites are present, they are always very small, and can be rodlets, minute platelets, corpuscle-like forms and their surface may be almost smooth, often appearing opalescent, or coarse and crystalline. A microscopic feature of the family, observed only in histological preparations, is that only the dorsal pair of mesenteries

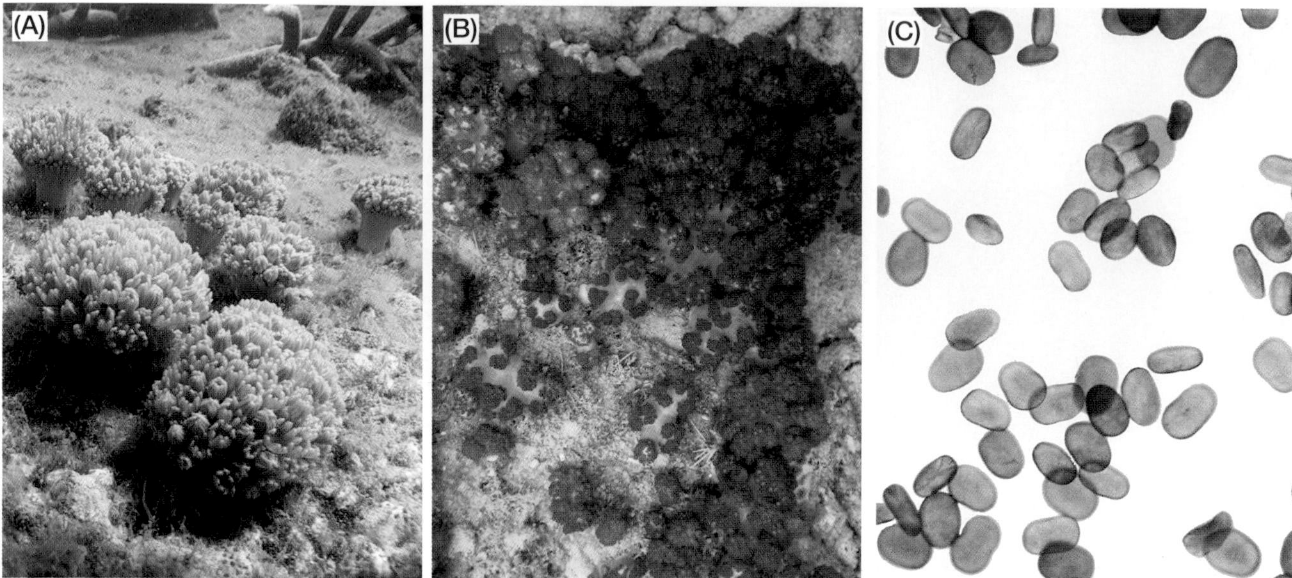

Fig. 23.16. *Xenia* sp. (family Xeniidae): **(A–B)**, colonies; **(C)** the small platelets representative of this genus (Photos: K. Fabricius).

retain their filaments in the adult polyps; the filaments on the other six mesenteries are absent or rudimentary.

Xenia (Fig. 23.16). Many species of the genus *Xenia* are not branched, but if they are it is only sparsely. Colonies have non-retractile polyps that are confined to the domed ends of the branches. Individual colonies tend to be small (often <5 cm in diameter), but often form larger super-colonies (colony clusters). Sclerites are minute platelets. The colour may be white, grey to yellow or brown and blue-green. Colonies with the last colour show an iridescence due to light being refracted by the sclerites' ultrastructure. *Xenia* are abundant in clear waters, but rarely found in inshore waters where they are dark brown. Similar genera: *Heteroxenia* and *Asterospicularia*.

Efflatounaria (Fig. 23.17). Species currently referred to as *Efflatounaria* probably represent an undescribed genus. They have firm yet contractile colonies with highly contractile polyps that can appear to have retracted. Daughter colonies are produced from 'runners' (stolonic outgrowths of terminal polyps). Sclerites are minute platelets or irregularly subspherical bodies. The colonies, which are commonly white, pink, yellow or blue, can be abundant in clear waters. The genus was actually based on a species that is soft and transparent and does not have any sclerites. Similar genus: *Cespitularia*.

Family Briareidae

Briareum is the only genus in the family. On the GBR, different species of *Briareum* grow as thin, smooth sheets or as groups of calyces up to 10 mm tall that are joined together, sometimes at several levels. Although colonies may look like encrusting soft corals, the division of the coenenchyme into an upper cortex and a basal medulla places them with certain groups of gorgonians that do not have a consolidated axis. The basal layer, or medulla, which is attached to the substratum is deep magenta due to the colour of the sclerites. The upper layer, or cortex, may be the same as the basal layer, brown, or almost white depending upon the amount of coloured sclerites it contains. Except for the most basal layer of the medulla, the sclerites are all spindles, sometimes branched, with low or tall, spiny wart-like outgrowths (tubercles) arranged in relatively distinct girdles. The most basal layer generally includes multiple branched, reticulate and fused forms with very tall, complex tubercles. Colonies encrust rock, or dead or live substrate. Species often resemble encrusting lobular soft

Fig. 23.17. *Efflatounaria* sp. (family Xeniidae): **(A)** colonies; **(B)** the small sclerites representative of this genus (Photos: *A*, K. Fabricius; *B*, P. Alderslade).

corals, but the lobes or bumps are where the thin colony is growing over a lumpy surface.

Briareum (Fig. 23.18). On the GBR, colonies form encrusting sheets, which in some species may form filo-pastry style colonies through successive layering. The polyps are fully retractile. Calyces are tall to exceedingly small. The sclerites are spindle-shaped and multiradiate, sometimes fused in clumps. Commonly the surface layer is beige-brown, but it can be magenta or almost white. The basal layer is magenta while the polyps may be iridescent green or yellow. Members of *Briareum* can be abundant on inshore reefs where they can grow to sheets >1 m in diameter. Tropical octocorals previously known as *Solenopodium* or *Pachyclavularia* should be called *Briareum*. Similar genus: *Rhytisma*.

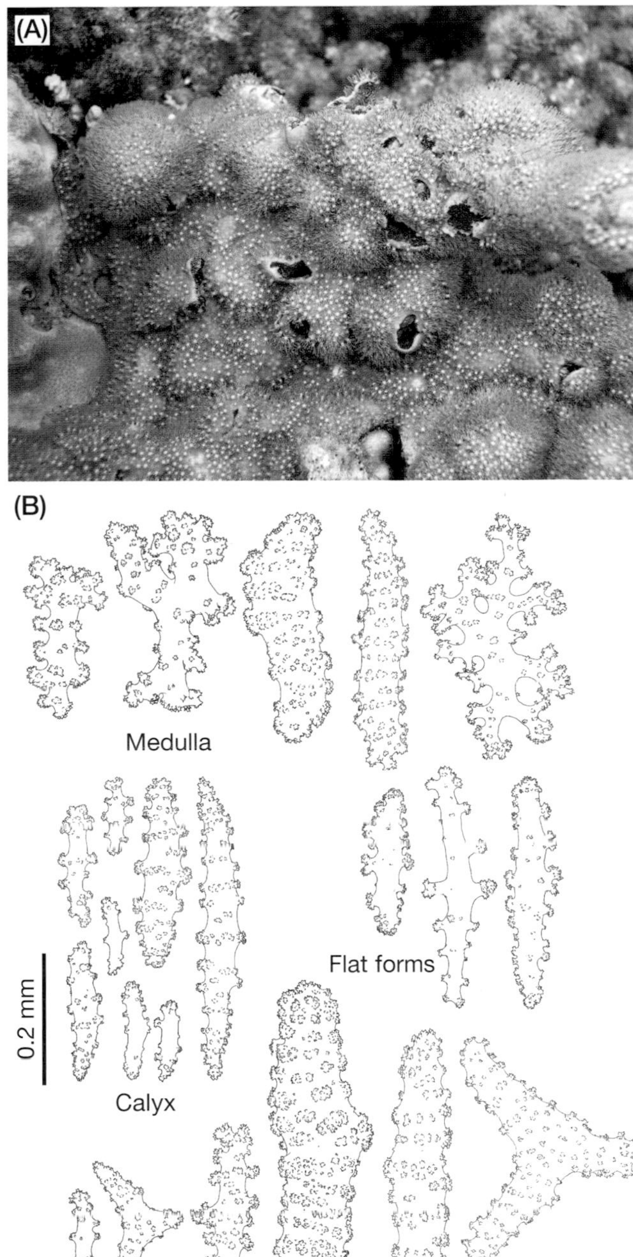

Fig. 23.18. *Briareum* sp. (family Briareidae): **(A)** colony; **(B)** some of the sclerites representative of this genus (Photo: K. Fabricius; illustration: P. Alderslade).

Family Subergorgiidae

In this family colonies are erect and either arborescent with free branches, or as net-like fans. The central medulla is relatively consolidated being formed from sclerites, often branched, that are

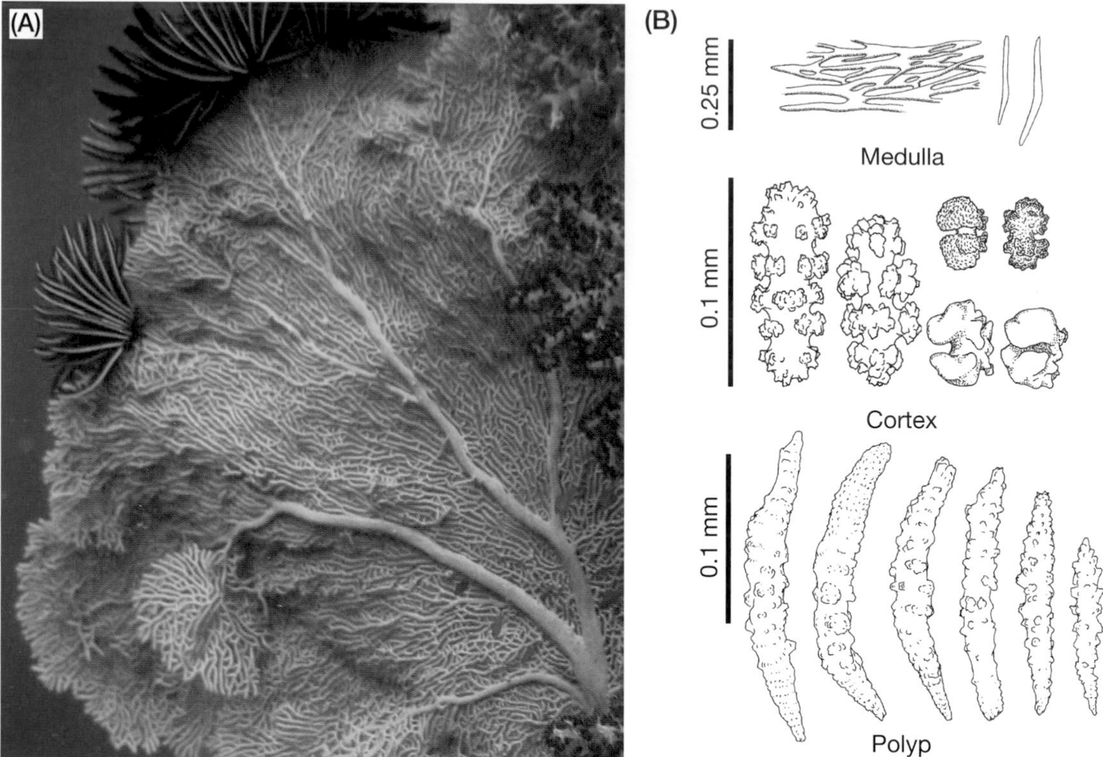

Fig. 23.19. *Annella* sp. (family Subergorgiidae): **(A)** colony; **(B)** some of the sclerites representative of this genus (Photo: K. Fabricius; illustration: P. Alderslade).

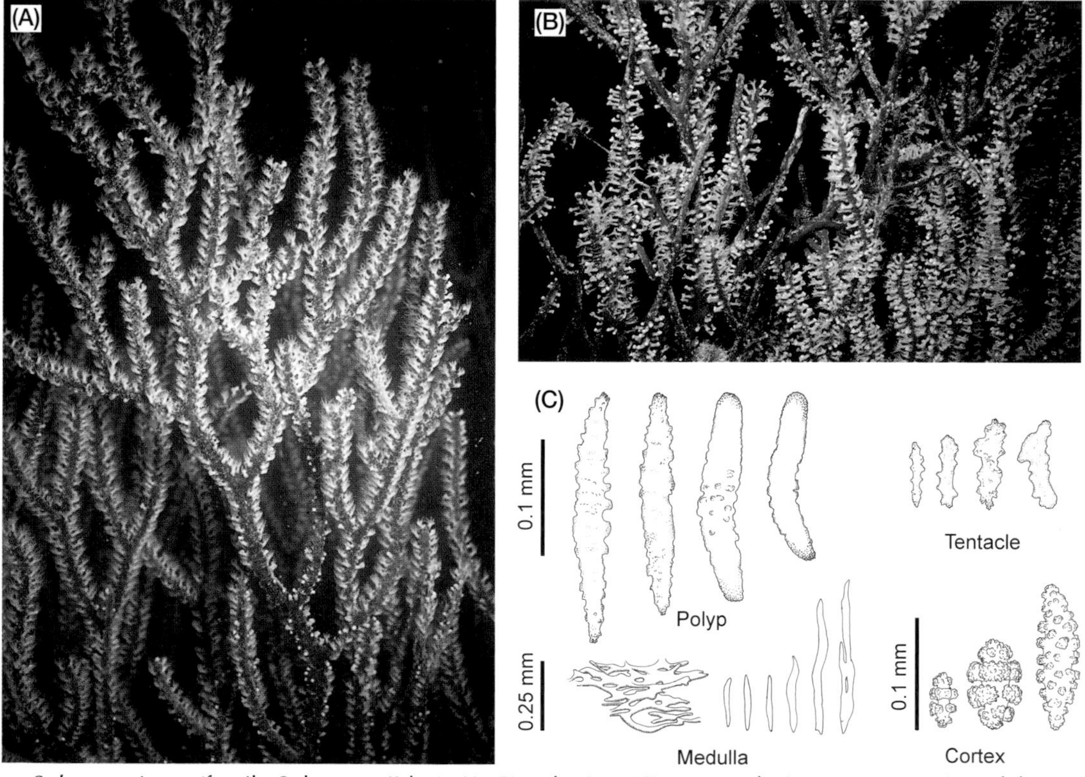

Fig. 23.20. *Subergorgia* sp. (family Subergorgiidae): **(A–B)** colonies; **(C)** some sclerites representative of this genus (Photo: K, Fabricius; illustration: P. Alderslade).

interlocked and partially fused and embedded in a tough matrix of gorgonin. There is a ring of longitudinal boundary canals directly outside the medulla separating it from the cortex, but there are virtually no canals running through the medulla. Axial sclerites are usually smooth, sinuous, and relatively long. Sclerites of the cortex are predominantly tuberculate spindles and ovals, and small, irregularly shaped dumb-bell-like sclerites often referred to as double-heads or double-wheels. Using just external features, species in this family are easily confused with those in families with no sclerites in the axis.

Annella (Fig. 23.19). Members of this genus grow as large reticulate fans up to 2 m in diameter. The axial skeleton comprises gorgonin and partially fused, smooth, sinuous sclerites. Sclerites of the cortex are double discs and spindles. The colour may be yellow, brown, orange or red. Similar genera: those that can form fans with a close mesh, such as *Villogorgia, Melithaea, Verrucella.*

Subergorgia (Fig. 23.20). One particular common species of this genus, *Subergorgia suberosa* may be encountered in many regions of the GBR, although it prefers more turbid inshore environments where individuals may be adorned with strings of detritus. They are usually branched dichotomously, more or less in one plane and colonies removed from the water have a shallow groove down the middle of both sides of the branches. The axis is like that of *Annella* while the sclerites of the outer cortex are ovals and spindles, many with the tubercles arranged in girdles. The colour is dark reddish brown to pale brown, with white polyps that are often expanded also during the day. Similar genera: *Dichotella, Melithaea.*

Family Melithaeidae

The Melithaeidae consists of fan-shaped or tangled bushy colonies in which the axial medulla is segmented, comprised of a series of short, rounded nodes alternating with longer, narrower internodes. The swollen axial nodes on the stem and main branches are generally conspicuous, while those at the points of branching within the fan may not be so obvious. With few exceptions, branching occurs at the spongy nodes that consist of small rod-like sclerites embedded in gorgonin. The internodes are formed from the same kind of sclerites cemented together with calcite. Colonies are quite fragile, breaking at node–internode joints. The sclerites are usually coloured, and besides the rods in the axis they include tuberculate clubs, leaf clubs, ovals and spindles with ridges or spines along one

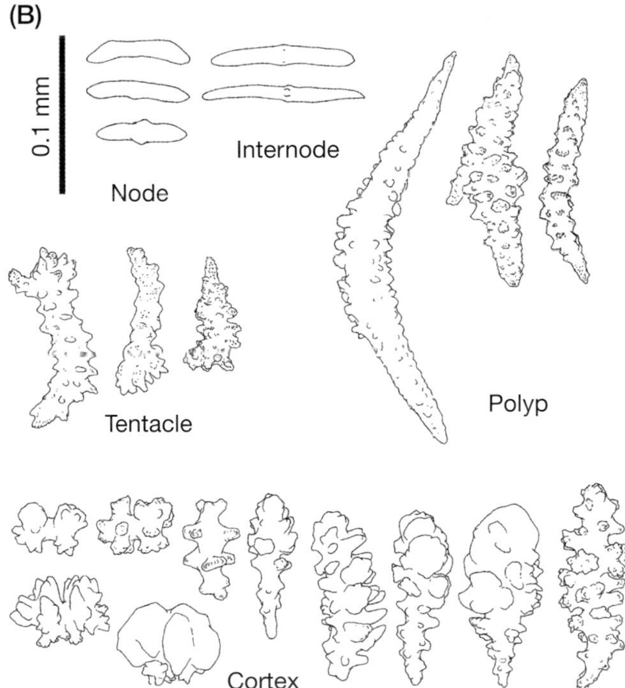

Fig. 23.21. Family Melithaeidae, *Melithaea*: **(A)** colony; **(B)** some of the sclerites representative of this genus (Photo: K. Fabricius; illustration: P. Alderslade).

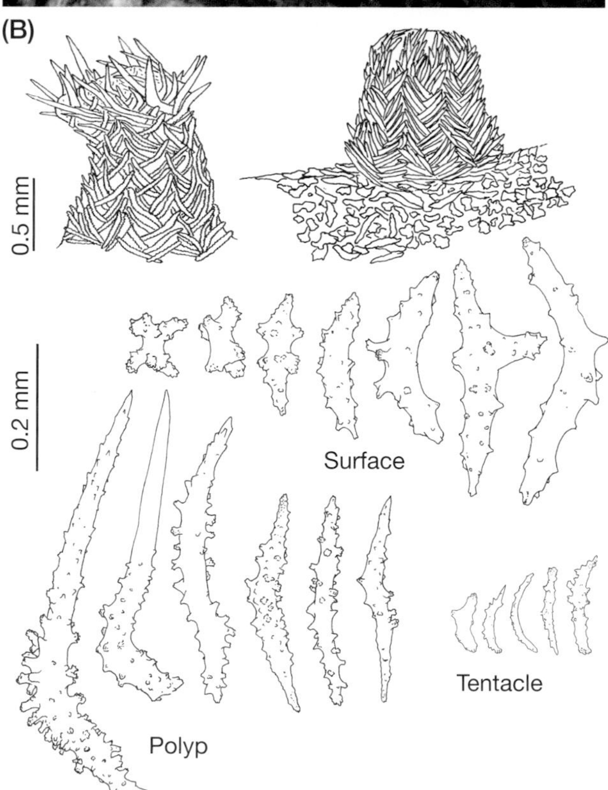

Fig. 23.22. *Acanthogorgia* sp. (family Acanthogorgiidae): **(A)** colony; **(B)** polyp drawings and some of the sclerites representative of this genus (Photo: K. Fabricius; illustration: P. Alderslade).

side and multirotulates (resembling two or more buns pressed together). The family contains five nominal genera that are probably just variations of the one genus, *Melithaea* (Fig. 23.21). The colour may be red, yellow or white. Melithaeidae are uncommon but widely distributed.

Family Acanthogorgiidae

Colonies are richly branched, fan-like, net-like, bushy, or untidy and tangled. The most noticeable features are the very conspicuous non-retractile polyps, which are completely covered with straight and curved needle- or spindle-shaped sclerites commonly arranged in eight double rows. Some species have a conspicuous crown of sharp spines around the top of the polyps. Sclerites on the branches are mostly tuberculate spindles, but tripods and capstans also occur. Colonies have a hard, black, purely horny axis with a wide, hollow, soft, cross-chambered central core. In some species, the coenenchyme is so thin that the black axis can be seen through it, which can be a clue to their identity, but in others it is quite thick, full of sclerites and opaque. In the latter species, the polyps can look like the calyces found in members of the family Plexauridae inside which a retractable polyp can be found, and so lead to a wrong identification.

Acanthogorgia (Fig. 23.22). Members of this genus usually form small fans with partial or full reticulation, but sometimes they are bushy. The axis is usually black and visible through the thin coenenchyme. Non-retractile polyps are often very tall, and covered in eight double rows of thin, bent spindles and the distal sclerites usually form a spiny crown around the polyp head. Coenenchymal sclerites may include small spindles, thorn-stars and capstans. The colour is bright, often multi-coloured. This genus is uncommon. Similar genera: all other members of the family; *Villogorgia*.

Family Plexauridae

Colonies in this family are bushy or fan-shaped (often net-like), and are sparsely to richly branched. They have a black to brown axis of gorgonin with a

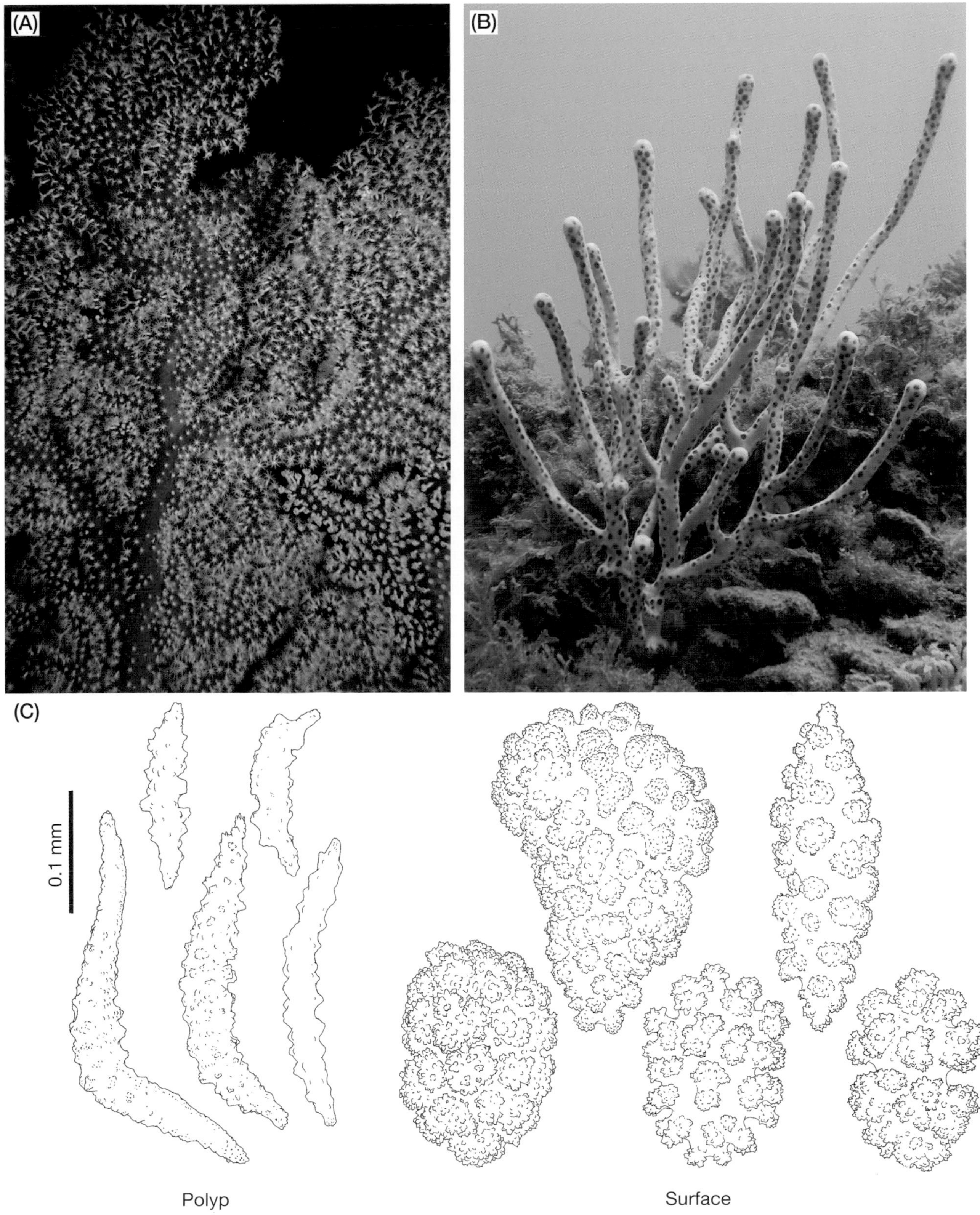

Fig. 23.23. *Euplexaura* sp. (family Plexauridae): **(A–B)** colonies; **(C)** some of the sclerites representative of this genus (Photos: K. Fabricius; illustration: P. Alderslade).

wide, hollow, soft, cross-chambered central core. Within the gorgonin are numerous microscopic spaces (loculi) filled with fibrous calcite. The polyps are retractile, often within prominent calyces, and are usually armed with large sclerites in a collaret and points arrangement. Sclerites are often quite large, longer than 0.3 mm, and sometimes as long as 5 mm. They are tuberculate, sometimes thorny, and the tubercles are rarely arranged in regular whorls. Sclerites called 'thorn-scales' often occur in the walls of calyces; these usually have a basal, spreading, root-like structure, and large, distal spines or blades that often protrude beyond the rim of the calyx. The sclerites in the coenenchyme come in a very wide range of forms.

Euplexaura (Fig. 23.23). The genus comprises planar or slightly bushy colonies, not richly branched, with branchlets usually arising at right angles and then bending upwards. Sclerites of the thick coenenchyme are plump spindles and spheroids. The colour may be white, grey or brown. This genus is uncommon.

Astrogorgia (Fig. 23.24). Colonies grow in one plane and are often densely branched. Polyps can be completely retracted into short or tall calyces, but the polyp head is often left sitting on the top of the calyx in preserved specimens. Polyp sclerites are spindle-like, usually numerous and arranged longitudinal often in eight groups or points. In tightly contracted polyps the lower point sclerites can tend to lay transversally resembling a collaret. The sclerites of the calyces and the branch surface are also spindles. In the calyces they are commonly arranged longitudinally, usually in eight double

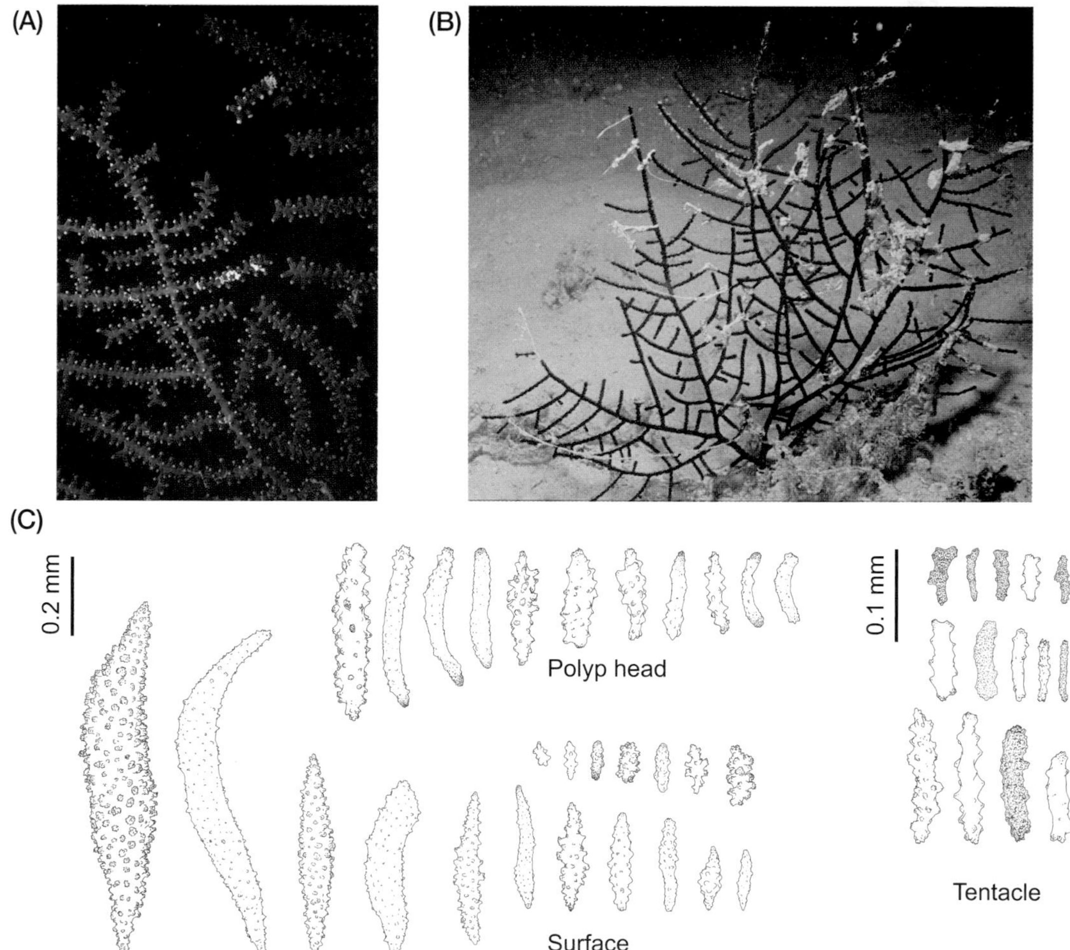

Fig. 23.24. *Astrogorgia* sp. (family Plexauridae): **(A–B)** colonies; **(C)** some of the sclerites representative of this genus. (Photos: A, K. Fabricius; B, L. Devantier; illustration: P. Alderslade).

rows, but can lie transversally. In branch surface they mostly lay along the branch, can be thin or thick, short or long, with the latter sometimes being longer then the branch is wide. The colour can be white, but such colours as reds, oranges, yellows, pinks and browns are far more common. Similar genera: *Muricella, Bebryce, Echinogorgia.*

Villogorgia (Fig. 23.25). Colonies are richly branched fans, often net-like, with polyps that are retractile into prominent calyces. The polyp sclerites are in a strong collaret and points arrangement (Fig. 23.2) and the tentacles contain curved scales commonly referred to as dragon wings. The sclerites of the calyces are thorn-scales that have a broad base, which fits the curve of the calyx, and a projecting upper part that maybe globular, but is usually thorny or leafy. Sclerites resembling the upper part of the thorn-scales, but modified as thorn-stars or thorn-spindle, occur in the coenenchyme of the branches. The colour may be brown,

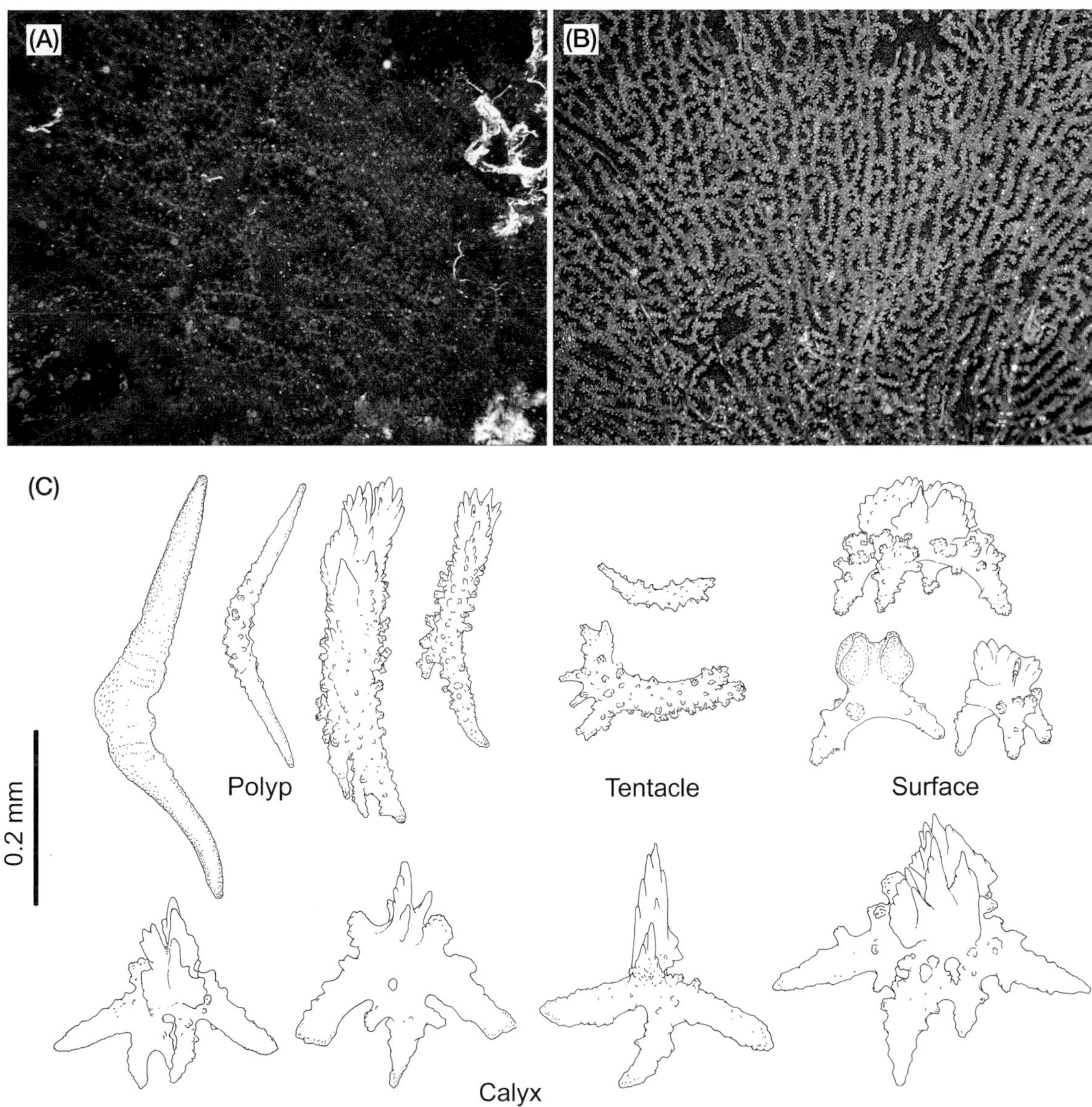

Fig. 23.25. *Villogorgia* sp. (family Plexauridae): **(A–B)** colonies; **(C)** some of the sclerites representative of this genus. (Photos: A, K. Fabricius; B, Queensland Museum (The photo of *Villogorgia compressa* G306181 was taken by Steve Cook on board the Southern Surveyor 7/9/1995, SW of Cape Jaubert); illustration: P. Alderslade).

orange-brown, dark red or yellow. This genus is uncommon but widespread. Similar genera: those that grow as finely branched fans like *Acanthogorgia, Echinogorgia.*

Echinogorgia (Fig. 23.26). Colonies grow in one plane and usually have short side branches that may join up to form net-like fans. The polyps are fully retractile into spiny calyces that may be low or very prominent. The polyp sclerites are arranged in a collaret and points arrangement and the tentacles contain curved scales commonly referred to as dragon wings. The sclerites of the calyces are thorn-scales that usually have three thorny or blade-like projections extending from a complex tuberculate base. The coenenchyme contains sclerites in various forms that generally consist of a complex tuberculate base from which arise thorns or blade-like processes. Large spindles with or without thorn-like projections on one side may randomly occur. The colour may be orange-red, yellow, red or white. Similar genera: those that grow as fans like *Astrogorgia, Menella, Paraplexaura.*

Family Gorgoniidae

Colony growth forms in this family are tree-shaped, bushy, pinnate, blade- or fan-shaped (sometimes net-like), and are sparsely to richly branched. They have a black to brown, horny axis, but in contrast to the plexaurid genera, the hollow, soft, cross-chambered central core is usually narrow, and the axial material surrounding the core is very dense, with little or no fibrous calcium carbonate present. Sclerites are generally small, usually less than 0.3 mm in length. Polyps are always retractile, sometimes into low mounds. The polyps may have no sclerites at all, but if present

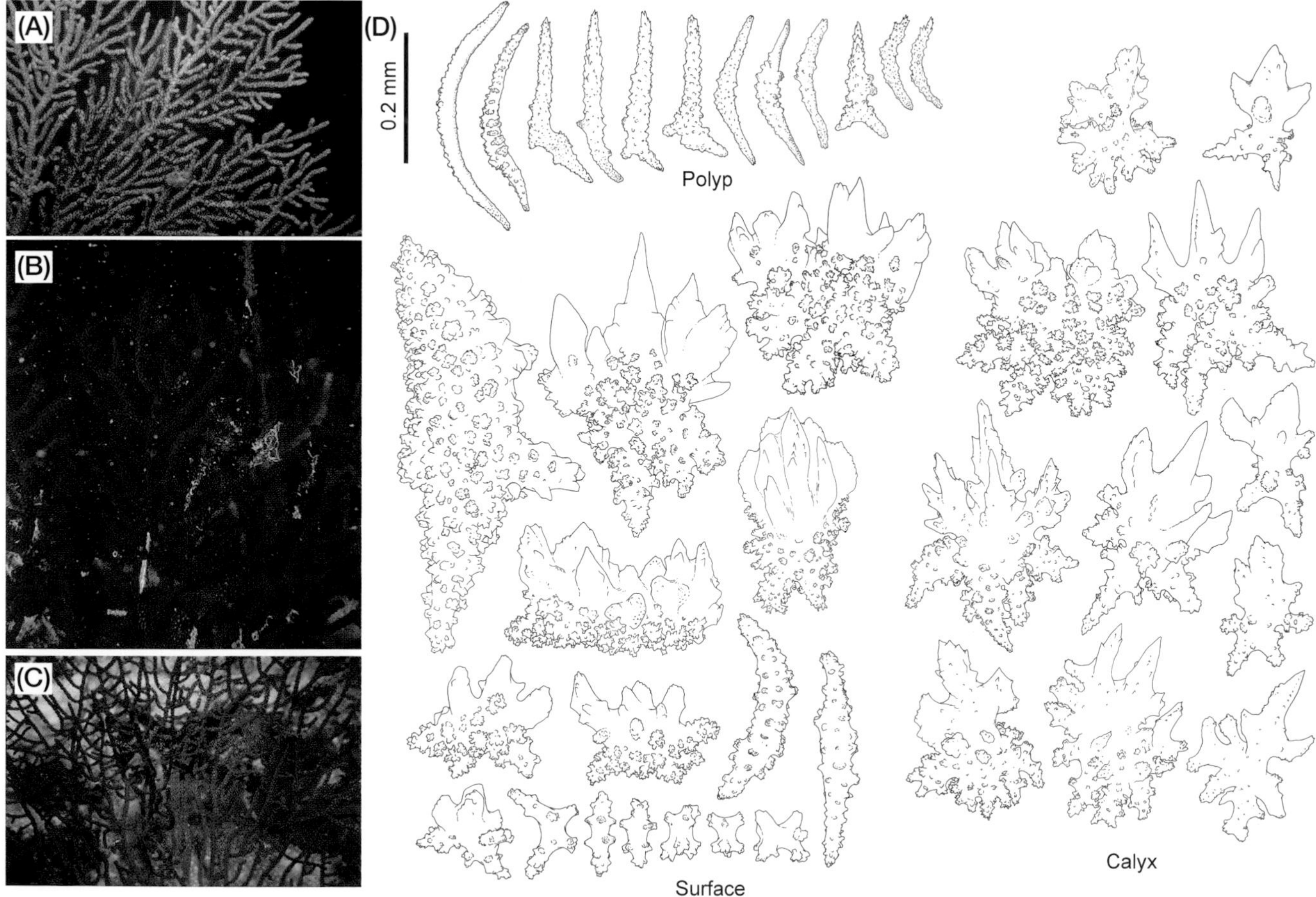

Fig. 23.26. *Echinogorgia* sp. (family Plexauridae): **(A–C)** colonies; **(D)** some of the sclerites representative of this genus (Photos: K. Fabricius; illustration: P. Alderslade).

Fig. 23.27. *Rumphella* sp. (family Gorgoniidae): **(A)** colony; **(B)** some of the sclerites representative of this genus (Photo: K. Fabricius; illustration: P. Alderslade).

they are generally small, flattened rodlets with scalloped edges. The sclerites in the rest of the colony are nearly always spindles with the tubercles arranged in whorls. Some species have curved, asymmetrically developed spindles.

Rumphella (Fig. 23.27). Colonies are bushy, often with a large, calcareous holdfast allowing them to grow in shallow wave exposed areas. Polyps are very small and retractile into the thick coenenchyme. Calyces are absent. Sclerites are symmetrical clubs and spindles with the tubercles in girdles. The colour may be brown to greenish-grey, contains zooxanthellae. *Rumphella* are uncommon, but widely distributed.

Family Ellisellidae

Colonies have a strongly calcified, continuous axis, and can be unbranched, loosely branched, or form broad, flat fans that may be net-like. It is very important to note the way a colony branches, because genera with the same sorts of sclerites are currently distinguished on growth form. Polyps are highly contractile, but not retractile. When contracted, they may fold over and lie against the branch surface, or just close up to form a small mound, which may look like a calyx. The major feature of this family is the characteristic form of the tiny sclerites, which are shaped like clubs, double heads and short spindles with distinctly separate, papillate tubercles. There can also be capstans with cone-like processes in the subsurface layer (Fig 23.30C).

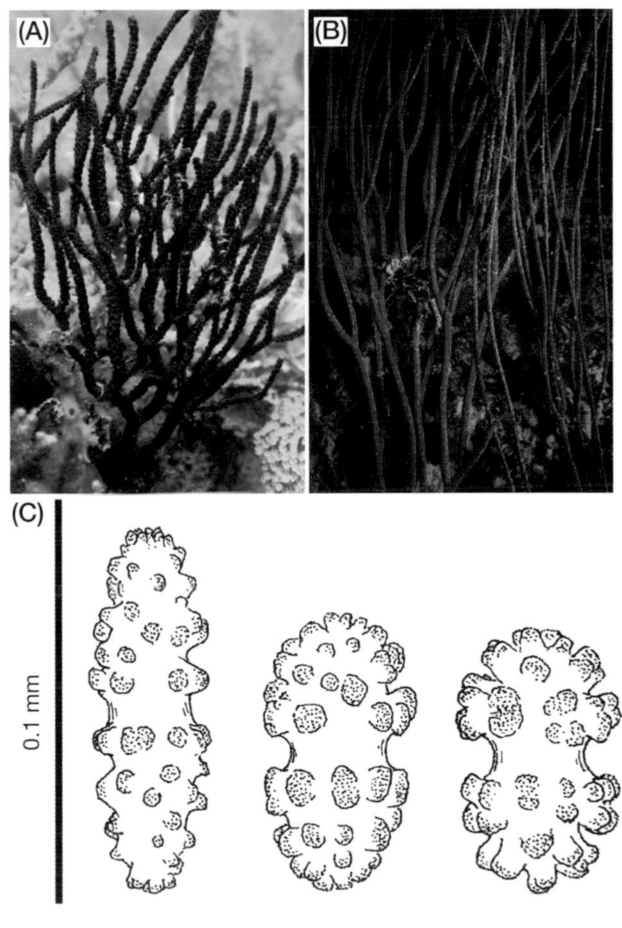

Fig. 23.28. *Ellisella* sp. (family Ellisellidae): **(A–B)** colonies; **(C)** some of the sclerites representative of this genus (Photos: K. Fabricius; illustration: P. Alderslade).

Ellisella (Fig. 23.28). Colonies are bushy, with whip-like branches, repeatedly branched in a dichotomous manner, generally <0.7 m in diameter. Polyps are usually folded up against the branch surface. Sclerites are double heads and waisted spindles. The colour may be red, yellow or white. *Ellisella* are uncommon but widely distributed.

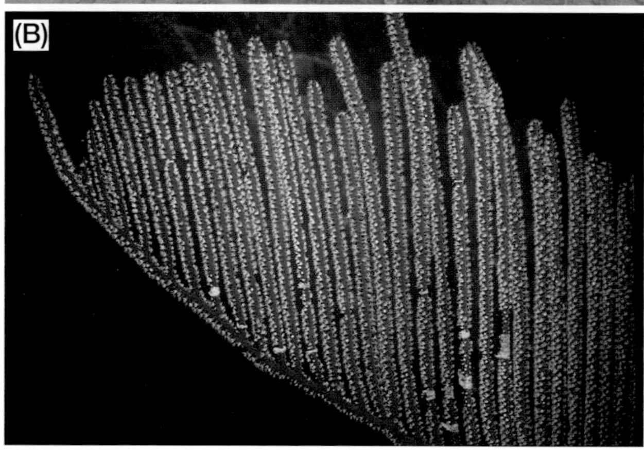

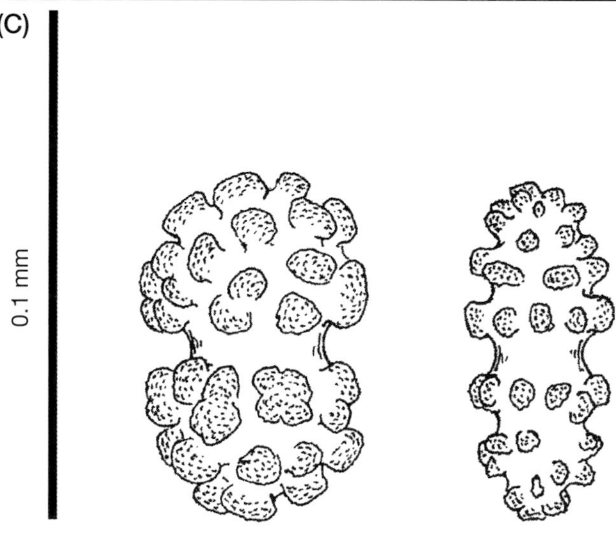

Fig. 23.29. *Ctenocella* sp. (family Ellisellidae): **(A–B)** colonies; **(C)** some of the sclerites representative of this genus (Photos: K. Fabricius; illustration: P. Alderslade).

Ctenocella (Fig. 23.29). Members of this genus have comb or lyre-shaped colonies that can grow to >1.5 m in size. Otherwise their characteristics are the much the same as those of *Ellisella*, although the waisted spindles are very short. They are usually red and generally occur in low light.

Junceella (Fig. 23.30). Unbranched, whip-like colonies up to >2 m in length occur in this genus. The polyps are usually folded up against the branch surface. Surface sclerites are clubs, subsurface sclerites are double stars. The colour may be white, cream, yellow or red. Some members of this genus can reproduce asexually by allowing the detached tip of the colony to fall to the substrate and so can be found in dense patches. They are quite common and widely distributed. Similar genus: *Viminella*.

Family Isididae

Colonies in this family have a distinctly segmented axis free of sclerites. Although the calcareous internodes are usually solid, in some species they are hollow, but this tubular nature is not to be confused with the soft, cross-chambered central core of the holaxonians. Internodes can be coloured, and although they can be quite smooth, they are often ornamented with prickles and ridges. The nodes that alternate with the internodes consist of pure gorgonin. Colonies can be whip-like, but are usually profusely branched, bushy or fan-like, but rarely net-like. Polyps can be retractile or non-retractile. If non-retractile, they are commonly covered with broad scales or narrow rods or needles. If retractile, the polyps either have no sclerites or are armed with spindles in a collaret and points arrangement. Sclerites in the rest of the colony can be of many forms. The majority of species in this family are deepwater inhabitants; *Isis hippuris* and *Jasminisis cavatica* are the only species found in shallow waters of the GBR.

Isis (Fig. 23.31). Members of this genus are planar to bushy, generally 20–40 cm tall. The axial skeleton consists of alternating horny nodes and non-spicular, opaque, solid calcium carbonate internodes. Polyps are small and retractile into a very thick

Fig. 23.30. *Junceella* sp. (family Ellisellidae): **(A–B)** colonies; **(C)** some of the sclerites representative of this genus (Photos: K. Fabricius; illustration: P. Alderslade).

Fig. 23.31. *Isis hippuris* (family Isididae): **(A–B)** colonies; **(C)** some of the sclerites representative of this genus (Photos: K. Fabricius; illustration: P. Alderslade).

coenenchyme containing abundant sclerites in the form of capstans and clubs. The colour may be bright yellow to green or brown. They are zooxanthellate. This genus is common in clear waters. Similar genera: none. By far the majority of the other genera in this family are quite different to *Isis*, and for quite a while it has been thought that the family is actually a mixture of several evolutionary lineages. This theory is now well supported by DNA analysis.

Further reading

Alderslade P (2001) Six new genera and six new species of soft coral, and some proposed familial and subfamilial changes within the Alcyonacea (Coelenterata: Octocorallia). *Bulletin of the Biological Society of Washington* **10**, 15–65.

Bayer FM (1981) Key to the genera of Octocorallia exclusive of Pennatulacea (Coelenterata: Anthozoa), with diagnoses of new taxa. *Proceedings of the Biological Society of Washington* **94**, 901–947.

Bridge TCL, Fabricius KE, Bongaerts P, Wallace CC, Muir P, Done TJ, *et al.* (2012) Diversity of Scleractinia and Octocorallia in the mesophotic zone of the Great Barrier Reef, Australia. *Coral Reefs* **31**, 179–189. doi:10.1007/s00338-011-0828-1

Dinesen ZD (1983) Patterns in the distribution of soft corals across the central Great Barrier Reef. *Coral Reefs* **1**, 229–236. doi:10.1007/BF00304420

Fabricius K, Alderslade P (2001) *Soft Corals and Sea Fans: A Comprehensive Guide to the Tropical Shallow Water Genera of the Central–West Pacific, the Indian Ocean and the Red Sea.* Australian Institute of Marine Science, Townsville.

Fabricius KE, De'ath G (2008) Photosynthetic symbionts and energy supply determine octocoral biodiversity in coral reefs. *Ecology* **89**, 3163–3173. doi:10.1890/08-0005.1

Fabricius KE, Alderslade P, Williams GC, Colin PL, Golbuu Y (2007) Octocorallia in Palau, Micronesia: effects of biogeography and coastal influences on local and regional biodiversity. In *Coral Reefs of Palau*. (Eds H Kayanne, M Omori, K Fabricius, E Verheij, P Colin, Y Golbuu and H Yurihira) pp. 79–91. Palau International Coral Reef Centre, Palau.

Goulet TL, LaJeunesse TC, Fabricius KE (2008) Symbiont specificity and bleaching susceptibility among soft corals in the 1998 Great Barrier Reef mass coral bleaching event. *Marine Biology* **154**, 795–804. doi:10.1007/s00227-008-0972-5

Grasshoff M (1999) The shallow water gorgonians of New Caledonia and adjacent islands (Coelenterata, Octocorallia). *Senckenbergiana Biologica* **78**, 1–121.

Hyman LH (1940) *The Invertebrates. Protozoa through Ctenophora*. McGraw-Hill, New York, USA.

McFadden CS, France SC, Sánchez JA, Alderslade P (2006) A molecular phylogenetic analysis of the Octocorallia (Cnidaria: Anthozoa) based on mitochondrial protein-coding sequences. *Molecular Phylogenetics and Evolution* **41**(3), 513–527. doi:10.1016/j.ympev.2006.06.010

McFadden CS, Sánchez JA, France SC (2010) Molecular phylogenetic insights into the evolution of Octocorallia: a review. *Integrative and Comparative Biology* **50**(3), 389–410. https://doi.org/10.1093/icb/icq056

Paulay G, Puglisi MP, Starmer JA (2003) The non-scleractinian Anthozoa (Cnidaria) of the Mariana Islands. *Micronesica* **35–36**, 138–155.

van Oppen MJH, Mieog JC, Sánchez CA, Fabricius KE (2005) Diversity of algal endosymbionts (zooxanthellae) in octocorals: the roles of geography and host relationships. *Molecular Ecology* **14**, 2403–2417. doi:10.1111/j.1365-294X.2005.02545.x

Verseveldt J (1977) Australian Octocorallia (Coelenterata). *Australian Journal of Marine and Freshwater Research* **28**, 171–240. doi:10.1071/MF9770171

Williams GC (1995) Living genera of sea pens (Coelenterata: Pennatulacea): illustrated key and synopses. *Zoological Journal of the Linnean Society* **113**, 93–140. doi:10.1111/j.1096-3642.1995.tb00929.x

Williams GC (2011) The global diversity of sea pens (Cnidaria: Octocorallia: Pennatulacea). *PLoS One* **6**, e22747. doi:10.1371/journal.pone.0022747

24

Worms

P. A. Hutchings

The common name 'worms' can refer to any of several groups of animals, including marine sea worms (includes polychaetes, sipunculans (peanut worms), echiurans (spoon worms) and marine earthworms), nemerteans (ribbon worms), myzostomes, nematodes, which may be free living or parasitic, and the parasitic worms the platyhelminths, which includes the cestodes (tapeworms), trematodes (liver flukes) and the free living turbellarians (flatworms). All these groups are abundant on the Great Barrier Reef (GBR), but the amount of information known about them varies according to the group. Each of these groups will now be discussed in terms of their diversity, where they occur on the reef, and their feeding and reproductive ecology. At the end of this chapter a series of references is provided, which will facilitate the identification of the groups and provide additional information about them.

Marine annelids

Marine annelids are one of the most diverse and abundant of these groups and were originally referred to as Polychaeta, a class within the phylum Annelida, together with Oligochaeta (earthworms) and the Hirudina (leeches), and this classification is still found in many textbooks. Recent studies (both morphological and molecular) on the annelids have failed to show that Polychaeta is monophyletic, and although the earthworms and leeches do form a single clade (Clitellata) within Polychaeta, their relationship to the polychaetes is still being debated. So today the term marine annelids is widely used rather than the term Polychaeta. Annelida currently includes all the traditional polychaete families as well as the leeches, earthworms and the siboglinids, sometimes known as 'beard worms', as well as the sipunculans and echiurans. Until recently, siboglinids were considered either as one or two phyla and included the Vestiminifera, which are restricted to cold and hot vents, and the Pogonophora, which are thin worms found in sediments. However, recent studies have shown that these worms are closely related to each other and each represents a group within the sea worms. To date, no examples of siboglinids have been found in the GBR, but some may occur in deeper waters off the outer barrier reef. Each of these groups will now be discussed.

Polychaetes

Polychaetes consist of two presegmental regions, the prostomium and peristomium (which may be fused), a segmented trunk and postsegmental pygidium (Fig. 24.1A). The body wall consists of circular and longitudinal muscle layers enclosing a body coelom, but in some groups the circular muscles are reduced. Usually they have a well-defined head with sensory and/or feeding appendages, followed by numerous body segments, which may be differentiated into thoracic and abdominal regions.

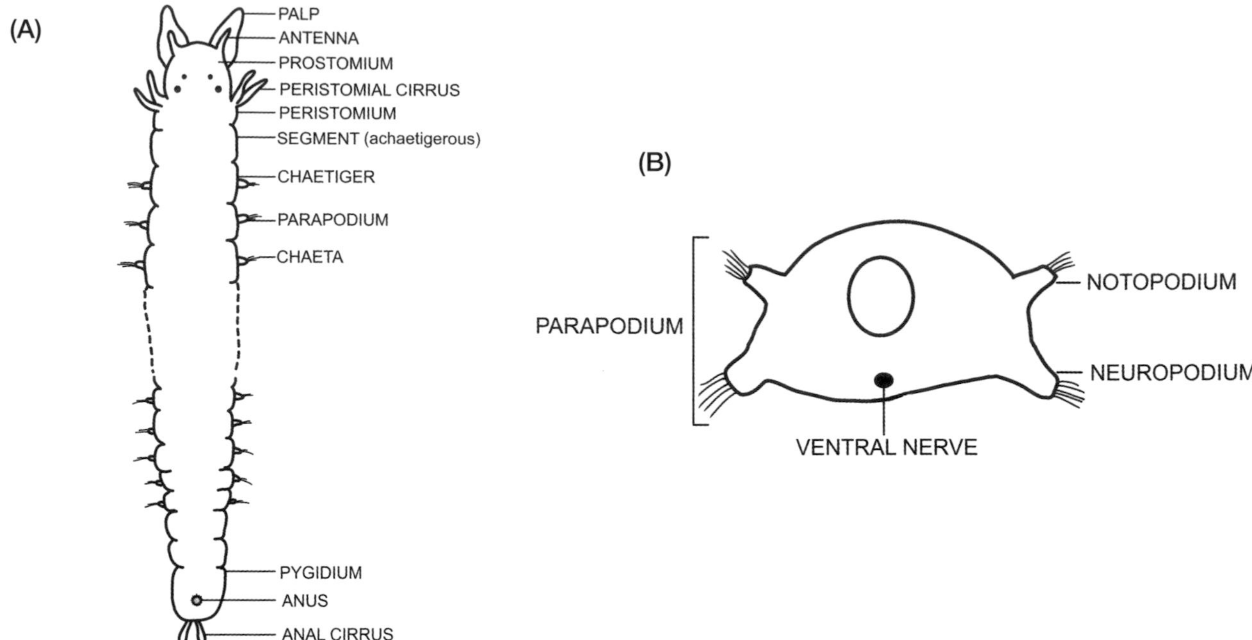

Fig. 24.1. **(A)** Stylised diagram showing major morphological characters of a generalised polychaete. **(B)** Cross-section of the body of a stylised polychaete, showing structure of the parapodia and location of the gut and ventral nerve cord (Source: After Fauchald 1977).

Fig. 24.2. Polychaetes. **(A) Family Nereididae:** *Perinereis helleri* with its pharynx everted showing paragnaths, an important diagnostic character for this family. Size: up to 4–5 cm in length. (Photo: H. Nguyen, lab of C. Glasby, Museum and Art Gallery Northern Territory, Darwin.) **(B) Family Amphinomidae:** *Chloeia flava* moving over the sediment. The chaetae easily break off and become embedded in fingers if the animal is picked up. Size: up to 100 mm in length (Photo: R. Steene). **(C) Family Amphinomidae:** *Eurythoe* spp. *complex*, 'fire worm'. These worms are commonly found under boulders intertidally and shallow subtidally. The dark red bushy branchiae are present adjacent to the parapodia all along body. Size: up to 60–80 mm in length (Photo: K. Atkinson). **(D) Family Amphinomidae:** *Pherecardia* sp. crawling over the substrate in search of food. It uses its eversible muscular pharynx to feed on sponges, anemones, hydroids, and so on. (Photo: R. Steene). **(E) Family Terebellidae:** *Reteterebella lirrf* deeply embedded within the coral substratum with highly extensile feeding tentacles spread out over the surface of the reef collecting food particles and moving them along the tentacles to the mouth (Photo: P. Hutchings). **(F) Family Serpulidae:** *Spirobranchus corniculatus*-complex illustrating the diversity of colours on a single live colony of *Porites*. This species settles on a damaged polyp, secretes a calcareous tube and encourages the coral to grow around the worm (Photo: R. Steene); **(G) Family Sabellidae:** *Megalomma interrupta*, a colourful fan worm. Each radiole has numerous fine filaments along its length that are used to strain the water passing through the crown. Particles pass down the axis of each radiole to the mouth, where they are eaten, used for tube construction or rejected. The crown is 50–100 mm in diameter (Photo: A. Semenov). **(H) Family Sabellidae:** *Sabellastarte* sp. This worm is fully extended from its muddy tube. The tentacular crown 50–100 mm in diameter is used in filter feeding and for respiration. (Photo: R. Steene.) **(I) Family Eunicidae:** *Marphysa* spp. complex, commonly known as blood worms, common in seagrass and muddy habitats adjacent to reefs, up to 200 mm in length. (Photo: K. Atkinson.) **(J) Family Eunicidae**: *Eunice aphroditois* emerges at night from its burrow to feed, keeping its jaws open for passing prey such as crustaceans and drift algae that are then grabbed and drawn down into the buccal cavity. They are sensitive to vibration and can rapidly withdraw into their burrows. They are found on shallow coral reefs and may reach 1–2 m in length (Photo: R. Steene). **(K) Family Phyllodocidae:** Phyllodocid found in coral sediments, with expanded dorsal cirri all along its body. Size: 4–5 cm in length (Photo: K. Atkinson). **(L) Family Phyllodocidae:** *Phyllodoce* sp. Mass spawning of these worms coincides with coral spawning. These worms normally live under rubble and swim up into the water column to spawn at around 2100 h (Photo: P. Hutchings).

Typically, each segment has a pair of parapodia with chaetae (bristles) (Fig. 24.1B). The diversity of life styles exhibited by polychaetes is often reflected by their morphology. For example, species that burrow through sediments, like the capitellids, tend to have few if any sensory or feeding appendages and reduced parapodia. In contrast, nereidids, which crawl actively over the substratum, have

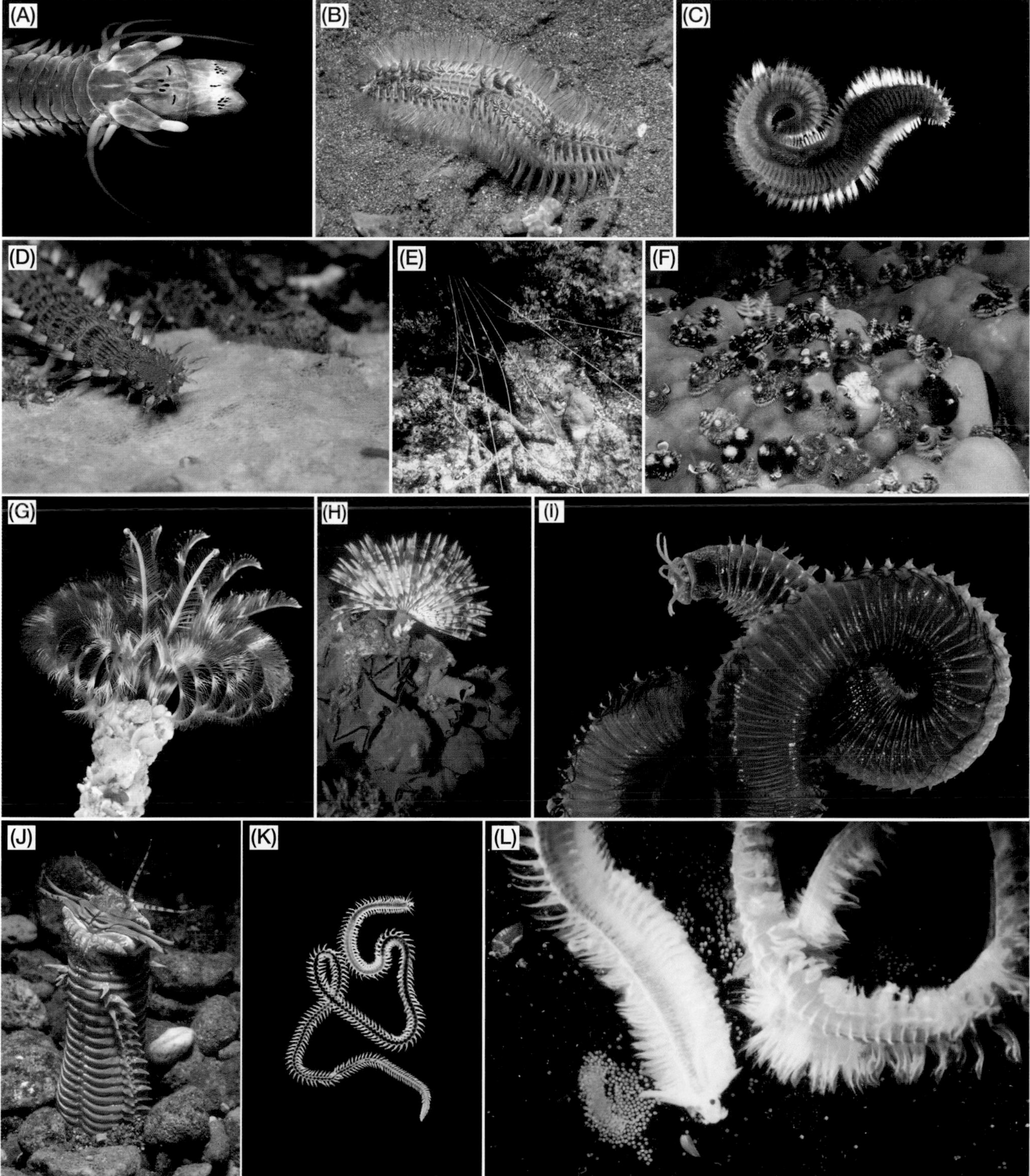

well-developed sensory appendages and parapodia (Fig. 24.2A) and also hesionids (Fig. 24.3F).

Traditionally, polychaetes consist of ~72 families grouped into clades whose composition, taxonomic rank and relationships are still being debated. Discussion of phylogenetic relationships, homologous characters, and phylogenetic relationships among polychaetes and other annelids is beyond the scope of this chapter but the references in 'Further reading' will provide interested readers with an introduction to the literature. Not only are the relationships within the polychaetes still being actively debated, but also their relationship to other invertebrate groups. For many years they were regarded as being closely related to the arthropods, based upon the presence of segmentation in both groups, but recently larval morphology and molecular analyses have shown they are more closely related to molluscs, brachiopods, phoronids and nemertines.

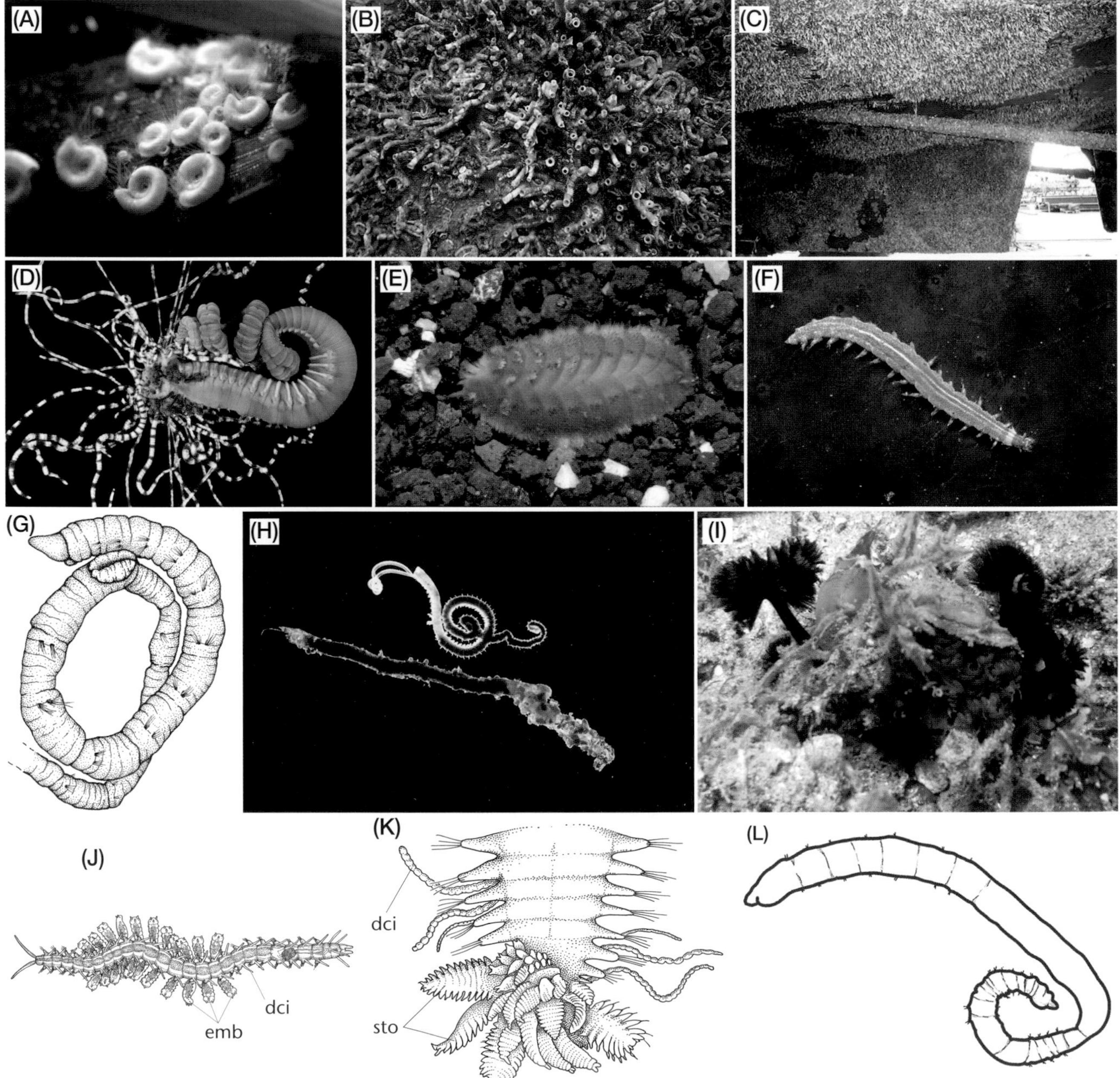

Polychaetes occur throughout the world in all habitats from the supralittoral to the deepest parts of the ocean. They are predominantly marine or estuarine but a few species occur in fresh water and moist terrestrial environments; most are free living although some are commensal or parasitic. Currently over 13 000 species are recognised worldwide and many more remain to be described. Many benthic marine environments are dominated by polychaetes, both in terms of number of species and numbers of individuals. In estuarine environments the diversity of polychaetes, like most invertebrates, is typically low, abundances are often high. Body size of polychaetes ranges from a few millimetres or less in length with few segments to worms many centimetres in length and with hundreds of segments.

Although polychaete families and many of the genera are known to occur worldwide, or have very wide distributions, it is at the species level that one finds restricted distributions. While no comprehensive survey of the polychaetes of the GBR has been undertaken, and our knowledge is mainly restricted to areas around field stations, we know that they are diverse and abundant both in inter-reefal sediments, and as borers or nestlers in dead coral substrate. Some families, such as the Serpulidae, always live in calcareous tubes firmly attached to hard substrates (Fig. 24.2F) or to other organisms such as algae (Fig. 24.3A), seagrasses, molluscs or crab carapaces. Some adult pelagic polychaetes are also found in the reefal plankton, as well as larval stages of many species. In addition, species are attached to floating debris, as commensals on the undersurface of holothurians and as fouling organisms on buoys and hulls of ships (Fig. 24.3B, C). Few polychaetes have common names and scientific names must be used. A recent workshop held at Lizard Island over 2 weeks yielded 91 new species and many new records, and only 24 families were studied, which highlights the amazing diversity awaiting discovery.

Many of the polychaetes living in reefal sediments are burrowers that break down organic matter in the sediment as it passes through their bodies, making it available to other organisms. They are abundant in seagrass beds and mangrove areas where large concentrations of organic matter accumulate from shed leaves. Such soft-bodied worms on intertidal reel flats are an important

Fig. 24.3. Polychaetes. **(A) Family Serpulidae:** *Eulaeospira* sp., firmly attached to a blade of seagrass. These small animals filter feed and they are able to seal off their tubes as one of the branchial filaments is modified to form a plug or an operculum. In some species a brood chamber is developed underneath the operculum (Photo: T. Macdonald). **(B) Family Serpulidae:** *Hydroides sanctaecrucis*. Close-up of calcareous tubes on a fouled yacht (Photo: J. Lewis). **(C) Family Serpulidae:** *Hydroides sanctaecrucis*. This is an introduced species that has colonised a yacht in a marina in Cairns. (Photo: J. Lewis.) **(D) Family Terebellidae:** *Lanice viridis*. This is a surface deposit feeder living under rocks or at the base of bommies. Size: 30–50 mm in length, excluding tentacles (Photo: A. Semenov). **(E) Family Polynoidae:** A scale worm, probably a species of *Iphione*, crawling over coarse sediment. Scale worms are active carnivores, catching prey using their eversible pharynx. Encrusting barnacles are present on some elytra in this specimen (Photo: R. Steene). **(F) Family Hesionidae:** *Hesione* sp. or *Leocrates* sp., crawling over the substrate. Hesionids are active carnivores that feed by everting the pharynx and sucking up their prey (Photo: R. Steene.) **(G) Family Questidae**: *Questa ersei*. Line drawing of anterior end of the animal, showing its simple body lacking parapodia, paired bundles of chaetae on all segments and the head reduced to a simple palpode with an eversible pharynx. Found living within sediment, it can reach lengths of up to 10 mm. (Figure reproduced with permission from Beesley PL, Ross GJB, Glasby CJ (Eds) (2000) *Fauna of Australia – Volume 4A: Polychaetes and Allies: The Southern Synthesis.* Australian Biological Resources Study/CSIRO Publishing). **(H) Family Chaetopteridae:** *Mesochaetopterus* sp. One is removed from its tube and the other is still in its flimsy tube made of sand grains. These worms are common in reefal sediments. The anterior palps are used for feeding. The enlarged parapodial lobes in the mid-body create water currents that pass through the tube and particles of food are collected. Size: up to 10–20 mm in length (Photo: K. Atkinson). **(I) Phoronid:** These animals are always associated with burrowing sea anemones (Photo: K. Attwood). **(J)** Externally brooded embryos in *Pionosyllis elegans* are attached to the ventral cirri by glandular secretions; dci= dorsal cirrus, emb= embryos (Source: after Pierantoni 1903). **(K)** Multiple male stolon production from the venter of posterior segments in *Trypanosyllus crosslandi*; dci= dorsal cirrus, sto= stolons (after Potts 1911, redrawn by K. Nolan). **(L)** Line drawings of a marine oligochaete (Illustration: K. Attwood).

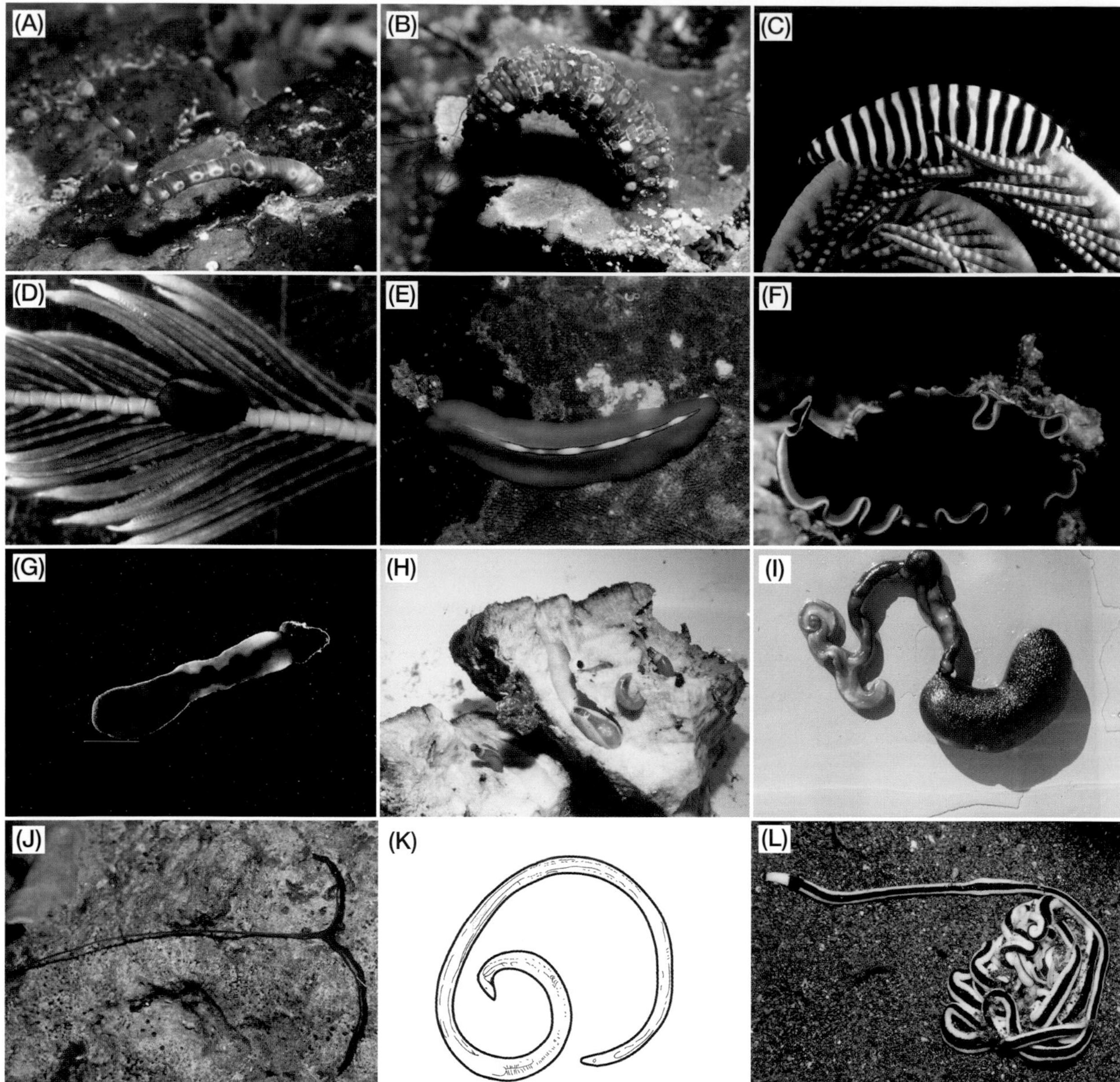

Figs 24.4. (A) Family Hirudinae: *Trachelobdella* sp. (Photo: R. Steene). **(B) Family Hirudinae:** *Stibarobdella macrothela* that feeds exclusively on elasmobranchs (Photo: R. Steene). **(C)** Myzostome, *Hypomyzostoma dodecephalcis* on its host *Zygometra elegans* (a crinoid) from Lizard Island (Photo: G. Rouse). **(D)** *Myzostoma bicorneon* on its host *Amphimetra tessellata* (a crinoid) from Lizard Island (Photo: G. Rouse). **(E)** Marine flatworm gliding over the coral (Photo: R. Steene). **(F)** Flatworm *Pseudobiceros hancocki* appears to be almost swimming over the reef (Photo: R. Steene). **(G)** Sipunculida, an unidentified species of *Aspidosiphon* (family Aspidosiphonidae), removed from its burrow deep within the coral substrate (Photo: K. Atkinson). **(H)** Sipunculans entombed in the coral (Photo: P. Hutchings). **(I)** Echuira extracted from its burrow, probably a species of *Bonellia* (Photo: Australian Museum collection). **(J)** Echuira, bifid tentacle of *Bonellia* spread out over the sediment at the base of a coral bommie at night (photo: R. Steene). **(K)** Line drawings of nematode (illustration: K. Attwood); **(L)** nemertean, *Baseodiscus hemprichii* (Photo: R. Steene).

food source for wading birds at low tide and for fish and crustaceans as they move over such flats at high tide. Other polychaete species found on the reef are active carnivores, such as the large fire worms commonly found underneath boulders or rubble at the base of reefs. Fire worms belong to the Amphinomidae (Fig. 24.2B, C), and they should be handled with care because their long barbed chaetae easily break off and lodge in skin and some people react to this. Their name reflects the burning sensation from this sting. This family is well represented in tropical waters, and floating pieces of pumice from a distant volcanic eruption are often covered with fire worms (and often with goose barnacles). Amphinomidae range from a few millimetres to several centimetres in length, with robust typically square shaped bodies, and most are active carnivores although others are omnivores or opportunistic feeders.

Long white tentacles can often be seen spread out over the reef – these belong to at least three species commonly referred to as the 'spaghetti worm' (*Reteterebella* spp.) belonging to the family Terebellidae. These tentacles collect food particles and when touched rapidly contract into their tubes (Fig. 24.2E), which are soft and flimsy, made of fine sediment cemented together with mucus. *Reteterebella queenslandia* are found in dense numbers on reef flats such as at Heron Island, whereas *R. lirrf* (Fig. 24.2E) occurs deep down in the coral. It seems likely that in both species these tentacles contain noxious compounds, because fish avoid them. Another common species of terebellid is *Lanice viridis* (Fig. 24.3D), which lives at the base of small bommies.

Conspicuous 'Christmas tree' (*Spirobranchus corniculatus* –complex) worms are common in live coral, especially the massive corals such as *Porites,* in a diverse range of colours (Fig. 24.2F) and they rapidly withdraw their branchial crowns in response to a shadow, closing the calcareous tube with a modified branchial filament which forms a plug (or operculum). Although these worms, which belong to the family Serpulidae, on first inspection appear to have bored into the live coral, they have not: rather they have stimulated the coral to grow around them. This happens when their pelagic larvae, which have spent several weeks in the plankton, settle on a damaged polyp on the living coral (to avoid being eaten), and rapidly secrete a thin calcareous tube. This stimulates the coral to grow around the tube and gradually the worm tube becomes firmly embedded in the coral. The branchial crown, when held upright in the water column, is able to strain the water column effectively and small particles of food are trapped in mucus along the ciliated filaments and these particles are transported along the axis of the filaments down to the mouth at the base of the crown where they are eaten. The crown is also important for respiration. On the GBR they are commonly, but mistakenly, referred to as *S. giganteus* but this species is actually restricted to the Caribbean. The Australian species belong to the *Spirobranchus corniculatus*-complex and some undescribed species occur. The genus has been widely reported from all coral reef areas of the world. The colour of the branchial crown, which can vary from yellow, blue, red and purple, is not a useful character to distinguish species, and the structure of the opening tube as well as chaetal characteristics are important. Many other species of serpulids are common on the reef and most have an operculum, the structure and ornamentation of which is critical for species and generic identification. Other genera of much smaller serpulids are common on algal fronds and on blades of seagrasses and form small calcareous tubes in a tight spiral, looking like a 'comma'. Once considered to belong to a separate family, the Spirorbidae (Fig. 24.3A), these now regarded as belonging to the family Serpulidae. These small worms also have an operculum and on some species a brood chamber develops on the undersurface where the fertilised eggs develop, producing miniature adults. They may breed continuously over several months. Serpulids are commonly transported across oceans by drift algae and as fouling organisms on the hulls of ships (Fig. 24.3B, C). Some species are gregarious and larvae are attracted to settle close to adults of the same species.

Another group of filter feeding polychaetes are the sabellids (family Sabellidae), or fan worms, which also have a branchial crown but are always found in muddy/sandy tubes and with no branchial filament modified to form an operculum (Figs 24.2G, H). A tremendous diversity is found ranging from species a few millimetres in length to large conspicuous fan worms with a branchial crown

2–3 cm in diameter. All these fan worms can retract rapidly back into their tubes in response to a passing shadow or vibration, but sometimes they are not quick enough and are eaten by a predatory fish. Providing the worm can rapidly contract the body wall to prevent loss of coelomic body fluids, they can regenerate the crown. Fan worms are found in sediments with the tubes anchored onto a small piece of hard substrate such as shell fragment, or they actively bore into coral substrate and line their burrow with a fine chitinous tube.

Underneath large boulders may be found species of eunicids, which may reach several centimetres in length, with numerous segments, often an iridescent epithelium and along the body on restricted segments, bright red, tufted branchiae are present (Fig. 24.2I). Another species of eunicid, *Eunice aphroditois,* is found in the low intertidal zone deeply embedded in the reef crest, and at early morning low tides they extend out of their burrows and graze on the algae (Fig. 24.2J). Some species of eunicids bore into dead coral substrate using both mechanical and acid secretion to excavate their burrows deep within the coral substrate, but maintaining an opening to the exterior through which they obtain their food, oxygen from the water column as well as discharging their gametes.

Another family, the nereidids or ragworms, are common on the reef, occurring in many habitats – in the sediments, associated with filamentous and encrusting algae, as nestlers in dead coral substrate, and in the infralittoral zone underneath decaying vegetation. They have an eversible pharynx with

Box 24.1. Asexual reproduction of polychaetes

Some species of polychaete can undertake asexual reproduction as well as sexual reproduction. *Dodecaceria* (family Cirratulidae) is common in dead coral substratum as it can bore into such habitats. An adult worm is capable of splitting into individual segments and each one develops a new head and tail to become a new individual. Another family that exhibits asexual reproduction is the Syllidae. In some cases, the worm develops a series of stolons, resembling railway carriages, with the most posterior one the next one to be released as a new individual (Fig. 24.3J); in other cases the stolons are proliferated off at angles to the adult worm (Fig. 24.3K).

Box 24.2. Mass spawning of polychaete species

In the late 1980s, when mass coral spawning was discovered, often polychaetes were also seen swimming in the water column and they too were spawning. Such mass spawning has been known and exploited by the Samoans for centuries. They could predict the night on which these sedentary eunicids, the 'Palolo' worms that live deep within the coral substrate, would swim up into the water column where their posterior ends would split open to release their gametes, triggered by the particular phase of the moon. The Samoans would collect these swimming sacs full of protein-rich eggs and sperm and dry them to produce a nutritious flour that they would make into biscuits or else trade with highlanders. On the GBR, a suite of species spawn at particular phases of the moon during the summer months including species of nereidids, amphinomids, syllids and phyllodocids (Fig. 24.2L). In these species the entire worm develops swimming chaetae and large eyes, and leaves the substratrum and swarms in the water column where the males secrete a pherome that stimulates the females to spawn. By the end of the night the beaches are littered with the remains of their bodies. All the gametes are released into the water column, providing fish and other predators with an amazing feast. Almost certainly eunicids spawn on the GBR like in Samoa, but to date this has not been documented. In all these cases, while water temperature and light levels are critical for coordinating spawning, other factors must be involved months before to ensure that the germinal epithelium of these worms proliferate the gametes and that there is sufficient time for the eggs and sperm to develop to maturity within the coelom. Laboratory studies have revealed that all these processes are controlled by an endocrine system that responds to external cues such as temperature and light.

well-developed jaws, which they use to collect prey and scavenge opportunistically (Fig. 24.2A). A closely related family, the predatory Phyllodocidae, is common on the reef and they can be abundant on reef flats at particular times of the year when the mature worms congregate on the surface to spawn and release their gametes, producing egg masses that are attached to pieces of algae or substrate, while other species spawn in the water column (Fig. 24.2L).

Scale worms are common underneath coral substrate and are characterised by a series of overlapping scales (or elytra) which cover the dorsum and which may be brightly coloured, with distinct pigment patterns and/or heavily ornamented (Fig. 24.3E). This group is rich in species on the reef and some live commensally, most often on holothurians and other echinoderms.

Another common group in reefal sediments are members of the family Chaetopteridae belonging to the genus *Mesochaetopterus* (Fig. 24.3H). These animals build sandy tubes that protrude from the surface of the sediment. The worm pumps water through the tubes, trapping particles that are used by the animal for tube construction or else eaten.

Another common family on the reef are the tiny syllids (Syllidae). These are well known among specialist because of their striking reproductive modes that imply strong changes not only in their morphology but also their behaviour. Some syllids are able to modify the posterior segments or produce new ones full of gametes that develop their own eyes and anterior appendages (Fig. 24.3J). These posterior segments with sensory structures are detached from the stock and become sexual units named 'stolons' that actively swim for spawning in the pelagic realm and, in some cases, brood eggs and embryos (Fig. 24.3K).

The distribution of these, and the many other kinds of polychaetes that occur on coral reefs, is largely dependent on the substratum, such as the size and type of sediment, presence of suitable substrate for the borers and nestlers and encrusting species. Factors such as exposure and water currents are also important for filter-feeding organisms. Species living in sediments need to have stable sediments so high energy beaches are typically low in number of species and individuals. Because all polychaetes are soft bodied, they need protection from predatory organisms, either by secreting a tube into which they can rapidly retract or habitats in which they burrow and avoid predation. A few species have developed anti predator strategies, and the numerous chaetae of amphinomids (Fig. 24.2B–D) may make them unpalatable to fish and other predators. However, fish, some molluscs and bird guts reveal that polychaetes are an important prey item for many species and even when buried in sediments they can still be eaten. Lacking an external skeleton or shell, they provide an easy to digest source of food.

Polychaetes exhibit an amazing diversity of reproductive modes, both sexual and asexual (see Boxes 24.1, 24.2). Although most polychaetes are dioecious (i.e. with separate males and females) some species are hermaphrodites, and others may be males at particular times of their life and then become females subsequently. Species may live for a few weeks or months to many years, some breed continuously over several months, whereas others are restricted to spawning on a single day. Some species actively mate, with the male fertilising the eggs as they are laid. Fertilised eggs are then placed under the dorsal scales (e.g. some polynoids), others are brooded within the adult tubes, or placed in a chamber below the operculum, others lay their fertilised eggs in brood capsules, which are then attached to the substrate where development occurs. In some cases, the developing embryos are cannibals, eating some of the other embryos, but in other cases the large yolky eggs have sufficient nutrients for them all to hatch as miniature adults. Other species are broadcast spawners, with gametes being released into the water column where fertilisation occurs (an example of broadcast spawning occurs on the GBR: see Box 24.2). In these cases, a free-swimming larvae, or trochophore, is produced, which spends anything from a few hours to many weeks in the plankton before settling and metamorphosing into an adult worm. Some larvae feed while in the plankton and may therefore be dispersed over great distances, whereas other species, which produce large yolky eggs, do not. In all cases, it is important that males and females mature at the same time, and that spawning is synchronised and this is achieved by a well-developed endocrine system.

Although not too many polychaetes will be seen when diving on the reef, closer inspection will usually reveal tentacles and expanded branchial crowns of sabellids and serpulids. At night more tentacles can be seen spreading out over the substrate or being held up in the water column fishing. Only when sediment samples are collected and sorted under the microscope will the diversity of the polychaetes be revealed, or if pieces of dead coral substrate are split open will all the borers and nestlers be seen. However, a permit is required to collect such samples (see Chapter 14).

Clitellates

Clitellates include oligochaetes (relatives of terrestrial earthworms) and the hirudineans (leeches). This group, like the polychaetes, is also a paraphyletic group which still needs to be resolved. Although oligochaetes and leeches are usually thought of as being terrestrial, both groups have many species in both freshwater and marine sediments. Few studies of clitellates have been carried out on the GBR but they are known to be common in marine sediments on the reef.

Oligochaetes are segmented worms without parapodia, with segmental bundles of simple hook-like or capillary chaetae. Species tend to be small, <1 cm in length, and thread-like (Fig. 24.3.L), often bright red in colour due to the blood pigment haemoglobin. They play an important role in these meiofaunal communities in the breakdown of particulate matter and they themselves are predated upon by many animals. The prostomium is simple, lacking any sensory appendages. They are hermaphrodites, and worms copulate and exchange gametes, with fertilised gametes being laid in cocoons secreted by the clitellum. The cocoons are poorly developed, containing both the male and female gonopores and deposited in the sediment. Miniature worms hatch out from the cocoons after variable periods of time. Some species undergo asexual reproduction by transverse division of the parent worm into two or more individuals. Marine oligochaetes feed on fine detritus, algae and other microorganisms using their eversible pharynx. The diversity of the oligochaete fauna is a good indication of the 'health' of the sediment.

Marine leeches (Hirudinea) are present on the GBR, while some are free living (Fig. 24.4A, B) others rely upon an intermittent supply of blood from marine vertebrates, including turtles, fish and elasmobranchs (sharks and rays). Some appear to be fairly host specific, whereas others are not. They use their anterior sucker to attach onto the host and inject an anticoagulant, which allows the leech to suck out the blood without it clotting. A blood-feeding leech will be two to three times larger after feeding than before. Leeches are typically dorsoventrally flattened and tapered at each end, with segments at both extremities modified to form suckers. Leeches are segmented, although external segmentation is obscured, and chaetae are absent. A clitellum is present and always formed by segments 9, 10 and 11, although only conspicuous during reproductive periods. Branchiae are present in the Piscicolidae, as lateral leaf-like or branching outgrowths. Species without branchiae respire through the skin. Like all clitellates, leeches are hermaphrodites and animals mate and exchange sperm, which is then stored and used as required to fertilise the eggs, which are laid in cocoons from which miniature adults hatch out after some weeks.

Sipunculans

Sipunculans are soft-bodied, unsegmented coelomate worm-like animals, and are commonly called peanut worms (Fig. 24.4G, H). They consist of a muscular trunk or body and an eversible introvert that bears the mouth at the anterior extremity. The introvert is highly elastic, capable of considerable extension, and at other times is partially or completely retracted within the body cavity. The mouth leads to a long, recurved, spirally wound alimentary canal, which lies within the coelom. Tentacles either surround the mouth or are associated with it. Chaetae are absent. One or two pairs of nephridia are present. The body wall is usually thick and often rubbery and has well-developed circular and longitudinal muscles and occasionally a thin layer of oblique or diagonal muscle is also present. The skin often contains pigmented papillae and the surface of the introvert may be covered in hooks. A caudal shield or calcareous knob or cone is often present around the anus. Sexes are separate and fertilisation is external and a free-swimming trochophore larva is produced.

Sipunculans respire through their body wall and tentacles, and cells in both the vascular and coelomic systems contain a red respiratory pigment, haemerythrin, which gives up its oxygen only at a very low partial pressure, much lower than commonly found in sea water. It has been suggested that it functions only in adverse situations, such as may occur on intertidal reef flats during low tides in the middle of the day during summer.

Sipunculans are abundant on the reef, although not often seen unless sediment samples are collected or reef rock broken off (Fig. 24.4H). Some species live in characteristic shaped burrows in reefal substrate or in tubes in the sediment, while others bore into solitary corals or inhabit empty gastropod shells. The solitary coral *Heteropsammia michelini* commonly contains the sipunculan *Aspidosiphon muelleri.* Rock boring species secrete an acid-like secretion and after settling as larvae metamorphose and bore into the substrate, forming a flask shaped burrow where they spend their lives while retaining a narrow passage to the exterior through which they obtain their oxygen and release their gametes. They are detritus feeders, using their tentacles and extensible introverts to collect algae, sediment and detritus. Sipunculans are selectively fed upon by some molluscs and by humans in selected localities in the tropical and subtropical Indo-Pacific.

Six families of sipunculans are known and numerous genera and species have been recorded from the GBR and many remain to be described but appear to have phylogeographic affinities with the Indo-Pacific. Increasingly, molecular studies are showing that sipunculan species have restricted distributions while genera are widely distributed.

Echiurans

Echiurans exhibit a developing nervous system which shows some traces which appear to be a reminiscence of segmented ancestors, coelomate, bilaterally symmetrical worm-like marine animals (Fig. 24.4I, J). They have an elongate muscular trunk and an anterior extensible proboscis. They are commonly referred to as spoon worms, because their proboscis when extended is used to collect sediment from around its burrow.

The muscular trunk typically bears numerous flat or swollen glandular and sensory papillae and is usually light to dark green, but some species are reddish-brown. A pair of golden-brown chaetae is usually present on the ventral surface of the trunk, just behind the mouth. The proboscis is usually flattened and ribbon-like, but may be fleshy and spatulate. It is highly extensible and contractile but cannot be withdrawn completely into the body cavity like sipunculans. The distal end of the proboscis is usually truncate or bifid. It is this proboscis that is typically seen on the reef, especially on night dives, and when touched it retracts and is smooth and slippery (Fig. 24.4J). The mouth of an echuiran is present anteroventrally at the base of the proboscis and the anus is at the posterior extremity of the trunk. Echuirans are detritus feeders, except for species of *Urechis,* which trap very fine particles by secreting a mucus net.

The sexes are separate, mature eggs and sperm pass out through the paired nephridia, of which there may be one to 10 pairs. Fertilisation is external, and a pelagic trochophore is produced. The only exception are species of *Bonellia* (Fig. 24.4I, J), which are common on the reef, where the male is minute and lives permanently within the female body.

Echiurans make burrows in sand and mud, live in the shells of molluscs and tests of sand dollars and in galleries made by other animals in the coral.

There is no evidence that they themselves are capable of burrowing into the coral. They also occur on tropical mud flats intertidally.

To date, only 13 species and subspecies of echiurans have been described from Australia and seven of these are endemic. Although at present on the GBR, it is not known how many species reside there.

Myzostomids

Myzostomids are a bizarre group of worm-like animals, which are all obligate symbionts mainly with crinoids and other echinoderms. They range from forms that are mobile and move freely over the host stealing food to others that are endoparasitic and live inside the host; some of the latter stimulate the host to form cysts or galls around them. Although myzostomids occur worldwide, they are commonest in reefal areas where crinoids are abundant. The body form of the mobile species varies from flattened oval to disc-like forms and some are ridged or have elongate extensions, which resemble the pinnules on the crinoid arms (Fig. 24.4C, D). Many are coloured to resemble their hosts and they may be quite difficult to spot on the crinoid. The mobile forms basically have a fused body and appear unsegmented, the head is not a distinct structure, with no eyes and the mouth is a ventral or terminal structure. If one dissects a specimen, the body is clearly segmented with the five segments separated by septa and with five pairs of appendages. The appendages have a stout emerging hook-shaped chaeta and are supported by a single aciculum. These structures are far less obvious in the endoparasitic species. They are usually protandric hermaphrodites and are initially functional males, which they then become simultaneous hermaphrodites at maturity. Fertilisation is internal and fertilised eggs are spawned into the water column and give rise to planktonic, non-feeding trochophores and they settle on a host within 5–8 days. The larvae have two bundles of long chaetae, which resemble those found in the larvae of some polychaete families. How the endoparasitic and the gall producing species reproduce is still not known.

Platyhelminthes

This group of unsegmented worms includes the free living turbellarians and three major parasitic groups. Turbellarians are mainly small and cryptic (although highly abundant), but the Polycladida are large (up to several centimetres long) and may have spectacular colouration.

Cestoda

The Cestoda (tapeworms, which are secondarily segmented) are at their most abundant as parasites in sharks and rays; on the GBR only rarely are individuals of these groups are without tapeworms. All tapeworms have complex multi-host life cycles that commence with an egg in the faeces of the final host. First intermediates are usually small crustaceans, although this has never been demonstrated on the GBR. Second intermediate hosts are a range of vertebrates, especially bony fishes, which are often heavily infected.

Monogeans

The Monogenea are ectoparasites of fishes, including both bony and cartilaginous fishes. Most species infect fish gills, but some infect fish skin and a few are internal parasites. The life cycle of monogeneans involves only a single host. Eggs are usually dropped by the adult worm and tiny ciliated larvae eventually hatch and must find another animal to infect; they may use complex host-finding behaviours to enhance this process. Typically, the host-specificity of monogeneans is high so that most species infect just one fish species.

Trematodes

The Trematoda (flukes) occur as sexual adults and are abundant as internal parasites in bony fishes, reptiles, birds and marine mammals; strangely they are not common in sharks and rays. Trematodes mainly infect the gut of their hosts, but some occur in the circulatory system, under the scales and in the muscle. Green turtles can host up to 12 species and 16 species have been recorded from dugongs. The life cycle of digenenean trematodes is complex, involving two, three or four hosts. The first intermediate host, in which asexual reproduction occurs, is almost always a mollusc. Most trematodes are transmitted to the final host by the ingestion of an intermediate host, but some are transmitted by direct penetration, direct attachment, or direct ingestion.

Combined, there are several hundred species of parasitic cestodes, monogeneans and trematodes known from GBR vertebrates and hundreds more await description. The observer can be confident that almost every individual fish, turtle, bird or marine mammal that is seen is infected by at least one species of platyhelminth. The effects of these parasites on the health of the hosts have been little studied. Certainly it is the general observation that seemingly perfectly healthy and reproducing vertebrates have these parasites without obvious deleterious effects. It is suspected, however, that parasite pathogenicity varies dramatically with parasite species and stage, and that under some circumstances they are indeed significantly pathogenic.

Free-living flatworms

Free-living flatworms are divided up into several groups, but they are all bilaterally symmetrical and unsegmented, with a body in which the organs are embedded in a solid cellular matrix of parenchymatous tissue rather than lying in a body cavity, the coelom, as occurs in the marine annelids. The gut is sac-like, unless it has been lost, and their nervous system has an anterior 'brain' and lateral nerve cords. The body is made up of three layers: the ectoderm from which the epidermis and nervous tissue develops; the endoderm from which the gut arises; and the mesoderm from which the muscles and other organs arise. These soft-bodied worms regulate their body fluids by a complex series of channels in which specialised cells the protonephridia beat and propel fluids to the exterior. Flatworms, especially those living on coral reefs, are often brightly coloured (Fig. 24.4E, F) and are sometimes confused with nudibranch molluscs. They can be easily separated because nudibranchs have a muscular foot, anterior rhinophores and posterior frilly gills, all of which are absent in flatworms. Flatworms are also usually much more fast moving than nudibranchs.

Flatworms are considered to be the first organisms that developed cephalisation, possessing an anterior region where light-sensitive eye-spots and chemosensory cells are gathered. These can be found in clusters on the head region or gathered around pseudotentacles or true tentacles in some species. Most free-living flatworms are predators of smaller invertebrates, using their muscular pharynx to swallow their prey. The pharynx can be either tubular or ruffled in nature, and the mouth pore is usually found on the anterior or middle region of the body. The animals glide on a sheet of mucus by the beating of the ciliary epidermis. Other species swim using well-developed dorsoventral muscles that allow the worm to send waves along the body and allow the worm to swim up into the water column. All flatworms can rapidly regenerate lost body tissues.

Flatworms occur on a variety of reefal habitats, and some live in close association with other invertebrates, especially species of echinoderms and soft corals. They are carnivores and may either ingest their prey whole or just remove bits at a time. They evert their pharynx and secrete enzymes that begin to digest the prey tissue and the partially digested material is sucked up into the gut. Target organisms include individual zooids of corals, bivalve molluscs and colonial ascidians. Because they lack an anus, any undigested particles must be ejected via the mouth. Nutrients diffuse into the body from the gut: there is no circulatory system.

Many of the species of flatworms on the reef are highly conspicuous and, as already mentioned, some can swim up into the water column by undulating their body margins, and yet they are almost universally ignored by predatory fish as their body tissue often contain toxins. The bright colours may also warn potential predators that they are poisonous to eat. Others mimic poisonous animals so that although they are not themselves poisonous the animal that they are mimicking is.

Although flatworms are hermaphrodites, they appear not to self fertilise, and when two mature individuals meet, one slowly glides over the other and raises the front half of the body. A penis armed with a hypodermic-like stylet is pushed into the body wall of the other worm and the sperm are injected into the animal, they then move through the body to find an egg in the oviduct where fertilisation occurs. In some other species the worms copulate tail to tail and the sperm are delivered into the female part of the reproductive system and the sperm may be released into the oviduct as required. Fertilised eggs are placed in a capsule, which is attached to the substratum until they hatch. Some species present direct development in which a miniature adult emerges from the egg and grows into adulthood, while some others show an indirect development through a larval stage.

Flatworms are abundant and diverse on the GBR, and range in size from very small (less than a 1 mm) to several centimetres in length. Colour patterns are useful in species identification, although these are often lost when the material is preserved. Identification from good quality photos can be possible although many species require histology studies for proper description.

Nematodes

Nematodes include both parasitic and free-living species burrowing in the sediments. They are characterised by shiny iridescent cuticle and an unsegmented body (Fig. 24.4K). In many species, males and females exhibit no sexual dimorphism and have streamlined bodies pointed at both ends; in other cases, males have complex copulatory organs. Free-living species are abundant in reefal sediments as a constituent of the meiofauna, and probably are extremely speciose but they have been poorly studied. Nematodes are important in recycling of nutrients in sediments and are an important food source for many other animals. Our knowledge of parasitic nematodes on the GBR is extremely limited but they can be found in the gut of many marine vertebrate species, although they may have been accidently ingested.

Nemerteans

Many species have long, thin unsegmented bodies, sometimes brightly coloured, and covered in sticky mucus. The head end is spade-shaped without any sensory appendages, although pigment spots may be visible in the surface epidermis. Their mouth is subterminal and they catch their prey by rapidly everting their proboscis. Only some of the large conspicuous species have been described from the GBR including *Baseodiscus hemprichii* (Fig. 24.4L) and many more remain to be described. They can be seen crawling over the substrate, but many of the smaller species live in the sediments. They appear to be unpalatable to many species and it has been suggested that their mucus sheath on which they glide contains unpalatable chemical compounds.

Phoronids

Phoronids represent a small group of unsegmented, sedentary worm-like animals that live in tubes (Fig. 24.3I) often attached to the tubes of boring sea anemones like species of *Cerianthus*. Superficially they resemble sabellid polychaetes, but on closer examination the inspection the retractile tentacles (lophophore) are ciliated and are arranged in a horseshoe shape around the mouth. They are hermaphrodites and gametes are released through their excretory ducts: the eggs drift up the tube and are trapped in the tentacles. After they have been fertilised, a pelagic larvae is released into the water column and they spend several weeks in the plankton before settling and metamorphosing into adults. Little is known about their diversity on the GBR, but there are only 11 species known worldwide.

Further reading

Polychaetes

Identification of polychaetes

Glasby CJ, Fauchald K (2003) POLIiKEY. An information system for polychaete families and higher taxa. Website. Version 2. Department of Environment and Energy, Canberra.

Fauchald K (1977) The polychaete worms, definitions and keys to the orders, families, and genera. *Natural History Museum of Los Angeles County, Science Series* **28**, 1–190.

Hutchings P, Kupriyanova E (2015) Coral reef-associated fauna of Lizard Island, Great Barrier Reef: polychaetes and allies. *Zootaxa* **4019**, 001–801.

Wilson R, Hutchings PA, Glasby CJ (2003) *Polychaetes: An Interactive Identification Guide.* CD. CSIRO Publishing, Melbourne.

General information on polychaete families, myzostomes, echuira and sipuncula

Beesley PL, Ross GJB, Glasby CJ (Eds) (2000) *Polychaetes & Allies: The Southern Synthesis. Fauna of Australia. Vol. 4A: Polychaeta, Myzostomida, Pogonophora, Echiura, Sipuncula.* CSIRO Publishing, Melbourne.

Rouse GW, Pleijel F (2001) *Polychaetes.* Oxford University Press, Oxford, UK.

Struck TH, Schult N, Kusen T, Hickman E, Bleidorn C, McHugh D, *et al.* (2007) Annelid phylogeny and the status of Sipuncula and Echiura. *BMC Evolutionary Biology* **7**, 57. doi:10.1186/1471-2148-7-57

Weigert A, Helm C, Meyer M, Nickel B, Arendt D, Hausdorf B, et al. (2014) Illuminating the base of the annelid tree using transcriptomics. *Molecular Biology and Evolution* **31**, 1391–1401. doi:10.1093/molbev/msu080

Flatworms

Laumer CE, Hejnol A, Giribet G (2015) Nuclear genomic signals of the 'microturbellarian' roots of platyhelminth evolutionary innovation. *eLife* **4**, e05503.

Egger B, Lapraz F, Tomiczek B, Müller S, Dessimoz C, Girstmair J, et al. (2015) A transcriptomic-phylogenomic analysis of the evolutionary relationships of flatworms. *Current Biology* **25**(10), 1347–1353. doi:10.1016/j.cub.2015.03.034

Newman L, Canon L (2003) *Marine Flatworms: The World of Polyclads.* CSIRO Publishing, Melbourne.

Newman L, Cannon L (2005) *Fabulous Flatworms: A Guide to Marine Polyclads*, CD. CSIRO Publishing, Melbourne.

Parasitic worms

Rohde K (Ed.) (2005) *Marine Parasitology.* CSIRO Publishing, Melbourne.

Nemerteans

Andrade SCS, Montenegro H, Strand M, Schwartz ML, Kajihara H, Norenburg JL, *et al.* (2014) A transcriptomic approach to ribbon worm systematics (Nemertea): resolving the Pilidiophora problem. *Molecular Biology and Evolution* **31**(12), 3206–3215. doi:10.1093/molbev/msu253

Gibson R (1983) Nemerteans of the Great Barrier Reef. 6. Enopla Hoplonemertea (Polystilifera: Reptantia). *Zoological Journal of the Linnean Society* **78**, 73–104. doi:10.1111/j.1096-3642.1983.tb00863.x

Gibson R (1997) Nemerteans (Phylum Nemertea). In *Marine Invertebrates of Southern Australia. Part III.* (Eds SA Shepherd and M Davies) pp. 905–974. South Australian Research and Development Institute, Adelaide.

25

Arthropods: crustaceans and pycnogonids

S. T. Ahyong

Crustaceans

Crustaceans are the most diverse marine arthropods, earning them the title 'insects of the sea'. Being arthropods, crustaceans have a tough, chitinous, segmented exoskeleton, but are distinguished from other arthropods by having two pairs of antennae at some stage in their life cycle (antennae are lost in adult barnacles). Crustaceans are primarily marine, but also occur on land and in fresh water. They range in size from microscopic copepods to the Japanese giant spider crab (*Macrocheira*) with a 4 m leg span. The best-known crustaceans are the crabs, shrimps and lobsters because of their large size and culinary qualities.

The crustacean body generally comprises three regions: the cephalon, thorax (or pereon) and abdomen (or pleon). Each region usually bears paired biramous limbs. The cephalon bears the compound eyes and two pairs of antennae. The limbs of the thorax form the mouthparts, and grasping and locomotory legs, and those of the abdomen typically form pleopods (swimmerets) and uropods. Many crustaceans breathe through gills located either at the base of the thoracic limbs (e.g. decapods) or abdominal limbs (e.g. stomatopods).

Crustaceans continue to grow throughout their lives, unlike insects, which stop growing at maturity. The hard exoskeleton, however, cannot stretch or expand to accommodate growth. To increase in size, crustaceans must periodically shed their old exoskeleton in a process known as moulting or ecdysis. In the lead-up to a moult, calcium and various other minerals are resorbed into the body from the old exoskeleton, beneath which a new, soft skin develops. The old exoskeleton 'loosens' at key locations and fractures along various defined points (called moult sutures). The animal extracts itself from the old exoskeleton, often within only a few minutes. Recently moulted animals, being soft and extremely vulnerable to predation, hide and remain inactive. At this time, calcium and other minerals from the old exoskeleton and surrounding sea water are slowly redeposited in the new soft exoskeleton. Within 24–48 h, the new exoskeleton hardens and the animal can resume normal activity.

Sexes are usually separate in crustaceans and most have planktonic larvae. Planktonic larvae pass through many stages and can spend months in the plankton before settling on the bottom. Crustacean larvae are a significant component of the zooplankton.

On the Great Barrier Reef (GBR), crustaceans are abundant in all habitats, with around 1300 species recorded. In general, back-reef sites on mid-shelf reefs have the highest overall crustacean diversity, but most habitats await detailed faunal study. The most conspicuous crustaceans are the comparatively large and usually colourful decapods (crabs, shrimps, lobsters) and stomatopods (mantis shrimps), but small-bodied, speciose groups such as peracarids, ostracods and copepods are abundant. The major crustacean groups to be encountered are discussed in the following sections.

Ostracoda

Ostracods (Fig. 25.1A), commonly called seed shrimps, have a bivalved carapace that fully encloses the body. They range in length from less than 1 mm to ~20 mm, but most species seen on the GBR are 3–5 mm long. Ostracods use their two pairs of antennae for locomotion, either swimming or 'creeping', and their thoracic limbs are reduced to two pairs. Ostracods are divided into two major groups: the Podocopa and Myodocopa. Podocopans are typically elongate and peanut shaped, whereas myodocopans are generally rounded or ovate with a notch in the margin. Almost any sediment or algal sample will contain ostracods; species may be detritivores, carnivores or scavengers. More than 40 species are known from Lizard Island. Elsewhere on the GBR, ostracods have been scarcely studied but many more species almost certainly await discovery.

Copepoda

Copepods are abundant on the GBR and form a major component of zooplankton. Most are no longer than a few millimetres, though some parasitic species reach 300 mm. Copepods lack a carapace, and most free-living species have a single median eye, a long pair of antennae that is held outwards, a broadened trunk area and slender abdomen. The most common copepods on the GBR are free-living forms: the compact harpacticoids and barrel-shaped calanoids and cyclopoids. Harpacticoids are generally benthic, living under rocks, in algae and sediment. Calanoids and cyclopoids are common in the plankton where they may form dense swarms. Free-living copepods feed variously on other zooplankton, diatoms and algae, and are an extremely important food source for larval fish and decapods. Parasitic copepods feed on the blood or tissue of their hosts. Examples include the siphonostomatoids and poecilostomatoids that attach to the skin of sharks and fish, and certain cyclopoids, which have become worm-like as internal invertebrate parasites.

Cirripedia

The sessile habit and interlocking shell plates of barnacles led early investigators to believe they were molluscs. Not until the 1830s, when barnacle larvae were studied, were they correctly recognised as crustaceans. Barnacles are more abundant in temperate waters than in the tropics, but nevertheless range throughout the GBR on all hard substrata, including the carapace of sea turtles and the shells of other crustaceans. More than 40 species and 20 genera of barnacles are presently known from GBR waters. The body of most barnacles is enclosed by a somewhat tubular or conical opened topped chamber formed by a series of interlocking calcareous plates – usually known as the shell. The animal is positioned on its back, and thrusts its modified, feathery thoracic limbs, known as cirri, out of the shell to strain food particles from the water.

Unlike most other crustaceans, barnacles are hermaphrodites, meaning that each animal possesses male and female reproductive organs. The majority of barnacles in the reproductive phase develop a greatly elongated penis (up to 10 times their body length) that is used to reach out of the shell to fertilise a receptive neighbour. The fertilised eggs develop into free-swimming larvae that eventually settle on the substratum close to others of their species, which are identified by their unique chemical signature.

Three major types of barnacles will be encountered on the reef: balanomorphs (Fig. 25.1B), lepadomorphs (Fig. 25.1C) and rhizocephalans. In balanomorph barnacles, the so-called acorn barnacles, the shell is firmly fixed to the substratum, be it rock, coral, mangroves or wharf piles. Common balanomorphs on the reef flat rocks and boulders include species of *Balanus*, *Chthamalus*, *Megabalanus*, *Tesseropora*, *Tetraclita* and *Tetraclitella*. Not all barnacles are fixed to the substratum by the shell, however. The lepadamorphs, the goose-necked barnacles, are attached to the substratum by a flexible stalk, with common genera including *Lepas* and *Ibla*. *Conchoderma virgatum* is unusual in using sea snakes as a substratum. In contrast to the balanomorphs and lepadomorphs that simply use the substratum as an anchor point, the rhizocephalan barnacles have an entirely different lifestyle as partially internal and partially external parasites of decapod crustaceans. Rhizocephalans have lost any resemblance to typical adult barnacles, including loss of the calcareous plates. A common

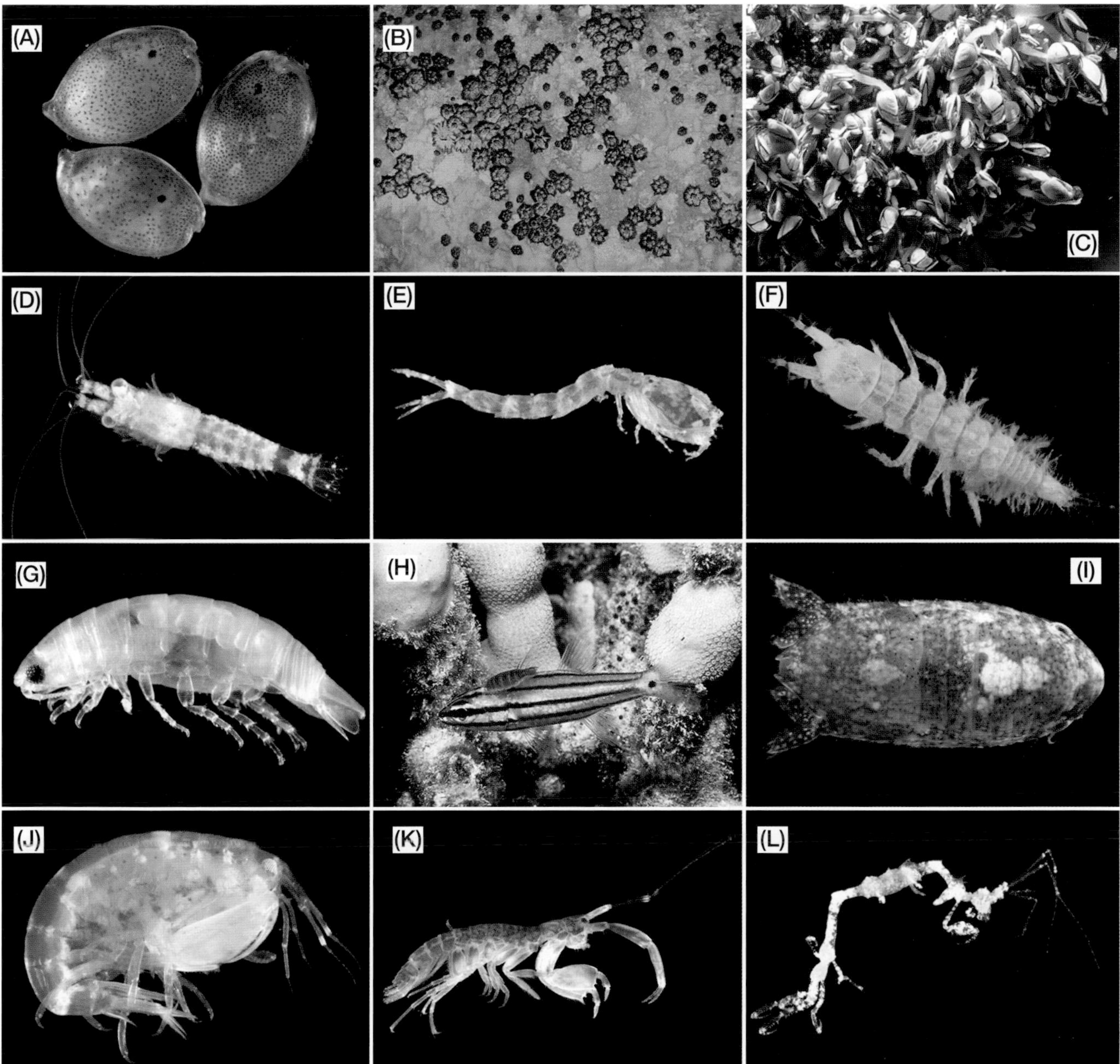

Fig. 25.1. **(A)** Ostracod, *Cypridinodes* sp.; **(B)** balanomorph barnacle, *Chthamalus* sp.; **(C)** lepadomorph barnacle, *Lepas anserifera*; **(D)** opossum shrimp (Mysida); **(E)** comma shrimp (Cumacea); **(F)** apseudomorph tanaid (Tanaidacea); **(G)** cirolanid isopod, *Cirolana curtensis*; **(H)** parasitic isopod, *Anilocra apogonae*, on fish; **(I)** sphaeromatid isopod, *Cerceis pravipalm*; **(J)** gammaridean amphipod, *Leucothoe* sp.; **(K)** senticaudatan amphipod, *Grandidierella* sp.; **(L)** senticaudatan caprelloid amphipod, *Metaprotella* sp. (Photos: *A, D–H, J–L,* R. Springthorpe, ©Australian Museum; *B, C,* ©GBRMPA; *I,* ©N. D. Pentcheff.)

rhizocephalan, *Sacculina*, infects crabs and is visible externally only as a fleshy mass beneath the abdomen of the crab. Rhizocephalans interfere with the reproductive cycle of crabs and through the effect of hormones, cause male crabs to slowly develop female characteristics, an effect known as 'parasitic castration'.

Peracarida

The peracarids are a large group of malacostracan crustaceans recognisable by the possession of the *lacinia mobilis* (a blade-like process on the mandible), and direct development of young, which are brooded in a pouch formed by lamellar outgrowths of the bases of the thoracic limbs (with the exception of the thermosbaenaceans, which brood their eggs under the carapace). Most peracarids are minute, generally not exceeding 10 mm in length (many exceptions exist, such as the giant deep-sea isopod, *Bathynomus*). Despite their small size, however, peracarids are highly speciose and abundant in most habitats. Of the nine generally recognised peracarid orders, the most important on the GBR are the Mysidacea, Cumacea, Tanaidacea, Isopoda and Amphipoda.

Mysidacea: opossum shrimps

The mysids (Fig. 25.1D) resemble juvenile decapod shrimps, and are usually less than 20 mm long. Though mysids superficially resemble decapod shrimps, an important distinction is the position of the statocyst (balance organ resembling a circular pore). The statocyst of mysids is located at the base of each branch of the tailfan, instead of the base of the first pair of antennae in decapods. Unlike other peracarids, mysidaceans have a well-developed carapace and stalked, instead of fixed, eyes. Most mysids are omnivorous suspension feeders, and form dense schools near the bottom or in the water column. Some species, however, have more particular associations, such as the red and white *Heteromysis harpaxoides*, which lives with hermit crabs inside the gastropod shell. More than 800 species in more than 120 genera occur worldwide, with around 50 species found on the GBR.

Cumacea: comma shrimps

The cumaceans (Fig. 25.1E) are distinctive because of their bulbous carapace and slender abdomen that terminates in a pair of stick-like uropods. They burrow in reef sediments or seagrass beds, and feed on detritus, algae and microbes. At night, cumaceans ascend into the water column to moult and mate. Of the ~1500 species of cumacean worldwide, around 40 species are known from Queensland, of which about 30 are known from the GBR.

Tanaidacea

The tanaids (Fig. 25.1F), usually less than 10 mm length, resemble a slender isopod (next section) with a pair of chelipeds (pincers). A unique feature of the tanaids is the fusion of the head with the first and second thoracic segments. Most tanaids fall into one of two major groups: the apseudomorphs, which have a somewhat flattened body; and tanaidomorphs, with a cylindrical body. Though seldom seen in the open, tanaids are abundant in reef sediments, seagrass, algae and hydrozoans where they burrow or build tubes from sand and fine particles. Despite their numerical abundance, however, the GBR tanaids are poorly known, with only about 10 of more than 900 known species formally recorded from the region.

Isopoda: slaters, pill bugs, fish lice

The isopods are a remarkably diverse group of peracarids that have colonised almost all habitats, from terrestrial leaf litter to abyssal depths. Isopods are characterised by fixed, immovable eyes, no carapace, seven pairs of thoracic legs, five pairs of pleopods and a pleotelson (formed by fusion of the last two body segments). Most isopods have the familiar, flattened, oval body shape. The largest isopod, the deep-sea *Bathynomus giganteus*, reaches 35 cm in length, but on the GBR, most species are less than 10 mm long. Some 250 species of isopod are known from the GBR, and most obvious are the mostly free-living flabelliferans, particularly species of the families Cirolanidae (Fig. 25.1G) and Sphaeromatidae (Fig. 25.1I). Cirolanids are efficient scavengers, emerging from the substratum to swim rapidly towards a food source such as a dead or maimed fish. They can also be voracious predators, sometimes biting humans. Sphaeromatids are perhaps less conspicuous than the cirolanids, but often have a highly sculptured pleotelson and trailing uropods. Some isopods vary markedly from the 'typical' form. These include the slender valviferans and anthurideans, and most notably the endoparasitic bopyroids that infect decapods. Bopyroids have lost many of the features of free-living isopods and, being internal parasites, are seldom seen. A rounded bulge on the surface of the carapace of the decapod host is the only external clue that a bopyroid lies beneath. Other parasitic isopods such as the cymothoids (Fig. 25.1H) infect the skin, mouth, and gills of fish, to which they cling with sharp, hooked 'feet'.

Amphipoda: sandhoppers, scuds and skeleton shrimps

The amphipods are the most speciose of the peracarids with more than 10 000 species worldwide. They can be recognised by their (usually) laterally compressed body, biramous antennules and three pairs of uropods. On the GBR, ~200 species are known from depths shallower than 50 m, of which 80% are yet to be formally named, and this all from near Lizard Island in the far northern section of the Reef. Thus, many more species can be expected from the GBR overall. Three of the four major suborders of Amphipoda occur on the reef (Gammaridea, Hyperiidea and Senticaudata). The majority of amphipods are gammarideans (Fig. 25.1J) or senticaudatans (Fig. 25.1K), commonly known as scuds and sandhoppers (for terrestrial species), and these are the most familiar form. Gammarideans and senticaudatans usually have a relatively deep, curved body that is obviously laterally compressed; they are the most accomplished swimmers of the Amphipoda, and can 'scuttle', 'skip' and 'flick' their bodies to escape capture. They live in almost every habitat on the GBR: among rubble, boulders, coral and algae, and sometimes in association with hydroids, ascidians and sponges. Some are predatory, some scavenge, and others graze on algae and detritus. Some of the most unusual senticaudatan amphipods are the caprelloids, commonly known as 'skeleton shrimps' (Fig. 25.1L), so named because of the very slender, stick-like body. Caprelloids can be seen standing upright with forelimbs spread out while clinging by their posterior legs to algae or hydroids. They crawl with a somewhat looping motion by alternately using their anterior and posterior limbs. Though appearing very different from other amphipods, caprelloids probably arose as a highly specialised lineage within the Senticaudata. The hyperiideans resemble gammarideans and senticaudatans with a swollen head and long slender legs. Unlike gammarideans, however, hyperiideans are seldom seen on the reef because they are oceanic creatures that live inside salps or on jellyfish.

Stomatopoda

The mantis shrimps, order Stomatopoda, are the most flamboyant crustaceans, and only distant relatives of the decapods. They are primarily shallow water tropical animals and are abundant throughout the GBR. They are active, aggressive predators and often vividly coloured. Adults range in size from less than 10 mm to almost 400 mm. The body is elongate, with a short, shield-like carapace that does not cover the last three thoracic somites. Unlike decapods, which either broadcast spawn or incubate the eggs on the pleopods, stomatopod females carry the egg mass with their mouthparts until hatching. The most characteristic feature of stomatopods, however, is the greatly enlarged raptorial forelimbs used to catch prey. The raptorial claw of stomatopods resembles that of the praying mantis (an insect), in which the last segment (dactylus) folds like a jackknife against the preceding segment (propodus). However, instead of grasping prey like the praying mantis, stomatopods strike violently with considerable speed and force. The raptorial strike of the stomatopod is one of the fastest known animal movements, being completed within five milliseconds. Obviously, large stomatopods must be handled with great care.

Stomatopods can be divided into two functional groups depending on how they use their raptorial claws during the strike. 'Spearers' (Fig. 25.2C, E, F) strike with an open dactylus, impaling prey – soft-bodied animals such as fish, cephalopods and shrimps – on a row of sharp spines. 'Smashers' (Fig. 25.2A, B, D) strike with the dactylus closed against the propodus. The outer edge of the dactylus is a calcified heel, used like a hammer to smash open hard-bodied prey such as crabs and snails. The power of the raptorial strike can be lethal – the force of the impact may approach that of a small calibre bullet. So, in 'smashers', the raptorial claws are used not only for hunting, but also to avoid

Fig. 25.2. **(A)** 'Smashing' mantis shrimp, *Gonodactylaceus graphurus*; **(B)** 'smashing' mantis shrimp, *Odontodactylus scyllarus,* carrying egg mass; **(C)** 'spearing' mantis shrimp, *Pseudosquilla ciliata*; **(B)** 'smashing' mantis shrimp, *Gonodactylus smithii*, showing threat display; **(E)** 'spearing' mantis shrimp, *Lysiosquillina maculata*; **(F)** 'spearing' mantis shrimp, *Pseudosquilla ciliata*; **(G)** red spot king prawn, *Melicertus longistylus*; **(H)** penaeid prawn, *Heteropenaeus longimanu*; **(I)** caridean shrimp, *Periclimenes* sp.; **(J)** snapping shrimp, *Synalpheus demani*; **(K)** hingebeak shrimp, *Rhynchocinetes durbanensis*; **(L)** caridean shrimp, *Periclimenes brevicarpalis*, on sea anemone. (Photos: *A–D, F,* ©R. Caldwell; *E,* ©S. Ahyong; *G,* ©G. Ahyong; *H, K, L,* ©GBRMPA; *I, J,* R. Springthorpe, ©Australian Museum).

unnecessary confrontations. Many 'smashers' have evolved elaborate threat displays that involve lunging and spreading the limbs to display a coloured 'meral spot' on the inner surface of the raptorial claw. The intensity of the display and brightness of the 'meral spot' varies between species, and gives a potential opponent some indication of the ferocity that it might encounter.

The powerful raptorial claws of stomatopods are complemented by finely tuned vision. Each eye can be independently rotated and is divided in half by a central band of ommatidia. This provides the stomatopod with binocular vision from each eye, and the central band of ommatidia provides many of the reef species with colour vision and polarising filters. The combination of excellent vision and powerful raptorial claws make the stomatopod an efficient predator of both slow and fast moving prey.

Almost 500 species of stomatopod are known worldwide and more than 90 occur in GBR waters. Many new species await scientific documentation, especially small species that live deep inside rock and coral crevices. Around Lizard Island, almost 30 species are known. The most common stomatopods among rubble and corals are 'smashers' of the superfamily Gonodactyloidea. Common gonodactyloid genera on the reef flat are *Gonodactylellus*, *Gonodactylus*, *Conodactylaceus*, *Chorisquilla* and *Haptosquilla*. *Haptosquilla glyptocercus* and *H. trispinosa*, both with striped antennae and not exceeding ~40 mm length can be seen on the reef flat peering out of circular holes in reef rock. 'Spearers', primarily members of the Squilloidea and Lysiosquilloidea, burrow in the sand and sediment of the reef flat and on the sea bed of the lagoon and between reefs. Several species of squilloid are commercially harvested, taken as trawl by-catch (e.g. *Harpiosquilla harpax* and *Oratosquillina* spp.). On the reef flat, the very large piscivore, *Lysiosquillina maculata* (up to 385 mm long), lives in mated pairs in deep, vertical burrows, with burrow entrances almost 100 mm across (Fig. 25.2E). *Lysiosquillina maculata* is the largest known stomatopod, with a body marked by bold, light and dark transverse bands. *Pseudosquilla ciliata*, another common spearer, burrows in seagrass beds or under rocks on the reef flat, and ranges in colour from bright yellow or dark green to mottled green and grey (Fig. 25.2C, F). On the reef slope, the number of species increases, and some of the most spectacularly coloured stomatopods, those of the 'smashing' genus *Odontodactylus*, can be seen walking out in the open (Fig. 25.2B).

Decapoda

The best known of all crustaceans are the decapods. They include the largest arthropods and many commercial species such as the mud crab (*Scylla serrata*), blue swimming crab (*Portunus armatus*, previously known as *P. pelagicus*), red-spot king prawn (*Melicertus longistylus*), Moreton Bay bug (*Thenus* spp.) and spiny lobsters (*Panulirus* spp.). Decapods have a well-developed carapace covering the thorax, and five (rarely four) pairs of legs, of which one or more pairs usually forms a cheliped. At least nine decapod infraorders are recognised: Dendrobranchiata (penaeidean prawns); Caridea (shrimps), Stenopodidea (coral shrimps); Polychelida (deep-sea blind lobsters); Achelata (spiny-crayfish); Astacidea (clawed lobsters and freshwater crayfish); Axiidea and Gebiidea (mud-shrimps, marine yabbies); Anomura (hermit crabs, squat lobsters, mole crabs); and Brachyura (true crabs). Members of each of these infraorders occur on the GBR, apart from polychelidans, which occur in deep water beyond the reef.

Dendrobranchiata: prawns

The dendrobranchiates, collectively known as prawns in Australia, are possibly the oldest lineage of the decapods. They have the familiar elongate body with a slender, muscular abdomen and three pairs of small chelipeds. Unlike other decapods, female prawns do not carry eggs until hatching, but spawn their fertilised eggs directly into the water column. Prawns are not generally common on corals or hard reef structures, though *Heteropenaeus longimanus* (Fig. 25.2H) and several species of *Metapenaeopsis*, *Sicyonia*, and *Acetes* have been recorded. Prawns are more abundant on the sediments of the lagoon and between reefs, where they support an important trawl fishery. One of the most important commercial species is the red-spot king prawn (Fig. 25.2G; *Melicertus longistylus*, previously known as *Penaeus longistylus*). Many commercial penaeids 'grow-out' in estuaries or nearshore habitats, and migrate offshore as adults, but red-spot king prawns spend their entire lives in the vicinity of the reef. Juveniles use shallow reef flats and lagoon areas as a nursery, and adults move to deeper parts of the lagoon to spawn.

Caridea: shrimps

The shrimps of the infraorder Caridea, like the prawns, are accomplished swimmers. Unlike prawns, however, carideans incubate their eggs beneath the abdomen, have only two pairs of chelipeds and have the pleuron (lateral plate) of the second abdominal segment overlapping both those of the first and third segments. More than 200 species of caridean shrimp are presently known from the GBR, the most common of which are members of the Palaemonidae and Alpheidae. Most GBR palaemonids are associates of other reef invertebrates, such as corals, anemones, nudibranchs, sponges, echinoderms and molluscs. Most reef palaemonids (Fig. 25.2I–L) tend to be delicate and retiring, and though common, must be searched for. One of the more common palaemonids is *Periclimenes brevicarpalis* (Fig. 25.2L), which lives under the protection of the tentacles of sea anemones. Here, *P. brevicarpalis* rolls and sways its body, which is transparent except for its several large white patches, and orange and purple highlights. Interestingly, few reef palaemonids have drab colouration: most are either brightly coloured or almost completely transparent. Alpheids, the snapping shrimps, are abundant throughout the GBR, but are more often heard than seen. Snapping shrimps live in burrows in sand, sponge, mud and coralline rock. The characteristic feature of the snapping shrimps is the greatly enlarged snapping claw, which, when snapped shut produces a loud, cracking sound with a localised shock wave sufficient to injure or kill small invertebrates and fish. Snapping shrimps usually live in pairs, but some form large social networks (e.g. *Synalpheus*; Fig. 25.2J), and many species of *Alpheus* cooperatively share a burrow with gobiid fish. Other conspicuous carideans on the reef include the cleaner shrimp, *Lysmata amboinensis*, the hinged-beak shrimps (*Rhynchocinetes durbanensis*; Fig. 25.2K) and the marbled shrimp (*Saron marmoratus*). One of the most unusual shrimps on the GBR is the harlequin shrimp (*Hymenocera picta*), which feeds exclusively on sea-stars. As a rule, caridean shrimps hide during the day and are best observed at night, when they emerge to feed.

Stenopodidea: coral shrimps

The stenopodidean shrimps are generally small, spiny shrimps characterised by having three pairs of chelipeds of which the third pair is the largest. Four species of stenopodidean are known from the GBR, of which the red and white banded coral shrimp, *Stenopus hispidus* (Fig. 25.3A), is the largest and most familiar, reaching ~75 mm in length. Banded coral shrimps live in mated pairs in deep, rock or coral crevices and on the underside of overhangs. They are also known as 'fish cleaners'. Pairs of *S. hispidus* attract fish to their 'cleaning station' by waving their long, white antennae. Large fish approach the cleaning station, where the shrimps 'pick-off' dead skin and parasites. Other stenopodideans on the GBR include other species of *Stenopus* and the 15 mm long *Microprosthema validum*.

Achelata: spiny and slipper lobsters

The spiny lobsters (Palinuridae) and slipper lobsters (Scyllaridae) are unusual decapods in that they lack chelipeds. They are common on the reef flat and slope where they hide in crevices by day and emerge at night to feed. Palinurids include the large, edible species such as the painted cray (*Panulirus versicolor*), ornate cray (*P. ornatus*) and spotted cray (*P. bispinosus* (Fig. 25.3B), often misidentified as *P. femoristraga*), for which a commercial fishery operates in northern reef waters between Cape York and Princess Charlotte Bay. Unlike their temperate water counterparts, species of *Panulirus* seldom enter traps, and are instead harvested by hand. Scyllarids are smaller than palinurids, and the antennae, rather than being long and whip-like, are formed into short flat plates. Whereas tropical palinurids are usually brightly coloured, scyllarids are often camouflaged to match their coralline surroundings. Scyllarids, such as *Scyllarides* and *Parribacus*, are common on the reef slope, among coral and rubble, whereas *Thenus* (Fig. 25.3C) and *Ibacus* (often called 'bugs') live on the muddy inter-reef sediments where they are commercially

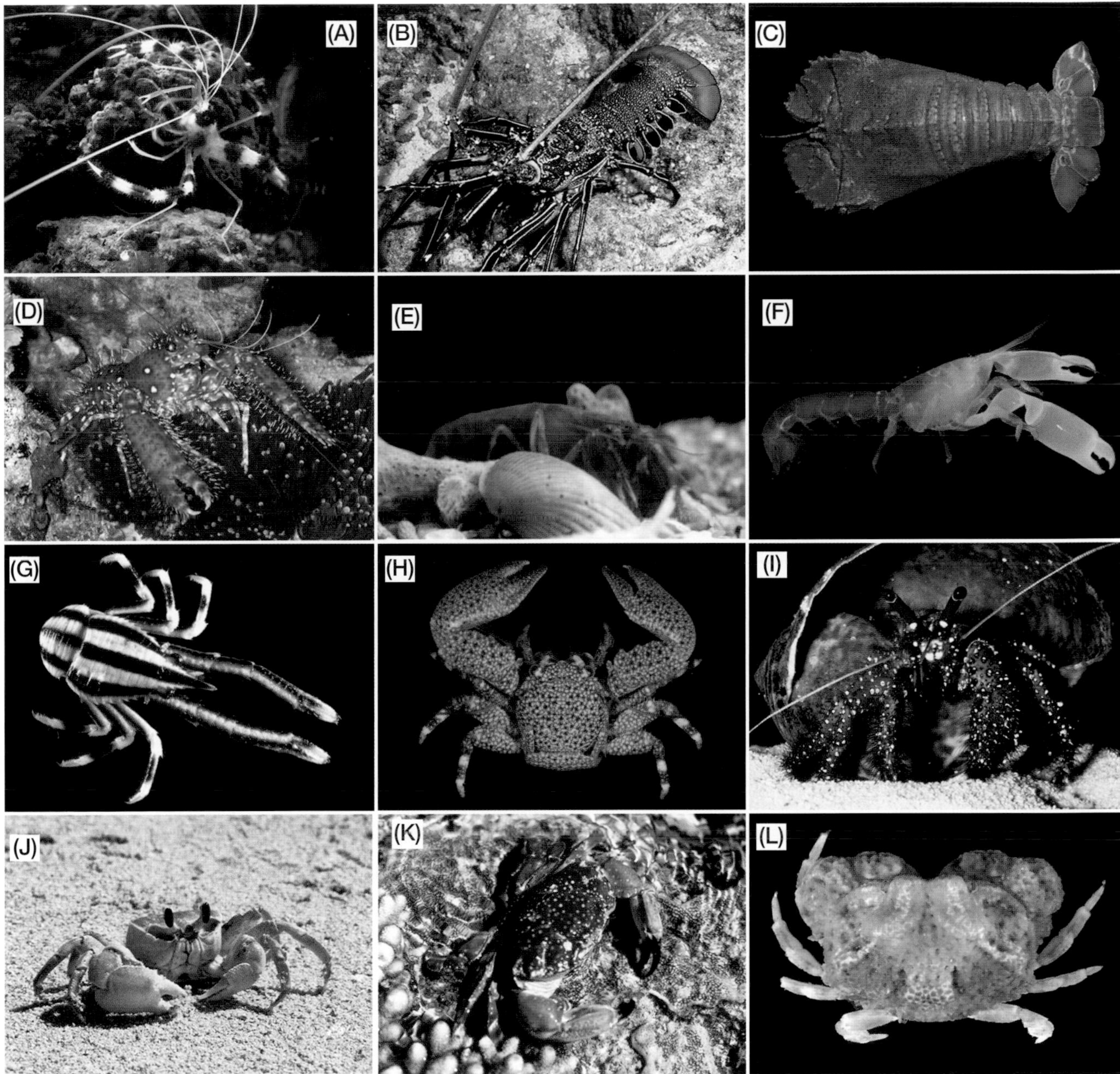

Fig. 25.3. **(A)** Banded coral shrimp, *Stenopus hispidus*; **(B)** spotted spiny crayfish, *Panulirus bispinosus*; **(C)** Moreton Bay bug, *Thenus australiensis*; **(D)** dwarf reef lobster, *Enoplometopus occidentalis*; **(E)** mud shrimp, *Strahlaxius glytocercus*; **(F)** ghost shrimp, *Axianassa heardi*; **(G)** squat lobster, *Allogalathea elegans*; **(H)** porcelain crab, *Petrolisthes* sp.; **(I)** hermit crab, *Dardanus megistos*; **(J)** ghost crab, *Ocypode cordimana*; **(K)** poisonous xanthid crab, *Lophozozymus erinnyes*; **(L)** rubble crab, *Actaeomorpha scruposa* (Photos: *A, D,* ©S. Ahyong; *B, I–K,* ©GBRMPA; *C,* ©G. Ahyong; *E,* ©C. Tudge; *F, H,* ©A. Anker; *G, L,* R. Springthorpe, ©Australian Museum).

trawled. An unusual feature of *Thenus* and *Ibacus* is their very flat carapace that extends sideways, covering their legs, which is probably an adaptation to living on level soft sediments.

Astacidea: clawed lobsters and freshwater crayfish

The Astacidea is the large group containing the marine clawed lobsters and freshwater crayfish. The most obvious feature of the astacideans, differentiating them from the palinurid lobsters, is the possession of at least one pair of enlarged, powerful chelipeds. In almost all cases, the large, first pair of chelipeds is followed by two pairs of smaller chelipeds. Freshwater crayfish, such as the yabby and marron (*Cherax*) are common on the mainland, and marine counterparts, such as scampi (*Metanephrops*, *Nephropsis* and *Thaumastochelopsis*) occur in deep, outer-shelf waters. However, only a few species of astacideans live on the reef itself: the dwarf reef lobsters of the genus *Enoplometopus* (Fig. 25.3D) and family Enoplometopidae. Enoplometopidae, containing ~20 species worldwide, possibly represents the last remnant of a lineage of lobsters that otherwise went extinct in the Lower Jurassic period. Species of *Enoplometopus* do not exceed 20 cm in length (more often less than 10 cm) but have vivid orange, red and purple colouration. Enoplometopids are very distinctive, but they are seldom seen because of their shy, nocturnal habit and lower reef slope habitat.

Axiidea and Gebiidea: mud shrimps, ghost shrimps, marine yabbies

Axiideans and gebiideans were until recently united under the Thalassinidea. Evolutionary studies, however, have shown that these shrimp-like decapods are composed two separate groups, Axiidea and Gebiidea, that are not closely related despite very similar general appearance. They typically have at least one pair of enlarged chelipeds, small eyes, a much-reduced first abdominal segment, but long abdomen, and tailfan. The carapace and abdominal exoskeleton of axiideans and gebiideans is usually thin, translucent and somewhat soft, so, not surprisingly, they rarely leave their burrows. Most burrow in sand and mud, but some burrow into sponges, live in holes or under coral rocks. Those burrowing in soft sediments, especially species of the Callianassidae (Axiidea) and Upogebiidae (Gebiidea), are important bioturbators, often creating tall mounds of excavated sand and mud around the burrow entrances. The species most likely to be seen on the reef flat is the orange axiidean, *Strahlaxius plectorhynchus*, living in vertical, rubble lined burrows on the inner reef flat. Other axiideans, *Strahlaxius glyptocercus* (Fig. 25.3E) and *Trypaea australiensis*, occur in muddier or more fine-grained habitats, especially closer to the mainland coast. The gebiidean *Axiannassa* lives in sand burrows on the reef flat (Fig. 25.3F). At least 38 species in nine families of axiidean and gebiidean shrimp are known from GBR and adjacent waters. Axiideans and gebiideans are usually difficult to capture because of their deep burrows, and no doubt many more species than presently recorded occur in the region.

Anomura: hermit crabs, squat lobsters, porcelain crabs and allies

The anomurans are the nearest relatives of the Brachyura, the true crabs, and include the hermit crabs (Paguroidea), squat lobsters and porcelain crabs (Galatheoidea), and mole crabs (Hippoidea). Unifying features of the Anomura include the very small, reduced last walking legs, and presence of an uncalcified groove along the sides of the carapace (the *linea anomurica*). The last pair of walking legs is so reduced that, at first sight, most anomurans appear to have only six, instead of eight, walking legs. The last pair of walking legs are usually folded under the abdomen or edges of the carapace, and used for grooming. Anomurans are abundant on the GBR, with the most obvious being the hermit crabs. Most hermit crabs have a soft, vulnerable abdomen, and usually use a gastropod shell for protection. Hermit crabs (Fig. 25.3I) are especially well adapted to occupying gastropod shells: the asymmetrical, coiled abdomen and asymmetry of walking legs and claws enable the animal to fit snugly inside. Species range from less than 10 mm to almost 300 mm in length. Common hermit crabs on the GBR fall into one of three major groups: the coenobitids, or terrestrial hermit crabs; the diogenids (the left-handed hermit crabs); and pagurids (the right-handed hermit crabs). Coenobitids, represented by *Coenobita* on the GBR, live on beaches above the tide line where they forage among flotsam and jetsam. To avoid desiccation, *Coenobita* carries water in its gastropod home and shelters in the shade or under driftwood during the hottest part of the day. Though they spend their entire adult life out of water, coenobitids must return to the sea to breed. Diogenids, with the left cheliped generally larger than the right, and the pagurids, with the opposite pattern, are common on all parts of the reef, from the intertidal beach rock to the reef flat, slope, and inter-reef areas. Hermit crabs are generally scavengers or herbivores, and large numbers can often be seen at night foraging on the reef flat or on the reef slope.

Other anomurans, such as the colourful squat lobsters and porcelain crabs, are common in most parts of the reef. Squat lobsters (Galatheidae and Munididae) resemble a small, flattened lobster with numerous transverse groves on the carapace, with the tail tucked under the body and chelipeds pointing forwards. Common squat lobsters include species of *Munida* that live in deeper, inter-reefal waters, species of *Galathea*, that live among corals, on bryozoans and sponges, and the conspicuously striped *Allogalathea elegans* (Fig. 25.3G) that lives in pairs on crinoids. Porcelain crabs (Porcellanidae) have a short, smooth or spiny carapace and chelipeds protruding sideways. Common porcellanid genera include the free-living *Petrolisthes* (Fig. 25.3H) and *Polyonyx*, and anemone associated *Neopetrolisthes*. Not more than ~30 species each of porcelain crabs and squat lobsters are presently formally recorded from GBR waters, although the actually numbers are certainly much higher.

The mole crabs, Hippoidea, are highly specialised for burrowing in sand. Their elongate bodies are flattened or oval in shape, and the legs are flattened into digging spades. They can burrow extremely rapidly and are difficult to detect, let alone capture. Most hippoids live in shallow water, but some live at depths beyond 200 m.

Brachyura: crabs

The Brachyura, the true crabs, are possibly the most successful of all decapods, with around 7000 described species in more than 100 families worldwide. Almost 500 species in ~40 families occur on the GBR. The shallow water crabs of the region are relatively well documented, though many new species continue to be discovered. The name Brachyura, means 'short-tail' and refers to the characteristic feature of crabs: a short abdomen that is tucked beneath the body. Crabs occur in every reef habitat, and some species occasionally enter the water column as hitchhikers on jellyfish or among floating algae or debris.

On beaches above the tideline, the fast running stalk-eyed ghost crab (*Ocypode*; Fig. 25.3J) is conspicuous, particularly early and late in the day. Ghost crabs scavenge and opportunistically prey on whatever they can find on the beach, including the occasional turtle hatchling. Though ghost crabs live high up on the beach, their burrows usually reach down to the water table. On the reef flat, the slow moving xanthoid crabs, such as *Liomera* (Fig. 25.4A), *Atergatis*, *Carpilius*, *Chlorodiella*, *Eriphia*, *Etisus*, and the poisonous *Lophozozymus* (Fig. 25.3K) are common in and under crevices of coral and rocks. Xanthoids are well adapted to living in and among corals and rock, and are the most abundant and diverse of coral reef crabs. The xanthoid body is typically compact, often covered in small spines, bumps, or bristles, and the legs are relatively short and stout. The claws of many xanthoids have dark fingertips, hence the common name 'dark fingered crabs'. The parthenopoids also occur among the rubble of the reef flat, and common genera include *Actaeomorpha* (Fig. 25.3L) and *Daldorfia*. Parthenopoids usually have a somewhat triangular or rectangular carapace, with the whole surface of the body and claws covered in tubercles knobs and pits, effectively simulating their rubbly habitat. Unlike the xanthoids and parthenopoids, the swimming crabs (family Portunidae) are fast moving and aggressive. The hind legs of portunids form swimming paddles and the other legs are flattened to assist swimming and digging. The most common swimming crabs on the reef flat are of the genus *Thalamita* (Fig. 25.4B, C), recognisable by their widely spaced eyes. Species of *Thalamita* are powerful and aggressive, and forage over open over sand, rubble and seagrass.

The mud crab, *Scylla serrata* (Fig. 25.4G), is the largest known swimming crab, and is recreationally and commercially fished throughout tropical Australia. It usually lives along the coast in

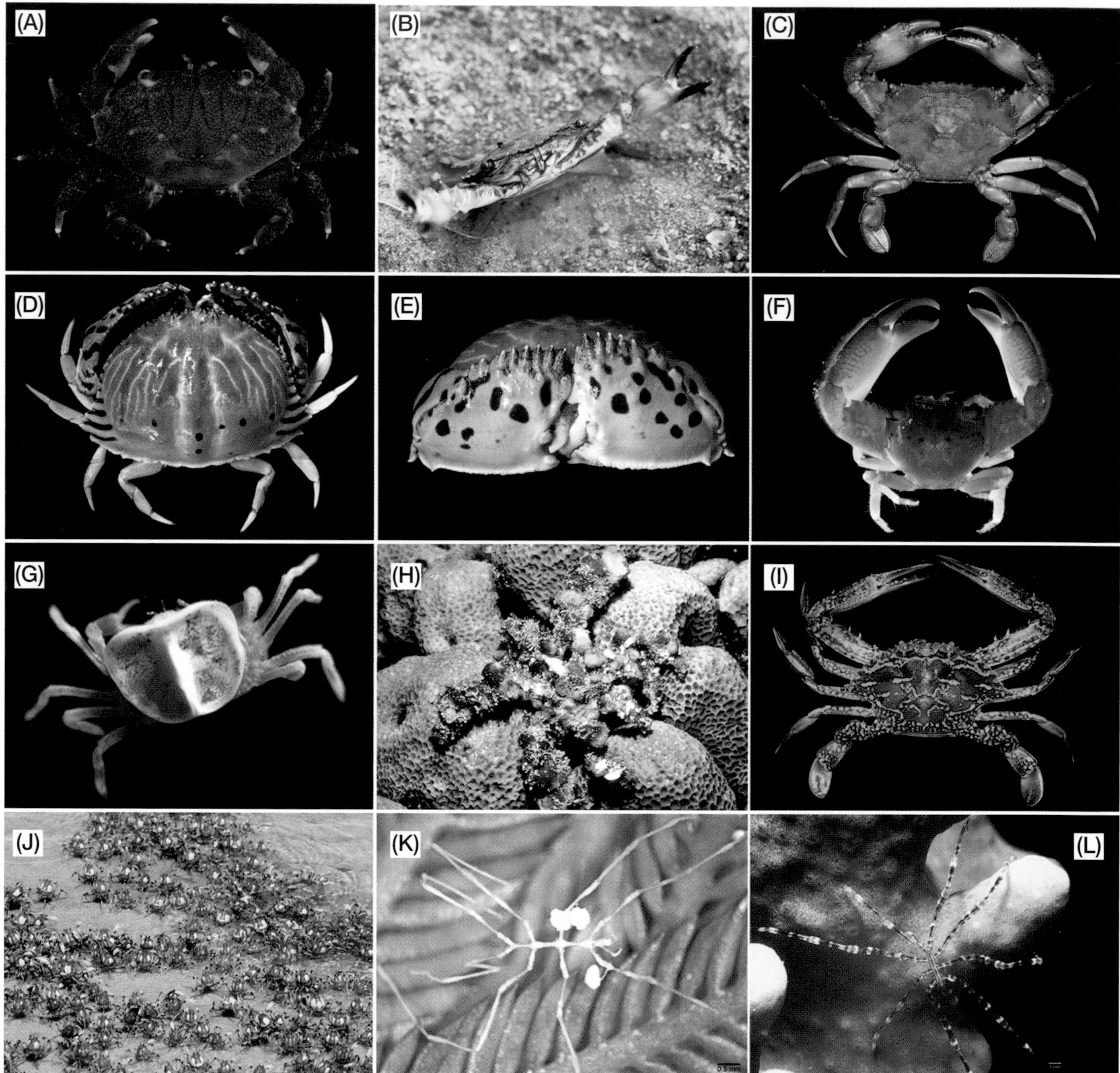

Fig. 25.4. **(A)** Xanthid crab, *Liomera bella*; **(B–C)** swimming crab, *Thalamita crenata*; **(D–E)** box crab, *Calappa lophos*; **(F)** coral crab, *Trapezia cymodoce*; **(G)** mud crab, *Scylla serrata*; **(H)** a well camouflaged spider crab (Majidae); **(I)** blue swimming crab, *Portunus armatus*; **(J)** soldier crabs, *Mictyris longicarpus*; **(K)** male sea spider carrying egg masses, *Anoplodactylus perissoporus*; **(L)** sea spider, *Endeis mollis*, on *Millepora* coral (Photos: A, ©A. Anker; ©Australian Museum; B, H, J, ©GBRMPA; C–E, I ©S. Ahyong; F, R. Springthorpe; G, I. Loch; © Australian Museum; K, L, ©C. Arango).

mangrove swamps, but also on the soft offshore sediments between reefs, and among mangroves that sometimes occur on or near reefs flats. The claws are particularly powerful and can cause serious injury if mishandled. A much smaller, more 'gentle' crab of the mangroves is the soldier crab (*Mictyris*; Fig. 25.4J), which forages in large numbers on receding tides. Soldier crabs have a somewhat spherical, bluish body and forward, instead of sideways, walk. They may march in the thousands, but at the first sign of danger, bury themselves in the sand and mud in a quick 'corkscrew' motion leaving only a small pock mark on the surface where they previously stood. Other conspicuous crabs on the open reef flat include the box crabs (*Calappa*; Fig. 25.4D, E). Box crabs hide, buried in the sand, and emerge at low tide to feed on molluscs, especially gastropods. Box crabs are often called 'shame-faced crabs' because the chelipeds are broad and high, almost entirely hiding the mouth and eyes. A peculiar feature of box crabs is a special 'peg-like' tooth on the side of the moveable finger of the right cheliped. The box crab uses this special tooth on the right cheliped like a can opener to literally 'peel' open gastropod shells.

Some crabs are very specific about their habitat. These include: the turtle-grass crab (*Caphyra rotundifrons*), which lives only in the green alga *Chlorodesmis*; the swimming crab (*Lissocarcinus orbicularis*), which lives on and inside holothurians; and coral crabs (*Trapezia* (Fig. 25.4F) and *Tetralia* of the families Trapeziidae and Tetraliidae, respectively), which live in the branches of scleractinian corals such as *Acropora* and *Pocillopora*. Trapeziids and tetraliids feed mostly on coral mucus, and though not more than 20 mm wide, will attempt to defend their host against intruders, including crown-of-thorns starfish (*Acanthaster* cf. *solaris*). The pinnotherid crabs, commonly called pea crabs, usually live in another type of host: in most cases, bivalve molluscs. Species of *Nepinnotheres* and *Arcotheres* are common in pearl oysters, the rare *Durckheimia lochi* lives in file shells (Limidae), and *Xanthasia murigera* lives in giant clams. Spider crabs (Majidae; Fig. 25.4H) and sponge crabs (Dromiidae) are common but usually overlooked because they move slowly and are well camouflaged. As their name implies, sponge crabs carry a sponge, held against the carapace by the last two pairs of legs. Spider crabs usually adorn themselves with algae, coral and sponge that are attached to velcro-like hooked setae; they are often known as 'decorator crabs'. Over time, the sponge, algae or coral grows over the surface of the crab to the extent that the crab can hardly be detected unless it moves. As a small, mobile piece of 'reef', other hitchhikers including worms, amphipods and even other small crabs can be found on large majids. Numerous species of crabs occur on the soft sediments between reefs, particularly swimming crabs of the genera *Charybdis* and *Portunus*. Among these is the edible blue swimming crab, *Portunus armatus* (previously known as *P. pelagicus*) (Fig. 25.4I) and mud crab, *Scylla serrata* (Fig. 25.4G).

Pycnogonida

Sea spider is an apt name for the pycnogonids (Fig. 25.4K, L), because they superficially resemble true spiders (Araneae), albeit with a more slender body and sometimes more than four pairs of legs (up to six pairs in some Antarctic species). The body of pycnogonids is small and slender. The head and thorax is fused to form the prosoma, and the abdomen (or opisthosoma) is small and unsegmented. Pycnogonids are also unusual in having the mouth placed at the end of a proboscis that may be as long as half the body. The pycnogonids are an ancient group dating back to the Devonian, probably representing the nearest relatives of the Chelicerata (spiders, mites, scorpions), or possibly the nearest relatives of the remaining Arthropoda. Around 1500 species of pycnogonid are known worldwide, with ~30 species in 17 genera known from the GBR and adjacent Queensland coast. Some deep-sea pycnogonids have a leg span of 750 mm, but species on the GBR are much smaller, at less than 20 mm across. All are cryptic and slow moving, so detection can be difficult. Pycnogonids prey on bryozoans, polychaetes, hydroids, zoanthids and even scleractinian corals.

Further reading

Arango C (2003) Sea spiders (Pycnogonida, Arthropoda) from the Great Barrier Reef, Australia: new species, new records and ecological annotations. *Journal of Natural History* **37**, 2723–2772. doi:10.1080/00222930210158771

Davie PJF (2002) *Crustacea: Malacostraca: Eucarida (Part 1): Phyllocarida, Hoplocarida, Eucarida. Zoological Catalogue of Australia 19.3A.* CSIRO Publishing, Melbourne.

Davie PJF (2002) *Crustacea: Malacostraca: Eucarida (Part 2): Decapoda – Anomura, Brachyura. Zoological Catalogue of Australia 19.3B.* CSIRO Publishing, Melbourne.

Jones D, Morgan G (2002) *A Field Guide to Crustaceans of Australian Waters.* 2nd edn. Reed Holland, Sydney.

Poore GCB (2004) *Marine Decapod Crustacea of Southern Australia: A Guide to Their Identification.* CSIRO Publishing, Melbourne.

Poupin J, Juncker M (2010) *A Guide to the Decapod Crustaceans of the South Pacific.* CRISP and SPC, Noumea, New Caledonia.

26
Molluscs

R. C. Willan

Introduction

Molluscs form a group of highly sophisticated invertebrates in terms of their morphology, having the greatest biodiversity of any phylum in the marine environment. This diversity is exceedingly high in the tropical waters of the Indo-Pacific Ocean, particularly in coral environments such as those of the Great Barrier Reef (GBR), where it has been estimated that ~60% of all the marine invertebrate species are molluscs. There are as many as 6000 species of molluscs on the GBR.

On the GBR, molluscs vary in size from the impressive metre-long giant clams (Tridacnidae) down to micromolluscs (Rissoidae, Barleeidae, Stenothyridae) and nudibranchs (Hedylopsidae, Microhedylidae) that are less than 1 mm long when full grown. Molluscs can exploit every possible kind of plant or animal as food, using numerous feeding strategies. Herbivores and carnivores among them use a specialised feeding organ (radula) to scrape food off the substratum into their mouth. Some species graze algae or sea sponges from rocks, while others are active hunters, detecting their prey either by smell (gastropods) or by acute eyesight (cephalopods). Many molluscs, particularly bivalves, feed on plankton suspended in the water by filtering out edible particles using mucous nets or gill sieves. Additionally, detrital feeding molluscs separate minute organic particles out from the sediments. In doing so, these detrital feeders recycle nutrients that would otherwise be locked up in the sediments.

Molluscs are linked into every food chain on the GBR, and they are integral for the very formation and destruction of the reef itself. Their shells serve as settlement sites for crust-forming red algae (Corallinaceae) that weld the hard corals together as they grow, thus building up the reef. Other molluscs (especially Mytilidae, Pholadidae and Gastrochaenidae) actively bore into live (and dead) coral, thus weakening it, and eventually causing its collapse. Molluscs are fundamental to the productivity of the GBR: their eggs are laid in tens of millions and the floating larval stages form a very important component of the zooplankton in the waters over the reef.

During the life of a mollusc, its shell serves as a living space for other invertebrates that also take advantage of the protection afforded by the rightful inhabitant. In doing so, they reward the mollusc by helping to defend the shell from attackers such as fishes and crabs. Examples of such commensal invertebrates are marine worms that live inside the shell and hydroids that live on the outside. After the mollusc that manufactured the shell has died, the shell continues to be recycled by other animals such as hermit crabs, tanaid crustaceans and peanutworms as their 'home' until it finally breaks down completely. One species of hermit crab, *Trizopagurus strigatus*, is specially adapted to live only in empty cone shells: its body, claws and legs are all flattened to fit inside the narrow-mouthed shell. During its decomposition the shell also serves as a site of attachment for a multitude of other invertebrates, general and specific, such as boring sponges, sea anemones, barnacles and foraminiferans (single celled animals enclosed within a tiny calcareous shell belonging to the phylum Protista; indicated by an arrow in Fig. 26.1).

Molluscs are present in all habitats on the GBR, from coral outcrops and rocky shores, to sandy lagoons, as well as in the water column – in the plankton that washes over the reef. The sheer abundance of some species such as those of the Littorinidae (periwinkles) can be staggeringly high. Some families (Pyramidellidae and Eulimidae) are almost all parasitic and members of several other families (Cancellariidae, Galeommatidae, Tergipedidae) and live commensally with a range of animals from other phyla.

One of the most significant aspects about the molluscs of the GBR is that there are no non-native (i.e. introduced or exotic) species among them; not a single one. This situation is unique in Australia today and it is one reason authorities need to be so wary when dealing with the shipping traffic traversing the GBR every day, because ships can easily transport non-native marine 'pests' (as they are termed in Australia) on their hulls, in ballast water and in internal seawater compartments called 'sea chests'.

Fig. 26.1. Example of micromolluscs (including triphorid gastropods, one indicated by the black arrow), the most abundant group of molluscs numerically on the Great Barrier Reef. This sample was sorted from a single bag of coral sand from Arlington Reef. Note also the foraminiferan indicated by the white arrow. (Photo: U. Weinreich.)

The larger gastropods and bivalves with external shells are the best known groups taxonomically on the GBR, so the level of accuracy of identification for them is relatively high compared with the smaller molluscs, particularly the micromolluscs, and those without shells.

Body plan

There is no standard molluscan shape and, indeed, there is no word in the English language that takes in the whole of the phylum Mollusca. The term 'mollusc' means soft-bodied and all molluscs lack an internal supporting skeleton. In its absence, there is no 'typical' shape that characterises the phylum, resulting in high variation in the shapes of molluscan animals. In fact, the shape of the molluscan animal is largely determined by the shape of the external shell it manufactures for protection. Molluscan shells are hard and three-layered, consisting of crystals of calcium carbonate (either calcite or aragonite) set in a matrix of protein material called conchiolin. An additional layer of a horny substance (periostracum) is laid over the exterior of the shell for extra protection or camouflage. Some gastropods also have a patch of this same substance on top of the foot (operculum), and this is used as a plug to seal the aperture when the animal withdraws into its shell.

The body of all molluscs consists of two main parts, the head/foot and the visceral mass. The former consists of a muscular foot in combination with the head with its array of sensory appendages. The visceral mass is essentially a sac containing the organs responsible for digestion, respiration, circulation and reproduction, all encased in a thin membrane (the mantle). The mantle produces the shell from its outer edge.

Not all molluscs have an external shell. Within each of the major subdivision of molluscs (classes), there are representatives with simplified and reduced shells. It seems that evolution of modern day molluscs has proceeded since the Palaeocene era (65 million years ago (Mya)) from forms with external shells to forms with reduced shells or no shell at all.

The Mollusca is a very ancient phylum and its members separated early in the Palaeozoic era (540 Mya) into several branches. Eight such branches (ranked as classes) exist today, with the ninth, Rostroconchia, being entirely extinct. The univalves (Gastropoda, Monoplacophora and Scaphopoda) have a single shell that encases the animal, or none at all. Although squid, cuttlefish and octopus (Cephalopoda) have no external shell, it is clear from their long-distant relative, the pearly nautilus, which is still surviving as a 'living fossil' on the GBR today that they evolved from ancestors with a single coiled shell. The bivalves (Bivalvia) have two shells, one on either side of the body, with an elastic ligament as a hinge at the top. The chitons (Polyplacophora) have eight shell plates roofing the body, all bound together by a leathery girdle. The gastroverms (Aplacophora and Solenogastres) are worm-like and apparently simple morphologically with their skin impregnated with tiny rods of calcium (spicules). All these classes, except the most primitive ones, the Monoplacophora, Aplacophora and Solenogastres, are represented on the GBR. In fact, the latter two classes probably do live on the GBR, but nobody has searched for them because they are inconspicuous and very difficult to locate.

In terms of their body plan, the chitons (Fig. 26.6A) are the most primitive and simplest molluscs one can encounter on the GBR. Chitons are dorsoventrally flattened, bilaterally symmetrical molluscs. There is a large flat foot ventrally and the body is covered dorsally with eight separate, usually articulating, shell plates bearing sensory aesthetes (simple light-detecting organs). The shell plates are surrounded by a muscular girdle that is either naked or covered with calcareous plates like a suit of armour. The head lacks eyes and tentacles. There are ~30 species of chitons on the GBR and they all live on hard substrata.

In terms of numbers of species, the most diverse molluscs on the GBR are the gastropods (Gastropoda) (Figs 26.2–26.6). There are ~4500 species of gastropods on the GBR. Gastropods have a multitude of different body forms, encompassing sea snails and slugs, limpets, sea hares and nudibranchs. Gastropods are united by the way in which the larva develops. In the free swimming larva (called the veliger), the cavity containing the gills faces backwards and ventrally, and the viscera are massed above the head/foot. At metamorphosis when the larva settles onto the bottom, an asymmetrical retractor muscle pulls the visceral mass through 180°. As a result, the mantle cavity becomes dorsal behind the head and faces forwards. The digestive system and its associated nerve fibres also become twisted during this dramatic process called torsion. In all gastropods but the true limpets, torsion is accompanied by spiral coiling of the shell as it grows. Gastropod shells are asymmetric because growth takes place in a clockwise direction, resulting in a shell with the aperture situated to the right of the longitudinal axis. Only a few gastropods (such as those of the family Triphoridae, one of which is arrowed in Fig. 26.1), consistently coil in an anticlockwise direction. Many slugs, both sea slugs and land slugs, have become (secondarily) bilaterally symmetrical externally, in an apparent reversal torsion, but internally their digestive, nervous and reproductive systems are all twisted as the result of this process in the larva.

The foot of gastropods has a flat sole for creeping, but it may be modified for swimming, grasping prey, digging or blocking the aperture of the shell. For shelled gastropods, the foot is last to be drawn into the most spacious final coil of the shell (the body whorl) and the aperture is finally plugged with the operculum: a horny or calcareous plate, borne on top of the foot. For shelled gastropods, the head/foot region, as the complex above the foot is known, extends from beneath the shell when the animal is active and the visceral mass is permanently contained within the upper coils of the shell (the spire) and is covered by the mantle. A large part of the visceral mass is occupied by the digestive gland that extends nearly to its apex and is part of the digestive system. The rest of this system consists of a mouth at the end of a snout or retractable proboscis, a feeding organ (radula) and jaws within the pharynx, salivary glands, oesophagus, stomach (that opens into the digestive gland), intestine, rectum and anus. Because of the process of torsion, faecal pellets are discharged over the head. The structure of the radula is a fundamental diagnostic feature used in the classification of gastropods.

Bivalves (Bivalvia) (Fig. 26.7) are laterally compressed, bilaterally symmetrical molluscs. There are ~1000 species of bivalves on the GBR. The shell consists of two valves joined dorsally by an elastic ligament and held together by two large muscles attached to both shell valves (adductor muscles). The mantle is often fused around the edges and extended posteriorly into retractable siphons. The head is simple, lacking eyes and tentacles, but there is a flap of tissue (labial palp) on either side of the mouth. The foot is axe-shaped and often bears a gland that secretes a bundle of threads (byssus) for attachment to the substratum. The foot is retracted into the shell by a special pedal retractor muscle. The gills are much enlarged and serve more for filter feeding than they do for respiration. The gut includes a complex stomach.

The watering pots (family Penicillidae) are the most bizarre and unusual of all bivalves on the GBR, albeit with only a few species. Juvenile watering pots, which are fully shelled, either burrow into the sediment or bore into soft calcareous rock immediately after settlement. As they grow, the original shell valves of the juvenile fuse to the anterior end of a sealed, calcareous tube that is permanently buried in the sand. The anterior end of this tube is swollen like a balloon and perforated by numerous tiny tubes, like the holes of a watering can. Watering pots have lost the posterior adductor and pedal retractor muscles and their anterior equivalents are vestigial, so the animal is only attached to its tube by retractor muscles arising from the pallial line and by an array of muscular papillae. A foot is present, but it is only small. In its place, an enlarged pedal disc acts as a hydraulic pump to bring water into the mantle cavity from the interstitial water surrounding the anterior tube through the perforations.

Tusk snails (Scaphopoda) are very elongate, cylindrical, bilaterally symmetrical molluscs. There

Fig. 26.2. Shells of representative taxa from the *Cymbiola* (*Cymbiolacca*) *pulchra* species group (Volutidae) (from left to right): *C. pulchra woolacottae* from Heron Island reef, 66.1 mm shell length; *C. houarti* from the Swain Reefs, 78.0 mm; *C. craecenta* from John Brewer Reef, 82.7 mm; *C. intruderi* from Halfmoon Reef, 71.5 mm; *C. pulchra excelsior* from Elusive Reef, 62.5 mm (Photos: *C. pulchra excelsior*, A. Limpus; the rest, MAGNT).

are ~25 species of scaphopods on the GBR. The mantle is fused mid-ventrally and the long tubular shell is open at both ends. The head bears a long snout and two groups of slender tentacles (captacula). The foot is cylindrical and pointed. Scaphopods have no gills, distinct blood vessels or auricles. They burrow in sediments and use the captacula to haul foraminiferans to the mouth.

Cephalopods (squid, cuttlefish, octopus) (Fig. 26.7E) are bilaterally symmetrical molluscs with a dorso-ventrally elongated body. There are ~35 described species of cephalopods on the GBR. In modern cephalopods the shell is internal and the visceral hump is covered by a muscular mantle,

Fig. 26.3. Triton's trumpet, *Charonia tritonis* (Ranellidae), *in situ*, about to devour a crown-of-thorns starfish, *Acanthaster* cf. *solaris* (Photo: GBRMPA).

Fig. 26.4. Triton's trumpet, *Charonia tritonis* (Ranellidae), *in situ*, devouring a large sea cucumber, *Bohadschia* sp., from its dorsal surface. Trumpet snails and balers often dine on large prey upside-down like this (Photo: C. Thomas).

Fig. 26.5. This geography cone snail, *Conus geographus* (Conidae), has stung a small demoiselle fish and is expanding the anterior end of its digestive system in preparation for ingesting the fish (Photo: U. Weinreich).

giving the body a rounded or streamlined shape. Speed, alertness and large body size are the key morphological features of cephalopods and they rival fishes in their locomotory prowess. They swim freely in the sea like fish, or move nimbly over the bottom. The head bears a pair of large, morphologically complex eyes. Cephalopods have prehensile arms around the head and the mouth lies at their centre. These muscular arms bear rows of suckers and have been described as 'super lips'. Squid and cuttlefish have two extra, retractile tentacles bearing suckers at their distal tips. Cephalopods have a specialised muscular organ, called a funnel, formed from part of the foot. Water from the mantle cavity is ejected via the funnel and is used as a means of jet propulsion in swimming.

Biogeography

The northern GBR is the region of the Australian continent where the greatest number of molluscan species exist today (~6000 species). Interestingly, however, it is also the region of the GBR with the lowest endemicity of molluscs, probably reflecting the youthfulness of its geological formation and the fact that most species have planktonic larvae and so can mix with populations of the same species elsewhere in the tropical Indo-Pacific Ocean. Considerably greater diversity of molluscs (between 10 000 and 12 000 marine species) exists in the 'Coral Triangle' region immediately to the north of the northern GBR. The diversity of molluscs does attenuate slightly towards the southern GBR, but total biodiversity is little changed because of the presence of long-range, endemic, temperate Australian taxa at the northern limits of their range. Here, the deeper water taxa (Turritellidae, Cypraeidae and Volutidae) have short-lived or direct-developing larvae and there are permanent populations in the cooler inter-reefal areas and channels deeper than 100 m. By contrast, the eastern Australian shallow-water endemic taxa that do occur on the GBR, such as the nudibranch *Goniobranchus splendidus* (Chromodorididae), have planktonic larvae and it is not known if their sporadic occurrences on the southern GBR represent breeding populations or adults

Fig. 26.6. **(A)** The chiton, *Cryptoplax larvaeformis* (Cryptoplacidae), *in situ*, emerging at night to graze on algae (Photo: GBRMPA); **(B)** ass's ear abalone, *Haliotis asinina* (Haliotidae), *in situ* showing the animal extended from its shell (Photo: GBRMPA); **(C)** Gilbert's top snail, *Jujubinus gilberti* (Trochidae), showing the animal extended from its shell (Photo: U. Weinreich); **(D)** strawberry top snail, *Clanculus margaritarius margaritarius* (Trochidae), showing the animal extended from its shell. Note the remarkable similarity of the pattern of the animal's foot to its shell (Photo: U. Weinreich); **(E)** sea hare, *Aplysia argus* (Aplysiidae), with its parapodia (upward-directed extensions from the foot) opened to show the mantle protecting the internal shell (Photo: G. Cobb); **(F)** tiger cowrie, *Cypraea tigris* (Cypraeidae), *in situ*, showing the animal extended from its shell (Photo: GBRMPA); **(G)** magnificent dorid nudibranch, *Chromodoris magnifica* (Chromodorididae), *in situ* (Photo: R. C. Willan); **(H)** much-desired aeolid nudibranch, *Coryphellina expotata* (Flabellinidae), *in situ*. Note the specialised defensive sacs (cnidosacs; one indicated by an arrow) at the ends of outgrowths (cerata) on the dorsal surface (Photo: R. C. Willan); **(I)** acorn dog whelk, *Nassarius glans* (Nassariidae), *in situ* showing the animal extended from its shell. (Photo: GBRMPA); **(J)** parasitic snail, *Balcis* sp. (Eulimidae), *in situ* on host crinoid (Photo: U. Weinreich); **(K)** this striated cone snail, *Conus striatus* (Conidae), has extended its proboscis (the narrower tube (arrowed) above the siphon) from its foregut in preparation for firing a toxin-loaded radular tooth into its prey (Photo: U. Weinreich).

resulting from chance northward-flowing larvae that have successfully grown to adulthood.

One cannot cover the biogeography of molluscs of the GBR without mentioning the remarkable and spectacularly beautiful group of endemic Australian baler shells known collectively as Heron Island volutes. The subgenus *Cymbiolacca* of the genus *Cymbiola* (Volutidae) contains a complex of ~20 allopatric species, subspecies and forms with very different shaped shells and colour patterns

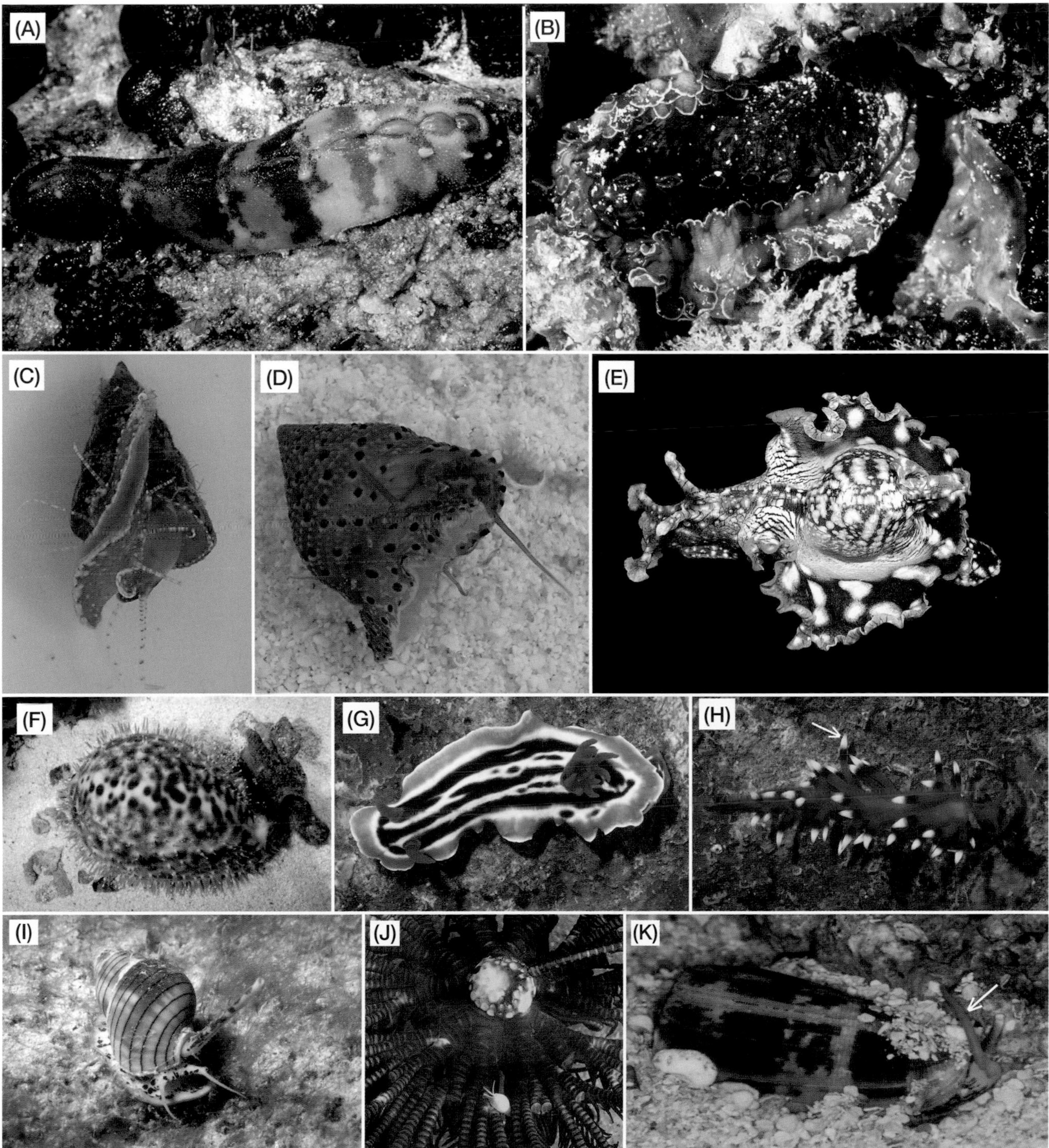

(Fig. 26.2) that live along the eastern Australian coast between 16°S and 32°S and includes the majority on the GBR (12 taxa alone occur in the section between the Swain Reefs and Fraser Island). Taxa of this assemblage occur from the Ribbon Reefs east of Cape Flattery in the north and extend continuously, in both shallow water on the coral reefs themselves and the deep water channels throughout the GBR, to Lady Elliot Island. South of Lady Elliott Island they are only found in deeper water (i.e. greater than 100 m). Interestingly, the deeper water taxa are ancestral and they have diverged less than the shallow water forms. Like all balers, these taxa of *Cymbiolacca* hatch from their egg capsules as crawl-away juveniles, so there is no genetic mixing between them that reinforces their distinctiveness. Although the evolution of these balers must have taken place only since the formation of the GBR in the Pleistocene era (2 Mya), we are a long way from understanding their evolutionary mosaic.

Feeding

Were it not for the combined efforts of all the gastropods and chitons grazing algae and algal sporelings from hard substrata, the whole GBR would be green with seaweed. Such herbivores have the foregut elaborated into a cuticularised rasping structure (radula) and have specialised ciliary fields in the stomach. Examples of algal grazing molluscs common on the GBR are snake skin chitons (Chitonidae), narrow-plated chitons (Cryptoplacidae) (Fig. 26.6A), limpets (Lottiidae, Patellidae, Siphonariidae), abalones (Haliotidae) (Fig. 26.6B), top and turban snails (Trochidae, Tegulidae, Turbinidae, Phasianellidae) (Fig. 26.6C, D), nerites (Neritidae), clusterwinkles (Planaxidae), periwinkles (Littorinidae), a few cowries (Cypraeidae), bubble snails (Bullidae, Haminoeidae), sap-suckers (Juliidae, Plakobranchidae, Limapontiidae, Polybranchiidae) and sea hares (Aplysiidae) (Fig. 26.6E).

Fig. 26.7. **(A)** Flashing file clam, *Ctenoides ales* (Limidae), showing the animal extended from its shell. (Photo: U. Weinreich); **(B)** the pedum oyster, *Pedum spondyloideum* (Pectinidae), *in situ*, is an unusual scallop that lives permanently buried in brain corals (Photo: J. G. Marshall); **(C)** lilac venus clam, *Callista erycina* (Veneridae), showing the animal extended from its shell (Photo: U. Weinreich); **(D)** giant clam, *Tridacna gigas* (Tridacnidae), *in situ* showing the brightly coloured mantle (Photo: GBRMPA); **(E)** cuttlefish, *Sepia* sp. (Sepiidae), *in situ*, showing the head, eyes, arms and funnel (Photo: GBRMPA).

Even more gastropods graze on encrusting animals than they do on plants. Animal-grazing is accomplished by the radula as in the algal grazers. Probably sponges serve as the main food group within this category of invertebrate 'meat' (if that term can be applied to such animal tissues). Examples of sponge-grazing molluscs common on the GBR are spikey chitons (Acanthochitonidae), keyhole and slit-limpets (Fissurellidae), top snails (Chilodontidae), cerithiopsids (Cerithiopsidae), triphoras (Triphoridae) (Fig. 26.1), the majority of cowries (Cypraeidae) (Fig. 26.6F), side-gilled sea slugs (Pleurobranchidae), and many nudibranchs (Hexabranchidae, Aegiridae, Dorididae, Discodorididae, Chromodorididae, Actinocyclidae, etc.) (Fig. 26.6G). But there are lots of other groups of invertebrates that constitute the food for grazing gastropods apart from sponges. Examples of hydroid-grazers are top snails (Calliostomatidae) and aeolid nudibranchs (Flabellinidae, Eubranchidae, Tergipedidae, Facelinidae). Examples of hard coral-grazers are coral snails (Coralliophilidae), wentletraps (Epitoniidae) and nudibranchs (Tergipedidae). Examples of grazers on sea anemones are some nudibranchs (Aeolidiidae). Examples of soft coral grazers are egg cowries (Ovulidae) and some nudibranchs (Tritoniidae). Examples of bryozoan-grazers are some nudibranchs (Polyceridae, Goniodorididae). Examples of kamptozoan-grazers are some nudibranchs (Goniodorididae). Examples of ascidian-grazers are lamellarias (Velutinidae), bean cowries (Triviidae) and some nudibranchs (Polyceridae, Goniodorididae).

The most remarkable of all these carnivorous gastropods are the aeolid nudibranchs (Fig. 26.6H), all of which feed on cnidarians. After ingestion, the tissue of their cnidarian prey is separated mechanically by cilia into digestible material and indigestible stinging cells (nematocysts), the latter being prevented from firing through copious volumes of mucus. These stinging cells are moved (carefully!) along fine branches of the digestive gland by other types of cilia to specialised sacs (cnidosacs; indicated by an arrow in Fig. 26.6H) at the ends of finger-like outgrowths from the dorsal surface of the body (cerata) that otherwise serve for respiration. When a predatory fish takes a bite out of an aeolid, the nematocyst-loaded cerata are quickly put into action and the nematocysts are fired off all at once, deterring the attacker. Aeolids provide the best example of animals using the defensive structures made by other organisms for their own defence (kleptoplasty).

The natural progression from grazing other species of animal is to feeding upon one's own kind, and indeed there are two families of nudibranchs (Polyceridae and Gymnodorididae) and one family of bubble snails (Agjajidae) that have members specialising for eating other sea slugs. This usually involves ingesting the prey's body whole.

Instead of grazing 'meat' unselectively, some gastropods have become highly specialised as parasites on specific tissues (such as the blood) of their hosts. The 'blood-suckers' are pyramidellids (Pyramidellidae) that feed on a range of invertebrate hosts, eulimids (Eulimidae) (Fig. 26.6J) that parasitise echinoderms (either as ectoparasites or as endoparasites living entirely in 'galls' inside the host), and margin snails (Marginellidae) and vampire snails (Colubrariidae) that suck the blood of sleeping fishes.

On the GBR, an enormous number of gastropods actively hunt down mobile prey. Those that prey specifically on marine worms are cone snails (Conidae, Terebridae), drupes (Muricidae), vase snails (Turbinellidae), bubble snails (Acteonidae, Aplustridae) and some nudibranchs (Vayssiereidae). Mitre and tulip snails (Mitridae, Costellariidae, Fasciolariidae) prey on peanut worms (sipunculans). Those that prey on crustaceans are harp snails (Harpidae), rock snails (Muricidae) and some nudibranchs (Tethydidae). Those that prey on echinoderms are helmet snails (Cassidae, Tonnidae) and trumpet snails (Ranellidae) (Figs 26.3, 26.4). Those that only prey on other gastropods are frog snails (Bursidae), balers (Volutidae) and cone snails (Conidae). Those that prey on bivalves are moon snails (Naticidae) and rock snails (Muricidae). Those that prey on fishes are basket snails

(Cancellariidae) and some cone snails (Conidae) (Figs 26.5, 26.6K).

Most species of dove snails (Columbellidae) are carnivorous, eating crustaceans and marine worms. However, a few species are scavengers and, remarkably, two species (one of which, *Euplica scripta*, occurs on the GBR) are facultative herbivores: that is, they occasionally eat algae in addition to their normal diet of crustaceans.

Balers (Volutidae) live on soft substrata, occasionally foraging on hard substrata. They hunt down other gastropods (such as top snails and turban snails, and even poisonous cone snails), capturing them by smothering them with their large foot. Having entrapped the prey, balers often then 'larder' them in a special pouch under the foot. This pouch may contain several gastropods, still alive, that have been hunted down during a night's foraging.

For an essentially bottom-dwelling group, the gastropods have a large number of species that, although living attached to reefal substrata, exploit the plankton in the water column above the reef for food. Examples of plankton-feeding gastropods are cap snails (Capulidae, Hipponicidae), slipper limpets (Calyptraeidae) and worm snails (Vermetidae, Siliquariidae). The worm snails (Vermetidae), the most interesting group of these plankton-feeding gastropods and the commonest encountered on the GBR, use a mucous trap to feed – a gland on their foot secretes copious quantities of sticky mucus that form thin threads streamed into the water as a feeding web. This web is hauled back into the mouth with the aid of the radula and the plankton stuck onto it is ingested. To maximise the time spent feeding, at least some species of worm snails are able to produce a second web concurrently with the ingestion of the first one.

The most agile and beautiful planktonic-feeding gastropods are those that live permanently in the water column, forming part of the macroplankton. The sea butterflies (Cavoliniidae, Limacinidae) have their foot modified into two enlargements anteriorly (the so-called 'wings') for swimming. Many species of sea butterflies use a transparent and delicate mucous 'fishing' web to strain other microscopic algal cells (i.e. diatoms and dinoflagellates) out of the plankton. Sea goddesses (Clionidae, Hydromylidae, Pneumodermatidae), another wholly planktonic group of gastropods, are active predators upon these sea butterflies, spotting them with their well-developed eyes. Sea goddesses capture sea butterflies with the suckered tentacles on their heads, attaching these tentacles to the inside of the aperture of the sea butterfly shell. The proboscis is then thrust into the shell of the sea butterfly and the body is ripped out whole.

There is an array of gastropods dedicated to scavenging dead animals. Indeed, one family of scavengers, the dog whelks (Nassariidae) (Fig. 26.6I), has ~30 species on the GBR. These different species have partitioned the habitats neatly between themselves: some only live intertidally, others in shallow lagoons, and yet others in fine sediments on outer slopes. Other scavenging gastropods are a few species of cowrie (Cypraeidae) and whelks (Buccinidae).

When plankton falls down onto the sea floor, it joins other organic and inorganic material deposited there. Several gastropods 'vacuum' up these nutritionally rich deposits from the surface of both hard and soft substrata. Examples of deposit-feeding gastropods are sand creepers (Cerithiidae), longbums (Potamididae), conchs and spider snails (Strombidae, Seraphsidae), carrier shells (Xenophoridae), onchidiids (Onchidiidae) and ear snails (Ellobiidae).

Though sedentary, bivalves use their extremely large and complex gills as filters to strain the plankton. Long cilia on the gills draw a powerful current of sea water into the mantle cavity (i.e. the space between the shell valves) where it is passed through the gill sieve formed by other types of cilia. A typical bivalve filters 30 to 60 times its own volume of water every hour. Common examples of filter-feeding bivalves on the GBR are mussels (Mytilidae), ark clams (Arcidae, Cucullaeidae, Noetidae), bittersweet clams (Glycymerididae), pearl oysters (Pteriidae), hammer oysters (Malleidae), mangrove oysters (Isognomonidae), pen shells (Pinnidae), file clams (Limidae) (Fig. 26.7A), oysters (Gryphaeidae, Ostreidae), kittens paws (Plicatulidae), saucer scallops and fan shells (Pectinidae) (Fig. 26.7B), thorny oysters (Spondylidae), jingle shells (Anomiidae), jewel boxes (Chamidae), yoyo clams (Galeommatidae), cardita clams (Carditidae), cockles (Cardiidae), giant clams (Tridacnidae), trough clams (Mactridae), razor clams (Solenidae, Pharidae), wedge clams (Donacidae), venus clams (Veneridae) (Fig. 26.7C), basket clams (Corbulidae), piddocks (Pholadidae) and watering pots (Penicillidae).

Giant clams (Tridacnidae) (Fig. 26.7D) lie upside down in depressions on the coral. Their mantle lobes are extensively developed and 'farm' dense colonies of microscopic dinoflagellates (*Symbiodinium* sp., popularly called zooxanthellae). These zooxanthellae appear to supply the majority of older clams' carbon requirements, most importantly glucose, through photosynthetic by-products. The largest family of bivalves numerically, the wafer clams (Tellinidae), all feed on organic particles deposited on the sea floor, thus recycling nutrients in the way of the deposit-feeding gastropods mentioned earlier. They have a very long and flexible inhalant siphon that 'vacuums' these deposits off the top layer of sediment. The ingested sediment is sucked into the mantle cavity where it is sorted by the enormous labial palps into edible particles and waste matter. Two other families of bivalves closely related to the wafer clams are also deposit feeders: the sunset clams (Psammobiidae) and semele clams (Semelidae).

The lucine clams (Lucinidae) are a particularly specialised family of bivalves whose shells are quite commonly encountered on the GBR. Lucines lack an inhalant siphon and have an inhalant opening instead for feeding and respiration. They construct an inhalant tube up through the sediment by the foot. The foot is highly extensible, with the tip in the form of a pointed bulb capable of secreting mucus that lines the inhalant tube. The gills have a rich, resident bacterial flora contained in large vacuoles; these bacteria undertake sulphide-oxidising reactions.

The cephalopods are all carnivorous, rivalling the fishes in their ability to see and hunt down active prey such as fishes and swimming crustaceans. At least two species of the ancient pearly nautilus (Nautilidae) live on the GBR. Common examples of free-swimming cephalopods are cuttlefish (Sepiidae) (Fig. 26.7E) and calamari squid (Loliginidae). These catch their prey by rapidly shooting out their pair of feeding tentacles. These tentacles move so fast that they are difficult to see. Dumpling squid (Sepiolidae) and octopuses (Octopodidae) spend most of their time on the sea floor, feeding on less active prey such as benthic crustaceans, bivalves and sleeping fishes. However, their ability to see and hunt down their prey is every bit as good as their free-swimming relatives. All octopuses have strong toxins that quickly immobilise their prey.

Reproduction

Most molluscan species have separate sexes and there are generally no external differences between males and females of the same species. However, the spider snail *Lambis lambis* (Strombidae) shows some dimorphism when the shells are fully grown, with females having larger shells with long upward-curved spines and males having smaller shells with short horizontally pointing spines. Other notable examples are the very small yoyo clams (Galeommatidae), where dwarf males live inside the mantle cavity of the female.

Many bivalves (most notably Ostreidae, Tridacnidae and Galeommatidae) change sex from male to female as they grow; that is, they are protandric hermaphrodites. All the sea slugs are simultaneous hermaphrodites, meaning that each individual has both male and female reproductive organs that are physiologically mature at the same time. However, they never fertilise themselves (although this is physically possible), and instead are able to mate with any other mature animal of the same species they encounter, both individuals acting as sperm donors and egg recipients at the time of copulation. The sea hares (Aplysiidae) (Fig. 26.6E) have an even more extreme kind of partnering where they form long mating chains; the individual at the rear of such a chain acts as a male to the one in front of it by delivering sperm. This second-to-last individual acts as a female by receiving the sperm and, simultaneously, as a male by delivering sperm to the partner in front of it, and so on through the chain. Indeed, there is one report in the literature of the animals at the front and rear of such a chain coming together to form a complete mating ring!

Some sea butterflies (Cavoliniidae) are believed to possess remarkable asexual reproduction. Under laboratory conditions, an unusual-looking skinny individual is formed by transverse splitting of the body of a normal individual. The new individual then detaches itself from its shell as a naked animal, transporting only gonads, and grows independently into a fully shelled separate individual.

Most molluscs simply shed their gametes (eggs and sperm) directly into the sea water where fertilisation occurs, but fertilisation is internal in some gastropods and all cephalopods. In those taxa with an intermediate stage, the resulting larva has a shell like a tiny cap, and lobes on its foot for swimming and capturing plankton for feeding. As mentioned earlier, a larva of this type, termed a veliger, is typical of all molluscs. Veligers drift with the currents until they come across the adult food source at which time they break off their swimming lobes and metamorphose into crawling juveniles. Those molluscs that have internal fertilisation lay their eggs in tough cases. Either veligers hatch from these cases and join the plankton, or a miniature version of the adult crawls out. In the first case, the female produces a relatively large number of small eggs, each with a small amount of yolk (lecithotrophic development), and this is the most common pattern found in warmer waters such as those of the GBR. In the second case, which is termed direct development, the female produces only a relatively few large eggs, each with a rich supply of yolk, and this occurs mainly in molluscs living in cold waters, although some examples are found on the GBR.

Reproduction reaches new heights in cephalopods, in terms of both mating and brood protection. The sexes are always separate. Many shallow water cephalopods come together in large numbers to reproduce. They have elaborate courtship displays with the male 'dancing' and rapidly changing its colour and pattern at the same time to impress the female. The males of some octopuses display specially enlarged suckers to females as a sign of sexual maturity, with males of one species on the GBR, *Octopus cyanea*, waving a raised and coiled modified arm tip in the direction of the female during its courtship display. The sperm duct of cephalopods has become extremely specialised for the manufacture of sperm bundles (spermatophores). Each spermatophore is a narrow, hard, torpedo-shaped tube containing a dense mass of sperm. Although a penis is used in some species to transfer spermatophores directly to the female, in most cephalopods, the male uses a specially modified arm (the hectocotylus) to transfer spermatophores from the terminal opening of the genital duct to the female (the cephalopod equivalent of copulation), sometimes even leaving his arm holding the bunch of spermatophores inside the female's mantle cavity. Males of other cephalopods bite small holes in the female's skin into which they insert the spermatophores. These implanted spermatophores resemble small, white, parasitic worms under the skin. Cephalopod eggs and embryos are different to those of all other molluscs. The eggs are comparatively large and yolky, and do not completely cleave after fertilisation so that the embryo is built up from a smaller disc of cells on the upper pole of the egg, and the larger part of the egg goes to form a yolk sac from which the young animal is nourished. Female octopuses brood their eggs until the embryos hatch, fiercely defending them from predators.

Although brood protection is best developed in octopuses, it is by no means unique to octopuses or just cephalopods. Females of several gastropods (Cypraeidae, Ranellidae and Facelinidae) remain with their spawn masses until the young hatch out and it is a common sight to see a female cowrie 'sitting on her eggs' when one turns over a dead coral slab on the GBR. It is one of the reasons people should always replace rocks and coral slabs if they turn them over; if not, the female and all the embryos she is guarding will die from predation or desiccation (if they are intertidal).

Females of a few bivalves (most notably the yoyo clams of the family Galeommatidae) take brood protection to the extreme by incubating their eggs, and even the developing embryos, inside their mantle cavities. These clams only produce a small number of embryos (up to a dozen) because of the considerable energy that needs to be invested in such maternal care.

Dangerous molluscs found on the Great Barrier Reef

Only a few molluscs living on the GBR are deadly to humans, and they cause harm accidentally by defending themselves.

The blue-ringed octopus, *Hapalochlaena lunulata*, has particularly potent saliva that quickly paralyses its prey (normally crustaceans). One component of the saliva is tetrodotoxin, which is produced by bacteria in the salivary glands, and to which the human nervous system is particularly susceptible. This chemical prevents messages coming from the brain from reaching the muscles, so that the human victim is paralysed. Interestingly, tetrodotoxin only paralyses voluntary muscles; involuntary muscles, such as the heart, the iris of the eye and the gut, continue to function normally. The victim scarcely feels the bite of a blue-ringed octopus, but after a short time has difficulty in breathing. This is followed by nausea and vomiting, complete cessation of breathing and collapse. The victim remains fully conscious and may die from lack of oxygen. There have been no deaths from blue-ringed octopuses on the GBR, but in 1954 a man in Darwin died within 2 h of being bitten by one.

A victim definitely feels a sharp pain if stung by a cone snail. There are actually more than 100 species of cone snails (*Conus*) in all habitats on the GBR, but only three of them living among live coral and coral rubble (*Conus geographus*, *C. tulipa* and *C. obscura*) are capable of killing humans. These cone snails are all fish-eating (piscivorous) species and have an enlarged aperture to the shell into which the dead fish can be hauled for digestion (Fig. 26.5). The sting is caused by the piercing of the skin by a single radular tooth resembling a tiny harpoon. The tooth, which is shot out forcefully from the tip of a long flexible proboscis, injects a powerful neurotoxin (contoxin) as it is driven into the flesh of the prey. Cone snails have an elaborate venom apparatus consisting of a muscular bulb and a tubular secretory duct opening into the mouth cavity. A modified proboscis that is tubular and muscular like an elephant's trunk (Fig. 26.5) actually fires out this radular tooth. A human victim of a sting from a cone snail can lose vision, or have hearing or speech affected, and may become partially or completely paralysed within half an hour. Indeed, there is one recorded death of a man on Hayman Island in 1935 from the sting of a *Conus geographus*. A sting by a species of cone snail other than these three may cause pain, swelling and discolouration of the area near the puncture, but not death. So one should walk carefully over the reef and not pick up a cone shell, even if it appears to be empty, because its animal may be fully retracted within the shell.

Future research

At present the most urgent need is for taxonomic research to establish what species of molluscs live on the GBR and where they live. At present, the figure of 6000 species is a best guess, and the number of adequately identified species is rapidly increasing as more and more groups of micromolluscs such as those shown in Fig. 26.1 are studied. An indication of the richness of these micromolluscs on the GBR can be obtained from just one family, the curious left-twisted triphoras (Triphoridae; a triphora shell is arrowed in Fig. 26.1), where a single sample of coral sand is known to have contained 80 species of triphoras, many of them undescribed scientifically.

Genetic technology has become an important tool for taxonomic research. Genetics provides a method to separate taxa that do not offer sufficient morphological characters to distinguish them or present a confusing array of morphological characters. The prime candidate waiting for genetic research would have to be the beautiful balers of the genus *Cymbiola* (*Cymbiolacca*) (Volutidae) mentioned earlier and shown in Fig. 26.2.

Despite the very large number of species of molluscs on the GBR, research on their behaviour and physiology has barely begun. As described earlier, we have some general knowledge about the feeding types of molluscs, but there has hardly been any research on individual species. For example, although we know some chitons have 'home' sites that they return to after feeding excursions, we do not know how individuals find their way 'home' or what makes a particularly good 'home' site. Similarly, we know that some members of the mussel genus *Lithophaga* (Mytilidae) burrow into coral using a chelating agent secreted by pallial glands, but we do not know what turns this agent off when individual mussels attain maturity and stop burrowing, or how they avoid being overgrown by living corals.

As the advent of scuba diving on the coral reefs has opened our eyes to the nudibranch fauna of the GBR, drift diving in mid-water has just started to reveal some astonishing information about the behaviour of the holoplanktonic molluscs that live there. For example, the very peculiar sea goddess *Hydromyles globulosa*, which is moderately common on the GBR, has its entire animal encased in a transparent and flexible cuticle. When threatened, an animal can retract completely into this cuticle and seal the slit-like opening with a fold in the cuticle, thus turning itself into a completely impervious gelatinous sphere.

If one investigates the body of any mollusc closely, one is likely to find it harbours numerous parasites (e.g. ciliated protists, nematodes, trematodes, isopod crustaceans), both externally and internally. For example, on Heron Island, the common clusterwinkle *Planaxis sulcatus* (Planaxidae) acts as an intermediate host for several trematodes, including *Austrobilharzia terrigalensis*, which is responsible for 'swimmer's itch'.

Nudibranchs have come to symbolise all the molluscs on the GBR because of their fragile bodies, bizarre shapes, bright colours and remarkable behaviours. Nowadays no diver can visit the GBR without taking at least some digital images of nudibranchs. Photography is certainly a form of 'collecting', albeit one that does not involve the removal of any live animals themselves, though damage is inadvertently caused to the reef by trampling, boat's anchors and incorrect weighting.

In the past, shell collectors received much criticism by outsiders for their activities. One hears anecdotes of the 'devastation' caused by shell collectors and one reads emotive passages targeting them, such as: 'By their thoughtless depredation the whole productivity of that reef flat and the natural habitat of thousands upon thousands of its inhabitants can literally be ruined in a matter of hours' (Bennett 1986: 126). In fact, such hysterical statements have never been tested on the GBR. Humans have collected molluscs for ages – for food, trade and decoration – and detailed studies on population ecology have shown many species are very resilient to collecting. Responsible shell collectors only take a few individuals and never remove immature specimens or females guarding eggs. Furthermore, they are aware of the ecological consequences of turning over dead coral slabs. It is the shell collectors who have provided us scientists with the specimens for our comparative studies and museum reference collections. In recognising shell collectors as the least threatening of a host of (natural and unnatural) processes that could affect the GBR – human trampling, bottom trawling, commercial fishing, collection for consumption by Aboriginal and non-Indigenous people, eutrophication from land-derived nutrients, cyclones, rising sea levels and increased ocean temperatures along with other effects of climate change – the Great Barrier Reef Marine Park Authority currently allows recreational collection of molluscs on certain reefs on the GBR and along the adjacent coast without a permit. This collecting is limited to five individuals of a species (i.e. a total that comprises living animals and/or their shells, but not more than five of any species) to be taken in one day. For all other purposes, a permit is required from GBRMPA (see Chapter 14).

References

Taxonomy of balers of the *Cymbiola pulchra* species group with hypotheses on their evolution

Bail P, Limpus A (1998) *Revision of* Cymbiola (Cymbiolacca) *from the East Australian Coast: The* 'pulchra *complex'*. Evolver Publications, Rome.

Bail P, Limpus A (2015) The eastern Australian *Cymbiola* Swainson, 1831 (Gastropoda: Volutidae). New revision of the subgenus *Cymbiolacca* Iredale, 1929. The "*pulchra* complex" 1998 updated. *Visaya* **6**(supplement), 1–94.

Detailed treatment of all mollusc families occurring in Australia

Beesley PL, Ross GJB, Wells AE (Eds) (1998) *Mollusca: The Southern Synthesis. Fauna of Australia. Vol. 5.* CSIRO Publishing, Melbourne.

Species present in Australian waters with geographical distributions

Gowlett-Holmes K (2001) Polyplacophora. In *Zoological Catalogue of Australia 17.2 Mollusca; Aplacophora, Polyplacophora, Scaphopoda, Cephalopoda*. (Eds AE Wells and WWK Houston) pp. 10–84. CSIRO Publishing, Melbourne.

Lamprell KL, Healy JM (1998) A revision of the Scaphopoda from Australian waters. *Records of the Australian Museum* **24**(Supplement), 1–189. doi:10.3853/j.0812-7387.24.1998.1267

Lu CC (2001) Cephalopoda. In *Zoological Catalogue of Australia 17.2 Mollusca; Aplacophora, Polyplacophora, Scaphopoda, Cephalopoda*. (Eds AE Wells and WWK Houston) pp. 129–308. CSIRO Publishing, Melbourne.

Detailed descriptions and illustrations of all the species from Heron Island plus a listing of species present on the GBR

Marshall JG, Willan RC (1999) *Nudibranchs of Heron Island, Great Barrier Reef: A Survey of the Opisthobranchia (Sea Slugs) of Heron and Wistari Reefs.* Backhuys Publishers, Leiden, Netherlands.

Source of quotation

Bennett I (1986) *The Great Barrier Reef.* Lansdown Press, Sydney.

27

Bryozoa

P. E. Bock and D. P. Gordon

Bryozoans are colonial animals, commonly called 'sea mats' or 'moss animals'. 'Lace coral' is used in Australia, particularly for the erect net-like forms, but this usage is discouraged, because in other regions this term is applied to stylasterid hydrocorals (phylum Cnidaria). The great majority of bryozoans are marine, with slightly over 100 species of freshwater species.

In the 1800s, the Bryozoa were held to comprise two very distinct groups, Ectoprocta (bryozoans in the strict sense) and Entoprocta (also called Kamptozoa, goblet worms and nodding animals), each of which was raised to phylum rank. Ectoprocta as a phylum name may be found in some publications. The International Bryozoology Association has formally adopted the name Bryozoa for the ectoprocts. Entoprocts may not be closely related to genuine bryozoans, but the evolutionary origins of both groups is very obscure. Only six species of entoprocts have been reported from the Great Barrier Reef (GBR), with many more to discover: they are not further discussed in this account. Phylogenetic studies demonstrate that bryozoans form a distinct group within the Lophotrochozoa, although relationships to other phyla such as the Brachiopoda or Phoronida are still uncertain.

Bryozoans are colonial, with the individual units known as 'zooids' (Box 27.1). Colonies can grow from a larva derived from sexual reproduction or through fragmentation and regeneration giving rise to daughter colonies (Box 27.2). Adult colonies range in size from less than a millimetre to more than a metre in diameter. Most are one to a few centimetres in width (if encrusting) or height (if erect). In the tropics, bryozoans are mostly dwarfed by scleractinian corals, but in cooler temperate waters they come into their own, with a few reaching colony diameters up to 15 m, as in Western Port Bay, Victoria, where the world's largest bryozoan colonies are found.

Box 27.1. How are they constructed?

Colonies are made up mostly of feeding zooids (autozooids), but bryozoans are justifiably famous for exhibiting a higher degree of polymorphism ('many forms') than almost any other invertebrate group, by having different kinds of non-feeding zooids. The most famous are avicularia, named by Charles Darwin, which have various roles (cleaning, defence, locomotion). Then there are kenozooids, simple chambers modified to serve as attachment rootlets (rhizoids, stalks), spines, stolons, space-fillers, or part of the support structure for large erect colonies. To feed, bryozoans evert a funnel-shaped plume of tentacles (the lophophore) through an opening (orifice) in the body wall that merely puckers inwards upon closure or is variously modified as a pleated collar, as a pair of stiff lip-like folds, or as a lid-like operculum. Retraction of the lophophore is achieved by contraction of long muscles. The eversion of the crown of tentacles requires a muscular method of increasing the internal pressure in the zooid, thus squeezing out the lophophore. There are several bodily arrangements to achieve the necessary redistribution of fluid.

Box 27.2. Reproduction

Usually zooids are hermaphroditic, their sperm maturing before their ova. Some bryozoan species have separate female and male zooids that differ in shape, size, and anatomy. Purely reproductive zooids may lack a gut and have a few tiny non-ciliated tentacles whose sole function is to release sperm. This surprising use of tentacles as vasa deferentia (sperm ducts) appears to be true for all bryozoans. The lophophore also serves as a gill for respiration, so it is a truly a multipurpose structure.

Fertilisation in bryozoans is internal. Sperm from one colony are captured by a recipient colony's tentacles, to which they first adhere then move downwards towards a duct (intertentacular organ) or, more commonly, a pore that allows entry into the body cavity. Fertilised eggs may be incubated in a special sac within the zooid, within a modified zooid (gonozooid or brood chamber), or, as in the majority of species, in a hood-like ooecium distal to the maternal orifice.

Phylum Bryozoa has an excellent fossil record from the Ordovician to the present. It is the only major phylum lacking indisputable records from the Cambrian: reports of Cambrian bryozoans exist, but these fossils are arguably not bryozoan. Current estimates of recorded species diversity for Holocene bryozoans is just under 6300, from a total phylum diversity of nearly 23 000 species. The phylum is most diverse in tropical shelf waters, but species are recorded down to abyssal depths, and bryozoans are important in polar waters. As current studies are proving an undiscovered further diversity of modern bryozoans, it is likely ultimate species diversity may be up to twice current estimates.

As the British biologist John Ryland noted, the most outstanding thing that can be said about research on GBR Bryozoa is its paucity, given the undoubted richness of the fauna, although the number of studies on particular taxa have been increasing in recent years. Ryland (1984)

Box 27.3. Collecting bryozoans

Erect bushy, stick-like, and lace-like bryozoans can easily be detached from their substratum. Encrusting bryozoans are harder to collect unless they are on pieces of shell, seaweed or coral. A thin blade can be used to lift or scrape some off the substratum. More resistant substrata may be broken using a hammer and cold chisel. A very useful tool is a high-speed rotating craft tool fitted with a small diamond blade. Details regarding permits required for collecting in the various zones of the GBR are provided in Chapter 14.

Preservation

Uncalcified and delicate bryozoans are best preserved in alcohol, usually 70% ethanol, but 90–100% ethanol is necessary for DNA studies. Freezing is an alternative where suitable facilities exist. The easiest way to treat most encrusting and robust erect species is to let them dry out. There is very little smell.

Preparation for study

Field determination of the bryozoan using a dissecting microscope is useful. Painting colonies with dilute food colouring helps to accentuate key morphological features. In some groups, a thick outer cuticle may obscure significant morphological details. For strongly calcified forms, laboratory study and illustration is generally made from material that has been bleached using diluted domestic bleach, removing cuticle and internal tissue. Scanning electron microscopy has proved essential for adequate characterisation of microscopic morphological details of bryozoan skeletons.

Identification

Information on bryozoans is scattered in monographs and scientific papers and there are very few popular guides (none for the GBR). However, the internet is a good source of information. The best entry point, especially for photographs of Australian bryozoans, is http://bryozoa.net and the two volumes of *Australian Bryozoa* (2018, CSIRO Publishing). The official website of the International Bryozoology Association is http://bryozoa.net/iba/.

Table 27.1. Bryozoan taxonomic diversity from the Great Barrier Reef Province

Taxon	No. of families	No. of genera	No. of species
Class Stenolaemata	10	13	16
Order Cyclostomata	10	13	16
Class Gymnolaemata	72	136	316
Order Ctenostomata	6	8	10
Order Cheilostomata	66	128	306
Totals	82	149	332

summarised the history of collecting from the 1840s to the 1970s and wrote a short account of the Bryozoa of the GBR for the *Coral Reef Handbook* based on his personal collecting. This collection, mostly from Heron Island, formed the basis of two important papers published with Peter Hayward (Ryland and Hayward 1992; Hayward and Ryland 1995). For information on collecting bryozoans, see Box 27.3.

Current known bryozoan diversity in the GBR Province from Torres Strait to the southern-most section is based on distributional information, species lists, and descriptions provided by 20 bryozoologists and two ecologists from 1852 to 2016. The tally from all sources, after accounting for synonyms, is 332 species for the entire GBR Province (see Table 27.1). This figure for the number of species is probably extremely conservative, especially given the geographic area covered by the GBR, as well as the wide variety of habitats and niches that can be occupied by the range of colony forms that bryozoans exhibit (see later section). Several bryozoan species from the GBR are illustrated in Fig. 27.1. The neighbouring New Caledonian bryofauna comprises 407 species, of which 178 species are found in the first 100 m, and 232 species ranging into or found only at depths greater than 100 m. Some 178 species were recently reported in the Solomon Islands fauna, of which 72 (40%) were new, giving an indication of the level of taxonomic novelty that tends to occur whenever a tropical island group is explored intensively and for the first time. Some 725 species are known for the Philippine-Indonesian region (i.e. the 'Coral Triangle'), which is the most biodiverse marine area globally. Even this region, however, has been explored relatively superficially, both in shallow water and at depth, and one can expect this tally to be doubled. Equally, one should expect more than 1000 bryozoan species within the boundaries of the entire GBR Province. While many GBR species are also found elsewhere in the Coral Triangle, several species clearly have limited distributions.

Bryozoans in the coral-reef environment

It is probably not generally appreciated that bryozoan diversity can be very high in tropical reef environments. Bryozoans are generally inconspicuous in reef habitats. Intertidal exposed surfaces are normally covered by the more successful competitors, including hard corals, soft corals, algae and sponges. Smaller blocks of reef rubble are capable of being moved by storm wave action, and these substrata support few bryozoans. The really large blocks and fixed coral ramparts provide space for a few species in well-lit waters, but once again bryozoans tend to lose the race to cover space.

Bryozoan associations with other biota may clearly be seen on the erect coral forms such as acroporids. The growing tips of the coral are occupied by living polyps. Further down a branch, where the corals are no longer living, the branch is encrusted by a mixture of algae, ascidians, sponges and bryozoans, while the interior of the branch is hollowed out by clionid sponges. This bioerosion opens up yet more microhabitats for mobile or encrusting organisms. In the case of blocky coral ramparts, cracks and cavities may be present. Examination of these shows few bryozoans near the surface, where light penetration is high. Excavating further shows increasing cover and diversity of bryozoans, as the lower light levels make competition from algae less significant.

Away from the physical reef habitats, the GBR Province contains a diverse range of bryozoan occurrences. The Seabed Biodiversity Project of 2003–2006, coordinated by the Australian Institute of Marine Science, sampled a large number of sites

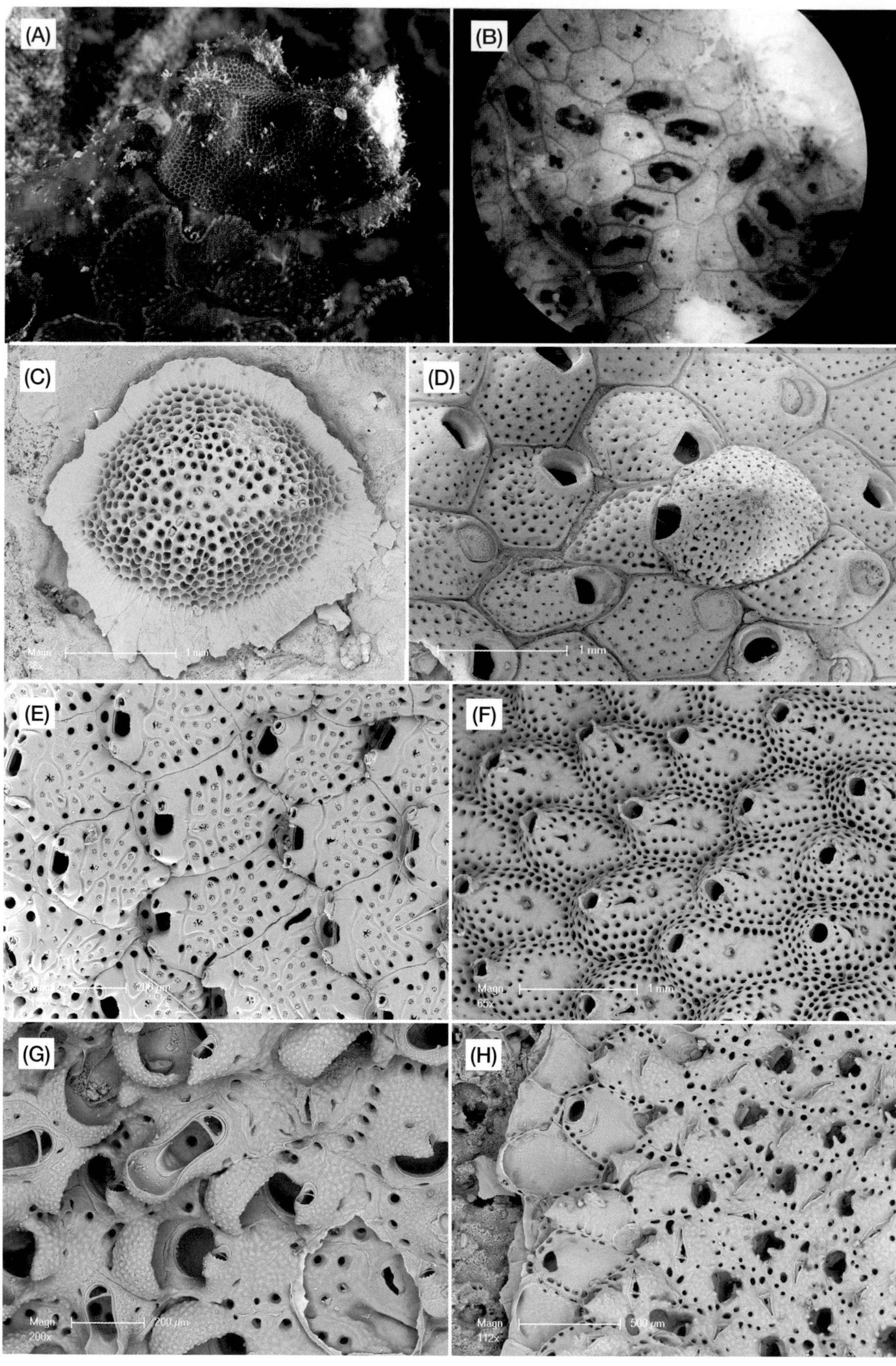
(A)
(B)
(C)
1 mm
(D)
1 mm
(E)
(F)
Magn
65x
1 mm
(G)
Magn
200x
200 µm
(H)
Magn
112x
500 µm

throughout the province. An extensive collection of samples has been stored as frozen material in the collections of the Museum of Tropical Queensland. A cursory examination of this material suggests that erect and flexible forms are an important part of the inter-reef habitats. Among the range of habitats, one of the significant examples is the 'sand-fauna' habitat, which is colonised by a small number of free-living colonies, such as species of *Selenaria*, whose cap-shaped colonies are capable of autonomous movement using peripheral avicularia mandibles. There are many groups that maintain their place by growing root or stem zooids anchored in the sand. Examples in this group are members of the order Cheilostomata in the genera *Sphaeropora*, *Conescharellina*, *Flabellopora*, *Siphonicytara*, and many from the family Candidae. A similar important group in the order Ctenostomata are from the genus *Amathia*. As these groups are not closely related, it is clear that the growth of rootlets has evolved many times independently.

Epiphytic bryozoans are another important group. In cooler water, bryozoans are often seen on a variety of red or brown algae. A distinctive set of bryozoans in tropical waters are those found growing on the genus *Sargassum*. These have weakly calcified joints or zooidal junctions to maintain flexibility. Another form found washed ashore are some epiplanktonic bryozoans, which colonise floating objects such as pumice, plant material or plastic. The genus *Jellyella* is the main member of this group. Bryozoans may also be found attached to swimming animals, such as turtles, although there are no such records from Australia.

Beyond the tropical shelf of Australia, the bryozoan fauna is poorly known. A few samples are available, but have not been studied in detail. The bryozoan slope fauna is quite distinct: the Australian material is related to other deeper water regions, such as New Caledonia and Indonesia.

Bryozoan growth forms

Bryozoans exhibit an impressive range of colony forms, a fact noted and much used by non-bryozoologists (as well as bryozoologists), particularly palaeontologists who have sought to correlate colony form with environment in fossil assemblages in which bryozoans have been found in rock-forming abundance. Colonies can be two-dimensional encrusters (the majority) or, through frontal budding and/or self-overgrowth, these can become mounded (as seen in species of *Celleporaria*) or erect. Hence, erect bryozoans may be firmly fixed to the substratum, but many are not, having only root-like attachments. Owing to the fact that rootlets can spread out, and in a few cases even become finely divided and attach to sand grains, some species are adapted to live on soft sediments. Some soft-sediment dwellers can be tiny conical forms, scarcely more than 1 mm in size and may have only a single anchoring rootlet. In some others, the rootlet is a robust tube that anchors a large 'platey' colony above the sediment. Bryozoan colonies may be thickly calcified, lightly calcified or (in the case of ctenostomes) uncalcified. If erect colonies are lightly calcified, they can bend in a current and are said to be flexibly erect (members of the families Bugulidae and Candidae). But well-calcified colonies can also be flexible if they are articulated by uncalcified joints (Cellariidae, Catenicellidae, Poricellariidae, Quadricellariidae, Savignyellidae, Tetraplariidae) or are basally rooted.

Fig. 27.1. **(A)** *Steginoporella crassa* (above) and *Iodictyum willeyi* (below). Lizard Island region (Photo G. Cranitch); **(B)** *Alcyonidium* species on plastic fouling panel. Lizard Island region. These uncalcified bryozoans are difficult to observe in the field; **(C)** *Disporella* species – a cyclostome. Yonge Reef, NQ; **(D)** *Macropora* species, showing ovicell and calcified opercula. North-west Island, Capricornia Group; **(E)** *Inversiula inversa*. The operculum hinges on the distal edge of the orifice. Heron Island region; **(F)** *Reptadeonella hystricosus*. The central pore is a spiramen for water transfer. Yonge Reef, NQ; **(G)** *Celleporaria* species. Showing ovicells, and avicularium, and a budding zooid. North Direction Island, NQ. **(H)** *Rhynchozoon haha*. Showing growing edge of colony, and progressive thickening of skeleton. Ovicells are present, but obscured, at lower right. Yonge Reef, NQ.

Apart from encrusting, fixed-erect, and flexible-erect colonies, there are two other categories. One is free-living, where the colony is not attached to the substratum and may be transported by waves or currents (families Lunulariidae, Otionellidae and Selenariidae). In at least one species, bristle-like avicularian mandibles around the periphery of the colony can move in a coordinated way to move the colony laterally, or to return it from an over-turned state to the normal convex-up position. The other distinctive category is that of shell-borer, found among several families of ctenostomes. None has yet been reported from Australia, although they have been collected from cooler water locations. Preliminary searching has not identified any from the GBR: detailed examination of shells of molluscs will almost certainly reveal some. Their small size, transparency, and cryptic habit render them easily overlooked: using ink or dye can help in making them visible in dried shells.

For the entire GBR Province, 55% of the bryozoan species are two-dimensional encrusters, 20% are flexibly erect and 15% are firm three-dimensional structures that are rigidly attached to the substratum. These categories of colonial morphology are very broad. They can be subdivided, and more than 20 such categories have been recognised and named (e.g. encrusting colonies may be runners, spots or sheets; erect colonies may be sticks, trees or fronds). Such categories based on colony form potentially allow for the possibility of paleoenvironmental analysis, but it must be said that correlations between morphology and environmental conditions in the literature have typically been inferential and not backed up by experimental or ecological observations based on living colonies. For this reason, a new, integrated classification scheme based on growth-habit characteristics and the disposition of modules (zooids) has been developed to allow a testable common ground for systematic comparison of character states among varied bryozoan growth habits. It would be fruitful applying this system to the varied range of colony forms represented in the GBR Province.

Bryozoans in an acidifying ocean

Bryozoan skeletal material is composed of calcite, aragonite, or a mixture of both mineralogies. This is related to taxonomy, but not in a simple way. About two-thirds of living bryozoans are made of calcite alone, including almost all from the class Stenolaemata. Of the rest, about half are made from aragonite alone. The bryozoans with bimineralic skeletons are of three types, including a group with both high-Mg and low-Mg calcite together. It is predicted that decreasing pH as a result of increasing atmospheric carbon dioxide will result in lower growth rates and dissolution of skeletal material. However, response to decreasing pH may be affected also by skeletal morphology and by the characters of the cuticular covering. It can be anticipated that some groups will be affected strongly while others will be able to adapt to growing with a smaller amount of calcification.

Further reading

Bock PE, Gordon DP (2013) Phylum Bryozoa Ehrenberg, 1831. *Zootaxa* **3703**, 67–74. doi:10.11646/zootaxa.3703.1.14

Cook PL, Bock PE, Gordon DP, Weaver HJ (Eds) (2018) *Australian Bryozoa*. 2 vols. CSIRO Publishing, Melbourne.

Gordon DP, Bock PE (2008) Bryozoa. In *The Great Barrier Reef: Biology, Environment and Management*. (Eds P Hutchings, M Kingsford and O Hoegh-Guldberg) pp. 290–295. CSIRO Publishing, Melbourne.

Hageman SJ, Bock PE, Bone Y, McGowran B (1998) Bryozoan growth habits: classification and analysis. *Journal of Paleontology* **72**, 418–436. doi:10.1017/S0022336000024161

Hayward PJ, Ryland JS (1995) Bryozoa from Heron Island, Great Barrier Reef, 2. *Memoirs of the Queensland Museum* **38**, 533–573.

Hyman LH (1959) *The Invertebrates: Smaller Coelomate Groups. Chaetognatha, Hemichordata, Pogonophora, Phoronida, Ectoprocta, Brachiopoda, Sipunculida, the Coelomate Bilateria*. McGraw-Hill, New York, USA.

Pitcher CR, Doherty P, Arnold P, Hooper J, Gribble N, Bartlett C, *et al.* (2007) 'Seabed biodiversity on the continental shelf of the Great Barrier Reef World Heritage Area'. AIMS/CSIRO/QM/QDPI CRC Reef Research Task Final Report. CSIRO Marine and Atmospheric Research, Brisbane.

Ryland JS (1974) Bryozoa in the Great Barrier Reef Province. In *Proceedings of the Second International Reef Symposium Volume 2*. (Eds AM Cameron, BM Campbell, AB Cribb, R

Endean, JS Jell, OA Jones, *et al.*) pp. 341–348. The Great Barrier Reef Committee, Brisbane.

Ryland JS (1984) Phylum Bryozoa – lace corals and their relatives. In *A Coral Reef Handbook*. Handbook Series No. 1, 2nd edn. (Eds P Mather and I Bennett) pp. 68–75. The Australian Coral Reef Society, Brisbane.

Ryland JS, Hayward PJ (1992) Bryozoa from Heron Island, Great Barrier Reef. *Memoirs of the Queensland Museum* **32**, 223–301.

Sebastian P, Cumming RL (2016) Three new species of *Calyptotheca* (Bryozoa: Lanceoporidae) from the Great Barrier Reef, tropical Australia. *Zootaxa* **4079**, 467–479. doi:10.11646/zootaxa.4079.4.6

Smith AM (2014) Growth and calcification of marine bryozoans in a changing ocean. *The Biological Bulletin* **226**, 203–210. doi:10.1086/BBLv226n3p203

Tilbrook KJ (2006) Cheilostomatous Bryozoa from the Solomon Islands. In *Santa Barbara Museum of Natural History Monographs 4: Studies in Biodiversity 3*. (Ed. HW Chaney) pp. 1–385. Santa Barbara Museum of Natural History, Santa Barbara, CA, USA.

Website

The Bryozoan Home Page: http://www.bryozoa.net. This includes lists of species by genus, and references arranged by year of publication. Selected lists, including a species checklist for the Great Barrier Reef, can be seen at http://www.bryozoa.net/checklists/.

28
Echinoderms

M. Byrne

Introduction

Echinoderms are a conspicuous and diverse component of the invertebrate fauna of the Great Barrier Reef (GBR) and have been reviewed in several taxonomic and biogeographic studies (see 'Further reading'). Most echinoderms from tropical Australia have a broad distribution in the Indo-Pacific Ocean. The 630 species of echinoderms recorded from the GBR are from the five classes: sea stars (Asteroidea) 137 species; brittle stars (Ophiuroidea) 166 species; sea urchins (Echinoidea) 110 species; sea cucumbers (Holothuroidea) 127 species; and feather stars (Crinoidea) 90 species. Echinoderms are a conspicuous and ecologically important component of reef communities. They are often the dominant organisms on the sea floor and this is particularly true of tropical Holothuroidea, the elongate sausage-shaped animals (e.g. *Holothuria*) seen on reef flats and sandy areas. Some of the more common echinoderm species on the GBR are illustrated in Figs 28.1–28.8.

The body symmetry of adult echinoderms is secondarily radial and based on pentamery. There are usually five radii (e.g. arms of sea stars), although multi-armed (six or more arms) asteroids and brittle stars are common and some burrowing sea urchins have four radii. Echinoderm classes are easily identified by their distinctive body profiles. Asteroids are star-shaped with the arms (five or more) tapering from the disc (e.g. *Linckia, Nardoa*) or a cushion-like pentagon shape lacking arms (e.g. *Culcita*) (Figs 28.1, 28.2A–28.2A–E). Holothuroids by contrast are elongate (Figs 28.2F–I, 28.3–28.5). Ophiuroids have a round central disc and slender flexible arms that are sharply set-off from the body (Figs 28.6, 28.7). Their body can have a simple (brittle stars or serpent stars) or branched (basket stars) profile. Echinoids have a rigid, globose body (test) covered by spines of varying length (Fig. 28.8A–F). Crinoids have an array of feather-like arms that range in number from five to several hundred (Fig. 28.8G–I). Some aspects of the biology of echinoderms, for instance reproduction, are general to the phylum, while other features such as feeding are class specific. Class specific features are dealt with in the following sections, with a focus on species that occur on the GBR.

Reproduction and life history

For the most part, there are no external morphological differences between males and females. Most echinoderms are sexual reproducers, spawning copious numbers of eggs. The crown-of-thorns starfish, *Acanthaster,* is estimated to release up to 200 million eggs per year (Box 28.1). Many echinoderms from the GBR spawn in the spring and summer. Mass spawning of echinoderms is common and many species spawn at the same time as the corals on the GBR. Release of gametes by erect sea cucumbers during summer evenings is a sight often seen by divers. Reproductive periodicity of some echinoderms along the GBR changes with latitude. *Archaster typicus*, a widespread and abundant sea star on muddy sand flats throughout the Indo-Pacific, is unusual in showing pairing of males and females during spawning. As the breeding season approaches, the male climbs on the female's aboral surface. This pairing behaviour undoubtedly enhances fertilisation success.

Most echinoderms are free spawners and have dispersive larvae. These larvae are beautiful and distinct for each class. In contrast to adults, the larvae have bilateral symmetry. Some need to feed in the plankton (planktotrophic larvae) and others are sustained by egg nutrients (lecithotrophic larvae). Some echinoderms, such as the asteroid *Aquilonastra byrneae* (Figs 28.2C), have benthic development and consequently lack a dispersive stage. Brooding echinoderms such as the asteroid *Cryptasterina hystera* (Fig. 28.2A) and the ophiuroid *Ophiopeza spinosa* (Fig. 28.7C) care for their young and give rise directly to juveniles. These two species are unusual in having a pelagic-type larva in the brood chamber and so have potential to brood and broadcast their young, but it is not known if they do so.

Asexual reproduction is also common in the echinoderms of the GBR and occurs in asteroids, ophiuroids and holothuroids and can generate very high densities. This involves the animals breaking

Fig. 28.1. Asteroidea: **(A)** *Acanthaster cf. solaris* (×0.07) (photo: J. Keesing); **(B)** *Linckia laevigata* (×0.25); **(C)** *Linckia guildingii* (×0.30); **(D)** *Linckia multifora* (×0.25); **(E)** *Nardoa novaecaledoniae* (×0.33); **(F)** *Culcita novaeguineae* adult (×0.18); **(G)** *Culcita novaeguineae* juvenile (×1.00) (photo: S. Walker); **(H)** *Echinaster luzonicus* (×0.40); **(I)** *Fromia milleporella* (×0.80). (Photos: M. Byrne, unless noted).

in half or fragmenting a part of their body. As a result, these echinoderms often have an unusual body profile (Figs 28.1D, 28.2D). This process is followed by regeneration of each portion to form a complete individual. On the GBR fission (splitting in half) is particularly common in brittle stars and sea cucumbers. Some small asterinid sea stars are fissiparous (Fig. 28.2D). *Holothuria atra* and *Stichopus chloronotus* are well known for asexual reproduction by fission. The asteroid *Linckia multifora* has an impressive capacity to propagate asexually and is often seen with arms at different stages of regeneration (Fig. 28.1D). Whole stars can regenerate from a single autotomised arm.

Echinoderm diversity

Asteroidea

Asteroids (Figs 28.1, 28.2A–E), although diverse, are generally not abundant on the GBR, with the exception of the spectacular outbreaks of *Acanthaster* (Fig. 28.1A) (Box 28.1). They occur in reef, soft sediment and rubble habitats. *Luidia* and *Astropecten* species are found in lagoon and inter-reefal sediment habitats. The Ophidiasteridae is a large family of tropical sea stars including the distinct blue sea star *Linckia laevigata* (Fig. 28.1B) and other common species *Fromia milleporella* (Fig. 28.1I), *Nardoa novaecaledoniae* (Fig. 28.1E) and *Ophidiaster granifer*. The Asterinidae (Fig. 28.2A–E) is a species-rich family that is particularly challenging with respect to species identification due to the presence of morphospecies complexes comprising several species. Several cryptic species have been identified in tropical Queensland based on differences in reproduction and development.

Echinaster luzonicus (Fig. 28.1H) is a common species in the GBR. This multi-armed species often has an irregular profile with unequal portions due to its propensity to autotomise distal portions of its arms followed by regeneration. The cushion star *Culcita novaeguineae* is also common (Fig. 28.1F, G).

Box 28.1. Family Acanthasteridae

Acanthaster* cf. *solaris

The crown-of-thorns starfish (COTS), *Acanthaster* (Fig. 28.1A), now known to be a species complex of at least four species across the Indo-Pacific, is probably the best studied asteroid in the world because of the effect that periodic outbreaks of this species have on coral reefs. The proposed name for the Pacific COTS is *A. solaris*, but this is still equivocal and requires further taxonomic work. We have comprehensive knowledge of the biology and ecology of *A.* cf *solaris*. This is an unusual asteroid in being a specialist corallivore and outbreaks result in marked decrease in live coral cover to below 1–5% in some reports. *Acanthaster* prefers acroporid and pocilloporid corals, but other corals are also consumed. The outbreak ends when coral prey is exhausted. Reef recovery following intense predation by COTS is variable, with some reefs not recovering for 10–15 years. The eggs contain saponins, a chemical toxic to fish and presumed to protect the eggs and embryos from predation.

There are three hypotheses on the causes of *Acanthaster* outbreaks: (1) the predator removal hypothesis proposes that overfishing of COTS predators (triton shells, fishes) influences the increase in numbers; (2) the natural phenomenon-larval resilience hypothesis; and (3) the terrestrial runoff hypothesis that proposes that nutrient runoff from the land leads to an increase in the phytoplankton food of the larvae, leading to enhanced recruitment. Although the nutrient hypothesis has received considerable traction, emphasising the need to control nutrient runoff from anthropogenic sources into the GBR lagoon (see Chapter 13), outbreaks also occur in remote areas away from anthropogenic sources. The larvae appear to be well adapted to low nutrient conditions with plastic growth that allows them to adjust to nutrient levels to optimise feeding. Overfishing of predators of juvenile *Acanthaster* is probably also important.

Although it was thought that COTS outbreaks might be due to a major pulse of recruitment, it now appears that they are comprised of individuals from multiple recruitment years.

Juvenile *Culcita* are flatter and have more distinct rays than the adult (Fig. 28.1G).

The feeding biology of sea stars varies from specific to general diets. Many species are predators on molluscs, sponges, corals and other echinoderms. Other species are surface grazers and scavengers. These asteroids predominantly feed off hard substrata such as coral rubble, eating sponges, microscopic organisms or coral mucus. *Culcita novaeguineae* feeds on corals, other invertebrates and benthic films. Coral mortality caused by individual *Culcita* is less than that caused by individual *Acanthaster*, but its selective feeding habits may influence the relative abundance of some coral species.

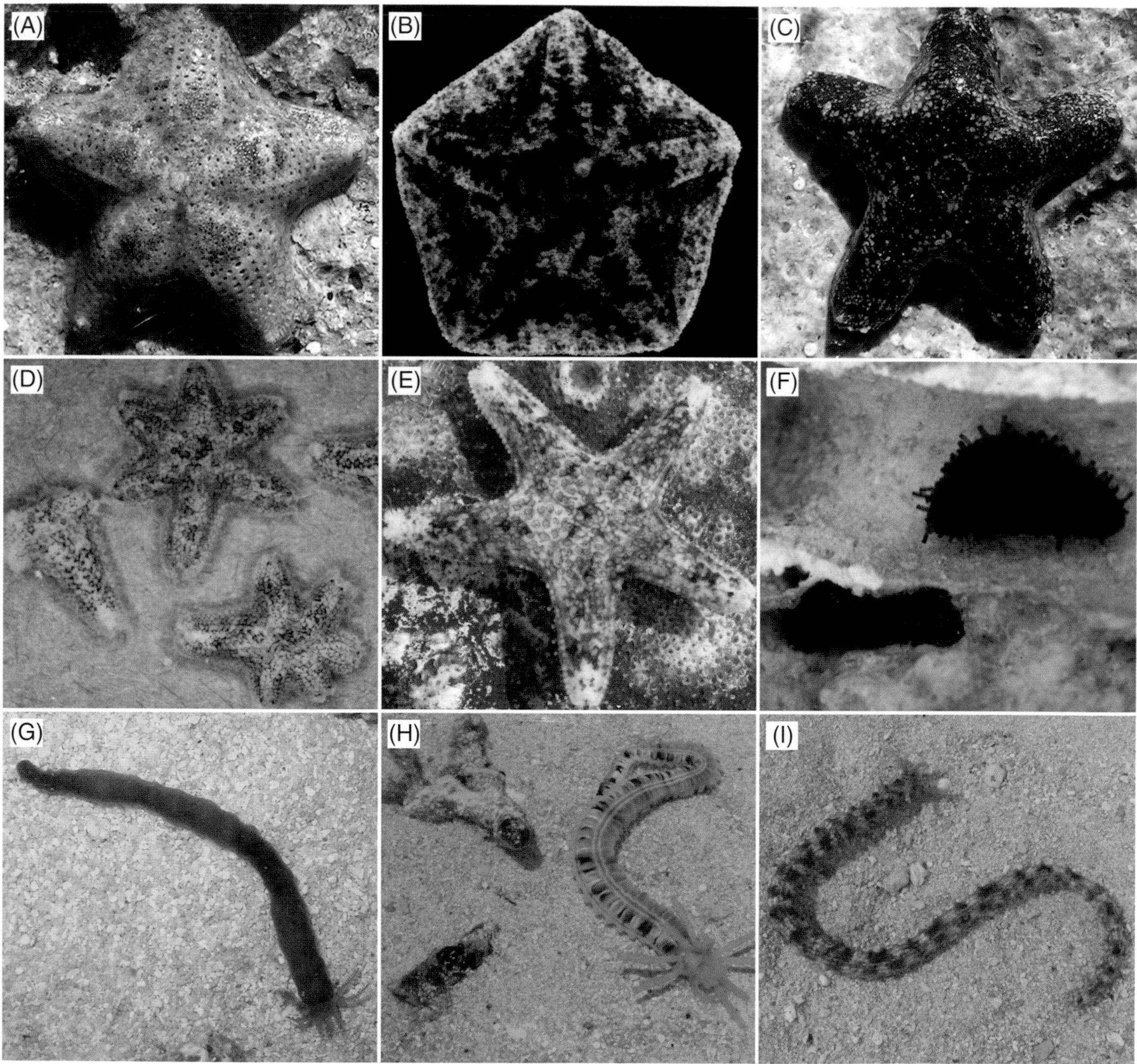

Fig. 28.2. Asteroidea: **(A)** *Cryptasterina hystera* (×1.67) (photo: from Byrne and Walker 2007); **(B)** *Cryptasterina pentagona* (×1.67) (photo: from Dartnall *et al.* 2003 ©Magnolia Press, reproduced with permission); **(C)** *Aquilonastra byrneae* (×1.67) (photo: from Byrne and Walker 2007); **(D)** *Ailsastra* sp. (× 0.30); **(E)** *Disasterina* sp. (×1.25). Holothuroidea. **(F)** *Afrocucumis africana* (×0.92); **(G)** *Polyplectana kefersteinii* (×1.10); **(H)** *Euapta godeffroyi* (×0.10); **(I)** *Synapta maculata* (×0.23). (Photos: M. Byrne, unless noted).

Holothuroidea

The Holothuroidea (Figs 28.2F–I, 28.3–28.5) are diverse and common all along the GBR and there are three major orders: (1) the Aspidochirotida, the common deposit feeders that graze on the surface of the sediment; (2) the Dendrochirotida, suspension feeders with an array of branching tentacles and; (3) the Apodida, which are also deposit feeders. Several surveys of holothuroid diversity on the GBR have been undertaken.

Aspidochirotids are by far the most abundant and conspicuous holothuroids on the GBR (Figs 28.3–28.5). They occur in a variety of habitats from the reef dwelling *Actinopyga* (surf red fish, Fig. 28.5F), to the intertidal and deep water aspidochirotids *Actinopyga*, *Bohadschia*, *Holothuria* and *Stichopus* species. The most abundant species on the GBR are capable of asexual and sexual reproduction. These include *H. atra* (Fig. 28.3A), *H. difficilis* (Fig. 28.3G), *H. edulis* (Fig. 28.3D), *H. hilla* (Fig. 28.4C) and *S. chloronotus* (Fig. 28.5A). They are fissiparous, splitting in half and subsequently regenerating the anterior and posterior halves to make two complete individuals. Although these species also reproduce sexually, spawning gametes and have dispersive larvae, it appears that fission is important in

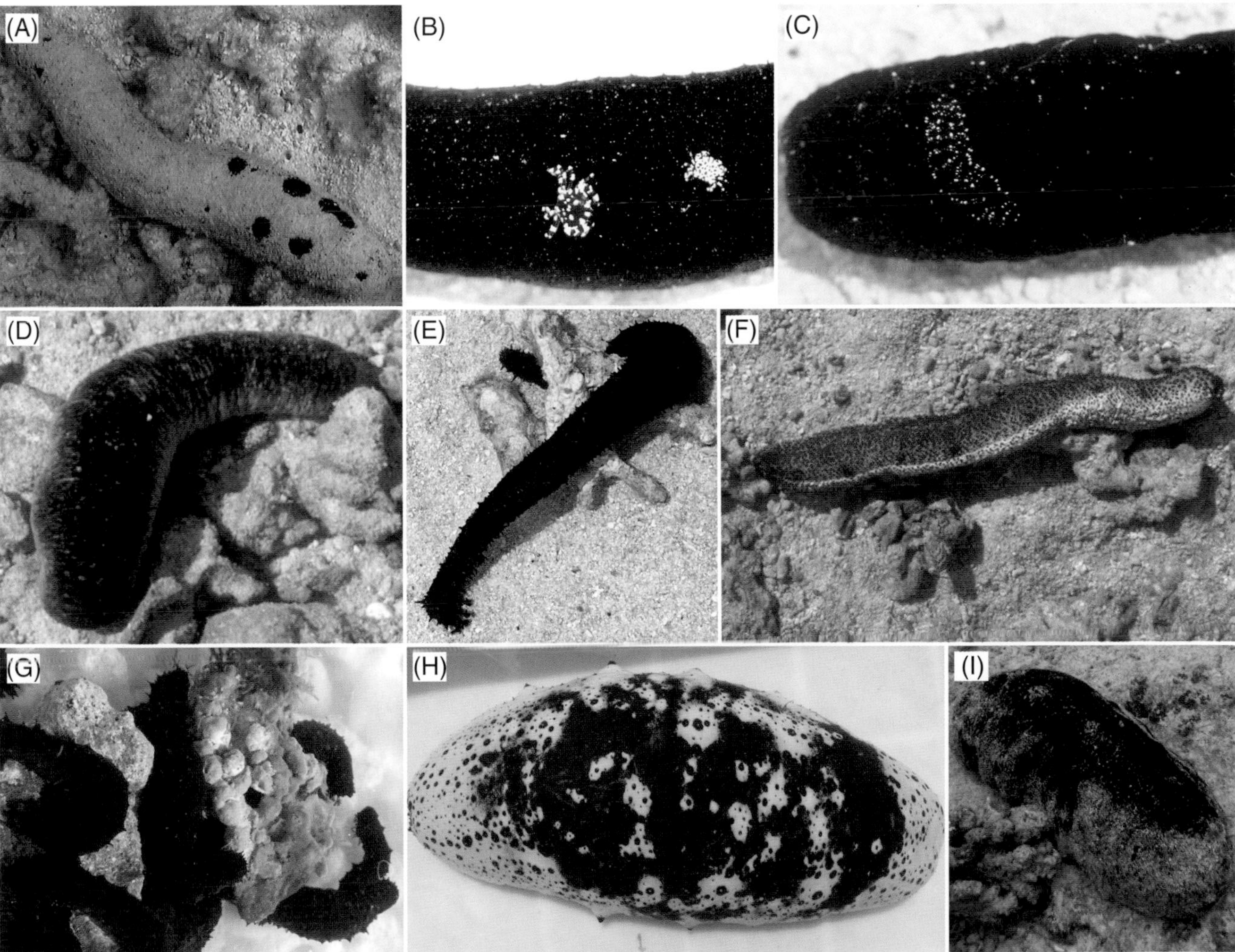

Fig. 28.3. Holothuroidea: **(A)** *Holothuria atra* (×0.27); **(B)** *Holothuria atra* with commensal crabs (× 0.50); **(C)** *Holothuria atra* with commensal worm (× 0.67); **(D)** *Holothuria edulis* (×0.38); **(E)** *Holothuria leucospilota* (×0.17); **(F)** *Holothuria dofleinii* (×0.21); **(G)** *Holothuria difficilis* (×0.63); **(H)** *Holothuria fuscogilva* (× 0.19); **(I)** *Holothuria whitmaei* (×0.15) (photo: H. Eriksson). (Photos: M. Byrne, unless noted).

maintaining the populations. Many species provide habitat to commensal crabs and scale worms. These are commonly seen on the body wall of *H. atra* (Fig. 28.3B, C). Parasitic eulimid snails are also commonly seen on the body surface of some holothuroids (Fig. 28.4F). These snails extend their proboscis through the body wall and into the coelom where they use nutrients in their host body fluids.

Aspidochirotids are deposit feeders and are prominent members of the biota of soft sediment environments. They feed with their spatula-like tentacles (Fig. 28.5G), scooping up sand, algal films and detritus. Aspidochirotids provide important ecosystem services by enhancing local productivity through their bioturbation and feeding/digestive activity. Burrowing species are particularly important in bioturbation of the nutrient-poor carbonate sediments that dominate much of the GBR. Taxonomically, these sea cucumbers have long been a challenge, with some aspidochirotids well characterised, but others being intractable to traditional taxonomy. Several commercial species of *Stichopus* and *Actinopyga* appear to be morphospecies complexes and the identity of species currently being fished needs to be assessed (Box 28.2). Synaptids (Fig. 28.2G–I) are less familiar because many are small and some are nocturnal. The suspension feeding dendrochirotids are less common. *Afrocucumis africana* (Fig. 28.2F), a small, dark purple dendrochirotid is often seen under boulders and rubble in the intertidal zone. Other dendrochirotids burrow in sediment or live in the reef infrastructure or algal turfs and so are rarely seen. Their tree-like dendritic tentacles that they use for filter feeding can be seen emerging from the substrate.

Ophiuroidea

The species diversity of ophiuroids on the GBR is impressive, with species in the families Ophiocomidae, Ophiotrichidae and Ophiodermatidae being well represented (Figs 28.6, 28.7). *Ophiocoma* and *Macrophiothrix* species are a diverse assemblage of large species. Several genera are in need of revision, with cryptic species evident. Ophiuroids are often common shoreward of coral habitats where many species are sympatric, aggregating under slabs of coral rubble and in crevices. Along the GBR these include *Ophiarachnella gorgonia* (Fig. 28.7B), *Ophiolepis elegans* (Fig. 28.7D), *Breviturma*

Box 28.2. Bêche-de-mer species

Approximately 15 species comprise the bêche-de-mer fisheries across northern Australia and elsewhere in the Indo-Pacific. Several Australian fisheries for the high-value species collapsed in short order, repeating the pattern seen elsewhere in the Indo-Pacific. As illustrated by the teatfish (*Holothuria* [*Microthele*] *whitmaei* and *H.* [*M.*] *fuscogilva*) complex, there are serious conservation concerns for populations of commercial species that have been on the decline for some time or have completely disappeared from some localities. Holothuroids are particularly susceptible to overfishing because of their limited mobility, poor recruitment and density dependent propagation. Local areas are quickly stripped of valuable species. Currently, fishers in Australia are moving on to less valuable or less accessible (remote areas, deeper waters) stock, repeating the 'fishing-down' pattern seen in other jurisdictions. Due to the perilous state of conservation of the bêche-de-mer species, 14 were recently listed as Endangered or Vulnerable to Extinction by the IUCN.

One of the first species targeted by fishers on the GBR was the high value species, the black teatfish (BTF) *Holothuria whitmaei*. This species is still listed incorrectly as *H. nobilis* in some guides, but this is the name of the Indian Ocean species. The GBR fishery for the BTF was short lived because the stock of this shallow water species was quickly depleted. Unfished reefs and green zones support high densities of bêche-de-mer species. The large populations of *H. whitmaei* at Raine Reef and other reserves are crucial for conservation of the species.

The general lack of juveniles in sea cucumber populations indicates that recruitment will be slow. The first principle of reproduction, gamete contact, is likely to be compromised by the current low densities of spawning individuals on some reefs.

dentata (Fig. 28.6A), *Ophiocoma erinaceus* (Fig. 28.6C), *Ophiocoma scolopendrina* (Fig. 28.6B), *Ophionereis porrecta* (Fig. 28.7F) and *Macrophiothrix* species (Fig. 28.7G, H). Ophiuroids also inhabit the reef infrastructure and soft sediments. *Ophiocoma scolopendrina* (Fig. 28.6B) is the most conspicuous ophiuroid in the Indo-Pacific. This species specialises in shallow habitats that are emersed at low tide where it reaches densities up to 320 individuals/m^2 and is often seen feeding on surface scum at low tide.

Ophiuroids have varied diets, with many being filter feeders, extending their arms from their hiding places in the reef infrastructure to feed. This is often at night. Burrowing amphiurid ophiuroids keep their disc below the surface of the sediment and extend their arms into the water column to filter feed. Ophiodermatids are predators and scavengers. *Ophiarachna incrassata* (Fig. 28.7A), a spectacular ophiodermatid, traps fishes under its arms and also feeds on carrion left behind by major predators. It has been seen scavenging turtle flesh left behind by tiger sharks.

Echinoidea

The Echinoidea includes the familiar sea urchins and irregular urchins (sand dollars and spatangoids). On the GBR, most sea urchins are cryptic during the day, with *Diadema savignyi* (Fig. 28.8A)

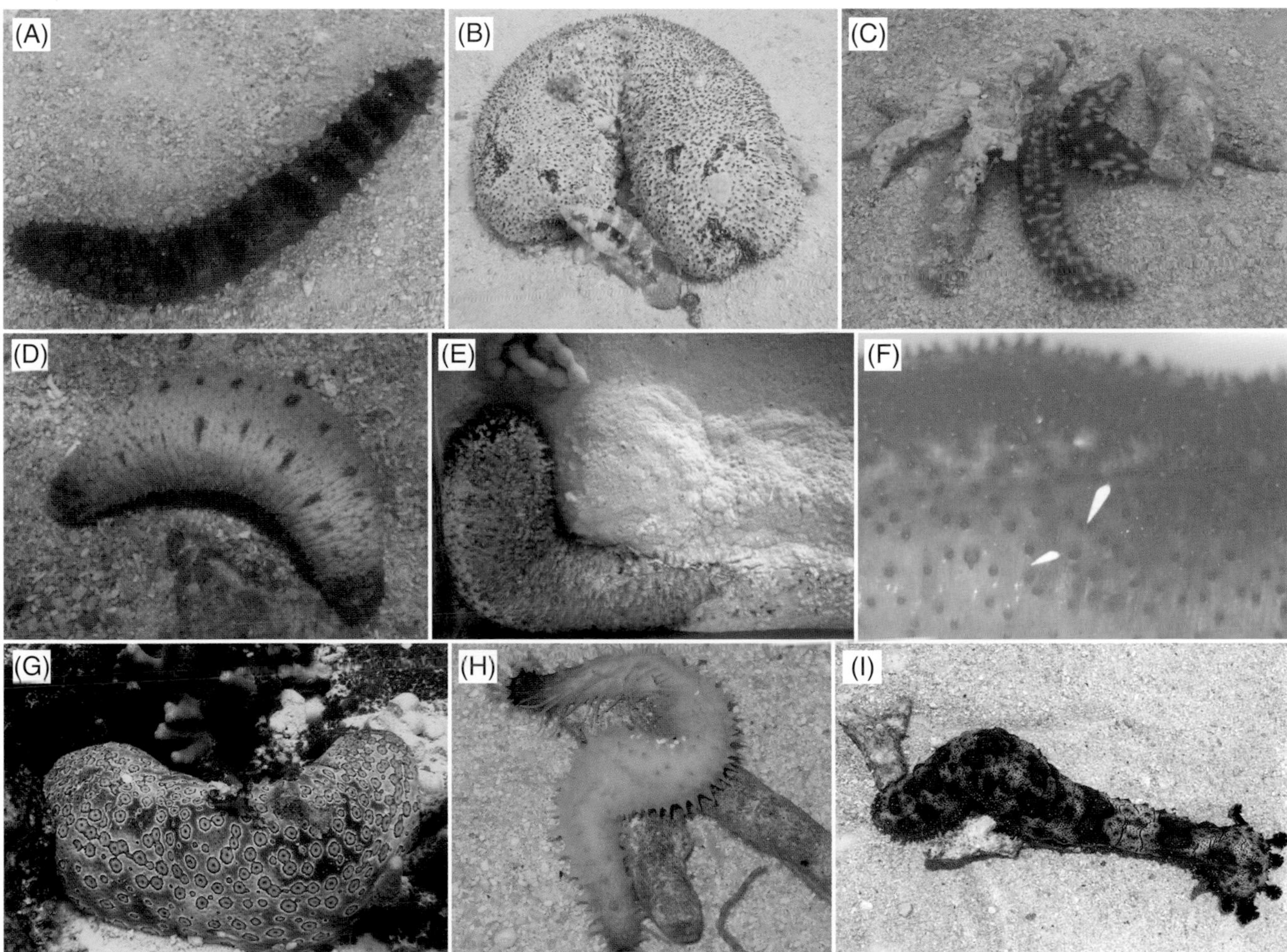

Fig. 28.4. Holothuroidea: **(A)** *Holothuria impatiens* (×0.23); **(B)** *Holothuria lessoni* (×0.20) (Photo: H. Eriksson); **(C)** *Holothuria hilla* (×0.14); **(D)** *Holothuria arenicola* (×0.25); **(E)** *Holothuria isuga* (×0.12); **(F)** *Holothuria isuga* with parasitic snails (×1.00); **(G)** *Bohadschia argus* (×0.17) (Photo: H. Eriksson); **(H)** *Labidodemas semperianum* (×0.40); **(I)** *Personothuria graeffei* (×0.16). (Photos: M. Byrne, unless noted).

and species in the *Echinometra* species complex (Fig. 28.8C) emerging at night. *Echinostrephus* excavates holes in coral to form a permanent residence from which its spines emerge. The commercial urchin *Tripneustes gratilla* (Fig. 28.8D) is common in seagrass areas. Common irregular echinoids include the sand dollars *Arachnoides placenta* and *Peronella lesueuri* that occur in abundance along the Queensland coast and the spatangoid urchin *Breynia australasiae* that burrows in sandy inter-reefal areas and can be locally abundant. The burrowing activity of echinoids, particularly by *Echinometra* species, is important in bioerosion of coral reef and intertidal habitats (see Chapter 9).

Most sea urchins are grazers, using their hard calcareous teeth to remove algae and encrusting organisms from the surface. As a member of the grazing guild on coral reefs, sea urchins contribute to keeping the biomass of algae low, thereby preventing overgrowth of algae over coral substrate. Sand dollars are suspension feeders, collecting particles from flow and transferring these by specialised tube feet down the food groves on the surface of the test to the mouth. Spatangoids have a secondary bilateral profile associated with their burrowing life style. They are deposit feeders removing organic matter from ingested sand as they propel themselves through sediments using their specialised 'rowing' spines.

Fig. 28.5. Holothuroidea: **(A)** *Stichopus chloronotus* (×0.15); **(B)** *Stichopus herrmanni* (×0.17); **(C)** *Stichopus monotuberculatus* (×0.20); **(D)** *Stichopus vastus* (×0.18); **(E)** *Thelenota ananas* (×0.14); **(F)** *Actinopyga* sp. (surf red fish) (×0.17) (photo: S. Walker); **(G)** *Actinopyga* tentacles (×0.50); **(H)** *Actinopyga echinites* (× 0.13); **(I)** *Actinopyga miliaris* (×0.11). (Photos: M. Byrne, unless noted).

Crinoidea

Most crinoids on the GBR are subtidal and a few are abundant in lower intertidal areas. Reefs along the GBR have a diverse assemblage of crinoid species. Some species occupy the surface of soft sediments. There are several taxonomic reviews and surveys of the crinoid fauna of the GBR (see 'Further reading'). The Comasteridae is a major family of tropical crinoids including *Clarkcomanthus* (Fig. 28.8G), *Comaster* (Fig. 28.8H), *Comatella*, *Comatula* and *Phanogenia* species. The Mariametridae also has several genera on the GBR including *Anneissia*, *Dichrometra*, *Lamprometra*, *Mariametra*, *Himermetra* (Fig. 28.8I) and *Stephanometra*.

Crinoids are locally abundant on the GBR, as shown in several investigations of their population ecology. They extend their feather-like arms into the water column and may or may not have cirri (basal claw-like appendages) to cling to the substrate. Those that lack cirri use their arms for attachment. Many crinoids are fully or partially concealed within the reef infrastructure during the day and emerge at night to feed. Other crinoids are fully exposed at all times. The cryptic behaviour of

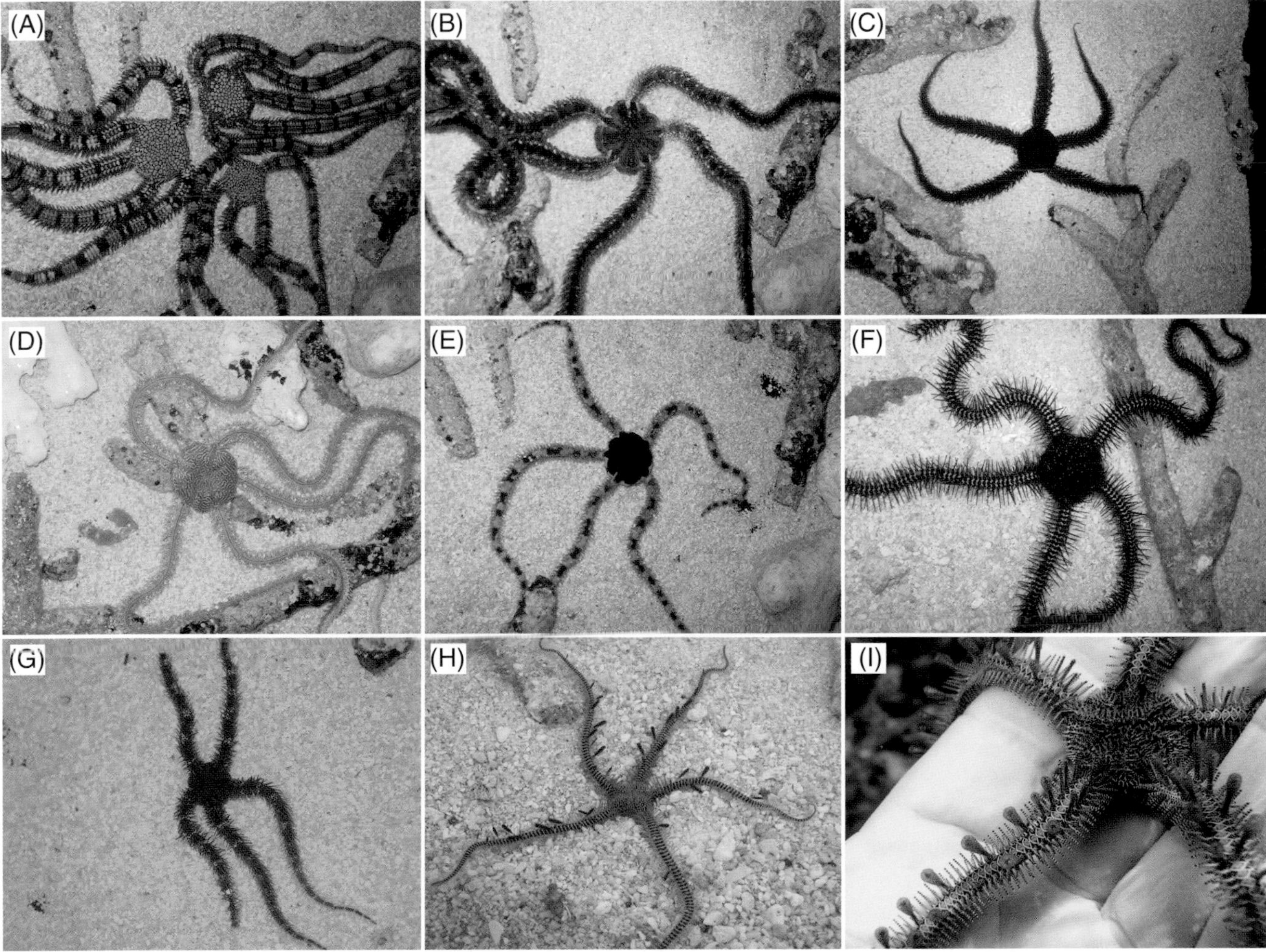

Fig. 28.6. Ophiuroidea: **(A)** *Breviturma dentata* (×0.50); **(B)** *Ophiocoma scolopendrina* (×0.50); **(C)** *Ophiocoma erinaceus* (×0.33); **(D)** *Ophiomastix pictum* (×0.38); **(E)** *Ophiomastix elegans* (×0.50); **(F)** *Ophiomastix janualis* (×0.38); **(G)** *Ophiomastix mixta* (×0.50); **(H)** *Ophiomastix caryophyllata* (×0.50); **(I)** *Ophiomastix annulosa* (×1.00) (Photo: S. Uthicke). (Photos: M. Byrne, unless noted).

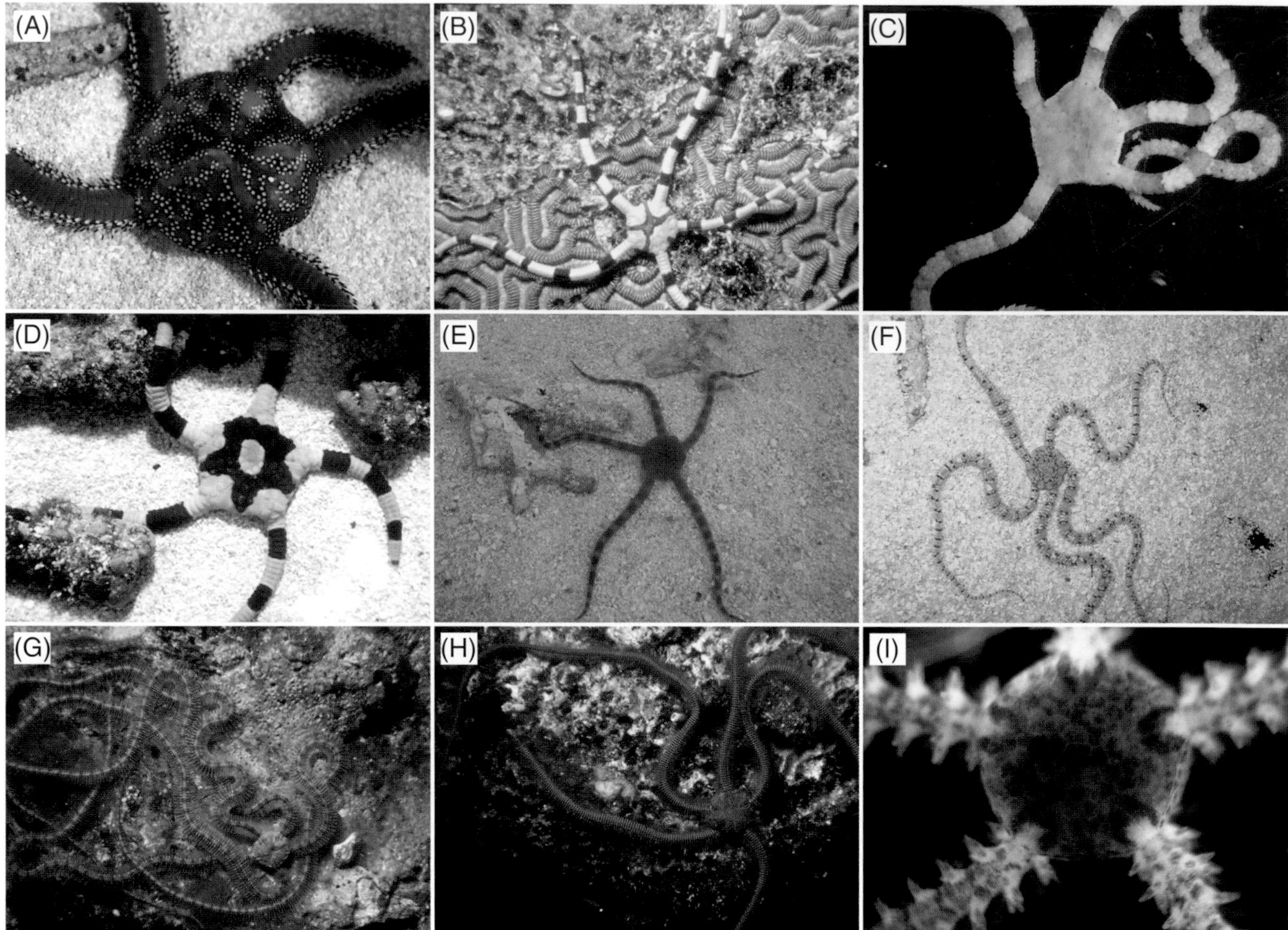

Fig. 28.7. Ophiuroidea: **(A)** *Ophiarachna incrassata* (×0.71); **(B)** *Ophiarachnella gorgonia* (×0.50) (Photo: J. Keesing); **(C)** *Ophiopeza spinosa* (×2.50); **(D)** *Ophiolepis elegans* (×0.63); **(E)** *Ophiomyxa australis* (×0.25); **(F)** *Ophionereis porrecta* (×0.50); **(G)** *Macrophiothrix lorioli* (×0.33) (Photo: J. Keesing); **(H)** *Macrophiothrix nereidina* (×0.33) (Photo: J. Keesing); **(I)** *Amphipholis squamata* (×8.33). (Photos: M. Byrne, unless noted).

crinoids may be to avoid predation by fishes, and several species have defensive chemicals that may deter predators. When disturbed from their perch, crinoids exhibit a distinctive swimming behaviour with coordinated movement of the arms.

Crinoids are filter feeders, extending their feather-like arms above the substrate to capture food particles on their small tube feet that line the arms. The food is captured from flow and so crinoids take up positions on the reef and adjust the arrangement of the arms to take advantage of the ambient water movement. Food particles are conveyed to the mouth in the centre of the disc on food tracts.

Any collecting of echinoderms requires a permit (see Chapter 14).

Further reading

General references

Balogh R, Wolfe K, Byrne M (2018) Gonad development and spawning of the vulnerable commercial sea cucumber, *Stichopus herrmanni*, in the southern Great Barrier Reef. *Journal of the Marine Biological Association of the United Kingdom*. doi.org/10.1017/S0025315418000061

Byrne M, O'Hara TD (Eds) (2017) *Australian Echinoderms: Biology Ecology and Evolution*. CSIRO Publishing, Melbourne.

Byrne M, Walker SJ (2007) Distribution and reproduction of intertidal species of *Aquilonastra* and *Cryptasterina*

Fig. 28.8. Echinoidea: **(A)** *Diadema savignyi* (×0.14) (Photo: A. Hoggett); **(B)** *Echinothrix calamaris* (×0.23) (Photo: S. Doo); **(C)** *Echinometra* species A (×0.54); **(D)** *Tripneustes gratilla* (×0.50); **(E)** *Mespilia globulus* (×1.00); **(F)** *Eucidaris metularia* (×1.00) (Photo: A. Hoggett). Crinoidea: **(G)** *Clarkcomanthus alternans* (×0.16) (Photo: A. Hoggett); **(H)** *Comaster schlegelii* (×0.20) (Photo: A. Hoggett); **(I)** *Himerometra robustipinna* (×0.10) (Photo: A. Hoggett).

(Asterinidae) from One Tree Reef, Southern Great Barrier Reef. *Bulletin of Marine Science* **81**, 209–218.

Byrne M, Cisternas P, O'Hara T (2008) Brooding of a pelagic-type larva in *Ophiopeza spinosa:* reproduction and development in a tropical ophiodermatid brittlestar. *Invertebrate Biology* **127**, 98–107. doi:10.1111/j.1744-7410.2007.00110.x

Clark AM, Rowe FWE (1971) *Monograph of Shallow-Water Indo-West Pacific Echinoderms*. Trustees of the British Museum (Natural History), London, UK.

Dartnall AJ, Byrne M, Collins J, Hart MW (2003) A new viviparous species of asterinid (Echinodermata, Asteroidea, Asterinidae) and a new genus to accommodate the species of pan-tropical exiguoid sea stars. *Zootaxa* **359**, 1–14. doi:10.11646/zootaxa.359.1.1

Messing CG, Meyer DL, Siebeck UE, Jermin LS, Vaney DI, Rouse GW (2006) A modern soft-bottom, shallow-water crinoid fauna (Echinodermata) from the Great Barrier Reef, Australia. *Coral Reefs* **25**, 164–168. doi:10.1007/s00338-005-0076-3

Rowe FWE, Gates J (1995) *Echinodermata. Zoological Catalogue of Australia. Vol. 33*. CSIRO Publishing, Melbourne.

Crown-of-thorns starfish

Haszprunar G, Vogler C, Wörheide G (2017) Persistant gaps in knowledge for naming and distinguishing multiple species of Crown-of-Thorns-Seastar in the *Acanthaster planci* species complex. *Diversity (Basel)* **9**, 22. doi:10.3390/d9020022

Pratchett MS, Cabelles C, Rivera-Posada JA, Sweatman HPA (2013) Limits to understanding and managing outbreaks of crown-of-thorns starfish (*Acanthaster* spp.). *Oceanography and Marine Biology - an Annual Review* **52**, 133–200.

Pratchett MS, Caballes CF, Wilmes JC, Matthews S, Mellin C, Sweatman HPA, *et al.* (2017) Thirty years of research on crown-of-thorns starfish (1986–2016): scientific advances and emerging opportunities. *Diversity* **9**(4) 1–49. doi:10.3390/d9040041

Uthicke S, Schaffelke B, Byrne M (2009) A boom or bust phylum? Ecological and evolutionary consequences of large population density. *Ecological Monographs* **79**, 3–24. doi:10.1890/07-2136.1

Wolfe K, Graba-Landry A, Dworjanyn SA, Byrne M (2017) Superstars: assessing nutrient thresholds for enhanced larval success of *Acanthaster planci*, a review of the evidence. *Marine Pollution Bulletin* **116**, 307–314. doi:10.1016/j.marpolbul.2016.12.079

Bêche-de-mer holothuroids

Eriksson A, Byrne M (2015) The sea cucumber fishery in Australia's Great Barrier Reef Marine Park follows global patterns of serial exploitation. *Fish and Fisheries* **16**, 329–341. doi:10.1111/faf.12059

Uthicke S, O'Hara TD, Byrne M (2004) Species composition and molecular phylogeny of the Indo-Pacific teatfish (Echinodermata: Holothuroidea) bêche-de-mer fishery. *Marine and Freshwater Research* **55**, 837–848. doi:10.1071/MF04226

Uthicke S, Welch D, Benzie JAH (2004) Slow growth and lack of recovery in overfished holothurians on the Great Barrier Reef: evidence from DNA fingerprints and repeated large-scale surveys. *Conservation Biology* **18**, 1395–1404. doi:10.1111/j.1523-1739.2004.00309.x

29

Tunicates (ascidians and their allies)

M. Ekins

Introduction

The Subphylum Tunicata include the planktonic thaliaceans and appendicularians, as well as the more widely known sessile ascidians. The Class Thaliacea includes the orders Doliolida, Salpida and Pyrosomatida. All thaliaceans share the ability to alternate between asexual and sexual reproduction with each generation. This reproductive strategy enables thaliaceans to build up to large blooms in the ocean (*thalia* = 'bloom' in Greek). All of the thaliaceans found so far in the Great Barrier Reef (GBR) are cosmopolitan, with no currently known endemic species.

Doliolids are solitary tunicates with a barrel-shaped body resembling a jet engine in shape and function, with a large incurrent mouth at one end and the excurrent at the other end (Fig. 29.1). Like ascidians, the doliolids produce a tadpole larval stage.

Salps are transparent and mostly colonial planktonic ascidians that usually range in size from short chains of less than 1 cm up to several centimetres (Fig. 29.2). In general, salps are found in these chains of asexual reproducing (budding) individuals. These chains reproduce sexually with other chains, then mature and release individuals that go on to produce asexual chains again. *Thetys vagina* is an unusually large species that can grow up to 30 cm long, which is usually found in small numbers in tropical waters, but occasionally is also found in colder waters (Fig. 29.3).

The Order Pyrosomatida, or purse salps, is represented by only eight species that are well known for their highly visible bioluminescent displays that are visible from space. The most common species *Pyrosoma atlanticum* (Fig. 29.4), can form massive swarms. This group also includes the diver's delight, Neptunes sword (*Pyrostremma spinosum*), which can be up to 20 m long. The purse salps are the only exclusively colonial salps. The colonies are tubular, enclosed at one end, with the sexual, hermaphroditic individuals (ascidiozooids) embedded in the wall. Each individual has their oral aperture outside of the tube drawing in water and expelling the filtered water into the colony's interior towards the open end of the tube, providing propulsion for the colony.

The Class Appendicularia is also known as larvaceans because they maintain the larval tadpole shape of tunicates all their life. They are divided into a trunk (body) and a tail that can be up to five times the length of the body. They secrete a protein mucus coating (the tunic) over the trunk forming a 'house' around the larvacean. The larvacean produces water currents by tail movements that filters food particles on the outside. The larvaceans discard and replace their houses often when the filters become clogged (Fig. 29.5).

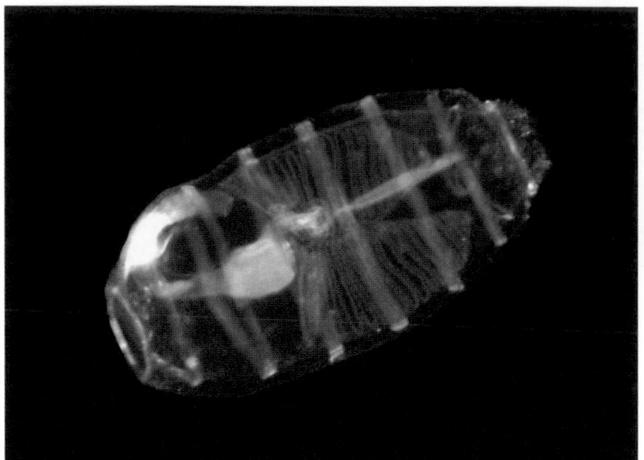

Fig. 29.1. The doliolid *Doliolum gegenbauri* (Photo: Lisa Gershwin).

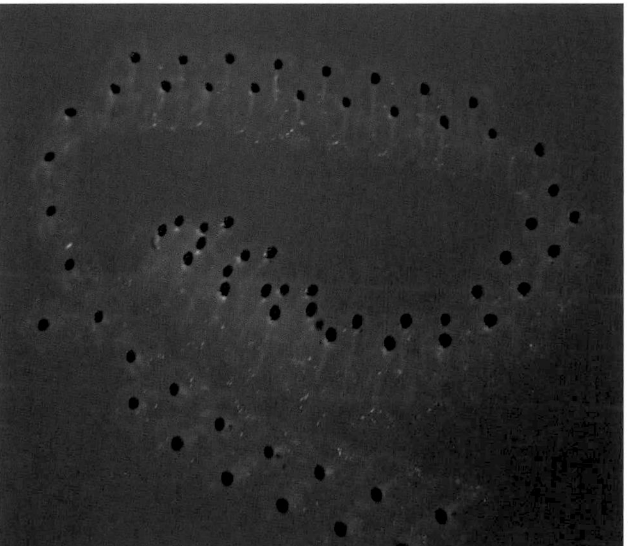

Fig. 29.2. Salps forming a colonial chain (Photo: Jon Cragg, Fish Rock Dive Centre).

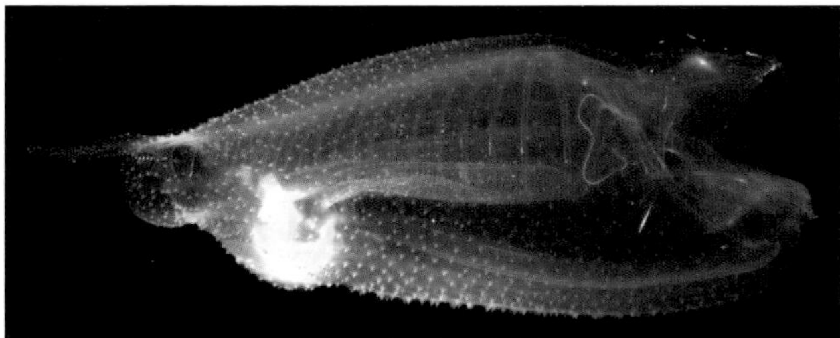

Fig. 29.3. The salp *Thetys vagina* in solitary phase (Photo: Merrick Ekins).

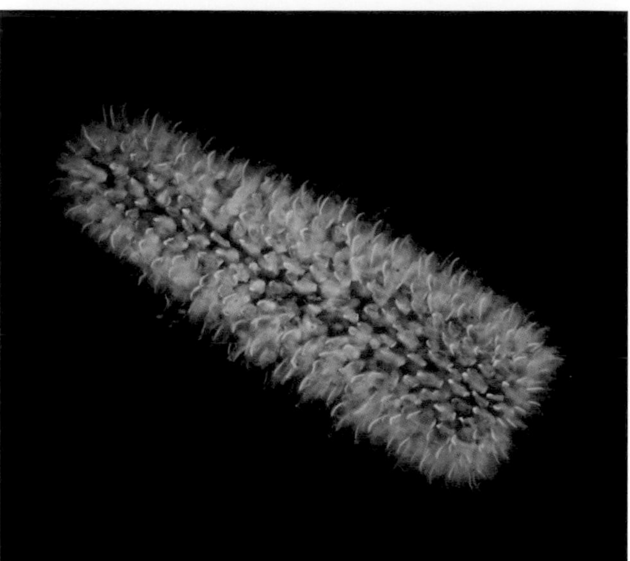

Fig. 29.4. The purse salp *Pyrosoma atlanticum* (Photo: Neville Coleman – QM images).

Fig. 29.5. Class Appendicularia better known as larvaceans, the top most larvacean with its 'house' (Photo: Julian Uribe-Palomino, CSIRO-IMOS).

The tunicates are currently a subphyla in Chordata, although they have been regarded as their own phylum. It was the Russian embryologist Kowalevsky, who recognised in 1866 the notochord and gill slits in the larval phase of ascidians and its similarities with frog tadpoles, leading to their placement in the evolutionary tree with chordates. As well as having a notochord, tunicates have a dorsal nerve cord, a perforated pharynx (like fish gills) and an endostyle (thyroid gland) secreting thyroxin (like vertebrates). It is the sessile colonial or solitary ascidian (Class Ascidiacea), commonly known as the sea squirt, that is most visible to visitors to the underwater realm. These colourful and

Fig. 29.6. A colourful collection of solitary and colonial ascidians competing for space (Photo: Merrick Ekins).

cryptic sessile ascidians found on the GBR will occupy the remainder of this chapter (Fig. 29.6).

Body plan

The ascidian body plan is illustrated in Fig. 29.7.

The ascidian body is enclosed in a tunic also known as the test. This tunic is composed of polysaccharides of tunicin similar to plant cellulose. It is extremely variable, being very soft and transparent in *Didemnid* species to tough and leathery and incorporating sand and algae as camouflage in the solitary Pyuridae. This external tunic provides protection against the environment and predators,

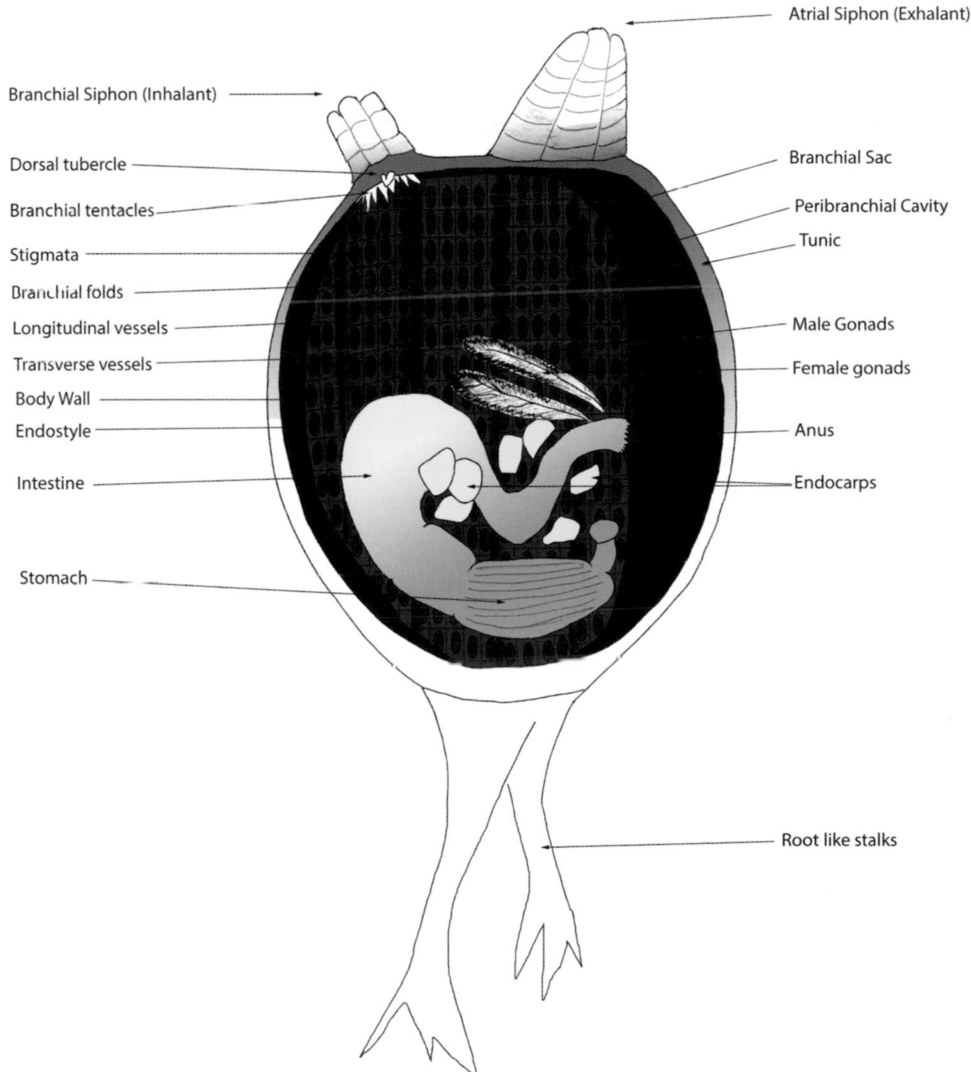

Fig. 29.7. The general structure of the ascidian *Cnemidocarpa stolonifera* (Drawing: Merrick Ekins).

but it also provides support that the muscles (as part of the body wall or mantle) can act upon to retract the siphons and internal organs when disturbed. This contraction of the body and the siphons causes water to be forcibly expelled from the ascidian, giving rise to the common name of sea squirts.

The branchial sac is the most obvious structure and occupies most of the cavity within the tunic. The branchial sac is attached to the base of the branchial or oral (inhalant) siphon at the top of the ascidian and separated by the oral tentacles. It is also attached to the mantle via the mid-ventral line, where the endostyle is located and thyroxin production occurs. The branchial sac has an intricate replicated transverse pattern of stigmata. Some species such as *Clavelina molluscensis* are very photogenic because their mantle is transparent revealing the striking patterns of the branchial sac (Fig. 29.8). These stigmata or slits in the branchial wall are bordered by cilia that draw the water into the ascidian and expel it out into the peribranchial cavity before exiting through the atrial or cloacal (exhalant) siphon. The branchial wall can be more complex with the addition of papillae and folding, all of which are very important characteristics for identification. If the pharyngeal wall has only simple stigmata, usually in longitudinal rows separated by transverse vessels, then these simple but elegant stigmata belong to the Order Aplousobranchia (*aplouso* = 'simple' in Greek), of which there are 13 families. The three largest and most important families of colonial ascidians are Didemnidae, Polyclinidae and Polycitoridae. If the stigmata have papillae or are joined by longitudinal vessels, they belong to the Order Phleobranchia (*phlebo* = 'vessel' in Greek). This order consists of nine families, the most common of which are Ascidiidae, Cionidae and Perophoridae. The stigmata in this order maybe more complex in shape: they vary from straight to curved in morphology. If the stigmata have so many longitudinal vessels that the branchial wall folds to increase the surface area, then the ascidian belongs to the Order Stolidobranchia (*stolido* = 'folded' in Greek). This order includes the solitary and some colonial ascidians of only three families: Styelidae, Pyuridae and Molgulidae. The shape and size of stigmata in this group varies.

Fig. 29.8. The transparent tunic of *Clavelina moluccensis* showing the branchial sac (Photo: Merrick Ekins).

Another group of organs important for identification is the digestive tract and its distinctive gut loop. The gut loop is located either adjacent to or below the branchial sac. The location of the gonads with respect to the gut loop is a handy way to separate ascidians into two different groups, which were old orders and no longer in use, but are still handy for identification. If the gonads are situated adjacent to the gut loop (i.e. just inside the thoracic wall), the ascidian belongs to the old group Pleurogona (*pleura* = membrane lining the thorax, surrounding the branchial sac in Greek). If the gonads are in the gut loop, the ascidian belongs to the group Enterogona (*entero* = relating to the intestine in Greek). The digestive tract is most visible as the tube-like oesophagus, generally round stomach and an intestine that is often divided. The faeces are usually pellet-like and ejected into the atrial cavity and removed via the cilia generated current out through the atrial siphon.

Ascidians are hermaphroditic and so have both ovaries and testes within the same individual, but they can be separated spatially or temporarily. Ascidians may have one combined gonad or many separated gonads. Location of the gonads in the ascidian body is another morphological characteristic used for identification: they can be found within the gut loop or in the peribranchial wall. Ascidians reproduce sexually, but they can also reproduce asexually via a variety of different methods. These include different ways of budding including lobulation, stolonic budding, strobilation, oesophageal–rectal and peribranchial budding. Ovoviviparous ascidians may also have pouches or chambers for brooding of young embryos. In these, and in oviparous species where sperm and eggs are both released into the sea, the fertilised eggs develop into a short-lived free-swimming larval stage. This stage usually only lasts from a few minutes to a few hours before the larva settles onto a hard substrate and undergoes metamorphosis. During metamorphosis the tail is rapidly reabsorbed and the animal rotates within the tunic and quickly develops into the shape we recognise as an ascidian.

Ascidians are fantastic filter feeders capable of filtering over 10 times their own volume in water in an hour. Some deepwater species have large siphons and reduced cilia to enable passive movement of water through the pharynx. There are also abyssal ascidians that, owing to the paucity of organic particles and bacteria in the water column, have abandoned the filtering behaviour that is characteristic for this group and have become carnivorous. The large mouth-like siphons serve as traps for marine organisms. Carnivory among ascidians is quite rare and most ascidians actively filter food from the water, with some species making use of the available sunlight by having an algal symbiont to assist in nutrition. Some of the colonial ascidians in the family Didemnidae are in symbiosis with the chlorophyll-containing cyanobacteria *Prochloron* to absorb sunlight for energy. Some colonies have internal structures within the colony, as well as in the test to accommodate the photosymbiont. *Lissoclinum bistratum* is an example of an ascidian with symbiotic *Prochloron*. The ascidian can regulate the amount of sunlight reaching the photosymbiont by altering the amount of spicules in the test. When part of the colony is in partial sun, say on side of the rock, it will appear white due to the large amount of white calcareous spicules that the ascidian has developed in the test. In direct sunlight, on top of the rock, the ascidian may appear pink from the carotenoid pigments. The same colony looks bright green when shaded under a rock, because the ascidian has reduced the amount of spicules in the test, so that sunlight falling on the photosymbiont is maximised. This species can form large ascidial mats on the reef flats of the northern GBR and other tropical regions. Ascidians usually prefer a more cryptic habitat to hide in. However, with the help of symbionts, these didemnids have access to a plentiful food supply and, with the high rates of asexual reproduction, have this ability to colonise large areas within the reef flat.

Diversity

The great diversity of species on the GBR is most likely due to its tropical location with proximity to the coral triangle and its highly biodiverse fauna. A large number of the tropical species are endemic to the GBR, but it also contains many species that have an Indo-Pacific tropical distribution.

The GBR contains four-fifths of known species of ascidians from Queensland and almost half of the known ascidians in Australia. On the reef flats, the colonial ascidians dominate, particularly the Didemnidae. Many of these are small and cryptic, hiding on the underside of coral rubble. The most common ascidian on the GBR is the solitary *Polycarpa pigmentata*, while the second most common ascidian is *Botrylloides leachi*, which is also the most prevalent colonial ascidian on the reef. Both are easily identified in the field: *Polycarpa pigmentata* by its blue siphon lining (Fig. 29.9) and *Botrylloides leachii* by its bright dramatic floral patterns (Fig. 29.10).

Ascidians come in every colour under the rainbow, with bright beautiful colours caused by pigments, symbiotic algae and spicules (white), microcrystals or diffraction. Generally, the colonial ascidians are the most colourful, while the majority

Fig. 29.9. *Polycarpa pigmentata* is often obscured by overgrowing epibionts, but revealed by its blue siphon linings (Photo: Merrick Ekins).

Fig. 29.11. *Polycarpa aurata* stands out from its surroundings with its bright contrasting yellow and purple colours (Photo: Merrick Ekins).

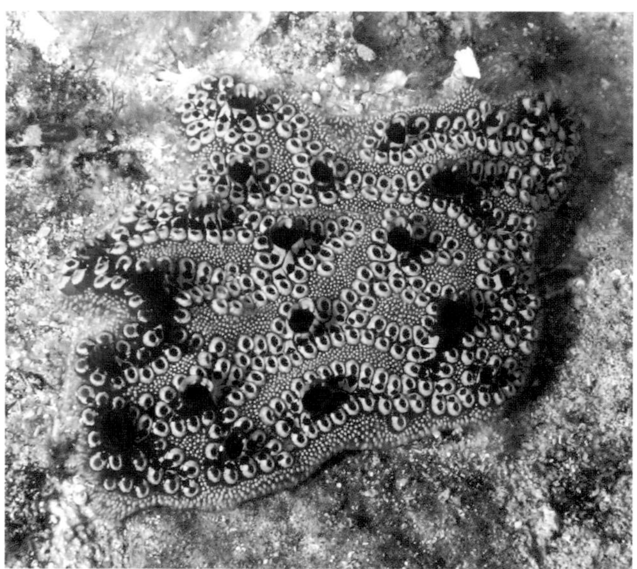

Fig. 29.10. *Botrylloides leachii* showing some of its dramatic floral patterns (Photo: Merrick Ekins).

of solitary species are drab or covered by epibionts, with the exception of *Polycarpa aurata* with its bright contrasting yellow and purple colours (Fig. 29.11). Some species, such as *Perophora namei* and *Pycnoclavella detorta*, are very photogenic with their transparent mantle (Figs 29.12, 29.13).

Many ascidians live as solitary individuals displaying an erect posture, and some are even

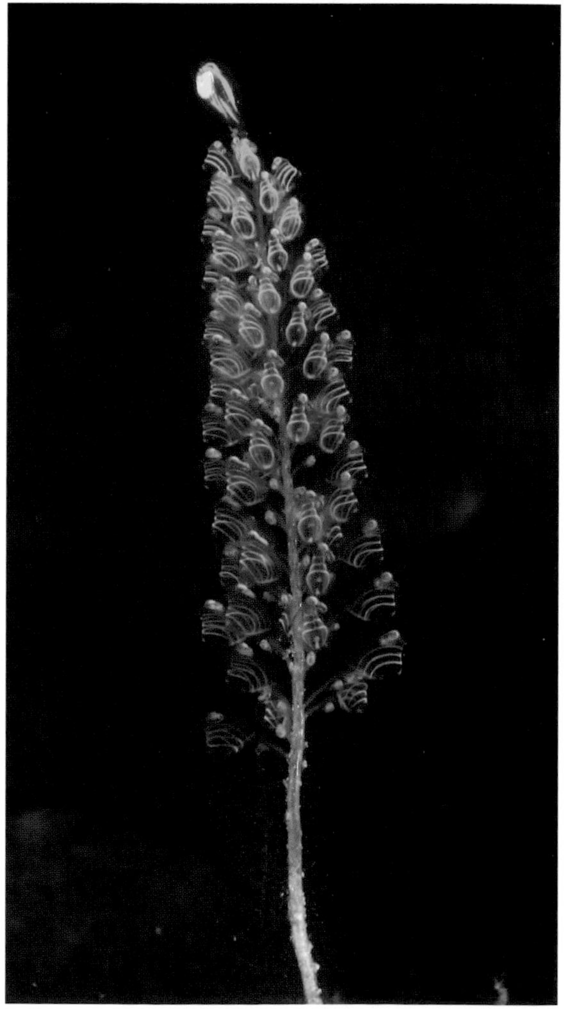

Fig. 29.12. The photogenic *Perophora namei* (Photo: Merrick Ekins).

Fig. 29.13. Another photogenic ascidian, *Pycnoclavella detorta* (Photo: Merrick Ekins).

stalked. Although they may often have transparent tunics, the majority of solitary ascidians have a camouflaged appearance because the tunic is often covered in epibionts and the ascidian is hidden in cryptic places. The name ascidian is derived from the Greek *askos*, meaning 'wine skin', as this is the shape of many solitary ascidians. A good example from the GBR is *Herdmania momus*, which is totally obscured by epibonts with the exception of its red siphon linings (Fig. 29.14). In contrast, *Polycarpa clavata* is a large ascidian with bright colours, and

Fig. 29.14. *Herdmania momus* totally obscured except for its siphons (Photo: Merrick Ekins).

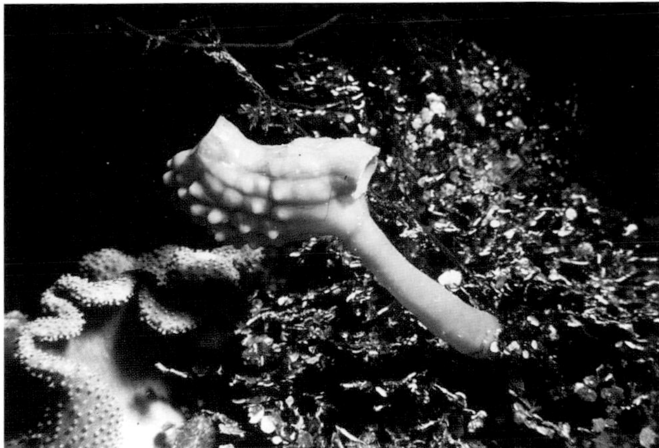

Fig. 29.15. *Polycarpa clavata* (Photo: Neville Coleman – QM Images).

projects via its stalk up above the substrate, which can make it very prominent on the reef (Fig. 29.15). Ascidians can attach to hard substrates such as rock, corals and other ascidians with adhesive papillae, but they can also have hair- or root-like processes anchoring them in soft sediments of sand and mud (Fig. 29.16). The ability of ascidians to grow and reproduce rapidly also enables them to overgrow many other plants and animals in the marine environment including algae, seagrasses, bryozoans, corals and sponges, which are often the same epibonts it may use as camoflague. Many other marine organisms (even fish) use the tunic as a convenient surface to grow on. Ascidians are also hosts to commensals such as amphipods, copepods

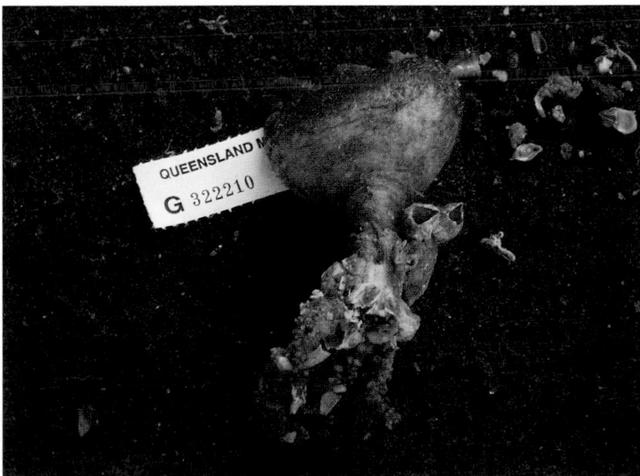

Fig. 29.16. *Cnemidocarpa stolonifera* with its root-like processes (Photo: Merrick Ekins).

and shrimps, which can commonly be found in the branchial sacs or atrial cavity making use of the ascidians as shelters from predation and currents. The tunics of solitary ascidians can also harbour bivalves that have burrowed into the tunic.

Solitary ascidians can occur together to give the appearance of a colonial nature when larvae settle together, such as *Polycarpa nigricans* on the GBR (Fig. 29.17). However, ascidians have adopted many other forms of colonial living. Stolonic colonies occur among certain ascidian families. Here the ascidian produces stolons that grow outwards from the pioneering individual. Under favourable conditions, these stolons produce asexual buds that grown into new individuals and the process continues. The ascidians can either grow tightly together so that the tunics fuse between individuals, or they can maintain a reasonable space between individuals. In stolonic colonies the founding individuals may be reabsorbed by the colony and reused at the budding growing edge of the colony, enabling it to migrate to a more favourable environment (Fig. 29.18). Another colonial arrangement is the formation of cluster colonies, where individuals are only partially attached. This cluster arrangement can simply be individuals joined by a shared thin membranous tunic on the substrate (Fig. 29.19). The next level of colonial integration as a cluster, includes those clusters with a shared and thicker tunic including their abdomens, but their thoraces are independent, such as the spectacular *Nephtheis fascicularis* (Fig. 29.20). After this level, all individuals are incorporated into a common tunic, forming large encrusting colonies, but each with their own inhalant and exhalant siphons. The ultimate stage of the colony formation is the incorporation into the common tunic and the joining of the waste systems, not unlike a town sewer system, to form the common cloacal apertures (Fig. 29.21). It is this colonial

Fig. 29.18. Stolonic colonies of *Symplegma brackenhielmi* (Photo: Neville Coleman – QM Images).

Fig. 29.17. Solitary *Polycarpa nigricans* giving the appearance of a colony (Photo: Merrick Ekins).

Fig. 29.19. *Clavelina arafuriensis* individuals cluster together sharing a tunic (Photo: Merrick Ekins).

Fig. 29.20. *Nephtheis fascicularis* cluster together and share thoraces (Photo: Merrick Ekins).

formation with associated clonal reproduction that enables certain ascidians to form large mats. Colonial ascidian species are about five times as numerous as solitary ones in both temperate and tropical waters.

Ascidians, like most other sessile inhabitants of the reef, have to compete for suitable substrate for attachment, especially when the larvae first settle. Some of the solitary ascidians simply become part of the substrate with epibionts growing all over them. Colonial ascidians on the other hand are more likely to overgrow other inhabitants of the reef. *Didemnids* and *Botryllus* species are particularly good at this and are known as possible invaders in many parts of the world. Ascidians, both solitary and colonial, are well known as fouling organisms in ports, on jetties, buoys, ship hulls, and so on, because these provide good surfaces for larval attachment. Harbours in particular provide sheltered waters that have a good mixing of turbid water that is often full of nutrients (and therefore bacterial food): an ideal environment for a shade-loving filter feeder. Harbours around the world generally have similar conditions, thus facilitating introductions. Harbours usually provide very different habitats from those of surrounding reefs, thus maintaining different ascidian faunas.

Fig. 29.21. *Botryllus anceps* showing its extensive common cloacal apertures (Photo: Merrick Ekins).

Impacts of climate change

The majority of ascidians are not sensitive to environmental stressors and will probably cope well with changes in the climate. The most likely impact of climate change will be a change in the reef structure and composition, and filter-feeding organisms such as ascidians will probably have a competitive advantage over reef-building corals. However, there will be a significant loss of biodiversity of ascidians. Those species with a narrow range of tolerances to abiotic conditions and with very specific habitats will not cope well with upcoming climate changes, while those species with a wide range of tolerances will dominate in the ecosystem, taking advantage of decreased competition. Intertidal ascidians will probably be the most stable ascidians, because they already cope with incredible temperature, moisture and salinity changes on a daily basis, not to mention the constant physical battering they receive from wave action. Bio-eroding ascidians will probably dominate, further adding to the stressed coral foundations of the reef.

For many invasive species, it will be difficult to ascertain the true origin because of the spread via trade around the world over at least the last 400 years. With rising sea temperatures, many 'outbreaks' or blooms of ascidians can be expected,

Fig. 29.22. *Eudistoma elongatum* in the seagrass before undergoing boom cycles (Photo: Merrick Ekins).

most likely in artifical settings such as oyster fisheries because these provide uniform habitats. These blooms will be a combination of new exotic species colonising as well as native species undergoing boom–bust cycles as suitable conditions become available. Whether these boom–bust cycles are due to increased water temperatures or higher nutrient loading is yet to be determined. *Eudistoma elongatum* is an example of a colonial ascidian on the east coast of Australia since at least 1960, that frequently (every 10 years or so) has a sudden population explosion on jetty pylons and becomes a visible mass of elongated drooping colonies (Fig. 29.22). This species is always living cryptically among seagrass beds, but is only noticed by people once it reaches critical mass in publicly visible areas.

As sea temperatures rise, marine organisms such as ascidians will tend to change distributions to higher latitudes. In Australia, there will be a range expansion as tropical species will be able to colonise the subtropical southern end of the GBR. Subtropical species will move further south to temperate waters. Temperate species from southern Australia may disappear because they cannot retreat to higher latitudes. Most likely, greater numbers of exotic species will find suitable habitat on the southern GBR as temperatures rise. With the increase in temperature, the entire ecological composition of the reef will change and ascidians will probably increase as a proportion of the community, because the area occupied by the corals will decrease, thus creating more habitat. This will also depend on the available food (any increase in nutrient in the water will favour the filter feeders) and on the mode of reproduction of the different ascidian species. Any changes to climate will also affect the plethora of commensals and parasites associated with ascidians.

Deepwater species were initially thought to be shielded from extreme climatic changes, but, because of the constancy of conditions in the abyss, a 1° change in temperature may be catastrophic for some species. Recent studies have shown that climate change has been responsible for changes in ocean currents, stratification and intensified upwelling has resulted in increased organic material being transported into deeper shelf waters, leading to an increase in respiration, hypoxia and, in some cases, the eruption of toxic gases such as methane and hydrogen sulphide from deep anoxic sediments. These events causing changes in respiration and nutrients have already been associated with mass mortalities among some deepwater benthic communities.

The oceans have already absorbed a third of the anthropogenic carbon dioxide. With the increase in ocean acidification there has also been an associated decrease in the concentration of carbonate ions in the sea. Shallow water species, such as the colonial Didemnidae, that use calcium carbonate spicules to reduce sunlight, will have difficulty producing spicules under ocean acidification, and

Box 29.1. *Didemnum molle*: an ascidian that walks to more favourable places

Didemnum molle is an exceptional species, which by the process of division; it slowly creeps up the branches of corals to find just the right amount of sunlight. It does this by withdrawing the tendrils of test from the lower base of the colony and adding new tendrils ahead of the colony, thereby 'walking' (albeit slowly) up the corals (Figs 29.23, 29.24 and 29.25). It requires sunlight for the photosynthetic symbiont *Prochloron*.

Fig. 29.23. *Didemnum molle* crawling up an *Acropora* coral (Photo: Merrick Ekins).

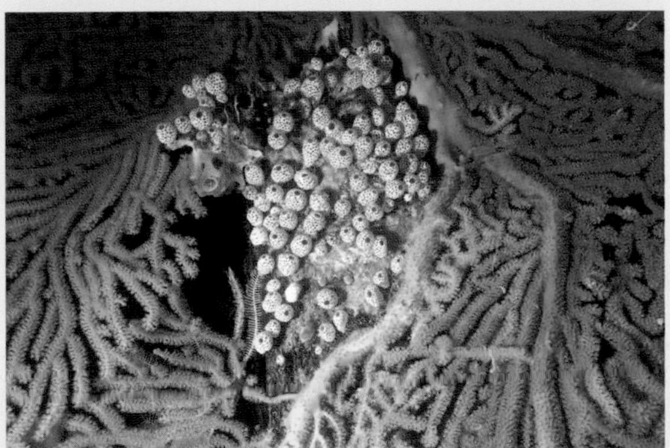

Fig. 29.24. Colonies of *Didemnum molle* occupying a sunny niche (Photo: Merrick Ekins).

Fig. 29.25. *Didemnum molle* showing its vast common cloacal cavity and the photosynthetic *Prochloron* (Photo: Merrick Ekins).

the dissolution of calcium carbonates. Ascidians that have symbiotic algae require sunlight, so are likely to be in shallower water, and those on reef flats are especially likely to be affected by rising sea temperatures.

Further reading

Bakun A, Weeks SJ (2004) Greenhouse gas build up, sardines, submarine eruptions and the possibility of abrupt degradation of intense marine upwelling ecosystems. *Ecology Letters* **7**(11), 1015–1023. doi:10.1111/j.1461-0248.2004.00665.x

Chan F, Barth JA, Lubchenco J, Kirincich A, Weeks H, Peterson WT, *et al.* (2008) Emergence of anoxia in the California Current large marine ecosystem. *Science* **319**(5865), 920. doi:10.1126/science.1149016

Diaz RJ, Rosenberg R (2008) Spreading dead zones and consequences for marine ecosystems. *Science* **321**(5891), 926–929. doi:10.1126/science.1156401

Fenaux R (1998) Anatomy and functional morphology of the appendicularia. In *The Biology of Pelagic Tunicates.* (Ed. Q Bone) pp. 25–34. Oxford University Press, Oxford, UK.

Gershwin L, Lewis M, Gowlett-Holmes K, Kloser R (2013) The Pelagic Tunicates. In *Pelagic Invertebrates of South-Eastern Australia: A Field Reference Guide.* CSIRO Marine and Atmospheric Research, Hobart.

Godeaux JEA (1989) Functions of the endostyle in the tunicates. *Bulletin of Marine Science* **45**(2), 228–242.

Godeaux JEA (1998) The relationships and systematics of the Thaliacea with keys for identification. In *The Biology of Pelagic Tunicates.* (Ed. Q Bone) pp. 272–274. Oxford University Press, Oxford, UK.

Godeaux J, Bone Q, Braconnet JC (1998) The anatomy of Thaliacea. In *The Biology of Tunicates.* (Ed. Q Bone) pp. 1–23. Oxford University Press, Oxford, UK.

Goodbody I (1974) The physiology of ascidians. *Advances in Marine Biology* **12**, 1–149.

Govindarajan AF, Bucklin A, Madin LP (2011) A molecular phylogeny of the Thaliacea. *Journal of Plankton Research* **33**(6), 843–853. doi:10.1093/plankt/fbq157

Hoegh-Guldberg O, Bruno JF (2010) The impact of climate change on the world's marine ecosystems. *Science* **328**(5985), 1523–1528. doi:10.1126/science.1189930

Kelmo F, Attrill MJ, Jones MB (2006) Mass mortality of coral reef ascidians following the 1997/1998 El Niño event. *Hydrobiologia* **555**, 231–240. doi:10.1007/s10750-005-1119-z

Kott P (1985) The Australian Ascidiacea Pt 1, Phlebobranchia and Stolidobranchia. *Memoirs of the Queensland Museum* **23**, 1–440.

Kott P (1990) The Australian Ascidiacea Pt 2, Aplousobranchia (1). *Memoirs of the Queensland Museum* **29**(1), 1–266.

Kott P (1992) The Australian Ascidiacea, Pt 3, Aplousobranchia (2). *Memoirs of the Queensland Museum* **32**(2), 377–620.

Kott P (2001) The Australian Ascidiacea Pt 4, Didemnidae. *Memoirs of the Queensland Museum* **47**(1), 1–410.

Kott P (2005) *Catalogue of Tunicata in Australian Waters.* Australian Biological Resources Study, Canberra.

Kott P (2008) Tunicata. In *The Great Barrier Reef: Biology, Environment and Management.* (Eds P Hutchings, M Kingsford and O Hoegh-Guldberg) pp. 308–326. CSIRO Publishing, Melbourne.

Kowalevsky A (1901) Les Hedylidés, étude anatomique. *Zapiski Imperatorskoi Akademii Nauk.* **12**, 1–32.

Lord J, Whitlatch R (2015) Predicting competitive shifts and responses to climate change based on latitudinal distributions of species assemblages. *Ecology* **96**(5), 1264–1274. doi:10.1890/14-0403.1

Monniot C, Monniot F, Laboute P (1991) *Coral Reef Ascidians of New Caledonia.* Editions de l'ORSTOM, Institut Français de Recherche Scientifique pour le Développement en Coopération, Paris.

Newcomb E, Pugh T (1975) Blue-green algae associated with ascidians of the Great Barrier Reef. *Nature* **253**, 533–534. doi:10.1038/253533a0

Pelejero C, Calvo E, Hoegh-Guldberg O (2010) Paleo-perspectives on ocean acidification. *Trends in Ecology & Evolution* **25**(6), 332–344. doi:10.1016/j.tree.2010.02.002

Petit JR, Jouzel J, Raynaud D, Barkov NI, Barnola JM, Basile I, *et al.* (1999) Climate and atmospheric history of the past 420,000 years from the Vostok ice core, Antarctica. *Nature* **399**(6735), 429–436. doi:10.1038/20859

Sahade R, Lagger C, Torre L, Momo F, Monien P, Schloss I, *et al.* (2015) Climate change and glacier retreat drive shifts in an Antarctic benthic ecosystem. *Science Advances* **1**(10), e1500050. doi:10.1126/sciadv.1500050

Shenkar N, Bronstein O, Loya Y (2008) Population dynamics of a coral reef ascidian in a deteriorating environment. *Marine Ecology Progress Series* **367**, 163–171. doi:10.3354/meps07579

Thompson H (1948) *Pelagic Tunicates of Australia.* Commonwealth Council for Scientific and Industrial Research Australia, Melbourne.

Van Soest RWM (1981) A monograph of the order Pyrosomatida (Tunicata, Thaliacea). *Journal of Plankton Research* **3**, 603–631. doi:10.1093/plankt/3.4.603

Weeks SJ, Currie B, Bakun A, Peard KR (2004) Hydrogen sulphide eruptions in the Atlantic Ocean off southern Africa: implications of a new view based on SeaWiFS satellite imagery. *Deep-sea Research. Part I, Oceanographic Research Papers* **51**(2), 153–172. doi:10.1016/j.dsr.2003.10.004

30

The fish assemblages of the Great Barrier Reef: their diversity and origin

J. H. Choat, B. C. Russell and A. Chin

Introduction

There are three critical features of coral reef fish faunas. First, they represent the most diverse assemblages of vertebrates on the planet. Second, this diversity may be seen at very local scales: hundreds of species co-occur within relatively small areas. Third, most species have broad geographical distributions. Thus, the observer moving across a small area of reef will encounter many different species of fish. Moving over relatively small areas (500 m^2) of reef habitat can reveal up to 100 species of reef fishes: far more vertebrates than would be encountered in any terrestrial habitat. If the scale of observation is increased to cover geographically distant reefs within the same ocean basin similar diversities may be encountered and in many instances the same species will be observed. However, species diversity and species identity do vary on geographic scales in ways that reflect the location and the evolutionary history of the reef habitat. In order to understand the processes that underlie this incredible diversity of fishes, we must also appreciate the forces that have modified the reef habitat over time.

This chapter provides an introduction to some of the important species groups and to the functional diversity of fishes on the Great Barrier Reef (GBR). It does not try to provide a catalogue or identification guide to all the different groups: there are many studies that do this (Box 30.1). Instead, it poses and answers some general questions about the GBR reef fish fauna. What is the relationship of the GBR fishes to other reef fish faunas? Do they have any unusual or unique features? Are there distinctive features of the GBR fauna that reflect the location, structure and history of the GBR itself? Therefore it describes not only the fishes but also the geological history and oceanographic processes that have contributed to the formation of their habitat: the largest and most complex reef structure in the world.

The chapter also provides examples of different groups of fishes that illustrate some of the important ecological features that help define the GBR and reflect its history. These examples provide an insight into some of the latest research initiatives on GBR fishes and identify future directions this may take.

Fishes – the most diverse vertebrates

Fishes are the largest and most diverse group of vertebrates on the planet. With an estimated 32 500 species, they constitute approximately half of all the known species of vertebrates. Moreover, the potential for discovering new species is far greater in fishes than for any of the other groups of vertebrates. Even after three centuries, the rate of

Box 30.1. Catalogues and key references for identification of Great Barrier Reef fishes

Carpenter KE, Niem VH (Eds) (1998) *FAO Species Identification Guide for Fishery Purposes. The Living Marine Resources of the Western Central Pacific. Vols 1–6*. FAO, Rome, Italy.

Eschmeyer WN (Ed.) (1998) *Catalog of Fishes*. Center for Biodiversity Research and Information, Special Publication 1. California Academy of Sciences, San Francisco, CA, USA, <www.calacademy.org/research/ichthyology/catalog/fishcatsearch.html>.

Froese R, Pauly D (Eds) (2007) FishBase. Website, <www.fishbase.org>.

Hoese DF, Bray DJ, Allen GR, Paxton J (2006) *Zoological Catalogue of Australia. Vol. 35: Fishes* (Eds PL Beesley and A Wells) ABRS, Canberra and CSIRO Publishing, Melbourne.

Lowe GR, Russell BC (1994) 'Additions and revisions to the checklist of fishes of the Capricorn-Bunker Group Great Barrier Reef Australia. Technical Memoir GBRMPA-TM-19'. Great Barrier Reef Marine Park Authority, Townsville.

Nelson JS (2006) *Fishes of the World*. 4th edn. John Wiley and Sons, New York, USA.

Randall JE, Allen GR, Steene RC (1990) *Fishes of the Great Barrier Reef and Coral Sea*. Crawford House Press, Bathurst.

Russell BC (1983) 'Annotated checklist of the coral reef fishes in the Capricorn–Bunker Group Great Barrier Reef Australia.' Special Publication No. 1. Great Barrier Reef Marine Park Authority, Townsville.

discovery of new species remains undiminished and fish experts estimate ~5000 marine species remain to be discovered and identified.

Although the diversity of fishes is relatively high, their distribution through the biosphere is taxonomically biased. For example, the 11 952 fishes recorded from freshwater environments are strongly represented by just three groups: the carps, characins and catfishes that constitute 65% of the diverse fauna. Similar biases occur in the marine environment. Coral reefs are dominated by a single but complex group of bony fishes: the Perciformes. These are perch-like fishes with spiny fin rays characterised by modifications to the feeding and locomotory apparatus that allow them to exploit a variety of food items ranging from microscopic sessile organisms to highly motile invertebrates and fishes.

The distribution of fish through the aquatic biosphere is also heavily biased. The majority of fish species occur in warm water with high local productivity, on shallow tropical marine reefs and in tropical lakes and streams. Tropical reefs make up less than 1% of the total marine habitat and freshwater species occupy less than 1% of the world's aquatic habitat. Thus, the greatest diversity of fishes occurs within a tiny fraction of the Earth's aquatic habitats, in shallow productive waters at low latitudes.

The diversity of marine reef fishes is reflected in the following figures. The total number of species is ~16 000, with 10 000 occurring in shallow tropical waters. The diversity of coral reef fishes has a strong geographical focus, with ~2400 species occurring in the Indonesian and Philippine archipelagos. The majority of these are Perciformes that have an extended evolutionary history to the start of the Cenozoic. Although some lineages were present in the Cretaceous period at 80–100 million years ago (Mya) the great morphological variety that we see in the present day was not established until the Eocene period, ~50 Mya.

The most speciose groups of present day coral reef fishes are gobies (family Gobiidae), wrasses (Labridae), groupers (Serranidae), damselfishes (Pomacentridae), cardinalfishes (Apogonidae) and blennies (Blenniidae). In addition, many groups such as butterflyfishes (Chaetodontidae), surgeonfishes (Acanthuridae), parrotfishes (Scaridae) and snappers (Lutjanidae) are also conspicuous and diverse in terms of species numbers, and with wrasses and damselfishes make up a majority of the individual fishes observed on coral reefs. In terms of size, coral reef fishes are highly skewed towards the lower end of the size range, with the majority of reef fishes being less than 20 cm in length. In addition, many reef fish species are locally rare, and a characteristic of fish communities on coral reefs is that they are dominated by a few very abundant species and numerous small rare species.

Coral reefs as fish habitats – an ecological and historical perspective

Coral reefs are biological formations resulting from constructive processes that produce calcium carbonate and erosive processes that reduce this to sediment. Because important metabolic processes of corals are dependent on light, coral reefs are shallow-water phenomena. Coral reefs are important fish habitats as they provide shelter, especially for the juveniles of many species. In shallow, clear water, coral reefs support complexes of small turfing algae capable of rapid growth and turnover. These algae trap organic detritus and provide sites for bacterial growth. These highly productive algal complexes and the associated detritus and microbes are a major source of readily accessible primary and secondary productivity for grazing animals. The calcium carbonate substrate is relatively soft and porous and subject to colonisation by boring organisms. These in turn provide habitats for microorganisms that represent an important protein source for those reef fishes that can excavate the substratum and process the material to extract invertebrates and living plants. The feeding activities of grazing and excavating fishes and the subsequent passage of ingested material through the alimentary tract produce a rain of fine sediment and enriched detrital material. The detritus serves as the primary food source for some of the most abundant groups of grazing fishes. Lastly, the currents in the vicinity of coral reefs form complex eddies and accumulate large volumes of planktonic and small nektonic organisms that serve as a food source for numerous species of plankton-feeding reef fishes that are preyed upon by larger pelagic predators.

Understanding reef fish faunas for the purposes of management and conservation requires an appreciation of how these faunas differ between reefs and regions. Reef history, location and environmental influences may all have profound effects on fish assemblages. The GBR has had a distinctive history that, in association with its regional location, has left its imprint on the fish fauna. The following sections will deal with the location of the GBR with respect to other large reef systems, its history and the geological processes as they relate to the present day fish assemblages and their habitats.

Great Barrier Reef fishes

Patterns in space

Some 1625 species, including trawl fishes, are recorded from the GBR, of which 1468 are coral reef species. This is a relatively high value and may be explained by two critical influences.

First, there is the geographical location of the reef and its setting in terms of the hydrodynamic environment. The position of the GBR relative to the Indonesian and Philippine archipelagos to the north and the close proximity to the western Pacific reefs has resulted in a diverse fauna of reef fishes. Colonisation by larval reef fish from sources to the north and west are reflected in the strong biogeographical affinities of the GBR fish fauna with widespread tropical Indo-West Pacific elements.

Second, there is the size, configuration and habitat structure of the reef. The GBR is exceptional in terms of its size (350 000 km^2), extending along the east Australian coastline from 10°S to 24°S, a distance of 1200 km (Fig. 30.1). The main structure of the reef terminates just south of the tropic of Capricorn. Beyond this there are several isolated coral reefs (Elizabeth and Middleton Reefs and Lord Howe Island) extending as far south as 31°30′S. These support faunas of coral reef fishes and also representatives of subtropical and temperate fish groups. A recent survey of Elizabeth and Middleton identified 322 species of reef-associated fishes including 26 groupers, 25 butterfly fishes, 21 benthic feeding damselfishes, 22 parrotfishes and 21 surgeonfishes. Some species are endemic to these southern regions. Similar data from the northern end of the reef (Lizard Island) identifies a major shift in diversity, with ~900 species of reef-associated fishes recorded. For the same groups, the diversities were: groupers 41, butterfly fishes 38, benthic feeding damselfishes 48, parrotfishes 25 and surgeonfishes 31. However, the southern reefs also harbour species with warm temperate water affinities including morwongs (Cheilodactylidae), ludericks or sea chubs (*Kyphosus, Girella*), large pomacentrids (*Parma*), wrasses (*Pseudolabrus*) and surgeonfishes (*Prionurus*). These species contribute to the overall diversity of the reef, providing a small but distinctive southern element to the reef fauna (Fig. 30.2).

There are also strong longitudinal trends in reef fish diversity. Comparison of equivalent habitats from inner coastal reefs and mid-shelf and outer barrier reefs at the central region of the GBR revealed the following differences in species diversities (inner versus outer reefs): butterfly fishes (7 versus 15); damselfishes (10 versus, 26); parrotfishes (8 versus 20); and surgeonfishes (4 versus 15). An important aspect of this distribution is that some inshore species have restricted distributions that reflect some of the unusual habitat features of inshore reefs (Fig. 30.3). The GBR maintains a very high diversity of fishes for two reasons. The first is its sheer size, comprising 2000 reefs spread over 350 000 km^2. This provides a massive target with a high probability of suitable habitats for larval fishes that may disperse from other regions to the north and east. The second is the variety of habitats, including inshore coastal reefs extending out to reefs exposed to fully oceanic conditions.

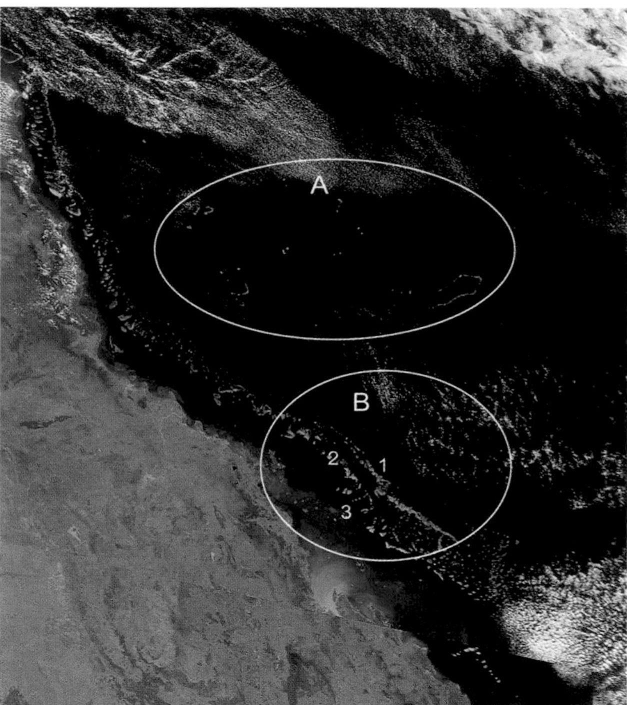

Fig. 30.1. A satellite image of the north-east Australian coast showing the Great Barrier Reef and the characteristic structure from the coast to the outer barrier with inner, mid- and outer-shelf reefs across a longitudinal gradient. Reefs of the Queensland Plateau represent habitats from which reef fishes could recolonise the GBR during periods of rising sea level. Area A: Coral sea reefs of the Queensland Plateau. Area B: structure of the GBR; 1, outer reefs exposed to oceanic influences; 2, mid-shelf reefs; 3, inner reefs and islands merging with mid-shelf reefs (Image by NASA).

The continuous nature of the reef, its latitudinal distribution from tropical to subtropical environments and the strong longitudinal gradient of habitat structure from the coasts to the Coral Sea are factors that underlie the notable diversity of reef fish species. In addition, the unique evolutionary history of the Australian coastal fauna has made an important contribution to the diversity of the present day GBR fish assemblage.

Patterns in time

Although reef fish faunas vary according to their location, the most profound and interesting changes are those that occur over time. Reef fish diversities vary in response to processes that occur at different timescales. These include the following.

- Intergeneration changes driven by the variation in the recruitment of juvenile fishes. Reef fish populations are subject to an open water dispersive stage before commencing life on the reef. This influences the number and type of fish that constitute the next generation of reef life.
- Decadal-scale changes associated with climatic variation such as that encountered in El Niño

Fig. 30.2. **(A)** The large surgeon fish *Prionurus maculatus* schooling on the Middleton reef crest. **(B)** The kyphosid *Kyphosus pacificus* that frequently schools with *P. maculatus* on the Middleton reef crest. Both species use fermentation to digest their primary food source: macroscopic algae. These species extend into the southern GBR. **(C)** The girellid fish *Girella cyanea* schooling with *K. pacificus*. *Girella cyanea* feeds on a higher proportion of animal material than the kyphosid (Photos: A. M. Ayling, Sea Research).

years, cyclonic activity and in some cases biological agents such as the crown-of-thorns starfish, *Acanthaster* (see Chapter 28). Pulses of temperature increase *Acanthaster* feeding, and destructive cyclones result in declines of living coral, with concomitant changes in the numbers of many small reef fishes.

- Century-scale changes associated with longer term shifts in ocean temperature or current systems as exemplified by the Little Ice Age (1200–1800 CE). Such climatic shifts will modify distributional patterns of fish, including migration into warmer sections of their range.
- Geological and oceanographic processes and long-term climatic trends that usually operate over thousands or millions of years. These include changes to the geographical location of reefs (plate tectonics) and drastic modification of sea level (glaciation cycles).

The most informative approach to understanding the present day GBR fish fauna is to summarise the history of the reef through geological time. A comprehensive description of the geological history of the GBR is provided in Chapters 2 and 3.

At the start of reef fish history, 55 Mya, much of eastern Australia lay well south of the tropics. Over the next 50 million years tectonic processes moved the Australian continent northwards with the northern boundary of the reef reaching present tropical latitudes ~25 Mya. However, the entire extent of the reef was not wholly tropical until

Fig. 30.3. **(A)** Inner reef feeding group of grazing fishes. The group consists of two species of parrotfishes *Scarus ghobban*, often found in non-reef environments and a predominantly inshore species *S. rivulatus*. Macroscopic algae is usually present at such sites. **(B)** A mixed school of grazing fishes on a mid-shelf reef. These diverse groups of grazing fishes are characteristic of mid-shelf reefs. The example consists of eight species of parrotfish and four species of surgeonfish. **(C)** Fishes of the exposed outer barrier reef crest. This group is dominated by the large browsing surgeonfish *Naso tonganus* and plankton feeding *Acanthurus mata*. (Photos: A, B, A. Lewis, Tevene'i Marine; C, JCU).

3 Mya. Three things are important about this historical pattern. First, unlike the continents of the Northern Hemisphere, the Australian coast was not subject to episodes of extensive glaciation during the mid to late Cenozoic. In contrast to the temperate coasts of North America and Europe, the southern Australian fish fauna underwent periods of extensive diversification, resulting in lineages of reef fish especially within the wrasses, morwongs and leather jackets (Monacanthidae) that are unique to the Southern Hemisphere. Some of these temperate-water groups have been able to penetrate tropical environments.

Second, over this period, temperatures of the surrounding oceans have varied substantially. From 60 to 45 Mya, the GBR was subject to water temperatures ranging from 9°C to 19°C, thus inhibiting coral reef formation. The combination of northward continental movement and increasing oceanic temperatures provided a period of 17 My to the present day when an increasing proportion of the reef enjoyed temperatures that permitted coral reef growth.

A third factor, however, further inhibited tropical reef formation. Glaciation cycles drive major fluctuations in sea level and 32 glaciations have been recorded over the last 1.8 My. Over the last 430 000 years, there has been an increase in the magnitude of cycles, resulting in four episodes of rapid sea level variation with maximum amplitudes of 120–140 m. Cycles of sea level fluctuation reduce and alter habitats that in turn modify reef fish populations. For the GBR, the most dramatic changes have occurred over the last 130 thousand years (ky). At that time sea levels were equivalent to those of the present day. Sea levels then declined in a series of steps to 125 m below present day levels 20 ky ago. There followed an abrupt rise to present levels commencing 16–18 ky ago.

Given the configuration of the GBR and association with the continental shelf over much of the previous 130 ky, there would have been no reef formation on the north-east Australian coast other than a fringing reef at the continental margin. The characteristic mid and outer-shelf reefs that harbour most of the fish species would have been non-existent. The best estimates of the rate of sea level rise place the age of the GBR in its present configuration as less than 7 ky old.

In summary, the present GBR is surprisingly young and has been subject to enormous changes over the last 200 ky. Over this period, characterised by cyclic changes in sea level, the reef has ceased to exist during low stands and then been reconstituted through the rising seas. In the periods of rising seas, the reef must have been recolonised by reef fishes from the reef systems to the east and north. The present-day configuration with the system of midshelf platform reefs is only 6–8 ky old, although the fish species themselves are far more ancient.

Functionally important groups of fishes on the GBR

Grazing fishes–parrotfishes and surgeon fishes

What defines a reef fish fauna? This debate has tended to focus on the taxonomic structure of reef fish faunas and their history. One emerging conclusion is that present day reef fish did not arise on coral reefs. The taxonomic debate has been influenced by the observation that the most abundant groups of coral reef fishes also have representatives in temperate waters. It may be easier to define reef fishes in terms of their functional attributes and ecological features. The strongest associations with coral reefs are seen in groups such as parrotfishes, surgeonfishes, butterflyfishes, blennies and many benthic-feeding damselfishes. This association has strong links with their nutritional ecology.

A study of these groups reveals many similar species that co-occur within small areas. This is especially true of the grazing fishes that characterise reef crests and flats. This begs the question as to how they share resources. To understand this, it is necessary to examine how space is used in foraging and feeding activities on reefs. In terms of fish real estate, reefs consist of contrasting areas of complex structure provided by living corals and extensive areas of carbonate rock and associated coral debris. Living corals are used extensively by smaller reef fishes and recruits of larger species as shelter. Extensive carbonate flats and fields of coral rubble support complexes of turfing and encrusting algae, fine sediment and detritus. It is the extensive areas of apparently bare calcareous substratum that provide a key to the diversity and dynamics of many types of reef fishes. If this substratum is examined microscopically then a complex 'tangled bank' harbouring a great variety of small plants and microorganisms, which (to quote Darwin) are 'different from each other, and dependent upon each other in so complex a manner' is revealed. Very large numbers of fishes graze on this 'bank', with the primary groups being parrotfishes, surgeonfishes and rabbitfishes (Siganidae).

Although these groups are usually classed as 'herbivores' their trophic biology is more complex than just eating living plants (Fig. 30.4). Many, including the most abundant species, feed on detrital material, bacterial aggregates and small invertebrates. Others, and especially some species of surgeonfishes, target turfing and filamentous algae, but most of the species that constitute the mixed foraging schools of grazing Perciformes that are so characteristic of reefs feed on detrital and bacterial aggregates and cannot be defined as herbivores. This is especially true of parrotfishes that have the capacity to remove calcareous material with scraping (Fig. 30.4A) or excavating (Fig. 30.4B) oral jaws to extract the microbial assemblages that occur within the porous upper substratum. Many species of surgeonfishes graze mainly on detritus and sediment (Fig. 30.4C). Moreover, analysis of feeding and foraging of smaller fish, including many damselfishes and most blennies, shows that they are also targeting detrital resources.

The foraging behaviour of abundant grazing fishes provides a consistent ecological signature for reef fish assemblages. The initial impression is one of uniform feeding by large multispecific schools of grazing fishes moving over tracts of reef. However, both feeding behaviour and the resources targeted are more complex than an initial impression suggests. Some species (exemplified by the aggressive surgeonfish *Acanthurus lineatus*) actively defend territories that support dense growths of turfing algae, their primary food source (Fig. 30.5A). Similar behaviour is exhibited by territorial damselfishes of the genus *Stegastes* (Fig. 30.5B) where defence of feeding territories against much larger grazers is common. In some reef habitats, such as reef crests, territories may cover up to 70% of the available reef substratum. Other groups of grazing fishes, especially most parrotfishes and many surgeonfishes, form mobile schools that graze over extensive areas, with feeding episodes continuously disrupted by territorial species.

The complexity of grazing behaviour is illustrated by the variety of diets, or the 'nutritional ecology' of the different species. The range

Fig. 30.4. **(A)** An example of a scraping parrotfish, *Scarus oviceps*, feeding is predominantly by removal of the top 2–3 mm of the calcareous substratum. **(B)** An example of an excavating parrotfish, *Chlorurus microrhinos*. Note the deep profile of the head that incorporates the massive oral jaws and associated musculature. Feeding is predominantly by excavating the calcareous substratum. **(C)** *Acanthurus olivaceus*, an abundant detrital feeding surgeonfish. Detritus and sediment are triturated in a muscular gizzard before digestion (Photos: A. Lewis, Tevene'i Marine).

includes: (1) acanthurids that feed on filamentous algae, processing food via acid digestion (e.g. *Acanthurus lineatus*); (2) species that concentrate on large brown algae, through bacterial fermentation (e.g. *Naso lituratus*); (3) species that selectively harvest smaller turfing algae and also use bacterial fermentation (e.g. *Zebrasoma scopas*); (4) species that feed on both algae and plankton (e.g. *Naso brevirostris, N. vlamingii*); (5) species that feed exclusively on detritus (e.g. *Ctenochaetus striatus*); (6) species that feed on mixtures of detritus, animal matter and algae (e.g. *Acanthurus blochii*); and (7) species that scrape or excavate calcareous surfaces to extract detritus, animal material and turfing algae

Fig. 30.5. **(A)** The highly aggressive surgeon fish *Acanthurus lineatus*. This species defends territories from grazing fishes and enhances the growth of red algae, the primary food source. **(B)** A territorial damselfish of the genus *Stegastes*, which employs active defence of an algal garden against grazing fishes (Photos: A. Lewis, Tevene'i Marine).

(e.g. *Scarus frenatus, Chlorurus microrhinos*). All of these groups may combine to form multi-species foraging schools that are frequently joined by rabbitfishes and species of goatfish (Mullidae) and wrasses that forage on small invertebrates disturbed by group feeding. At the upper end of the size spectrum of grazing fishes are the very large excavating parrotfishes (*Bolbometopon, Chlorurus*) that are capable of removing hundreds of tonnes of solid calcareous material per year and redistributing it as sediment on reef faces.

These groups have traditionally been classified as herbivores, placed at the bottom of the food chain and attributed the dual roles of enhancing coral growth through removal of algae and acting as a major conduit of carbon through reef systems by linking plant production to carnivores that consume the abundant herbivorous fish. The removal of algae by grazing fish has been demonstrated experimentally. However, the widespread consumption of bacteria, small invertebrates and a cosmopolitan mix of detrital material demonstrates that food chains and the flow of carbon through coral reef communities are likely to be far more complex than usually considered in trophic schemes. The apparent uniformity of grazing activity in mixed schools masks a much greater diversity of feeding behaviour, dietary targets and food processing than is currently recognised.

One of the best examples of the diversity of feeding behaviour in 'herbivorous' fishes is provided by an analysis of the evolutionary relationship in the surgeonfish genus *Naso* (unicorn fishes) carried out using gene sequences extracted from mitochondrial DNA. This provides a means of constructing a robust picture of the evolutionary relations among the 19 currently recognised species. This group is of interest on account of their morphological variation, ranging from slow moving reef grazers and plankton feeders closely associated with shallow reef habitats, often with large frontal horns or head-bumps, to rapidly swimming pelagic tuna-like species. These species have a wide variety of diets including benthic algae, benthic invertebrates, macroplankton and small rapidly swimming nekton such as small fishes (Fig. 30.6). The unicorn fishes have a long evolutionary history, with closely related genera occurring as well-preserved fossils in Eocene (50 Mya) shallow water marine deposits. The purpose of the evolutionary analysis was to determine the evolutionary pathways of the different feeding modes. Were the herbivorous species basal to the other groups or did they emerge much later? Is plant feeding restricted to certain evolutionary groups? What is the basis of the extraordinary morphological diversity of the unicorn fishes?

The answers (Box 30.2) were surprising. Several well-supported monophyletic evolutionary groupings (clades) were identified. The ancestral group was represented by the small, pelagic, fast swimming species of the *Naso minor* complex, not the more abundant reef-associated species. Particular types of feeding behaviour and nutritional ecology (algal grazing, plankton feeding, pelagic foraging on nekton, fermentative digestion) did not occur in cohesive evolutionary groupings but were scattered through the different clades. The trademark morphology of the group, the extended frontal horn, occurred in unrelated species in different clades (e.g. *Naso annulatus* and *Naso brevirostris*). This was surprising, because such distinctive morphologies had strongly suggested natural groupings in past studies. The basic message is one of greater evolutionary flexibility in terms of feeding behaviour and morphology than anticipated from ecological studies. The large groups of grazing fishes that characterise the GBR are therefore deserving of more detailed study.

Predatory guilds of reef fishes

It is a natural progression from grazing fishes to the predatory fishes that inhabit the reef. Many species are piscivores (fish eaters) and tend to concentrate on parrotfishes, damselfishes and near-reef pelagic groups such as fusiliers (Caesionidae). Not surprisingly, species with well-developed dorsal and anal fin spines (butterflyfishes) and with caudal knives (surgeonfishes) are avoided by predators and occur very infrequently in the stomach contents. There are numerous different types of predatory fishes including pelagic species such as mackerels (Scombridae) and large trevallies (Carangidae) but the dominant predators on coral reefs are usually groupers. These are dominated by four genera, *Epinephelus, Cephalopholis, Plectropomus* and *Variola,* and may be partitioned into ecological and foraging groupings. The largest genus is *Epinephelus,* with 27 species recorded from

Box 30.2. Evolutionary relationships in foraging and feeding modes in the genus *Naso* (unicorn fishes)

Figure 30.6 illustrates the evolutionary relationships among the different species of the genus *Naso* in the form of a phylogenetic tree based on sequences from mitochondrial and nuclear genes. Members of the genus display a variety of behaviours including pelagic foraging in the water column, benthic foraging over the reef surface and a mixture of pelagic and benthic foraging modes. Species that swim slowly in open water develop extended frontal horns; fast swimming pelagic foragers have a tuna-like morphology. Different foraging modes are associated with different feeding patterns: macroplankton, benthic animal material, brown algae or mixtures of green and red algae. Some species feed on both planktonic and benthic animals; others feed on algae and plankton. The phylogenetic reconstruction shows that the two most speciose evolutionary groupings or clades (1 and 2) each show examples of the different foraging modes. The distinctive morphologies such as frontal horns and tuna-like body shapes have developed independently in the different clades. The ancestral groups are represented by small pelagic species in the subgenus *Axinurus*. The genus is a relatively ancient widely distributed group of reef fishes with the distribution ranges: IPO, Indo-Pacific; PO, Pacific; IO, Indian Ocean, WIO, West Indian Ocean indicated. The numbers represent the robustness of the tree structure using Bayesian, MP and ML analyses.

Further details are available in Klanten *et al.* (2004) *Molecular Phylogenetics and Evolution* **32**, 221–235.

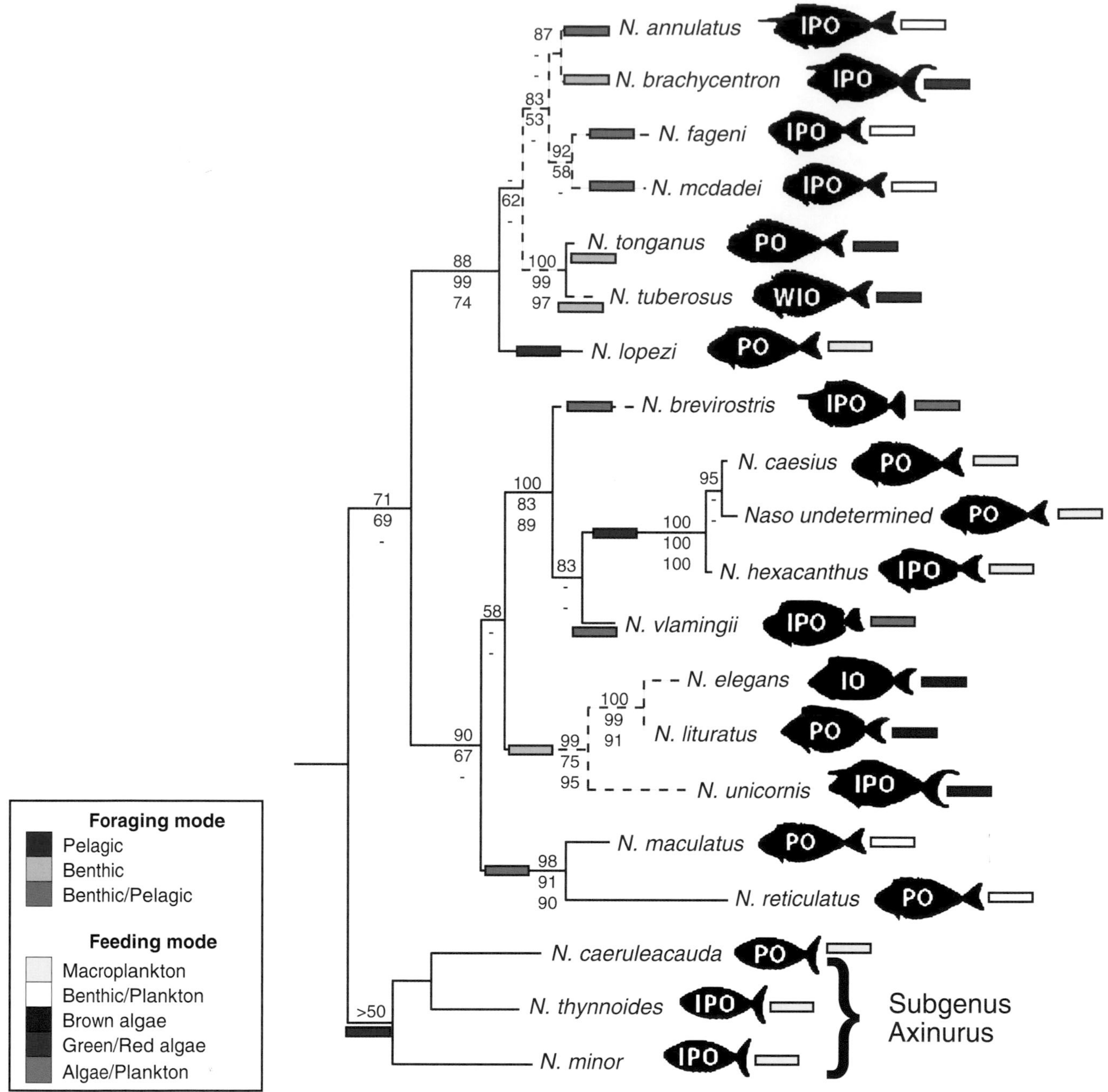

Fig. 30.6. Evolutionary relationships in foraging and feeding patterns in *Naso* (Illustration: S. Klanten JCU).

the GBR, followed by *Cephalopholis*, 10; *Plectropomus*, five and *Variola*, two.

Epinephelus is notable in that the species cover a very wide size, ranging from *E. lanceolatus*, which reaches 2.5 m total length (TL), to small cryptic species such as *E. merra* at 0.25 m TL. The genus may be ecologically subdivided by size, with large species between 0.6–2.0 m TL often being mobile species but usually associated with bottom cover. Smaller species are more numerous than the larger species, with GBR grouper assemblages being dominated by fish 15–25 cm TL. Size is related to foraging activities. Small groupers, including members of the genus *Cephalopholis*, small species of *Epinephelus* and juveniles of the larger species are invariably cryptic, associated with the shelter of coral growths and debris. The larger species of *Epinephelus* are more mobile and may move considerable distances when foraging, but are strongly associated with complex reef structures and usually seek shelter when disturbed. Members of the genus are usually ambush predators preying on other reef fish. In contrast to the genera *Cephalopholis* and *Epinephelus,* members of the other two genera, *Plectropomus* and *Variola,* have a roving habit and usually forage above the substratum, especially on reef fronts, slopes and deeper reef bases. For this reason, they are more visible to the observer than other groupers. There are two unifying features of groupers. First, juveniles are invariably secretive and associated with areas of high cover. For this reason, it is difficult to estimate recruitment patterns in reef groupers and to obtain information on the early growth stages. Second, members of this family frequently aggregate to spawn, with some species of *Epinephelus* reaching very high local densities at spawning sites. Other genera such as *Plectropomus* also form spawning aggregations, but with numerous local groups with relatively low numbers of fish.

The GBR grouper fauna has some unusual features. Most reef areas are dominated by smaller cryptic species, with the larger mobile species being comparatively rare, even in areas protected from fishing. Consequently, observers see few groupers during normal activities. For example, the dominant groupers on relatively undisturbed reefs in the southern Seychelles were *Cephalopholis urodeta* and *Epinephelus fasciatus*: two small species that made up 70% of the grouper fauna. Larger species, such as *E. polyphekadion* and *E. fuscoguttatus,* occurred at less than one individual per 1000 m^2. Although four small species of *Cephalopholis* made up 47% of the GBR grouper fauna, two species of *Plectropomus* (*P. leopardus* and *P. laevis*) made up 33% of the fauna. The presence of these large, actively foraging predators on reef crests and reef fronts are a unique feature of the GBR predator fauna. By comparison, this genus was very rare in the southern Seychelles. This is not a reflection of fishing because other groupers of great commercial importance were recorded in high numbers from the monitoring sites. This simply reflects the natural distribution of *Plectropomus;* a genus confined mainly to the south-west Pacific and the Indo-Australian region. The ecological equivalent was *Variola louti,* a species that shows a similar above-bottom foraging habit to *Plectropomus,* although it made up only 6% of the serranid fauna. The most abundant populations of *Plectropomus* occur on the Australian plate and especially on the north-eastern Australian coasts and reefs where they also support extensive fisheries. This high concentration of roving and highly visible predators is very much an Australian phenomenon.

Predation is intense on coral reefs and the influence of predators such as groupers on reef fishes may be manifested in behavioural and morphological responses in the prey species. Many reef fishes have evolved colour patterns and behaviours as a means of avoiding predation, or in some cases to enhance their ability as predators. The most common way of avoiding predation is through protective resemblance or camouflage, where the fish closely resembles a part of a substratum, a plant, or a sedentary animal such as a sponge or soft coral to avoid predation (Fig. 30.7; Box 30.3).

The recent history of the GBR – how has this influenced the fish fauna?

The complex structure of the GBR gives an impression of stability and permanence. The tracts of coastal fringing reefs, mid-shelf platform reefs and ribbon reefs of the outer barrier provide a parallel series of distinctive habitats extending along a north–south axis for 1000 km. The different habitats support distinctive fish communities. At several localities along the reef the habitats intergrade, especially in the vicinity of large coastal islands (Fig. 30.1). At these localities, elements of the inshore and mid-shelf reef assemblages may mix.

The geological history of the GBR shows that this structural and biological partitioning of the reef has occurred very recently, and that sea level

Fig. 30.7. Mimicry in Great Barrier Reef fishes: **(A)** noxious tetradontid *Canthigaster valentini*; **(B)** monocanthid *Paraleuteres prionurus* that mimics *C. valentini*; **(C)** poison fang blenny *Meiacanthus lineatus*; **(D)** *Scolopsis bilineatus* that mimics *M. lineatus*; **(E)** *Petroscirtes fallax* that also mimics *M. lineatus* (Photos: A, B, D, Gerry Allen; C, E, Roger Steene).

changes associated with glacial cycles have profoundly modified Australian shallow water reef systems. For the last 130 ky, sea levels have fallen on a global basis from levels that were similar to those of the present day to a low of 140 m below present levels only 20 ky. The biogeographical consequences of this were profound, with isolation of the north-east Australian coast from the west due to the closure of the Torres Strait from 116 ky to 30 ky, and then by the emergence of the exposed coasts of Papua New Guinea to the north and east. Rapid sea level rises linked to the termination of the last glacial period 19 ky ago flooded continental shelves so that reef habitats increased from 50 000 to 225 000 km^2 at present. Thus, over the last 6000 years the north-eastern Australian coast has been transformed from a shelf habitat dominated by fringing reefs to one supporting the extensive and ecologically diverse system of the present day GBR. One consequence of the recent emergence of a new reef system is that the majority of the fish fauna must be the product of a rapid colonisation process from external sources. Where did the present day fish fauna come from and are there any evolutionary and ecological signatures of this colonisation process?

To explore this question, consider two of the most prominent members of the reef fauna: the bar cheeked coral trout (*Plectropomus maculatus*), characteristic of inshore reefs; and the leopard coral trout (*Plectropomus leopardus*) common on mid-shelf reefs. The geographical source of these populations and their recent evolutionary history was investigated by examining and comparing sequences of molecules amplified from mitochondrial genes. Such sequences are passed on through female parents and, because they are inherited, they provide a method for disentangling patterns of evolutionary descent in the sampled population. They also provide a species identification code and allow assessment of the degree of relatedness among groups of species.

An analysis of mitochondrial sequences of GBR *Plectropomus* species was carried out to establish the pattern and degree of relatedness of three species (*P. leopardus*, *P. maculatus* and *P. laevis*) that are very similar in appearance and colour pattern (Fig. 30.8). *Plectropomus leopardus* achieves high abundances on the Australian plate, including New Caledonia, but abundances decline through the reef at lower latitudes, then increase in higher northern latitudes, which suggests an anti-tropical distribution.

Box 30.3. Disguise, defence and aggression

Mimicry is a special kind of resemblance that involves co-evolution of colour and morphology, and even behaviour, to enhance the deception. Mimicry among coral reef fishes, once thought to be rare, appears to be a general and widespread phenomenon, with ~100 cases now reported. Many of the known cases of interspecific mimicry in fishes involve one or more species of the family Blenniidae. Mimicry also appears to be particularly important during juvenile stages, with more than 25% of mimic species losing their mimic colouration when they outgrow their models and become less vulnerable to predation.

Most of the cases of interspecific mimicry reported so far can be classified as Batesian, Müllerian or aggressive mimicry. Batesian mimicry is the resemblance of a harmless or palatable species to a harmful or unpalatable one. An example of Batesian mimicry among GBR fishes is that between the noxious *Cathigaster valentini* (Fig. 30.7A) and the triggerfish *Paraleuteres prionurus* (Fig. 30.7B). In Müllerian mimicry, both species possess some undesirable qualities. This type of mimicry appears to be rare among fishes, although it may contribute to the mimetic complexes involving members of the blenniid tribe Nemophini. Aggressive mimicry is the resemblance of a predatory species to a harmless or non-predatory form. Aggressive mimicry is the most prevalent type of mimicry in coral reef fishes, constituting about half of all known cases. An example among GBR fishes is the aggressive mimic blenny *Aspidontus taeniatus,* which closely resembles the colour and behaviour of the cleaner wrasse *Labroides dimidiatus,* and uses this deceit to closely approach and bite pieces from unsuspecting prey fishes.

In some cases, where two or more species of fishes are involved in a mimetic complex, elements of all three types of mimicry may be present. The spatial distribution of mimics also appears to be limited by that of their model species, although some mimic different models or different colour morphs of the same model in different habitats or in different parts of their range. For example, the juvenile coral bream *Scolopsis bilineatus* (Fig. 30.7D) and the blenny *Petroscirtes fallax* (Fig. 30.7E) both mimic the yellow and black-striped poison-fang blenny *Meiacanthus lineatus* (Fig. 30.7C) on the GBR, but in Fiji, where an all yellow morph of *M. lineatus* occurs, it is mimicked by an unusual yellow colour form of the coral bream and also by a yellow form of the aggressive sabretooth blenny *Plagiotremus laudandus,* which preys on the soft tissue of other fishes. These examples suggest a high degree of phenotypic plasticity in mimetic colouration and little genetic differentiation among different mimics of the same species. For further details see the following papers:

Randall JE (2005) A review of mimicry in marine fishes. *Zoological Studies* **44**(3), 299–328.

Moland E, Eagle JV, Jones GP (2005) Ecology and evolution of mimicry in coral reef fishes. *Oceanography and Marine Biology: An Annual Review* **43**, 455–482.

Analysis of the geographical structure of *P. leopardus* populations includes some surprises. Eastern and western Australian populations are distinct, which is a reflection of the long period of closure of the Torres Strait and the sparse reef environment in the present day Arafura Sea. In fact, the closest relatives of the west Australian population appear to be from Taiwan. The GBR populations had their strongest affinities with New Caledonian fish, and the analysis of larval migration patterns strongly suggests an east to west gene flow. This provides a key to the question of the rapid colonisation of the GBR reef over the last 6–7 ky. During the period of low sea level stands, the Coral Sea was characterised by large shallow areas of actively growing reefs – the Queensland and Marion plateaus. These are now inundated and well below the level of active reef growth and are now represented by only scattered groups of Coral Sea reefs and islands. However, during periods of low sea level, driven by glaciation cycles over the last 500 ky, these reefs, and those of New Caledonia further to the east, would have served as recruitment sources when rising sea levels provided the opportunity for recolonisation of the GBR.

One of the most important messages is that the process of colonisation of the newly forming GBR and the partitioning of species into different habitats happened over a very short period. Are there any genetic signatures of these events? Genetic analysis of the relationships among species of *Plectropomus* shows that *P. leopardus, P. maculatus* and

P. laevis are closely related, something that is reflected in their colour patterns. On the west coast of Australia *P. leopardus* and *P. maculatus* form distinct monophyletic groups or clades. However, on the east coast *P. maculatus* has the same mitochondrial genetic signature or haplotype as *P. leopardus*. Using mitochondrial genes, *Plectropomus maculatus* from the GBR cannot be distinguished from *P. leopardus*, which is in striking contrast to the pattern observed in Western Australia. The most parsimonious explanation for these distinctive coastal patterns is that on the east coast the species have a history of hybridisation, with male *P. maculatus* joining spawning groups of *P. leopardus* (Fig. 30.8; Box 30.4). The opportunity for the mixing and overlap of the species populations on the east coast was greatly enhanced by the turbulent history of the reef, with very rapid episodes of colonisation of new reef structures and sorting among the newly formed habitats. In contrast, the coast of Western Australia has had a far more stable history, with limited effects of sea level change and with populations of each species clearly partitioned between coastal and oceanic offshore reefs.

The distinctive genetic structures of *P. leopardus* and *P. maculatus* populations on the GBR, with evidence of past episodes of hybridisation, are in effect a signature of a distinctive geological and evolutionary history of this reef system. It is highly probable that other species will show similar evidence

Box 30.4. Genetic structure of *Plectropomus* populations

The genetic structure of *Plectropomus* on the east and west coasts of Australia (Fig. 30.8). The diagram represents a phylogenetic tree based on molecular sequences from the D-loop region of the mitochondrial (mt) genome of *P. leopardus* and *P. maculatus* collected from the tropical coasts of eastern and western Australia. Sequences from *P. laevis* constitute an outgroup. The colours represent mt sequences characteristic of each species (blue, *P. leopardus;* green, *P. maculatus*).

Plectropomus leopardus sequences grouped as a single clade (A), with populations from the two coasts occurring as two genetically distinct sister clades. *Plectropomus maculatus* populations from the west coast form a distinct clade (B). However, on the east coast *P. leopardus* and *P. maculatus* do not separate into species-specific clades and are genetically indistinguishable with the mitochondrial marker used in this study. The failure of the mitochondrial marker to distinguish the two species is attributable to interspecific hybridisation between these species on the GBR. The structure of the phylogenetic tree is strongly supported in ML and Bayesian analyses. Further details of this study are provided in:
L. van Herwerden *et al.* (2006) Contrasting patterns of genetic structure in two species of the coral trout *Plectropomus* (Serranidae) from east and west Australia: introgressive hybridisation or ancestral polymorphisms. *Molecular Phylogenetics and Evolution* **41**, 420–435.

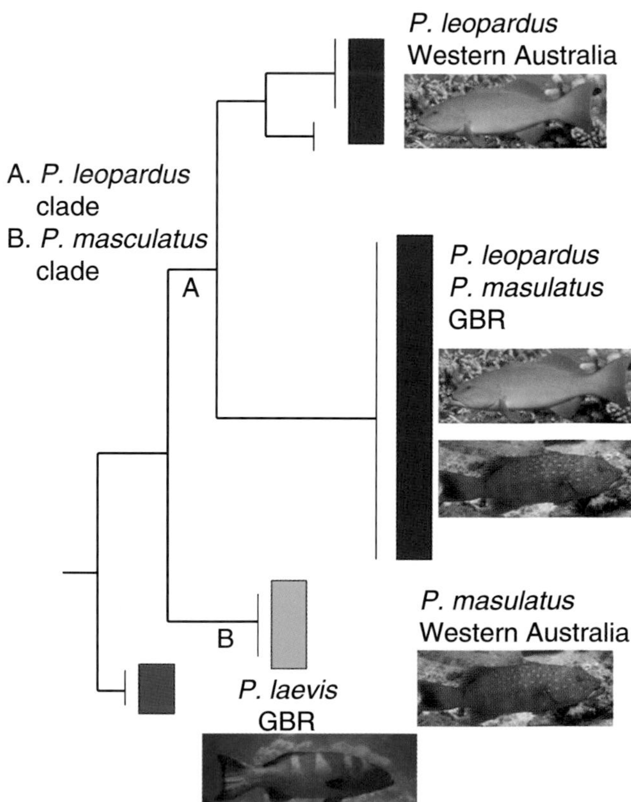

Fig. 30.8. Genetic structure of *Plectropomus* populations (Illustration: L. van Herwerden, JCU).

Box 30.5. Sharks and rays of the Great Barrier Reef World Heritage Area

The chondrichthyan fishes (sharks, rays and chimaeras – henceforth called sharks and rays) of the Great Barrier Reef (GBR) have important social, cultural, economic and ecological values and roles. With 135 species recorded from the World Heritage Area, the GBR has one of the world's highest levels of chondrichthyan diversity and includes 10 skate species, 82 shark species, 39 ray species and four chimaeras. These species have a diverse range of ecological traits and habitat use patterns, ranging from use of shallow estuarine habitats and coral reefs, to pelagic habitats and the deep waters off the continental shelf. Some species such as the large-tooth sawfish (*Pristis pristis*) also use freshwater habitats in coastal rivers and floodplains. Similarly, these species have a wide range of life history traits. The silvertip shark (*Carcharhinus albimarginatus*) grows to 2.75 m and reaches sexual maturity at a late age and large size, making it vulnerable to fishing pressure. In contrast, the Australian sharpnose shark (*Rhizoprionodon taylori*) grows to only 67 cm, matures at 40–45 cm and has high fecundity. With a population doubling time of 2.5 years, this species is very resilient to fishing pressure. This diversity is reflected in the varying conservation status assessments for these GBR sharks and rays. Species such as sawfishes (*Pristis* spp.), speartooth sharks (*Glyphis* spp.) and the grey nurse shark (*Carcharias taurus*) are at high risk of extinction and of high conservation concern, while other species such as the Australian sharpnose shark are abundant and widely distributed.

Sharks and rays are iconic species in the GBR. With many species being large and charismatic, the public have a heightened awareness of and interest in sharks and rays, although this interest also extends to debates regarding balancing swimmer risks and conservation concerns. Sharks and rays are also iconic species for the GBR's largest economic activity – tourism. Sharks and rays are valuable 'living attractions' that tourists want to see. These interests mainly focus on reef sharks such as the grey reef shark (*Carcharhinus amblyrhynchos*), blacktip reef shark (*Carcharhinus melanopterus*) and whitetip reef shark (*Triaenodon obesus*), and in the southern GBR, on mobulid rays. Traditional Owners of GBR sea country also have important cultural connections with sharks and rays, and these animals feature widely in art, stories and Indigenous astronomy (see Fig. 30.9). In some Indigenous communities, stingrays are an important food source and hunting provides a way to continue cultural practices and maintain Indigenous knowledge.

In ecological terms, most GBR sharks and rays are meso-predators or top-predators, and only a very few, such as the tiger shark (*Galeocerdo cuvier*) are apex predators. In these roles, they are purported to perform important ecological roles in applying predation pressure or behavioural modification upon prey species, which maintains ecosystem function and balance. Stingrays may also play important roles in habitat formation and bioturbation. Although ecological theory suggests that sharks play important top-down trophic controls and cascades throughout marine ecosystems, evidence demonstrating these effects is equivocal and some studies have been shown to have serious flaws.

The main pressures facing GBR sharks and rays is fishing (see Chapter 10). Most of the GBR shark and ray catch is taken in the commercial inshore net fishery, which is a multispecies fishery targeting mainly barramundi (*Lates calcarifer*), mackerels (*Scombermorus* spp.) and salmon (Polynemidae), although some fishers specifically target sharks such as the Australian blacktip (*Carcharhinus tilstoni*), common blacktip (*C. limbatus*), spot-tail shark (*C. sorrah*), and spinnershark, (*C. brevipinna*). Current harvest levels appear to be sustainable for many of these species, but there are concerns about the level of harvest of Australian blacktip and hammerheads. The inshore net fishery is complex (see Chapter 10), with a wide range of species taken and a variety of management measures, and the diversity and spatial scale of the fishery make it challenging to monitor, manage and assess. In contrast, the reef line fishery takes a relatively narrow range of sharks. Although the reef line fishery takes a relatively minor component of the total GBR shark catch, landings consist of fewer species: namely the grey reef shark, blacktip reef shark and whitetip reef shark. Research suggests significant declines in reef sharks had occurred before 1989, with shark density and abundance much lower in areas subjected to fishing. Sharks and rays are also taken as by-catch in trawl fisheries, with the majority of by-catch being rays. The risk posed by trawling varies between species, with 11 of 33 assessed shark and ray species categorised as being at 'high risk'. Some sharks are also taken by commercial and Indigenous fishers. Although the limited information available suggests that the impact from these activities is low, more information is needed about the catch and sustainability of sharks and rays in these two fisheries.

Aside from fisheries, sharks and rays can also be impacted by habitat loss and degradation, disturbance, and environmental changes. Some species are heavily reliant upon specific habitats, and loss or degradation of these habitats could have serious impacts. Climate change may also affect some sharks and rays, particularly freshwater and estuarine, and reef-associated species.

Since 2004, a raft of management changes have been introduced that are likely to benefit GBR sharks and rays. The rezoning of the GBR Marine Park substantially increased the area of 'no-take' marine reserves, which is likely to provide some protection for sharks and rays, although the level of protection varies between species. Aside from rezoning, fisheries restructures have significantly reduced fishing levels and new regulations have been introduced (see Chapter 10). Unless a commercial fisher has a specific 'S' (shark and ray fishery) endorsement, a net fisher can possess only 10 sharks and a line fisher can possess only four. Shark finning (where fins are removed and the carcass is discarded) is illegal, and sharks of particular concern must be landed with fins and tail attached. An annual total allowable catch of 480 tonnes has also been applied to sharks and rays in the GBR World Heritage Area. Lastly, several shark and ray species such as sawfishes, the grey nurse shark, white shark (*Carcharodon carcharias*) and the speartooth shark are protected species, and other species such as some reef sharks have reduced catch limits. Although there is some evidence that collectively these measures may be contributing to recovery of some species, the long-term sustainability of sharks and rays in the GBR is not yet assured. These improved fisheries management measures must be properly resourced and implemented, and their effects on sharks and rays evaluated over coming years to ensure existing measures are working. Furthermore, the cumulative effects of impacts from fishing, habitat loss and climate change, are difficult to predict, and fisheries management needs to be supplemented by measures to protect GBR habitats and environmental quality.

Fig. 30.9. Sharks and rays have a multitude of important social and cultural values and in some GBR communities, this significance has persisted through tens of thousands of years (Image: A. Chin).

of a reef system subject to rapid structural and biological change.

The future for reef fish research on the GBR

Given the diversity of the fishes and the ecological complexity of the GBR, there are a multitude of novel research issues to follow up. Much of what we hear about the GBR concerns issues of habitat disturbance and loss, climate change and the consequences of overfishing. These are legitimate issues and the global picture of the status of coral reefs demonstrates that, although the GBR is relatively healthy, the consequences of natural and anthropogenic disturbances can be manifested very quickly. The most effective way to evaluate the potential influences of disturbance and exploitation is through an understanding of the biology of the organisms themselves. Reef animals, and especially the fishes, vary in the way they respond to stress and environmental change. It is difficult to generalise from one group to another and the best approach is to understand the biological mechanisms, both evolutionary and ecological, that underlies the way different groups respond to change.

Three areas of future studies of GBR fishes that will yield the type of information required are discussed in the following sections. This list is not exhaustive or prioritised, but it captures some of the excitement of ongoing reef fish research.

Depth distributions of reef fishes

How deep does the GBR go? Using baited video cameras it is now possible to obtain visual samples of the diversity of reef fishes to a depth of 200 m, below the level of active coral growth. These surveys have revealed that several species that were considered to be shallow water fishes extend to 100 m depth. This work requires a re-evaluation of the habitat association for some common reef species and also identifies species that are able to extend their depth range in response to shallow water disturbances, such as sudden increases in temperature or fishing effects.

Genetic analysis of population structures

Is there a consistent genetic signature associated with the very recent colonisation of the GBR by reef fishes including further examples of hybridisation between related species? Surveys of reef fish genetic structure are also important with respect to the detection of 'cryptic' species. However, this carries with it an important caveat: it is unwise to use a genetic criterion such as differences in mitochondrial sequences to differentiate fish species, as the *Plectropomus* example shows. Two things are essential for the confirmation of cryptic species. At least two lines of evidence such as morphological and/or structural distinctions as well as genetic differences are required. More importantly, it is critical to have the capacity to correctly identify reef fishes in the field. This capacity is being eroded as taxonomic research in our museums winds down. Collecting tissues from reef fishes for such schemes as the Barcode of Life without proper taxonomic quality control is a recipe for disaster.

Multi-scale analysis of life history features

After a slow start, there are several study programs investigating the age-based life history features of coral reef fishes using age information extracted from reef fish otoliths. This is important for several reasons, including the analysis of stock structure. Many species of reef fish show significant differences in growth and mortality rates between localities. This is very conspicuous in GBR fishes. Such studies are valuable when linked to analyses of genetic structure over the same scale because they provide insights into the mechanisms underlying spatial differences in demography. Preliminary work suggests that reef fishes show great plasticity with respect to growth rates, reproductive outputs and age-structure, but it is unclear to what extent this has a genetic basis. When examined on a broader geographic scale, GBR reef fishes show some unique demographic features, including far greater life spans than those in adjacent reef systems although the significance of this is not clear at present.

References and further reading

General

Allen GR (2007) Conservation hotspots of biodiversity and endemism for Indo-Pacific coral reef fishes. *Aquatic Conservation* **18**, 541–556. doi:10.1002/aqc.880

Sale PF (Ed.) (1991) *The Ecology of Fishes on Coral Reefs*. Academic Press, San Diego, CA, USA.

Sale PF (Ed.) (2006) *Coral Reef Fishes. Dynamics and Diversity in a Coral Reef Ecosystem*. Academic Press, San Diego, CA, USA.

Sharks and rays

Chin A, Kyne PM (2007) Vulnerability of chondrichthyan fishes of the Great Barrier Reef to climate change. In *Climate Change and the Great Barrier Reef: A Vulnerability Assessment*. (Eds JE Johnson, PA Marshall) pp. 393–426. Great Barrier Reef Marine Park Authority and Australian Greenhouse Office, Townsville.

Chin A, Heupel MR, Simpfendorfer CA, Tobin AJ (2016a) Population organisation in reef sharks: new variations in coastal habitat use by mobile marine predators. *Marine Ecology Progress Series* **544**, 197–211. doi:10.3354/meps11545

Chin A, Kyne PM, White WT, Hillcoat S (2016b) *Checklist of the Chondrichthyan fishes (sharks, rays and chimaeras) of the Great Barrier Reef World Heritage Area*. Shark Search Indo-Pacific, James Cook University, Townsville.

Espinoza M, Cappo M, Heupel MR, Tobin AJ, Simpfendorfer CA (2014) Quantifying shark distribution patterns and species-habitat associations: implications of marine park zoning. *PLoS One* **9**, e106885. doi:10.1371/journal.pone.0106885

Grubbs RD, Carlson JK, Romine JG, Curtis TH, McElroy WD, McCandless CT, *et al.* (2016) Critical assessment and ramifications of a purported marine trophic cascade. *Scientific Reports* **6**, 20970. doi:10.1038/srep20970

Harry AV, Macbeth WG, Gutteridge AN, Simpfendorfer CA (2011) The life histories of endangered hammerhead sharks (Carcharhiniformes, Sphyrnidae) from the east coast of Australia. *Journal of Fish Biology* **78**, 2026–2051. doi:10.1111/j.1095-8649.2011.02992.x

Harry AV, Saunders RJ, Smart JJ, Yates PM, Simpfendorfer CA, Tobin AJ (2016) Assessment of a data-limited, multi-species shark fishery in the Great Barrier Reef Marine Park and south-east Queensland. *Fisheries Research* **177**, 104–115. doi:10.1016/j.fishres.2015.12.008

Heupel MR, Simpfendorfer CA, Espinoza M, Smoothey AF, Tobin A, Peddemors V (2015) Conservation challenges of sharks with continental scale migrations. *Frontiers in Marine Science* **2**, 12. doi:10.3389/fmars.2015.00012

Heupel MR, Williams AJ, Welch DJ, Ballagh A, Mapstone BD, Carlos G, *et al.* (2009) Effects of fishing on tropical reef associated shark populations on the Great Barrier Reef. *Fisheries Research* **95**, 350–361. doi:10.1016/j.fishres.2008.10.005

Last PR, Stevens JD (2009) *Sharks and Rays of Australia*. CSIRO Publishing, Melbourne.

Lynch AMJ, Sutton SG, Simpfendorfer CA (2010) Implications of recreational fishing for elasmobranch conservation in the Great Barrier Reef Marine Park. *Aquatic Conservation* **20**, 312–318. doi:10.1002/aqc.1056

O'Shea OR, Thums M, van Keulen M, Meekan M (2012) Bioturbation by stingrays at Ningaloo Reef, Western Australia. *Marine and Freshwater Research* **63**, 189–197. doi:10.1071/MF11180

Pears RJ, Morison AK, Jebreen EJ, Dunning MC, Pitcher CR, Courtney AJ, *et al.* (2012) 'Ecological risk assessment of the East Coast Otter Trawl Fishery in the Great Barrier Reef Marine Park: Technical report'. Great Barrier Reef Marine Park Authority, Townsville.

Queensland Primary Industries and Fisheries (2009) Guidelines for commercial operators in the East Coast Inshore Finfish Fishery, PR09_4406. Department of Employment, Economic Development and Innovation, Brisbane.

Rizzari JR, Frisch AJ, Hoey AS, McCormick MI (2014) Not worth the risk: apex predators suppress herbivory on coral reefs. *Oikos* **123**, 829–836. doi:10.1111/oik.01318

Robbins WD, Hisano M, Connolly SR, Choat JH (2006) Ongoing collapse of coral-reef shark populations. *Current Biology* **16**, 2314–2319. doi:10.1016/j.cub.2006.09.044

Roff G, Doropoulos C, Rogers A, Bozec Y-M, Krueck NC, Aurellado E, *et al.* (2016) The ecological role of sharks on coral reefs. *Trends in Ecology & Evolution* **31**, 395–407. doi:10.1016/j.tree.2016.02.014

Simpfendorfer CA (1999) Mortality estimates and demographic analysis for the Australian sharpnose shark, *Rhizoprionodon taylori*, from northern Australia. *Fishery Bulletin* **97**, 978–986.

Smart JJ, Chin A, Baje L, Tobin AJ, Simpfendorfer CA, White WT (2017) Life history of the silvertip shark *Carcharhinus albimarginatus* from Papua New Guinea. *Coral Reefs* **36**, 577–588.

Stevens JD, Bonfil R, Dulvy NK, Walker PA (2000) The effects of fishing on sharks, rays, and chimaeras (chondrichthyans), and the implications for marine ecosystems. *ICES Journal of Marine Science* **57**, 476–494. doi:10.1006/jmsc.2000.0724

Stoeckl N, Birtles A, Farr M, Mangott A, Curnock M, Valentine P (2010) Live-aboard dive boats in the Great Barrier Reef: regional economic impact and the relative values of their target marine species. *Tourism Economics* **16**, 995–1018. doi:10.5367/te.2010.0005

Whatmough S, Van Putten I, Chin A (2011) From hunters to nature observers: a record of 53 years of diver attitudes towards sharks and rays and marine protected areas. *Marine and Freshwater Research* **62**, 755–763. doi:10.1071/MF10142

31

Marine mammals and reptiles

H. Marsh, H. Heatwole and V. Lukoschek

Introduction

The Great Barrier Reef World Heritage Area supports a diverse marine mammal and reptile fauna. Our knowledge of their distributions suggests that more than 30 species of marine mammals spend at least part of their lives in the region (Fig. 31.1). Most of these species are members of the order Cetacea (whales and dolphins). The region also supports globally significant populations of one member of the order Sirenia (sea cows): the dugong, *Dugong dugon*. Both cetaceans and sirenians spend their entire lives in the water. There is one report of a vagrant leopard seal (*Hydrurga leptonyx*) from Heron Island.

Six species of sea turtles occur in the waters of the Great Barrier Reef (GBR). Five of these belong to the family Cheloniidae: green turtle (*Chelonia mydas*) (Fig. 31.2A); flatback (*Natator depressus*); olive ridley (*Lepidochelys olivacea*); hawksbill (*Eretmochelys imbricata*) (Fig. 31.2B); and loggerhead (*Caretta caretta*) (Fig. 31.2C). The remaining species, the leatherback or luth (*Dermochelys coriacea*), is a member of the family Dermochelidae. Sea turtles come out of the water to lay their eggs. Australian waters also support 32 species of true sea snakes, of which 13 species maintain permanent breeding populations in the GBR region. Like marine mammals, sea sakes spend all their lives in the water.

Land mammals and reptiles are known from the islands and cays of the GBR but are beyond the scope of this chapter.

Marine mammals

There are two major groups (suborders) of cetaceans, the Mysticeti or baleen whales and the Odontoceti or toothed whales (Fig. 31.1). Mysticete whales lack teeth and have baleen plates hanging from roof of their mouths to catch krill or small fish. Toothed whales generally capture their prey, typically fish or squid, one at a time. Their most usual dentition is a large number (up to 100) of peg-like teeth, all similar in form in each of the upper and lower jaws.

Each cetacean suborder contains several families, which in turn contain one or more genera. Two baleen whales – the humpback (*Megaptera novaengliae*) and the dwarf minke, (*Balaenoptera acutorostrata*) – are commonly seen in the GBR region during the winter and are important tourist attractions. The coastal odontocetes include the Australian humpback dolphin (*Sousa sahulensis*) (Fig. 31.1) and Australian snubfin dolphin (*Orcaella heinsohni*) (Fig. 31.1) (the only cetaceans endemic to northern Australia and Papua New Guinea waters).

Other cetacean species such as killer whales and common dolphins are known to occur in the GBR region but are rarely reported. The rare Omura's whale was sighted on the GBR in 2016, the first confirmed sighting of this elusive whale in Australian waters. Other species have never been seen alive in the region but are known from strandings on the

Dolphins & Small Whales of Northern Australia

IDENTIFICATION GUIDE

SMALL WHALES & LARGE DOLPHINS (OCEANIC)
From 5 to 10 metres (16-30 feet) long

RELATIVELY SMALL FIN SET FAR BACK ON BODY

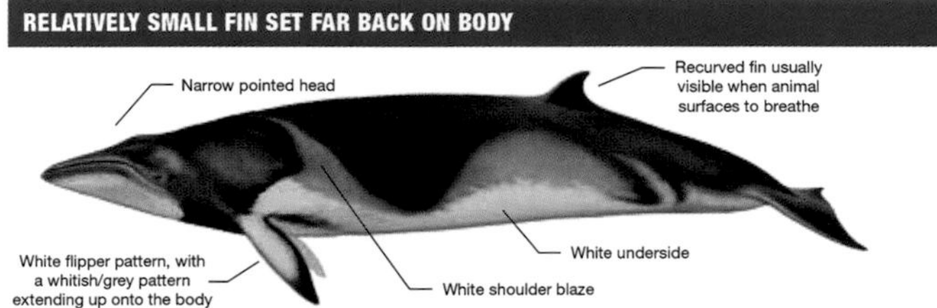

DWARF MINKE WHALE *Balaenoptera acutorostrata*

Size: up to 7.8m | Baleen Plates: 231-285 pairs of short, white/cream colored plates
Small, sleek, dark body. Usually seen singly or in small groups. Dwarf minke whale tends to be inquisitive and circle boats. The Antarctic minke whale, *Balaenoptera bonaerensis* may be sighted from May-August.

CONSPICUOUS BLACK AND WHITE MARKINGS

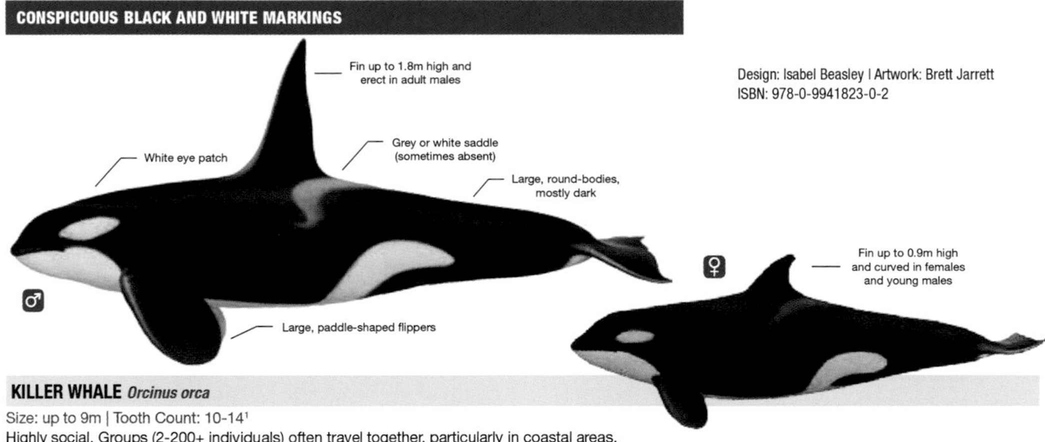

Design: Isabel Beasley | Artwork: Brett Jarrett
ISBN: 978-0-9941823-0-2

KILLER WHALE *Orcinus orca*

Size: up to 9m | Tooth Count: 10-14[1]
Highly social. Groups (2-200+ individuals) often travel together, particularly in coastal areas.
Fast swimmers with unpredictable movements and behaviour, sometimes inquisitive and approachable.

RECURVED FIN SET MID-BACK ON BODY

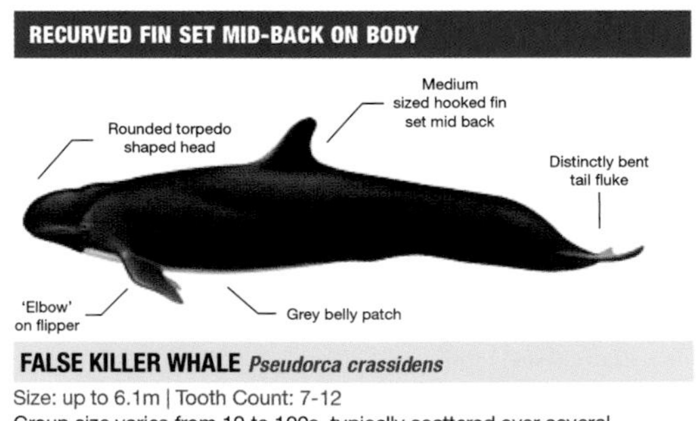

FALSE KILLER WHALE *Pseudorca crassidens*

Size: up to 6.1m | Tooth Count: 7-12
Group size varies from 10 to 100s, typically scattered over several kilometres. Fast-moving and energetic, very acrobatic and frequently bow-ride.

FIN CLOSER TO HEAD THAN TAIL

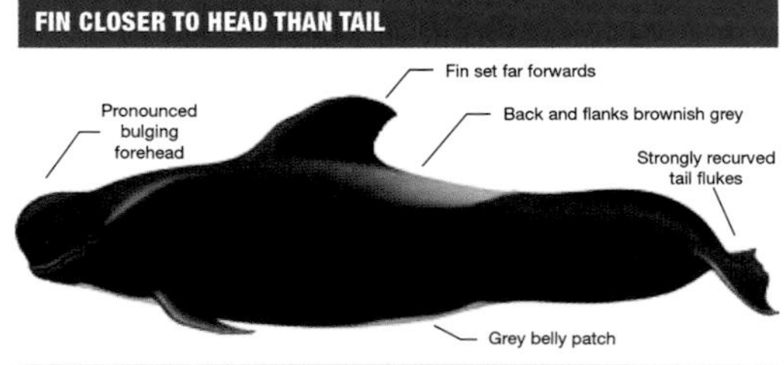

SHORT-FINNED PILOT WHALE *Globicephala macrorhynchus*

Size: up to 7.2m | Tooth Count: 7-9
Pilot whales form large groups. There are two species, not easily distinguishable at sea. The short-finned pilot whale, *Globicephala macrorhynchus* is most likely to be sighted around Northern Australia.

[1] Tooth counts are per tooth row, with typically 4 tooth rows. Some species have no upper tooth rows. Tooth counts based on *Jefferson, T.A.J., Webber, M. & Pitman, R. 2007. Marine Mammals of the World: A Comprehensive Guide to their Identification.*

Fig. 31.1. Identification guide to the medium-sized and smaller marine mammals most likely to the seen in the Great Barrier Reef region. Not all animals from the region are illustrated (Design: Isabel Beasley; artwork: Brett Jarrett).

BEAKED DOLPHINS
Less than 5 metres (16 feet) long

MEDIUM BEAK

COASTAL: DISTINCT MELON ON FOREHEAD, LARGE FIN

High, hooked fin
Spinal blaze
Curved mouthline
Subtle spotting on belly

INDO-PACIFIC BOTTLENOSE DOLPHIN *Tursiops aduncus*

Size: up to 2.6m | Tooth Count: 21-29

Generally playful, approachable and inquisitive. Occasionally bow-ride, although groups can also be shy and less playful. Distinguished from other inshore dolphins by their high, curved dorsal fin.

LONG BEAK

COASTAL: LOW, WIDE DORSAL FIN

Dark cape
Low dorsal fin with whitish/pink patches on top of dorsal fin (older adults)
Long beak with whitish/pink patches (older adults)
Thick tail stock

AUSTRALIAN HUMPBACK DOLPHIN *Sousa sahulensis*

Size: Up to 2.7m | Tooth Count: 31-35

Slow moving coastal species that rarely bow-ride or leap out of water. Usually seen in small groups of up to six animals. Mottled light to dark grey in colour, with patches of whitish/pink on dorsal fin and head in older adults.

OCEANIC: LARGE FIN & LARGE GROUP SIZE

COMMON BOTTLENOSE DOLPHIN *Tursiops truncatus*

Size: up to 3.8m | Tooth Count: 18-27

Large, strong swimmer, that spends a lot of time at the surface. Group size from 10 to 100s. Very playful, frequently bow-ride, with numerous associated acrobatics.

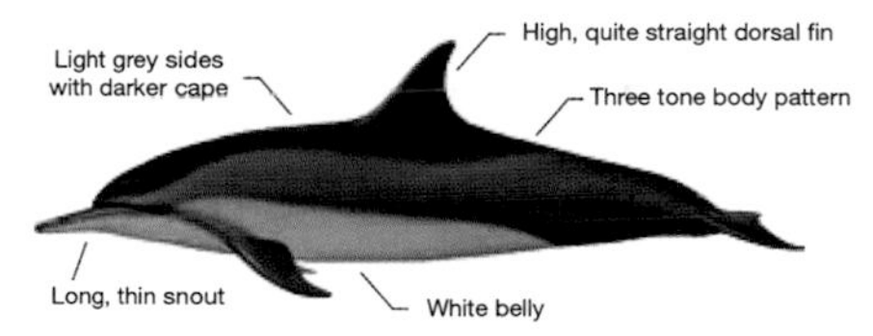

SPINNER DOLPHIN *Stenella longirostris*

Size: up to 2m | Tooth Count: 40-62

Some groups rest in shallow coastal waters and oceanic islands during the day, however most groups primarily oceanic. Very small, fast, highly acrobatic dolphin which can 'spin' when leaping.

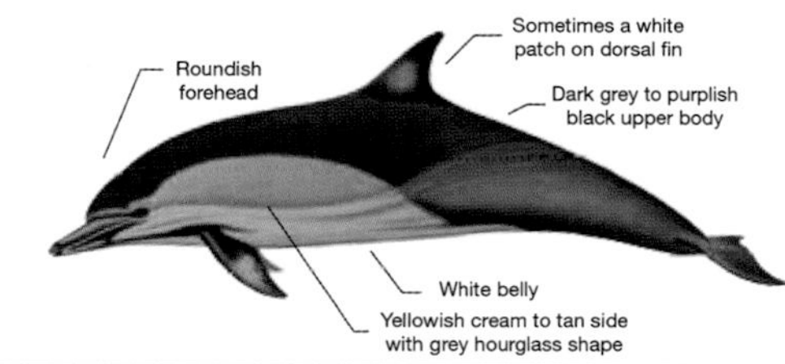

SHORT-BEAKED COMMON DOLPHIN *Delphinus delphis*

Size: up to 2.4m | Tooth Count: 41-57

Distinct 'hourglass' colour pattern on sides. Usually seen in open ocean, often in large schools.

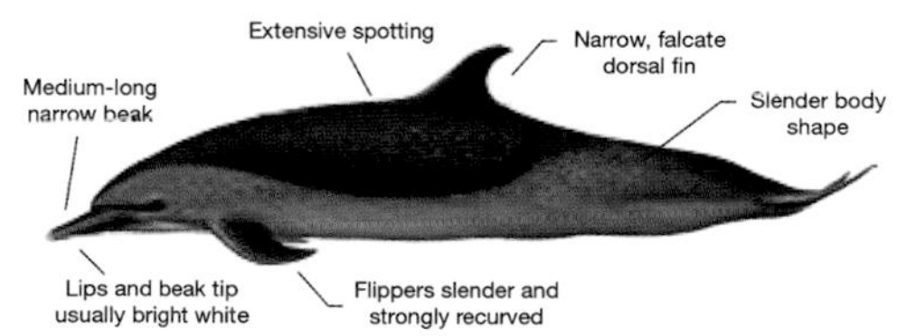

PANTROPICAL SPOTTED DOLPHIN *Stenella attenuata*

Size: up to 2.5m | Tooth Count: 34-48

Extensive spotting over body is visible when close to individuals, young animals unspotted (some populations little or no spotting visible). Fast and acrobatic, frequently bow rides. Lips and beak tip usually white.

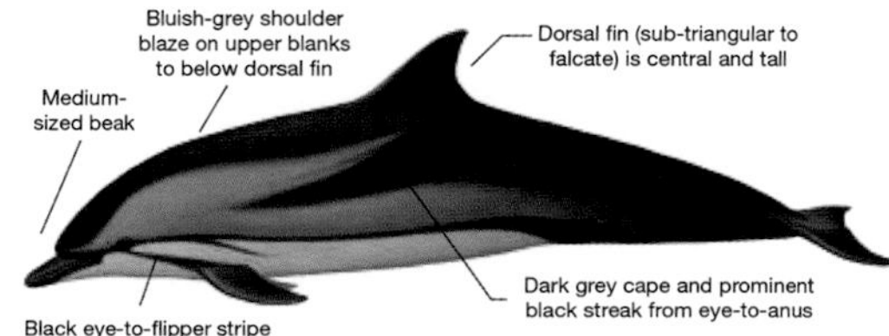

STRIPED DOLPHIN *Stenella coeruleoalba*

Size: up to 2.7m | Tooth Count: 40-55

Fast, energetic swimmer, bow-rides in some areas. Black eye-to-flipper stripe and shoulder blaze the main distinguishing features.

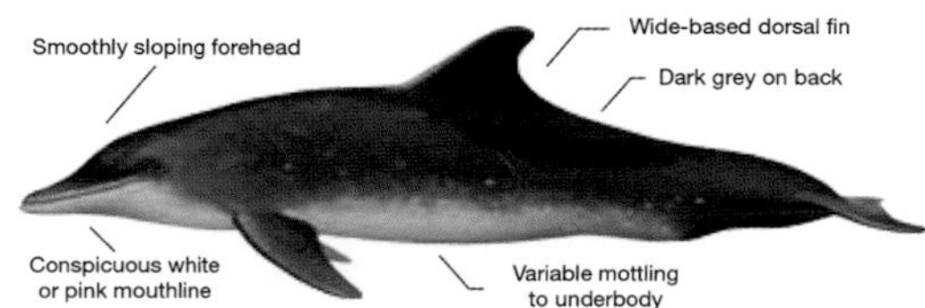

ROUGH-TOOTHED DOLPHIN *Steno bredanensis*

Size: up to 2.7m | Tooth Count: 19-28

Usually sluggish and does not porpoise out of the water when traveling. Often swims shoulder-to-shoulder with others in group, with beak and tip of chin out of water. Occasionally leaps out of water. Sloping forehead the main distinguishing feature.

Fig. 31.1. (Continued)

BLUNT HEADED DOLPHINS & DUGONG
Less than 5 metres (16 feet) long

LARGE DORSAL FIN

OCEANIC: LARGE BODY SIZE

RISSO'S DOLPHIN ***Grampus griseus***

Size: up to 4m | Tooth Count: 2-7 (lower), 0 (upper)
Often active at the surface, forming lines when hunting. Group size of 5-50+. Often extensive scarring over body.

OCEANIC: SMALL BODY SIZE

MELON-HEADED WHALE ***Peponocephala electra***

Size: up to 2.8m | Tooth Count: 20-25
Fast swimmer, porpoising regularly in low leaps with much spray, highly gregarious. Dark cape dips down below dorsal fin. Flippers have pointed tips. White chin/lips.

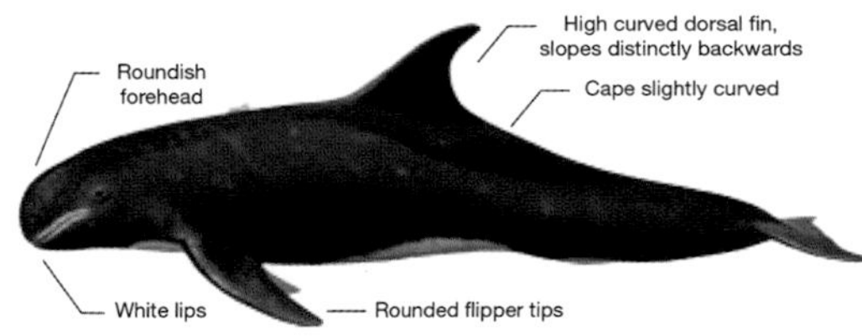

PYGMY KILLER WHALE ***Feresa attenuata***

Size: up to 2.8m | Tooth Count: 8-13
Commonly sighted in small groups of 6-10 individuals, which roll synchronously shoulder-to-shoulder. Often a slow sluggish swimmer. Dark cape sweeps down slightly near dorsal fin. Flippers have rounded tips. White chin/lips.

OCEANIC: SMALL BODY SIZE & VERY SHORT BEAK

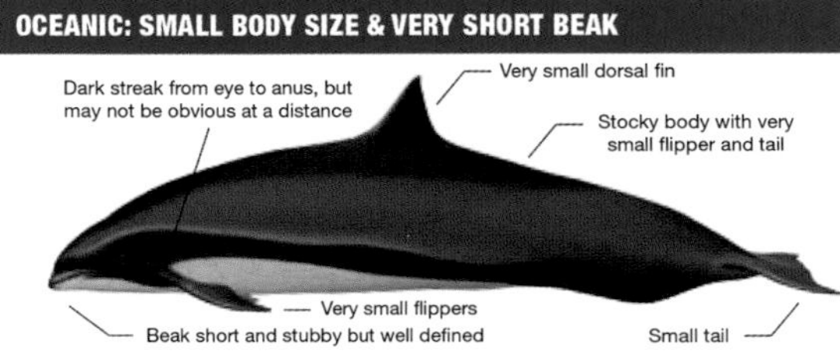

FRASER'S DOLPHIN ***Lagenodelphis hosei***

Size: up to 2.7m | Tooth Count: 38-44
Usually seen in open ocean off the continental shelf. Fast, aggressive swimmers with a lot of splashing. Commonly in large groups of 100s to even 1000s.

SMALL DORSAL FIN

COASTAL: ROUND HEAD & SHY

AUSTRALIAN SNUBFIN DOLPHIN ***Orcaella heinsohni***

Size: up to 2.3m | Tooth Count: 11-22
Occurs singly or in small groups, sometimes up to 20 individuals. Commonly sighted near mangrove, river habitats. Shy and does not bow-ride, keeps low on the surface.

OCEANIC: LOGGING AT SURFACE

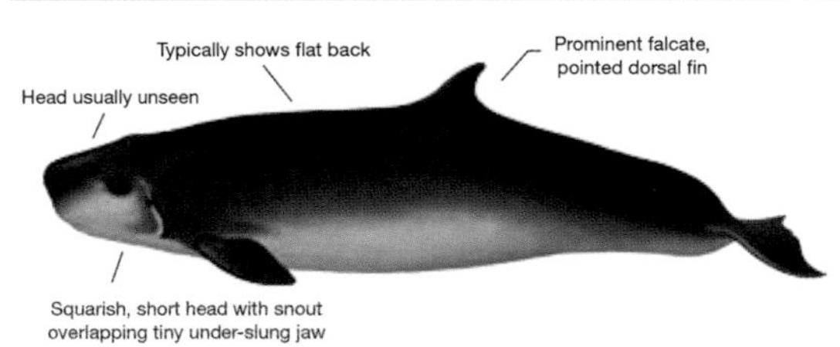

DWARF SPERM WHALE ***Kogia sima***

Size: up to 2.7m | Tooth Count: 7-13 (lower), 3 (upper)
Smaller size with tall, pointed dorsal fin distinguishes the dwarf from pygmy sperm whale. Inconspicuous surfacing similar to pygmy sperm whale.

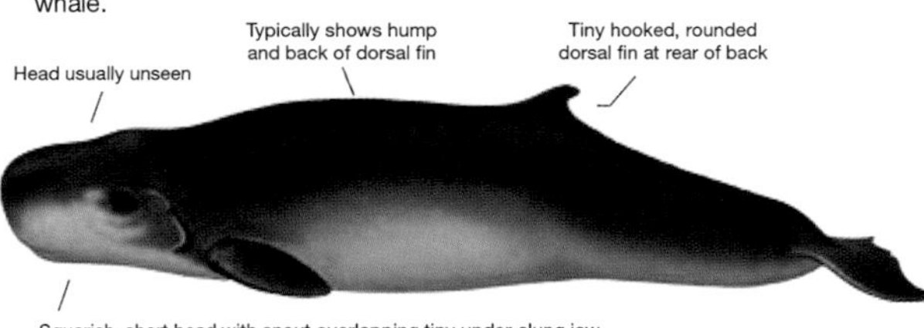

PYGMY SPERM WHALE ***Kogia breviceps***

Size: up to 3.5m | Tooth Count: 10-16 (lower), 0 (upper)
Difficult to detect unless sea conditions calm. Typically rises to surface and floats motionless with blowhole and back out of water, then sinks below surface without rolling.

DUGONG ***Dugong dugon***

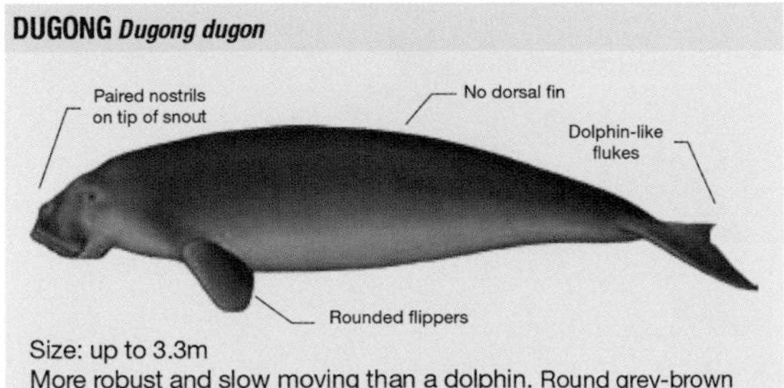

Size: up to 3.3m
More robust and slow moving than a dolphin. Round grey-brown body, usually seen singly or in small aggregations, although large groups of 100's can sometimes be seen.

Fig. 31.1. (Continued)

Coastal species identification

AUSTRALIAN SNUBFIN DOLPHIN
Orcaella heinsohni

Fin: Small, triangular, rounded tip
Colour: Brownish to grey
Face: Blunt, no snout
Length: Up to 2.3m

Habitat & Behaviour:
Endemic to Australia, occur in coastal waters within 10km of coastline and less than 15m depth, very site attached, usually found within 20km of estuary mouths. Feed and socialise in just 1-2m of water in estuaries (which are their critical habitat). Often found in groups of 5-6 dolphins. May spit water as a method of herding fish.

AUSTRALIAN HUMPBACK DOLPHIN
Sousa sahulensis

Fin: Low & wide, triangular, pointed tip
Colour: Pale grey, adults have pink patches on snout & dorsal fin
Face: Long, pointed snout
Length: Up to 2.7m

Habitat & Behaviour:
Usually seen in coastal waters less than 15m deep (but up to 30m), feed in or near estuaries, in water depths of 2-5m (slightly deeper than snubfins). Usually found in small groups of 2-5 animals, but groups of up to 10-30 animals have also been reported. Known to follow trawlers as a method of hunting fish.

INDO-PACIFIC BOTTLENOSE DOLPHIN
Tursiops aduncus

Fin: Tall & curved
Colour: Light to dark grey
Face: Distinct forehead, medium length snout
Length: Up to 2.6m

Habitat & Behaviour:
Indo-Pacific bottlenose dolphins can be found inshore in shallow coastal waters including estuaries, mangroves, beaches and bays. Very approachable, occasionally bow-ride and exhibit aerial activity. Found in groups of 1-10, but can occur in groups of 20s in some areas.

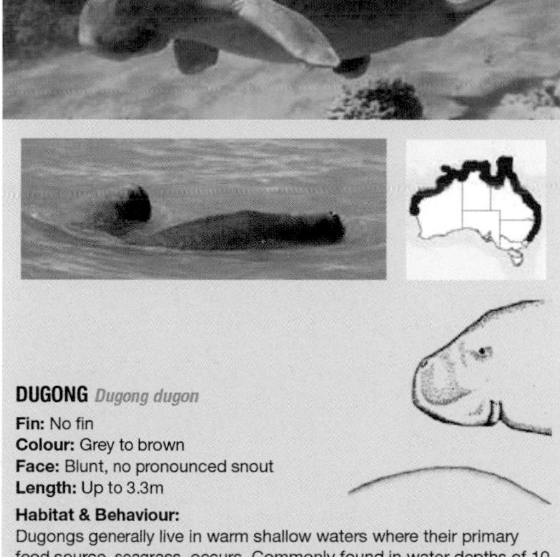

DUGONG *Dugong dugon*

Fin: No fin
Colour: Grey to brown
Face: Blunt, no pronounced snout
Length: Up to 3.3m

Habitat & Behaviour:
Dugongs generally live in warm shallow waters where their primary food source, seagrass, occurs. Commonly found in water depths of 10 metres or less. Seagrass beds form part of their critical habitat. Known to travel distances of up to 600km. Usually shy and avoid boats.

Drawings - Owen Li - Petrified Pencils

Fig. 31.1. (Continued)

adjacent Queensland coast. One species, Longman's beaked whale, (*Indopacetus pacificus*) was considered to be the world's rarest cetacean until recently. A skull found in Mackay, Queensland, was the basis for the initial description of this species.

The populations of great whales that occur in the GBR region were severely depleted by commercial whaling operations in the 20th century. The decline is best documented for humpback whales but presumably also affected the GBR stocks of other baleen whales. The population of humpback

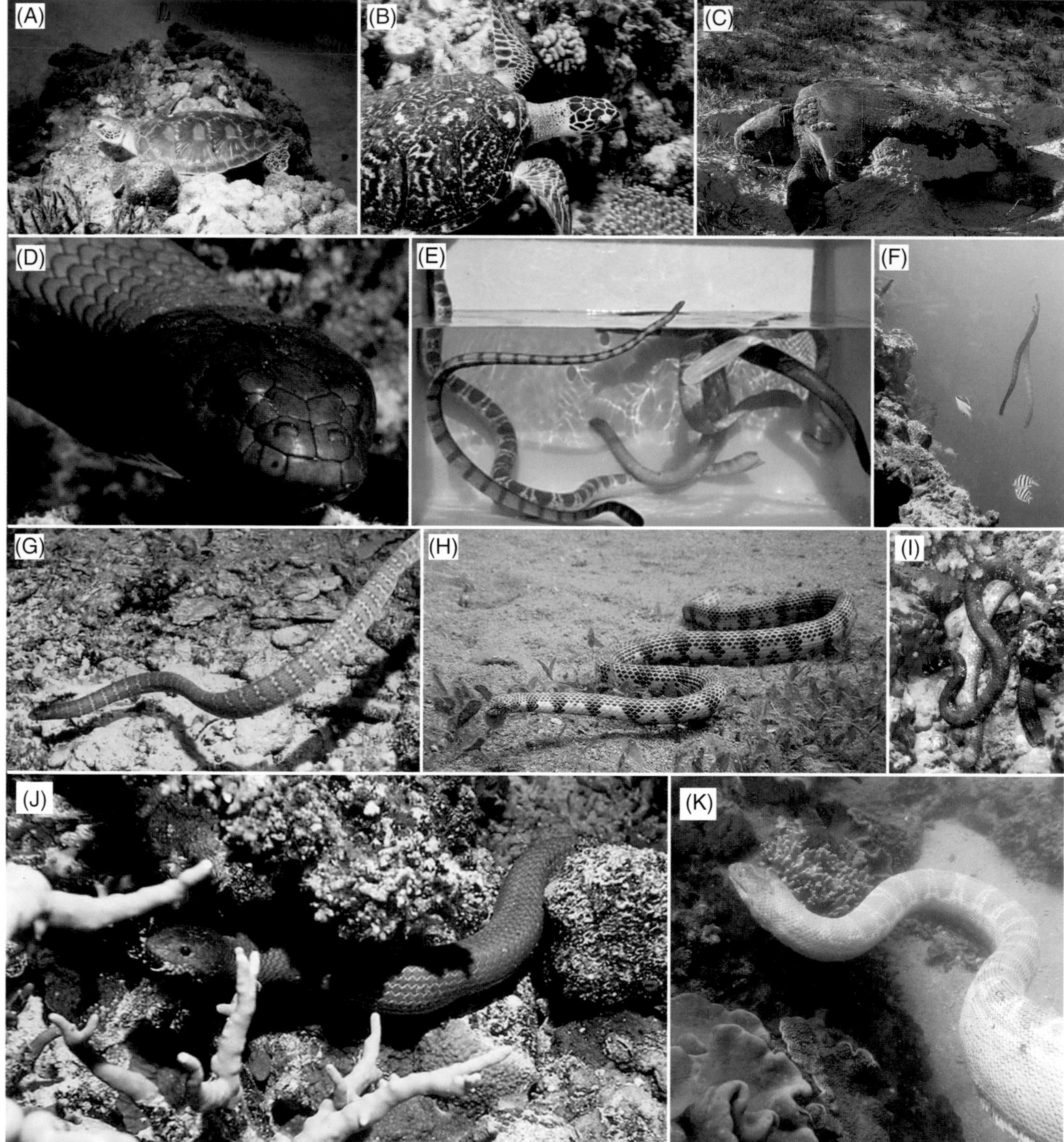

whales was reduced to a few hundred individuals in the 1960s but has increased rapidly, especially since the cessation of illegal Russian whaling. The most recent population estimate (for 2010) was more than 14 000 animals with a long-term rate of increase of between 10% and 11% annually. Thus the current (2018) population is likely to be much larger than the 2010 estimate.

Whales are frequently encountered by visitors to the GBR, especially between May and September. Humpback whales are sighted throughout the region; minke whales are mainly seen from Cairns north to Lizard Island. The outer shelf region from Ribbon Reef 10 (near Lizard Island) south along the Ribbon Reefs to Agincourt Reef is the main focus for minke whale tourism, which mainly occurs in June and July. People are allowed to swim with dwarf minke whales from permitted vessels, but only if the whales initiate the encounter. Regulations govern people's behaviour during such encounters.

The dugong (Fig. 31.1) is of great cultural and dietary value to Indigenous Australians living adjacent to the GBR region. The importance of the populations of dugongs occurring in the GBR was part of the rationale for the region's World Heritage listing. Dugongs are legally harvested by traditional owners as a Native Title Right. Several initiatives assist Indigenous hunters (Traditional Owners) to manage their harvests within sustainable limits.

The most serious contemporary threats to marine mammals in the GBR are extreme weather events and incidental capture in fishing gear. An unknown number of inshore dolphins and dugongs are killed each years in commercial gill-nets. The number killed has presumably been reduced by the declaration of Dugong Protection Areas in 1997 and the rezoning of the marine park in 2003, both of which reduced gill-netting activities in key dugong habitats.

Sea turtles

Each of the species of marine turtle that occurs in the GBR has its own feeding and nesting grounds, but there is broad overlap among species. In the non-breeding season, individuals of all species disperse widely to feeding grounds within and well beyond the region. Turtles nesting on or near the GBR may go as far as Indonesia, Papua New Guinea (PNG), the Solomon Is, Vanuatu, New Caledonia and Fiji. Only the flatback is considered endemic to Australia because it nests only on the Australian mainland, but some individual flatbacks migrate to Indonesian and PNG waters to feed.

Green turtles (Fig. 31.2A) nest on the islands of the Capricorn-Bunker and Swains Reefs and on the

Fig. 31.2. **(A)** Green turtle (*Chelonia mydas*) (photo: GBRMPA); **(B)** hawksbill turtle (*Eretmochelys imbricata*) (Photo: GBRMPA); **(C)** loggerhead sea turtle (*Caretta caretta*) digging a nest pit on Heron Island, Great Barrier Reef (Photo: H. Heatwole); **(D)** head of olive sea snake (*Aipysurus laevis)* showing the two semicircular nostril valves (closed) (Photo: V. Lukoschek); **(E)** various sea snake species in an aquarium illustrating the differences in size, shape and colour pattern within the group. On the left with saddle-blotches is *Hydrophis stokesii* and the banded sea snake is a different species of *Hydrophis* (note that the head is small and protruding above the water and the paddle-shaped tail is large and submerged). The three sea snakes on the right are *Aipysurus laevis* (note their marked differences in colour). Colour patterns vary greatly within many species, especially from one locality to another (Photo: H. Heatwole.); **(F)** two turtleheaded sea snakes, *Emydocephalus annulatus*, courting. The male is the smaller (dark) snake pursing the larger female. There is enormous colour pattern variation in both sexes of this species, ranging from entirely black (melanistic) to very pale salmon (Photo: M. Beger); **(G)** juvenile olive sea snake, *Aipysurus laevis*, showing strongly banded pattern of this species in its first year of life (Photo: V. Lukoschek); **(H)** *Aipysurus duboisii* swimming over a seagrass bed in sandy habitat adjacent to a coral reef (Photo: V. Lukoschek); **(I)** two turtleheaded sea snakes, *Emydocephalus annulatus*, courting. The male is the smaller snake on top of the larger female. (Photo: M. Carvolth); **(J)** adult olive sea snake, *Aipysurus laevis*, showing typical colour pattern (olive head and uniform grey to brown body) found on the Great Barrier Reef (Photo: V. Lukoschek); **(K)** *Hydrophis stokesii*, the bulkiest of all sea snake species, may reach 2 m in length and has a large head and an exceptionally wide gape. Its fangs are long and it is one of the few species capable of penetrating a wet suit (Photo: Australian Insistitute of Marine Science).

cays of the outer barrier north of Princess Charlotte Bay. Loggerheads (Fig. 31.2C) nest on the cays of the Swain Reefs and Capricorn-Bunker islands as well as on beaches such as Mon Repos on the adjacent Queensland coast. Hawksbills (Fig. 31.2B) lay their eggs mostly on the cays of the inner shelf north of Princess Charlotte Bay. Flatbacks nest on the inshore continental islands and mainland beaches, mostly between Bundaberg and Ayr. The olive ridley does not nest in the GBR region and no leatherback turtle nesting has been seen in the GBR since 1996. All species also live and nest well beyond the GBR, and indeed all except the flatback are distributed worldwide in subtropical and tropical waters. The leatherback also ranges into cooler waters.

Female turtles return repeatedly to lay their eggs on beaches in the region from which they hatched. In the GBR region, sea turtles aggregate and begin to court near nesting beaches in spring (September to November) and females come ashore and dig pits in the sand in which they lay their eggs between October and March. Eggs incubate in the sand and the temperature of the nest determines the sex of the hatchlings: cooler nests produce males and warmer nests produce females. Hatchlings emerge in summer and autumn (January to May) and go to sea where they move with currents and feed on macroscopic plankton. After 5–10 years, they move into their traditional feeding grounds and adopt the adult diet.

Adult green turtles feed mainly on algae and seagrass on coral reefs and inshore seagrass flats. The loggerhead eats mainly molluscs and crabs in sandy lagoons on the reef or in inshore bays and estuaries. The olive ridley consumes small crabs in inter-reefal areas. Hawksbills eat algae, seagrasses, sponges, ascidians, bryozoans and molluscs, mainly on coral reefs. The flatback feeds on soft corals and other soft-bodied animals on soft bottoms in inter-reefal areas. The leatherback eats jellyfish and salps, mainly in deeper water away from reefs.

Several threatening processes affect marine turtles at all stages of their life cycles and in different parts of their geographic ranges. Because of this, marine turtle conservation requires a multifaceted approach and international cooperation. Throughout the world, key threatening processes include climate change, the loss of nesting habitat on key nesting beaches, artificial lights at nesting beaches that attract hatchlings towards the light and away from the ocean, direct harvest of eggs for human consumption, increased predation on nests by feral animals (such as wild pigs), direct harvest of adults for consumption of meat and for 'tortoiseshell', tourism, boat strike, ingestion and/or entanglement in marine debris, and incidental capture in fisheries gear.

In the GBR region (as with many other places), numerous conservation initiatives are in place, which aim to reduce or eliminate these impacts. These initiatives include the protection of key nesting beaches and important inter-nesting and foraging habitats, assisting Indigenous hunters (Traditional Owners) to manage their harvests within sustainable limits, increasing awareness of the need for boats to 'go slow in key turtle habitats, increasing awareness of the impacts of pollution litter in the marine environment, and a legal requirement that turtle exclusion devices (TEDs) be used by trawl fishers at all times. Ecotourism ventures in Queensland allow visitors the opportunity to observe females dig their nests and lay eggs (Fig. 31.2C) and to participate in the release of hatchlings: these activities aim to educate people about the fascinating lives of marine turtles.

Raine Island, near the remote northern tip of the GBR is the breeding ground for one of the world's largest populations of green turtles. During the nesting season, up to 60 000 female green turtles swim thousands of kilometres from their feeding grounds to this tiny island to lay their eggs. Despite the island's remoteness and high levels of protection, hatchling production from this rookery has been low for many years. The Raine Island Recovery Project is restoring nesting habitat on the island and reducing the mortality of nesting turtles.

Sea snakes

True sea snakes are a diverse group of live-bearing predatory marine reptiles that arose less than 7 million years ago (Mya). True sea snakes evolved within the live-bearing group of Australia's terrestrial venomous snakes, and the two groups comprise the subfamily Hydrophiinae, which together with Asia's terrestrial venomous subfamily Elapinae comprise the family Elapidae, defined by having fixed front fangs. Hydrophiine sea snakes differ from their terrestrial relatives by numerous marine adaptations, including smaller belly scales, a propulsive paddle-shaped tail, nostril valves that close from the inside to prevent water entering the lungs while submerged (Fig. 31.2D) and a salt-excreting gland located beneath the tongue sheath.

As reptiles, sea snakes need to surface at regular intervals to breathe air, but they are extraordinary divers able to descend to depths >100 m and remain submerged for 2 h or more, during which time they supplement their oxygen and carbon dioxide exchange by cutaneous respiration. The other group of venomous marine snakes (elapids) is the amphibious egg-laying sea kraits, comprising eight species in the genus *Laticauda,* that use terrestrial environments to oviposit, rest, digest and mate but forage at sea. Although sea kraits are abundant in New Guinea and many south Pacific Islands, they have no breeding populations on the GBR or elsewhere in Australia, but they sometimes appear in Australian waters as waifs.

Hydrophiine sea snakes essentially comprise two evolutionary lineages. The GBR has relatively low species richness of the *Aipysurus* lineage, with only four widespread species, *A. laevis, A. duboisii* and *A. mosaicus* and *Emydocephalus annulatus,* occurring in its waters (Table 31.1, Fig. 31.2 E–J). Nine members of the *Hydrophis* lineage occur in the GBR including six of the species endemic to Australasia (Table 31.1). No sea snake species is endemic to the GBR. Within the GBR, there tends to be a progressive decrease in species richness from north to south. Nonetheless, some southern reefs, particularly in the Swains and Keppel Island regions, are known to have high sea snake abundance, particularly of the reef-associated species, *A. laevis* and *E. annulatus.*

Aipysurus group species are typically associated with coral (and rocky) reefs and species' distributions tend to mirror the patchy distributions of these habitats. By contrast, the *Hydrophis* group species typically occur in inter-reefal areas with sandy or muddy bottoms, or in estuarine habitats. However, there is considerable overlap in the habitats used by species from each group. For example, *A. laevis* and *A. duboisii* are commonly found on coral reefs but also occur in inter-reefal habitats on the GBR and *A. mosaicus* is predominantly an inter-reefal species. By contrast, *E. annulatus* from the *Aipysurus* group is exclusively associated with coral reefs on the GBR. Only a few *Hydrophis* group species are occasionally found on coral reefs, most notably *H. stokesii,* whereas the remaining species are found exclusively in inter-reefal or estuarine habitats. The one exception is *H. platurus,* a pelagic species that lives at the surface, predominantly associated with slicks.

The diets of sea snake species range from generalist to extremely specialised. *Aipysurus eydouxii* and *E. annulatus* have the most specialised diets and eat only fish eggs, while *H. peronii* feeds principally on goby fish, and several *Hydrophis* species are eel-specialists: *H. elegans* is such a species. Other species, such as *A. duboisii, A. laevis, H. curtus, H. major* and *H. platurus,* are more general in their diet and eat fish from a variety of families (some even take an occasional invertebrate). The diets of the other sea snake species from the GBR are poorly known (Table 31.1).

Most sea snake species feed on the bottom. They poke their heads into burrows and crevices and flick out the tongue to capture odours whereby they identify their prey. Most species do not attempt to catch fish in open water. The one exception is *Hydrophis platurus,* which catches its prey while floating on the surface. It slowly bends its

Table 31.1. Sea snake fauna of the Great Barrier Reef and Coral Sea and the habitats where they most commonly occur

Species	Habitat	Diet	Water depth
***Aipysurus* Group**			
Aipysurus duboisii[†]	Coral reefs, over seagrass or sandy habitats	Generalist	Variable
Aipysurus mosaicus[†#]	Muddy and sandy habitats; rivers and estuaries	Fish eggs	Shallow
Aipysurus laevis[†]	Coral reefs, rocky coasts and soft sediments	Generalist	Variable
Emydocephalus annulatus	Coral reefs	Fish eggs	Shallow
Hydrophis* Group			
Hydrophis curtus	Muddy, sandy and seagrass habitats in turbid waters	Generalist	Variable
Hydrophis elegans[†]	Sandy, muddy and seagrass habitats; estuaries	Eels	Shallow
Hydrophis kingii[†]	Inter-reefal	Unknown	Deep
Hydrophis major[†]	Inter-reefal over sandy and muddy habitats	Generalist	Variable
Hydrophis mcdowelli[†]	Inter-reefal over sandy and muddy habitats	Eels	Variable
Hydrophis ocellatus[†]	Inter-reefal over sandy and muddy habitats	Unknown	Deep
Hydrophis peronei	Inter-reefal and coral reefs	Eels	Deep
Hydrophis platurus	Pelagic	Fish	Surface
Hydrophis stokesii	Coral reefs, sandy and muddy habitats; estuaries	Generalist	Variable
Hydrophis zweifelli[†#]	Estuaries and creeks; muddy habitats	Unknown	Shallow

† Species endemic to Australasia (including New Guinea, New Caledonia and islands and reefs in the Coral Sea).

* *Hydrophis* group species in the GBR were previously thought represent six genera but recent molecular data indicate they all belong to the genus *Hydrophis*.

\# *Aipysurus mosaicus* was previously thought to belong to the South-East Asian species, *A. eydouxi,* while *H. zweifelli* was previously thought to belong to the south-east Asian species, *H. schistosa*.

head towards vibrations of the water made by small fish and captures its prey using a fast lateral strike, sometimes accompanied by backward swimming. All species manipulate their prey until they reach the head and prey items are swallowed head first.

While no species of sea snake occurring in GBR waters is currently listed as threatened with extinction, unexplained declines have been documented on protected reefs for *A. laevis* and *E. annulatus* in the southern GBR. Genetic and mark–recapture studies have shown that *Aipysurus* group species have highly restricted dispersal over multiple spatial scales, including between adjacent reefs. Low connectivity among populations, combined with the intrinsically patchy distributions of *Aipysurus* group species, and their relatively low reproductive outputs, means that declines in abundance or local extinctions are unlikely to be reversed by dispersal from adjacent reefs over timeframes relevant to conservation.

In order to implement targeted and effective conservation strategies, specific threatening processes impacting sea snakes need to be identified. Possible threatening processes include: disease, particularly as increased sea surface temperatures can promote the spread and prevalence of pathogens and increase host disease susceptibility; invasive species; recruitment failure; seismic surveys; and pollution, including oil spills. Inter-reefal species comprise a significant component of commercial trawl by-catch on the GBR, and considerable effort is being made to reduce this impact by developing and evaluating the effectiveness of by-catch reduction devices for sea snakes. Protection of intertidal habitats, which are too shallow for trawl fishing and important refugia for juvenile sea snakes and gravid females, may help mitigate the impacts of trawling. The ongoing degradation of GBR coral reefs is unlikely to be beneficial for sea snakes.

Identifying marine mammals and reptiles at sea

Identifying marine mammals and reptiles at sea can be extremely difficult, especially if only a small portion of an animal's body is seen fleetingly as it surfaces to breathe. High-quality photographs and videos of whales or dolphins breaching (leaping from the water) or underwater (all species of marine mammals and reptiles) with most or all of the body in the picture, or of stranded animals (alive or dead) are important aids to identification and can be used by researchers to identify species, and even individuals of some species. Sightings and photographs of marine mammals and reptiles can be uploaded using the 'Eye and the Reef' app (http://www.gbrmpa.gov.au/managing-the-reef/how-the-reefs-managed/eye-on-the-reef/report-sightings).

Identifying stranded animals is also difficult, especially when they are decomposing. All stranded marine mammals and reptiles should be reported through the stranding hotline at 1300 264 625.

Figure 31.1 is a sighting sheet that will help identify some of the species of marine mammals most likely to be seen in the GBR region; the species list is not exhaustive. There are several good identification guides that provide more detail (see 'Further reading'). There is still much to learn about the distribution of marine mammals and reptiles in the GBR and the reported distribution maps for some species are probably inaccurate. A single report, or a small number of reports, of stranded individuals does not necessarily reflect the normal distribution and range of a species.

Further reading

Baker AN (1999) *Whales and Dolphins of New Zealand and Australia: An Identification Guide.* Victoria University Press, Wellington, New Zealand.

Bryden MM, Marsh H, Shaughnessy P (1998) *Dugongs, Whales, Dolphins and Seals: A Guide to the Sea Mammals of Australia.* Allen & Unwin, Sydney.

Cogger H (2014) *Reptiles and Amphibians of Australia.* 7th edn. CSIRO Publishing, Melbourne.

Commonwealth of Australia (2014) *National Dugong and Turtle Protection Plan 2014–2017.* Department of the Environment and Energy, Canberra, <http://www.environment.gov.au/system/files/resources/19f5c341-2705-4c48-b7b3-f66a52517e16/files/dugong-turtle-protection-plan.pdf>.

Commonwealth of Australia (2017) *Recovery Plan for Marine Turtles in Australia.* Department of the Environment and Energy, Canberra, <http://www.environment.gov.au/marine/publications/recovery-plan-marine-turtles-australia-2017>.

Elfes C, Livingstone SR, Lane A, Lukoschek V, Sanders KL, Courtney AJ, *et al.* (2013) Fascinating and forgotten: the conservation status of marine elapid snakes. *Herpetological Conservation and Biology* **8**, 37–52.

GBRMPA (2007) *Operational Policy on Whale and Dolphin Conservation in the Great Barrier Reef Marine Park.* Great Barrier Reef Marine Park Authority, Townsville, <http://elibrary.gbrmpa.gov.au/jspui/bitstream/11017/409/1/Operational-policy-on-whale-and-dolphin-conservation-GBRMP.pdf>.

Geraci JR, Lounsbury VJ (2005) *Marine Mammals Ashore: A Field Guide for Strandings.* 2nd edn. National Aquarium in Baltimore, Baltimore, MD, USA.

Heatwole H (1999) *Sea Snakes.* University of New South Wales Press, Sydney.

Heatwole H, Cogger H (1994) Sea snakes of Australia. In *Sea Snake Toxicology.* (Ed. P Gopalakrishnakone) pp. 167–205. Singapore University Press, Singapore.

Lukoschek V, Heatwole H, Grech A, Burns G, Marsh H (2007) Distribution of two species of marine snakes, *Aipysurus laevis* and *Emydocephalus annulatus,* in the southern Great Barrier Reef: metapopulation dynamics, marine protected areas and conservation. *Coral Reefs* **26**, 291–307. doi:10.1007/s00338-006-0192-8

Lukoschek V, Waycott M, Marsh H (2007) Phylogeographic structure of the olive sea snake, *Aipysurus laevis* (Hydrophiinae) indicates recent Pleistocene range expansion but low contemporary gene flow. *Molecular Ecology* **16**, 3406–3422. doi:10.1111/j.1365-294X.2007.03392.x

Lutz PL, Musick JA (Eds) (1996) *The Biology of Sea Turtles.* Vol. I. CRC Press, Boca Raton, FL, USA.

Lutz PL, Musick JA, Wyneken J (Eds) (2003) *The Biology of Sea Turtles.* Vol. II. CRC Press, Boca Raton, FL, USA.

Marsh H, O'Shea TJ, Reynolds JE, III (2011) *The Ecology and Conservation of Sirenia: Dugongs and Manatees*. Cambridge University Press, Cambridge, UK.

Perrin W, Wursig B, Thewissen J (Eds) (2009) *Encyclopedia of Marine Mammals*. 2nd edn. Academic Press, Los Angeles, CA, USA.

Woinarski J, Burbidge A, Harrison P (2014) *The Action Plan for Australian Mammals 2012*. CSIRO Publishing, Melbourne.

Wyneken J, Lohman KJ, Musick JA (Eds) (2013) *The Biology of Sea Turtles*. Vol. III. CRC Press, Boca Raton, FL, USA.

32

Seabirds

B. C. Congdon

General life-history characteristics

Seabirds are highly visible, charismatic predators in marine ecosystems that feed primarily or exclusively at sea. They have a range of relatively unique life-history characteristics associated with this marine lifestyle. Many of these characteristics are directly linked with having to forage over large distances to obtain sufficient food to breed.

In general, seabirds are long-lived species that have deferred sexual maturity, small clutch sizes, slow chick growth rates and extended fledgling periods. For example, crested terns (*Sterna bergii*, Fig. 32.1A) may live 18–20 years, sooty terns (*S. fuscata*, Fig. 32.1B) up to 32 years and larger species such as boobies and frigatebirds (Fig. 32.1C) even longer. Most seabird species do not become sexually mature or return to breed for between 5 to 12 years after fledging and the nestling period for some species, such as frigatebirds, can be up to 6 months long.

When nesting, care by both parents is usually required to successfully rear chicks. Following hatching, adults take turns brooding chicks until they are able to regulate their own body temperature. After this, chicks are left alone at the colony while both parents forage for food. At this stage the chicks of some ground-nesting tern species are mobile and gather together in crèches for protection from predators. Chicks of other species such as the bridled tern (*S. anaethetus*, Fig. 32.1D) usually remain concealed at, or near the nest site.

Chicks of many seabirds also accumulate extensive body fat reserves to buffer themselves against long periods between adult feeding visits. Greatest mortality occurs during early life-history stages, such as eggs, nestlings and in the post-fledging and pre-reproductive periods. Predation levels in seabird colonies are often significant, and any disturbance by humans usually leads to increased levels of predation on exposed eggs or chicks.

Most seabirds breed on relatively remote islands, different species tending to nest in specific habitat types. Seabirds of the Great Barrier Reef (GBR) prefer coral cays to other island types. Most nest in dense colonies on the ground using either bare scrape nests or nests built among low lying vegetation. Other species use rocky crevices or burrows dug among vegetation dense enough to stabilise the soil structure and still others nest in trees on platform nests built of leaves or sticks (Table 32.1).

Distribution and abundance on the GBR

Australia's GBR supports a significant and diverse seabird community. Twenty different species (Table 32.1), comprising 1.7 million individuals and more than 25% of Australia's tropical seabirds, nest within the GBR region. This includes greater than 50% of Australia's roseate terns (*S. dougallii*), lesser crested terns (*S. bengalensis*, Fig. 32.1E), black-naped terns (*S. sumatrana*, Fig. 32.1F) and black noddies

Fig. 32.1. Seabird of the Great Barrier reef: **(A)** crested tern; **(B)** sooty tern; **(C)** least frigatebird; **(D)** bridled tern; **(E)** lesser crested tern; **(F)** black-naped tern; **(G)** black noddy; **(H)** wedge-tailed shearwater; **(I)** brown booby; **(J)** masked booby; **(K)** common noddy. (Photos: A, B, D, E, F, K, C. Devney; C, H, J, B. Congdon; G, I, D. Peck.)

(*Anous minutus*, Fig. 32.1G), and ~25% of the wedge-tailed shearwater (*Puffinus pacificus*, Fig. 32.1H), brown booby (*Sula leucogaster*, Fig. 32.1I), masked booby (*S. dactylatra*, Fig. 32.1J) and red-tailed tropicbirds (*Phaeton rubricauda*).

In general, seabirds are colonial nesters. Throughout the GBR there are 56 major and 20 minor seabird colonies (Fig. 32.2). Most major colonies are located in either the far northern, northern or far southern regions of the GBR. Raine Island in the far northern sector is one of the largest and most significant tropical seabird breeding sites in Australia. Of the 24 seabird species recorded as breeding in Queensland, 14 breed at Raine Island. Michaelmas Cay in the northern section of the GBR is a tropical seabird colony rated as the second most important nesting site in the GBR. The island constitutes a major nesting site for sooty terns and common noddies (*A. stolidus*, Fig. 32.1K), as well as for crested and lesser crested terns. The islands of the Swain reefs in the far south-eastern region of the GBR also constitute one of six core seabird breeding areas.

Finally, the Capricorn-Bunker group of islands in the far southern GBR contains nationally and internationally significant seabird breeding

Table 32.1. Seabird species/approximate population estimates for the Great Barrier Reef only/nesting habitat/foraging guild (modified from Hulsman *et al.* 1997)

Species		GBR population	Nesting habitat	Foraging guild
Herald petrel	*Pterodroma arminjoniana*	>26	Ground: scrape	Pelagic
Wedge-tailed shearwater	*Puffinus pacificus*	560 000	Burrow	Pelagic
Australian pelican	*Pelicanus conspicillatus*	1024	Ground: scrape	Coastal
Red-footed booby	*Sula sula*	172	Trees/shrubs/ground	Pelagic
Brown booby	*Sula leucogaster*	18 500	Ground: low grass/scrape	Offshore
Masked booby	*Sula dactylatra*	1100	Ground: scrape	Pelagic
Great frigatebird	*Fregata minor*	20	Trees/ground	Pelagic
Least frigatebird	*Fregata ariel*	2500	Shrubs/ground	Pelagic
Red-tailed tropicbird	*Paethon rubricauda*	100	Holes/crevices/tree roots	Pelagic
Caspian tern	*Sterna caspia*	70	Ground: low grass/scrape	Coastal
Roseate tern	*Sterna dougallii*	6000	Ground: scrape	Inshore
Black-naped tern	*Sterna sumatrana*	3900	Ground: scrape	Inshore
Little tern	*Sterna albifrons*	51 (up to 1000)	Ground: scrape	Coastal
Sooty tern	*Sterna fuscata*	48 000	Ground: low grass/scrape	Pelagic
Bridled tern	*Sterna anaethetus*	13 900	Ground: low grass/scrape	Offshore
Crested tern	*Sterna bergii*	26 000	Ground: scrape	Inshore
Lesser crested tern	*Sterna bengalensis*	6300	Ground: scrape	Inshore
Common noddy	*Anous stolidus*	46 000	Ground: low grass/scrape	Offshore
Black noddy	*Anous minutus*	300 000	Trees	Offshore
Silver Gull	*Larus novaehollandiae*	750	Ground: low grass/scrape	Coastal

populations. This island group supports the Pacific Ocean's largest breeding colony of wedge-tailed shearwaters and over 97% of the black noddy populations of the GBR. The Capricornia Cays also contain ~75% of the seabird biomass of the GBR World Heritage Area.

Many seabird species that breed within the GBR and in adjacent areas are considered migratory and/or threatened species and are listed under the Australian *Environmental Protection and Biodiversity Conservation Act 1999* (EPBC Act) in a variety of categories. Many are also variously protected under

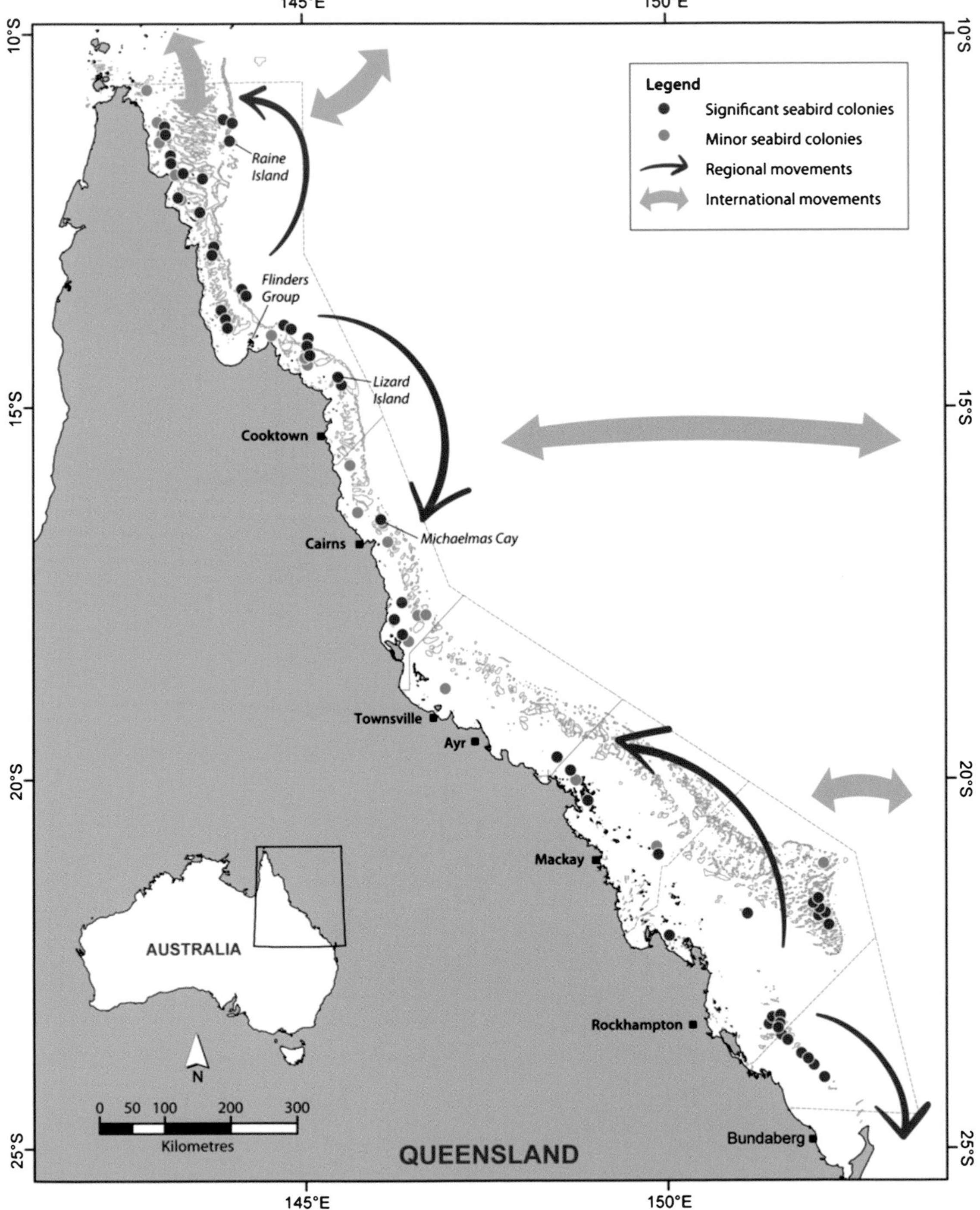

Fig. 32.2. Seabird colony distribution on the Great Barrier Reef (Map: Johanna Johnson).

international agreements such as: the China–Australia Migratory Bird Agreement (CAMBA), Japan–Australia Migratory Bird Agreement (JAMBA) and the Convention on the Conservation of Migratory Species of Wild Animals (Bonn Convention 1979). Additionally, the GBR region hosts migrating populations of some Northern Hemisphere breeding species such as common tern (*Sterna hirundo*) and much of the Asian population of roseate tern (*Sterna dougallii*).

Foraging modes

Seabirds of the GBR can be classified into four major foraging guilds: coastal, inshore, offshore and open-ocean or pelagic. Species in each of these guilds travel different distances to access food and are away from young for different periods of time. Offshore and pelagic species may travel very long distances to obtain food (see Box 32.1) and, because of this, usually regurgitate food for young. Inshore foragers travel much shorter distances, forage in and around reefs adjacent to colonies and carry whole fish in their bill to feed young. Foraging mode also affects both the number of eggs laid and chick growth rates, with more pelagic foraging species having single egg clutches and much slower growing chicks.

Inshore, offshore and pelagic foragers show consistent differences in the size of breeding colonies. Inshore foragers usually have small, more densely packed colonies of up to a few thousand birds, while pelagic species can have enormous colonies spread over multiple closely spaced islands. For example, >500 000 wedge-tailed shearwaters breed in the 13 islands of the Capricorn and Bunker groups.

Such high population abundance means that where large seabird colonies occur the quantity of marine resources consumed by seabirds is significant. Such high consumption rates also mean that seabirds play several important functional roles in marine ecosystems. Most seabird prey items are either pelagic fish or squid. Therefore, seabirds play a major role in transferring nutrients from offshore and pelagic areas to islands and reefs. For example, black noddies deposit an estimated 4 tonnes of guano on Heron Island per annum. Some species of seabirds also play an important role in seed dispersal and in distributing organic matter into lower parts of the developing soil profile (e.g. burrow-nesting species such as shearwaters).

Seabirds can forage alone, but most forage in large mixed species aggregations or flocks. The likely reason for this is that flocking offers advantages in both finding and accessing prey. Different species forage in different ways. Species that capture prey within a metre or so of the ocean surface usually just dip their beaks underwater after landing, or while still in the air. Other species that forage to depths of up to 20 m may plunge dive at speed using just the momentum of the dive to carry them to prey, or they may extend the pursuit of prey underwater using slight movements of their wings. A third group more actively pursues prey by swimming underwater using their wings or feet for propulsion. Regardless of foraging mode, most seabirds are also heavily reliant on the activity of subsurface predators to drive prey within reach. For this reason, the presence and activity of large predatory fishes such as tuna and mackerel, as well as other subsurface predators, are extremely important to seabirds. This is particularly so in the relatively resource poor foraging environment of the tropics.

The location and types of food used by a single seabird species can also differ significantly between the breeding and non-breeding periods and between different life-history stages such as adults and offspring or sexes, during breeding. This is particularly true for offshore and pelagic foraging species. Partitioning resources in this way allows these species to expand the resource base available to them and so breed in what would otherwise be poor or marginal habitat. However, being reliant on food availability in many different locations also makes them particularly susceptible to disturbance at any one of these sites (see Box 32.1).

Timing of breeding

The timing of seabird breeding activity on the GBR is complex and can vary with location, year and/or species. In general, both northern and southern colonies tend to have specific seasonal peaks in activity. Most breeding peaks occur over summer from October through to April each year, but for some species, and especially in northern colonies, breeding activity can occur at all times of the year. Different species may also use the same areas at different times of year. For example, masked booby breeding in the Swains Reefs peaks in the austral winter, while brown booby breeding at the same locations peaks during the austral summer.

Threats to seabirds

There are a range of identified anthropogenic threats to the long-term population viability of seabirds of the GBR. By far the biggest contemporary threat is climate change and its associated impacts on food supply and loss of breeding habitat. The total dependence of seabirds on marine resources makes them key upper trophic level predators in marine systems. This means that seabird demographics and reproductive parameters are strongly impacted by, and closely reflect, changing food availability and oceanographic conditions. For this reason, they are widely considered important indicators of marine ecosystem health. Therefore, understanding how seabird population dynamics and reproductive ecology are impacted by changing oceanography leads directly to important insights into the potential future impacts of climate change: not only on seabirds, but also on other functionally important components of reef ecosystems in general (see Box 32.2).

Other important identified threats to seabirds include, but are not limited to: commercial fishing, breeding habitat destruction, the introduction of exotic plants and animals, and direct disturbance by visitors to islands. Multiple forms of pollution, particularly plastics and water quality degradation, are also becoming increasing important threats to the GBR system as a whole, including to seabirds.

Non-marine birds

Many other terrestrial, coastal and migratory bird species also use the continental islands and cays of the GBR. The numbers and types are too extensive to detail here, but these bird communities often resemble those occurring in similar habitats on the mainland. Some of the more common non-migratory species include the buff-banded rail (*Gallirallus philippensis*), Australian pelican (*Pelicanus*

Box 32.1. Critical foraging locations for wedge-tailed shearwaters of the GBR

High-resolution tracking of individual wedge-tailed shearwaters (Fig. 32.3A) has shown that birds breeding on Heron Island rely on food obtained from three distinctly different foraging areas. When not breeding they are trans-equatorial migrants that fly over 6000 km from breeding sites to overwinter in Micronesia (Fig. 32.3B). When breeding adults obtain food for their chick within 300 km of Heron Island, in five distinct foraging zones (Fig. 32.3C). Prey availability in these near-colony zones is generally poor and is driven by the dynamics of the 'Capricorn Eddy', whose location and intensity is in turn determined by the dynamics of the East Australian Current as it flows along the edge of the GBR (Fig. 32.3D).

Poor food availability near the colony means that adults cannot simultaneously feed both themselves and their chick using locally caught prey. Because of this, adults use a specialised dual-foraging strategy. During near-colony trips they sacrifice their own body condition by feeding nearly everything they obtain to their chick. Then, when their own reserves are low, they perform an extended trip of up to 1000 km to distant foraging grounds in the Coral and Tasman Seas (Fig. 32.3E) to quickly regain condition, while their partner takes over chick feeding duties. At these distant locations, adults forage with tuna and other large subsurface predators (Fig. 32.3F, G). These large predatory fish drive baitfish towards the sea surface when feeding, making them available to foraging seabirds.

The conservation significance of such key foraging refuges cannot be overstated. Most are outside the GBR management zone and currently have no special protected status. The breeding success of shearwaters and other seabird species on a regional scale is likely to be totally dependent on the continued stability of these sites and the top-predator interactions that occur at them.

conspicillatus), eastern reef egret (*Egretta sacra)*, pied (Torresian) imperial pigeon (*Ducula bicolor*), sooty oyster catcher (*Haematopus fuliginosus*) and beach stone-curlew (*Esacus neglectus*). Three species of fish-eating raptors also inhabit islands in the GBR: the osprey (*Pandion haliaetus*), brahminy kite (*Haliastur indus*) and white-bellied sea eagle (*Haliaeetus leucogaster*).

The eastern reef egret or eastern reef heron is common throughout the GBR and adjacent

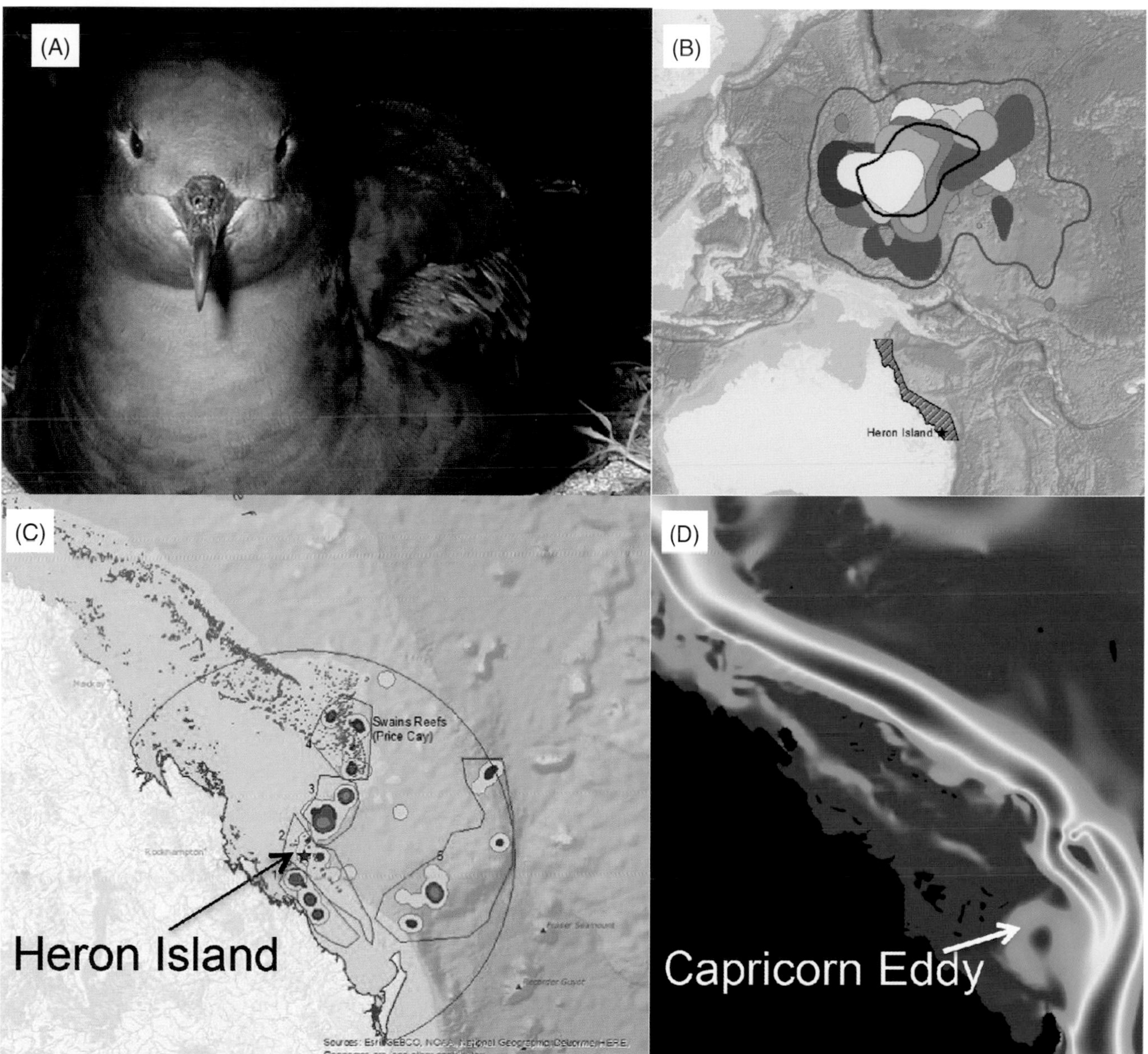

Fig. 32.3. **(A)** adult wedge-tailed shearwater; **(B)** wedged-tailed shearwater non-breeding foraging locations; **(C)** wedged-tailed shearwater near-colony foraging locations used by adults in the Capricorn-Bunker Island group to supply food to chicks during breeding; **(D)** the position of the 'Capricorn Eddy' adjacent to the East Australian Current at the southern end of the Great Barrier Reef (image courtesy of C. Steinberg); **(E)** 'at-distance' foraging locations used by wedge-tailed shearwater adults for self provisioning during incubation and breeding; **(F)** facilitated foraging of wedge-tailed shearwaters with subsurface predatory fish; **(G)** overlap of wedge-tailed shearwater foraging locations with predicted tuna distribution and abundance, for shearwaters breeding at Heron and Lord Howe Islands 2015 (Photos and original images: A, B. Congdon; B, C, E, F. McDuie; D, C. Steinberg; G, M. Miller).

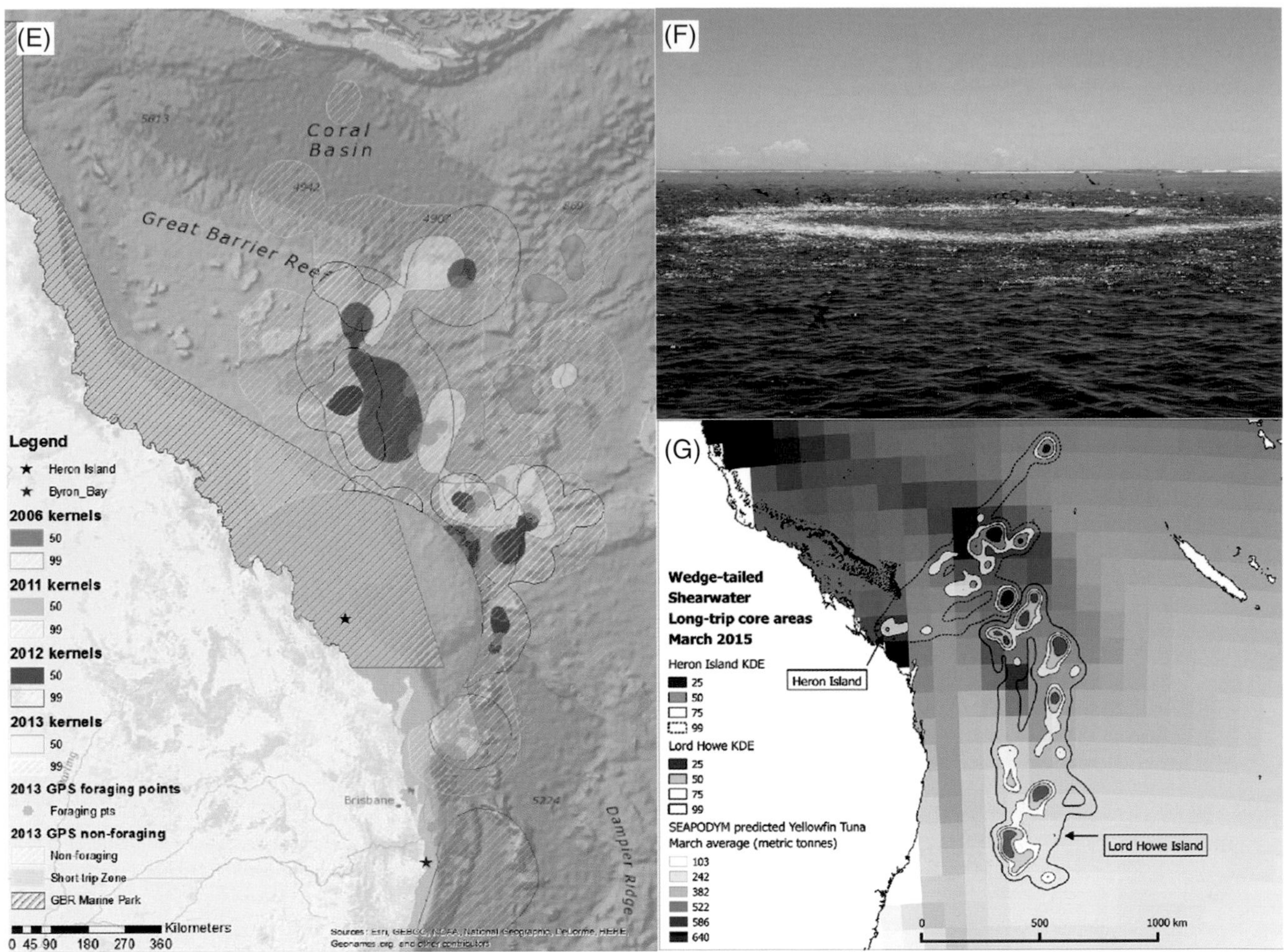

Fig. 32.3. (Continued)

Box 32.2. Seabirds predict the weather: seabirds and climate change on the GBR

Increases in sea-surface temperature (SST) and the intensity of El Niño events negatively impact many seabird species breeding on the Great Barrier Reef (GBR). During chick rearing, these phenomena cause both short-term and seasonal-scale reductions in food supply (Fig. 32.4A) that result in increased adult foraging ranges and effort, poor chick growth and survivorship and even seasonal reproductive collapse at some colonies.

Changes in El Niño intensity have now also been linked to decreased breeding participation at one of the most important seabird breeding sites on the GBR, Michaelmas Cay (Fig. 32.4C). The numbers of sooty terns and common noddies breeding at this location in any one year can actually predict El Niño intensity 9–12 months into the future! This means that whether adults are able to obtain the food they need to reach breeding condition is determined by precursory, El Niño driven changes in the ocean that occur up to 12 months ahead of recorded El Niño events.

Combined, such findings demonstrate that predicted increases in SST and changes to the intensity or frequency of El Niño events are likely to have serious detrimental impacts on tropical seabird populations throughout the GBR and in adjacent areas of the Coral Sea. Importantly, current evidence suggests that at least some species at significant GBR breeding colonies may already be in decline.

Whether seabirds can cope with these climate stressors in the future is unknown. However, much will depend on the effective identification and management of additional anthropogenic threats to nesting sites and foraging resources, as well as on our ability to adaptively modify conservation and fisheries management strategies in response.

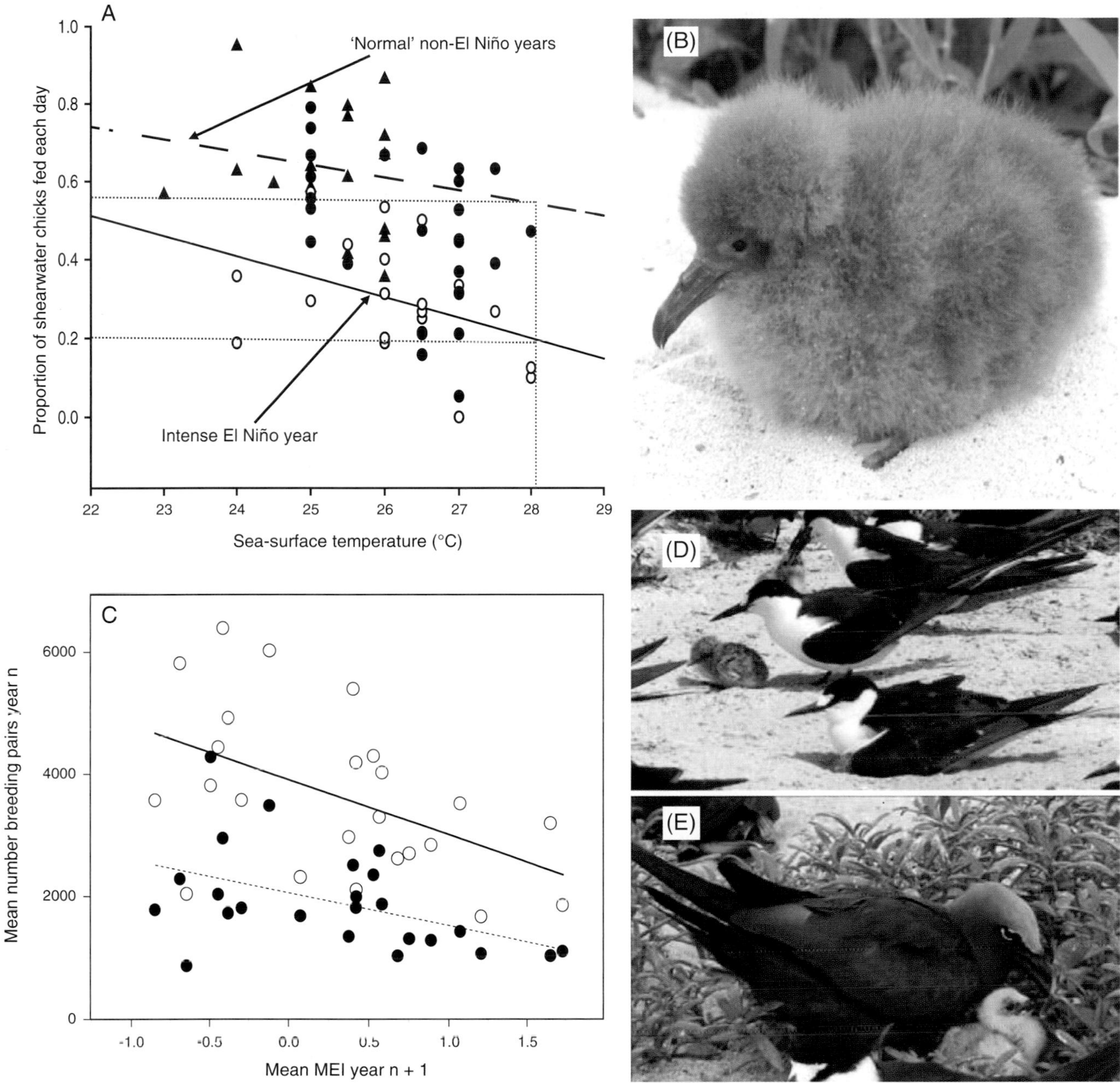

Fig. 32.4. **(A)** The relationship between day-to-day variation in sea surface temperature (SST), El Niño intensity and the proportion of chicks fed each day, for wedge-tailed shearwaters breeding on Heron Island. Note that increasing SSTs reduce food availability to chicks in all years, and that during intense El Niño conditions the proportion of chicks fed decreases by a further ~35% for equivalent SSTs (red dashed lines). **(B)** wedge-tailed shearwater chick. **(C)** Relationship between the number of breeding pairs of pelagic foraging terns and annual averages of the multivariate El Niño index (MEI). The average annual number of breeding pairs counted from 1984 to 2001 on Michaelmas Cay (Great Barrier Reef, Australia) for the sooty tern (open circles, solid line) and common noddy (solid circles, dotted line) in each year (n) is shown relative to the mean annual MEI in the following year (n+1). **(D)** Sooty tern. **(E)** Common noddy (Photos and original images: A,B, D. Peck; C–E, C. Devney).

mainland. The species has an unusual, non-sexual colour dimorphism, with different individuals having either entirely white or slate grey plumage. Globally, the dark form is more common in temperate areas while the white form is more abundant in the tropics. The reason for the colour dimorphism is unknown. The two colour forms randomly interbreed and both white and dark offspring can be found in the same brood. When nesting on coral cays, reef egrets hunt both day and night at low tide using small territories they establish and maintain on the reef flat.

The buff-banded rail is widespread and secretive on the Australian mainland, but is also a highly dispersive and an obvious reef island coloniser, reaching many of the heavily vegetated cays. It is an omnivorous scavenger that feeds on small vertebrates and invertebrates as well as seeds, fallen fruit and other vegetable matter that it collects from within a defended territory. The species is highly susceptible to the impacts of introduced predators or competitors and often disappears from islands where these threats occur.

The Torresian imperial pigeon (*Ducula spilorrhoa*) is one of the most numerous terrestrial birds on the GBR. It is a migratory species that visits coastal reef islands and the adjacent mainland south to about Mackay during summer. It then returns to Papua New Guinea during the winter. Greatest numbers of pigeons occur between Cooktown and Cape York, with some colonies in this area reaching greater than ten thousand pairs.

Other less numerous but important terrestrial subspecies on the GBR include the Capricorn silvereye (*Zosterops lateralis chlorocephalus*) and the Capricorn yellow chat (*Epthianura crocea macgregori*). The Capricorn white-eye is the only endemic bird breeding exclusively on islands of GBR. It is restricted to the wooded cays of the southern Capricorn-Bunker region. Genetic research suggests that it may be a distinct species. The yellow chat is the most vulnerable terrestrial bird species breeding in the GBR. This Australian east coast subspecies was believed to be extinct until a small number were discovered on a coastal island of the southern GBR island in the 1990s. The species currently has an extremely restricted distribution and total population numbers remain low (~300). It is listed as 'critically endangered'.

Large numbers of migratory wader species also spend their non-breeding season among the islands of the GBR. These birds breed in northern Asia and Alaska and migrate thousands of kilometres along the East Asian–Australasian Flyway to non-breeding sites in Australia and New Zealand. They are most commonly found in coastal estuaries and tidal mudflats but many species also form large groupings on coral cays and other islands. Common species include: greater (*Charadrius leschenaultii*) and lesser (*Charadrius mongolus*) sand plovers, turnstones (*Arenaria interpres*), golden plover (*Pluvialis apricaria*), grey-tailed tattler (*Tringa brevipes*), red-necked stint (*Calidris ruficollis*) and bar-tailed godwit (*Limosa lapponica*).

Further reading

Chambers LE, Devney CA, Congdon BC, Dunlop N, Woehler EJ, Dann P (2011) Observed and predicted impacts of climate on Australian seabirds. *Emu-Austral Ornithology* **111**, 235–251. doi:10.1071/MU10033

Congdon BC, Krockenberger AK, Smithers BV (2005) Dual-foraging and co-ordinated provisioning in a tropical Procellariiform, the wedge-tailed shearwater. *Marine Ecology Progress Series* **301**, 293–301. doi:10.3354/meps301293

Devney CA, Short M, Congdon BC (2009) Sensitivity of tropical seabirds to El Niño precursors. *Ecology* **90**, 1175–1183. doi:10.1890/08-0634.1

Devney CA, Short M, Congdon BC (2009) Cyclonic and anthropogenic influences on terns breeding at Michaelmas Cay, Great Barrier Reef. *Australian Wildlife Research* **36**, 368–378. doi:10.1071/WR08142

Dyer PK, O'Neill P, Hulsman K (2005) Breeding numbers and population trends of Wedge-tailed Shearwater (*Puffinus pacificus*) and Black Noddy (*Anous minutus*) in the Capricornia Cays, southern Great Barrier Reef. *Emu* **105**, 249–257. doi:10.1071/MU04011

Erwin CA, Congdon BC (2007) Day-to-day variation in sea-surface temperature negatively impacts sooty tern (*Sterna fuscata*) foraging success on the Great Barrier Reef, Australia. *Marine Ecology Progress Series* **331**, 255–266. doi:10.3354/meps331255

Higgins PJ, Davies SJJF (1996) *Handbook of Australian New Zealand and Antarctic Birds*. Vol. 3. Oxford University Press, Melbourne.

Hulsman K, O'Neill P, Stokes T (1997) Current status and trends of seabirds on the Great Barrier Reef. In *State of the Great Barrier Reef World Heritage Area Workshop, Townsville, Queensland, Australia, 27–29 November 1995.* (Eds D Wachenfeld, J Oliver and K Davis) pp. 259–282. Great Barrier Reef Marine Park Authority, Townsville, <http://hdl.handle.net/11017/259.

McDuie F, Congdon BC (2016) Trans-equatorial migration and non-breeding habitat of tropical shearwaters: implications for modelling pelagic Important Bird Areas. *Marine Ecology Progress Series* **550**, 219–234. doi:10.3354/meps11713

McDuie F, Weeks SJ, Miller MGR, Congdon BC (2015) Breeding tropical shearwaters use 'at-distance' foraging sites when self-provisioning. *Marine Ornithology* **43**, 123–129.

Miller MGR, Carlile N, Phillips JS, McDuie F, Congdon BC (2018) Importance of tropical tuna for seabird foraging over a marine productivity gradient. *Marine Ecology Progress Series* **586**, 233–249. https://doi.org/10.3354/meps12376

Miller MGR, Silva FRO, Macovsky-Capuska GE, Congdon BC (2018) Sexual segregation in tropical seabirds: drivers of sex-specific foraging in the Brown Booby *Sula leucogaster*. *Journal of Ornithology* **159**, 425–437.

O'Neill P, Minton C, Ozaki K, White R (2005) Three populations of non-breeding Roseate Terns (*Sterna dougallii*) in the Swain Reefs, southern Great Barrier Reef. *Emu* **105**, 57–66. doi:10.1071/MU03044

Peck DR, Smithers BV, Krockenberger AK, Congdon BC (2004) Sea surface temperature constrains wedge-tailed shearwater foraging success within breeding seasons. *Marine Ecology Progress Series* **281**, 259–266. doi:10.3354/meps281259

Weeks SJC, Steinberg C, Congdon BC (2013) Oceanography and seabird foraging: within-season impacts of increasing sea surface temperature on the Great Barrier Reef. *Marine Ecology Progress Series* **490**, 247–254. doi:10.3354/meps10398

Epilogue and acknowledgements

The Great Barrier Reef (GBR) has been an important part of Australia's identity and has become an international icon. So much has happened since we published the first edition of this book. The concern about degradation of the GBR has encircled the globe. Reports of record levels of mass bleaching and mortality have had special traction worldwide and the increased frequency of major bleaching events has concerned people that our precious reef will be lost as predicted almost 20 years ago. Climate change and a related cascade of impacts can no longer be ignored. Furthermore, other types of impacts have not abated including managing runoff form the land, fisheries, continued coastal development and the activities of mining companies continue to threaten the reef.

Australia has the oldest coral reef society in the world, namely the Australian Coral Reef Society (ACRS), which was founded in 1992. This Society began as a Great Barrier Reef Committee. The Committee instigated the first major expedition on the GBR 1927–1928. We are pleased that members of the Society have continued to provide leading roles in describing the patterns of biodiversity and impacts on reefs, as well as the processes that underlie the patterns. The engagement of stakeholders and their connected research and learning initiatives with others worldwide is critical for the GBR and the future of all coral reefs.

We, as scientists and conservationists, have been motivated by the desire to ensure that we continue to enhance our knowledge and understanding of the GBR. The Society sponsored this book and most of the contributing experts have been, or are, members. As a result, all proceeds from the sale of this book go to ACRS for the promotion and support of further efforts to understand and manage coral reefs not only in Australia but worldwide, especially through postgraduate students. The details for how to join ACRS can be found at its website http://www.australiancoralreefsociety.org/.

There are many people who have contributed to this book by contributing text or images. The Australian Museum, James Cook University and the University of Queensland have supported the editors in undertaking the task of bringing this revised edition of the book to completion and support research programs on the GBR. The 48 experts that provided their time for free to be able to compile a fascinating series of chapters, and the editors are enormously grateful for their contributions. We thank CSIRO Publishing for their attention to detail in the final stages. It is our hope that the book will be a useful starting point to any study of coral reefs, both in Australia and abroad.

Shane Ahyong gratefully acknowledges Glenn Ahyong (Canberra), Claudia Arango (Queensland Museum), Arthur Anker (Universidade Federal de Goiás, Brazil), Roy Caldwell (University of California, Berkeley), Roger Springthorpe and Ian Loch (both Australian Museum), N. Dean Pentcheff (Los Angeles), Chris Tudge (American University, Washington DC) and the Great Barrier Reef Marine Park Authority for the use of photographs.

Tom Bridge would like to thank Stefan Williams and Oscar Pizarro and the Australian Centre for Field Robotics, University of Sydney, and Rob Beaman from James Cook University. Without their expertise, we would still know very little about mesophotic coral ecosystems in the GBR. He also acknowledges funding support from the Australian Research Council. Jody Webster acknowledges funding from the ARC Discovery grant program and Emma Kennedy for providing the high-resolution images of the *Halimeda* bioherms. Tiffany Sih would like to thank M.J. Kingsford and M. Cappo.

Phil Bock and Dennis Gordon wish to thank the Australian Institute of Marine Science for participation in the Census of Coral Reefs, and, in New Zealand, NIWA. Gary Cranitch from the Queensland Museum generously gave permission to use the colour photo. Phil Bock would like to thank Robyn Cumming for constructive comments.

Maria Byrne gives thanks to many colleagues for permission to use their images. Thomas Prowse and Paulina Selvakumaraswamy assisted with the images.

Andrew Chin would like to thank Peter Kind who reviewed his chapter.

Brad Congdon would like to thank Carol Devney, Fiona McDuie, Mark Miller, Darren Peck, Craig Steinberg and Scarla Weeks for their contributions to both the research and final chapter. This research was funded by the Australian Research Council (ARC), the Marine and Tropical Sciences Research Facility (MTSRF), the Great Barrier Reef Marine Park Authority (GBRMPA), the National Environmental Research Program (NERP), the Holdsworth Wildlife Research Endowment, the Lord Howe Island Board, Birdlife Australia and Birds Queensland.

Jon Day acknowledges that the RAP and the rezoning process resulted from the involvement and support of virtually the entire GBRMPA staff, as well as many external researchers, other experts, many thousands of local users, and the wider public, who were all concerned for the future of the GBR; their collective efforts therefore need to be acknowledged. Particular thanks also to Mick Bishop, the late Max Day, Kirstin Dobbs, Leanne Fernandes and Andrew Skeat, Richard Kenchington for comments on drafts of his chapter.

Guillermo Diaz-Pulido acknowledges the funding support from the Australian Research Council and the Great Barrier Reef Foundation for funding.

Merrick Ekins would like to acknowledge Julian Uribe-Palomino, Neville Coleman, Lisa Gershwin and Jon Cragg for supplying some of the images in Chapter 29 and Noa Shenkar for reviewing an early draft.

Dennis Gordon and Phillip Bock thank the New Zealand Foundation for Research, Science and Technology (Contract C01X0502) for funding (Dennis Gordon), and Karen Gowlett-Holmes (CSIRO) for permission to reproduce a photograph.

Ove Hoegh-Guldberg thanks artist Diana Kleine, and members of the Coral Reef Ecosystems Laboratory and ARC Centre for Excellent in Coral Reef Studies at the University of Queensland, as well as the ARC Laureate program, and Morgan Pratchett for his comments on Chapter 12.

John Hooper would like to thank Natural Products Discovery Griffith University, and the GBR Seabed Biodiversity Project consortium (CRC Reef, AIMS, QDPI, QM and CSIRO Marine and Atmospheric Research) for their significant sponsorship of research on Porifera of the reefs and inter-reef regions, respectively, of the GBR. Without this funding, we would still know very little about the tropical Australasian sponge faunas.

Pat Hutchings would like to thank Kathy Atkinson, Kate Attwood, David Bellwood, Karen Gowlett-Holmes, Jim Johnson, Brendan Kelaher, Tara Macdonald, Ashley Miskelly, Huy Nguyen, Greg Rouse, Alexander Semenov, Roger Steene, Lyle Vail and Dave Wachenfeld for allowing her to use their superb photos, Tom Cribb for commenting on the section on platyhelminths, Jorge Rodriguez Monter for commenting on the flat worms section, Chris Glasby, Robin Wilson for commenting on the polychaetes. Maite Aguado and Christophe Bleidorn for commenting on the current status of our understanding of worm phylogeny. Pat and Mike would like to thank Roy Caldwell, University of California, Berkeley for his image of the stomatopod used in Chapter 15.

Mike Kingsford would like to thank AIMS, GBRMPA and CSIRO for allowing us to use images from their archives, especially in chapters 4 and 5. Further, under creative commons, we acknowledge NOAA as an important source of data. Phil Munday and Jon Knott kindly provided images for Chapter 5. Mike Kingsford and David McKinnon would like to thank lain Suthers for constructive comments on drafts of the heavily revised plankton chapter. Lachlan McKinna and Norman Kuring (NASA) are also thanked for providing and interpreting remote sensing images, Fiona Martin (visualizingscience.com) for infographics, Samantha Talbot née Duggan (AIMS) for processing plankton data, Murray Logan (AIMS) for statistical advice, and Fabien Lombard (Observatoire Océanographique de Villefranche), Kirsten Heimann (JCU), Doug Capone (University of Southern California), Hans Paerl (UNC) and Kennedy Wolfe (University of Sydney) for advice and their images. Images tell different stories to different people and the collective of image providers have greatly enhanced the effectiveness of this book. Mike thanks all his co-authors for contributing to chapters that portray the complexity, beauty and vulnerabilities of the Great Barrier Reef.

Helene Marsh thanks Rebecca Fong for assistance with the illustrations that were based on line drawings by Geoff Kelly and a concept diagram developed with the assistance of the late Peter Arnold. Isabel Beasely, Alistair Birtles, Mike Noad, Guido Parra, Steve van Dyck and anonymous referees provided valuable comments on various drafts of the manuscript. Photographs were provided by Guido Parra.

Helene Marsh, Hal Heatwole and Vimoksalehi Lukoschek thank the contributors to the figures in Chapter 31 listed in the figure legends and the reviewers, especially Mark Hamann, who provided valuable comments on various drafts of the manuscript.

John Pandolfi and Russell Kelley acknowledge their thanks to Robin Beaman for use of the two photos he supplied, to David Hopley for his thoughtful review, to Jody Webster for discussions, and to the ARC Centre of Excellence for Coral Reef Studies for funding.

Roland Pitcher, Peter Doherty and Tara Anderson acknowledge that the Great Barrier Reef Seabed Biodiversity Project was a collaboration between the Australian Institute of Marine Science (AIMS), the Commonwealth Scientific and Industrial Research Organisation (CSIRO), Queensland Department of Primary Industries & Fisheries (QDPI&F), and the Queensland Museum (QM); funded by the CRC Reef Research Centre (CRCReef), the Fisheries Research and Development Corporation (FRDC), and the National Oceans Office (NOO) of the Department of Environment and Water Resources. We also thank the multi-agency teams and the crews of the RV *Lady Basten* (AIMS) and FRV *Gwendoline May* (QDPI&F) that contributed to the success of the fieldwork; the research agencies AIMS, CSIRO, QM, QDPI&F for providing support to the project; and all the project's team members without whose valuable efforts the project would not have been possible.

Morgan Pratchett would like to thank Hugh Sweatman for reviewing his chapter.

Carden Wallace and Andrea Crowther thank: Paul Muir for preparing and revising plates for two chapters and providing advice and input throughout, Michel Pichon for access to unpublished compilations of new coral nomenclature and identification advice, Zoe Richards for excellent review and advice on Chapter 22, Wayne Napier for artwork in Chapter 21. Great Barrier Reef Marine Park Authority, Peter Harrison and Roger Steene, as well as Paul Muir of Queensland Museum who gave permission to publish images. We thank also the countless colleagues whose high-quality research and data compilation, as well as custodianship of online databases, expand our knowledge of reef Cnidaria on an almost daily basis.

Richard Willan acknowledges Gary Cobb (Sunshine Coast), Julie Jones (GBRMPA, Townsville), Allan Limpus (Bundaberg), Julie Marshall (Melbourne), Craig Thomas (Kailu, Hawaii) and David Wachenfeld (GBRMPA, Townsville) for the use of an image. John Collins (James Cook University, Townsville) is acknowledged for helpful comments on an earlier draft of the mollusc chapter.

Paul Southgate (James Cook University, Townsville) provided information on aquaculture of giant clams, and Uwe Weinreich (Cairns) is thanked for the use of images and for information on feeding in Bursidae.

Finally, the editors, Pat Hutchings, Mike Kingsford and Ove Hoegh-Guldberg would like to thank their respective institutions, the Australian Museum, James Cook University and the University of Queensland, for their support and encouragement in the preparation of this second edition. Without this support, the book could not have been produced. We also thank all the authors who contributed and the organisations and individuals who supplied photographs. This has certainly been a team effort of the Australian coral reef community.

Further reading

Hill D (1985) The Great Barrier Reef Committee, 1922–82. Part 11: The last three decades. *Historical Records of Australian Science* **6**(2), 195–221. doi:10.1071/HR9850620195

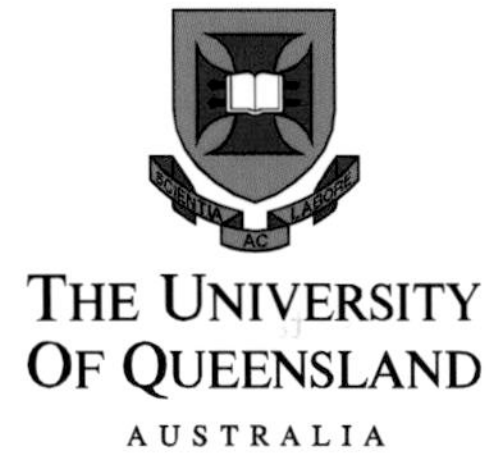

Index

Note: Bold page numbers refer to illustrations.

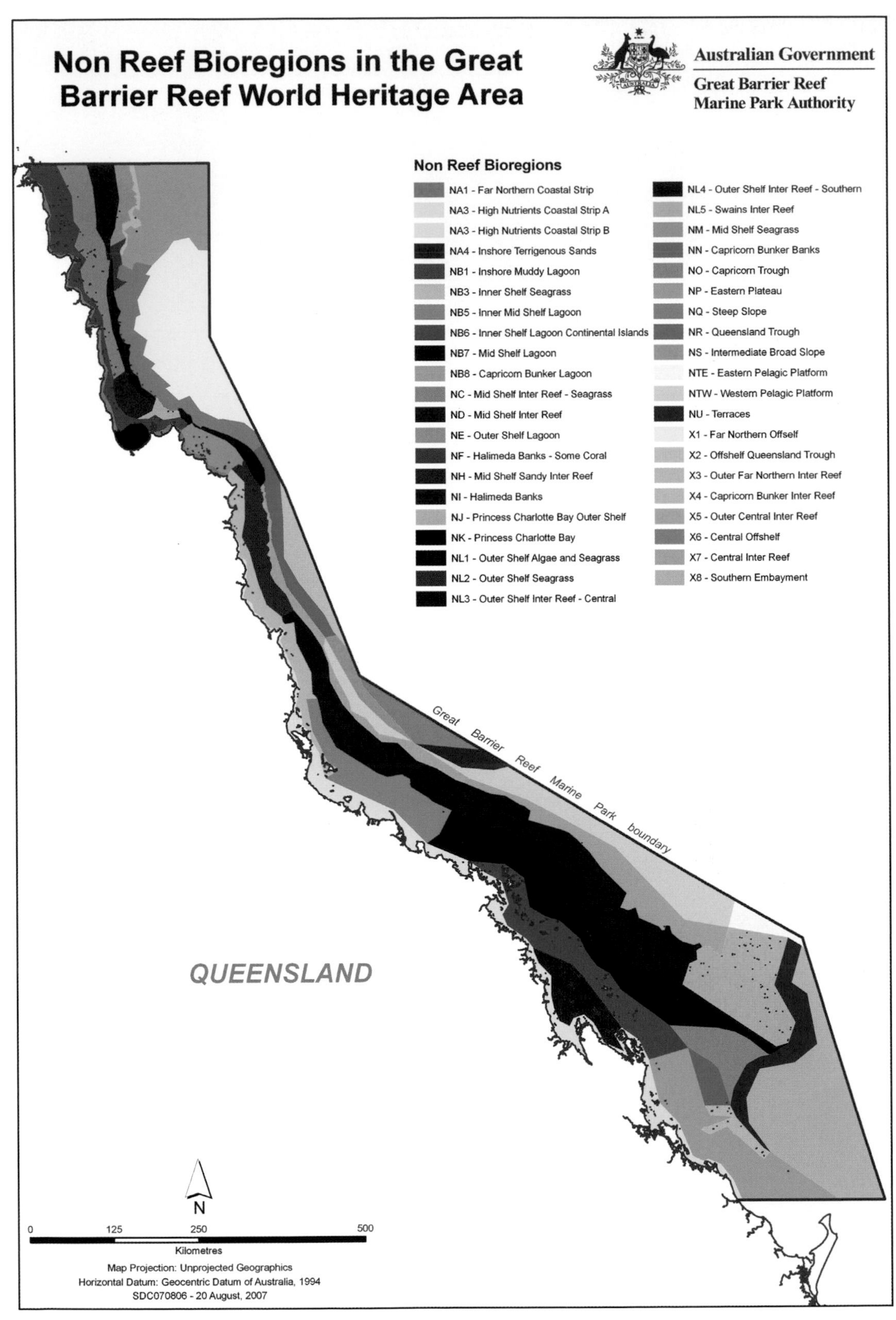

Non-reef bioregions in the Great Barrier Reef World Heritage Area. Each of these is characterised by a unique suite of biota and physical characteristics. (Map: Courtesy of the Spatial Data Centre, Great Barrier Reef Marine Park Authority © Commonwealth of Australia (GBRMPA)).